THERMODYNAMIC PROPERTIES OF GASES

THERMODYNAMIC PROPERTIES
OF GASES

FOURTH EDITION, REVISED
In coordination with the National Service
for Standard Reference Data of the USSR

S. L. Rivkin

English Edition Editor

D. D. Wagman

⬤HEMISPHERE PUBLISHING CORPORATION
A member of the Taylor & Francis Group

New York   Washington   Philadelphia   London

THERMODYNAMIC PROPERTIES OF GASES: Fourth Edition, Revised

Translated by **J. I. Ghojel.**

1 2 3 4 5 6 7 8 9 0   B R B R   8 9 8

This book was set in Times Roman by Hemisphere Publishing Corporation. The editor was Victoria Danahy; and the typesetter was Janis Durbin.
Braun-Brumfield, Inc.was printer and binder.
Cover design by Tammy Marshall.

Library of Congress Cataloging-in-Publication Data

Rivkin, S. L. (Solomon Lazarevich)
  Thermodynamic properties of gases.

  Translation of: Termodinamicheskie svoĭstva gazov.
  Bibliography: p.
  Includes index.
  1. Gases—Thermal properties—Tables. 2. Thermo—
dynamics—Tables. I. Wagman, Donald D. II. Title.
QC165.5.R5813 1988   530.4'3   88-16248
ISBN 0-89116-750-1

DISTRIBUTION OUTSIDE NORTH AMERICA:
ISBN 3-540-18885-1 Springer-Verlag Berlin

CONTENTS

FOREWORD   vii

NOTATIONS AND UNITS
OF PHYSICAL QUANTITIES   ix

**I. TABLES OF THERMODYNAMIC
PROPERTIES OF GASES**   1

Thermodynamic principles   1
Description of the tables   3
Region of applicability of the tables   5
Examples of use of the tables   9

**II. THERMAL DIAGRAM FOR AIR AND
PRODUCTS OF COMBUSTION
OF A GIVEN FUEL**   17

Method of constructing the diagram   17
Tables of caloric quantities for gases   20
Example of diagram construction   25
Example of diagram use   28

III. TABLES OF THERMODYNAMIC PROPERTIES OF VARIOUS GASES   35

Tables III.1, III.2. Air   36
Tables III.3, III.4. Nitrogen   65
Tables III.5, III.6. Atmospheric nitrogen   94
Tables III.7, III.8. Oxygen   123
Tables III.9, III.10. Carbon dioxide   152
Tables III.11, III.12. Water vapor   181
Tables III.13, III.14. Carbon monoxide   210
Tables III.15, III.16. Hydrogen   239
Table III.17. Helium   268

FOREWORD

The first edition of the present book included detailed tables of thermodynamic properties of a number of technically important gases, compiled in the old system of units [1]. In the second, revised, and expanded edition [2] the tables were given in the International System of Units (SI).

The method of construction of the tables has received wide acclaim. The description of the tables was included in a textbook on technical thermodynamics [3], and the tabular calculation method of thermal processes in gases is expounded as part of the thermodynamic course in a number of educational institutions. The suggested tables for the calculation of thermal processes in ideal gases are recommended for use in a "collection of problems on technical thermodynamics" [4].

The ever-increasing application of computers in thermal engineering calculations was taken into account in the third edition of the book [7]. The temperature dependence of constant pressure specific heat of some gases was approximated by the method of least squares with a high degree of accuracy. Values of constant pressure specific heat calculated using the approximating equations deviated from the exact values obtained directly from spectroscopic data by less than several hundredths of a percent. The expressions for the temperature dependence of other thermodynamic quantities (enthalpy, internal energy, and so on) were obtained on the basis of known thermodynamic relations, and data on thermodynamic properties of gases calculated by equations presented in the book. The tables are compiled with such a temperature interval as to render interpolation unnecessary in the majority of technical calculations, which obviously simplifies these calculations. A number of revisions were introduced in the third edition. In particular, along with the approximate calculation of the isentropic process for gaseous mixtures, its exact calculation was presented, taking into consideration the remarks in reference [5].

The method of construction of thermal diagrams on the basis of tables for the products of combustion of a specified fuel has been retained. This method is useful for numerous calculations of processes with combustion products of a given fuel with an arbitrary excess of air. These diagrams, which are equivalent to the usual T-S diagrams, can be easily constructed on any convenient scale to ensure high accuracy for the calculation. Furthermore, they are more exact than the universal diagram [6] and do not require the introduction of correction factors. An example of constructing a diagram for the combustion products of underground gasification gas with excess air is presented, and examples of diagram applications are discussed.

The proposed equations for thermodynamic quantities are applicable to computer-aided thermal engineering calculations. Due to the relative simplicity of these equations, their utilization does not require a large on-line computer memory and long computer time.

In this edition of the book, the data of the third edition have been revised. The data presented in the book in accordance with GOST 8.310-78 pertain to the category of informational data.

Comments on the book may be addressed to 113114, Moscow, M-114, Shlyuzovaya nab., 10, Energoatomizdat.

NOTATIONS AND UNITS
OF PHYSICAL QUANTITIES

Temperature	t, °C; T, K
Molecular mass	μ, kg/kmol
Enthalpy	h, kJ/kg; H or μh, kJ/kmol
Internal energy	u, kJ/kg; U or μu, kJ/kmol
Entropy	s, kJ/(kg·K); S or μs, kJ/(kmol·K)
Constant-pressure heat capacity (specific heat)	c_p, kJ/(kg·K); C_p or μc_p, kJ/(kmol·K)
Constant-volume heat capacity (specific heat)	c_v, kJ/(kg·K); C_v or μc_v, kJ/(kmol·K)
Relative pressure	π, dimensionless quantity
Relative volume	θ, dimensionless quantity

In the following tables, equations, and figures, commas have been used instead of decimal points.

I. TABLES OF THERMODYNAMIC PROPERTIES OF GASES

THERMODYNAMIC PRINCIPLES OF TABLE COMPILING

In the present tables, it is assumed that the gases obey the equation of state for ideal gases

$$pV = RT \tag{1}$$

over the entire range of tabular data.

The error in this assumption, generally speaking, increases with decreasing temperature and increasing pressure. Practically, in thermal engineering calculations we frequently deal with elevated pressures in the region of high temperatures and with low pressures in the region of relatively low temperatures. In other words, in practical thermal engineering we usually deal with parameters of state that are favorable from the standpoint of accuracy of the assumed equation of state. Since the internal energy and enthalpy of a gas obeying Eq. (1) depend only on temperature, the tables can be compiled in terms of only one parameter, which greatly simplifies the calculations.

We write the principal equation of thermodynamics in the form

$$Tds = dh - vdp \tag{2}$$

or

$$Tds = c_p dT - vdp \tag{2a}$$

For an isentropic process

$$0 = c_p dT - vdp \tag{3}$$

By substituting v in the last equation by its value from Eq. (1) and separating the variables, we obtain

$$\frac{dp}{p} = \frac{\mu c_p}{R} \frac{dT}{T} \tag{3a}$$

and following integration

$$\ln \frac{p}{p_0} = \frac{\mu}{R} \int_{T_0}^{T} \frac{c_p}{T} dT \tag{4}$$

Denoting

$$\int_{T_0}^{T} \frac{c_p}{T} dT$$

by s^0, we rewrite the last expression as

$$\ln \frac{p}{p_0} = \frac{\mu s^0}{R} \tag{5}$$

If we adopt some temperature T_0 as the base temperature, then the pressure ratio $p/p_0 = \pi_0$, which is called relative pressure below, is determined as a function of temperature irrespective of the value of entropy.

The corresponding ratios of specific volumes are also functions of temperature and are determined by

$$\ln \frac{v}{v_0} = -\frac{\mu}{R} \int_{T_0}^{T} \frac{c_v dT}{T} \tag{6}$$

or

$$\theta_0 = \frac{v}{v_0} = \frac{T}{T_0} \frac{p_0}{p} \tag{7}$$

Equation (2) can be written in this form

$$ds = c_p \frac{dT}{T} - \frac{R}{\mu} \frac{dp}{p} \tag{8}$$

The value of entropy can be determined by the expression

$$s = \int_{T_0}^{T} \frac{c_p dT}{T} - \frac{R}{\mu} \ln p = s^0 - \frac{R}{\mu} \ln p \tag{9}$$

Here, entropy is calculated relative to an arbitrarily selected zero value at base temperature T_0 and pressure equal to unity.

The change of entropy between the two states is equal to

$$s_2 - s_1 = s_2^0 - s_1^0 + \frac{R}{\mu} \ln \frac{p_2}{p_1} \tag{10}$$

By analyzing Eqs. (5), (9), and (10), we can establish that the ratio of pressures p_1 and p_2 in an isentropic process, taking place from temperature T_1 to temperature T_2, is equal to the ratio of the relative pressures π_{01} and π_{02} corresponding to temperatures t_1 and t_2:

$$\left(\frac{p_1}{p_2} \right)_{s=\text{const}} = \frac{\pi_{01}}{\pi_{02}} \tag{11}$$

Similarly,

$$\left(\frac{v_1}{v_2} \right)_{s=\text{const}} = \frac{\theta_{01}}{\theta_{02}} \tag{12}$$

DESCRIPTION OF THE TABLES

The tables in this book were compiled on the basis of the relations presented above. It turned out to be expedient to give detailed tables for each component of gas mixtures in order to conduct the calculations of the processes with both individual gases and with various gas mixtures, including the combustion products of fuels of arbitrary composition. In this case, the data for gas mixtures are determined from the data for the components according to a simple mixing rule (additivity).

When analyzing the processes with combustion products we practically have to use four tables (for N_2, CO_2, O_2, and H_2O). A separate table is compiled for air. Also presented is a table of the thermodynamic properties of helium, which can be regarded as a prospective working medium in gas turbines intended for nuclear power stations.

At the basis of the calculation of the tables of thermodynamic properties of gases lie the latest values of specific heat calculated from spectroscopic data taking into account new values of the fundamental physical constants, and for water vapor and carbon dioxide—taking into account the centrifugal deformation of the molecules with their rotation.

Due to the growing application of computers in thermal engineering calculations, it is found expedient to approximate the specific heat of gases by analytical expressions and use these expressions for compiling the tables.

The values of specific heats are approximated in the computer by the least squares method in the form of a polynomial of the absolute temperature

$$\mu c_p = C_p = \sum_{n=-1}^{n-7} a_n^c \left(\frac{T}{1000} \right)^n \tag{13}$$

On the basis of the known thermodynamic relations and expression (13) for specific heats, we find the coefficients in the expressions for enthalpy and entropy

$$\mu h = H = \int c_p \, dT = \sum_{n=0}^{n-8} a_n^h \left(\frac{T}{1000} \right)^n + b_h \ln \frac{T}{1000} \tag{14}$$

$$\mu s^0 = S^0 = \int \frac{c_p}{T} \, dT = \sum_{n=-1}^{n-7} a_n^s \left(\frac{T}{1000} \right)^n + b_s \ln \frac{T}{1000} \tag{15}$$

The values of the coefficients of equations (13), (14), and (15), used in compiling the tables, are given in tables I.1, I.2, and I.3.

Tables III.1 to III.17 contain values for the thermodynamic properties of air, nitrogen, "atmospheric nitrogen" (i.e., air with the oxygen removed), oxygen, water vapor, carbon dioxide, carbon monoxide, hydrogen, and helium. For each of the above gases except helium data are presented in two tables. The first table contains values for the constant-pressure specific heats c_p, in kJ/(kg·K), and $\mu c_p = C_p$, in kJ/(kmol·K), and for the constant-volume specific heats c_v, in kJ/(kg·K), and $\mu c_v = C_v$, in kJ/(kmol·K). The last column of the table gives the adiabatic index k $= c_p/c_v$. The values are given at 25 degree intervals from $-50\,°C$ (223.15 K) to $200\,°C$ (473.15 K) and then at 50 degree intervals to $1500\,°C$ (1773.15 K).

The second table contains values for enthalpy h, in kJ/kg, the internal energy u, in kJ/kg, the dimensionless quantities relative pressure π_0 and relative volume Θ_0 and the entropy s^0 in kJ/(kg·K):

$$s^0 = \int_{T_0}^{T} \frac{c_p dT}{T} \tag{16}$$

Also, since calculations on a per mole basis are often more convenient, particularly with gaseous mixtures, this table also contains values for $\mu h = H$, in kJ/kmol, $\mu u = U$, in kJ/kmol, and $\mu s^0 = S^0$, in kJ/(kmol·K). The quantities are given at one degree intervals from $-50\,°C$ (223.15 K) to $1500\,°C$, (1773.15 K).

For helium gas, the values of c_p (5.193 kJ/(kg·K)), C_p (20.786 kJ/(kmol K), c_v (3.116 kJ/(kg·K) and C_v (12.472 kJ/(kmol·K)) are constant over the range of temperatures considered here, so only the table of values for enthalpy, internal energy, entropy, and relative pressures and volumes is given (Table III.17).

The absolute zero temperature is taken as a base temperature. The values of enthalpy and

$$s^0 = \int_{T_0}^{T} \frac{c_p dT}{T}$$

at temperature 298.15 °C are taken as standard values. The value of the universal gas constant is taken to be equal to R $= 8.31441$ kJ/(kmol·K).

In compiling the tables for air and atmospheric nitrogen, we assume the following volume composition:

For air		For atmospheric nitrogen	
Nitrogen	0.7803	Nitrogen	0.9876
Oxygen	0.2099	Argon	0.0119
Argon	0.0094	Carbon dioxide	0.0004
Hydrogen	0.0001	Hydrogen	0.0001
Carbon dioxide	0.0003		

REGION OF APPLICABILITY OF THE TABLES

The behavior of real gases and gaseous mixtures does not strictly obey the equation of state of ideal gases, which was assumed when compiling the present tables. Therefore, it is important to give an estimate of the accuracy of the

Table I.1 Coefficients of equations

Coefficient	Air	N$_2$	N$_2'$	O$_2$
a^c_{-1}	0	0	0	0
a^c_0	$+2,9438265 \cdot 10^1$	$+2,8298404 \cdot 10^1$	$+2,8151964 \cdot 10^1$	$+3,3051759 \cdot 10^1$
a^c_1	$-1,6108220 \cdot 10^0$	$+1,2689906 \cdot 10^1$	$+1,3197123 \cdot 10^1$	$-4,1834166 \cdot 10^1$
a^c_2	$-1,1991744 \cdot 10^1$	$-7,2418092 \cdot 10^1$	$-7,4482113 \cdot 10^1$	$+1,4802410 \cdot 10^2$
a^c_3	$+6,8828384 \cdot 10^1$	$+1,8536290 \cdot 10^2$	$+1,8998363 \cdot 10^2$	$-2,0502229 \cdot 10^2$
a^c_4	$-9,8239929 \cdot 10^1$	$-2,2042323 \cdot 10^2$	$-2,2661680 \cdot 10^2$	$+1,4536800 \cdot 10^2$
a^c_5	$+6,4883505 \cdot 10^1$	$+1,3735517 \cdot 10^2$	$+1,4204175 \cdot 10^2$	$-5,2290720 \cdot 10^1$
a^c_6	$-2,0909380 \cdot 10^1$	$-4,3809407 \cdot 10^1$	$-4,5640429 \cdot 10^1$	$+7,5770768 \cdot 10^0$
a^c_7	$+2,6652402 \cdot 10^0$	$+5,6619528 \cdot 10^0$	$+5,9487537 \cdot 10^0$	0

proposed tables in various regions of state parameters. For such an estimate, it is logical to use an equation of state for real gases which reflects as far as possible the behavior of these gases in the region of state parameters of interest to us.

Analysis of numerous experimental data on the compressibility of pure gases shows that, under moderate pressures (approximately up to 5 MPa) and temperatures remote from the critical, the isotherms in the coordinate system $p - pv$ are practically linear within a wide range of temperatures (Fig. 1).

The equation of state of the gas for this region can be written as

$$pv = \frac{R}{\mu} T + Bp \tag{17}$$

where B is a function of temperature, the so-called second virial coefficient. Knowing the temperature dependence of the second virial coefficient B for the given gas, the pressure dependence of the caloric quantities for this gas can be determined. It is easily shown that the effect of pressure on enthalpy of the gas will be linear, i.e.,

$$h - h_0 = B_1 p \tag{18}$$

$$\mu c_p = C_p = \sum_{n=-1}^{n=7} a_n^c \left(\frac{T}{1000}\right)^n$$

CO_2	H_2O	CO	H_2
0	$+7{,}3147600 \cdot 10^{-1}$	0	$-1{,}7412059 \cdot 10^0$
$+1{,}7640049 \cdot 10^1$	$+2{,}7885805 \cdot 10^1$	$+2{,}9161791 \cdot 10^1$	$+4{,}0010590 \cdot 10^1$
$+9{,}3726944 \cdot 10^1$	$+8{,}4430197 \cdot 10^0$	$+4{,}1350040 \cdot 10^0$	$-2{,}3707024 \cdot 10^1$
$-1{,}3037466 \cdot 10^2$	$+1{,}1985297 \cdot 10^1$	$-4{,}3774782 \cdot 10^1$	$+1{,}9219936 \cdot 10^1$
$+1{,}5397055 \cdot 10^2$	$-1{,}6092233 \cdot 10^1$	$+1{,}4623050 \cdot 10^2$	$+1{,}4567966 \cdot 10^0$
$-1{,}3999603 \cdot 10^2$	$+1{,}3636273 \cdot 10^1$	$-1{,}9532958 \cdot 10^2$	$-8{,}5065648 \cdot 10^0$
$+8{,}3151862 \cdot 10^1$	$-6{,}4729000 \cdot 10^0$	$+1{,}3090545 \cdot 10^2$	$+4{,}1386238 \cdot 10^0$
$-2{,}7578508 \cdot 10^1$	$+1{,}1891256 \cdot 10^0$	$-4{,}4090211 \cdot 10^1$	$-6{,}6190591 \cdot 10^{-1}$
$+3{,}8298136 \cdot 10^0$	0	$+5{,}9600455 \cdot 10^0$	0

where h is the enthalpy of the real gas; h_0 is the enthalpy of the gas in the ideal-gas state (when $p \to 0$); and B_1 is the temperature function:

$$B_1 = B - T \frac{dB}{dT} \tag{19}$$

Following the determination of the effect of pressure on enthalpy of the gas thus, we estimate the region of pressures and temperatures in which this effect will be felt to a certain degree. By specifying the allowable error of the tables at 0.5% ($h - h_0 = 0.005\,h_0$), i.e., a value of the same order as the error of water vapor tables, we can find from Eq. (18) for individual gases the limit pressure, as a function of temperature, at which the error of the tables does not exceed the indicated value.

Figure 2 shows the values of the limit pressures, plotted on the ordinate, for nitrogen, oxygen, carbon dioxide, and water vapor as a function of temperature. An example of using this graph is also shown in the figure.

For a gaseous mixture, an equation of state in the form of (17) can be used as a first approximation, and the specific volume of the mixture can be determined from the mixing rule in terms of the specific volumes of the components.

Table I.2 Coefficients of equations

Coefficient	Air	N_2	N_2	O_2
a_0^h	$-5,4200000 \cdot 10^1$	$+2,5450000 \cdot 10^1$	$+4,0100000 \cdot 10^1$	$-3,0030000 \cdot 10^2$
a_1^h	$+2,9438265 \cdot 10^4$	$+2,8298404 \cdot 10^4$	$+2,8151964 \cdot 10^4$	$+3,3051759 \cdot 10^4$
a_2^h	$-8,0541099 \cdot 10^2$	$+6,3449526 \cdot 10^3$	$+6,5985613 \cdot 10^3$	$-2,0917083 \cdot 10^4$
a_3^h	$-3,9972481 \cdot 10^3$	$-2,4139364 \cdot 10^4$	$-2,4827371 \cdot 10^4$	$+4,9341367 \cdot 10^4$
a_4^h	$+1,7207096 \cdot 10^4$	$+4,6340725 \cdot 10^4$	$+4,7495900 \cdot 10^4$	$-5,1255572 \cdot 10^4$
a_5^h	$-1,9647986 \cdot 10^4$	$-4,4084647 \cdot 10^4$	$-4,5323358 \cdot 10^4$	$+2,9073599 \cdot 10^4$
a_6^h	$+1,0813917 \cdot 10^4$	$+2,2892528 \cdot 10^4$	$+2,3673626 \cdot 10^4$	$-8,7151202 \cdot 10^3$
a_7^h	$-2,9870543 \cdot 10^3$	$-6,2584868 \cdot 10^3$	$-6,5200710 \cdot 10^3$	$+1,0824395 \cdot 10^3$
a_8^h	$+3,3315502 \cdot 10^2$	$+7,0774408 \cdot 10^2$	$+7,4359419 \cdot 10^2$	0
b_h	0	0	0	0

In this case, the effect of pressure on the enthalpy of the gaseous mixture will also be linear:

$$(h - h_0)_{\text{mix}} = B_{1\,\text{mix}} p \tag{20}$$

$B_{1\,\text{mix}}$ being determined from the values of B_1 of the components in conformity with the additivity rule. Thus, the limit pressure, for which the error of the tables does not exceed the stipulated 0.5%, can also be estimated for mixtures.

Calculations based on these relations indicate that the proposed tables are sufficiently accurate in the pressure region up to 2.5–3 MPa for all practically used temperatures in gas turbine engineering.

It is pertinent to mention that comparison of the heat drop obtained by the existing, reliable, and experimentally substantiated tables of water vapor and carbon dioxide with the corresponding values obtained from the present tables has shown their good agreement (within the bounds of the estimated region of applicability of the tables). At pressures up to the limits shown in Fig. 2, the actual adiabatic heat drop never differed from the heat drop obtained from the

$$\mu h = H = \sum_{n=0}^{n=8} a_n^h \left(\frac{T}{1000} \right)^n + b_h \ln \frac{T}{1000}$$

CO_2	H_2O	CO	H_2
$+8,4740000 \cdot 10^2$	$+2,0129470 \cdot 10^3$	$-3,2540000 \cdot 10^1$	$-4,6847200 \cdot 10^3$
$+1,7640049 \cdot 10^4$	$+2,7885805 \cdot 10^4$	$+2,9161791 \cdot 10^4$	$+4,0010590 \cdot 10^4$
$+4,6863472 \cdot 10^4$	$+4,2215098 \cdot 10^3$	$+2,0675020 \cdot 10^3$	$-1,1853512 \cdot 10^4$
$-4,3458218 \cdot 10^4$	$+3,9950990 \cdot 10^3$	$-1,4591594 \cdot 10^4$	$+6,4066451 \cdot 10^3$
$+3,8492636 \cdot 10^4$	$-4,0230580 \cdot 10^3$	$+3,6557624 \cdot 10^4$	$+3,6419916 \cdot 10^2$
$-2,7999206 \cdot 10^4$	$+2,7272546 \cdot 10^3$	$-3,9065915 \cdot 10^4$	$-1,7013130 \cdot 10^3$
$+1,3858643 \cdot 10^4$	$-1,0788167 \cdot 10^3$	$+2,1817574 \cdot 10^4$	$+6,8977065 \cdot 10^2$
$-3,9397868 \cdot 10^3$	$+1,6987509 \cdot 10^2$	$-6,2986014 \cdot 10^3$	$-9,4557986 \cdot 10^1$
$+4,7872667 \cdot 10^2$	0	$+7,4500568 \cdot 10^2$	0
0	$+7,3147600 \cdot 10^2$	0	$-1,7412059 \cdot 10^3$

tables by more than 0.5%, which is a good verification of the reliability of the adopted method of estimating the region of applicability of the tables. The estimates of accuracy were conducted for triatomic gases due to the fact that they are characterized by the largest deviation from the equation of state for ideal gases in the temperature region under consideration.

EXAMPLES OF USE OF THE TABLES

Example 1. Compression of air in a compressor. Air is sucked into the compressor at pressure $p_1 = 0.1$ MPa and temperature of 20 °C and compressed to pressure 0.4 MPa. Determine the theoretical work of the adiabatic compression, actual compression work for an internal relative efficiency of the compressor $\eta_{oi} = 0.85$, temperature of air at the end of compression, and entropy change.

Table I.3 Coefficients of equations

Coefficient	Air	N,	N$_2$	O,
a^s_{-1}	0	0	0	0
a^s_0	$+2,3017630 \cdot 10^2$	$+2,2391890 \cdot 10^2$	$+2,2322190 \cdot 10^2$	$+2,5249950 \cdot 10^2$
a^s_1	$-1,6108220 \cdot 10^0$	$+1,2689906 \cdot 10^1$	$+1,3197123 \cdot 10^1$	$-4,1834166 \cdot 10^1$
a^s_2	$-5,9958719 \cdot 10^0$	$-3,6209046 \cdot 10^1$	$-3,7241056 \cdot 10^1$	$+7,4012048 \cdot 10^1$
a^s_3	$+2,2942794 \cdot 10^1$	$+6,1787635 \cdot 10^1$	$+6,3327875 \cdot 10^1$	$-6,8340764 \cdot 10^1$
a^s_4	$-2,4559982 \cdot 10^1$	$-5,5105807 \cdot 10^1$	$-5,6654199 \cdot 10^1$	$+3,6342000 \cdot 10^1$
a^s_5	$+1,2976701 \cdot 10^1$	$+2,7471033 \cdot 10^1$	$+2,8408351 \cdot 10^1$	$-1,0458144 \cdot 10^1$
a^s_6	$-3,4848967 \cdot 10^0$	$-7,3015678 \cdot 10^0$	$-7,6067499 \cdot 10^0$	$+1,2628462 \cdot 10^0$
a^s_7	$+3,8074860 \cdot 10^{-1}$	$+8,0885040 \cdot 10^{-1}$	$+8,4982197 \cdot 10^{-1}$	0
b_s	$+2,9438205 \cdot 10^1$	$+2,8298404 \cdot 10^1$	$+2,8151964 \cdot 10^1$	$+3,3051759 \cdot 10^1$

Solution. From Table III.2 for $t_1 = 20\,°C$ we find $\pi_{01} = 1.2784$, $h_1 = 293.39$ kJ/kg and $s^0_1 = 6.6812$ kJ/(kg·K), where subscript 1 is ascribed to the data of the sucked air. For the determination of the final state of the air following adiabatic compression, we find the relative pressure corresponding to the state of the gas at the end of compression

$$\frac{p_2}{p_1} = \frac{0,4}{0,1} = 4 \qquad \pi_{02s} = 4 \cdot 1,2784 = 5,1136$$

From Table III.2 by interpolation we determine the values $h_{2s} = 436.48$ kJ/kg, $s^{\theta}_{2s} = 7.0792$ kJ/kg·K), and $t_{2s} = 161.70\,°C$ corresponding to π_{02s}.
Theoretical compression work

$$\Delta h_{ad} = h_{2s} - h_1 = 436,48 - 293,39 = 143,1 \text{ kJ/kg}$$

Actual compression work when $\eta_{oh} = 0.85$

$$\Delta h_{act} = h_2 - h_1 = \frac{143,1}{0.85} = 168,3 \text{ kJ/kg}$$

$$\mu s^0 = S^0 = \sum_{n=-1}^{n=7} a_n^s \left(\frac{T}{1000}\right)^n + b_s \ln \frac{T}{1000}$$

CO₂	H₂O	CO	H₂
0	$-7{,}3147600 \cdot 10^{-1}$	0	$+1{,}7412059 \cdot 10^0$
$+2{,}1171710 \cdot 10^2$	$+2{,}2197830 \cdot 10^2$	$+2{,}3294050 \cdot 10^2$	$+1{,}7938530 \cdot 10^2$
$+9{,}3726944 \cdot 10^1$	$+8{,}4430197 \cdot 10^0$	$+4{,}1350040 \cdot 10^0$	$-2{,}3707024 \cdot 10^1$
$-6{,}5187329 \cdot 10^1$	$+5{,}9926485 \cdot 10^0$	$-2{,}1887391 \cdot 10^1$	$+9{,}6099679 \cdot 10^0$
$+5{,}1323515 \cdot 10^1$	$-5{,}3640779 \cdot 10^0$	$+4{,}8743499 \cdot 10^1$	$+4{,}8559888 \cdot 10^{-1}$
$-3{,}4999006 \cdot 10^1$	$+3{,}4089868 \cdot 10^0$	$-4{,}8832394 \cdot 10^1$	$-2{,}1266412 \cdot 10^0$
$+1{,}6630372 \cdot 10^1$	$-1{,}2945800 \cdot 10^0$	$+2{,}6181090 \cdot 10^1$	$+8{,}2772475 \cdot 10^{-1}$
$-4{,}5964181 \cdot 10^0$	$+1{,}4818760 \cdot 10^{-1}$	$-7{,}3483677 \cdot 10^0$	$-1{,}1031765 \cdot 10^{-1}$
$+5{,}4711620 \cdot 10^{-1}$	0	$+8{,}5143506 \cdot 10^{-1}$	0
$+1{,}7640049 \cdot 10^1$	$+2{,}7885805 \cdot 10^{-1}$	$+2{,}4161791 \cdot 10^1$	$+4{,}0010590 \cdot 10^1$

We now find the state parameters of air at the end of the actual compression process.

The enthalpy of air at the end of compression

$$h_2 = h_1 + \Delta h_{\text{act}} = 293{,}39 + 168{,}3 = 461{,}7 \text{ kJ/kg}$$

From Table III.2 we find

$$t_2 = 186{,}44 \ °C \ (h_2 = 461{,}7 \text{ kJ/kg})$$

$$s_2^0 = 7{,}1357 \text{ kJ/kg}$$

The change of entropy can be determined from the relation

$$s_2 - s_1 = -\frac{R}{\mu} \ln \frac{p_2}{p_1} + s_2^0 - s_1^0$$

or

$$s_2 - s_1 = s_2^0 - s_{2s}^0$$

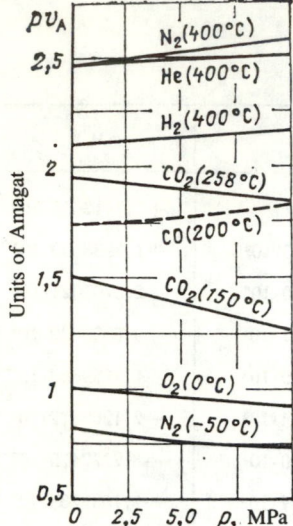

Figure 1

$$s_2 - s_1 = 7{,}1357 - 7{,}0792 = 0{,}0565 \text{ kJ/(kg·K)}$$

Example 2. Gaseous product of fuel combustion having the following composition in volume fractions:

$r \, (CO_2) = 7{,}5\% \qquad r \, (O_2) = 10\%$

$r \, (N_2) = 76{,}5\% \qquad$ and $\qquad r(H_2O) = 6{,}0\%$

does work by expanding from $p_1 = 0.5$ MPa to $p_2 = 0.1$ MPa, corresponding to temperatures $t_1 = 500\,°C$ and $t_2 = 290\,°C$.

Determine the expansion work and internal relative efficiency of the expansion process.

Solution. From Table III.2–III.16 we determine the enthalpy of the mixture prior to expansion corresponding to the initial temperature $t_1 = 500\,°C$:

$H_1 \, (CO_2) = 30804 \text{ kJ/kmol} \qquad \pi_{01} \, (CO_2) = 22.422$

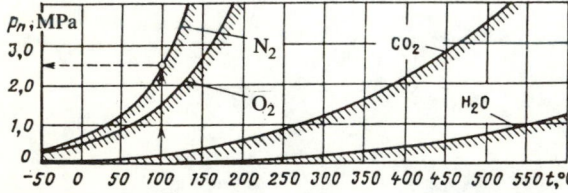

Figure 2

$H_1 \; (H_2O) = 26868 \; kJ/kmol$ $\pi_{01} \; (H_2O) = 41.340$

$H_1 \; (O_2) = 23588 \; kJ/kmol$ $\pi_{01} \; (O_2) = 181.13$

$H_1 \; (N_2) = 22867 \; kJ/kmol$ $\pi_{01} \; (N_2) = 30.393$

$H_{mix} = 0.075 \times 30804 + 0.060 \times 26868 + 0.10 \times 23588 + 0.765$

$\times \; 22867 = 23774 \; kJ/kmol$

We now determine from the tables the values of the enthalpy at the end of the adiabatic expansion for each component taken separately:

$\pi_{02s}(CO_2) = 22.422/5 = 4.4844$ $H_{2s} = 21732 \; kJ/kmol$

    $t_{2s} = 315.43 \,°C$

$\pi_{02s}(H_2O) = 41.340/5 = 8.268$ $H_{2s} = 18170 \; kJ/kmol$

    $t_{2s} = 264.91 \,°C$

$\pi_{02s}(O_2) = 181.13/5 = 36.226$ $H_{2s} = 15093 \; kJ/kmol$

    $t_{2s} = 238.15 \,°C$

$\pi_{02s}(N_2) = 30.393/5 = 6.0786$ $H_{2s} = 14486 \; kJ/kmol$

    $t_{2s} = 223.90 \,°C$

Approximately, $H_{2s \, mix} = 21732 \times 0.075 + 18170 \times 0.060 + 15093 \times 0.100 + 14486 \times 0.765 = 15311 \; kJ/kmol$.
    The theoretical work of expansion

$$\Delta H_T = H_{1 \, mix} - H_{2s \, mix} = 23774 - 15311 = 8463 \; kJ/kmol$$

We calculate the actual work of expansion, given the temperature of the gas mixture at the end of the expansion. For this, we find the enthalpies of the components at temperature $t_2 = 290 \,°C$ (at the end of the actual expansion process) from Tables III.4, III.8, III.10, and III.12, and multiply them by their corresponding mole fractions:

$H_2 \; (CO_2) = 20544 \times 0.075 = 1541 \; kJ/kmol$

$H_2 \; (H_2O) = 19068 \times 0.060 = 1144 \; kJ/kmol$

H_2 (O_2) = 16725 × 0.100 = 1672 kJ/kmol

H_2 (N_2) = 16452 × 0.765 = 12586 kJ/kmol

Using the simple mixing rule

$H_{2s\,mix}$ = 16943 kJ/kmol

The actual work of expansion

$\Delta H_{act} = H_{1s\,mix} - H_{2s\,mix}$ = 23774 − 16943 = +6831 kJ/kmol

The relative efficiency of the expansion process

$\eta_{0h} = \Delta H_{act}/\Delta H_T$ = 6831/8463 = 0.807

For a more exact solution of the problem, we must determine the value of S^0 corresponding to the state of the gas mixture at the end of the isentropic process. This value is determined from Eq. (10):

$S_{2s}^0 = S_1^0 - R \ln (p_1/p_2)$

From the tables for the individual components we find S_1^0 for the mixture at 500 °C using the mixing rule:

$S_1^0(CO_2)$ = 255.59 kJ/(kmol·K) $S_1^0(O_2)$ = 234.68 kJ/(kmol·K)

$S_1^0(H_2O)$ = 222.39 kJ/(kmol·K) $S_1^0(N_2)$ = 219.83 kJ/(kmol·K)

$S_{1\,mix}^0$ = 255.59 × 0.075 + 222.39 × 0.060 + 234.68 × 0.100

+ 219.83 × 0.765 = 224.15 kJ(kmol·K)

Thus at the end of the expansion process

$S_{2\,mix}^0$ = 224.15 − 8.31441 × ln 5 = 210.77 kJ/(kmol·K)

The problem is now reduced to determining the temperature of the gaseous mixture which corresponds to the above value of $S_{2\,mix}^0$. For this purpose we first assume a final temperature of the expansion process close to the final temperature of the principal component—in this case nitrogen (see the approximate solution above). For convenience in interpolation, we round the approximate temperature to the nearest ten degrees upward, considering that the final temperatures of the other gases upon separate expansion are higher than that for nitrogen. Thus we assume $t_2' = 230\,°C$. From the tables for the components we find for this temperature, using the mixing rule

$$S_{2\,mix}^{0'} = 235.02 \times 0.075 + 206.63 \times 0.06 + 220.79 \times 0.10$$

$$+ 206.81 \times 0.765 = 210.31 \text{ kJ/(kmol·K)}$$

Since the resulting value $S_2^{0'} < S_2^0$ the second trial temperature is assumed as 10 degrees higher, i.e., $t_{2s}'' = 240\,°C$. For this temperature we find $S_{2\,mix}'' = 210.93$ kJ/(kmol·K). We now find by linear interpolation that the temperature of the gas mixture at the end of the expansion process is

$$t_2 = 237.4\,°C$$

and at this temperature the molar enthalpy of the mixture at the end of the expansion is

$$H_{2s\,mix} = 18141.7 \times 0.075 + 17194.0 \times 0.06 + 15069.8 \times 0.10$$

$$+ 14886.0 \times 0.765 = 15287.0 \text{ kJ/kmol}$$

The theoretical expansion work (in this more exact calculation) equals $\Delta H_T = H_{1\,mix} - H_{2s\,mix} = 23774 - 15287 = 8487$ kJ/kmol. This is 24 kJ/kmol or 0.28 % greater than the value obtained in the more approximate calculation.

The internal relative efficiency of the expansion process

$$\eta_{0h} = 6831/8487 = 0.805$$

i.e., about 0.25 % less than in the first calculation.

II. THERMAL DIAGRAM FOR AIR AND PRODUCTS OF COMBUSTION OF A GIVEN FUEL

METHOD OF CONSTRUCTING THE DIAGRAM

In practical calculations, it is often more convenient to use not only tables, but also diagrams. Such diagrams can be constructed using the same relations that were applied in the compilation of the tables.

For the construction of the diagram for air (Fig. 3), temperature is plotted on the horizontal axis (for example, in degrees Celsius), and on the vertical axis enthalpy H and the logarithm of the relative pressure lg π_0. Then we plot the curves of the temperature dependent quantities H, lg π_0, and RT for a mol of air. From such a diagram, one can make the same calculations as from H, S-diagrams for air, the former being superior due to its simplicity: instead of a large number of curves plotted on the H, S-diagrams, we have here two main curves and one auxiliary straight line.

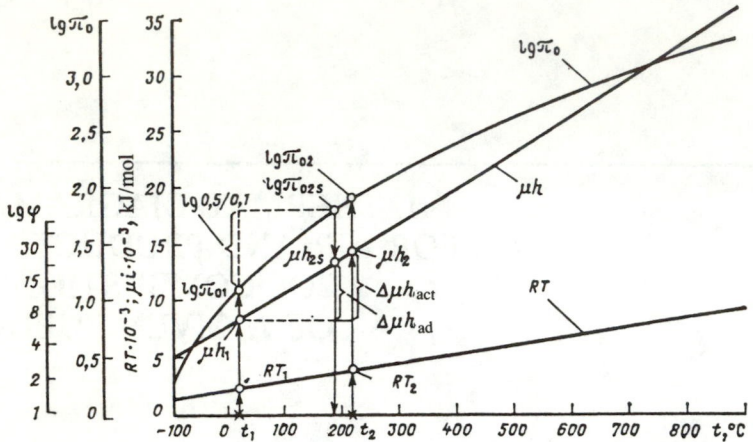

Figure 3

For the determination of the adiabatic heat drop only an auxiliary scale, on which the logarithms of the degree of pressure change in the adiabatic process $\lg \varphi$ is plotted, is required. It must be noted that the curve of the logarithm of relative pressures $\lg \pi_0$ can be used for the determination of entropy change in the real process. As a matter of fact, for the isentropic process Eq. (5) is true:

$$\ln \pi_0 = \frac{\mu s^0}{R} \quad \text{and} \quad \lg \pi_0 = \frac{\mu s^0}{R} \lg e = \frac{S^0}{R} \lg e$$

where e is the base of natural logarithms; $\lg e = 0.434294$; and R is the molar gas constant equal to 8.31441 kJ/(kmol·K).

Consequently

$$S^0 = \frac{R}{\lg e} \lg \pi_0 = \frac{8{,}31441}{0{,}434294} \lg \pi_0 = 19{,}1447 \lg \pi_0 \tag{21}$$

For the isentropic process we have from Eq. (10)

$$0 = R \ln \frac{p_2}{p_1} + S^0_{2s} - S^0_1 \tag{22}$$

where S^0_{2s} is the value of S^0 at a gas temperature corresponding to the end of the isentropic process; and S^0_1 is the value of S^0 at a gas temperature corresponding to the beginning of the isentropic process.

For a real adiabatic process taking place under the same pressures p_1 and p_2 and at the same initial temperature

$$S_2 - S_1 = R \ln \frac{p_2}{p_1} + S_2^0 - S_1^0 \tag{23}$$

where S_2^0 is the value of S^0 at a gas temperature corresponding to the end of the real process.

Subtracting Eq. (22) from Eq. (23), we obtain

$$S_2 - S_1 = S_2^0 - S_{2s}^0 \tag{24}$$

i.e., the change in entropy in the real process is equal to the difference of the values of S^0 corresponding to the gas temperature at the end of the real and isentropic processes (for the same pressure ratios).

Following the substitution of the values of S^0 from Eq. (21), the last equality assumes the form

$$S_2 - S_1 = 19{,}1447 \,(\lg \pi_{02} - \lg_{02s}^h) \tag{25}$$

i.e., change of entropy is determined in terms of the values of the logarithm of relative pressure.

For the purpose of constructing a similar diagram for the products of combustion of a given fuel, one must calculate the volume composition of the products of combustion from the composition of this fuel at an excess air coefficient $\alpha = 1$. Then, using the values of enthalpy and the logarithm of the relative pressure for components N_2, CO_2, H_2O, and O_2, which are presented in Tables II.1–II.6, one must determine $H = \mu h$ and $\lg \pi_0$ from the mixing rule for the products of the computed composition at various temperatures with an interval of 50–100 °C.

Given the values of enthalpy and the logarithm of the relative pressure for air (Table II.1) and for the combustion products of a fuel without excess air (Table II.7), one can plot on one diagram the curves $H = \mu h$ and $\lg \pi_0$ as a function of temperature for these two gaseous mixtures (air and combustion products)[1].

We shall now consider the gaseous products of combustion of a specified fuel with any excess air as a mixture of gaseous products of combustion of the fuel at $\alpha = 1$ and moist air. In this case, assuming the values of α (e.g., 2, 3, 4, 5), one can determine the volume fraction of this combustion product at $\alpha = 1$ and of air in the given mixture. Such a computation is very simple.

Once we determine the theoretical amount of air necessary for the combustion of fuel L_0 and the volume of water vapor in this amount of air and the total volume of combustion products V^0 for $\alpha = 1$, it is easy to find the composition of gases for any value of α:

---

[1]When constructing the diagram, it is more convenient to plot on scale $\lg \pi_0$ the value $\lg \pi_0 - 9$.

TABLES OF CALORIC QUANTITIES OF GASES

Table II.1 Air (MW = 28.97)

i	T	h	u	s^0	H	U	S^0	$\lg \pi_0$
−50	223,15	223,11	159,07	6,4054	6463,5	4608,2	185,564	9,6930
0	273,15	273,23	194,84	6,6080	7915,4	5644,4	191,435	9,9997
50	323,15	323,42	230,68	6,7768	9369,4	6682,7	196,323	10,2550
100	373,15	373,79	266,70	6,9217	10 828,6	7726,2	200,522	10,4743
150	423,15	424,43	302,99	7,0491	12 295,8	8777,7	204,211	10,6670
200	473,15	475,44	339,65	7,1630	13 773,6	9839,7	207,512	10,8394
250	523,15	526,90	376,75	7,2664	15 264,2	10 914,6	210,506	10,9959
300	573,15	578,85	414,36	7,3612	16 769,3	12 004,0	213,254	11,1394
400	673,15	684,45	491,26	7,5310	19 828,5	14 231,8	218,172	11,3963
500	773,15	792,43	570,54	7,6805	22 956,6	16 528,5	222,504	11,6226
600	873,15	902,78	652,19	7,8147	26 153,6	18 894,0	226,392	11,8256
700	973,15	1015,35	736,07	7,9367	29 414,8	21 323,8	229,927	12,0103
800	1073,15	1129,91	821,92	8,0488	32 733,5	23 811,1	233,173	12,1798
900	1173,15	1246,19	909,50	8,1524	36 102,2	26 348,4	236,174	12,3366
1000	1273,15	1363,97	998,58	8,2487	39 514,3	28 929,1	238,965	12,4824
1100	1373,15	1483,06	1088,96	8,3387	42 964,2	31 547,5	241,573	12,6186
1200	1473,15	1603,31	1180,52	8,4233	46 447,8	34 199,7	244,022	12,7466
1300	1573,15	1724,60	1273,10	8,5029	49 961,7	36 882,2	246,330	12,8671
1400	1673,15	1846,84	1366,65	8,5783	53 503,1	39 592,2	248,512	12,9811
1500	1773,15	1969,92	1461,02	8,6497	57 068,7	42 326,4	250,582	13,0892

Table II.2 Oxygen (MW = 32.0)

t	T	h	u	s^0	H	U	S^0	$\lg \pi_0$
−50	223,15	202,18	144,20	6,1438	6469,8	4614,5	196,600	10,2695
0	273,15	247,75	176,78	6,3280	7928,1	5657,1	202,496	10,5774
50	323,15	293,60	209,64	6,4821	9395,1	6708,4	207,428	10,8350
100	373,15	339,95	242,99	6,6155	10 878,3	7775,8	211,694	11,0579
150	423,15	386,95	277,01	6,7336	12 382,4	8864,3	215,477	11,2555
200	473,15	434,70	311,77	6,8403	13 910,4	9976,6	218,889	11,4337
250	523,15	483,24	347,31	6,9378	15 463,6	11 114,1	222,009	11,5967
300	573,15	532,57	383,65	7,0278	17 042,2	12 276,9	224,891	11,7472
400	673,15	633,50	458,60	7,1901	20 271,9	14 675,2	230,083	12,0184
500	773,15	737,14	536,26	7,3336	23 588,4	17 160,3	234,675	12,2583
600	873,15	843,05	616,19	7,4624	26 977,7	19 718,2	238,797	12,4736
700	973,15	950,83	697,98	7,5792	30 426,4	22 335,3	242,536	12,6689
800	1073,15	1060,12	781,30	7,6861	33 924,0	25 001,6	245,957	12,8476
900	1173,15	1170,71	865,91	7,7847	37 462,8	27 709,0	249,109	13,0123
1000	1273,15	1282,43	951,64	7,8760	41 037,7	30 452,5	252,033	13,1650
1100	1373,15	1395,15	1038,38	7,9613	44 644,8	33 228,2	254,761	13,3075
1200	1473,15	1508,77	1126,01	8,0411	48 280,5	36 032,5	257,316	13,4410
1300	1573,15	1623,17	12 14,43	8,1163	51 941,5	38 862,0	259,721	13,5666
1400	1673,15	1738,29	1303,56	8,1872	55 625,2	41 714,3	261,991	13,6852
1500	1773,15	1854,16	1393,45	8,2545	59 333,1	44 590,9	264,143	13,7977

Table II.3 Carbon dioxide ($\mu = 44.01$)

t	T	h	u	s^0	H	U	S^0	$\lg \pi_0$
−50	223,15	152,60	110,44	4,6220	6716,0	4860,7	203,420	10,6257
0	273,15	192,08	140,48	4,7815	8453,7	6182,6	210,437	10,9922
50	323,15	234,27	173,22	4,9232	10 310,5	7623,8	216,675	11,3181
100	373,15	278,90	208,40	5,0515	12 274,5	9172,0	222,322	11,6130
150	423,15	325,71	245,78	5,1692	14 335,0	10 816,8	227,502	11,8836
200	473,15	374,52	285,13	5,2782	16 482,9	12 549,0	232,298	12,1341
250	523,15	425,12	326,29	5,3798	18 710,0	14 360,4	236,771	12,3678
300	573,15	477,36	369,08	5,4752	21 009,1	16 243,8	240,967	12,5870
400	673,15	586,17	459,00	5,6500	25 797,7	20 201,0	248,663	12,9890
500	773,15	699,92	553,86	5,8075	30 804,0	24 375,9	255,594	13,3510
600	873,15	817,77	652,81	5,9508	35 990,9	28 731,4	261,901	13,6805
700	973,15	939,04	755,19	6,0823	41 328,3	33 237,3	267,687	13,9827
800	1073,15	1063,18	860,44	6,2037	46 791,7	37 869,3	273,030	14,2618
900	1173,15	1189,74	968,11	6,3164	52 361,9	42 608,1	277,992	14,5210
1000	1273,15	1318,38	1077,86	6,4216	58 023,3	47 438,1	282,623	14,7629
1100	1373,15	1448,81	1189,39	6,5203	63 763,6	52 347,0	286,963	14,9896
1200	1473,15	1580,78	1302,47	6,6130	69 571,9	57 323,8	291,046	15,2028
1300	1573,15	1714,08	1416,88	6,7006	75 438,4	62 358,9	294,898	15,4041
1400	1673,15	1848,50	1532,41	6,7834	81 354,4	67 443,5	298,544	15,5945
1500	1773,15	1983,92	1648,94	6,8620	87 314,4	72 572,1	302,004	15,7752

Table II.4 Water vapor

t	T	h	u	s^0	H	U	S^0	$\lg \pi_0$
−50	223,15	409,89	306,91	9,9359	7384,5	5529,2	179,005	9,3504
0	273,15	502,80	376,74	10,3116	9058,4	6787,4	185,773	9,7039
50	323,15	596,03	446,90	10,6250	10 738,2	8051,4	191,420	9,9989
100	373,15	690,04	517,84	10,8954	12 431,8	9329,4	196,292	10,2534
150	423,15	785,11	589,83	11,1345	14 144,6	10 626,4	200,599	10,4784
200	473,15	881,43	663,07	11,3496	15 879,8	11 945,9	204,475	10,6808
250	523,15	979,14	737,71	11,5459	17 640,1	13 290,5	208,011	10,8655
300	573,15	1078,33	813,83	11,7270	19 427,3	14 662,0	211,273	11,0359
400	673,15	1281,50	970,84	12,0536	23 087,4	17 490,7	217,157	11,3432
500	773,15	1491,34	1134,53	12,3441	26 868,0	20 439,9	222,391	11,6167
600	873,15	1708,12	1305,16	12,6077	30 773,5	23 514,0	227,140	11,8647
700	973,15	1931,95	1482,84	12,8503	34 806,1	26 715,1	231,511	12,0931
800	1073,15	2162,85	1667,59	13,0761	38 966,0	30 043,6	235,579	12,3056
900	1173,15	2400,72	1859,31	13,2880	43 251,4	33 497,6	239,397	12,5050
1000	1273,15	2645,36	2057,80	13,4881	47 658,8	37 073,6	243,001	12,6932
1100	1373,15	2896,46	2262,75	13,6779	52 182,6	40 765,9	246,421	12,8719
1200	1473,15	3153,62	2473,76	13,8587	56 815,5	44 567,5	249,678	13,0420
1300	1573,15	3416,38	2690,38	14,0312	61 549,5	48 470,0	252,787	13,2044
1400	1673,15	3684,29	2912,13	14,1963	66 376,1	52 465,2	255,761	13,3597
1500	1773,15	3956,95	3138,64	14,3546	71 288,3	56 546,0	258,612	13,5087

Table II.5 Atmospheric nitrogen ($\mu = 28.15$)

t	T	h	u	s^0	H	U	S^0	$\lg \pi_0$
−50	223,15	229,85	163,94	6,4890	6470,4	4615,0	182,665	9,5416
0	273,15	281,38	200,70	6,6973	7920,8	5649,7	188,530	9,8479
50	323,15	332,91	237,47	6,8706	9371,5	6684,8	193,407	10,1027
100	373,15	384,53	274,31	7,0191	10 824,4	7722,0	197,587	10,3210
150	423,15	436,30	311,32	7,1493	12 281,7	8763,6	201,252	10,5125
200	473,15	488,32	348,57	7,2655	13 746,1	9812,2	204,523	10,6833
250	523,15	540,68	386,17	7,3707	15 220,1	10 870,6	207,484	10,8380
300	573,15	593,47	424,18	7,4670	16 706,1	11 940,8	210,197	10,9797
400	673,15	700,57	501,75	7,6392	19 721,0	14 124,3	215,044	11,2329
500	773,15	809,97	581,6!	7,7907	22 800,8	16 372,6	219,309	11,4556
600	873,15	921,78	663,89	7,9267	25 948,0	18 688,5	223,136	11,6556
700	973,15	1035,88	748,45	8,0504	29 160,2	21 069,2	226,618	11,8375
800	1073,15	1152,09	835,12	8,1640	32 431,3	23 508,9	229,818	12,0046
900	1173,15	1270,15	923,64	8,2692	35 754,6	26 000,8	232,778	12,1592
1000	1273,15	1389,83	1013,79	8,3671	39 123,7	28 538,5	235,534	12,3032
1100	1373,15	1510,95	1105,38	8,4587	42 533,3	31 116,6	238,112	12,4378
1200	1473,15	1633,35	1198,24	8,5447	45 978,8	33 730,8	240,534	12,5643
1300	1573,15	1756,88	1292,24	8,6258	49 456,2	36 376,8	242,817	12,6836
1400	1673,15	1881,39	1387,21	8,7026	52 961,2	39 050,3	244,977	12,7965
1500	1773,15	2006,76	1483,04	8,7753	56 490,2	41 747,9	247,026	12,9036

Table II.6 Carbon monoxide ($\mu = 28.011$)

t	T	h	u	s^0	H	U	S^0	$\lg \pi_0$
−50	223,15	231,60	165,36	6,7640	6487,3	4632,0	189,467	9,8969
0	273,15	283,54	202,47	6,9741	7942,4	5671,3	195,351	10,2042
50	323,15	335,52	239,60	7,1488	9398,2	6711,5	200,245	10,4599
100	373,15	387,63	276,86	7,2987	10 857,9	7755,5	204,445	10,6792
150	423,15	439,99	314,38	7,4304	12 324,7	8806,5	208,134	10,8719
200	473,15	492,72	352,27	7,5482	13 801,6	9867,8	211,433	11,0442
250	523,15	545,91	390,62	7,6551	15 291,5	10 941,9	214,426	11,2006
300	573,15	599,63	429,50	7,7531	16 796,3	12 031,0	217,173	11,3441
400	673,15	708,91	509,10	7,9288	19 857,4	14 260,6	222,094	11,6011
500	773,15	820,80	591,31	8,0837	22 991,5	16 563,4	226,434	11,8278
600	873,15	935,28	676,10	8,2230	26 187,1	18 938,5	230,333	12,0315
700	973,15	1052,14	763,27	8,3496	29 471,5	21 380,6	233,882	12,2169
800	1073,15	1171,11	852,56	8,4660	32 803,9	23 881,5	237,141	12,3871
900	1173,15	1291,89	943,66	8,5736	36 187,1	26 433,3	240,155	12,5446
1000	1273,15	1414,24	1036,32	8,6737	39 614,2	29 028,9	242,958	12,6910
1100	1373,15	1537,95	1130,35	8,7672	43 079,6	31 663,0	245,578	12,8278
1200	1473,15	1662,88	1225,59	8,8550	46 578,8	34 330,8	248,038	12,9563
1300	1573,15	1788,85	1321,88	8,9378	50 107,4	37 028,0	250,355	13,0774
1400	1673,15	1915,71	1419,06	9,0159	53 661,0	39 750,1	252,545	13,1918
1500	1773,15	2043,35	1517,01	9,0900	57 236,3	42 494,0	254,621	13,3002

Table II.7 Composition and caloric values of the combustion products of some fuels at ($\alpha = 1$)

t	Blast-furnace gas Composition of combustion products at $\alpha = 1$: $r(CO_2) = 0{,}2301$; $r(H_2O) = 0{,}0609$; $r(N_2) = 0{,}7090$		Underground gasification gas Composition of combustion products at $\alpha = 1$: $r(CO_2) = 0{,}1701$; $r(H_2O) = 0{,}1211$; $r(N_2) = 0{,}7085$		Fuel oil Composition of combustion products at $\alpha = 1$: $r(CO_2) = 0{,}1263$; $r(H_2O) = 0{,}1484$; $r(N_2') = 0{,}7253$		Saratov gas Composition of combustion products at $\alpha = 1$: $r(CO_2) = 0{,}0951$; $r(H_2O) = 0{,}1961$; $r(N_2') = 0{,}7088$		Moist air (0.01 water in volume)	
	$\lg \pi_0$	H, kJ/kmol	$\lg \pi_0$	H, kJ/kmol	$\lg \pi_0$	H, kJ/kmol	$\lg \pi_0$	H, kJ/kmol	$\lg \pi_0$	H, kJ/kmol
−50	9,7781	6584	9,7014	6626	9,6483	6640	9,6058	6676	9,6895	6473
0	10,1015	8112	10,0241	8150	9,9696	8158	9,9274	8195	9,9965	7927
50	10,3753	9668	10,2960	9696	10,2396	9692	10,1971	9728	10,2523	9383
100	10,6134	11 252	10,5317	11 263	10,4729	11 243	10,4297	11 275	10,4720	10 803
150	10,8248	12 863	10,7404	12 853	10,6790	12 814	10,6349	12 838	10,6650	11 895
200	11,0151	14 500	10,9280	14 466	10,8617	14 405	10,8189	14 420	10,8377	13 793
300	11,3504	17 853	11,2573	17 760	11,1880	17 646	11,1408	17 641	11,1382	16 795
400	11,6406	21 312	11,5419	21 152	11,4676	20 978	11,4183	20 948	11,3956	19 861
500	11,8981	24 877	11,7941	24 644	11,7150	24 405	11,6638	24 348	11,6223	22 996
600	12,1302	28 537	12,0213	28 228	11,9378	27 921	11,8850	27 836	11,8259	26 200
700	12,3422	32 289	12,2290	31 902	12,1416	31 524	12,0871	31 412	12,0109	29 468
800	12,5375	36 111	12,4203	35 646	12,3292	35 197	12,2734	35 058	12,1808	32 794
900	12,7186	40 002	12,5978	39 462	12,5034	38 940	12,4463	38 778	12,3380	36 173
1000	12,8867	43 955	12,7628	43 340	12,6653	42 747	12,6074	42 562	12,4842	39 594
1100	13,0455	47 964	12,9187	47 277	12,8184	46 612	12,7597	46 410	12,6209	43 056
1200	13,1944	52 016	13,0651	51 260	12,9622	50 524	12,9029	50 302	12,7492	46 951
1300	13,3350	56 111	13,2035	55 290	13,0982	54 484	13,0384	54 247	12,8702	50 077
1400	13,4681	60 243	13,3345	59 357	13,2270	58 482	13,1667	58 232	12,9846	53 632
1500	13,5918	64 403	13,4593	63 455	13,3496	62 511	13,2891	62 251	13,0931	57 210

23

$$r_g = \frac{100 V^0}{V^0 + (L_0 + 0{,}124 L_0 d_w / 100)(\alpha - 1)} , \quad \%$$

$$r_{air} = 100 - r_g, \quad \%$$

where r_g is the volume fraction of combustion products for $\alpha = 1$; r_{air} is the volume fraction of air; and d_w is the quantity of water vapor in grams per normal cubic meter of air (in accordance with standard calculations of boiler assemblies it is assumed that $d_w = 8$ g/m^3).

If the volume fraction of air in the combustion products of the fuel is taken as a parameter, then the values of enthalpy and the logarithm of the relative pressure of the given mixture are determined by linear interpolation between the values of these values for moist air and for the combustion products of the fuel at $\alpha = 1$, i.e.,

$$H_{1\,mix} = H_{air} + (H_{\alpha - 1} - H_{air})(1 - r_{air})$$

In other words, if the intervals between the enthalpy and the lg π_0 curves for combustion products when $\alpha = 1$ and for moist air are divided into n equal parts, and the corresponding points are connected by smooth curves, then the obtained curves will correspond to the combustion products containing $1/n$, $2/n$, $3/n$, . . . volume fractions of air.

Practically, it is enough to divide the intervals into 10 parts for the interpolation to be quite reliable. In order to select curve H and curve lg π_0 corresponding to the combustion products with the specified excess air, it suffices to plot on the diagram the curve of dependence of r_g on α calculated by the foregoing method.

For the determination of the molecular mass μ and density ρ of a gaseous combustion product at normal conditions and for any value of α, two straight lines are plotted: μ and ρ.

As an example, presented in Table II.7 are the values of enthalpy and logarithm of the relative pressure calculated by the mixing rule, for the combustion products of a number of fuels at $\alpha = 1$.

A diagram similar to the one described can be constructed for the combustion products of any fuel (solid, liquid, and gaseous) in any scale that ensures high accuracy of the calculation. The use of such a diagram is justified by the exceptional simplicity of its construction, high accuracy, and convenience of use in practical calculations. The diagram can be recommended in cases when numerous calculations are conducted of processes with the combustion products of a specified fuel.

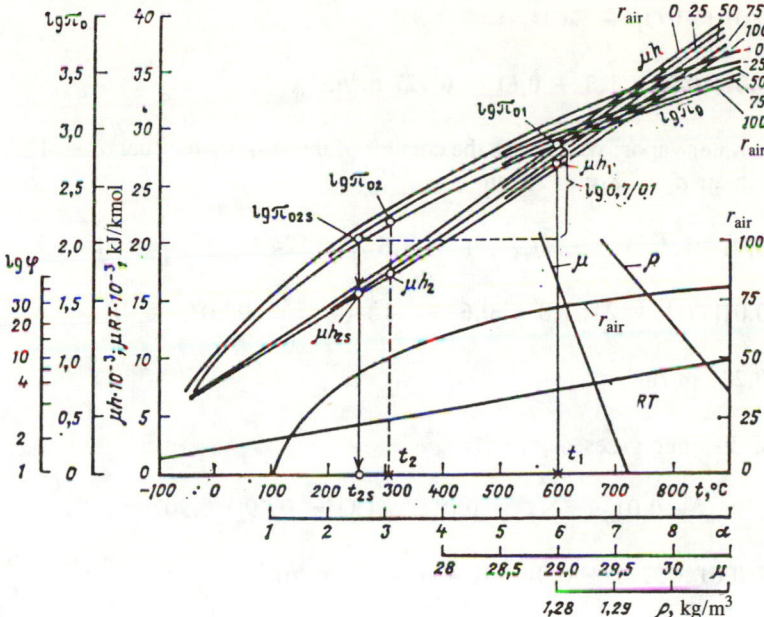

Figure 4

EXAMPLE OF DIAGRAM CONSTRUCTION

Let it be required to construct a diagram (Fig. 4) for the products of combustion of underground gasification gas having the following volume composition: $H_2S = 0.6\%$; $CO_2 = 10.3\%$; $O_2 = 0.2\%$; $CO = 18.4\%$; $H_2 = 11.1\%$; $CH_2 = 1.8\%$, and $N_2' = 57.6\%$ (N_2' is atmospheric nitrogen).

    1. The theoretical amount of oxygen needed for the combustion of 1 m^3 of the gas

$$V_{O_2} = 0.5\, V_{H_2} + 0.5\, V_{CO} + (m + n/4)V_{C_m H_n} + 1.5\, V_{H_2S}$$

Since air contains 21% oxygen and the fuel gas contains 0.2% oxygen the volume of air required for the combustion of the gas is

$$L_0 = (1/21)(0.5 \times 11.1 + 0.5 \times 18.4 + 2.0 \times 1.8 + 1.5 \times 0.6 - 0.2)$$

$$= 0.907\ \text{m}^3_{\text{air}}/\text{m}^3$$

    2. Volume composition of combustion products when $\alpha = 1$: volume of polyatomic gases (without water vapor)

$$V_{RO_2} = 0,01 [\%CO_2 + \%CH_4 + \%H_2S]$$

$$= 0,01(10,3 + 1,8 + 0,6) = 0,127 \text{ m}^3/\text{m}^3$$

volume of water vapor: we assume the content of moisture in the fuel $d_f = 47.5$ g/m^3 and in air $d_a = 8$ g/m^3, then

$$V_{H_2O} = 0,01 [\%H_2 + 2 \cdot \%CH_4 + \%H_2S + 0,124 (d_f + \alpha L_0 d_w)]$$

$$= 0,01[11,1 + 2 \cdot 1,8 + 0,6 + 0,124(47,5 + 0,907 \cdot 8,0)]$$

$$= 0,221 \text{ m}^3/\text{m}^3$$

volume of diatomic gases

$$V_{R_2} = 0,79L_0 + 0,01 \cdot \%N_2' + 0,01 \cdot \%CO = 0,79 \cdot 0,907$$

$$+ 0,01 \cdot 57,6 + 0,01 \cdot 18,4 = 1,477 \text{ m}^3/\text{m}^3$$

total volume of combustion products

$$V^0 = \Sigma V = 0,127 + 0,221 + 1,477 = 1,825 \text{ m}^3/\text{m}^3$$

The volume composition of the combustion products when $\alpha = 1$:

$$r (RO_2) = \frac{0,127}{1,825} = 0,070$$

$$r (H_2O) = \frac{0,221}{1,825} = 0,121$$

$$r (R_2) = \frac{1,477}{1,825} = 0,809$$

3. From the obtained composition and values of enthalpy and the logarithm of relative pressure of individual components, presented in Tables II.2–II.6, we determine the values of enthalpy and the logarithm of relative pressure for combustion products for various temperatures with an interval 50–100°C and $\alpha = 1$ (we assume that RO$_2$ represents carbon dioxide, since the content of SO$_2$ in the gas is insignificant).

For this purpose we use the mixing rule:

$$H_{\alpha - 1} = H(CO_2) \cdot r (RO_2) + H (H_2O)$$

$$\cdot \; r\,(H_2O) \; + \; H(N_2') \cdot r\,(R_2), \; kJ/kmol$$

$$(\lg \pi_0)_{a-1} \; = \; \lg \pi_0 \,(CO_2) \cdot r\,(RO_2)$$

$$+ \; \lg \pi_0 \,(H_2O) \cdot r(H_2O) \; + \; \lg \pi_0(N_2') \cdot r\,(R_2)$$

4. We can now construct the curves of enthalpy and the logarithm of relative pressure for the combustion products when $\alpha = 1$ and for air as a function of temperature plotted on the horizontal axis of the diagram. The interval between the enthalpy curves are divided into 10 (or 20) parts and the corresponding division points are connected with smooth curves. These curves will correspond to the content of 90, 80, 70, . . . , 10% (or 95, 90, 85, 80, . . . , 5%) of air by volume (if the reference point is the curve for air).

Similarly, intermediate curves of the logarithm of relative pressure[2] can be plotted.

5. For convenience of diagram use, the following must be plotted:

a) straight line RT as a function of temperature, for which it suffices to determine the value of RT [$R = 8.31441$ kJ/kmol·K for two or three temperatures];

b) the curve of dependence of r_a on α. To construct this curve we must find the values of r_a for $\alpha = 2, 3, \ldots, 15$ from formula

$$r_a \; = \; \left[1 \; - \; \frac{V^0}{V^0 \; + \; (L_0 \; + \; 0{,}124 L_0 d_w/100)(\alpha \; - \; 1)} \right] \cdot 100$$

$$\approx \; \left[1 \; - \; \frac{V^0}{V^0 \; + \; 1{,}01 L_0(\alpha \; - \; 1)} \right] \cdot 100\%$$

For example, when $\alpha = 6$, for underground gasification gas

$$r_a \; = \; 100 \left[1 \; - \; \frac{1{,}825}{1{,}825 \; + \; 1{,}01 \cdot 0{,}907 \, (6 \; - \; 1)} \right] \; = \; 71{,}5\%$$

From the obtained values of r_a, we must construct in a convenient scale a curve as a function of α plotting α on the horizontal axis (see the diagram for underground gasification gas presented in Fig. 4);

c) the curve or scale of the logarithms of quantities $\lg \varphi$ must be plotted in the same scale as for the logarithms of relative pressures;

d) two straight lines for the determination of the molecular mass μ and density of the gaseous product of combustion ascribed to the normal conditions ρ, kg/m^3.

For the construction of these straight lines, we must determine μ of the

[2]Since $\ln \pi_0 = s^0/R$, the described diagram is equivalent to the T, S-diagram for one mol of an ideal gas.

combustion products when $\alpha = 1$, from the composition found above, using formula

$$\mu_{\text{com.p}} = 44{,}01r \ (CO_2) + 18{,}016r \ (H_2O) + 28{,}15r(N_2')$$

(when calculating the data for the construction of the diagram, we specify the values of H and $\lg \pi_0$ for atmospheric nitrogen, the molecular mass of which $\mu = 28.15$).

In our case

$$\mu_{\text{com.p}} = 44{,}01 \cdot 0{,}070 + 18{,}016 \cdot 0{,}121 + 28{,}15 \cdot 0{,}809 = 28{,}03$$

the molecular mass of humid air $\mu = 28.86$.

We plot on a convenient scale on the horizontal axis the values of the molecular masses and mark two points: one corresponding to the molecular mass of air when $r_a = 100\%$, and the other corresponding to the molecular mass of the combustion products computed above for $r_a = 0$ ($\alpha = 1$). By connecting these points by a straight line, we can determine the molecular mass of the combustion products for any r_a, i.e., for any value of α.

The curve for ρ is similarly plotted.

We plot the values of ρ on the horizontal axis in a convenient scale and mark two points: one corresponding to ρ of air ($\rho_a = \mu_a/22.4146 = 28.86/22.4146 = 1.288$ kg/m^3), and the other corresponding to ρ of the combustion products for $\alpha = 1$ ($\rho_{\text{com.p}} = 28.03/22.4146 = 1.251$ kg/m^3).

By connecting these points, we obtain a straight line for the determination of ρ for combustion products of the given fuel at any value of excess air.

It is pertinent to mention that the diagram for the products of combustion of a given fuel at any value of excess air can be constructed not for 1 kmol but for 1 kg of gas. For this we must only determine the enthalpy per 1 kg of combustion products for $\alpha = 1$ and 1 kg of moist air (divide the molar enthalpy by the corresponding molecular mass). Also, as a parameter we adopt not the volume content of air in the combustion products at a given excess air but its mass fraction. The logarithm of the relative pressure does not change, and the construction of the diagram stays the same as for a mol of gas.

EXAMPLES OF DIAGRAM USE

Example 1. Air is sucked into a compressor at pressure $p_1 = 0.1$ MPa and temperature $t_1 = 20\,°C$ and compressed to pressure 0.5 MPa. Determine the theoretical compression work and internal relative efficiency if the air temperature at the end of compression is equal to $t_2 = 220\,°C$.

Solution. The calculation path is shown in Fig. 3.

1. From the enthalpy curves $\mu h = H$ of the diagram we obtain the molar enthalpy of air corresponding to temperature $t_1 = 20\,^\circ C$:

$H_1 = 8500$ kJ/kmol

On curve $\lg \pi_0$ we find point $\lg \pi_{01}$ corresponding to the same temperature.

2. From this point, a vertical section is drawn upwards corresponding to the degree of pressure change $\lg p_2/p_1 = \lg 0.5/0.1$, and by drawing a horizontal line from the end of this section we find the value $\lg \pi_{02s}$ on curve $\lg \pi_0$, enthalpy H_{2s} on curve $\mu h = H$, and then the temperature of the air at the end of the adiabatic process t_{2s}:

$t_{2s} = 190\,^\circ C \qquad H_{2s} = 13480$ kJ/kmol

3. The theoretical work

$\Delta H_{ad} = 13480 - 8500 = 4980$ kJ/kg

4. Enthalpy of air at the end of the actual process when $t_2 = 220\,^\circ C$,

$H_2 = 14360$ kJ/kmol

5. Actual work

$DH_{act} = 14360 - 8500 = 5860$ kJ/kg

6. The internal relative efficiency of the compression process

$$\eta_{oh} = \frac{4980}{5860} = 0{,}848$$

Example 2. From the conditions of the previous example, determine the change in internal energy of air in an actual process and the molar volumes of air at the beginning and end of compression.

Solution. Change in internal energy

$\Delta U = \Delta H - (RT_2 - RT_1)$

From line RT plotted in the diagram, we determine the values of RT_2 and RT_1 corresponding to the temperature of air at the beginning and end of the process:

$RT_2 = 4100$ kJ/kmol

$RT_1 = 2450$ kJ/kmol

$\Delta U + 5860 - (4100 - 2450) = 4210$ kJ/kmol

The molar volume of air, m^3/kmol, at the beginning of the compression process

$V_1 = RT_1/p_1$

and at the end of the compression process

$V_2 = RT_2/p_2$

(here, RT must be in J/kmol, and the pressure in Pa).
The molar volumes:

$$V_1 = \frac{2450 \cdot 10^3}{0.1 \cdot 10^6} = 24.5 \ m^3/\text{kmol}$$

$$V_2 = \frac{4100 \cdot 10^3}{0.5 \cdot 10^6} = 8.20 \ m^3/\text{kmol}$$

Example 3. Determine the theoretical expansion work of the combustion products of underground gasification gas for $\alpha = 7$ having a temperature of $t_1 = 600\,°C$ at pressure $p_1 = 0.7$ MPa if the final pressure p_2 is equal to 0.1 MPa. Find the density of the gas ascribed to the normal conditions at the indicated α.

Solution. The calculation path is shown in Fig. 4.
From curve r_a we determine r_a corresponding to value $\alpha = 7$:

$r_a = 75\%$

We find in the diagram curves μh and lg π_0 corresponding to the value of r_a closest to the obtained value; for temperature $600\,°C$ we find by interpolation

$H_1 = \mu h_1 = 26670$ kJ/mol   lg $\pi_{01} = 2.870$

From point lg π_{01} downwards we draw a section equal to lg p_1/p_2, i.e., lg 0.7/0.1, taken from the scale of lg φ.
From the end of this section along the horizontal, we find point lg $\pi_{02s} = 2.025$ and then find the gas temperature corresponding to this value $t_{2s} =$

253 °C at the end of expansion and the enthalpy of the gas

$$H_{2s} = \mu h_{2s} = 15560 \text{ kJ/kmol}$$

The theoretical expansion work of one mol of gas

$$\Delta H_{ad} = H_1 - H_{2s} = 26670 - 15560 = 11110 \text{ kJ/kmol}$$

We find the molecular mass of the gas and its density ascribed to the normal conditions. From the straight lines μ and ρ we find for $\alpha = 7$:

$$\mu = 29.06 \qquad \rho = 1.296 \text{ kg/m}^3$$

The theoretical expansion work of 1 kg of gas

$$\Delta h_{ad} = 11210/29.06 = 385.8 \text{ kJ/kg}$$

Example 4. From the conditions of the last example, determine the internal relative efficiency and entropy change in the process if the temperature at the end of the actual expansion process $t_2 = 305 °C$.

Solution. For temperature $t_2 = 305 °C$ we find the enthalpy at the end of the actual expansion process

$$H_2 = \mu h_2 = 17230 \text{ kJ/kmol}$$

and the logarithm of relative pressure

$$\lg \pi_{02} = 2.180$$

Then, the actual work is

$$\Delta H_{act} = H_1 - H_2 = 26670 - 17230 = 9440 \text{ kJ/kmol}$$

and the efficiency

$$\eta_{oh} = 9440/11110 = 0.850$$

The change in entropy is determined from the found values of $\lg \pi_0$ as follows:

$$S_2 - S_1 = \mu s_2 - \mu s_1 = 19{,}1448 \, (\lg \pi_{02} - \lg \pi_{02s})$$

$$= 19{,}1448 \, (2{,}180 - 2{,}025) = 2{,}967 \text{ kJ(kmol} \cdot \text{K)}$$

REFERENCES

1. Rivkin, C. L. Tables of thermodynamic properties of gases. Moscow-Leningrad, Gosenergoizdat, 1953.
2. Rivkin, C. L. Thermodynamic properties of gases. Moscow-Leningrad, Energiya, 1964.
3. Litvin, A. M. Technical thermodynamics. Moscow-Leningrad, Gosenergoizdat, 1958, 1963.
4. Andrianova, T. N., Dzampov, B. V., Zubarev, V. N., and Remizov, C. A. Collected problems in technical thermodynamics. Moscow, Energiya, 1971.
5. Kalafati, D. D. Analysis of calculation methods of the adiabatic process of gaseous mixtures from tables and diagrams. Izv. vuzov. Ser. Energetika, 1965, No. 8.
6. Rivkin, C. L. Thermodynamic properties of air and products of combustion of fuels. Moscow-Leningrad: Gosenergoizdat, 1954, 1962.
7. Rivkin, C. L. Thermodynamic properties of gases. Moscow, Energiya, 1973.

NOTES TO USERS OF THE TABLES*

Users of these tables may notice some slight numerical inconsistencies between the Tables II.1–II.6 and III.1–III.16. Because of the recent unfortunate death of Professor Rivkin we have been unable to establish all of the causes for these variations. However, the nature and magnitude of these discrepancies are such that they cannot introduce any significant error in the thermomechanical calculations of the type for which these tables are designed.

Tables II.1–II.6 constitute a set of skeleton tables for six of the gases considered here. They have been calculated from the coefficients given in Table I.1–I.3, as are the corresponding tables in Chapter III. However, slightly different values of the universal gas constant R and molecular weights have been used in the calculations of the related values in the two sets. The values in the more detailed tables of Chapter III are to be preferred.

In Table III.1 the value of the molecular mass used to convert molar values (calculated by Eqs. (13)–(15)) to values per kilogram ($\mu = 28.97$) differs from

*Added for the English language edition by D. D. Wagman.

that used in Table III.2 ($\mu = 28.96$). Similarly, in Table III.3 $\mu = 28.016$, whereas in Table III.4 $\mu = 28.0134$.

The values of lg π_0 given in Tables II.1–II.6 are calculated from the relation

$$\lg \pi_0 = S^0/19.144$$

In Tables III.1–III.17 the values of π_0 are calculated as follows: The value of lg π_0 is calculated by the relation

$$\lg \pi_0 = S^0/19.1447$$

The characteristic of the logarithm is then reduced by a convenient integer (5 for helium, 6 for hydrogen, 10 for air, nitrogen, atmospheric nitrogen, oxygen, water vapor, and carbon monoxide, and 12 for carbon dioxide), and the value of π_0 is then calculated from the reduced logarithm. This scaling is done for convenience in the use of the numbers in the calculations, and again introduces no error in the calculations.

In Table III.17 the values given for the entropy, enthalpy, and internal energy for helium differ significantly from the currently accepted values. However, the temperature dependence of these quantities and of the relative pressure π_0 are all correct. Hence no problem will arise from the use of the values in Table III.17 for the types of calculations described here.

Note: In the following tables, commas have been used instead of decimal points.

III. TABLES OF THERMODYNAMIC PROPERTIES OF VARIOUS GASES

Table III.1 Air ($\mu = 28.970$)

t	T	c_p	μc_p	c_v	μc_v	$k = \dfrac{c_p}{c_v}$
−50	223,15	1,0020	29,026	0,7150	20,712	1,401
−25	248,15	1,0023	29,036	0,7153	20,722	1,401
0	273,15	1,0028	29,050	0,7158	20,736	1,401
25	298,15	1,0038	29,079	0,7168	20,765	1,400
50	323,15	1,0053	29,123	0,7183	20,809	1,400
75	348,15	1,0073	29,181	0,7203	20,867	1,398
100	373,15	1,0098	29,255	0,7228	20,941	1,397
125	398,15	1,0128	29,342	0,7259	21,028	1,395
150	423,15	1,0163	29,442	0,7293	21,128	1,394
175	448,15	1,0202	29,554	0,7332	21,240	1,391
200	473,15	1,0244	29,677	0,7374	21,363	1,389
250	523,15	1,0339	29,952	0,7469	21,638	1,384
300	573,15	1,0445	30,260	0,7575	21,946	1,379
350	623,15	1,0559	30,589	0,7689	22,275	1,373
400	673,15	1,0678	30,933	0,7808	22,619	1,368
450	723,15	1,0798	31,282	0,7928	22,968	1,362
500	773,15	1,0918	31,630	0,8048	23,316	1,357
550	823,15	1,1036	31,972	0,8166	23,658	1,352
600	873,15	1,1150	32,301	0,8280	23,987	1,347
650	923,15	1,1258	32,615	0,8388	24,301	1,342
700	973,15	1,1361	32,912	0,8491	24,598	1,338
750	1023,15	1,1457	33,190	0,8587	24,876	1,334
800	1073,15	1,1546	33,449	0,8676	25,135	1,331
850	1123,15	1,1629	33,690	0,8760	25,376	1,328
900	1173,15	1,1707	33,914	0,8837	25,600	1,325
950	1223,15	1,1779	34,123	0,8909	25,809	1,322
1000	1273,15	1,1846	34,318	0,8976	26,004	1,320
1050	1323,15	1,1909	34,501	0,9039	26,187	1,318
1100	1373,15	1,1969	34,673	0,9099	26,359	1,315
1150	1423,15	1,2025	34,837	0,9155	26,523	1,314
1200	1473,15	1,2079	34,993	0,9209	26,679	1,312
1250	1523,15	1,2130	35,140	0,9260	26,826	1,310
1300	1573,15	1,2179	35,282	0,9309	26,968	1,308
1350	1623,15	1,2225	35,415	0,9355	27,101	1,307
1400	1673,15	1,2268	35,540	0,9398	27,226	1,305
1450	1723,15	1,2309	35,658	0,9439	27,344	1,304
1500	1773,15	1,2347	35,768	0,9477	27,454	1,303

Table III.2 Air ($\mu = 28.960$)

t	T	h	u	π_0	θ_0	s^0	H	U	S
−50	223,15	223,19	159,12	0,4929	12997	6,4076	6463,5	4608,1	185,564
−49	224,15	224,19	159,84	0,5006	12853	6,4121	6492,5	4628,8	185,694
−48	225,15	225,19	160,55	0,5085	12711	6,4166	6521,5	4649,6	185,823
−47	226,15	226,19	161,27	0,5164	12572	6,4210	6550,6	4670,3	185,952
−46	227,15	227,20	161,98	0,5244	12434	6,4254	6579,6	4691,0	186,080
−45	228,15	228,20	162,70	0,5325	12299	6,4298	6608,7	4711,7	186,208
−44	229,15	229,20	163,41	0,5407	12166	6,4342	6637,7	4732,4	186,335
−43	230,15	230,20	164,13	0,5490	12034	6,4386	6666,7	4753,2	186,461
−42	231,15	231,21	164,84	0,5574	11905	6,4429	6695,8	4773,9	186,587
−41	232,15	232,21	165,56	0,5658	11778	6,4473	6724,8	4794,6	186,712
−40	233,15	233,21	166,27	0,5744	11652	6,4516	6753,8	4815,3	186,837
−39	234,15	234,21	166,99	0,5830	11528	6,4559	6782,9	4836,0	186,962
−38	235,15	235,22	167,71	0,5918	11407	6,4601	6811,9	4856,8	187,085
−37	236,15	236,22	168,42	0,6006	11287	6,4644	6840,9	4877,5	187,208
−36	237,15	237,22	169,14	0,6096	11168	6,4686	6870,0	4898,2	187,331
−35	238,15	238,23	169,85	0,6186	11052	6,4728	6899,0	4918,9	187,453
−34	239,15	239,23	170,57	0,6277	10937	6,4770	6928,0	4939,6	187,575
−33	240,15	240,23	171,28	0,6369	10824	6,4812	6957,1	4960,4	187,696
−32	241,15	241,23	172,00	0,6472	10712	6,4854	6986,1	4981,1	187,817
−31	242,15	242,24	172,71	0,6556	10603	6,4895	7015,1	5001,8	187,937
−30	243,15	243,24	173,43	0,6651	10494	6,4937	7044,2	5022,5	188,057
−29	244,15	244,24	174,15	0,6747	10387	6,4978	7073,2	5043,2	188,176
−28	245,15	245,24	174,86	0,6844	10282	6,5019	7102,2	5064,0	188,294
−27	246,15	246,25	175,58	0,6942	10178	6,5060	7131,3	5084,7	188,413
−26	247,15	247,25	176,29	0,7041	10076	6,5100	7160,3	5105,4	188,530
−25	248,15	248,25	177,01	0,7141	9975,5	6,5141	7189,3	5126,1	188,648
−24	249,15	249,25	177,72	0,7242	9876,1	6,5181	7218,4	5146,8	188,764
−23	250,15	250,26	178,44	0,7344	9778,0	6,5221	7247,4	5167,6	188,881
−22	251,15	251,26	179,15	0,7447	9681,2	6,5261	7276,5	5188,3	188,997
−21	252,15	252,26	179,87	0,7552	9585,8	6,5301	7305,5	5209,0	189,112
−20	253,15	253,26	180,58	0,7657	9491,7	6,5341	7334,5	5229,7	189,227
−19	254,15	254,27	181,30	0,7763	9398,9	6,5380	7363,6	5250,5	189,341
−18	255,15	255,27	182,02	0,7870	9307,3	6,5420	7392,6	5271,2	189,455
−17	256,15	256,27	182,73	0,7978	9217,0	6,5459	7421,6	5291,9	189,569
−16	257,15	257,27	183,45	0,8088	9128,0	6,5498	7450,7	5312,6	189,682
−15	258,15	258,28	184,16	0,8198	9040,1	6,5537	7479,7	5333,4	189,795
−14	259,15	259,28	184,88	0,8309	8953,4	6,5576	7508,8	5354,1	189,907
−13	260,15	260,28	185,59	0,8422	8867,8	6,5614	7537,8	5374,8	190,019
−12	261,15	261,29	186,31	0,8536	8783,4	6,5653	7566,8	5395,5	190,130
−11	262,15	262,29	187,03	0,8650	8700,1	6,5691	7595,9	5416,3	190,241
−10	263,15	263,29	187,74	0,8766	8617,9	6,5729	7624,9	5437,0	190,352
−9	264,15	264,29	188,46	0,8883	8536,8	6,5767	7654,0	5457,7	190,462
−8	265,15	265,30	189,17	0,9001	8456,8	6,5805	7683,0	5478,4	190,572
−7	266,15	266,30	189,89	0,9120	8377,8	6,5843	7712,1	5499,2	190,681
−6	267,15	267,30	190,60	0,9240	8299,8	6,5881	7741,1	5519,9	190,790
−5	268,15	268,31	191,32	0,9362	8222,9	6,5918	7770,1	5540,6	190,899
−4	269,15	269,31	192,04	0,9484	8146,9	6,5955	7799,2	5561,4	191,007
−3	270,15	270,31	192,75	0,9608	8071,9	6,5993	7828,2	5582,1	191,114
−2	271,15	271,32	193,47	0,9733	7997,9	6,6030	7857,3	5602,8	191,222
−1	272,15	272,32	194,18	0,9859	7924,8	6,6067	7886,3	5623,6	191,329
0	273,15	273,32	194,90	0,9986	7852,6	6,6103	7915,4	5644,3	191,435
1	274,15	274,32	195,62	1,0114	7781,4	6,6140	7944,4	5665,0	191,541
2	275,15	275,33	196,33	1,0244	7711,1	6,6176	7973,5	5685,8	191,647
3	276,15	276,33	197,05	1,0375	7641,6	6,6213	8002,5	5706,5	191,753
4	277,15	277,33	197,76	1,0506	7573,0	6,6249	8031,6	5727,2	191,858

Air (*Continued*)

t	T	h	u	π_0	θ_0	s^0	H	U	S
5	278,15	278,34	198,48	1,0640	7505,3	6,6285	8060,6	5748,0	191,962
6	279,15	279,34	199,20	1,0774	7438,4	6,6321	8089,7	5768,7	192,066
7	280,15	280,34	199,91	1,0909	7372,3	6,6357	8118,8	5789,5	192,170
8	281,15	281,35	200,63	1,1046	7307,1	6,6393	8147,8	5810,2	192,274
9	282,15	282,35	201,35	1,1184	7242,6	6,6429	8176,9	5831,0	192,377
10	283,15	283,35	202,06	1,1323	7179,0	6,6464	8205,9	5851,7	192,480
11	284,15	284,36	202,78	1,1464	7116,1	6,6499	8235,0	5872,4	192,582
12	285,15	285,36	203,49	1,1605	7054,0	6,6535	8264,0	5893,2	192,684
13	286,15	286,36	204,21	1,1748	6992,7	6,6570	8293,1	5913,9	192,786
14	287,15	287,37	204,93	1,1892	6932,0	6,6605	8322,2	5934,7	192,888
15	288,15	288,37	205,64	1,2038	6872,2	6,6640	8351,2	5955,4	192,989
16	289,15	289,38	206,36	1,2184	6813,0	6,6674	8380,3	5976,2	193,089
17	290,15	290,38	207,08	1,2332	6754,5	6,6709	8409,4	5996,9	193,190
18	291,15	291,38	207,79	1,2481	6696,8	6,6744	8438,4	6017,7	193,290
19	292,15	292,39	208,51	1,2632	6639,7	6,6778	8467,5	6038,5	193,389
20	293,15	293,39	209,23	1,2784	6583,3	6,6812	8496,6	6059,2	193,489
21	294,15	294,39	209,94	1,2937	6527,6	6,6847	8525,7	6080,0	193,588
22	295,15	295,40	210,66	1,3091	6472,5	6,6881	8554,7	6100,7	193,686
23	296,15	296,40	211,38	1,3247	6418,1	6,6915	8583,8	6121,5	193,785
24	297,15	297,41	212,09	1,3404	6364,3	6,6948	8612,9	6142,2	193,883
25	298,15	298,41	212,81	1,3563	6311,1	6,6982	8642,0	6163,0	193,980
26	299,15	299,41	213,53	1,3722	6258,5	6,7016	8671,0	6183,8	194,078
27	300,15	300,42	214,25	1,3884	6206,6	6,7049	8700,1	6204,5	194,175
28	301,15	301,42	214,96	1,4046	6155,2	6,7083	8729,2	6225,3	194,272
29	302,15	302,43	215,68	1,4210	6104,5	6,7116	8758,3	6246,1	194,368
30	303,15	303,43	216,40	1,4375	6054,3	6,7149	8787,4	6266,9	194,464
31	304,15	304,44	217,11	1,4542	6004,7	6,7182	8816,5	6287,6	194,560
32	305,15	305,44	217,83	1,4710	5955,6	6,7215	8845,5	6308,4	194,655
33	306,15	306,44	218,55	1,4879	5907,1	6,7248	8874,6	6329,2	194,750
34	307,15	307,45	219,27	1,5050	5859,2	6,7281	8903,7	6350,0	194,845
35	308,15	308,45	219,98	1,5222	5811,8	6,7314	8932,8	6370,7	194,940
36	309,15	309,46	220,70	1,5395	5764,9	6,7346	8961,9	6391,5	195,034
37	310,15	310,46	221,42	1,5570	5718,6	6,7379	8991,0	6412,3	195,128
38	311,15	311,47	222,14	1,5747	5672,7	6,7411	9020,1	6433,1	195,222
39	312,15	312,47	222,85	1,5925	5627,4	6,7443	9049,2	6453,9	195,315
40	313,15	313,48	223,57	1,6104	5582,6	6,7475	9078,3	6474,7	195,408
41	314,15	314,48	224,29	1,6285	5538,3	6,7507	9107,4	6495,4	195,501
42	315,15	315,49	225,01	1,6467	5494,4	6,7539	9136,5	6516,2	195,594
43	316,15	316,49	225,73	1,6651	5451,1	6,7571	9165,6	6537,0	195,686
44	317,15	317,50	226,44	1,6836	5408,2	6,7603	9194,7	6557,8	195,778
45	318,15	318,50	227,16	1,7022	5365,7	6,7634	9223,9	6578,6	195,869
46	319,15	319,51	227,88	1,7210	5323,8	6,7666	9253,0	6599,4	195,961
47	320,15	320,51	228,60	1,7400	5282,3	6,7697	9282,1	6620,2	196,052
48	321,15	321,52	229,32	1,7591	5241,2	6,7729	9311,2	6641,0	196,143
49	322,15	322,52	230,04	1,7784	5200,6	6,7760	9340,3	6661,8	196,233
50	323,15	323,53	230,75	1,7978	5160,4	6,7791	9369,4	6682,6	196,323
51	324,15	324,54	231,47	1,8173	5120,7	6,7822	9398,6	6703,4	196,413
52	325,15	325,54	232,19	1,8371	5081,3	6,7853	9427,7	6724,3	196,503
53	326,15	326,55	232,91	1,8569	5042,4	6,7884	9456,8	6745,1	196,593
54	327,15	327,55	233,63	1,8770	5003,9	6,7915	9485,9	6765,9	196,682
55	328,15	328,56	234,35	1,8971	4965,8	6,7946	9515,1	6786,7	196,771
56	329,15	329,57	235,07	1,9175	4928,1	6,7976	9544,2	6807,5	196,859
57	330,15	330,57	235,79	1,9380	4890,8	6,8007	9573,3	6828,3	196,948
58	331,15	331,58	236,50	1,9586	4853,9	6,8037	9602,5	6849,2	197,036
59	332,15	332,58	237,22	1,9794	4817,4	6,8068	9631,6	6870,0	197,124

38

t	T	h	u	π_0	θ_0	$s^\bullet$	H	U	S
60	333,15	333,59	237,94	2,0004	4781,3	6,8098	9660,8	6890,8	197,211
61	334,15	334,60	238,66	2,0215	4745,5	6,8128	9689,9	6911,7	197,299
62	335,15	335,60	239,38	2,0428	4710,1	6,8158	9719,1	6932,5	197,386
63	336,15	336,61	240,10	2,0643	4675,1	6,8188	9748,2	6953,3	197,473
64	337,15	337,62	240,82	2,0859	4640,4	6,8218	9777,4	6974,2	197,559
65	338,15	338,62	241,54	2,1076	4606,1	6,8248	9806,5	6995,0	197,646
66	339,15	339,63	242,26	2,1296	4572,1	6,8277	9835,7	7015,8	197,732
67	340,15	340,64	242,98	2,1517	4538,5	6,8307	9864,8	7036,7	197,817
68	341,15	341,64	243,70	2,1740	4505,2	6,8337	9894,0	7057,5	197,903
69	342,15	342,65	244,42	2,1964	4472,2	6,8366	9923,2	7078,4	197,988
70	343,15	343,66	245,14	2,2190	4439,6	6,8396	9952,3	7099,2	198,074
71	344,15	344,57	245,86	2,2418	4407,3	6,8425	9981,5	7120,1	198,158
72	345,15	345,67	246,58	2,2647	4375,4	6,8454	10010,7	7141,0	198,243
73	346,15	346,68	247,30	2,2878	4343,7	6,8483	10039,9	7161,8	198,328
74	347,15	347,69	248,02	2,3111	4312,4	6,8512	10069,0	7182,7	198,412
75	348,15	348,70	248,74	2,3345	4281,4	6,8541	10098,2	7203,5	198,496
76	349,15	349,70	249,46	2,3582	4250,7	6,8570	10127,4	7224,4	198,579
77	350,15	350,71	250,18	2,3820	4220,3	6,8599	10156,6	7245,3	198,663
78	351,15	351,72	250,90	2,4059	4190,2	6,8628	10185,8	7266,2	198,746
79	352,15	352,73	251,62	2,4301	4160,3	6,8656	10215,0	7287,0	198,829
80	353,15	353,73	252,35	2,4544	4130,8	6,8685	10244,1	7307,9	198,912
81	354,15	354,74	253,07	2,4789	4101,6	6,8714	10273,3	7328,8	198,994
82	355,15	355,75	253,79	2,5035	4072,6	6,8742	10302,5	7349,7	199,077
83	356,15	356,76	254,51	2,5284	4044,0	6,8770	10331,7	7370,6	199,159
84	357,15	357,77	255,23	2,5534	4015,6	6,8799	10361,0	7391,5	199,241
85	358,15	358,78	255,95	2,5786	3987,5	6,8827	10390,2	7412,4	199,322
86	359,15	359,78	256,67	2,6040	3959,6	6,8855	10419,4	7433,2	199,404
87	360,15	360,79	257,39	2,6296	3932,1	6,8883	10448,6	7454,1	199,485
88	361,15	361,80	258,12	2,6553	3904,7	6,8911	10477,8	7475,0	199,566
89	362,15	362,81	258,84	2,6812	3877,7	6,8939	10507,0	7496,0	199,647
90	363,15	363,82	259,56	2,7074	3850,9	6,8967	10536,2	7516,9	199,727
91	364,15	364,83	260,28	2,7336	3824,4	6,8994	10565,5	7537,8	199,808
92	365,15	365,84	261,00	2,7601	3798,1	6,9022	10594,7	7558,7	199,888
93	366,15	366,85	261,73	2,7868	3772,0	6,9050	10623,9	7579,6	199,968
94	367,15	367,86	262,45	2,8137	3746,2	6,9077	10653,2	7600,5	200,048
95	368,15	368,87	263,17	2,8407	3720,7	6,9105	10682,4	7621,4	200,127
96	369,15	369,88	263,89	2,8679	3695,4	6,9132	10711,6	7642,4	200,206
97	370,15	370,89	264,62	2,8953	3670,3	6,9159	10740,9	7663,3	200,286
98	371,15	371,90	265,34	2,9229	3645,4	6,9187	10770,1	7684,2	200,364
99	372,15	372,91	266,06	2,9507	3620,8	6,9214	10799,4	7705,2	200,443
100	373,15	373,92	266,79	2,9787	3596,4	6,9241	10828,6	7726,1	200,522
101	374,15	374,93	267,51	3,0069	3572,3	6,9268	10857,9	7747,0	200,600
102	375,15	375,94	268,23	3,0353	3548,3	6,9295	10887,1	7768,0	200,678
103	376,15	376,95	268,96	3,0639	3524,6	6,9322	10916,4	7788,9	200,756
104	377,15	377,96	269,68	3,0926	3501,1	6,9349	10945,7	7809,9	200,834
105	378,15	378,97	270,40	3,1216	3477,8	6,9375	10974,9	7830,8	200,911
106	379,15	379,98	271,13	3,1508	3454,7	6,9402	11004,2	7851,8	200,988
107	380,15	380,99	271,85	3,1801	3431,9	6,9429	11033,5	7872,8	201,066
108	381,15	382,00	272,57	3,2097	3409,2	6,9455	11062,8	7893,7	201,143
109	382,15	383,01	273,30	3,2394	3386,8	6,9482	11092,1	7914,7	201,219
110	383,15	384,02	274,02	3,2694	3364,5	6,9508	11121,3	7935,7	201,296
111	384,15	385,04	274,75	3,2996	3342,5	6,9535	11150,6	7956,6	201,372
112	385,15	386,05	275,47	3,3299	3320,6	6,9561	11179,9	7977,6	201,448
113	386,15	387,06	276,20	3,3605	3298,9	6,9587	11209,2	7998,6	201,524
114	387,15	388,07	276,92	3,3912	3277,5	6,9613	11238,5	8019,6	201,600

t	T	h	u	π_0	θ_0	s^0	H	U	S
115	388,15	389,08	277,64	3,4222	3256,2	6,9639	11267,8	8040,6	201,676
116	389,15	390,09	278,37	3,4534	3235,1	6,9665	11297,1	8061,6	201,751
117	390,15	391,11	279,09	3,4848	3214,2	6,9691	11326,4	8082,6	201,826
118	391,15	392,12	279,82	3,5164	3193,5	6,9717	11355,8	8103,6	201,901
119	392,15	393,13	280,54	3,5482	3173,0	6,9743	11385,1	8124,6	201,976
120	393,15	394,14	281,27	3,5802	3152,6	6,9769	11414,4	8145,6	202,051
121	394,15	395,16	282,00	3,6124	3132,5	6,9795	11443,7	8166,6	202,125
122	395,15	396,17	282,72	3,6448	3112,5	6,9820	11473,0	8187,6	202,200
123	396,15	397,18	283,45	3,6775	3092,6	6,9846	11502,4	8208,6	202,274
124	397,15	398,19	284,17	3,7104	3073,0	6,9871	11531,7	8229,6	202,348
125	398,15	399,21	284,90	3,7434	3053,5	6,9897	11561,1	8250,7	202,421
126	399,15	400,22	285,62	3,7767	3034,2	6,9922	11590,4	8271,7	202,495
127	400,15	401,23	286,35	3,8102	3015,0	6,9948	11619,7	8292,7	202,569
128	401,15	402,25	287,08	3,8439	2996,1	6,9973	11649,1	8313,8	202,642
129	402,15	403,26	287,80	3,8779	2977,2	6,9998	11678,4	8334,8	202,715
130	403,15	404,28	288,53	3,9120	2958,6	7,0023	11707,8	8355,9	202,788
131	404,15	405,29	289,26	3,9464	2940,1	7,0049	11737,2	8376,9	202,861
132	405,15	406,30	289,98	3,9810	2921,8	7,0074	11766,5	8398,0	202,933
133	406,15	407,32	290,71	4,0158	2903,6	7,0099	11795,9	8419,0	203,006
134	407,15	408,33	291,44	4,0509	2885,5	7,0124	11825,3	8440,1	203,078
135	408,15	409,35	292,17	4,0861	2867,7	7,0148	11854,7	8461,1	203,150
136	409,15	410,36	292,89	4,1216	2849,9	7,0173	11884,0	8482,2	203,222
137	410,15	411,38	293,62	4,1573	2832,4	7,0198	11913,4	8503,3	203,293
138	411,15	412,39	294,35	4,1933	2814,9	7,0223	11942,8	8524,3	203,365
139	412,15	413,41	295,08	4,2294	2797,6	7,0247	11972,2	8545,4	203,436
140	413,15	414,42	295,80	4,2658	2780,5	7,0272	12001,6	8566,5	203,508
141	414,15	415,44	296,53	4,3025	2763,5	7,0297	12031,0	8587,6	203,579
142	415,15	416,45	297,26	4,3393	2746,7	7,0321	12060,4	8608,7	203,650
143	416,15	417,47	297,99	4,3764	2729,9	7,0345	12089,8	8629,8	203,720
144	417,15	418,48	298,72	4,4137	2713,4	7,0370	12119,2	8650,9	203,791
145	418,15	419,50	299,45	4,4513	2696,9	7,0394	12148,7	8672,0	203,861
146	419,15	420,51	300,18	4,4891	2680,6	7,0418	12178,1	8693,1	203,932
147	420,15	421,53	300,90	4,5271	2664,4	7,0443	12207,5	8714,2	204,002
148	421,15	422,55	301,63	4,5653	2648,4	7,0467	12236,9	8735,3	204,072
149	422,15	423,56	302,36	4,6038	2632,5	7,0491	12266,4	8756,4	204,142
150	423,15	424,58	303,09	4,6426	2616,7	7,0515	12295,8	8777,6	204,211
151	424,15	425,60	303,82	4,6815	2601,1	7,0539	12325,3	8798,7	204,281
152	425,15	426,61	304,55	4,7207	2585,5	7,0563	12354,7	8819,8	204,350
153	426,15	427,63	305,28	4,7602	2570,2	7,0587	12384,2	8841,0	204,419
154	427,15	428,65	306,01	4,7999	2554,9	7,0611	12413,6	8862,1	204,488
155	428,15	429,66	306,74	4,8398	2539,7	7,0634	12443,1	8883,3	204,557
156	429,15	430,68	307,47	4,8800	2524,7	7,0658	12472,5	8904,4	204,626
157	430,15	431,70	308,20	4,9204	2509,8	7,0682	12502,0	8925,6	204,695
158	431,15	432,72	308,93	4,9611	2495,0	7,0705	12531,5	8946,7	204,763
159	432,15	433,73	309,66	5,0020	2480,3	7,0729	12561,0	8967,9	204,831
160	433,15	434,75	310,40	5,0432	2465,8	7,0753	12590,4	8989,1	204,899
161	434,15	435,77	311,13	5,0846	2451,3	7,0776	12619,9	9010,2	204,967
162	435,15	436,79	311,86	5,1263	2437,0	7,0799	12649,4	9031,4	205,035
163	436,15	437,81	312,59	5,1682	2422,8	7,0823	12678,9	9052,6	205,103
164	437,15	438,83	313,32	5,2103	2408,7	7,0846	12708,4	9073,8	205,171
165	438,15	439,85	314,05	5,2528	2394,7	7,0869	12737,9	9095,0	205,238
166	439,15	440,86	314,78	5,2954	2380,9	7,0893	12767,4	9116,2	205,305
167	440,15	441,88	315,52	5,3384	2367,1	7,0916	12796,9	9137,4	205,372
168	441,15	442,90	316,25	5,3815	2353,4	7,0939	12826,5	9158,6	205,439
169	442,15	443,92	316,98	5,4250	2339,9	7,0962	12856,0	9179,8	205,506

Air (*Continued*)

t	T	h	u	π_0	θ_0	s^0	H	U	S
170	443,15	444,94	317,71	5,4687	2326,4	7,0985	12885,5	9201,0	205,573
171	444,15	445,96	318,45	5,5126	2313,1	7,1008	12915,0	9222,2	205,639
172	445,15	446,98	319,18	5,5569	2299,8	7,1031	12944,6	9243,4	205,706
173	446,15	448,00	319,91	5,6013	2286,7	7,1054	12974,1	9264,7	205,772
174	447,15	449,02	320,65	5,6461	2273,7	7,1077	13003,7	9285,9	205,838
175	448,15	450,04	321,38	5,6911	2260,7	7,1100	13033,2	9307,1	205,904
176	449,15	451,06	322,11	5,7364	2247,9	7,1122	13062,8	9328,4	205,970
177	450,15	452,08	322,85	5,7819	2235,1	7,1145	13092,3	9349,6	206,036
178	451,15	453,10	323,58	5,8277	2222,5	7,1168	13121,9	9370,9	206,102
179	452,15	454,13	324,31	5,8738	2210,0	7,1190	13151,5	9392,1	206,167
180	453,15	455,15	325,05	5,9201	2197,5	7,1213	13181,1	9413,4	206,232
181	454,15	456,17	325,78	5,9667	2185,2	7,1235	13210,6	9434,6	206,298
182	455,15	457,19	326,52	6,0136	2172,9	7,1258	13240,2	9455,9	206,363
183	456,15	458,21	327,25	6,0608	2160,7	7,1280	13269,8	9477,2	206,428
184	457,15	459,23	327,99	6,1082	2148,7	7,1303	13299,4	9498,5	206,492
185	458,15	460,26	328,72	6,1559	2136,7	7,1325	13329,0	9519,8	206,557
186	459,15	461,28	329,46	6,2039	2124,8	7,1347	13358,6	9541,0	206,622
187	460,15	462,30	330,19	6,2521	2113,0	7,1369	13388,2	9562,3	206,686
188	461,15	463,32	330,93	6,3006	2101,2	7,1392	13417,8	9583,6	206,750
189	462,15	464,35	331,66	6,3494	2089,6	7,1414	13447,4	9604,9	206,815
190	463,15	465,37	332,40	6,3985	2078,1	7,1436	13477,1	9626,2	206,879
191	464,15	466,39	333,13	6,4479	2066,6	7,1458	13506,7	9647,6	206,942
192	465,15	467,41	333,87	6,4976	2055,2	7,1480	13536,3	9668,9	207,006
193	466,15	468,44	334,61	6,5475	2044,0	7,1502	13566,0	9690,2	207,070
194	467,15	469,46	335,34	6,5977	2032,7	7,1524	13595,6	9711,5	207,133
195	468,15	470,49	336,08	6,6482	2021,6	7,1546	13625,3	9732,9	207,197
196	469,15	471,51	336,82	6,6990	2010,6	7,1568	13654,9	9754,2	207,260
197	470,15	472,53	337,55	6,7501	1999,6	7,1590	13684,6	9775,6	207,323
198	471,15	473,56	338,29	6,8014	1988,7	7,1611	13714,2	9796,9	207,386
199	472,15	474,58	339,03	6,8531	1977,9	7,1633	13743,9	9818,3	207,449
200	473,15	475,61	339,77	6,9050	1967,2	7,1655	13773,6	9839,6	207,512
201	474,15	476,63	340,50	6,9573	1956,6	7,1676	13803,3	9861,0	207,575
202	475,15	477,66	341,24	7,0098	1946,0	7,1698	13832,9	9882,4	207,637
203	476,15	478,68	341,98	7,0626	1935,5	7,1719	13862,6	9903,7	207,700
204	477,15	479,71	342,72	7,1157	1925,1	7,1741	13892,3	9925,1	207,762
205	478,15	480,73	343,46	7,1691	1914,8	7,1762	13922,0	9946,5	207,824
206	479,15	481,76	344,19	7,2229	1904,5	7,1784	13951,7	9967,9	207,886
207	480,15	482,78	344,93	7,2769	1894,3	7,1805	13981,4	9989,3	207,948
208	481,15	483,81	345,67	7,3312	1884,2	7,1827	14011,2	10010,7	208,010
209	482,15	484,84	346,41	7,3858	1874,1	7,1848	14040,9	10032,1	208,072
210	483,15	485,86	347,15	7,4407	1864,2	7,1869	14070,6	10053,5	208,133
211	484,15	486,89	347,89	7,4959	1854,3	7,1890	14100,3	10074,9	208,195
212	485,15	487,92	348,63	7,5514	1844,4	7,1912	14130,1	10096,3	208,256
213	486,15	488,94	349,37	7,6073	1834,7	7,1933	14159,8	10117,8	208,317
214	487,15	489,97	350,11	7,6634	1825,0	7,1954	14189,6	10139,2	208,378
215	488,15	491,00	350,85	7,7199	1815,4	7,1975	14219,3	10160,6	208,439
216	489,15	492,03	351,59	7,7766	1805,8	7,1996	14249,1	10182,1	208,500
217	490,15	493,05	352,33	7,8337	1796,3	7,2017	14278,8	10203,5	208,561
218	491,15	494,08	353,07	7,8910	1786,9	7,2038	14308,6	10225,0	208,622
219	492,15	495,11	353,81	7,9487	1777,5	7,2059	14338,4	10246,4	208,682
220	493,15	496,14	354,55	8,0067	1768,2	7,2080	14368,2	10267,9	208,743
221	494,15	497,17	355,30	8,0650	1759,0	7,2101	14397,9	10289,4	208,803
222	495,15	498,20	356,04	8,1237	1749,9	7,2121	14427,7	10310,9	208,863
223	496,15	499,22	356,78	8,1826	1740,8	7,2142	14457,5	10332,3	208,923
224	497,15	500,25	357,52	8,2419	1731,7	7,2163	14487,3	10353,8	208,983

Air (*Continued*)

t	T	h	u	π_n	θ_n	$s°$	H	U	S
225	498,15	501,28	358,26	8,3015	1722,8	7,2183	14517,1	10375,3	209,043
226	499,15	502,31	359,01	8,3614	1713,8	7,2204	14547,0	10396,8	209,103
227	500,15	503,34	359,75	8,4216	1705,0	7,2225	14576,8	10418,3	209,163
228	501,15	504,37	360,49	8,4822	1696,2	7,2245	14606,6	10439,8	209,222
229	502,15	505,40	361,23	8,5431	1687,5	7,2266	14636,4	10461,3	209,282
230	503,15	506,43	361,98	8,6043	1678,8	7,2286	14666,3	10482,9	209,341
231	504,15	507,46	362,72	8,6658	1670,2	7,2307	14696,1	10504,4	209,400
232	505,15	508,49	363,46	8,7277	1661,7	7,2327	14725,9	10525,9	209,460
233	506,15	509,52	364,21	8,7898	1653,2	7,2348	14755,8	10547,5	209,519
234	507,15	510,55	364,95	8,8524	1644,7	7,2368	14785,7	10569,0	209,578
235	508,15	511,59	365,70	8,9152	1636,4	7,2388	14815,5	10590,6	209,636
236	509,15	512,62	366,44	8,9784	1628,0	7,2409	14845,4	10612,1	209,695
237	510,15	513,65	367,18	9,0419	1619,8	7,2429	14875,3	10633,7	209,754
238	511,15	514,68	367,93	9,1058	1611,6	7,2449	14905,1	10655,3	209,812
239	512,15	515,71	368,67	9,1700	1603,4	7,2469	14935,0	10676,8	209,871
240	513,15	516,74	369,42	9,2345	1595,3	7,2489	14964,9	10698,4	209,929
241	514,15	517,78	370,16	9,2994	1587,3	7,2509	14994,8	10720,0	209,987
242	515,15	518,81	370,91	9,3646	1579,3	7,2529	15024,7	10741,6	210,045
243	516,15	519,84	371,66	9,4302	1571,4	7,2549	15054,6	10763,1	210,103
244	517,15	520,88	372,40	9,4961	1563,5	7,2569	15084,5	10784,7	210,161
245	518,15	521,91	373,15	9,5623	1555,6	7,2589	15114,5	10806,4	210,219
246	519,15	522,94	373,89	9,6289	1547,9	7,2609	15144,4	10828,0	210,277
247	520,15	523,98	374,64	9,6958	1540,1	7,2629	15174,3	10849,6	210,334
248	521,15	525,01	375,39	9,7631	1532,5	7,2649	15204,3	10871,2	210,392
249	522,15	526,04	376,13	9,8308	1524,8	7,2669	15234,2	10892,8	210,449
250	523,15	527,08	376,88	9,8988	1517,3	7,2689	15264,2	10914,5	210,506
251	524,15	528,11	377,63	9,9671	1509,7	7,2708	15294,1	10936,1	210,564
252	525,15	529,15	378,38	10,035	1502,3	7,2728	15324,1	10957,8	210,621
253	526,15	530,18	379,12	10,104	1494,8	7,2748	15354,0	10979,4	210,678
254	527,15	531,22	379,87	10,174	1487,5	7,2767	15384,0	11001,1	210,735
255	528,15	532,25	380,62	10,244	1480,1	7,2787	15414,0	11022,7	210,791
256	529,15	533,29	381,37	10,314	1472,9	7,2807	15444,0	11044,4	210,848
257	530,15	534,32	382,12	10,384	1465,6	7,2826	15474,0	11066,1	210,905
258	531,15	535,36	382,86	10,455	1458,4	7,2846	15504,0	11087,8	210,961
259	532,15	536,39	383,61	10,526	1451,3	7,2865	15534,0	11109,4	211,018
260	533,15	537,43	384,36	10,598	1444,2	7,2885	15564,0	11131,1	211,074
261	534,15	538,47	385,11	10,670	1437,2	7,2904	15594,0	11152,8	211,130
262	535,15	539,50	385,86	10,742	1430,2	7,2924	15624,0	11174,5	211,187
263	536,15	540,54	386,61	10,815	1423,2	7,2943	15654,0	11196,3	211,243
264	537,15	541,58	387,36	10,888	1416,3	7,2962	15684,1	11218,0	211,299
265	538,15	542,61	388,11	10,961	1409,4	7,2981	15714,1	11239,7	211,354
266	539,15	543,65	388,86	11,035	1402,6	7,3001	15744,1	11261,4	211,410
267	540,15	544,69	389,61	11,109	1395,8	7,3020	15774,2	11283,2	211,466
268	541,15	545,73	390,36	11,184	1389,1	7,3039	15804,3	11304,9	211,521
269	542,15	546,77	391,11	11,259	1382,4	7,3058	15834,3	11326,7	211,577
270	543,15	547,80	391,87	11,334	1375,8	7,3077	15864,4	11348,4	211,632
271	544,15	548,84	392,62	11,409	1369,2	7,3097	15894,5	11370,2	211,688
272	545,15	549,88	393,37	11,486	1362,6	7,3116	15924,5	11391,9	211,743
273	546,15	550,92	394,12	11,562	1356,1	7,3135	15954,6	11413,7	211,798
274	547,15	551,96	394,87	11,639	1349,6	7,3154	15984,7	11435,5	211,853
275	548,15	553,00	395,62	11,716	1343,1	7,3173	16014,8	11457,3	211,908
276	549,15	554,04	396,38	11,794	1336,7	7,3192	16044,9	11479,1	211,963
277	550,15	555,08	397,13	11,871	1330,4	7,3211	16075,0	11500,9	212,018
278	551,15	556,12	397,88	11,950	1324,1	7,3229	16105,2	11522,7	212,072
279	552,15	557,16	398,64	12,029	1317,8	7,3248	16135,3	11544,5	212,127

Air (*Continued*)

t	T	h	u	π_v	θ_p	s^u	H	U	S
280	553,15	558,20	399,39	12,108	1311,5	7,3267	16165,4	11566,3	212,182
281	554,15	559,24	400,14	12,187	1305,3	7,3286	16195,5	11588,1	212,236
282	555,15	560,28	400,90	12,267	1299,2	7,3305	16225,7	11609,9	212,290
283	556,15	561,32	401,65	12,348	1293,0	7,3323	16255,8	11631,8	212,345
284	557,15	562,36	402,40	12,428	1287,0	7,3342	16286,0	11653,6	212,399
285	558,15	563,40	403,16	12,509	1280,9	7,3361	16316,2	11675,5	212,453
286	559,15	564,44	403,91	12,591	1274,9	7,3379	16346,3	11697,3	212,507
287	560,15	565,49	404,67	12,673	1268,9	7,3398	16376,5	11719,2	212,561
288	561,15	566,53	405,42	12,755	1263,0	7,3417	16406,7	11741,0	212,615
289	562,15	567,57	406,18	12,838	1257,1	7,3435	16436,9	11762,9	212,668
290	563,15	568,61	406,93	12,921	1251,2	7,3454	16467,1	11784,8	212,722
291	564,15	569,66	407,69	13,004	1245,4	7,3472	16497,3	11806,7	212,776
292	565,15	570,70	408,45	13,088	1239,6	7,3491	16527,5	11828,6	212,829
293	566,15	571,74	409,20	13,173	1233,8	7,3509	16557,7	11850,5	212,882
294	567,15	572,79	409,96	13,258	1228,1	7,3528	16587,9	11872,4	212,936
295	568,15	573,83	410,71	13,343	1222,4	7,3546	16618,1	11894,3	212,989
296	569,15	574,87	411,47	13,428	1216,8	7,3564	16648,3	11916,2	213,042
297	570,15	575,92	412,23	13,514	1211,1	7,3583	16678,6	11938,1	213,095
298	571,15	576,96	412,99	13,601	1205,6	7,3601	16708,8	11960,0	213,148
299	572,15	578,01	413,74	13,688	1200,0	7,3619	16739,1	11982,0	213,201
300	573,15	579,05	414,50	13,775	1194,5	7,3637	16769,3	12003,9	213,254
301	574,15	580,10	415,26	13,862	1189,0	7,3656	16799,6	12025,9	213,307
302	575,15	581,14	416,02	13,951	1183,6	7,3674	16829,9	12047,8	213,359
303	576,15	582,19	416,77	14,039	1178,1	7,3692	16860,1	12069,8	213,412
304	577,15	583,23	417,53	14,128	1172,7	7,3710	16890,4	12091,8	213,465
305	578,15	584,28	418,29	14,217	1167,4	7,3728	16920,7	12113,7	213,517
306	579,15	585,32	419,05	14,307	1162,1	7,3746	16951,0	12135,7	213,569
307	580,15	586,37	419,81	14,398	1156,8	7,3764	16981,3	12157,7	213,622
308	581,15	587,42	420,57	14,488	1151,5	7,3782	17011,6	12179,7	213,674
309	582,15	588,46	421,33	14,579	1146,3	7,3800	17041,9	12201,7	213,726
310	583,15	589,51	422,09	14,671	1141,1	7,3818	17072,2	12223,7	213,778
311	584,15	590,56	422,85	14,763	1135,9	7,3836	17102,6	12245,7	213,830
312	585,15	591,61	423,61	14,855	1130,8	7,3854	17132,9	12267,7	213,882
313	586,15	592,65	424,37	14,948	1125,7	7,3872	17163,2	12289,8	213,934
314	587,15	593,70	425,13	15,041	1120,6	7,3890	17193,6	12311,8	213,985
315	588,15	594,75	425,89	15,135	1115,6	7,3908	17223,9	12333,8	214,037
316	589,15	595,80	426,65	15,229	1110,6	7,3926	17254,3	12355,9	214,089
317	590,15	596,85	427,41	15,324	1105,6	7,3943	17284,7	12377,9	214,140
318	591,15	597,90	428,18	15,419	1100,6	7,3961	17315,0	12400,0	214,192
319	592,15	598,94	428,94	15,514	1095,7	7,3979	17345,4	12422,0	214,243
320	593,15	599,99	429,70	15,610	1090,8	7,3997	17375,8	12444,1	214,294
321	594,15	601,04	430,46	15,707	1085,9	7,4014	17406,2	12466,2	214,345
322	595,15	602,09	431,23	15,804	1081,1	7,4032	17436,6	12488,3	214,396
323	596,15	603,14	431,99	15,901	1076,3	7,4050	17467,0	12510,4	214,447
324	597,15	604,19	432,75	15,999	1071,5	7,4067	17497,4	12532,5	214,498
325	598,15	605,24	433,51	16,097	1066,7	7,4085	17527,8	12554,6	214,549
326	599,15	606,29	434,28	16,196	1062,0	7,4102	17558,3	12576,7	214,600
327	600,15	607,34	435,04	16,295	1057,3	7,4120	17588,7	12598,8	214,651
328	601,15	608,40	435,81	16,395	1052,6	7,4137	17619,1	12620,9	214,702
329	602,15	609,45	436,57	16,495	1048,0	7,4155	17649,6	12643,1	214,752
330	603,15	610,50	437,33	16,595	1043,4	7,4172	17680,0	12665,2	214,803
331	604,15	611,55	438,10	16,696	1038,8	7,4190	17710,5	12687,3	214,853
332	605,15	612,60	438,86	16,798	1034,2	7,4207	17741,0	12709,5	214,904
333	606,15	613,65	439,63	16,900	1029,7	7,4224	17771,4	12731,6	214,954
334	607,15	614,71	440,39	17,002	1025,2	7,4242	17801,9	12753,8	215,004

t	T	h	u	π₀	θ₀	s⁰	H	U	S
335	608,15	615,76	441,16	17,105	1020,7	7,4259	17832,4	12776,0	215,054
336	609,15	616,81	441,93	17,208	1016,2	7,4276	17862,9	12798,2	215,104
337	610,15	617,87	442,69	17,312	1011,8	7,4294	17893,4	12820,3	215,154
338	611,15	618,92	443,46	17,417	1007,4	7,4311	17923,9	12842,5	215,204
339	612,15	619,97	444,22	17,521	1003,0	7,4328	17954,4	12864,7	215,254
340	613,15	621,03	444,99	17,627	998,65	7,4345	17984,9	12886,9	215,304
341	614,15	622,08	445,76	17,732	994,32	7,4363	18015,4	12909,1	215,354
342	615,15	623,13	446,52	17,839	990,00	7,4380	18046,0	12931,4	215,403
343	616,15	624,19	447,29	17,945	985,71	7,4397	18076,5	12953,6	215,453
344	617,15	625,24	448,06	18,053	981,45	7,4414	18107,1	12975,8	215,503
345	618,15	626,30	448,83	18,160	977,21	7,4431	18137,6	12998,1	215,552
346	619,15	627,35	449,60	18,269	972,99	7,4448	18168,2	13020,3	215,601
347	620,15	628,41	450,36	18,377	968,80	7,4465	18198,7	13042,5	215,651
348	621,15	629,46	451,13	18,487	964,63	7,4482	18229,3	13064,8	215,700
349	622,15	630,52	451,90	18,596	960,48	7,4499	18259,9	13087,1	215,749
350	623,15	631,58	452,67	18,706	956,36	7,4516	18290,5	13109,3	215,798
351	624,15	632,63	453,44	18,817	952,26	7,4533	18321,1	13131,6	215,847
352	625,15	633,69	454,21	18,928	948,18	7,4550	18351,7	13153,9	215,896
353	626,15	634,75	454,98	19,040	944,13	7,4567	18382,3	13176,2	215,945
354	627,15	635,80	455,75	19,152	940,09	7,4584	18412,9	13198,5	215,994
355	628,15	636,86	456,52	19,265	936,08	7,4600	18443,5	13220,8	216,043
356	629,15	637,92	457,29	19,378	932,10	7,4617	18474,1	13243,1	216,092
357	630,15	638,98	458,06	19,492	928,13	7,4634	18504,8	13265,4	216,140
358	631,15	640,03	458,83	19,606	924,19	7,4651	18535,4	13287,8	216,189
359	632,15	641,09	459,60	19,721	920,27	7,4668	18566,0	13310,1	216,237
360	633,15	642,15	460,37	19,836	916,37	7,4684	18596,7	13332,4	216,286
361	634,15	643,21	461,15	19,952	912,49	7,4701	18627,4	13354,8	216,334
362	635,15	644,27	461,92	20,068	908,63	7,4718	18658,0	13377,1	216,383
363	636,15	645,33	462,69	20,185	904,80	7,4734	18688,7	13399,5	216,431
364	637,15	646,39	463,46	20,302	900,98	7,4751	18719,4	13421,9	216,479
365	638,15	647,45	464,23	20,420	897,19	7,4768	18750,1	13444,2	216,527
366	639,15	648,51	465,01	20,539	893,41	7,4784	18780,8	13466,6	216,575
367	640,15	649,57	465,78	20,658	889,66	7,4801	18811,5	13489,0	216,623
368	641,15	650,63	466,55	20,777	885,93	7,4817	18842,2	13511,4	216,671
369	642,15	651,69	467,33	20,897	882,22	7,4834	18872,9	13533,8	216,719
370	643,15	652,75	468,10	21,017	878,52	7,4850	18903,6	13556,2	216,767
371	644,15	653,81	468,87	21,138	874,85	7,4867	18934,3	13578,6	216,815
372	645,15	654,87	469,65	21,260	871,20	7,4883	18965,1	13601,0	216,862
373	646,15	655,93	470,42	21,382	867,57	7,4900	18995,8	13623,5	216,910
374	647,15	656,99	471,20	21,505	863,95	7,4916	19026,6	13645,9	216,957
375	648,15	658,06	471,97	21,628	860,36	7,4933	19057,3	13668,3	217,005
376	649,15	659,12	472,75	21,752	856,79	7,4949	19088,1	13690,8	217,052
377	650,15	660,18	473,52	21,876	853,23	7,4965	19118,9	13713,2	217,100
378	651,15	661,24	474,30	22,001	849,70	7,4982	19149,6	13735,7	217,147
379	652,15	662,31	475,07	22,126	846,18	7,4998	19180,4	13758,2	217,194
380	653,15	663,37	475,85	22,252	842,68	7,5014	19211,2	13780,6	217,241
381	654,15	664,43	476,63	22,378	839,20	7,5031	19242,0	13803,1	217,289
382	655,15	665,50	477,40	22,505	835,74	7,5047	19272,8	13825,6	217,336
383	656,15	666,56	478,18	22,633	832,30	7,5063	19303,6	13848,1	217,383
384	657,15	667,63	478,96	22,761	828,88	7,5079	19334,4	13870,6	217,430
385	658,15	668,69	479,74	22,890	825,47	7,5095	19365,3	13893,1	217,476
386	659,15	669,75	480,51	23,019	822,09	7,5112	19396,1	13915,6	217,523
387	660,15	670,82	481,29	23,149	818,72	7,5128	19426,9	13938,2	217,570
388	661,15	671,88	482,07	23,279	815,37	7,5144	19457,8	13960,7	217,617
389	662,15	672,95	482,85	23,410	812,03	7,5160	19488,6	13983,2	217,663

t	T	h	u	π_0	θ_0	s^0	H	U	S
390	663,15	674,02	483,63	23,542	808,72	7,5176	19519,5	14005,8	217,710
391	664,15	675,08	484,40	23,674	805,42	7,5192	19550,4	14028,3	217,756
392	665,15	676,15	485,18	23,806	802,14	7,5208	19581,2	14050,9	217,803
393	666,15	677,21	485,96	23,940	798,87	7,5224	19612,1	14073,5	217,849
394	667,15	678,28	486,74	24,073	795,63	7,5240	19643,0	14096,0	217,896
395	668,15	679,35	487,52	24,208	792,40	7,5256	19673,9	14118,6	217,942
396	669,15	680,41	488,30	24,343	789,18	7,5272	19704,8	14141,2	217,988
397	670,15	681,48	489,08	24,478	785,99	7,5288	19735,7	14163,8	218,034
398	671,15	682,55	489,86	24,614	782,81	7,5304	19766,6	14186,4	218,080
399	672,15	683,62	490,64	24,751	779,64	7,5320	19797,5	14209,0	218,126
400	673,15	684,68	491,42	24,888	776,50	7,5336	19828,5	14231,6	218,172
401	674,15	685,75	492,20	25,026	773,37	7,5352	19859,4	14254,2	218,218
402	675,15	686,82	492,99	25,164	770,25	7,5367	19890,3	14276,9	218,264
403	676,15	687,89	493,77	25,303	767,16	7,5383	19921,3	14299,5	218,310
404	677,15	688,96	494,55	25,443	764,08	7,5399	19952,3	14322,2	218,356
405	678,15	690,03	495,33	25,583	761,01	7,5415	19983,2	14344,8	218,401
406	679,15	691,10	496,11	25,724	757,96	7,5431	20014,2	14367,5	218,447
407	680,15	692,17	496,90	25,866	754,93	7,5446	20045,2	14390,1	218,493
408	681,15	693,24	497,68	26,008	751,91	7,5462	20076,2	14412,8	218,538
409	682,15	694,31	498,46	26,150	748,91	7,5478	20107,1	14435,5	218,584
410	683,15	695,38	499,25	26,293	745,92	7,5493	20138,1	14458,2	218,629
411	684,15	696,45	500,03	26,437	742,95	7,5509	20169,1	14480,8	218,674
412	685,15	697,52	500,81	26,582	739,99	7,5525	20200,2	14503,5	218,720
413	686,15	698,59	501,60	26,727	737,05	7,5540	20231,2	14526,2	218,765
414	687,15	699,66	502,38	26,872	734,12	7,5556	20262,2	14549,0	218,810
415	688,15	700,73	503,17	27,019	731,21	7,5572	20293,2	14571,7	218,855
416	689,15	701,81	503,95	27,165	728,32	7,5587	20324,3	14594,4	218,900
417	690,15	702,88	504,74	27,313	725,43	7,5603	20355,3	14617,1	218,945
418	691,15	703,95	505,52	27,461	722,57	7,5618	20386,4	14639,9	218,990
419	692,15	705,02	506,31	27,610	719,71	7,5634	20417,4	14662,6	219,035
420	693,15	706,10	507,09	27,759	716,88	7,5649	20448,5	14685,4	219,080
421	694,15	707,17	507,88	27,909	714,05	7,5665	20479,6	14708,1	219,125
422	695,15	708,24	508,66	28,060	711,24	7,5680	20510,7	14730,9	219,170
423	696,15	709,31	509,45	28,211	708,45	7,5696	20541,8	14753,7	219,214
424	697,15	710,39	510,24	28,363	705,67	7,5711	20572,9	14776,5	219,259
425	698,15	711,46	511,02	28,515	702,90	7,5726	20604,0	14799,3	219,303
426	699,15	712,54	511,81	28,668	700,15	7,5742	20635,1	14822,1	219,348
427	700,15	713,61	512,60	28,822	697,41	7,5757	20666,2	14844,9	219,392
428	701,15	714,69	513,39	28,977	694,68	7,5772	20697,3	14867,7	219,437
429	702,15	715,76	514,17	29,132	691,97	7,5788	20728,4	14890,5	219,481
430	703,15	716,84	514,97	29,287	689,27	7,5803	20759,6	14913,3	219,526
431	704,15	717,91	515,75	29,444	686,59	7,5818	20790,7	14936,1	219,570
432	705,15	718,99	516,54	29,601	683,92	7,5834	20821,9	14959,0	219,614
433	706,15	720,06	517,33	29,758	681,26	7,5849	20853,0	14981,8	219,658
434	707,15	721,14	518,12	29,917	678,61	7,5864	20884,2	15004,7	219,702
435	708,15	722,22	518,91	30,075	675,98	7,5879	20915,4	15027,5	219,746
436	709,15	723,29	519,70	30,235	673,36	7,5894	20946,6	15050,4	219,790
437	710,15	724,37	520,49	30,395	670,76	7,5910	20977,7	15073,3	219,834
438	711,15	725,45	521,28	30,556	668,17	7,5925	21008,9	15096,2	219,878
439	712,15	726,52	522,07	30,718	665,59	7,5940	21040,1	15119,0	219,922
440	713,15	727,60	522,86	30,880	663,02	7,5955	21071,4	15141,9	219,966
441	714,15	728,68	523,65	31,043	660,47	7,5970	21102,6	15161,8	220,010
442	715,15	729,76	524,44	31,206	657,92	7,5985	21133,8	15187,7	220,053
443	716,15	730,84	525,23	31,371	655,40	7,6000	21165,0	15210,7	220,097
444	717,15	731,91	526,02	31,536	652,88	7,6015	21196,3	15233,6	220,140

Air (*Continued*)

t	T	h	u	π_0	θ_0	s^0	H	U	S
445	718,15	732,99	526,81	31,701	650,38	7,6030	21227,5	15256,5	220,184
446	719,15	734,07	527,61	31,867	647,88	7,6045	21258,8	15279,4	220,227
447	720,15	735,15	528,40	32,034	645,40	7,6060	21290,0	15302,4	220,271
448	721,15	736,23	529,19	32,202	642,94	7,6075	21321,3	15325,3	220,314
449	722,15	737,31	529,98	32,370	640,48	7,6090	21352,5	15348,3	220,358
450	723,15	738,39	530,78	32,539	638,04	7,6105	21383,8	15371,3	220,401
451	724,15	739,47	531,57	32,709	635,61	7,6120	21415,1	15394,2	220,444
452	725,15	740,55	532,36	32,879	633,19	7,6135	21446,4	15417,2	220,487
453	726,15	741,63	533,16	33,050	630,78	7,6150	21477,7	15440,2	220,530
454	727,15	742,71	533,95	33,222	628,38	7,6165	21509,0	15463,2	220,574
455	728,15	743,80	534,74	33,394	626,00	7·6180	21540,3	15486,2	220,617
456	729,15	744,88	535,54	33,567	623,63	7,6195	21571,6	15509,2	220,660
457	730,15	745,96	536,33	33,741	621,27	7,6209	21603,0	15532,2	220,703
458	731,15	747,04	537,13	33,915	618,92	7,6224	21634,3	15555,2	220,745
459	732,15	748,12	537,92	34,091	616,58	7,6239	21665,6	15578,2	220,788
460	733,15	749,21	538,72	34,267	614,25	7,6254	21697,0	15601,3	220,831
461	734,15	750,29	539,51	34,443	611,93	7,6269	21728,3	15624,3	220,874
462	735,15	751,37	540,31	34,620	609,63	7,6283	21759,7	15647,4	220,916
463	736,15	752,45	541,11	34,798	607,34	7,6298	21791,1	15670,4	220,959
464	737,15	753,54	541,90	34,977	605,05	7,6313	21822,5	15693,5	221,002
465	738,15	754,62	542,70	35,157	602,78	7,6327	21853,8	15716,6	221,044
466	739,15	755,71	543,50	35,337	600,52	7,6342	21885,2	15739,6	221,087
467	740,15	756,79	544,29	35,518	598,27	7,6357	21916,6	15762,7	221,129
468	741,15	757,87	545,09	35,699	596,03	7,6371	21948,0	15785,8	221,172
469	742,15	758,96	545,89	35,882	593,80	7,6386	21949,4	15808,9	221,214
470	743,15	760,04	546,69	36,065	591,59	7,6401	22010,9	15832,0	221,256
471	744,15	761,13	547,48	36,248	589,38	7,6415	22042,3	15855,1	221,298
472	745,15	762,21	548,28	36,433	587,18	7,6430	22073,7	15878,2	221,341
473	746,15	763,30	549,08	36,618	585,00	7,6444	22105,2	15901,4	221,383
474	747,15	764,39	549,88	36,804	582,82	7,6459	22136,6	15924,5	221,425
475	748,15	765,47	550,68	36,991	580,66	7,6473	22168,1	15947,6	221,467
476	749,15	766,56	551,48	37,178	578,50	7,6488	22199,5	15970,8	221,509
477	750,15	767,64	552,28	37,366	576,36	7,6502	22231,0	15993,9	221,551
478	751,15	768,73	553,08	37,555	574,22	7,6517	22262,5	16017,1	221,593
479	752,15	769,82	553,88	37,745	572,10	7,6531	22293,9	16040,3	221,635
480	753,15	770,91	554,68	37,935	569,98	7,6546	22325,4	16063,4	221,677
481	754,15	771,99	555,48	38,127	567,88	7,6560	22356,9	16086,6	221,718
482	755,15	773,08	556,28	38,318	565,78	7,6575	22388,4	16109,8	221,760
483	756,15	774,17	557,08	38,511	563,70	7,6589	22419,9	16133,0	221,802
484	757,15	775,26	557,88	38,704	561,62	7,6603	22451,4	16156,2	221,844
485	758,15	776,35	558,68	38,899	559,56	7,6618	22483,0	16179,4	221,885
486	759,15	777,43	559,48	39,094	557,50	7,6632	22514,5	16202,6	221,927
487	760,15	778,52	560,28	39,289	555,45	7,6647	22546,0	16225,8	221,968
488	761,15	779,61	561,09	39,486	553,42	7,6661	22577,6	16249,1	222,010
489	762,15	780,70	561,89	39,683	551,39	7,6675	22609,1	16272,3	222,051
490	763,15	781,79	562,69	39,881	549,37	7,6689	22640,7	16295,5	222,093
491	764,15	782,88	563,49	40,080	547,37	7,6704	22672,3	16318,8	222,134
492	765,15	783,97	564,30	40,279	545,37	7,6718	22703,8	16342,1	222,175
493	766,15	785,06	565,10	40,480	543,38	7,6732	22735,4	16365,3	222,216
494	767,15	786,15	565,90	40,681	541,40	7,6746	22767,0	16388,6	222,258
495	768,15	787,24	566,71	40,882	539,43	7,6761	22798,6	16411,9	222,299
496	769,15	788,33	567,51	41,085	537,47	7,6775	22830,2	16435,2	222,340
497	770,15	789,43	568,32	41,289	535,51	7,6789	22861,8	16458,4	222,381
498	771,15	790,52	569,12	41,493	533,57	7,6803	22893,4	16481,7	222,422
499	772,15	791,61	569,93	41,698	531,63	7,6817	22925,0	16505,0	222,463

t	T	h	u	π_n	θ_o	s^o	H	U	S
500	773,15	792,70	570,73	41,904	529,71	7,6831	22959,6	16528,4	222,504
501	774,15	793,79	571,54	42,110	527,79	7,6846	22988,3	16551,7	222,545
502	775,15	794,89	572,34	42,317	525,88	7,6860	23019,9	16575,0	222,586
503	776,15	795,98	573,15	42,526	523,98	7,6874	23051,6	16598,3	222,626
504	777,15	797,07	573,95	42,735	522,09	7,6888	23083,2	16621,7	222,667
505	778,15	798,17	574,76	42,944	520,21	7,6902	23114,9	16645,0	222,708
506	779,15	799,26	575,57	43,155	518,34	7,6916	23146,6	16668,4	222,749
507	780,15	800,35	576,37	43,366	516,47	7,6930	23178,2	16691,7	222,789
508	781,15	801,45	577,18	43,579	514,62	7,6944	23209,9	16715,1	222,830
509	782,15	802,54	577,99	43,792	512,77	7,6958	23241,6	16738,5	222,870
510	783,15	803,64	578,79	44,005	510,93	7,6972	23273,3	16761,9	222,911
511	784,15	804,73	579,60	44,220	509,10	7,6986	23305,0	16785,3	222,951
512	785,15	805,83	580,41	44,436	507,28	7,7000	23336,7	16808,6	222,992
513	786,15	806,92	581,22	44,652	505,46	7,7014	23368,4	16832,0	223,032
514	787,15	808,02	582,03	44,869	503,66	7,7028	23400,1	16855,5	223,072
515	788,15	809,11	582,83	45,087	501,86	7,7042	23431,9	16878,9	223,113
516	789,15	810,21	583,64	45,306	500,07	7,7056	23463,6	16902,3	223,153
517	790,15	811,30	584,45	45,525	498,29	7,7069	23495,4	16925,7	223,193
518	791,15	812,40	585,26	45,746	496,52	7,7083	23527,1	16949,2	223,233
519	792,15	813,50	586,07	45,967	494,75	7,7097	23558,9	16972,6	223,273
520	793,15	814,59	586,88	46,189	492,99	7,7111	23590,6	16996,1	223,313
521	794,15	815,69	587,69	46,412	491,24	7,7125	23622,4	17019,5	223,353
522	795,15	816,79	588,50	46,636	489,50	7,7139	23654,2	17043,0	223,393
523	796,15	817,89	589,31	46,860	487,77	7,7152	23686,0	17066,4	223,433
524	797,15	818,98	590,12	47,086	486,04	7,7166	23717,8	17089,9	223,473
525	798,15	820,08	590,93	47,312	484,33	7,7180	23749,6	17113,4	223,513
526	799,15	821,18	591,74	47,539	482,61	7,7194	23781,4	17136,9	223,553
527	800,15	822,28	592,56	47,767	480,91	7,7207	23813,2	17160,4	223,593
528	801,15	823,38	593,37	47,996	479,22	7,7221	23845,0	17183,9	223,633
529	802,15	824,48	594,18	48,226	477,53	7,7235	23876,8	17207,4	223,672
530	803,15	825,57	594,99	48,456	475,85	7,7249	23908,6	17230,9	223,712
531	804,15	826,67	595,80	48,688	474,18	7,7262	23940,5	17254,5	223,751
532	805,15	827,77	596,62	48,920	472,51	7,7276	23972,3	17278,0	223,791
533	806,15	828,87	597,43	49,153	470,85	7,7290	24004,2	17301,5	223,831
534	807,15	829,97	598,24	49,387	469,20	7,7303	24036,0	17325,1	223,870
535	808,15	831,07	599,05	49,622	467,56	7,7317	24067,9	17348,6	223,910
536	809,15	832,18	599,87	49,858	465,93	7,7330	24099,8	17372,2	223,949
537	810,15	833,28	600,68	50,095	464,30	7,7344	24131,7	17395,7	223,988
538	811,15	834,38	601,50	50,332	462,68	7,7358	24163,6	17419,3	224,028
539	812,15	835,48	602,31	50,571	461,06	7,7371	24195,4	17442,9	224,067
540	813,15	836,58	603,12	50,810	459,46	7,7385	24227,4	17466,5	224,106
541	814,15	837,68	603,94	51,050	457,86	7,7398	24259,3	17490,1	224,145
542	815,15	838,78	604,75	51,291	456,26	7,7412	24291,2	17513,7	224,185
543	816,15	839,89	605,57	51,533	454,68	7,7425	24323,1	17537,3	224,224
544	817,15	840,99	606,38	51,776	453,10	7,7439	24355,0	17560,9	224,263
545	818,15	842,09	607,20	52,020	451,53	7,7452	24387,0	17584,5	224,302
546	819,15	843,19	608,02	52,265	449,96	7,7466	24418,9	17608,1	224,341
547	820,15	844,30	608,83	52,510	448,41	7,7479	24450,8	17631,8	224,380
548	821,15	845,40	609,65	52,757	446,86	7,7493	24482,8	17655,4	224,419
549	822,15	846,50	610,47	53,004	445,31	7,7506	24514,8	17679,1	224,458
550	823,15	847,61	611,28	53,253	443,77	7,7520	24546,7	17702,7	224,497
551	824,15	848,71	612,10	53,502	442,24	7,7533	24578,7	17726,4	224,535
552	825,15	849,82	612,92	53,752	440,72	7,7546	24610,7	17750,0	224,574
553	826,15	850,92	613,73	54,004	439,20	7,7560	24642,7	17773,7	224,613
554	827,15	852,03	614,55	54,255	437,69	7,7573	24674,7	17797,4	224,652

t	T	h	u	π₀	θ₀	s°	H	U	S
555	828,15	853,13	615,37	54,508	436,19	7,7586	24706,7	17821,1	224,690
556	829,15	854,24	616,19	54,762	434,69	7,7600	24738,7	17844,8	224,729
557	830,15	855,34	617,01	55,016	433,20	7,7613	24770,7	17868,5	224,768
558	831,15	856,45	617,82	55,272	431,72	7,7626	24802,7	17892,2	224,806
559	832,15	857,55	618,64	55,529	430,24	7,7640	24834,7	17915,9	224,845
560	833,15	858,66	619,46	55,786	428,77	7,7653	24866,8	17939,6	224,883
561	834,15	859,77	620,28	56,045	427,30	7,7666	24898,8	17963,4	224,922
562	835,15	860,87	621,10	56,304	425,84	7,7680	24930,9	17987,1	224,960
563	836,15	861,98	621,92	56,565	424,39	7,7693	24962,9	18010,8	224,998
564	837,15	863,09	622,74	56,826	422,94	7,7706	24995,0	18034,6	225,037
565	838,15	864,19	623,56	57,088	421,50	7,7719	25027,1	18058,3	225,075
566	839,15	865,30	624,38	57,351	420,07	7,7732	25059,1	18082,1	225,113
567	840,15	866,41	625,20	57,616	418,64	7,7746	25091,2	18105,9	225,151
568	841,15	867,52	626,02	57,881	417,22	7,7759	25123,3	18129,6	225,189
569	842,15	868,63	626,84	58,147	415,80	7,7772	25155,4	18153,4	225,228
570	843,15	869,73	627,67	58,414	414,39	7,7785	25187,5	18177,2	225,266
571	844,15	870,84	628,49	58,682	412,99	7,7798	25219,6	18201,0	225,304
572	845,15	871,95	629,31	58,951	411,59	7,7811	25251,7	18224,8	225,342
573	846,15	873,06	630,13	59,221	410,20	7,7825	25283,8	18248,6	225,380
574	847,15	874,17	630,95	59,492	408,82	7,7838	25316,0	18272,4	225,418
575	848,15	875,28	631,78	59,763	407,44	7,7851	25348,1	18296,2	225,456
576	849,15	876,39	632,60	60,036	406,06	7,7864	25380,2	18320,1	225,494
577	850,15	877,50	633,42	60,310	404,70	7,7877	25412,4	18343,9	225,531
578	851,15	878,61	634,24	60,585	403,33	7,7890	25444,5	18367,7	225,569
579	852,15	879,72	635,07	60,861	401,98	7,7903	25476,7	18391,6	225,607
580	853,15	880,83	635,89	61,137	400,63	7,7916	25508,9	18415,4	225,645
581	854,15	881,94	636,72	61,415	399,28	7,7929	25541,0	18439,3	225,682
582	855,15	883,05	637,54	61,694	397,95	7,7942	25573,2	18463,2	225,720
583	856,15	884,16	638,36	61,974	396,61	7,7955	25605,4	18487,0	225,758
584	857,15	885,28	639,19	62,255	395,28	7,7968	25637,6	18510,9	225,795
585	858,15	886,39	640,01	62,536	393,96	7,7981	25669,8	18534,8	225,833
586	859,15	887,50	640,84	62,819	392,65	7,7994	25702,0	18558,7	225,870
587	860,15	888,61	641,66	63,103	391,34	7,8007	25734,2	18582,6	225,908
588	861,15	889,73	642,49	63,388	390,03	7,8020	25766,4	18606,5	225,945
589	862,15	890,84	643,32	63,673	388,73	7,8033	25798,7	18630,4	225,983
590	863,15	891,95	644,14	63,960	387,44	7,8046	25830,9	18654,3	226,020
591	864,15	893,06	644,97	64,248	386,15	7,8058	25863,1	18678,2	226,057
592	865,15	894,18	645,79	64,537	384,86	7,8071	25895,4	18702,2	226,095
593	866,15	895,29	646,61	64,827	383,59	7,8084	25927,6	18726,1	226,132
594	867,15	896,41	647,45	65,118	382,31	7,8097	25959,9	18750,1	226,169
595	868,15	897,52	648,27	65,409	381,05	7,8110	25992,2	18774,0	226,206
596	869,15	898,63	649,10	65,702	379,78	7,8123	26024,4	18798,0	226,243
597	870,15	899,75	649,93	65,996	378,53	7,8136	26056,7	18821,9	226,280
598	871,15	900,86	650,76	66,291	377,28	7,8148	26089,0	18845,9	226,318
599	872,15	901,98	651,58	66,587	376,03	7,8161	26121,3	18869,9	226,355
600	873,15	903,09	652,41	66,884	374,79	7,8174	26153,6	18893,9	226,392
601	874,15	904,21	653,24	67,183	373,55	7,8187	26185,9	18917,9	226,429
602	875,15	905,32	654,07	67,482	372,32	7,8199	26218,2	18941,8	226,466
603	876,15	906,44	654,90	67,782	371,10	7,8212	26250,5	18965,9	226,502
604	877,15	907,56	655,73	68,083	369,88	7,8225	26282,8	18989,9	226,539
605	878,15	908,67	656,56	68,385	368,66	7,8238	26315,2	19013,9	226,576
606	879,15	909,79	657,39	68,689	367,45	7,8250	26347,5	19037,9	226,613
607	880,15	910,91	658,22	68,993	366,25	7,8263	26379,9	19061,9	226,650
608	881,15	912,02	659,05	69,299	365,05	7,8276	26412,2	19086,0	226,686
609	882,15	913,14	659,88	69,605	363,85	7,8288	26444,6	19110,0	226,723

Air (*Continued*)

t	T	h	u	π_0	θ_0	s^0	H	U	S
610	883,15	914,26	660,71	69,913	362,66	7,8301	26476,9	19134,0	226,760
611	884,15	915,38	661,54	70,222	361,48	7,8314	26509,3	19158,1	226,796
612	885,15	916,49	662,37	70,531	360,30	7,8326	26541,7	19182,2	226,833
613	886,15	917,61	663,20	70,842	359,12	7,8339	26574,0	19206,2	226,870
614	887,15	918,73	664,03	71,154	357,95	7,8352	26606,4	19230,3	226,906
615	888,15	919,85	664,86	71,467	356,78	7,8364	26638,8	19254,4	226,943
616	889,15	920,97	665,69	71,781	355,62	7,8377	26671,2	19278,5	226,979
617	890,15	922,09	666,52	72,096	354,47	7,8389	26703,6	19302,6	227,015
618	891,15	923,21	667,36	72,413	353,31	7,8402	26736,0	19326,7	227,052
619	892,15	924,33	668,19	72,730	352,17	7,8414	26768,5	19350,8	227,087
620	893,15	925,45	669,02	73,048	351,02	7,8427	26800,9	19374,9	227,125
621	894,15	926,56	669,85	73,368	349,89	7,8440	26833,3	19399,0	227,161
622	895,15	927,68	670,69	73,689	348,75	7,8452	26865,8	19423,1	227,197
623	896,15	928,81	671,52	74,010	347,63	7,8465	26898,2	19447,2	227,233
624	897,15	929,93	672,35	74,333	346,50	7,8477	26930,7	19471,4	227,270
625	898,15	931,05	673,19	74,657	345,38	7,8490	26963,1	19495,5	227,306
626	899,15	932,17	674,02	74,982	344,27	7,8502	26995,6	19519,7	227,342
627	900,15	933,29	674,86	75,308	343,16	7,8514	27028,0	19543,8	227,378
628	901,15	934,41	675,69	75,636	342,05	7,8527	27060,5	19568,0	227,414
629	902,15	935,53	676,52	75,964	340,95	7,8539	27093,0	19592,2	227,450
630	903,15	936,65	677,36	76,294	339,86	7,8552	27125,5	19616,3	227,486
631	904,15	937,78	678,19	76,624	338,76	7,8564	27158,0	19640,5	227,522
632	905,15	938,90	679,03	76,956	337,68	7,8577	27190,5	19664,7	227,558
633	906,15	940,02	679,86	77,289	336,59	7,8589	27223,0	19688,9	227,594
634	907,15	941,14	680,70	77,623	335,51	7,8601	27255,5	19713,1	227,630
635	908,15	942,27	681,54	77,958	334,44	7,8614	27288,0	19737,3	227,665
636	909,15	943,39	682,37	78,295	333,37	7,8626	27320,5	19761,5	227,701
637	910,15	944,51	683,21	78,632	332,30	7,8638	27353,1	19785,7	227,737
638	911,15	945,64	684,04	78,971	331,24	7,8651	27385,6	19809,9	227,773
639	912,15	946,76	684,88	79,311	330,19	7,8663	27418,2	19834,2	227,808
640	913,15	947,88	685,72	79,652	329,13	7,8675	27450,7	19858,4	227,844
641	914,15	949,01	686,56	79,994	328,08	7,8688	27483,3	19882,7	227,880
642	915,15	950,13	687,39	80,337	327,04	7,8700	27515,8	19906,9	227,915
643	916,15	951,26	688,23	80,681	326,00	7,8712	27548,4	19931,2	227,951
644	917,15	952,38	689,07	81,027	324,96	7,8725	27581,0	19955,4	227,986
645	918,15	953,51	689,91	81,374	323,93	7,8737	27613,6	19979,7	228,022
646	919,15	954,63	690,74	81,722	322,90	7,8749	27646,1	20004,0	228,057
647	920,15	955,76	691,58	82,071	321,88	7,8761	27678,7	20028,2	228,093
648	921,15	956,88	692,42	82,421	320,86	7,8774	27711,3	20052,5	228,128
649	922,15	958,01	693,26	82,773	319,84	7,8786	27743,9	20076,8	228,164
650	923,15	959,14	694,10	83,125	318,83	7,8798	27776,6	20101,1	228,199
651	924,15	960,26	694,94	83,479	317,83	7,8810	27809,2	20125,4	228,234
652	925,15	961,39	695,78	83,834	316,82	7,8822	27841,8	20149,7	228,270
653	926,15	962,51	696,62	84,190	315,82	7,8835	27874,4	20174,0	228,305
654	927,15	963,64	697,46	84,548	314,83	7,8847	27907,1	20198,4	228,340
655	928,15	964,77	698,30	84,906	313,84	7,8859	27939,7	20222,7	228,375
656	929,15	965,90	699,14	85,266	312,85	7,8871	27972,4	20247,0	228,410
657	930,15	967,02	699,98	85,627	311,86	7,8883	28005,0	20271,4	228,446
658	931,15	968,15	700,82	85,989	310,88	7,8895	28037,7	20295,7	228,481
659	932,15	969,28	701,66	86,353	309,91	7,8907	28070,3	20320,1	228,516
660	933,15	970,41	702,50	86,717	308,94	7,8919	28103,0	20344,4	228,551
661	934,15	971,54	703,34	87,083	307,97	7,8932	28135,7	20368,8	228,586
662	935,15	972,66	704,18	87,450	307,00	7,8944	28168,4	20393,2	228,621
663	936,15	973,79	705,03	87,818	306,04	7,8956	28201,1	20417,5	228,656
664	937,15	974,92	705,87	88,188	305,09	7,8968	28233,8	20441,9	228,691

Air (*Continued*)

t	T	h	u	π_0	θ_0	s^0	H	U	S
665	938,15	976,05	706,71	88,558	304,13	7,8980	28266,5	20466,3	228,725
666	939,15	977,18	707,55	88,930	303,19	7,8992	28299,2	20490,7	228,760
667	940,15	978,31	708,39	89,304	302,24	7,9004	28331,9	20515,1	228,795
668	941,15	979,44	709,24	89,678	301,30	7,9016	28364,6	20539,5	228,830
669	942,15	980,57	710,08	90,054	300,36	7,9028	28397,3	20563,9	228,865
670	943,15	981,70	710,92	90,431	299,43	7,9040	28430,1	20588,3	228,899
671	944,15	982,83	711,77	90,809	298,49	7,9052	28462,8	20612,8	228,934
672	945,15	983,96	712,61	91,188	297,57	7,9064	28495,6	20637,2	228,969
673	946,15	985,09	713,45	91,569	296,64	7,9076	28528,3	20661,6	229,003
674	947,15	986,22	714,30	91,951	295,72	7,9088	28561,1	20686,1	229,038
675	948,15	987,36	715,14	92,334	294,81	7,9100	28593,8	20710,5	229,073
676	949,15	988,49	715,99	92,718	293,90	7,9112	28626,6	20735,0	229,107
677	950,15	989,62	716,83	93,104	292,99	7,9123	28659,4	20759,4	229,142
678	951,15	990,75	717,62	93,491	292,08	7,9135	28692,1	20783,9	229,176
679	952,15	991,88	718,52	93,879	291,18	7,9147	28724,9	20808,4	229,211
680	953,15	993,02	719,37	94,268	290,28	7,9159	28757,7	20832,8	229,245
681	954,15	994,15	720,21	94,659	289,39	7,9171	28790,5	20857,3	229,279
682	955,15	995,28	721,06	95,051	288,49	7,9183	28823,3	20881,8	229,314
683	956,15	996,41	721,90	95,444	287,61	7,9195	28856,1	20906,3	229,348
684	957,15	997,55	722,75	95,839	286,72	7,9207	28889,0	20930,8	229,382
685	958,15	998,68	723,60	96,235	285,84	7,9218	28921,8	20955,3	229,417
686	959,15	999,81	724,44	96,632	284,96	7,9230	28954,6	20979,8	229,451
687	960,15	1000,95	725,29	97,031	284,09	7,9242	28987,4	21004,4	229,485
688	961,15	1002,08	726,14	97,430	283,22	7,9254	29020,3	21028,9	229,519
689	962,15	1003,22	726,98	97,831	282,35	7,9266	29053,1	21053,4	229,553
690	963,15	1004,35	727,83	98,234	281,49	7,9277	29086,0	21077,9	229,588
691	964,15	1005,48	728,68	98,637	280,62	7,9289	29118,8	21102,5	229,622
692	965,15	1006,62	729,52	99,042	279,77	7,9301	29151,7	21127,0	229,656
693	966,15	1007,75	730,37	99,449	278,91	7,9313	29184,6	21151,6	229,690
694	967,15	1008,89	731,22	99,856	278,06	7,9324	29217,4	21176,2	229,724
695	968,15	1010,02	732,07	100,26	277,21	7,9336	29250,3	21200,7	229,758
696	969,15	1011,16	732,92	100,67	276,37	7,9348	29283,2	21225,3	229,792
697	970,15	1012,30	733,77	101,08	275,53	7,9360	29316,1	21249,9	229,826
698	971,15	1013,43	734,62	101,50	274,69	7,9371	29349,0	21274,5	229,859
699	972,15	1014,57	735,46	101,91	273,86	7,9383	29381,9	21299,0	229,893
700	973,15	1015,70	736,31	102,33	273,02	7,9395	29414,8	21323,6	229,927
701	974,15	1016,84	737,16	102,74	272,20	7,9406	29447,7	21348,2	229,961
702	975,15	1017,98	738,01	103,16	271,37	7,9418	29480,6	21372,8	229,995
703	976,15	1019,11	738,86	103,58	270,55	7,9430	29513,6	21397,5	230,028
704	977,15	1020,25	739,71	104,00	269,73	7,9441	29546,5	21422,1	230,062
705	978,15	1021,39	740,56	104,42	268,91	7,9453	29579,4	21446,7	230,096
706	979,15	1022,53	741,41	104,85	268,10	7,9465	29612,4	21471,3	230,130
707	980,15	1023,66	742,26	105,27	267,29	7,9476	29645,3	21496,0	230,163
708	981,15	1024,80	743,11	105,70	266,48	7,9488	29678,3	21520,6	230,197
709	982,15	1025,94	743,97	106,13	265,68	7,9499	29711,2	21545,2	230,230
710	983,15	1027,08	744,82	106,56	264,88	7,9511	29744,2	21569,9	230,264
711	984,15	1028,22	745,67	106,99	264,08	7,9523	29777,2	21594,6	230,297
712	985,15	1029,36	746,52	107,42	263,29	7,9534	29810,2	21619,2	230,331
713	986,15	1030,50	747,37	107,85	262,50	7,9546	29843,1	21643,9	230,364
714	987,15	1031,63	748,22	108,29	261,71	7,9557	29876,1	21668,6	230,398
715	988,15	1032,77	749,08	108,72	260,92	7,9569	29909,1	21693,2	230,431
716	989,15	1033,91	749,93	109,16	260,14	7,9580	29942,1	21717,9	230,465
717	990,15	1035,05	750,78	109,60	259,36	7,9592	29975,1	21742,6	230,498
718	991,15	1036,19	751,63	110,04	258,59	7,9603	30008,1	21767,3	230,531
719	992,15	1037,33	752,49	110,48	257,81	7,9615	30041,2	21792,0	230,565

t	T	h	u	π_0	θ_v	s^0	H	U	S
720	993,15	1038,47	753,34	110,92	257,04	7,9626	30074,2	21816,7	230,598
721	994,15	1039,61	754,19	111,37	256,27	7,9638	30107,2	21841,4	230,631
722	995,15	1040,75	755,05	111,81	255,51	7,9649	30140,2	21866,2	230,664
723	996,15	1041,90	755,90	112,26	254,75	7,9661	30173,3	21890,9	230,697
724	997,15	1043,04	756,75	112,71	253,99	7,9672	30206,3	21915,6	230,731
725	998,15	1044,18	757,61	113,16	253,23	7,9684	30239,4	21940,3	230,764
726	999,15	1045,32	758,46	113,61	252,48	7,9695	30272,4	21965,1	230,797
727	1000,15	1046,46	759,32	114,06	251,73	7,9706	30305,5	21989,8	230,830
728	1001,15	1047,60	760,17	114,52	250,98	7,9718	30338,6	22014,6	230,863
729	1002,15	1048,74	761,03	114,97	250,23	7,9729	30371,6	22039,3	230,896
730	1003,15	1049,89	761,88	115,43	249,49	7,9741	30404,7	22064,1	230,929
731	1004,15	1051,03	762,74	115,89	248,75	7,9752	30437,8	22088,9	230,962
732	1005,15	1052,17	763,59	116,35	248,02	7,9763	30470,9	22113,7	230,995
733	1006,15	1053,31	764,45	116,81	247,28	7,9775	30504,0	22138,4	231,028
734	1007,15	1054,46	765,30	117,27	246,55	7,9786	30537,1	22163,2	231,061
735	1008,15	1055,60	766,16	117,74	245,82	7,9797	30570,2	22188,0	231,094
736	1009,15	1056,74	767,02	118,20	245,10	7,9809	30603,3	22212,8	231,126
737	1010,15	1057,89	767,87	118,67	244,37	7,9820	30636,4	22237,6	231,159
738	1011,15	1059,03	768,73	119,14	243,65	7,9831	30669,5	22262,4	231,192
739	1012,15	1060,17	769,59	119,61	242,93	7,9843	30702,7	22287,2	231,225
740	1013,15	1061,32	770,44	120,08	242,22	7,9854	30735,8	22312,1	231,257
741	1014,15	1062,46	771,30	120,55	241,51	7,9865	30768,9	22336,9	231,290
742	1015,15	1063,61	772,16	121,03	240,80	7,9877	30802,1	22361,7	231,323
743	1016,15	1064,75	773,02	121,50	240,09	7,9888	30835,2	22386,5	231,355
744	1017,15	1065,90	773,87	121,98	239,39	7,9899	30868,4	22411,4	231,388
745	1018,15	1067,04	774,73	122,46	238,68	7,9910	30901,5	22436,2	231,421
746	1019,15	1068,19	775,59	122,94	237,98	7,9922	30934,7	22461,1	231,453
747	1020,15	1069,33	776,45	123,42	237,29	7,9933	30967,9	22485,9	231,486
748	1021,15	1070,48	777,31	123,91	236,59	7,9944	31001,1	22510,8	231,518
749	1022,15	1071,62	778,17	124,39	235,90	7,9955	31034,2	22535,7	231,551
750	1023,15	1072,77	779,02	124,88	235,21	7,9967	31067,4	22560,5	231,583
751	1024,15	1073,92	779,88	125,37	234,53	7,9978	31100,6	22585,4	231,616
752	1025,15	1075,06	780,74	125,86	233,84	7,9989	31133,8	22610,3	231,648
753	1026,15	1076,21	781,60	126,35	233,16	8,0000	31167,0	22635,2	231,680
754	1027,15	1077,36	782,46	126,84	232,48	8,0011	31200,2	22660,1	231,713
755	1028,15	1078,50	783,32	127,33	231,81	8,0022	31233,4	22685,0	231,745
756	1029,15	1079,65	784,18	127,83	231,13	8,0034	31266,7	22709,9	231,777
757	1030,15	1080,80	785,04	128,33	230,46	8,0045	31299,9	22734,8	231,810
758	1031,15	1081,94	785,90	128,82	229,79	8,0056	31333,1	22759,7	231,842
759	1032,15	1083,09	786,76	129,32	229,12	8,0067	31366,4	22784,6	231,874
760	1033,15	1084,24	787,62	129,83	228,46	8,0078	61399,6	22809,6	231,906
761	1034,15	1085,39	788,48	130,33	227,80	8,0089	31432,8	22834,5	231,938
762	1035,15	1086,54	789,34	130,83	227,14	8,0100	31466,1	22859,4	231,971
763	1036,15	1087,68	790,21	131,34	226,48	8,0111	31499,3	22884,4	232,003
764	1037,15	1088,83	791,07	131,85	225,83	8,0122	31532,6	22909,3	232,035
765	1038,15	1089,98	791,93	132,36	225,17	8,0134	31565,9	22934,3	232,067
766	1039,15	1091,13	792,79	132,87	224,52	8,0145	31599,1	22959,2	232,099
767	1040,15	1092,28	793,65	133,38	223,88	8,0156	31632,4	22984,2	232,131
768	1041,15	1093,43	794,52	133,89	223,23	8,0167	31665,7	23009,2	232,163
769	1042,15	1094,58	795,38	134,41	222,59	8,0178	31699,0	23034,1	232,195
770	1043,15	1095,73	796,24	134,93	221,95	8,0189	31732,3	23059,1	232,227
771	1044,15	1096,88	797,10	135,45	221,31	8,0200	31765,6	23084,1	232,259
772	1045,15	1098,03	797,97	135,97	220,67	8,0211	31798,9	23109,1	232,290
773	1046,15	1099,18	798,83	136,49	220,04	8,0222	31832,2	23134,1	232,322
774	1047,15	1100,33	799,69	137,01	219,41	8,0233	31865,5	23159,1	232,354

Air (*Continued*)

t	T	h	u	$\pi_\circ$	$\theta_\circ$	s°	H	U	S
775	1048,15	1101,48	800,56	137,54	218,78	8,0244	31898,8	23184,1	232,386
776	1049,15	1102,63	801,42	138,06	218,15	8,0255	31932,2	23209,1	232,418
777	1050,15	1103,78	802,28	138,59	217,53	8,0266	31965,5	23234,1	232,449
778	1051,15	1104,93	803,15	139,12	216,91	8,0277	31998,8	23259,1	232,481
779	1052,15	1106,08	804,01	139,65	216,29	8,0288	32032,2	23284,2	232,513
780	1053,15	1107,23	804,88	140,19	215,67	8,0299	32065,5	23309,2	232,545
781	1054,15	1108,39	805,74	140,72	215,05	8,0309	32098,9	23334,2	232,576
782	1055,15	1109,54	806,60	141,26	214,54	8,0320	32132,2	23359,3	232,608
783	1056,15	1110,69	807,47	141,80	213,83	8,0331	32165,6	23384,3	232,639
784	1057,15	1111,84	808,33	142,34	213,22	8,0342	32198,9	23409,4	232,671
785	1058,15	1112,99	809,20	142,88	212,61	8,0353	32232,3	23434,4	232,703
786	1059,15	1114,15	810,06	143,42	212,01	8,0364	32265,7	23459,5	232,734
787	1060,15	1115,30	810,93	143,96	211,41	8,0375	32299,1	23484,5	232,766
788	1061,15	1116,45	811,80	144,51	210,81	8,0386	32332,5	23509,6	232,797
789	1062,15	1117,61	812,66	145,06	210,21	8,0397	32365,8	23534,7	232,829
790	1063,15	1118,76	813,53	145,61	209,61	8,0407	32399,2	23559,8	232,860
791	1064,15	1119,91	814,39	146,16	209,02	8,0418	32432,6	23584,9	232,891
792	1065,15	1121,07	815,26	146,71	208,43	8,0429	32466,0	23610,0	232,923
793	1066,15	1122,22	816,13	147,26	207,84	8,0440	32499,5	23635,1	232,954
794	1067,15	1123,37	816,99	147,82	207,25	8,0451	32532,9	23660,2	232,985
795	1068,15	1124,53	817,86	148,38	206,67	8,0462	32566,3	23685,3	233,017
796	1069,15	1125,68	818,73	148,94	206,08	8,0472	32599,7	23710,4	233,048
797	1070,15	1126,84	819,60	149,50	205,50	8,0483	32633,2	23735,5	233,079
798	1071,15	1127,99	820,46	150,06	204,92	8,0494	32666,6	23760,6	233,111
799	1072,15	1129,14	821,33	150,63	204,35	8,0505	32700,0	23785,7	233,142
800	1073,15	1130,30	822,20	151,19	203,77	8,0516	32733,5	23810,9	233,173
801	1074,15	1131,45	823,07	151,76	203,20	8,0526	32766,9	23836,0	233,204
802	1075,15	1132,61	823,93	152,33	202,63	8,0537	32800,4	23861,2	233,235
803	1076,15	1133,77	824,80	152,90	202,06	8,0548	32833,9	23886,3	233,266
804	1077,15	1134,92	825,67	153,47	201,49	8,0558	32867,3	23911,5	233,297
805	1078,15	1136,08	826,54	154,05	200,93	8,0569	32900,8	23936,6	233,328
806	1079,15	1137,23	827,41	154,62	200,36	8,0580	32934,3	23961,8	233,359
807	1080,15	1138,39	828,28	155,20	199,80	8,0591	32967,8	23986,9	233,390
808	1081,15	1139,55	829,15	155,78	199,24	8,0601	33001,2	24012,1	233,421
809	1082,15	1140,70	830,02	156,36	198,69	8,0612	33034,7	24037,3	233,452
810	1083,15	1141,86	830,89	156,94	198,13	8,0623	33068,2	24062,5	233,483
811	1084,15	1143,02	831,76	157,53	197,58	8,0633	33101,7	24087,7	233,514
812	1085,15	1144,17	832,63	158,12	197,03	8,0644	33135,2	24112,9	233,545
813	1086,15	1145,33	833,50	158,70	196,48	8,0655	33168,7	24138,0	233,576
814	1087,15	1146,49	834,37	159,29	195,93	8,0665	33202,3	24163,2	233,607
815	1088,15	1147,64	835,24	159,88	195,38	8,0676	33235,8	24188,5	233,638
816	1089,15	1148,80	836,11	160,48	194,84	8,0687	33269,3	24213,7	233,669
817	1090,15	1149,96	836,98	161,07	194,30	8,0697	33302,8	24238,9	233,699
818	1091,15	1151,12	837,85	161,67	193,76	8,0708	33336,4	24264,1	233,730
819	1092,15	1152,28	838,72	162,27	193,22	8,0718	33369,9	24289,3	233,761
820	1093,15	1153,43	839,59	162,87	192,69	8,0729	33403,5	24314,6	233,791
821	1094,15	1154,59	840,46	163,47	192,15	8,0740	33437,0	24339,8	233,822
822	1095,15	1155,75	841,33	164,07	191,62	8,0750	33470,6	24365,0	233,853
823	1096,15	1156,91	842,21	164,68	191,09	8,0761	33504,1	24390,3	233,883
824	1097,15	1158,07	843,08	165,29	190,56	8,0771	33537,7	24415,5	233,914
825	1098,15	1159,23	843,95	165,90	190,03	8,0782	33571,3	24440,8	233,945
826	1099,15	1160,39	844,82	166,51	189,51	8,0793	33604,8	24466,0	233,975
827	1100,15	1161,55	845,69	167,12	188,99	8,0803	33638,4	24491,3	234,006
828	1101,15	1162,71	846,57	167,73	188,47	8,0814	33672,0	24516,6	234,036
829	1102,15	1163,87	847,44	168,35	187,95	8,0824	33705,6	24541,9	234,067

Air (*Continued*)

t	T	h	u	π_θ	θ_θ	s°	H	U	S
830	1103,15	1165,03	848,31	168,97	187,43	8,0835	33739,2	24567,1	234,097
831	1104,15	1166,19	849,19	169,59	186,91	8,0845	33772,8	24592,4	234,128
832	1105,15	1167,35	850,06	170,21	186,40	8,0856	33806,4	24617,7	234,158
833	1106,15	1168,51	850,93	170,83	185,89	8,0866	33840,0	24643,0	234,188
834	1107,15	1169,67	851,81	171,46	185,38	8,0877	33873,6	24668,3	234,219
835	1108,15	1170,83	852,68	172,09	184,87	8,0887	33907,2	24693,6	234,249
836	1109,15	1171,99	853,55	172,72	184,36	8,0898	33940,8	24718,9	234,279
837	1110,15	1173,15	854,43	173,35	183,86	8,0908	33974,5	24744,2	234,310
838	1111,15	1174,31	855,30	173,98	183,35	8,0919	34008,1	24769,5	234,340
839	1112,15	1175,47	856,18	174,61	182,85	8,0929	34041,7	24794,9	234,370
840	1113,15	1176,64	857,05	175,25	182,35	8,0939	34075,4	24820,2	234,401
841	1114,15	1177,80	857,93	175,89	181,85	8,0950	34109,0	24845,5	234,431
842	1115,15	1178,96	858,80	176,53	181,36	8,0960	34142,7	24870,9	234,461
843	1116,15	1180,12	859,68	177,17	180,86	8,0971	34176,3	24896,2	234,491
844	1117,15	1181,28	860,55	177,81	180,37	8,0981	34210,0	24921,5	234,521
845	1118,15	1182,45	861,43	178,46	179,88	8,0992	34243,7	24946,9	234,551
846	1119,15	1183,61	862,30	179,10	179,39	8,1002	34277,3	24972,3	234,581
847	1120,15	1184,77	863,18	179,75	178,90	8,1012	34311,0	24997,6	234,612
848	1121,15	1185,93	864,05	180,40	178,41	8,1023	34344,7	25023,0	234,642
849	1122,15	1187,10	864,93	181,06	177,93	8,1033	34378,4	25048,3	234,672
850	1123,15	1188,26	865,81	181,71	177,44	8,1043	34412,0	25073,7	234,702
851	1124,15	1189,42	866,68	182,37	176,96	8,1054	34445,7	25099,1	234,732
852	1125,15	1190,59	867,56	183,03	176,48	8,1064	34479,4	25124,5	234,762
853	1126,15	1191,75	868,43	183,69	176,01	8,1074	34513,1	25149,9	234,792
854	1127,15	1192,92	869,31	184,35	175,53	8,1085	34546,8	25175,3	234,821
855	1128,15	1194,08	870,19	185,01	175,05	8,1095	34580,6	25200,7	234,851
856	1129,15	1195,24	871,07	185,68	174,58	8,1105	34614,3	25226,1	234,881
857	1130,15	1196,41	871,94	186,35	174,11	8,1116	34648,0	25251,5	234,911
858	1131,15	1197,57	872,82	187,02	173,64	8,1126	34681,7	25276,9	234,941
859	1132,15	1198,74	873,70	187,69	173,17	8,1136	34715,4	25302,3	234,971
860	1133,15	1199,90	874,58	188,36	172,70	8,1147	34749,2	25327,7	235,001
861	1134,15	1201,07	875,45	189,04	172,24	8,1157	34782,9	25353,1	235,030
862	1135,15	1202,23	876,33	189,71	171,78	8,1167	34816,7	25378,6	235,060
863	1136,15	1203,40	877,21	190,39	171,31	8,1177	34850,4	25404,0	235,090
864	1137,15	1204,56	878,09	191,08	170,85	8,1188	34884,2	25429,4	235,119
865	1138,15	1205,73	878,97	191,76	170,39	8,1198	34917,9	25454,9	235,149
866	1139,15	1206,90	879,85	192,44	169,94	8,1208	34951,7	25480,3	235,179
867	1140,15	1208,06	880,72	193,13	169,48	8,1218	34985,4	25505,8	235,208
868	1141,15	1209,23	881,60	193,82	169,03	8,1229	35019,2	25531,2	235,238
869	1142,15	1210,39	882,48	194,51	168,57	8,1239	35053,0	25556,7	235,268
870	1143,15	1211,56	883,36	195,20	168,12	8,1249	35086,8	25582,2	235,297
871	1144,15	1212,73	884,24	195,90	167,67	8,1259	35120,6	25607,6	235,327
872	1145,15	1213,89	885,12	196,59	167,22	8,1269	35154,3	25633,1	235,356
873	1146,15	1215,06	886,00	197,29	166,78	8,1280	35188,1	25658,6	235,386
874	1147,15	1216,23	886,88	197,99	166,33	8,1290	35221,9	25684,1	235,415
875	1148,15	1217,39	887,76	198,70	165,89	8,1300	35255,7	25709,6	235,445
876	1149,15	1218,56	888,64	199,40	165,45	8,1310	35289,5	25735,0	235,474
877	1150,15	1219,73	889,52	200,11	165,01	8,1320	35323,4	25760,5	235,503
878	1151,15	1220,90	890,40	200,82	164,57	8,1330	35357,2	25786,0	235,533
879	1152,15	1222,06	891,28	201,53	164,13	8,1341	35391,0	25811,5	235,562
880	1153,15	1223,23	892,16	202,24	163,69	8,1351	35424,8	25837,1	235,592
881	1154,15	1224,40	893,04	202,95	163,26	8,1361	35458,6	25862,6	235,621
882	1155,15	1225,57	893,93	203,67	162,82	8,1371	35492,5	25888,1	235,650
883	1156,15	1226,74	894,81	204,39	162,39	8,1381	35526,3	25913,6	235,679
884	1157,15	1227,91	895,69	205,11	161,96	8,1391	35560,2	25939,1	235,709

t	T	h	u	π_0	θ_0	s^0	H	U	S
885	1158,15	1229,07	896,57	205,83	161,53	8,1401	35594,0	25964,7	235,738
886	1159,15	1230,24	897,45	206,56	161,10	8,1411	35627,9	25990,2	235,767
887	1160,15	1231,41	898,33	207,28	160,68	8,1421	35661,7	26015,8	235,796
888	1161,15	1232,58	899,22	208,01	160,25	8,1431	35695,6	26041,3	235,826
889	1162,15	1233,75	900,10	208,74	159,83	8,1442	35729,4	26066,8	235,855
890	1163,15	1234,92	900,98	209,48	159,41	8,1452	35763,3	26092,4	235,884
891	1164,15	1236,09	901,86	210,21	158,99	8,1462	35797,2	26118,0	235,913
892	1165,15	1237,26	902,75	210,95	158,57	8,1472	35831,1	26143,5	235,942
893	1166,15	1238,43	903,63	211,69	158,15	8,1482	35864,9	26169,1	235,971
894	1167,15	1239,60	904,51	212,43	157,73	8,1492	35898,8	26194,7	236,000
895	1168,15	1240,77	905,39	213,17	157,32	8,1502	35932,7	26220,2	236,029
896	1169,15	1241,94	906,28	213,91	156,91	8,1512	35966,6	26245,8	236,058
897	1170,15	1243,11	907,16	214,66	156,49	8,1522	36000,5	26271,4	236,087
898	1171,15	1244,28	908,05	215,41	156,08	8,1532	36034,4	26297,0	236,116
899	1172,15	1245,45	908,93	216,16	155,67	8,1542	36068,3	26322,6	236,145
900	1173,15	1246,62	909,81	216,91	155,27	8,1552	36102,2	26348,2	236,174
901	1174,15	1247,80	910,70	217,67	154,86	8,1562	36136,2	26373,8	236,203
902	1175,15	1248,97	911,58	218,43	154,45	8,1572	36170,1	26399,4	236,232
903	1176,15	1250,14	912,47	219,19	154,05	8,1582	36204,0	26425,0	236,261
904	1177,15	1251,31	913,35	219,95	153,65	8,1592	36237,9	26450,6	236,289
905	1178,15	1252,48	914,23	220,71	153,24	8,1602	36271,9	26476,2	236,318
906	1179,15	1253,65	915,12	221,48	152,84	8,1612	36305,8	26501,9	236,347
907	1180,15	1254,83	916,00	222,25	152,45	8,1621	36339,7	26527,5	236,376
908	1181,15	1256,00	916,89	223,02	152,05	8,1631	36373,7	26553,1	236,405
909	1182,15	1257,17	917,77	223,79	151,65	8,1641	36407,6	26578,8	236,433
910	1183,15	1258,34	918,66	224,56	151,26	8,1651	36441,6	26604,4	236,462
911	1184,15	1259,51	919,55	225,34	150,86	8,1661	36475,6	26630,0	236,491
912	1185,15	1260,69	920,43	226,12	150,47	8,1671	36509,5	26655,7	236,519
913	1186,15	1261,86	921,32	226,90	150,08	8,1681	36543,5	26681,3	236,548
914	1187,15	1263,03	922,20	227,68	149,69	8,1691	36577,5	26707,0	236,577
915	1188,15	1264,21	923,09	228,46	149,30	8,1701	36611,4	26732,7	236,605
916	1189,15	1265,38	923,98	229,25	148,91	8,1711	36645,4	26758,3	236,634
917	1190,15	1266,55	924,86	230,04	148,53	8,1720	36679,4	26784,0	236,662
918	1191,15	1267,73	925,75	230,83	148,14	8,1730	36713,4	26809,7	236,691
919	1192,15	1268,90	926,64	231,62	147,76	8,1740	36747,4	26835,4	236,719
920	1193,15	1270,08	927,52	232,42	147,38	8,1750	36781,4	26861,0	236,748
921	1194,15	1271,25	928,41	233,22	147,00	8,1760	36815,4	26886,7	236,776
922	1195,15	1272,42	929,30	234,02	146,62	8,1770	36849,4	26912,4	236,805
923	1196,15	1273,60	930,18	234,82	146,24	8,1779	36883,4	26938,1	236,833
924	1197,15	1274,77	931,07	235,62	145,86	8,1789	36917,4	26963,8	236,862
925	1198,15	1275,95	931,96	236,43	145,48	8,1799	36951,4	26989,5	236,890
926	1199,15	1277,12	932,85	237,24	145,11	8,1809	36985,4	27015,2	236,919
927	1200,15	1278,30	933,73	238,05	144,74	8,1819	37019,5	27040,9	236,947
928	1201,15	1279,47	934,62	238,86	144,36	8,1828	37053,5	27066,7	236,975
929	1202,15	1280,65	935,51	239,68	143,99	8,1838	37087,5	27092,4	237,004
930	1203,15	1281,82	936,40	240,49	143,62	8,1848	37121,6	27118,1	237,032
931	1204,15	1283,00	937,29	241,31	143,25	8,1858	37155,6	27143,8	237,060
932	1205,15	1284,17	938,18	242,14	142,89	8,1868	37189,7	27169,6	237,088
933	1206,15	1285,35	939,06	242,96	142,52	8,1877	37223,7	27195,3	237,117
934	1207,15	1286,53	939,95	243,79	142,15	8,1887	37257,8	27221,0	237,145
935	1208,15	1287,70	940,84	244,61	141,79	8,1897	37291,8	27246,8	237,173
936	1209,15	1288,88	941,73	245,44	141,43	8,1907	37325,9	27272,5	237,201
937	1210,15	1290,05	942,62	246,28	141,07	8,1916	37360,0	27298,3	237,229
938	1211,15	1291,23	943,51	247,11	140,71	8,1926	37394,0	27324,0	237,258
939	1212,15	1292,41	944,40	247,95	140,35	8,1936	37428,1	27349,8	237,286

r	T	h	u	π.	θ.	s°	H	U	S
940	1213,15	1293,58	945,29	248,79	139,99	8,1945	37462,2	27375,6	237,314
941	1214,15	1294,76	946,18	249,63	139,63	8,1955	37496,3	27401,3	237,342
942	1215,15	1295,94	947,07	250,48	139,28	8,1965	37530,4	27427,1	237,370
943	1216,15	1297,12	947,96	251,32	138,92	8,1974	37564,5	27452,9	237,398
944	1217,15	1298,29	948,85	252,17	138,57	8,1984	37598,6	27478,7	237,426
945	1218,15	1299,47	949,74	253,02	138,21	8,1994	37632,7	27504,5	237,454
946	1219,15	1300,65	950,63	253,87	137,86	8,2003	37666,8	27530,3	237,482
947	1220,15	1301,83	951,52	254,73	137,51	8,2013	37700,9	27556,0	237,510
948	1221,15	1303,00	952,41	255,59	137,16	8,2023	37735,0	27581,8	237,538
949	1222,15	1304,18	953,30	256,45	136,82	8,2032	37769,1	27607,6	237,566
950	1223,15	1305,36	954,19	257,31	136,47	8,2042	37803,2	27633,5	237,594
951	1224,15	1306,54	955,09	258,17	136,12	8,2052	37837,4	27659,3	237,622
952	1225,15	1307,72	955,98	259,04	135,78	8,2061	37871,5	27685,1	237,650
953	1226,15	1308,90	956,87	259,91	135,43	8,2071	37905,6	27710,9	237,677
954	1227,15	1310,07	957,76	260,78	135,09	8,2081	37939,8	27736,7	237,705
955	1228,15	1311,25	958,65	261,65	134,75	8,2090	37973,9	27762,5	237,733
956	1229,15	1312,43	959,54	262,53	134,41	8,2100	38008,0	27788,4	237,761
957	1230,15	1313,61	960,44	263,41	134,07	8,2109	38042,2	27814,2	237,789
958	1231,15	1314,79	961,33	264,29	133,73	8,2119	38076,3	27840,1	237,816
959	1232,15	1315,97	962,22	265,17	133,40	8,2128	38110,5	27865,9	237,844
960	1233,15	1317,15	963,11	266,06	133,06	8,2138	38144,7	27891,7	237,872
961	1234,15	1318,33	964,01	266,95	132,73	8,2148	38178,8	27917,6	237,900
962	1235,15	1319,51	964,90	267,84	132,39	8,2157	38213,0	27943,4	237,927
963	1236,15	1320,69	965,79	268,73	132,06	8,2167	38247,2	27969,3	237,955
964	1237,15	1321,87	966,68	269,62	131,73	8,2176	38281,3	27995,2	237,982
965	1238,15	1323,05	967,58	270,52	131,40	8,2186	38315,5	28021,0	238,010
966	1239,15	1324,23	968,47	271,42	131,07	8,2195	38349,7	28046,9	238,038
967	1240,15	1325,41	969,36	272,32	130,74	8,2205	38383,9	28072,8	238,065
968	1241,15	1326,59	970,26	273,23	130,41	8,2214	38418,1	28098,7	238,093
969	1242,15	1327,77	971,15	274,13	130,08	8,2224	38452,3	28124,5	238,120
970	1243,15	1328,95	972,05	275,04	129,76	8,2233	38486,5	28150,4	238,148
971	1244,15	1330,13	972,94	275,95	129,43	8,2243	38520,7	28176,3	238,175
972	1245,15	1331,32	973,83	276,87	129,11	8,2252	38554,9	28202,2	238,203
973	1246,15	1332,50	974,73	277,78	128,79	8,2262	38589,1	28228,1	238,230
974	1247,15	1333,68	975,62	278,70	128,47	8,2271	38623,3	28254,0	238,258
975	1248,15	1334,86	976,52	279,62	128,15	8,2281	38657,5	28279,9	238,285
976	1249,15	1336,04	977,41	280,54	127,83	8,2290	38691,8	28305,8	238,313
977	1250,15	1337,22	978,31	281,47	127,51	8,2300	38726,0	28331,7	238,340
978	1251,15	1338,41	979,20	282,40	127,19	8,2309	38760,2	28357,7	238,367
979	1252,15	1339,59	980,10	283,33	126,87	8,2319	38794,5	28383,6	238,395
980	1253,15	1340,77	980,99	284,26	126,56	8,2328	38828,7	28409,5	238,422
981	1254,15	1341,95	981,89	285,20	126,24	8,2337	38862,9	28435,4	238,449
982	1255,15	1343,14	982,78	286,14	125,93	8,2347	38897,2	28461,4	238,477
983	1256,15	1344,32	983,68	287,08	125,62	8,2356	38931,4	28487,3	238,504
984	1257,15	1345,50	984,57	288,02	125,31	8,2366	38965,7	28513,2	238,531
985	1258,15	1346,68	985,47	288,96	125,00	8,2375	39000,0	28539,2	238,558
986	1259,15	1347,87	986,37	289,91	124,69	8,2385	39034,2	28565,1	238,586
987	1260,15	1349,05	987,26	290,86	124,38	8,2394	39068,5	28591,1	238,613
988	1261,15	1350,23	988,16	291,81	124,07	8,2403	39102,8	28617,0	238,640
989	1262,15	1351,42	989,05	292,77	123,76	8,2413	39137,0	28643,0	238,667
990	1263,15	1352,60	989,95	293,73	123,46	8,2422	39171,3	28669,0	238,694
991	1264,15	1353,78	990,85	294,69	123,15	8,2431	39205,6	28694,9	238,722
992	1265,15	1354,97	991,74	295,65	122,85	8,2441	39239,9	28720,9	238,749
993	1266,15	1356,15	992,64	296,61	122,55	8,2450	39274,2	28746,9	238,776
994	1267,15	1357,34	993,54	297,58	122,24	8,2460	39308,5	28772,9	238,803

t	T	h	u	π_0	θ_0	s°	H	U	S
995	1268,15	1358,52	994,43	298,55	121,94	8,2469	39342,8	28708,8	238,830
996	1269,15	1359,70	995,33	299,52	121,64	8,2478	39377,1	28824,8	238,857
997	1270,15	1360,89	996,23	300,50	121,34	8,2488	39411,4	28850,8	238,884
998	1271,15	1362,07	997,13	301,48	121,05	8,2497	39445,7	28876,8	238,911
999	1272,15	1363,26	998,03	302,46	120,75	8,2506	39480,0	28902,8	238,938
1000	1273,15	1364,44	998,92	303,44	120,45	8,2515	39514,3	28928,8	238,965
1001	1274,15	1365,63	999,82	304,42	120,16	8,2525	39548,6	28954,8	238,992
1002	1275,15	1366,81	1000,72	305,41	119,86	8,2534	39582,9	28980,8	239,019
1003	1276,15	1368,00	1001,62	306,40	119,57	8,2543	39617,3	29006,8	239,046
1004	1277,15	1369,19	1002,52	307,39	119,28	8,2553	39651,6	29032,8	239,073
1005	1278,15	1370,37	1003,41	308,39	118,98	8,2562	39685,9	29058,9	239,099
1006	1279,15	1371,56	1004,31	309,39	118,69	8,2571	39720,3	29084,9	239,126
1007	1280,15	1372,74	1005,21	310,38	118,40	8,2580	39754,6	29110,9	239,153
1008	1281,15	1373,93	1006,11	311,39	118,11	8,2590	39789,0	29137,0	239,180
1009	1282,15	1375,11	1007,01	312,40	117,83	8,2599	39823,3	29163,0	239,207
1010	1283,15	1376,30	1007,91	313,40	117,54	8,2608	39857,7	29189,0	239,233
1011	1284,15	1377,49	1008,81	314,41	117,25	8,2617	39892,0	29215,1	239,260
1012	1285,15	1378,67	1009,71	315,43	116,97	8,2627	39926,4	29241,1	239,287
1013	1286,15	1379,86	1010,61	316,44	116,68	8,2636	39960,7	29267,2	239,314
1014	1287,15	1381,05	1011,51	317,46	116,40	8,2645	39995,1	29293,2	239,340
1015	1288,15	1382,23	1012,41	318,48	116,12	8,2654	40029,5	29319,3	239,367
1016	1289,15	1383,42	1013,31	319,51	115,83	8,2664	40063,9	29345,3	239,394
1017	1290,15	1384,61	1014,21	320,53	115,55	3,2673	40098,2	29371,4	239,420
1018	1291,15	1385,79	1015,11	321,56	115,27	8,2682	40132,6	29397,5	239,447
1019	1292,15	1386,98	1016,01	322,59	114,99	8,2691	40167,0	29423,5	239,474
1020	1293,15	1388,17	1016,91	323,63	114,71	8,2700	40201,4	29449,6	239,500
1021	1294,15	1389,36	1017,81	324,66	114,44	8,2710	40235,8	29475,7	239,527
1022	1295,15	1390,55	1018,71	325,70	114,16	8,2719	40270,2	29501,8	239,553
1023	1296,15	1391,73	1019,61	326,74	113,88	8,2728	40304,6	29527,9	239,580
1024	1297,15	1392,92	1020,51	327,79	113,61	8,2737	40339,0	29554,0	239,607
1025	1298,15	1394,11	1021,41	328,83	113,33	8,2746	40373,4	29580,1	239,633
1026	1299,15	1395,30	1022,31	329,88	113,06	8,2755	40407,8	29606,2	239,660
1027	1300,15	1396,49	1023,21	330,94	112,79	8,2765	40442,2	29632,3	239,686
1028	1301,15	1397,67	1024,11	331,99	112,51	8,2774	40476,7	29658,4	239,713
1029	1302,15	1398,86	1025,02	333,05	112,24	8,2783	40511,1	29684,5	239,739
1030	1303,15	1400,05	1025,92	334,11	111,97	8,2792	40545,5	29710,6	239,765
1031	1304,15	1401,24	1026,82	335,17	111,70	8,2801	40579,9	29736,7	239,792
1032	1305,15	1402,43	1027,72	336,24	111,43	8,2810	40614,4	29762,8	239,818
1033	1306,15	1403,62	1028,62	337,31	111,17	8,2819	40648,8	29788,9	239,845
1034	1307,15	1404,81	1029,53	338,38	110,90	8,2828	40683,3	29815,1	239,871
1035	1308,15	1406,00	1030,43	339,45	110,63	8,2837	40717,7	29841,2	239,897
1036	1309,15	1407,19	1031,33	340,53	110,37	8,2847	40752,1	29867,3	239,924
1037	1310,15	1408,38	1032,23	341,61	110,10	8,2856	40786,6	29893,5	239,950
1038	1311,15	1409,57	1033,14	342,69	109,84	8,2865	40821,1	29919,6	239,976
1039	1312,15	1410,76	1034,04	343,77	109,58	8,2874	40855,5	29945,8	240,002
1040	1313,15	1411,95	1034,94	344,86	109,31	8,2883	40890,0	29971,9	240,029
1041	1314,15	1413,14	1035,84	345,95	109,05	8,2892	40924,4	29998,1	240,055
1042	1315,15	1414,33	1036,75	347,04	108,79	8,2901	40958,9	30024,2	240,081
1043	1316,15	1415,52	1037,65	348,14	108,53	8,2910	40993,4	30050,4	240,107
1044	1317,15	1416,71	1038,55	349,24	108,27	8,2919	41027,9	30076,5	240,134
1045	318,15	1417,90	1039,46	350,34	108,02	8,2928	41062,3	30102,7	240,160
1046	319,15	1419,09	1040,36	351,44	107,76	8,2937	41096,8	30128,9	240,186
1047	320,15	1420,28	1041,27	352,55	107,50	8,2946	41131,3	30155,0	240,212
1048	321,15	1421,47	1042,17	353,66	107,24	8,2955	41165,8	30181,2	240,238
1049	322,15	1422,66	1043,07	354,77	106,99	8,2964	41200,3	30207,4	240,264

t	T	h	u	π₀	θ₀	s°	H	U	S
1050	1323,15	1423,85	1043,98	355,88	106,74	8,2973	41234,8	30233,6	240,290
1051	1324,15	1425,05	1044,88	357,00	106,48	8,2982	41269,3	30259,8	240,316
1052	1325,15	1426,24	1045,79	358,12	106,23	8,2991	41303,8	30286,0	240,342
1053	1326,15	1427,43	1046,69	359,24	105,98	8,3000	41338,3	30312,2	240,368
1054	1327,15	1428,62	1047,60	360,37	105,72	8,3009	41372,8	30338,4	240,395
1055	1328,15	1429,81	1048,50	361,50	105,47	8,3018	41407,4	30364,6	240,421
1056	1329,15	1431,00	1049,41	362,63	105,22	8,3027	41441,9	30390,8	240,446
1057	1330,15	1432,20	1050,31	363,76	104,98	8,3036	41476,4	30417,0	240,472
1058	1331,15	1433,39	1051,22	364,90	104,73	8,3045	41510,9	30443,2	240,498
1059	1332,15	1434,58	1052,12	366,04	104,48	8,3054	41545,5	30469,4	240,524
1060	1333,15	1435,77	1053,03	367,18	104,23	8,3063	41580,0	30495,6	240,550
1061	1334,15	1436,97	1053,93	368,33	103,99	8,3072	41614,5	30521,9	240,576
1062	1335,15	1438,16	1054,84	369,48	103,74	8,3081	41649,1	30548,1	240,602
1063	1336,15	1439,35	1055,74	370,63	103,50	8,3090	41683,6	30574,3	240,628
1064	1337,15	1440,54	1056,65	371,78	103,25	8,3099	41718,2	30600,5	240,654
1065	1338,15	1441,74	1057,55	372,94	103,01	8,3108	41752,7	30626,8	240,680
1066	1339,15	1442,93	1058,46	374,10	102,77	8,3116	41787,3	30653,0	240,705
1067	1340,15	1444,12	1059,37	375,26	102,52	8,3125	41821,8	30679,3	240,731
1068	1341,15	1445,32	1060,27	376,43	102,28	8,3134	41856,4	30705,5	240,757
1069	1342,15	1446,51	1061,18	377,60	102,04	8,3143	41891,0	30731,8	240,783
1070	1343,15	1447,70	1062,09	378,77	101,80	8,3152	41925,5	30758,0	240,808
1071	1344,15	1448,90	1062,99	379,94	101,56	8,3161	41960,1	30784,3	240,834
1072	1345,15	1450,09	1063,90	381,12	101,33	8,3170	41994,7	30810,5	240,860
1073	1346,15	1451,29	1064,81	382,30	101,09	8,3179	42029,2	30836,8	240,886
1074	1347,15	1452,48	1065,71	383,48	100,85	8,3188	42063,8	30863,1	240,911
1075	1348,15	1453,67	1066,62	384,67	100,61	8,3196	42098,4	30889,3	240,937
1076	1349,15	1454,87	1067,53	385,85	100,38	8,3205	42133,0	30915,6	240,963
1077	1350,15	1456,06	1068,44	387,05	100,14	8,3214	42167,6	30941,9	240,988
1078	1351,15	1457,26	1069,34	388,24	99,915	8,3223	42202,2	30968,2	241,014
1079	1352,15	1458,45	1070,25	389,44	99,681	8,3232	42236,8	30994,5	241,039
1080	1353,15	1459,65	1071,16	390,64	99,448	8,3241	42271,4	31020,8	241,065
1081	1354,15	1460,84	1072,07	391,84	99,216	8,3250	42306,0	31047,1	241,091
1082	1355,15	1462,04	1072,97	393,05	98,985	8,3258	42340,6	31073,3	241,116
1083	1356,15	1463,23	1073,88	394,25	98,754	8,3267	42375,2	31099,6	241,142
1084	1357,15	1464,43	1074,79	395,47	98,524	8,3276	42409,9	31126,0	241,167
1085	1358,15	1465,62	1075,70	396,68	98,295	8,3285	42444,5	31152,3	241,193
1086	1359,15	1466,82	1076,61	397,90	98,066	8,3294	42479,1	31178,6	241,218
1087	1360,15	1468,02	1077,52	399,12	97,838	8,3302	42513,7	31204,9	241,244
1088	1361,15	1469,21	1078,43	400,34	97,611	8,3311	42548,4	31231,2	241,269
1089	1362,15	1470,41	1079,33	401,57	97,384	8,3320	42583,0	31257,5	241,295
1090	1363,15	1471,60	1080,24	402,80	97,158	8,3329	42617,6	31283,8	241,320
1091	1364,15	1472,80	1081,15	404,03	96,933	8,3337	42652,3	31310,2	241,345
1092	1365,15	1474,00	1082,06	405,27	96,708	8,3346	42686,9	31336,5	241,371
1093	1366,15	1475,19	1082,97	406,51	96,484	8,3355	42721,6	31362,8	241,396
1094	1367,15	1476,39	1083,88	407,75	96,261	8,3364	42756,2	31389,2	241,421
1095	1368,15	1477,59	1084,79	408,99	96,038	8,3373	42790,9	31415,5	241,447
1096	1369,15	1478,78	1085,70	410,24	95,816	8,3381	42825,5	31441,9	241,472
1097	1370,15	1479,98	1086,61	411,49	95,595	8,3390	42860,2	31468,2	241,497
1098	1371,15	1481,18	1087,52	412,74	95,374	8,3399	42894,9	31494,5	241,523
1099	1372,15	1482,37	1088,43	414,00	95,154	8,3407	42929,5	31520,9	241,548
1100	1373,15	1483,57	1089,34	415,26	94,934	8,3416	42964,2	31547,3	241,573
1101	1374,15	1484,77	1090,25	416,52	94,715	8,3425	42998,9	31573,6	241,598
1102	1375,15	1485,96	1091,16	417,79	94,497	8,3434	43033,5	31600,0	241,624
1103	1376,15	1487,16	1092,07	419,06	94,280	8,3442	43068,2	31626,4	241,649
1104	1377,15	1488,36	1092,98	420,33	94,063	8,3451	43102,9	31652,7	241,674

Air (*Continued*)

t	T	h	u	π_0	θ_0	s^0	H	U	S
1105	1378,15	1489,56	1093,89	421,60	93,846	8,3460	43137,6	31679,1	241,699
1106	1379,15	1490,76	1094,80	422,88	93,631	8,3468	43172,3	31705,5	241,724
1107	1380,15	1491,95	1095,71	424,16	93,416	8,3477	43207,0	31731,9	241,750
1108	1381,15	1493,15	1096,62	425,45	93,201	8,3486	43241,7	31758,2	241,775
1109	1382,15	1494,35	1097,54	426,73	92,987	8,3494	43276,4	31784,6	241,800
1110	1383,15	1495,55	1098,45	428,02	92,774	8,3503	43311,1	31811,0	241,825
1111	1384,15	1496,75	1099,36	429,32	92,561	8,3512	43345,8	31837,4	241,850
1112	1385,15	1497,95	1100,27	430,61	92,349	8,3520	43380,5	31863,8	241,875
1113	1386,15	1499,14	1101,18	431,91	92,138	8,3529	43415,2	31890,2	241,900
1114	1387,15	1500,34	1102,09	433,22	91,927	8,3538	43449,9	31916,6	241,925
1115	1388,15	1501,54	1103,00	434,52	91,717	8,3546	43484,7	31943,0	241,950
1116	1389,15	1502,74	1103,92	435,83	91,508	8,3555	43519,4	31969,4	241,975
1117	1390,15	1503,94	1104,83	437,14	91,299	8,3564	43554,1	31995,8	242,000
1118	1391,15	1505,14	1105,74	438,46	91,090	8,3572	43588,9	32022,3	242,025
1119	1392,15	1506,34	1106,65	439,78	90,882	8,3581	43623,6	32048,7	242,050
1120	1393,15	1507,54	1107,57	441,10	90,675	8,3589	43658,3	32075,1	242,075
1121	1394,15	1508,74	1108,48	442,42	90,469	8,3598	43693,1	32101,5	242,100
1122	1395,15	1509,94	1109,39	443,75	90,263	8,3607	43727,8	32128,0	242,125
1123	1396,15	1511,14	1110,30	445,08	90,057	8,3615	43762,6	32154,4	242,150
1124	1397,15	1512,34	1111,22	446,41	89,853	8,3624	43797,3	32180,8	242,175
1125	1398,15	1513,54	1112,13	447,75	89,648	8,3632	43832,1	32207,3	242,200
1126	1399,15	1514,74	1113,04	449,09	89,445	8,3641	43866,8	32233,7	242,224
1127	1400,15	1515,94	1113,96	450,44	89,242	8,3650	43901,6	32260,2	242,249
1128	1401,15	1517,14	1114,87	451,78	89,039	8,3658	43936,3	32286,6	242,274
1129	1402,15	1518,34	1115,78	453,13	88,837	8,3667	43971,1	32313,1	242,299
1130	1403,15	1519,54	1116,70	454,48	88,636	8,3675	44005,9	32339,5	242,324
1131	1404,15	1520,74	1117,61	455,84	88,435	8,3684	44040,7	32366,0	242,348
1132	1405,15	1521,94	1118,52	457,20	88,235	8,3692	44075,4	32392,4	242,373
1133	1406,15	1523,14	1119,44	458,56	88,036	8,3701	44110,2	32418,9	242,398
1134	1407,15	1524,34	1120,35	459,93	87,837	8,3709	44145,0	32445,4	242,423
1135	1408,15	1525,55	1121,27	461,30	87,638	8,3718	44179,8	32471,9	242,447
1136	1409,15	1526,75	1122,18	462,67	87,440	8,3727	44214,6	32498,3	242,472
1137	1410,15	1527,95	1123,09	464,05	87,243	8,3735	44249,4	32524,8	242,497
1138	1411,15	1529,15	1124,01	465,42	87,046	8,3744	44284,2	32551,3	242,521
1139	1412,15	1530,35	1124,92	466,81	86,850	8,3752	44319,0	32577,8	242,546
1140	1413,15	1531,55	1125,84	468,19	86,654	8,3761	44353,8	32604,3	242,571
1141	1414,15	1532,75	1126,75	469,58	86,459	8,3769	44388,6	32630,8	242,595
1142	1415,15	1533,96	1127,67	470,97	86,265	8,3778	44423,4	32657,3	242,620
1143	1416,15	1535,16	1128,58	472,37	86,071	8,3786	44458,2	32683,8	242,645
1144	1417,15	1536,36	1129,50	473,77	85,877	8,3795	44493,0	32710,3	242,669
1145	1418,15	1537,56	1130,41	475,17	85,684	8,3803	44527,8	32736,8	242,694
1146	1419,15	1538,77	1131,33	476,57	85,492	8,3812	44562,7	32763,3	242,718
1147	1420,15	1539,97	1132,24	477,98	85,300	8,3820	44597,5	32789,8	242,743
1148	1421,15	1541,17	1133,16	479,39	85,109	8,3828	44632,3	32816,3	242,767
1149	1422,15	1542,37	1134,07	480,81	84,918	8,3837	44667,1	32842,8	242,792
1150	1423,15	1543,58	1134,99	482,22	84,728	8,3845	44702,0	32869,3	242,816
1151	1424,15	1544,78	1135,91	483,65	84,538	8,3854	44736,8	32895,9	242,841
1152	1425,15	1545,98	1136,82	485,07	84,349	8,3862	44771,7	32922,4	242,865
1153	1426,15	1547,19	1137,74	486,50	84,161	8,3871	44806,5	32948,9	242,890
1154	1427,15	1548,39	1138,65	487,93	83,973	8,3879	44841,4	32975,4	242,914
1155	1428,15	1549,59	1139,57	489,37	83,785	8,3888	44876,2	33002,0	242,938
1156	1429,15	1550,80	1140,49	490,80	83,598	8,3896	44911,1	33028,5	242,963
1157	1430,15	1552,00	1141,40	492,24	83,412	8,3904	44945,9	33055,1	242,987
1158	1431,15	1553,20	1142,32	493,69	83,226	8,3913	44980,8	33081,6	243,012
1159	1432,15	1554,41	1143,24	495,14	83,040	8,3921	45015,6	33108,2	243,036

t	T	h	u	π_0	θ_u	s^0	H	U	S
1160	1433,15	1555,61	1144,15	496,59	82,855	8,3930	45050,5	33134,7	243,060
1161	1434,15	1556,82	1145,07	498,04	82,671	8,3938	45085,4	33161,3	243,085
1162	1435,15	1558,02	1145,99	499,50	82,487	8,3946	45120,3	33187,8	243,109
1163	1436,15	1559,22	1146,91	500,96	82,304	8,3955	45155,1	33214,4	243,133
1164	1437,15	1560,43	1147,82	502,43	82,121	8,3963	45190,0	33241,0	243,158
1165	1438,15	1561,63	1148,74	503,90	81,939	8,3972	45224,9	33267,5	243,182
1166	1439,15	1562,84	1149,66	505,37	81,757	8,3980	45259,8	33294,1	243,206
1167	1440,15	1564,04	1150,58	506,84	81,575	8,3988	45294,7	33320,7	243,230
1168	1441,15	1565,25	1151,49	508,32	81,395	8,3997	45329,6	33347,2	243,255
1169	1442,15	1566,45	1152,41	509,80	81,214	8,4005	45364,5	33373,8	243,279
1170	1443,15	1567,66	1153,33	511,29	81,035	8,4013	45399,4	33400,4	243,303
1171	1444,15	1568,86	1154,25	512,78	80,855	8,4022	45434,3	33427,0	243,327
1172	1445,15	1570,07	1155,17	514,27	80,676	8,4030	45469,2	33453,6	243,351
1173	1446,15	1571,27	1156,08	515,77	80,498	8,4038	45504,1	33480,2	243,375
1174	1447,15	1572,48	1157,00	517,27	80,320	8,4047	45539,0	33506,8	243,400
1175	1448,15	1573,68	1157,92	518,77	80,143	8,4055	45573,9	33533,4	243,424
1176	1449,15	1574,89	1158,84	520,27	79,966	8,4063	45608,8	33560,0	243,448
1177	1450,15	1576,10	1159,76	521,78	79,790	8,4072	45643,7	33586,6	243,472
1178	1451,15	1577,30	1160,68	523,30	79,614	8,4080	45678,7	33613,2	243,496
1179	1452,15	1578,51	1161,60	524,81	79,439	8,4088	45713,6	33639,8	243,520
1180	1453,15	1579,71	1162,51	526,33	79,264	8,4097	45748,5	33666,4	243,544
1181	1454,15	1580,92	1163,43	527,86	79,089	8,4105	45783,4	33693,0	243,568
1182	1455,15	1582,13	1164,35	529,38	78,916	8,4113	45818,4	33719,7	243,592
1183	1456,15	1583,33	1165,27	530,91	78,742	8,4122	45853,3	33746,3	243,616
1184	1457,15	1584,54	1166,19	532,45	78,569	8,4130	45888,3	33772,9	243,640
1185	1458,15	1585,75	1167,11	533,99	78,397	8,4138	45923,2	33799,6	243,664
1186	1459,15	1586,95	1168,03	535,53	78,225	8,4146	45958,2	33826,2	243,688
1187	1460,15	1588,16	1168,95	537,07	78,053	8,4155	45993,1	33852,8	243,712
1188	1461,15	1589,37	1169,87	538,62	77,882	8,4163	46028,1	33879,5	243,736
1189	1462,15	1590,57	1170,79	540,17	77,712	8,4171	46063,0	33906,1	243,760
1190	1463,15	1591,78	1171,71	541,73	77,542	8,4179	46098,0	33932,8	243,784
1191	1464,15	1592,99	1172,63	543,29	77,372	8,4188	46132,9	33959,4	243,808
1192	1465,15	1594,20	1173,55	544,85	77,203	8,4196	46167,9	33986,1	243,831
1193	1466,15	1595,40	1174,47	546,41	77,034	8,4204	46202,9	34012,7	243,855
1194	1467,15	1596,61	1175,39	547,98	76,866	8,4212	46237,9	34039,4	243,879
1195	1468,15	1597,82	1176,31	549,56	76,698	8,4221	46272,8	34066,0	243,903
1196	1469,15	1599,03	1177,23	551,13	76,531	8,4229	46307,8	34092,7	243,927
1197	1470,15	1600,23	1178,15	552,71	76,364	8,4237	46342,8	34119,4	243,951
1198	1471,15	1601,44	1179,08	554,30	76,198	8,4245	46377,8	34146,0	243,974
1199	1472,15	1602,65	1180,00	555,88	76,032	8,4254	46412,8	34172,7	243,998
1200	1473,15	1603,86	1180,92	557,47	75,866	8,4262	46447,8	34199,4	244,022
1201	1474,15	1605,07	1181,84	559,07	75,701	8,4270	46482,7	34226,1	244,046
1202	1475,15	1606,28	1182,76	560,67	75,537	8,4278	46517,7	34252,7	244,069
1203	1476,15	1607,48	1183,68	562,27	75,373	8,4286	46552,7	34279,4	244,093
1204	1477,15	1608,69	1184,60	563,87	75,209	8,4294	46587,7	34306,1	244,117
1205	1478,15	1609,90	1185,53	565,48	75,046	8,4303	46622,8	34332,8	244,141
1206	1479,15	1611,11	1186,45	567,10	74,883	8,4311	46657,8	34359,5	244,164
1207	1480,15	1612,32	1187,37	568,71	74,721	8,4319	46692,8	34386,2	244,188
1208	1481,15	1613,53	1188,29	570,33	74,559	8,4327	46727,8	34412,9	244,211
1209	1482,15	1614,74	1189,21	571,95	74,397	8,4335	46762,8	34439,6	244,235
1210	1483,15	1615,95	1190,14	573,58	74,236	8,4343	46797,8	34466,3	244,259
1211	1484,15	1617,16	1191,06	575,21	74,076	8,4352	46832,9	34493,0	244,282
1212	1485,15	1618,37	1191,98	576,85	73,916	8,4360	46867,9	34519,7	244,306
1213	1486,15	1619,58	1192,90	578,49	73,756	8,4368	46902,9	34546,5	244,330
1214	1487,15	1620,79	1193,82	580,13	73,597	8,4376	46937,9	34573,2	244,353

Air (*Continued*)

t	T	h	u	π_0	θ_0	s^θ	H	U	S
1215	1488,15	1622,00	1194,75	581,77	73,438	8,4384	46973,0	34599,9	244,377
1216	1489,15	1623,21	1195,67	583,42	73,279	8,4392	47008,0	34626,6	244,400
1217	1490,15	1624,42	1196,59	585,08	73,121	8,4400	47043,1	34653,3	244,424
1218	1491,15	1625,63	1197,52	586,73	72,964	8,4409	47078,1	34680,1	244,447
1219	1492,15	1626,84	1198,44	588,39	72,807	8,4417	47113,2	34706,8	244,471
1220	1493,15	1628,05	1199,36	590,06	72,650	8,4425	47148,2	34733,5	244,494
1221	1494,15	1629,26	1200,29	591,72	72,494	8,4433	47183,3	34760,3	244,518
1222	1495,15	1630,47	1201,21	593,40	72,338	8,4441	47218,3	34787,0	244,541
1223	1496,15	1631,68	1202,13	595,07	72,183	8,4449	47253,4	34813,8	244,565
1224	1497,15	1632,89	1203,06	596,75	72,028	8,4457	47288,4	34840,5	244,588
1225	1498,15	1634,10	1203,98	598,43	71,873	8,4465	47323,5	34867,3	244,611
1226	1499,15	1635,31	1204,90	600,12	71,719	8,4473	47358,6	34894,0	244,635
1227	1500,15	1636,52	1205,83	601,81	71,565	8,4481	47393,7	34920,8	244,658
1228	1501,15	1637,73	1206,75	603,51	71,412	8,4489	47428,7	34947,5	244,682
1229	1502,15	1638,94	1207,68	605,20	71,259	8,4498	47463,8	34974,3	244,705
1230	1503,15	1640,15	1208,60	606,90	71,106	8,4506	47498,9	35001,1	244,728
1231	1504,15	1641,37	1209,53	608,61	70,954	8,4514	47534,0	35027,8	244,752
1232	1505,15	1642,58	1210,45	610,32	70,803	8,4522	47569,1	35054,6	244,775
1233	1506,15	1643,79	1211,37	612,03	70,651	8,4530	47604,1	35081,4	244,798
1234	1507,15	1645,00	1212,30	613,75	70,500	8,4538	47639,2	35108,2	244,822
1235	1508,15	1646,21	1213,22	615,47	70,350	8,4546	47674,3	35135,0	244,845
1236	1509,15	1647,43	1214,15	617,20	70,200	8,4554	47709,4	35161,7	244,868
1237	1510,15	1648,64	1215,07	618,92	70,050	8,4562	47744,5	35188,5	244,891
1238	1511,15	1649,85	1216,00	620,66	69,901	8,4570	47779,6	35215,3	244,915
1239	1512,15	1651,06	1216,92	622,39	69,752	8,4578	47814,7	35242,1	244,938
1240	1513,15	1652,27	1217,85	624,13	69,604	8,4586	47849,9	35268,9	244,961
1241	1514,15	1653,49	1218,77	625,88	69,456	8,4594	47885,0	35295,7	244,984
1242	1515,15	1654,70	1219,70	627,62	69,308	8,4602	47920,1	35322,5	245,007
1243	1516,15	1655,91	1220,63	629,37	69,161	8,4610	47955,2	35349,3	245,031
1244	1517,15	1657,12	1221,55	631,13	69,014	8,4618	47990,3	35376,1	245,054
1245	1518,15	1658,34	1222,48	632,89	68,867	8,4626	48025,4	35402,9	245,077
1246	1519,15	1659,55	1223,40	634,65	68,721	8,4634	48060,6	35429,7	245,100
1247	1520,15	1660,76	1224,33	636,42	68,576	8,4642	48095,7	35456,6	245,123
1248	1521,15	1661,98	1225,25	638,19	68,430	8,4650	48130,8	35483,4	245,146
1249	1522,15	1663,19	1226,18	639,97	68,285	8,4658	48166,0	35510,2	245,169
1250	1523,15	1664,40	1227,11	641,74	68,141	8,4666	48201,1	35537,0	245,192
1251	1524,15	1665,62	1228,03	643,53	67,997	8,4674	48236,3	35563,9	245,215
1252	1525,15	1666,83	1228,96	645,31	67,853	8,4682	48271,4	35590,7	245,238
1253	1526,15	1668,04	1229,89	647,10	67,709	8,4690	48306,6	35617,5	245,262
1254	1527,15	1669,26	1230,81	648,90	67,566	8,4698	48341,7	35644,4	245,285
1255	1528,15	1670,47	1231,74	650,70	67,424	8,4706	48376,9	35671,2	245,308
1256	1529,15	1671,69	1232,67	652,50	67,282	8,4714	48412,0	35698,0	245,331
1257	1530,15	1672,90	1233,59	654,31	67,140	8,4722	48447,2	35724,9	245,354
1258	1531,15	1674,11	1234,52	656,12	66,998	8,4729	48482,3	35751,7	245,377
1259	1532,15	1675,33	1235,45	657,93	66,857	8,4737	48517,5	35778,6	245,399
1260	1533,15	1676,54	1236,38	659,75	66,716	8,4745	48552,7	35805,4	245,422
1261	1534,15	1677,76	1237,30	661,57	66,576	8,4753	48587,8	35832,3	245,445
1262	1535,15	1678,97	1238,23	663,40	66,436	8,4761	48623,0	35859,1	245,468
1263	1536,15	1680,19	1239,16	665,23	66,296	8,4769	48658,2	35886,0	245,491
1264	1537,15	1681,40	1240,09	667,06	66,157	8,4777	48693,4	35912,9	245,514
1265	1538,15	1682,62	1241,01	668,90	66,018	8,4785	48728,6	35939,7	245,537
1266	1539,15	1683,83	1241,94	670,74	65,880	8,4793	48763,7	35966,6	245,560
1267	1540,15	1685,05	1242,87	672,59	65,742	8,4801	48798,9	35993,5	245,583
1268	1541,15	1686,26	1243,80	674,44	65,604	8,4809	48834,1	36020,4	245,606
1269	1542,15	1687,48	1244,73	676,29	65,466	8,4816	48869,3	36047,2	245,628

Air (Continued)

t	T	h	u	π_0	θ_0	s^0	H	U	S
1270	1543,15	1688,69	1245,65	678,15	65,329	8,4824	48904,5	36074,1	245,651
1271	1544,15	1689,91	1246,58	680,01	65,193	8,4832	48939,7	36101,0	245,674
1272	1545,15	1691,12	1247,51	681,88	65,056	8,4840	48974,9	36127,9	245,697
1273	1546,15	1692,34	1248,44	683,75	64,920	8,4848	49010,1	36154,8	245,720
1274	1547,15	1693,55	1249,37	685,63	64,785	8,4856	49045,3	36181,7	245,742
1275	1548,15	1694,77	1250,30	687,50	64,649	8,4864	49080,5	36208,6	245,765
1276	1549,15	1695,99	1251,23	689,39	64,514	8,4871	49115,7	36235,5	245,788
1277	1550,15	1697,20	1252,15	691,27	64,380	8,4879	49151,0	36262,4	245,810
1278	1551,15	1698,42	1253,08	693,17	64,246	8,4887	49186,2	36289,3	245,833
1279	1552,15	1699,63	1254,01	695,06	64,112	8,4895	49221,4	36316,2	245,856
1280	1553,15	1700,85	1254,94	696,96	63,978	8,4903	49256,6	36343,1	245,879
1281	1554,15	1702,07	1255,87	698,86	63,845	8,4911	49291,9	36370,0	245,901
1282	1555,15	1703,28	1256,80	700,77	63,712	8,4918	49327,1	36396,9	245,924
1283	1556,15	1704,50	1257,73	702,68	63,580	8,4926	49362,3	36423,8	245,947
1284	1557,15	1705,72	1258,66	704,60	63,448	8,4934	49397,6	36450,8	245,969
1285	1558,15	1706,93	1259,59	706,52	63,316	8,4942	49432,8	36477,7	245,992
1286	1559,15	1708,15	1260,52	708,44	63,185	8,4950	49468,0	36504,6	246,014
1287	1560,15	1709,37	1261,45	710,37	63,053	8,4958	49503,3	36531,6	246,037
1288	1561,15	1710,58	1262,38	712,30	62,923	8,4965	49538,5	36558,5	246,060
1289	1562,15	1711,80	1263,31	714,24	62,792	8,4973	49573,8	36585,4	246,082
1290	1563,15	1713,02	1264,24	716,18	62,662	8,4981	49609,0	36612,4	246,105
1291	1564,15	1714,24	1265,17	718,12	62,533	8,4989	49644,3	36639,3	246,127
1292	1565,15	1715,45	1266,10	720,07	62,403	8,4996	49679,5	36666,2	246,150
1293	1566,15	1716,67	1267,03	722,02	62,274	8,5004	49714,8	36693,2	246,172
1294	1567,15	1717,89	1267,96	723,98	62,145	8,5012	49750,1	36720,1	246,195
1295	1568,15	1719,11	1268,89	725,94	62,017	8,5020	49785,3	36747,1	246,217
1296	1569,15	1720,32	1269,82	727,91	61,889	8,5028	49820,6	36774,0	246,240
1297	1570,15	1721,54	1270,75	729,88	61,761	8,5035	49855,9	36801,0	246,262
1298	1571,15	1722,76	1271,68	731,85	61,634	8,5043	49891,1	36828,0	246,285
1299	1572,15	1723,98	1272,61	733,83	61,507	8,5051	49926,4	36854,9	246,307
1300	1573,15	1725,20	1273,55	735,81	61,380	8,5059	49961,7	36881,9	246,330
1301	1574,15	1726,42	1274,48	737,80	61,254	8,5066	49997,0	36908,9	246,352
1302	1575,15	1727,63	1275,41	739,79	61,128	8,5074	50032,3	36935,8	246,374
1303	1576,15	1728,85	1276,34	741,79	61,002	8,5082	50067,6	36962,8	246,397
1304	1577,15	1730,07	1277,27	743,79	60,877	8,5090	50102,9	36989,8	246,419
1305	1578,15	1731,29	1278,20	745,79	60,751	8,5097	50138,1	37016,8	246,442
1306	1579,15	1732,51	1279,13	747,80	60,627	8,5105	50173,4	37043,7	246,464
1307	1580,15	1733,73	1280,07	749,81	60,502	8,5113	50208,7	37070,7	246,486
1308	1581,15	1734,95	1281,00	751,83	60,378	8,5120	50244,0	37097,7	246,509
1309	1582,15	1736,17	1281,93	753,85	60,254	8,5128	50279,4	37124,7	246,531
1310	1583,15	1737,38	1282,86	755,88	60,131	8,5136	50314,7	37151,7	246,553
1311	1584,15	1738,60	1283,79	757,91	60,008	8,5144	50350,0	37178,7	246,576
1312	1585,15	1739,82	1284,73	759,94	59,885	8,5151	50385,3	37205,7	246,598
1313	1586,15	1741,04	1285,66	761,98	59,762	8,5159	50420,6	37232,7	246,620
1314	1587,15	1742,26	1286,59	764,02	59,640	8,5167	50455,9	37259,7	246,642
1315	1588,15	1743,48	1287,52	766,07	59,518	8,5174	50491,2	37286,7	246,665
1316	1589,15	1744,70	1288,46	768,12	59,397	8,5182	50526,6	37313,7	246,687
1317	1590,15	1745,92	1289,39	770,18	59,275	8,5190	50561,9	37340,7	246,709
1318	1591,15	1747,14	1290,32	772,24	59,154	8,5197	50597,2	37367,7	246,731
1319	1592,15	1748,36	1291,26	774,30	59,034	8,5205	50632,5	37394,8	246,754
1320	1593,15	1749,58	1292,19	776,37	58,913	8,5213	50667,9	37421,8	246,776
1321	1594,15	1750,80	1293,12	778,44	58,793	8,5220	50703,2	37448,8	246,798
1322	1595,15	1752,02	1294,05	780,52	58,674	8,5228	50738,6	37475,8	246,820
1323	1596,15	1753,24	1294,99	782,60	58,554	8,5236	50773,9	37502,9	246,842
1324	1597,15	1754,46	1295,92	784,69	58,435	8,5243	50809,2	37529,9	246,864

Air (*Continued*)

t	T	h	u	π_0	θ_0	s^0	H	U	S
1325	1598,15	1755,68	1296,85	786,78	58,316	8,5251	50844,6	37556,9	246,886
1326	1599,15	1756,90	1297,79	788,87	58,198	8,5258	50879,9	37584,0	246,909
1327	1600,15	1758,12	1298,72	790,97	58,080	8,5266	50915,3	37611,0	246,931
1328	1601,15	1759,35	1299,66	793,08	57,962	8,5274	50950,7	37638,0	246,953
1329	1602,15	1760,57	1300,59	795,19	57,844	8,5281	50986,0	37665,1	246,975
1330	1603,15	1761,79	1301,52	797,30	57,727	8,5289	51021,4	37692,1	246,997
1331	1604,15	1763,01	1302,46	799,42	57,610	8,5297	51056,7	37719,2	247,019
1332	1605,15	1764,23	1303,39	801,54	57,493	8,5304	51092,1	37746,2	247,041
1333	1606,15	1765,45	1304,33	803,67	57,377	8,5312	51127,5	37773,3	247,063
1334	1607,15	1766,67	1305,26	805,80	57,261	8,5319	51162,8	37800,3	247,085
1335	1608,15	1767,89	1306,19	807,93	57,145	8,8327	51198,2	37827,4	247,107
1336	1609,15	1769,12	1307,13	810,07	57,029	8,5335	51233,6	37854,5	247,129
1337	1610,15	1770,34	1308,06	812,22	56,914	8,5342	51269,0	37881,5	247,151
1338	1611,15	1771,56	1309,00	814,37	56,799	8,5350	51304,4	37908,6	247,173
1339	1612,15	1772,78	1309,93	816,52	56,685	8,5357	51339,7	37935,7	247,195
1340	1613,15	1774,00	1310,87	818,68	56,570	8,5365	51375,1	37962,7	247,217
1341	1614,15	1775,23	1311,80	820,84	56,456	8,5373	51410,5	37989,8	247,239
1342	1615,15	1776,45	1312,74	823,01	56,343	8,5380	51445,9	38016,9	247,261
1343	1616,15	1777,67	1313,67	825,18	56,229	8,5388	51481,3	38044,0	247,283
1344	1617,15	1778,89	1314,61	827,35	56,116	8,5395	51516,7	38071,1	247,305
1345	1618,15	1780,11	1315,54	829,53	56,003	8,5403	51552,1	38098,1	247,326
1346	1619,15	1781,34	1316,48	831,72	55,890	8,5410	51587,5	38125,2	247,348
1347	1620,15	1782,56	1317,41	833,91	55,778	8,5418	51622,9	38152,3	247,370
1348	1621,15	1783,78	1318,35	836,10	55,666	8,5425	51658,3	38179,4	247,392
1349	1622,15	1785,00	1319,29	838,30	55,554	8,5433	51693,7	38206,5	247,414
1350	1623,15	1786,23	1320,22	840,50	55,443	8,5440	51729,1	38233,6	247,436
1351	1624,15	1787,45	1321,16	842,71	55,332	8,5448	51764,6	38260,7	247,457
1352	1625,15	1788,67	1322,09	844,93	55,221	8,5456	51800,0	38287,8	247,479
1353	1626,15	1789,90	1323,03	847,14	55,110	8,5463	51835,4	38314,9	247,501
1354	1627,15	1791,12	1323,97	849,36	55,000	8,5471	51870,8	38342,0	247,523
1355	1628,15	1792,34	1324,90	851,59	54,890	8,5478	51906,3	38369,1	247,545
1356	1629,15	1793,57	1325,84	853,82	54,780	8,5486	51941,7	38396,3	247,566
1357	1630,15	1794,79	1326,77	856,06	54,670	8,5493	51977,1	38423,4	247,588
1358	1631,15	1796,01	1327,71	858,30	54,561	8,5501	52012,5	38450,5	247,610
1359	1632,15	1797,24	1328,65	860,54	54,452	8,5508	52048,0	38477,6	247,632
1360	1633,15	1798,46	1329,58	862,79	54,343	8,5516	52083,4	38504,7	247,653
1361	1634,15	1799,68	1330,52	865,05	54,235	8,5523	52118,9	38531,9	247,675
1362	1635,15	1800,91	1331,46	867,31	54,127	8,5531	52154,3	38559,0	247,697
1363	1636,15	1802,13	1332,39	869,57	54,019	8,5538	52189,8	38586,1	247,718
1364	1637,15	1803,36	1333,33	871,84	53,911	8,5546	52225,2	38613,3	247,740
1365	1638,15	1804,58	1334,27	874,11	53,804	8,5553	52260,7	38640,4	247,762
1366	1639,15	1805,81	1335,21	876,39	53,697	8,5561	52296,1	38667,5	247,783
1367	1640,15	1807,03	1336,14	878,67	53,590	8,5568	52331,6	38694,7	247,805
1368	1641,15	1808,25	1337,08	880,96	53,484	8,5575	52367,0	38721,8	247,827
1369	1642,15	1809,48	1338,02	883,25	53,377	8,5583	52402,5	38749,0	247,848
1370	1643,15	1810,70	1338,95	885,55	53,271	8,5590	52438,0	38776,1	247,870
1371	1644,15	1811,93	1339,89	887,85	53,166	8,5598	52473,4	38803,3	247,891
1372	1645,15	1813,15	1340,83	890,15	53,060	8,5605	52508,9	38830,4	247,913
1373	1646,15	1814,38	1341,77	892,46	52,955	8,5613	52544,4	38857,6	247,934
1374	1647,15	1815,60	1342,71	894,78	52,850	8,5620	52579,8	38884,8	247,956
1375	1648,15	1816,83	1343,64	897,10	52,745	8,5628	52615,3	38911,9	247,977
1376	1649,15	1818,05	1344,58	899,42	52,641	8,5635	52650,8	38939,1	247,999
1377	1650,15	1819,28	1345,52	901,75	52,537	8,5642	52686,3	38966,3	248,020
1378	1651,15	1820,50	1346,46	904,09	52,433	8,5650	52721,8	38993,4	248,042
1379	1652,15	1821,73	1347,40	906,43	52,329	8,5657	52757,3	39020,6	248,063

Air (*Continued*)

t	T	h	u	π_0	θ_0	s^0	H	U	S
1380	1653,15	1822,95	1348,33	908.77	52,226	8,5665	52792,7	39047,8	248,085
1381	1654,15	1824,18	1349,27	911,12	52,123	8,5672	52828,2	39075,0	248.106
1382	1655,15	1825,41	1350,21	913,47	52,020	8,5680	52863,7	39102,1	248,128
1383	1656,15	1826,63	1351,15	915.83	51,917	8,5687	52899,2	39129,3	248,149
1384	1657,15	1827,86	1352,09	918,20	51,815	8,5694	52934,7	39156,5	248,171
1385	1658,15	1829,08	1353,03	920,56	51,713	8,5702	52970,2	39183,7	248,192
1386	1659,15	1830,31	1353,97	922,94	51,611	8,5709	53005,7	39210,9	248,214
1387	1660,15	1831,53	1354,91	925,32	51,509	8,5716	53041,2	39238,1	248,235
1388	1661,15	1832,76	1355,84	927,70	51,408	8,5724	53076,8	39265,3	248,256
1389	1662,15	1833,99	1356,78	930,09	51,307	8,5731	53112,3	39292,5	248,278
1390	1663,15	1835,21	1357,72	932,48	51,206	8,5739	53147,8	39319,7	248,299
1391	1664,15	1836,44	1358,66	934,88	51,105	8,5746	53183,3	39346,9	248,320
1392	1665,15	1837,67	1359,60	937,28	51,005	8,5753	53218,8	39374,1	248,342
1393	1666,15	1838,89	1360,54	939,69	50,905	8,5761	53254,3	39401,3	248,363
1394	1667,15	1840,12	1361,48	942,10	50,805	8,5768	53289,9	39428,5	248,384
1395	1668,15	1841,35	1362,42	944,51	50,705	8,5775	53325,4	39455,7	248,406
1396	1669,15	1842,47	1363,36	946,94	50,606	8,5783	53360,9	39482,9	248,427
1397	1670,15	1843,80	1364,30	949,36	50,507	8,5790	53396,4	39510,1	248,448
1398	1671,15	1845,03	1365,24	951,80	50,408	8,5797	53432,0	39537,4	248,470
1399	1672,15	1846,25	1366,18	954,23	50,309	8,5805	53467,5	39564,6	248,491
1400	1673,15	1847,48	1367,12	956,67	50,211	8,5812	53503,1	39591,8	248,512
1401	1674,15	1848,71	1368,06	959,12	50,113	8,5820	53538,6	39619,0	248,533
1402	1675,15	1849,94	1369,00	961,57	50,015	8,5827	53574,1	39646,3	248,555
1403	1676,15	1851,16	1369,94	964,03	49,917	8,5834	53609,7	39673,5	248,576
1404	1677,15	1852,39	1370,88	966,49	49,820	8,5841	53645,2	39700,7	248,597
1405	1678,15	1853,62	1371,82	968,96	49,722	8,5849	53680,8	39728,0	248,618
1406	1679,15	1854,85	1372,76	971,43	49,626	8,5856	53716,3	39755,2	248,639
1407	1680,15	1856,07	1373,70	973,90	49,529	8,5863	53751,9	39782,4	248,660
1408	1681,15	1857,30	1374,64	976,39	49,432	8,5871	53787,5	39809,7	248,682
1409	1682,15	1858,53	1375,58	978,87	49,336	8,5878	53823,0	39836,9	248,703
1410	1683,15	1859,76	1376,53	981,36	49,240	8,5885	53858,6	39864,2	248,724
1411	1684,15	1860,99	1377,47	983,86	49,144	8,5893	53894,1	39891,4	248,745
1412	1685,15	1862,21	1378,41	986,36	49,049	8,5900	53929,7	39918,7	248,766
1413	1686,15	1863,44	1379,35	988,87	48,954	8,5907	53965,3	39945,9	248,787
1414	1687,15	1864,67	1380,29	991,38	48,858	8,5914	54000,9	39973,2	248,808
1415	1688,15	1865,90	1381,23	993,90	48,764	8,5922	54036,4	40000,5	248,829
1416	1689,15	1867,13	1382,17	996,42	48,669	8,5929	54072,0	40027,7	248,850
1417	1690,15	1868,36	1383,11	998,94	48,575	8,5936	54107,6	40055,0	248,872
1418	1691,15	1869,58	1384,06	1001,4	48,481	8,5944	54143,2	40082,3	248,893
1419	1692,15	1870,81	1385,00	1004,0	48,387	8,5951	54178,8	40109,5	248,914
1420	1693,15	1872,04	1385,94	1006,5	48,293	8,5958	54214,3	40136,8	248,935
1421	1694,15	1873,27	1386,88	1009,1	48,200	8,5965	54249,9	40164,1	248,956
1422	1695,15	1874,50	1387,82	1011,6	48,106	8,5973	54285,5	40191,4	248,977
1423	1696,15	1875,73	1388,76	1014,2	48,013	8,5980	54321,1	40218,6	248,998
1424	1697,15	1876,96	1389,71	1016,7	47,921	8,5987	54356,7	40245,9	249,019
1425	1698,15	1878,19	1390,65	1019,3	47,828	8,5994	54392,3	40273,2	249,040
1426	1699,15	1879,42	1391,59	1021,9	47,736	8,6002	54427,9	40300,5	249,061
1427	1700,15	1880,65	1392,53	1024,4	47,644	8,6009	54463,5	40327,8	249,082
1428	1701,15	1881,88	1393,48	1027,0	47,552	8,6016	54499,1	40355,1	249,102
1429	1702,15	1883,11	1394,42	1029,6	47,460	8,6023	54534,7	40382,4	249,123
1430	1703,15	1884,34	1395,36	1032,2	47,369	8,6030	54570,3	40409,7	249,144
1431	1704,15	1885,56	1396,30	1034,8	47,277	8,6038	54606,0	40437,0	249,165
1432	1705,15	1886,79	1397,25	1037,4	47,186	8,6045	54641,6	40464,3	249,186
1433	1706,15	1888,02	1398,19	1040,0	47,096	8,6052	54677,2	40491,6	249,207
1434	1707,15	1889,25	1399,13	1042,6	47,005	8,6059	54712,8	40518,9	249,228
1435	1708,15	1890,48	1400,08	1045,3	46,915	8,6067	54748,4	40546,2	249,249
1436	1709,15	1891,71	1401,02	1047,9	46,825	8,6074	54784,1	40573,5	249,270
1437	1710,15	1892,94	1401,96	1050,5	46,735	8,6081	54819,7	40600,8	249,290
1438	1711,15	1894,18	1402,90	1053,1	46,645	8,6088	54855,3	40628,1	249,311
1439	1712,15	1895,41	1403,85	1055,8	46,556	8,6095	54890,9	40655,4	249,332

Air (*Continued*)

t	T	h	u	π_0	θ_0	s^0	H	U	S
1440	1713,15	1896,64	1404,79	1058,4	46,466	8,6102	54926,6	40682,7	249,353
1441	1714,15	1897,87	1405,73	1061,1	46,377	8,6110	54962,2	40710,1	249,374
1442	1715,15	1899,10	1406,68	1063,7	46,289	8,6117	54997,9	40737,4	249,394
1443	1716,15	1900,33	1407,62	1066,4	46,200	8,6124	55033,5	40764,7	249,415
1444	1717,15	1901,56	1408,57	1069,1	46,112	8,6131	55069,1	40792,0	249,436
1445	1718,15	1902,79	1409,51	1071,7	46,024	8,6138	55104,8	40819,4	249,457
1446	1719,15	1904,02	1410,45	1074,4	45,936	8,6146	55140,4	40846,7	249,477
1447	1720,15	1905,25	1411,40	1077,1	45,848	8,6153	55176,1	40874,0	249,498
1448	1721,15	1906,48	1412,34	1079,8	45,760	8,6160	55211,7	40901,4	249,519
1449	1722,15	1907,71	1413,28	1082,5	45,673	8,6167	55247,4	40928,7	249,540
1450	1723,15	1908,94	1414,23	1085,2	45,586	8,6174	55283,0	40956,1	249,560
1451	1724,15	1910,18	1415,17	1087,9	45,499	8,6181	55318,7	40983,4	249,581
1452	1725,15	1911,41	1416,12	1090,6	45,412	8,6188	55354,4	11010,8	249,602
1453	1726,15	1912,64	1417,06	1093,3	45,326	8,6196	55390,0	41038,1	249,622
1454	1727,15	1913,87	1418,01	1096,0	45,240	8,6203	55425,7	41065,5	249,643
1455	1728,15	1915,10	1418,95	1098,7	45,153	8,6210	55461,4	41092,8	249,664
1456	1729,15	1916,33	1419,90	1101,5	45,068	8,6217	55497,0	41120,2	249,684
1457	1730,15	1917,57	1420,84	1104,2	44,982	8,6224	55532,7	41147,5	249,705
1458	1731,15	1918,80	1421,78	1107,0	44,897	8,6231	55568,4	41174,9	249,726
1459	1732,15	1920,03	1422,73	1109,7	44,811	8,6238	55604,1	41202,2	249,746
1460	1733,15	1921,26	1423,67	1112,5	44,726	8,6245	55639,7	41229,6	249,767
1461	1734,15	1922,49	1424,62	1115,2	44,641	8,6253	55675,4	41257,0	249,787
1462	1735,15	1923,73	1425,56	1118,0	44,557	8,6260	55711,1	41284,3	249,808
1463	1736,15	1924,96	1426,51	1120,7	44,472	8,6267	55746,8	41311,7	249,828
1464	1737,15	1926,19	1427,45	1123,5	44,388	8,6274	55782,5	41339,1	249,849
1465	1738,15	1927,42	1428,40	1126,3	44,304	8,6281	55818,2	41366,5	249,870
1466	1739,15	1928,66	1429,35	1129,1	44,220	8,6288	55853,9	41393,8	249,890
1467	1740,15	1929,89	1430,29	1131,9	44,137	8,6295	55889,5	41421,2	249,911
1468	1741,15	1931,12	1431,24	1134,7	44,053	8,6302	55925,2	41448,6	249,931
1469	1742,15	1932,35	1432,18	1137,5	43,970	8,6309	55960,9	41476,0	249,952
1470	1743,15	1933,59	1433,13	1140,3	43,887	8,6316	55996,6	41503,4	249,972
1471	1744,15	1934,82	1434,07	1143,1	43,804	8,6323	56032,3	41530,8	249,993
1472	1745,15	1936,05	1435,02	1145,9	43,721	8,6330	56068,1	41558,2	250,013
1473	1746,15	1937,28	1435,97	1148,7	43,639	8,6338	56103,8	41585,6	250,033
1474	1747,15	1938,52	1436,91	1151,6	43,557	8,6345	56139,5	41613,0	250,054
1475	1748,15	1939,75	1437,86	1154,4	43,475	8,6352	56175,2	41640,3	250,074
1476	1749,15	1940,98	1438,80	1157,2	43,393	8,6359	56210,9	41667,7	250,095
1477	1750,15	1942,22	1439,75	1160,1	43,311	8,6366	56246,6	41695,2	250,115
1478	1751,15	1943,45	1440,70	1162,9	43,230	8,6373	56282,3	41722,6	250,136
1479	1752,15	1944,68	1441,64	1165,8	43,148	8,6380	56318,1	41750,0	250,156
1480	1753,15	1945,92	1442,59	1168,6	43,067	8,6387	56353,8	41777,4	250,176
1481	1754,15	1947,15	1443,54	1171,5	42,986	8,6394	56389,5	41804,8	250,197
1482	1755,15	1948,39	1444,48	1174,4	42,906	8,6401	56425,2	41832,2	250,217
1483	1756,15	1949,62	1445,43	1177,3	42,825	8,6408	56461,0	41859,6	250,237
1484	1757,15	1950,85	1446,38	1180,1	42,745	8,6415	56496,7	41887,0	250,258
1485	1758,15	1952,09	1447,32	1183,0	42,665	8,6422	56532,4	41914,4	250,278
1486	1759,15	1953,32	1448,27	1185,9	42,585	8,6429	56568,2	41941,9	250,298
1487	1760,15	1954,55	1449,22	1188,8	42,505	8,6436	56603,9	41969,3	250,319
1488	1761,15	1955,79	1450,16	1191,7	42,426	8,6443	56639,6	41996,7	250,339
1489	1762,15	1957,02	1451,11	1194,6	42,346	8,6450	56675,4	42024,1	250,359
1490	1763,15	1958,26	1452,06	1197,6	42,267	8,6457	56711,1	42051,6	250,380
1491	1764,15	1959,49	1453,00	1200,5	42,188	8,6464	56746,9	42079,0	250,400
1492	1765,15	1960,73	1453,95	1203,4	42,109	8,6471	56782,6	42106,4	250,420
1493	1766,15	1961,96	1454,90	1206,3	42,031	8,6478	56818,4	42133,9	250,440
1494	1767,15	1963,20	1455,85	1209,3	41,952	8,6485	56854,1	42161,3	250,461
1495	1768,15	1964,43	1456,79	1212,2	41,874	8,6492	56889,9	42188,8	250,48₁
1496	1769,15	1965,66	1457,74	1215,2	41,796	8,6499	56925,6	42216,2	250,50₁
1497	1770,15	1966,90	1458,69	1218,1	41,718	8,6506	56961,4	42243,7	250,52₁
1498	1771,15	1968,13	1459,64	1221,1	41,640	8,6513	56997,2	42271,1	250,54₁
1499	1772,15	1969,37	1460,59	1224,1	41,563	8,6520	57032,9	42298,6	250,56₂
1500	1773,15	1970,60	1461,53	1227,0	41,485	8,6527	57068,7	42326,0	250,58₂

Table III.3 Nitrogen ($\mu = 28.016$)

t	T	c_p	μc_p	c_v	μc_v	$k = \dfrac{c_p}{c_v}$
−50	223,15	1,0388	29,103	0,7420	20,789	1,400
−25	248,15	1,0388	29,103	0,7420	20,789	1,400
0	273,15	1,0388	29,103	0,7420	20,789	1,400
25	298,15	1,0390	29,109	0,7422	20,795	1,400
50	323,15	1,0396	29,124	0,7428	20,810	1,400
75	348,15	1,0405	29,150	0,7437	20,836	1,399
100	373,15	1,0419	29,189	0,7451	20,875	1,398
125	398,15	1,0437	29,240	0,7469	20,926	1,397
150	423,15	1,0460	29,304	0,7492	20,990	1,396
175	448,15	1,0488	29,382	0,7520	21,068	1,395
200	473,15	1,0520	29,473	0,7552	21,159	1,393
250	523,15	1,0598	29,692	0,7631	21,378	1,389
300	573,15	1,0691	29,953	0,7724	21,639	1,384
350	623,15	1,0797	30,249	0,7830	21,935	1,379
400	673,15	1,0912	30,570	0,7944	22,256	1,374
450	723,15	1,1032	30,907	0,8064	22,593	1,368
500	773,15	1,1154	31,250	0,8187	22,936	1,362
550	823,15	1,1277	31,593	0,8309	23,279	1,357
600	873,15	1,1396	31,928	0,8429	23,614	1,352
650	923,15	1,1512	32,251	0,8544	23,937	1,347
700	973,15	1,1622	32,559	0,8654	24,245	1,343
750	1023,15	1,1725	32,850	0,8758	24,536	1,339
800	1073,15	1,1823	33,123	0,8855	24,809	1,335
850	1123,15	1,1914	33,378	0,8946	25,064	1,332
900	1173,15	1,1999	33,617	0,9032	25,303	1,328
950	1223,15	1,2079	33,841	0,9112	25,527	1,326
1000	1273,15	1,2154	34,051	0,9186	25,737	1,323
1050	1323,15	1,2225	34,249	0,9257	25,935	1,321
1100	1373,15	1,2292	34,436	0,9324	26,122	1,318
1150	1423,15	1,2354	34,612	0,9387	26,298	1,316
1200	1473,15	1,2414	34,779	0,9446	26,465	1,314
1250	1523,15	1,2470	34,935	0,9502	26,621	1,312
1300	1573,15	1,2521	35,080	0,9554	26,766	1,311
1350	1623,15	1,2569	35,214	0,9602	26,900	1,309
1400	1673,15	1,2613	35,337	0,9646	27,023	1,308
1450	1723,15	1,2654	35,453	0,9687	27,139	1,306
1500	1773,15	1,2694	35,565	0,9727	27,251	1,305

Table III.4 Nitrogen ($\mu = 28.0134$)

t	T	h	u	π_0	θ_0	s^0	H	U	S
−50	223,15	231,36	165,13	0,3650	18145	6,5350	6481,1	4625,8	183,067
−49	224,15	232,40	165,87	0,3707	17943	6,5396	6510,2	4646,6	183,197
−48	225,15	233,44	166,61	0,3765	17744	6,5442	6539,3	4667,4	183,327
−47	226,15	234,48	167,35	0,3824	17549	6,5488	6568,5	4688,1	183,456
−46	227,15	235,51	168,10	0,3884	17356	6,5534	6597,6	4708,9	183,584
−45	228,15	236,55	168,84	0,3944	17166	6,5580	6626,7	4729,7	183,712
−44	229,15	237,59	169,58	0,4005	16980	6,5625	6655,8	4750,5	183,839
−43	230,15	238,63	170,32	0,4066	16796	6,5671	6684,9	4771,3	183,966
−42	231,15	239,67	171,06	0,4129	16615	6,5716	6714,0	4792,1	184,092
−41	232,15	240,71	171,81	0,4192	16436	6,5761	6743,1	4812,9	184,218
−40	233,15	241,75	172,55	0,4255	16261	6,5805	6772,2	4833,7	184,343
−39	234,15	242,79	173,29	0,4319	16087	6,5850	6801,3	4854,5	184,467
−38	235,15	243,83	174,03	0,4384	15917	6,5894	6830,4	4875,3	184,591
−37	236,15	244,87	174,78	0,4450	15749	6,5938	6859,5	4896,1	184,715
−36	237,15	245,90	175,52	0,4516	15583	6,5982	6888,6	4916,9	184,838
−35	238,15	246,94	176,26	0,4583	15420	6,6026	6917,7	4937,7	184,960
−34	239,15	247,98	177,00	0,4651	15259	6,6069	6946,8	4958,4	185,082
−33	240,15	249,02	177,74	0,4719	15101	6,6113	6975,9	4979,2	185,204
−32	241,15	250,06	178,49	0,4789	14945	6,6156	7005,0	5000,0	185,325
−31	242,15	251,10	179,23	0,4858	14791	6,6199	7034,1	5020,8	185,445
−30	243,15	252,14	179,97	0,4929	14639	6,6242	7063,3	5041,6	185,565
−29	244,15	253,18	180,71	0,5000	14490	6,6284	7092,4	5062,4	185,685
−28	245,15	254,22	181,46	0,5072	14343	6,6327	7121,5	5083,2	185,804
−27	246,15	255,26	182,20	0,5145	14197	6,6369	7150,6	5104,0	185,922
−26	247,15	256,29	182,94	0,5219	14054	6,6411	7179,7	5124,8	186,040
−25	248,15	257,33	183,68	0,5293	13913	6,6453	7208,8	5145,6	186,158
−24	249,15	258,37	184,42	0,5368	13774	6,6495	7237,9	5166,3	186,275
−23	250,15	259,41	185,17	0,5444	13637	6,6536	7267,0	5187,1	186,391
−22	251,15	260,45	185,91	0,5520	13501	6,6578	7296,1	5207,9	186,507
−21	252,15	261,49	186,65	0,5598	13368	6,6619	7325,2	5228,7	186,623
−20	253,15	262,53	187,39	0,5676	13236	6,6660	7354,3	5249,5	186,738
−19	254,15	263,57	188,13	0,5755	13106	6,6701	7383,4	5270,3	186,853
−18	255,15	264,61	188,88	0,5834	12978	6,6742	7412,5	5291,1	186,967
−17	256,15	265,64	189,62	0,5915	12852	6,6783	7441,6	5311,9	187,081
−16	257,15	266,68	190,36	0,5996	12727	6,6823	7470,7	5332,7	187,194
−15	258,15	267,72	191,10	0,6078	12604	6,6863	7499,8	5353,4	187,307
−14	259,15	268,76	191,85	0,6161	12483	6,6904	7528,9	5374,2	187,420
−13	260,15	269,80	192,59	0,6245	12363	6,6944	7558,0	5395,0	187,532
−12	261,15	270,84	193,33	0,6329	12245	6,6984	7587,1	5415,8	187,644
−11	262,15	271,88	194,07	0,6414	12129	6,7023	7616,2	5436,6	187,755
−10	263,15	272,92	194,81	0,6500	12014	6,7063	7645,3	5457,4	187,866
−9	264,15	273,96	195,56	0,6587	11901	6,7102	7674,4	5478,2	187,976
−8	265,15	274,99	196,30	0,6675	11789	6,7141	7703,5	5499,0	188,086
−7	266,15	276,03	197,04	0,6763	11678	6,7181	7732,6	5519,8	188,196
−6	267,15	277,07	197.78	0,6853	11569	6,7220	7761,7	5540,5	188,305
−5	268,15	278,11	198,52	0,6943	11462	6,7258	7790,8	5561,3	188,413
−4	269,15	279,15	199,27	0,7034	11355	6,7297	7819,9	5582,1	188,522
−3	270,15	280,19	200,01	0,7126	11251	6,7336	7849,0	5602,9	188,630
−2	271,15	281,23	200,75	0,7219	11147	6,7374	7878,2	5623,7	188,737
−1	272,15	282,27	201,49	0,7312	11045	6,7412	7907,3	5644,5	188,844
0	273,15	283,31	202,23	0,7407	10944	6,7450	7936,4	5665,3	188,951
1	274,15	284,34	202,98	0,7502	10845	6,7488	7965,5	5686,1	189,057
2	275,15	285,38	203,72	0,7598	10746	6,7526	7994,6	5706,9	189,163
3	276,15	286,42	204,46	0,7696	10649	6,7564	8023,7	5727,6	189,269
4	277,15	287,46	205,20	0,7794	10554	6,7601	8052,8	5748,4	189,374

Nitrogen (*Continued*)

t	T	h	u	π₀	θᵥ	s°	H	U	S
5	278,15	288,50	205,95	0,7892	10459	6,7639	8081,9	5769,2	189,479
6	279,15	289,54	206,69	0,7992	10365	6,7676	8111,0	5790,0	189,583
7	280,15	290,58	207,43	0,8093	10273	6,7713	8140,1	5810,8	189,688
8	281,15	291,62	208,17	0,8194	10182	6,7750	8169,2	5831,6	189,791
9	282,15	292,66	208,91	0,8297	10092	6,7787	8198,3	5852,4	189,895
10	283,15	293,70	209,66	0,8400	10003	6,7824	8227,4	5873,2	189,998
11	284,15	294,73	210,40	0,8505	9915,8	6,7860	8256,5	5894,0	190,100
12	285,15	295,77	211,14	0,8610	9829,1	6,7897	8285,6	5914,8	190,202
13	286,15	296,81	211,88	0,8716	9743,4	6,7933	8314,7	5935,5	190,304
14	287,15	297,85	212,62	0,8823	9658,8	6,7970	8343,8	5956,3	190,406
15	288,15	298,89	213,37	0,8931	9575,2	6,8006	8372,9	5977,1	190,507
16	289,15	299,93	214,11	0,9040	9492,6	6,8042	8402,0	5997,9	190,608
17	290,15	300,97	214,85	0,9150	9411,0	6,8078	8431,1	6018,7	190,708
18	291,15	302,01	215,59	0,9261	9330,4	6,8113	8460,2	6039,5	190,808
19	292,15	303,05	216,34	0,9373	9250,7	6,8149	8489,4	6060,3	190,908
20	293,15	304,09	217,08	0,9486	9172,0	6,8184	8518,5	6081,1	191,008
21	294,15	305,12	217,82	0,9599	9094,2	6,8220	8547,6	6101,9	191,107
22	295,15	306,16	218,56	0,9714	9017,3	6,8255	8576,7	6122,7	191,206
23	296,15	307,20	219,30	0,9830	8941,4	6,8290	8605,8	6143,5	191,304
24	297,15	308,24	220,05	0,9947	8866,3	6,8325	8634,9	6164,3	191,402
25	298,15	309,28	220,79	1,0064	8792,1	6,8360	8664,0	6185,1	191,500
26	299,15	310,32	221,53	1,0183	8718,8	6,8395	8693,1	6205,9	191,598
27	300,15	311,36	222,27	1,0303	8646,3	6,8430	8722,2	6226,7	191,695
28	301,15	312,40	223,02	1,0423	8574,7	6,8464	8751,3	6247,4	191,791
29	302,15	313,44	223,76	1,0545	8503,9	6,8499	8780,4	6268,2	191,888
30	303,15	314,48	224,50	1,0668	8433,9	6,8533	8809,6	6289,0	191,984
31	304,15	315,52	225,24	1,0791	8364,7	6,8567	8838,7	6309,8	192,080
32	305,15	316,55	225,99	1,0916	8296,3	6,8601	8867,8	6330,6	192,176
33	306,15	317,59	226,73	1,1042	8228,7	6,8635	8896,9	6351,4	192,271
34	307,15	318,63	227,47	1,1169	8161,8	6,8669	8926,0	6372,2	192,366
35	308,15	319,67	228,21	1,1297	8095,7	6,8703	8955,1	6393,0	192,460
36	309,15	320,71	228,96	1,1426	8030,4	6,8737	8984,2	6413,8	192,555
37	310,15	321,75	229,70	1,1556	7965,8	6,8770	9013,3	6434,6	192,649
38	311,15	322,79	230,44	1,1687	7901,9	6,8804	9042,5	6455,4	192,743
39	312,15	323,83	231,18	1,1819	7838,7	6,8837	9071,6	6476,2	192,836
40	313,15	324,87	231,93	1,1952	7776,2	6,8870	9100,7	6497,0	192,929
41	314,15	325,91	232,67	1,2086	7714,4	6,8903	9129,8	6517,8	193,022
42	315,15	326,95	233,41	1,2221	7653,3	6,8936	9158,9	6538,6	193,114
43	316,15	327,99	234,15	1,2358	7592,9	6,8969	9188,1	6559,5	193,207
44	317,15	329,03	234,90	1,2495	7533,1	6,9002	9217,2	6580,3	193,299
45	318,15	330,07	235,64	1,2634	7474,0	6,9035	9246,3	6601,1	193,390
46	319,15	331,11	236,38	1,2773	7415,5	6,9068	9275,4	6621,9	193,482
47	320,15	332,15	237,12	1,2914	7357,7	6,9100	9304,5	6642,7	193,573
48	321,15	333,19	237,87	1,3056	7300,5	6,9133	9333,7	6663,5	193,664
49	322,15	334,23	238,61	1,3199	7243,9	6,9165	9362,8	6684,3	193,754
50	323,15	335,26	239,35	1,3343	7187,9	6,9197	9391,9	6705,1	193,844
51	324,15	336,30	240,10	1,3488	7132,6	6,9229	9421,0	6725,9	193,934
52	325,15	337,34	240,84	1,3634	7077,8	6,9261	9450,2	6746,7	194,024
53	326,15	338,38	241,58	1,3782	7023,6	6,9293	9479,3	6767,5	194,114
54	327,15	339,42	242,33	1,3930	6970,0	6,9325	9508,4	6788,3	194,203
55	328,15	340,46	243,07	1,4080	6916,9	6,9357	9537,5	6809,2	194,292
56	329,15	341,50	243,81	1,4231	6864,4	6,9388	9566,7	6830,0	194,380
57	330,15	342,54	244,55	1,4383	6812,5	6,9420	9595,8	6850,8	194,469
58	331,15	343,58	245,30	1,4536	6761,1	6,9451	9624,9	6871,6	194,557
59	332,15	344,62	246,04	1,4691	6710,3	6,9483	9654,1	6892,4	194,645

Nitrogen (*Continued*)

t	T	h	u	π_0	θ_0	s^0	H	U	S
60	333,15	345,66	246,78	1,4846	6659,9	6,9514	9683,2	6913,2	194,732
61	334,15	346,70	247,53	1,5003	6610,1	6,9545	9712,3	6934,1	194,820
62	335,15	347,74	248,27	1,5161	6560,9	6,9576	9741,5	6954,9	194,907
63	336,15	348,78	249,01	1,5320	6512,1	6,9607	9770,6	6975,7	194,993
64	337,15	349,82	249,76	1,5480	6463,8	6,9638	9799,7	6996,5	195,080
65	338,15	350,86	250,50	1,5642	6416,1	6,9669	9828,9	7017,4	195,166
66	339,15	351,90	251,24	1,5805	6368,8	6,9700	9858,0	7038,2	195,252
67	340,15	352,94	251,99	1,5969	6322,0	6,9730	9887,2	7059,0	195,338
68	341,15	353,98	252,73	1,6134	6275,7	6,9761	9916,3	7079,8	195,424
69	342,15	355,02	253,47	1,6300	6229,8	6,9791	9945,4	7100,7	195,509
70	343,15	356,06	254,22	1,6468	6184,4	6,9822	9974,6	7121,5	195,594
71	344,15	357,11	254,96	1,6637	6139,5	6,9852	10003,7	7142,3	195,679
72	345,15	358,15	255,70	1,6807	6095,0	6,9882	10032,9	7163,2	195,763
73	346,15	359,19	256,45	1,6978	6051,0	6,9912	10062,0	7184,0	195,848
74	347,15	360,23	257,19	1,7151	6007,4	6,9942	10091,2	7204,8	195,932
75	348,15	361,27	257,94	1,7324	5964,3	6,9972	10120,3	7225,7	196,016
76	349,15	362,31	258,68	1,7500	5921,6	7,0002	10149,5	7246,5	196,099
77	350,15	363,35	259,42	1,7676	5879,3	7,0032	10178,6	7267,3	196,183
78	351,15	364,39	260,17	1,7854	5837,4	7,0061	10207,8	7288,2	196,266
79	352,15	365,43	260,91	1,8032	5795,9	7,0091	10236,9	7309,0	196,349
80	353,15	366,47	261,66	1,8213	5754,9	7,0120	10266,1	7329,9	196,431
81	354,15	367,51	262,40	1,8394	5714,2	7,0150	10295,2	7350,7	196,514
82	355,15	368,55	263,14	1,8577	5674,0	7,0179	10324,4	7371,5	196,596
83	356,15	369,59	263,89	1,8761	5634,1	7,0209	10353,6	7392,4	196,678
84	357,15	370,63	264,63	1,8946	5594,7	7,0238	10382,7	7413,2	196,760
85	358,15	371,68	265,38	1,9133	5555,6	7,0267	10411,9	7434,1	196,841
86	359,15	372,72	266,12	1,9321	5516,9	7,0296	10441,1	7454,9	196,923
87	360,15	373,76	266,86	1,9511	5478,5	7,0325	10470,2	7475,8	197,004
88	361,15	374,80	267,61	1,9701	5440,5	7,0354	10499,4	7496,6	197,085
89	362,15	375,84	268,35	1,9893	5402,9	7,0382	10528,6	7517,5	197,165
90	363,15	376,88	269,10	2,0087	5365,7	7,0411	10557,7	7538,4	197,246
91	364,15	377,92	269,84	2,0282	5328,8	7,0440	10586,9	7559,2	197,326
92	365,15	378,96	270,59	2,0478	5292,3	7,0468	10616,1	7580,1	197,406
93	366,15	380,01	271,33	2,0675	5256,1	7,0497	10645,3	7600,9	197,486
94	367,15	381,05	272,08	2,0874	5220,2	7,0525	10674,4	7621,8	197,565
95	368,15	382,09	272,82	2,1074	5184,7	7,0554	10703,6	7642,7	197,645
96	369,15	383,13	273,57	2,1276	5149,5	7,0582	10732,8	7663,5	197,724
97	370,15	384,17	274,31	2,1479	5114,7	7,0610	10762,0	7684,4	197,803
98	371,15	385,21	275,06	2,1683	5080,2	7,0638	10791,2	7705,3	197,881
99	372,15	386,26	275,80	2,1889	5046,0	7,0666	10820,3	7726,1	197,960
100	373,15	387,30	276,55	2,2096	5012,1	7,0694	10849,5	7747,0	198,038
101	374,15	388,34	277,29	2,2305	4978,5	7,0722	10878,7	7767,9	198,116
102	375,15	389,38	278,04	2,2515	4945,3	7,0750	10907,9	7788,8	198,194
103	376,15	390,42	278,78	2,2726	4912,3	7,0778	10937,1	7809,6	198,272
104	377,15	391,47	279,53	2,2939	4879,7	7,0805	10966,3	7830,5	198,350
105	378,15	392,51	280,27	2,3153	4847,3	7,0833	10995,5	7851,4	198,427
106	379,15	393,55	281,02	2,3369	4815,3	7,0860	11024,7	7872,3	198,504
107	380,15	394,59	281,76	2,3586	4783,5	7,0888	11053,9	7893,2	198,581
108	381,15	395,64	282,51	2,3805	4752,1	7,0915	11083,1	7914,1	198,658
109	382,15	396,68	283,26	2,4025	4720,9	7,0943	11112,3	7935,0	198,734
110	383,15	397,72	284,00	2,4247	4690,0	7,0970	11141,5	7955,8	198,811
111	384,15	398,76	284,75	2,4470	4659,4	7,0997	11170,7	7976,7	198,887
112	385,15	399,81	285,49	2,4694	4629,0	7,1024	11199,9	7997,6	198,963
113	386,15	400,85	286,24	2,4920	4598,9	7,1051	11229,1	8018,5	199,038
114	387,15	401,89	286,99	2,5148	4569,1	7,1078	11258,4	8039,4	199,114

t	T	h	u	π_o	θ_o	g^o	H	U	S
115	388,15	402,93	287,73	2,5377	4539,6	7,1105	11287,6	8060,3	199,189
116	389,15	403,98	288,48	2,5607	4510,3	7,1132	11316,8	8081,2	199,264
117	390,15	405,02	289,22	2,5839	4481,3	7,1159	11346,0	8102,1	199,339
118	391,15	406,06	289,97	2,6073	4452,6	7,1185	11375,2	8123,1	199,414
119	392,15	407,11	290,72	2,6308	4424,1	7,1212	11404,5	8144,0	199,489
120	393,15	408,15	291,46	2,6544	4395,8	7,1239	11433,7	8164,9	199,563
121	394,15	409,19	292,21	2,6782	4367,8	7,1265	11462,9	8185,8	199,638
122	395,15	410,24	292,96	2,7022	4340,1	7,1291	11492,2	8206,7	199,712
123	396,15	411,28	293,70	2,7263	4312,5	7,1318	11521,4	8227,6	199,786
124	397,15	412,32	294,45	2,7506	4285,3	7,1344	11550,6	8248,6	199,859
125	398,15	413,37	295,20	2,7750	4258,2	7,1370	11579,9	8269,5	199,933
126	399,15	414,41	295,94	2,7996	4231,4	7,1397	11609,1	8290,4	200,006
127	400,15	415,46	296,69	2,8244	4204,9	7,1423	11638,3	8311,3	200,079
128	401,15	416,50	297,44	2,8493	4178,5	7,1449	11667,6	8332,3	200,152
129	402,15	417,54	298,19	2,8744	4152,4	7,1475	11696,8	8353,2	200,225
130	403,15	418,59	298,93	2,8996	4126,5	7,1501	11726,1	8374,1	200,298
131	404,15	419,63	299,68	2,9250	4100,9	7,1527	11755,3	8395,1	200,370
132	405,15	420,68	300,43	2,9505	4075,4	7,1552	11784,6	8416,0	200,443
133	406,15	421,72	301,18	2,9762	4050,2	7,1578	11813,9	8437,0	200,515
134	407,15	422,77	301,92	3,0021	4025,2	7,1604	11843,1	8457,9	200,587
135	408,15	423,81	302,67	3,0281	4000,4	7,1629	11872,4	8478,9	200,658
136	409,15	424,86	303,42	3,0543	3975,8	7,1655	11901,6	8499,8	200,730
137	410,15	425,90	304,17	3,0807	3951,4	7,1680	11930,9	8520,8	200,801
138	411,15	426,95	304,92	3,1072	3927,2	7,1706	11960,2	8541,7	200,873
139	412,15	427,99	305,66	3,1339	3903,2	7,1731	11989,5	8562,7	200,944
140	413,15	429,04	306,41	3,1607	3879,5	7,1757	12018,7	8583,6	201,015
141	414,15	430,08	307,16	3,1878	3855,9	7,1782	12048,0	8604,6	201,086
142	415,15	431,13	307,91	3,2150	3832,5	7,1807	12077,3	8625,6	201,156
143	416,15	432,17	308,66	3,2423	3809,3	7,1832	12106,6	8646,5	201,227
144	417,15	433,22	309,41	3,2698	3786,3	7,1857	12135,9	8667,5	201,297
145	418,15	434,26	310,15	3,2975	3763,5	7,1882	12165,1	8688,5	201,367
146	419,15	435,31	310,90	3,3254	3740,9	7,1907	12194,4	8709,5	201,437
147	420,15	436,35	311,65	3,3534	3718,5	7,1932	12223,7	8730,4	201,507
148	421,15	437,40	312,40	3,3817	3696,3	7,1957	12253,0	8751,4	201,576
149	422,15	438,44	313,15	3,4100	3674,2	7,1982	12282,3	8772,4	201,646
150	423,15	439,49	313,90	3,4386	3652,3	7,2007	12311,6	8793,4	201,715
151	424,15	440,54	314,65	3,4673	3630,6	7,2031	12340,9	8814,4	201,784
152	425,15	441,58	315,40	3,4962	3609,1	7,2056	12370,3	8835,4	201,853
153	426,15	442,63	316,15	3,5253	3587,7	7,2081	12399,6	8856,4	201,922
154	427,15	443,68	316,90	3,5546	3566,6	7,2105	12428,9	8877,4	201,991
155	428,15	444,72	317,65	3,5840	3545,6	7,2130	12458,2	8898,4	202,060
156	429,15	445,77	318,40	3,6136	3524,7	7,2154	12487,5	8919,4	202,128
157	430,15	446,82	319,15	3,6434	3504,1	7,2178	12516,8	8940,4	202,196
158	431,15	447,86	319,90	3,6733	3483,5	7,2203	12546,2	8961,4	202,264
159	432,15	448,91	320,65	3,7035	3463,2	7,2227	12575,5	8982,4	202,332
160	433,15	449,96	321,40	3,7338	3443,0	7,2251	12604,8	9003,4	202,400
161	434,15	451,00	322,15	3,7643	3423,0	7,2275	12634,2	9024,5	202,468
162	435,15	452,05	322,90	3,7950	3403,2	7,2299	12663,5	9045,5	202,535
163	436,15	453,10	323,65	3,8259	3383,5	7,2323	12692,8	9066,5	202,603
164	437,15	454,15	324,40	3,8569	3363,9	7,2347	12722,2	9087,5	202,670
165	438,15	455,19	325,15	3,8881	3344,5	7,2371	12751,5	9108,6	202,737
166	439,15	456,24	325,90	3,9196	3325,3	7,2395	12780,9	9129,6	202,804
167	440,15	457,29	326,65	3,9512	3306,2	7,2419	12810,2	9150,7	202,871
168	441,15	458,34	327,40	3,9829	3287,3	7,2443	12839,6	9171,7	202,937
169	442,15	459,39	328,16	4,0149	3268,5	7,2467	12869,0	9192,7	203,004

Nitrogen (*Continued*)

t	T	h	u	π_0	θ_0	s^0	H	U	S
170	443,15	460,43	328,91	4,0471	3249,8	7,2490	12898,3	9213,8	203,070
171	444,15	461,48	329,66	4,0794	3231,3	7,2514	12927,7	9234,8	203,136
172	445,15	462,53	330,41	4,1120	3213,0	7,2538	12957,1	9255,9	203,202
173	446,15	463,58	331,16	4,1447	3194,8	7,2561	12986,4	9277,0	203,268
174	447,15	464,63	331,91	4,1776	3176,7	7,2585	13015,8	9298,0	203,334
175	448,15	465,68	332,67	4,2107	3158,8	7,2608	13045,2	9319,1	203,400
176	449,15	466,73	333,42	4,2440	3141,0	7,2631	13074,6	9340,2	203,465
177	450,15	467,77	334,17	4,2775	3123,3	7,2655	13104,0	9361,2	203,530
178	451,15	468,82	334,92	4,3112	3105,8	7,2678	13133,4	9382,3	203,596
179	452,15	469,87	335,67	4,3451	3088,4	7,2701	13162,8	9403,4	203,661
180	453,15	470,92	336,43	4,3792	3071,2	7,2724	13192,1	9424,5	203,726
181	454,15	471,97	337,18	4,4134	3054,1	7,2747	13221,5	9445,6	203,790
182	455,15	473,02	337,93	4,4479	3037,1	7,2771	13251,0	9466,7	203,855
183	456,15	474,07	338,69	4,4826	3020,2	7,2794	13280,4	9487,7	203,920
184	457,15	475,12	339,44	4,5174	3003,5	7,2817	13309,8	9508,8	203,984
185	458,15	476,17	340,19	4,5525	2986,9	7,2840	13339,2	9529,9	204,048
186	459,15	477,22	340,95	4,5877	2970,4	7,2862	13368,6	9551,0	204,112
187	460,15	478,27	341,70	4,6232	2954,0	7,2885	13398,0	9572,2	204,177
188	461,15	479,32	342,45	4,6588	2937,8	7,2908	13427,5	9593,3	204,240
189	462,15	480,37	343,21	4,6947	2921,7	7,2931	13456,9	9614,4	204,304
190	463,15	481,42	343,96	4,7308	2905,7	7,2954	13486,3	9635,5	204,368
191	464,15	482,47	344,71	4,7670	2889,8	7,2976	13515,8	9656,6	204,431
192	465,15	483,53	345,47	4,8035	2874,0	7,2999	13545,2	9677,8	204,495
193	466,15	484,58	346,22	4,8402	2858,4	7,3021	13574,6	9698,9	204,558
194	467,15	485,63	346,98	4,8770	2842,9	7,3044	13604,1	9720,0	204,621
195	468,15	486,68	347,73	4,9141	2827,4	7,3066	13633,5	9741,2	204,684
196	469,15	487,73	348,49	4,9514	2812,1	7,3089	13663,0	9762,3	204,747
197	470,15	488,78	349,24	4,9889	2797,0	7,3111	13692,5	9783,4	204,810
198	471,15	489,83	350,00	5,0266	2781,9	7,3134	13721,9	9804,6	204,872
199	472,15	490,89	350,75	5,0645	2766,9	7,3156	13751,4	9825,7	204,935
200	473,15	491,94	351,51	5,1027	2752,1	7,3178	13780,9	9846,9	204,997
201	474,15	492,99	352,26	5,1410	2737,3	7,3200	13810,3	9868,1	205,059
202	475,15	494,04	353,02	5,1795	2722,7	7,3223	13839,8	9889,2	205,121
203	476,15	495,10	353,77	5,2183	2708,1	7,3245	13869,3	9910,4	205,183
204	477,15	496,15	354,53	5,2573	2693,7	7,3267	13898,8	9931,6	205,245
205	478,15	497,20	355,29	5,2964	2679,4	7,3289	13928,3	9952,7	205,307
206	479,15	498,25	356,04	5,3358	2665,2	7,3311	13957,8	9973,9	205,368
207	480,15	499,31	356,80	5,3755	2651,0	7,3333	13987,3	9995,1	205,430
208	481,15	500,36	357,55	5,4153	2637,0	7,3355	14016,8	10016,3	205,491
209	482,15	501,41	358,31	5,4553	2623,1	7,3377	14046,3	10037,5	205,553
210	483,15	502,47	359,07	5,4956	2609,3	7,3398	14075,8	10058,7	205,614
211	484,15	503,52	359,82	5,5361	2595,6	7,3420	14105,3	10079,9	205,675
212	485,15	504,57	360,58	5,5768	2581,9	7,3442	14134,8	10101,1	205,736
213	486,15	505,63	361,34	5,6177	2568,4	7,3464	14164,4	10122,3	205,796
214	487,15	506,68	362,10	5,6589	2555,0	7,3485	14193,9	10143,5	205,857
215	488,15	507,74	362,85	5,7002	2541,6	7,3507	14223,4	10164,7	205,918
216	489,15	508,79	363,61	5,7418	2528,4	7,3528	14252,9	10186,0	205,978
217	490,15	509,84	364,37	5,7836	2515,2	7,3550	14282,5	10207,2	206,039
218	491,15	510,90	365,13	5,8257	2502,2	7,3571	14312,0	10228,4	206,099
219	492,15	511,95	365,88	5,8679	2489,2	7,3593	14341,6	10249,6	206,159
220	493,15	513,01	366,64	5,9104	2476,4	7,3614	14371,1	10270,9	206,219
221	494,15	514,06	367,40	5,9532	2463,6	7,3636	14400,7	10292,1	206,279
222	495,15	515,12	368,16	5,9961	2450,9	7,3657	14430,3	10313,4	206,338
223	496,15	516,17	368,92	6,0393	2438,3	7,3678	14459,8	10334,6	206,398
224	497,15	517,23	369,68	6,0827	2425,8	7,3700	14489,4	10355,9	206,458

Nitrogen (*Continued*)

t	T	h	u	π_0	θ_0	s^0	H	U	S
225	498,15	518,29	370,43	6,1263	2413,3	7,3721	14519,0	10377,1	206,517
226	499,15	519,34	371,19	6,1702	2401,0	7,3742	14548,5	10398,4	206,576
227	500,15	520,40	371,95	6,2143	2388,7	7,3763	14578,1	10419,7	206,636
228	501,15	521,45	372,71	6,2586	2376,5	7,3784	14607,7	10440,9	206,695
229	502,15	522,51	372,47	6,3032	2364,4	7,3805	14637,3	10462,2	206,754
230	503,15	523,57	374,23	6,3480	2352,4	7,3826	14666,9	10483,5	206,813
231	504,15	524,62	374,99	6,3930	2340,5	7,3847	14696,5	10504,8	206,871
232	505,15	525,68	375,75	6,4383	2328,7	7,3868	14726,1	10526,1	206,930
233	506,15	526,74	376,51	6,4838	2316,9	7,3889	14755,7	10547,4	206,989
234	507,15	527,79	377,27	6,5295	2305,2	7,3910	14785,3	10568,7	207,047
235	508,15	528,85	378,03	6,5755	2293,6	7,3931	14815,0	10590,0	207,105
236	509,15	529,91	378,79	6,6217	2282,1	7,3952	14844,6	10611,3	207,164
237	510,15	530,97	379,55	6,6682	2270,6	7,3972	14874,2	10632,6	207,222
238	511,15	532,03	380,32	6,7149	2259,2	7,3993	14903,8	10653,9	207,280
239	512,15	533,08	381,08	6,7618	2248,0	7,4014	14933,5	10675,2	207,338
240	513,15	534,14	381,84	6,8090	2236,7	7,4034	14963,1	10696,6	207,396
241	514,15	535,20	382,60	6,8564	2225,6	7,4055	14992,8	10717,9	207,453
242	515,15	536,26	383,36	6,9041	2214,5	7,4076	15022,4	10739,2	207,511
243	516,15	537,32	384,12	6,9520	2203,5	7,4096	15052,1	10760,6	207,568
244	517,15	538,38	384,88	7,0002	2192,6	7,4117	15081,7	10781,9	207,626
245	518,15	539,43	385,65	7,0486	2181,7	7,4137	15111,4	10803,3	207,683
246	519,15	540,49	386,41	7,0973	2171,0	7,4157	15141,1	10824,6	207,740
247	520,15	541,55	387,17	7,1462	2160,3	7,4178	15170,7	10846,0	207,797
248	521,15	542,61	387,93	7,1954	2149,6	7,4198	15200,4	10867,4	207,854
249	522,15	543,67	388,70	7,2448	2139,1	7,4219	15230,1	10888,7	207,911
250	523,15	544,73	389,46	7,2944	2128,6	7,4239	15259,8	10910,1	207,968
251	524,15	545,79	390,22	7,3444	2118,1	7,4259	15289,5	10931,5	208,025
252	525,15	546,85	390,99	7,3945	2107,8	7,4279	15319,2	10952,9	208,081
253	526,15	547,91	391,75	7,4450	2097,5	7,4299	15348,9	10974,3	208,138
254	527,15	548,97	392,51	7,4957	2087,3	7,4320	15378,6	10995,7	208,194
255	528,15	550,03	393,28	7,5466	2077,1	7,4340	15408,3	11017,1	208,251
256	529,15	551,09	394,04	7,5978	2067,0	7,4360	15438,0	11038,5	208,307
257	530,15	552,16	394,81	7,6493	2057,0	7,4380	15467,8	11059,9	208,363
258	531,15	553,22	395,57	7,7010	2047,0	7,4400	15497,5	11081,3	208,419
259	532,15	554,28	396,34	7,7529	2037,1	7,4420	15527,2	11102,7	208,475
260	533,15	555,34	397,10	7,8052	2027,3	7,4440	15557,0	11124,1	208,531
261	534,15	556,40	397,87	7,8577	2017,5	7,4460	15586,7	11145,6	208,586
262	535,15	557,46	398,63	7,9104	2007,8	7,4479	15616,4	11167,0	208,642
263	536,15	558,53	399,40	7,9635	1998,2	7,4499	15646,2	11188,4	208,698
264	537,15	559,59	400,16	8,0168	1988,6	7,4519	15676,0	11209,9	208,753
265	538,15	560,65	400,93	8,0703	1979,1	7,4539	15705,7	11231,3	208,808
266	539,15	561,71	401,69	8,1241	1969,6	7,4559	15735,5	11252,8	208,864
267	540,15	562,78	402,46	8,1782	1960,2	7,4578	15765,3	11274,2	208,919
268	541,15	563,84	403,22	8,2326	1950,9	7,4598	15795,0	11295,7	208,974
269	542,15	564,90	403,99	8,2872	1941,6	7,4618	15824,8	11317,2	209,029
270	543,15	565,97	404,76	8,3421	1932,4	7,4637	15854,6	11338,6	209,084
271	544,15	567,03	405,52	8,3973	1923,2	7,4657	15884,4	11360,1	209,139
272	545,15	568,09	406,29	8,4527	1914,1	7,4676	15914,2	11381,6	209,193
273	546,15	569,16	407,06	8,5084	1905,1	7,4696	15944,0	11403,1	209,248
274	547,15	570,22	407,83	8,5644	1896,1	7,4715	15973,8	11424,6	209,303
275	548,15	571,29	408,59	8,6207	1887,2	7,4735	16003,6	11446,1	209,357
276	549,15	572,35	409,36	8,6772	1878,3	7,4754	16033,5	11467,6	209,411
277	550,15	573,41	410,13	8,7340	1869,5	7,4773	16063,3	11489,1	209,466
278	551,15	574,48	410,90	8,7911	1860,7	7,4793	16093,1	11510,6	209,520
279	552,15	575,54	411,67	8,8485	1852,0	7,4812	16123,0	11532,2	209,574

Nitrogen (*Continued*)

t	T	h	u	π_0	θ_0	s^θ	H	U	S
280	553,15	576,61	412,43	8,9062	1843,3	7,4831	16152,8	11553,7	209,628
281	554,15	577,68	413,20	8,9641	1834,7	7,4851	16182,6	11575,2	209,682
282	555,15	578,74	413,97	9,0223	1826,2	7,4870	16212,5	11596,7	209,736
283	556,15	579,81	414,74	9,0808	1817,7	7,4889	16242,4	11618,3	209,789
284	557,15	580,87	415,51	9,1396	1809,2	7,4908	16272,2	11639,8	209,843
285	558,15	581,94	416,28	9,1986	1800,9	7,4927	16302,1	11661,4	209,897
286	559,15	583,01	417,05	9,2580	1792,5	7,4946	16332,0	11683,0	209,950
287	560,15	584,07	417,82	9,3176	1784,2	7,4965	16361,8	11704,5	210,003
288	561,15	585,14	418,59	9,3776	1776,0	7,4984	16391,7	11726,1	210,057
289	562,15	586,21	419,36	9,4378	1767,8	7,5003	16421,6	11747,7	210,110
290	563,15	587,27	420,13	9,4983	1759,7	7,5022	16451,5	11769,2	210,163
291	564,15	588,34	420,90	9,5591	1751,6	7,5041	16481,4	11790,8	210,216
292	565,15	589,41	421,67	9,6202	1743,5	7,5060	16511,3	11812,4	210,269
293	566,15	590,48	422,44	9,6815	1735,6	7,5079	16541,2	11834,0	210,322
294	567,15	591,54	423,21	9,7432	1727,6	7,5098	16571,1	11855,6	210,375
295	568,15	592,61	423,98	9,8052	1719,7	7,5117	16601,1	11877,2	210,427
296	569,15	593,68	424,76	9,8675	1711,9	7,5136	16631,0	11898,8	210,480
297	570,15	594,75	425,53	9,9300	1704,1	7,5154	16660,9	11920,5	210,533
298	571,15	595,82	426,30	9,9929	1696,3	7,5173	16690,9	11942,1	210,585
299	572,15	596,89	427,07	10,056	1688,6	7,5192	16720,8	11963,7	210,637
300	573,15	597,96	427,84	10,119	1681,0	7,5210	16750,8	11985,4	210,690
301	574,15	599,02	428,62	10,183	1673,4	7,5229	16780,7	12007,0	210,742
302	575,15	600,09	429,39	10,247	1665,8	7,5248	16810,7	12028,6	210,794
303	576,15	601,16	430,16	10,311	1658,3	7,5266	16840,6	12050,3	210,846
304	577,15	602,23	430,94	10,376	1650,8	7,5285	16870,6	12072,0	210,898
305	578,15	603,30	431,71	10,441	1643,4	7,5303	16900,6	12093,6	210,950
306	579,15	604,37	432,48	10,506	1636,0	7,5322	16930,6	12115,3	211,002
307	580,15	605,44	433,26	10,572	1628,6	7,5340	16960,6	12137,0	211,054
308	581,15	606,52	434,03	10,638	1621,3	7,5359	16990,6	12158,6	211,105
309	582,15	607,59	434,80	10,704	1614,1	7,5377	17020,6	12180,3	211,157
310	583,15	608,66	435,58	10,770	1606,9	7,5395	17050,6	12202,0	211,208
311	584,15	609,73	436,35	10,837	1599,7	7,5414	17080,6	12223,7	211,260
312	585,15	610,80	437,13	10,904	1592,6	7,5432	17110,6	12245,4	211,311
313	586,15	611,87	437,90	10,972	1585,5	7,5450	17140,6	12267,1	211,362
314	587,15	612,94	438,68	11,039	1578,5	7,5469	17170,7	12288,9	211,414
315	588,15	614,02	439,45	11,108	1571,5	7,5487	17200,7	12310,6	211,465
316	589,15	615,09	440,23	11,176	1564,5	7,5505	17230,7	12332,3	211,516
317	590,15	616,16	441,00	11,245	1557,6	7,5523	17260,8	12354,0	211,567
318	591,15	617,23	441,78	11,314	1550,7	7,5542	17290,8	12375,8	211,618
319	592,15	618,31	442,56	11,383	1543,9	7,5560	17320,9	12397,5	211,668
320	593,15	619,38	443,33	11,453	1537,1	7,5578	17351,0	12419,3	211,719
321	594,15	620,45	444,11	11,523	1530,3	7,5596	17381,0	12441,0	211,770
322	595,15	621,53	444,89	11,593	1523,6	7,5614	17411,1	12462,9	211,820
323	596,15	622,60	445,66	11,664	1516,9	7,5632	17441,2	12484,6	211,871
324	597,15	623,68	446,44	11,735	1510,2	7,5650	17471,3	12506,3	211,921
325	598,15	624,75	447,22	11,806	1503,6	7,5668	17501,4	12528,1	211,972
326	599,15	625,82	448,00	11,878	1497,1	7,5686	17531,5	12549,9	212,022
327	600,15	626,90	448,77	11,949	1490,5	7,5704	17561,6	12571,7	212,072
328	601,15	627,97	449,55	12,022	1484,1	7,5722	17591,7	12593,5	212,122
329	602,15	629,05	450,33	12,094	1477,6	7,5740	17621,8	12615,3	212,172
330	603,15	630,13	451,11	12,167	1471,2	7,5757	17651,9	12637,1	212,222
331	604,15	631,20	451,89	12,241	1464,8	7,5775	17682,1	12658,9	212,272
332	605,15	632,28	452,67	12,314	1458,5	7,5793	17712,2	12680,7	212,322
333	606,15	633,35	453,45	12,388	1452,1	7,5811	17742,4	12702,6	212,372
334	607,15	634,43	454,23	12,462	1445,9	7,5829	17772,5	12724,4	212,422

t	T	h	u	π_0	θ_0	s^0	H	U	S
335	608,15	635,51	455,01	12,537	1439,6	7,5846	17802,7	12746,2	212,471
336	609,15	636,58	455,79	12,612	1433,4	7,5864	17832,8	12768,1	212,521
337	610,15	637,66	456,57	12,687	1427,3	7,5882	17863,0	12789,9	212,570
338	611,15	638,74	457,35	12,763	1421,1	7,5899	17893,2	12811,8	212,620
339	612,15	639,81	458,13	12,839	1415,0	7,5917	17923,3	12833,7	212,669
340	613,15	640,89	458,91	12,915	1409,0	7,5934	17953,5	12855,5	212,718
341	614,15	641,97	459,69	12,992	1402,9	7,5952	17983,7	12877,4	212,767
342	615,15	643,05	460,47	13,069	1397,0	7,5970	18013,9	12899,3	212,817
343	616,15	644,12	461,25	13,146	1391,0	7,5987	18044,1	12921,2	212,866
344	617,15	645,20	462,03	13,224	1385,1	7,6005	18074,3	12943,1	212,915
345	618,15	646,28	462,81	13,302	1379,2	7,6022	18104,5	12965,0	212,964
346	619,15	647,36	463,60	13,380	1373,3	7,6039	18134,8	12986,9	213,012
347	620,15	648,44	464,38	13,459	1367,5	7,6057	18165,0	13008,8	213,061
348	621,15	649,52	465,16	13,538	1361,7	7,6074	18195,2	13030,7	213,110
349	622,15	650,60	465,94	13,617	1355,9	7,6092	18225,5	13052,6	213,159
350	623,15	651,68	466,73	13,697	1350,2	7,6109	18255,7	13074,6	213,207
351	624,15	652,76	467,51	13,777	1344,5	7,6126	18286,0	13096,5	213,256
352	625,15	653,84	468,29	13,858	1338,8	7,6144	18316,2	13118,5	213,304
353	626,15	654,92	469,08	13,939	1333,2	7,6161	18346,5	13140,4	213,352
354	627,15	656,00	469,86	14,020	1327,6	7,6178	18376,8	13162,4	213,401
355	628,15	657,08	470,64	14,102	1322,0	7,6195	18407,0	13184,3	213,449
356	629,15	658,16	471,43	14,184	1316,5	7,6213	18437,3	13206,3	213,497
357	630,15	659,24	472,21	14,266	1310,9	7,6230	18467,6	13228,3	213,545
358	631,15	660,32	473,00	14,348	1305,5	7,6247	18497,9	13250,3	213,593
359	632,15	661,40	473,78	14,432	1300,0	7,6264	18528,2	13272,2	213,641
360	633,15	662,49	474,57	14,515	1294,6	7,6281	18558,5	13294,2	213,689
361	634,15	663,57	475,35	14,599	1289,2	7,6298	18588,8	13316,2	213,737
362	635,15	664,65	476,14	14,683	1283,8	7,6315	18619,1	13338,2	213,785
363	636,15	665,73	476,92	14,767	1278,5	7,6332	18649,5	13360,3	213,832
364	637,15	666,82	477,71	14,852	1273,2	7,6349	18679,8	13382,3	213,880
365	638,15	667,90	478,50	14,937	1267,9	7,6366	18710,1	13404,3	213,928
366	639,15	668,98	479,28	15,023	1262,6	7,6383	18740,5	13426,3	213,975
367	640,15	670,07	480,07	15,109	1257,4	7,6400	18770,8	13448,4	214,023
368	641,15	671,15	480,86	15,195	1252,2	7,6417	18801,2	13470,4	214,070
369	642,15	672,23	481,64	15,282	1247,1	7,6434	18831,6	13492,5	214,117
370	643,15	673,32	482,43	15,369	1241,9	7,6451	18861,9	13514,5	214,165
371	644,15	674,40	483,22	15,457	1236,8	7,6468	18892,3	13536,6	214,212
372	645,15	675,49	484,01	15,545	1231,7	7,6484	18922,7	13558,7	214,259
373	646,15	676,57	484,79	15,633	1226,7	7,6501	18953,1	13580,7	214,306
374	647,15	677,66	485,58	15,722	1221,6	7,6518	18983,5	13602,8	214,353
375	648,15	678,74	486,37	15,811	1216,6	7,6535	19013,9	13624,9	214,400
376	649,15	679,83	487,16	15,900	1211,7	7,6552	19044,3	13647,0	214,447
377	650,15	680,91	487,95	15,990	1206,7	7,6568	19074,7	13669,1	214,494
378	651,15	682,00	488,74	16,080	1201,8	7,6585	19105,2	13691,2	214,540
379	652,15	683,09	489,53	16,170	1196,9	7,6602	19135,6	13713,3	214,587
380	653,15	684,17	490,32	16,261	1192,0	7,6618	19166,0	13735,5	214,634
381	654,15	685,26	491,11	16,353	1187,2	7,6635	19196,5	13757,6	214,680
382	655,15	686,35	491,90	16,445	1182,4	7,6651	19226,9	13779,7	214,727
383	656,15	687,43	492,69	16,537	1177,6	7,6668	19257,4	13801,9	214,773
384	657,15	688,52	493,48	16,629	1172,8	7,6685	19287,8	13824,0	214,820
385	658,15	689,61	494,27	16,722	1168,1	7,6701	19318,3	13846,2	214,866
386	659,15	690,70	495,06	16,815	1163,4	7,6718	19348,8	13868,3	214,912
387	660,15	691,79	495,85	16,909	1158,7	7,6734	19379,3	13890,5	214,959
388	661,15	692,87	496,64	17,003	1154,0	7,6751	19409,7	13912,7	215,005
389	662,15	693,96	497,44	17,098	1149,3	7,6767	19440,2	13934,9	215,051

Nitrogen (Continued)

t	T	h	u	π_0	θ_0	s^0	H	U	S
390	663,15	695,05	498,23	17,193	1144,7	7,6784	19470,7	13957,0	215,097
391	664,15	696,14	499,02	17,288	1140,1	7,6800	19501,2	13979,2	215,143
392	665,15	697,23	499,81	17,384	1135,6	7,6816	19531,8	14001,4	215,189
393	666,15	698,32	500,60	17,480	1131,0	7,6833	19562,3	14023,6	215,235
394	667,15	699,41	501,40	17,576	1126,5	7,6849	19592,8	14045,9	215,280
395	668,15	700,50	502,19	17,673	1122,0	7,6865	19623,3	14068,1	215,326
396	669,15	701,59	502,98	17,771	1117,5	7,6882	19653,9	14090,3	215,372
397	670,15	702,68	503,78	17,868	1113,1	7,6898	19684,4	14112,5	215,417
398	671,15	703,77	504,57	17,967	1108,6	7,6914	19715,0	14134,8	215,463
399	672,15	704,86	505,37	18,065	1104,2	7,6930	19745,5	14157,0	215,508
400	673,15	705,95	506,16	18,164	1099,8	7,6947	19776,1	14179,3	215,554
401	674,15	707,04	506,95	18,264	1095,5	7,6963	19806,7	14201,5	215,599
402	675,15	708,13	507,75	18,364	1091,1	7,6979	19837,3	14223,8	215,645
403	676,15	709,23	508,54	18,464	1086,8	7,6995	19867,9	14246,1	215,690
404	677,15	710,32	509,34	18,564	1082,5	7,7011	19898,4	14268,3	215,735
405	678,15	711,41	510,14	18,666	1078,2	7,7028	19929,0	14290,6	215,780
406	679,15	712,50	510,93	18,767	1074,0	7,7044	19959,7	14312,9	215,825
407	680,15	713,60	511,73	18,869	1069,8	7,7060	19990,3	14335,2	215,870
408	681,15	714,69	512,52	18,971	1065,6	7,7076	20020,9	14357,5	215,915
409	682,15	715,78	513,32	19,074	1061,4	7,7092	20051,5	14379,8	215,960
410	683,15	716,88	514,12	19,177	1057,2	7,7108	20082,1	14402,2	216,005
411	684,15	717,97	514,91	19,281	1053,1	7,7124	20112,8	14424,5	216,050
412	685,15	719,06	515,71	19,385	1048,9	7,7140	20143,4	14446,8	216,095
413	686,15	720,16	516,51	19,490	1044,8	7,7156	20174,1	14469,2	216,139
414	687,15	721,25	517,31	19,595	1040,8	7,7172	20204,7	14491,5	216,184
415	688,15	722,35	518,10	19,700	1036,7	7,7188	20235,4	14513,9	216,229
416	689,15	723,44	518,90	19,806	1032,7	7,7203	20266,1	14536,2	216,273
417	690,15	724,54	519,70	19,912	1028,6	7,7219	20296,8	14558,6	216,318
418	691,15	725,63	520,50	20,019	1024,6	7,7235	20327,5	14581,0	216,362
419	692,15	726,73	521,30	20,126	1020,6	7,7251	20358,2	14603,3	216,407
420	693,15	727,82	522,10	20,234	1016,7	7,7267	20388,9	14625,7	216,451
421	694,15	728,92	522,90	20,342	1012,7	7,7283	20419,6	14648,1	216,495
422	695,15	730,02	523,70	20,450	1008,8	7,7298	20450,3	14670,5	216,539
423	696,15	731,11	524,50	20,559	1004,9	7,7314	20481,0	14692,9	216,583
424	697,15	732,21	525,30	20,668	1001,0	7,7330	20511,7	14715,3	216,628
425	698,15	733,31	526,10	20,778	997,23	7,7346	20542,5	14737,7	216,672
426	699,15	734,41	526,90	20,888	993,39	7,7361	20573,2	14760,2	216,716
427	700,15	735,50	527,70	20,999	989,56	7,7377	20603,9	14782,6	216,760
428	701,15	736,60	528,50	21,110	985,76	7,7393	20634,7	14805,0	216,803
429	702,15	737,70	529,30	21,222	981,97	7,7408	20665,5	14827,5	216,847
430	703,15	738,80	530,10	21,334	978,21	7,7424	20696,2	14849,9	216,891
431	704,15	739,90	530,90	21,446	974,46	7,7440	20727,0	14872,4	216,935
432	705,15	740,99	531,71	21,559	970,73	7,7455	20757,8	14894,9	216,979
433	706,15	742,09	532,51	21,673	967,02	7,7471	20788,6	14917,3	217,022
434	707,15	743,19	533,31	21,787	963,33	7,7486	20819,4	14939,8	217,066
435	708,15	744,29	534,11	21,901	959,65	7,7502	20850,2	14962,3	217,109
436	709,15	745,39	534,92	22,016	955,99	7,7517	20881,0	14984,8	217,153
437	710,15	746,49	535,72	22,131	952,36	7,7533	20911,8	15007,3	217,196
438	711,15	747,59	536,52	22,247	948,73	7,7548	20942,6	15029,8	217,240
439	712,15	748,69	537,33	22,363	945,13	7,7564	20973,4	15052,3	217,283
440	713,15	749,79	538,13	22,480	941,55	7,7579	21004,3	15074,9	217,326
441	714,15	750,89	538,93	22,597	937,98	7,7595	21035,1	15097,4	217,369
442	715,15	752,00	539,74	22,715	934,43	7,7610	21066,0	15119,9	217,413
443	716,15	753,10	540,54	22,833	930,89	7,7626	21096,8	15142,5	217,456
444	717,15	754,20	541,35	22,951	927,38	7,7641	21127,7	15165,0	217,499

Nitrogen (*Continued*)

t	T	h	u	π_0	θ_0	s^0	H	U	S
445	718,15	755,30	542,15	23,070	923,88	7,7656	21158,6	15187,6	217,542
446	719,15	756,40	542,96	23,190	920,40	7,7672	21189,4	15210,1	217,585
447	720,15	757,51	543,76	23,310	916,93	7,7687	21220,3	15232,7	217,628
448	721,15	758,61	544,57	23,430	913,48	7,7702	21251,2	15255,3	217,670
449	722,15	759,71	545,38	23,551	910,05	7,7718	21282,1	15277,8	217,713
450	723,15	760,81	546,18	23,673	906,64	7,7733	21313,0	15300,4	217,756
451	724,15	761,92	546,99	23,795	903,24	7,7748	21343,9	15323,0	217,799
452	725,15	763,02	547,80	23,917	899,86	7,7763	21374,8	15345,6	217,841
453	726,15	764,13	548,60	24,040	896,49	7,7779	21405,8	15368,2	217,884
454	727,15	765,23	549,41	24,163	893,14	7,7794	21436,7	15390,9	217,927
455	728,15	766,33	550,22	24,287	889,81	7,7809	21467,6	15413,5	217,969
456	729,15	767,44	551,03	24,412	886,49	7,7824	21498,6	15436,1	218,012
457	730,15	768,54	551,83	24,537	883,19	7,7839	21529,5	15458,8	218,054
458	731,15	769,65	552,64	24,662	879,90	7,7854	21560,5	15481,4	218,096
159	732,15	770,75	553,45	24,788	876,63	7,7869	21591,4	15504,0	218,139
460	733,15	771,86	554,26	24,914	873,38	7,7885	21622,4	15526,7	218,181
461	734,15	772,97	555,07	25,041	870,14	7,7900	21653,4	15549,4	218,223
462	735,15	774,07	555,88	25,168	866,91	7,7915	21684,4	15572,0	218,265
463	736,15	775,18	556,69	25,296	863,71	7,7930	21715,4	15594,7	218,308
464	737,15	776,28	557,50	25,425	860,51	7,7945	21746,4	15617,4	218,350
465	738,15	777,39	558,31	25,553	857,34	7,7960	21777,4	15640,1	218,392
466	739,15	778,50	559,12	25,683	854,17	7,7975	21808,4	15662,8	218,434
467	740,15	779,61	559,93	25,813	851,02	7,7990	21839,4	15685,5	218,476
468	741,15	780,71	560,74	25,943	847,89	7,8005	21870,4	15708,2	218,517
469	742,15	781,82	561,55	26,074	844,77	7,8020	21901,5	15730,9	218,559
470	743,15	782,93	562,36	26,205	841,67	7,8034	21932,5	15753,7	218,601
471	744,15	784,04	563,17	26,337	838,58	7,8049	21963,6	15776,4	218,643
472	745,15	785,15	563,98	26,470	835,51	7,8064	21994,6	15799,1	218,685
473	746,15	786,25	564,80	26,603	832,45	7,8079	22025,7	15821,9	218,726
474	747,15	787,36	565,61	26,736	829,40	7,8094	22056,7	15844,6	218,768
475	748,15	788,47	566,42	26,870	826,37	7,8109	22087,8	15867,4	218,809
476	749,15	789,58	567,23	27,005	823,35	7,8124	22118,9	15890,2	218,851
477	750,15	790,69	568,05	27,140	820,35	7,8138	22150,0	15912,9	218,892
478	751,15	791,80	568,86	27,275	817,36	7,8153	22181,1	15935,7	218,934
479	752,15	792,91	569,67	27,411	814,39	7,8168	22212,2	15958,5	218,975
480	753,15	794,02	570,49	27,548	811,43	7,8183	22243,3	15981,3	219,016
481	754,15	795,13	571,30	27,685	808,48	7,8197	22274,4	16004,1	219,058
482	755,15	796,25	572,12	27,823	805,54	7,8212	22305,5	16026,9	219,099
483	756,15	797,36	572,93	27,961	802,62	7,8227	22336,7	16049,7	219,140
484	757,15	798,47	573,74	28,100	799,72	7,8242	22367,8	16072,5	219,181
485	758,15	799,58	574,56	28,239	796,82	7,8256	22398,9	16095,4	219,222
486	759,15	800,69	575,37	28,379	793,95	7,8271	22430,1	16118,2	219,264
487	760,15	801,80	576,19	28,519	791,08	7,8286	22461,2	16141,1	219,305
488	761,15	802,92	577,01	28,660	788,23	7,8300	22492,4	16163,9	219,346
489	762,15	804,03	577,82	28,801	785,39	7,8315	22523,6	16186,8	219,386
490	763,15	805,14	578,64	28,943	782,56	7,8329	22554,8	16209,6	219,427
491	764,15	806,26	579,45	29,086	779,75	7,8344	22585,9	16232,5	219,468
492	765,15	807,37	580,27	29,229	776,94	7,8359	22617,1	16255,4	219,509
493	766,15	808,48	581,09	29,372	774,16	7,8373	22648,3	16278,3	219,550
494	767,15	809,60	581,91	29,517	771,38	7,8388	22679,5	16301,1	219,590
495	768,15	810,71	582,72	29,661	768,62	7,8402	22710,8	16324,0	219,631
496	769,15	811,82	583,54	29,807	765,87	7,8417	22742,0	16346,9	219,672
497	770,15	812,94	584,36	29,952	763,13	7,8431	22773,2	16369,9	219,712
498	771,15	814,05	585,18	30,099	760,41	7,8446	22804,4	16392,8	219,753
499	772,15	815,17	585,99	30,246	757,69	7,8460	22835,7	16415,7	219,793

Nitrogen (*Continued*)

t	T	h	u	π_0	θ_0	s^0	H	U	S
500	773,15	816,29	586,81	30,393	754,99	7,8474	22866,9	16438,6	219,834
501	774,15	817,40	587,63	30,541	752,31	7,8489	22898,2	16461,6	219,874
502	775,15	818.52	588,45	30,690	749,63	7,8503	22929,4	16484,5	219,914
503	776,15	819,63	589,27	30,839	746,97	7,8518	22960,7	16507,5	219,955
504	777,15	820,75	590,09	30,989	744,32	7,8532	22992,0	16530,4	219.995
505	778,15	821,87	590,91	31,139	741,68	7,8546	23023,3	16553,4	220,035
506	779,15	822,98	591,73	31,290	739,05	7,8561	23054,5	16576,4	220,075
507	780,15	824,10	592,55	31,441	736,44	7,8575	23085,8	16599,4	220,116
508	781,15	825,22	593,37	31,593	733,83	7,8589	23117,1	16622,3	220,156
509	782,15	826,34	594,19	31,746	731,24	7,8604	23148,5	16645,3	220,196
510	783,15	827,45	595,01	31,899	728,66	7,8618	23179,8	16668,3	220,236
511	784,15	828,57	595,83	32,053	726,09	7,8632	23211,1	16691,3	220,276
512	785,15	829,69	596,66	32,207	723,53	7,8647	23242,4	16714,4	220,316
513	786,15	830,81	597.48	32,362	720,99	7,8661	23273,8	16737,4	220,356
514	787,15	831,93	598,30	32,517	718,45	7,8675	23305,1	16760,4	220,395
515	788,15	833,05	599,12	32,673	715,93	7,8689	23336,5	16783,4	220,435
516	789,15	834,17	599,94	32,830	713,42	7,8703	23367,8	16806,5	220,475
517	790,15	835 28	600,77	32,987	710,92	7,8718	23399,2	16829,5	220,515
518	791,15	836,40	601,59	33,145	708,43	7,8732	23430,5	16852,6	220,554
519	792,15	837,52	602,41	33,303	705,95	7,8746	23461,9	16875,7	220,594
520	793,15	838,65	603,24	33,462	703,49	7,8760	23493,3	16898,7	220,634
521	794,15	839,77	604,06	33,622	701,03	7,8774	23524,7	16921,8	220,673
522	795,15	840,89	604,89	33,782	698,59	7,8788	23556,1	16944,9	220,713
523	796,15	842,01	605,71	33,943	696,15	7,8802	23587,5	16968,0	220,752
524	797,15	843.13	606,53	34,104	693,73	7,8816	23618,9	16991,1	220,792
525	798,15	844,25	607,36	34,266	691,32	7,8830	23650,3	17014,2	220,831
526	799,15	845,37	608.18	34,429	688,91	7,8845	23681,8	17037,3	220,870
527	800,15	846,49	609.01	34,592	686,52	7,8859	23713,2	17060,4	220,910
528	801,15	847,62	609,83	34,756	684,14	7,8873	23744,6	17083,5	220,949
529	802,15	848,74	610,66	34,920	681,77	7,8887	23776,1	17106,7	220,988
530	803,15	849,86	611,49	35,085	679,41	7,8901	23807,5	17129,8	221,027
531	804,15	850,99	612,31	35,250	677,06	7,8915	23839,0	17153,0	221,066
532	805,15	852,11	613,14	35,417	674,72	7,8928	23870,5	17176,1	221,105
533	806,15	853,23	613,97	35,583	672,39	7,8942	23901,9	17199,3	221,145
534	807,15	854,36	614,79	35,751	670,08	7,8956	23933,4	17222,4	221,184
535	808,15	855,48	615,62	35,919	667,77	7,8970	23964,9	17245,6	221,223
536	809,15	856,60	616,45	36,088	665,47	7,8984	23996,4	17268,8	221,262
537	810,15	857,73	617,27	36,257	663,18	7,8998	24027,9	17292,0	221,300
538	811,15	858,85	618,10	36,427	660,90	7,9012	24059,4	17315,2	221,339
539	812,15	859,98	618,93	36,597	658,63	7,9026	24090,9	17338,4	221,378
540	813,15	861,10	619,76	36,768	656,38	7,9040	24122,4	17361,6	221,417
541	814,15	862,23	620,59	36,940	654,13	7,9053	24154,0	17384,8	221,456
542	815,15	863,35	621,42	37,113	651,89	7,9067	24185,5	17408,0	221,494
543	816,15	864,48	622,25	37,286	649,66	7,9081	24217,0	17431,2	221,533
544	817,15	865,61	623,08	37,459	647,44	7,9095	24248,6	17454,5	221,572
545	818,15	866,73	623,91	37,633	645,23	7,9109	24280,1	17477,7	221,610
546	819,15	867,86	624,74	37,808	643,03	7,9122	24311,7	17501,0	221,649
547	820,15	868,99	625,57	37,984	640,84	7,9136	24343,3	17524,2	221,687
548	821,15	870,11	626,40	38,160	638,66	7,9150	24374,8	17547,5	221,726
549	822,15	871,24	627,23	38,337	636,49	7,9164	24406,4	17570,7	221,764
550	823,15	872,37	628,06	38,514	634,33	7,9177	24438,0	17594,0	221,803
551	824,15	873,50	628,89	38,693	632,17	7,9191	24469,6	17617,3	221,841
552	825,15	874,62	629,72	38,871	630,03	7,9205	24501,2	17640,6	221,879
553	826,15	875,75	630,55	39,051	627,90	7,9218	24532,8	17663,9	221,918
554	827,15	876,88	631,38	39,231	625,77	7,9232	24564,4	17687,2	221,956

t	T	h	u	π_0	θ_0	s^0	H	U	S
555	828,15	878,01	632,21	39,411	623,65	7,9246	24596,1	17710,5	221,994
556	829,15	879,14	633,05	39,593	621,55	7,9259	24627,7	17733,8	222,032
557	830,15	880,27	633,88	39,775	619,45	7,9273	24659,3	17757,1	222,070
558	831,15	881,40	634,71	39,957	617,36	7,9287	24691,0	17780,5	222,108
559	832,15	882,53	635,55	40,141	615,28	7,9300	24722,6	17803,8	222,147
560	833,15	883,66	636,38	40,325	613,21	7,9314	24754,3	17827,1	222,185
561	834,15	884,79	637,21	40,509	611,15	7,9327	24785,9	17850,5	222,223
562	835,15	885,92	638,05	40,695	609,09	7,9341	24817,6	17873,8	222,260
563	836,15	887,05	638,88	40,881	607,05	7,9354	24849,3	17897,2	222,298
564	837,15	888,18	639,71	41,067	605,01	7,9368	24881,0	17920,6	222,336
565	838,15	889,31	640,55	41,255	602,99	7,9381	24912,7	17943,9	222,374
566	839,15	890,44	641,38	41,443	600,97	7,9395	24944,4	17967,3	222,412
567	840,15	891,58	642,22	41,631	598,96	7,9408	24976,1	17990,7	222,450
568	841,15	892,71	643,05	41,820	596,95	7,9422	25007,8	18014,1	222,487
569	842,15	893,84	643,89	42,010	594,96	7,9435	25039,5	18037,5	222,525
570	843,15	894,97	644,72	42,201	592,98	7,9449	25071,2	18060,9	222,563
571	844,15	896,11	645,56	42,392	591,00	7,9462	25103,0	18084,4	222,600
572	845,15	897,24	646,40	42,584	589,03	7,9475	25134,7	18107,8	222,638
573	846,15	898,37	647,23	42,777	587,07	7,9489	25166,4	18131,2	222,675
574	847,15	899,51	648,07	42,971	585,12	7,9502	25198,2	18154,6	222,713
575	848,15	900,64	648,91	43,165	583,18	7,9516	25230,0	18178,1	222,750
576	849,15	901,77	649,74	43,359	581,24	7,9529	25261,7	18201,5	222,788
577	850,15	902,91	650,58	43,555	579,32	7,9542	25293,5	18225,0	222,825
578	851,15	904,04	651,42	43,751	577,40	7,9556	25325,3	18248,5	222,863
579	852,15	905,18	652,26	43,948	575,49	7,9569	25357,1	18271,9	222,900
580	853,15	906,31	653,09	44,145	573,58	7,9582	25388,8	18295,4	222,937
581	854,15	907,45	653,93	44,344	571,69	7,9596	25420,6	18318,9	222,974
582	855,15	908,58	654,77	44,543	569,80	7,9609	25452,4	18342,4	223,012
583	856,15	909,72	655,61	44,742	567,92	7,9622	25484,3	18365,9	223,049
584	857,15	910,85	656,45	44,942	566,05	7,9635	25516,1	18389,4	223,086
585	858,15	911,99	657,29	45,143	564,19	7,9649	25547,9	18412,9	223,123
586	859,15	913,12	658,13	45,345	562,33	7,9662	25579,7	18436,4	223,160
587	860,15	914,26	658,97	45,548	560,49	7,9675	25611,6	18459,9	223,197
588	861,15	915,40	659,81	45,751	558,65	7,9688	25643,4	18483,5	223,234
589	862,15	916,54	660,65	45,955	556,81	7,9702	25675,3	18507,0	223,271
590	863,15	917,67	661,49	46,159	554,99	7,9715	25707,1	18530,5	223,308
591	864,15	918,81	662,33	46,365	553,17	7,9728	25739,0	18554,1	223,345
592	865,15	919,95	663,17	46,571	551,36	7,9741	25770,9	18577,7	223,382
593	866,15	921,09	664,01	46,777	549,56	7,9754	25802,7	18601,2	223,419
594	867,15	922,22	664,85	46,985	547,77	7,9767	25834,6	18624,8	223,455
595	868,15	923,36	665,69	47,193	545,98	7,9780	25866,5	18648,4	223,492
596	869,15	924,50	666,54	47,402	544,20	7,9794	25898,4	18672,0	223,529
597	870,15	925,64	667,38	47,611	542,43	7,9807	25930,3	18695,5	223,566
598	871,15	926,78	668,22	47,822	540,66	7,9820	25962,2	18719,1	223,602
599	872,15	927,92	669,06	48,033	538,90	7,9833	25994,2	18742,7	223,639
600	873,15	929,06	669,91	48,245	537,15	7,9846	26026,1	18766,4	223,675
601	874,15	930,20	670,75	48,457	535,41	7,9859	26058,0	18790,0	223,712
602	875,15	931,34	671,59	48,671	533,67	7,9872	26089,9	18813,6	223,749
603	876,15	932,48	672,44	48,885	531,94	7,9885	26121,9	18837,2	223,785
604	877,15	933,62	673,28	49,099	530,22	7,9898	26153,8	18860,9	223,821
605	878,15	934,76	674,12	49,315	528,50	7,9911	26185,8	18884,5	223,858
606	879,15	935,90	674,97	49,531	526,80	7,9924	26217,8	18908,2	223,894
607	880,15	937,04	675,81	49,748	525,09	7,9937	26249,7	18931,8	223,931
608	881,15	938,18	676,66	49,966	523,40	7,9950	26281,7	18955,5	223,967
609	882,15	939,33	677,50	50,184	521,71	7,9963	26313,7	18979,1	224,003

Nitrogen (*Continued*)

t	T	h	u	π_0	θ_0	s^0	H	U	S
610	883,15	940,47	678,35	50,404	520,03	7,9976	26345,7	19002,8	224,039
611	884,15	941,61	679,19	50,624	518,36	7,9989	26377,7	19026,5	224,076
612	885,15	942,75	680,04	50,844	516,69	8,0002	26409,7	19050,2	224,112
613	886,15	943,89	680,88	51,066	515,03	8,0015	26441,7	19073,9	224,148
614	887,15	945,04	681,73	51,288	513,38	8,0027	26473,7	19097,6	224,184
615	888,15	946,18	682,58	51,511	511,73	8,0040	26505,7	19121,3	224,220
616	889,15	947,32	683,42	51,735	510,09	8,0053	26537,8	19145,0	224,256
617	890,15	948,47	684,27	51,959	508,46	8,0066	26569,8	19168,7	224,292
618	891,15	949,61	685,12	52,185	506,83	8,0079	26601,8	19192,5	224,328
619	892,15	950,76	685,96	52,411	505,21	8,0092	26633,9	19216,2	224,364
620	893,15	951,90	686,81	52,638	503,60	8,0105	26666,0	19239,9	224,400
621	894,15	953,04	687,66	52,865	501,99	8,0117	26698,0	19263,7	224,436
622	895,15	954,19	688,51	53,094	500,39	8,0130	26730,1	19287,4	224,472
623	896,15	955,33	689,36	53,323	498,80	8,0143	26762,2	19311,2	224,508
624	897,15	956,48	690,20	53,553	497,21	8,0156	26794,2	19335,0	224,543
625	898,15	957,62	691,05	53,784	495,63	8,0168	26826,3	19358,7	224,579
626	899,15	958,77	691,90	54,015	494,05	8,0181	26858,4	19382,5	224,615
627	900,15	959,92	692,75	54,247	492,49	8,0194	26890,5	19406,3	224,650
628	901,15	961,06	693,60	54,481	490,92	8,0207	26922,6	19430,1	224,686
629	902,15	962,21	694,45	54,714	489,37	8,0219	26954,7	19453,9	224,722
630	903,15	963,36	695,30	54,949	487,82	8,0232	26986,9	19477,7	224,757
631	904,15	964,50	696,15	55,185	486,27	8,0245	27019,0	19501,5	224,793
632	905,15	965,65	697,00	55,421	484,74	8,0257	27051,1	19525,3	224,828
633	906,15	966,80	697,85	55,658	483,20	8,0270	27083,3	19549,2	224,864
634	907,15	967,94	698,70	55,896	481,68	8,0283	27115,4	19573,0	224,899
635	908,15	969,09	699,55	56,134	480,16	8,0295	27147,6	19596,8	224,935
636	909,15	970,24	700,40	56,374	478,65	8,0308	27179,7	19620,7	224,970
637	910,15	971,39	701,25	56,614	477,14	8,0321	27211,9	19644,5	225,006
638	911,15	972,54	702,11	56,855	475,64	8,0333	27244,1	19668,4	225,041
639	912,15	973,69	702,96	57,097	474,14	8,0346	27276,2	19692,3	225,076
640	913,15	974,83	703,81	57,340	472,65	8,0358	27308,4	19716,1	225,111
641	914,15	975,98	704,66	57,583	471,17	8,0371	27340,6	19740,0	225,147
642	915,15	977,13	705,52	57,828	469,69	8,0384	27372,8	19763,9	225,182
643	916,15	978,28	706,37	58,073	468,22	8,0396	27405,0	19787,8	225,217
644	917,15	979,43	707,22	58,319	466,76	8,0409	27437,2	19811,7	225,252
645	918,15	980,58	708,07	58,565	465,30	8,0421	27469,4	19835,6	225,287
646	919,15	981,73	708,93	58,813	463,84	8,0434	27501,7	19859,5	225,322
647	920,15	982,88	709,78	59,061	462,39	8,0446	27533,9	19883,4	225,357
648	921,15	984,03	710,64	59,311	460,95	8,0459	27566,1	19907,3	225,392
649	922,15	985,18	711,49	59,561	459,51	8,0471	27598,4	19931,2	225,427
650	923,15	986,34	712,34	59,812	458,08	8,0484	27630,6	19955,2	225,462
651	924,15	987,49	713,20	60,063	456,66	8,0496	27662,9	19979,1	225,497
652	925,15	988,64	714,05	60,316	455,24	8,0509	27695,1	20003,1	225,532
653	926,15	989,79	714,91	60,569	453,82	8,0521	27727,4	20027,0	225,567
654	927,15	990,94	715,76	60,824	452,41	8,0534	27759,7	20051,0	225,602
655	928,15	992,10	716,62	61,079	451,01	8,0546	27792,0	20074,9	225,637
656	929,15	993,25	717,47	61,335	449,6!	8,0558	27824,2	20098,9	225,671
657	930,15	994,40	718,33	61,591	448,22	8,0571	27856,5	20122,9	225,706
658	931,15	995,55	719,19	61,849	446,83	8,0583	27888,8	20146,9	225,741
659	932,15	996,71	720,04	62,108	445,45	8,0596	27921,1	20170,9	225,776
660	933,15	997,86	720,90	62,367	444,07	8,0608	27953,4	20194,9	225,810
661	934,15	999,01	721,76	62,627	442,70	8,0620	27985,8	20218,9	225,845
662	935,15	1000,17	722,61	62,888	441,34	8,0633	28018,1	20242,9	225,879
663	936,15	1001,32	723,47	63,150	439,98	8,0645	28050,4	20266,9	225,914
664	937,15	1002,48	724,33	63,413	438,62	8,0657	28082,8	20290,9	225,948

Nitrogen (*Continued*)

t	T	h	u	π_0	θ_0	s^0	H	U	S
665	938,15	1003,63	725,19	63,676	437,27	8,0670	28115,1	20314,9	225,983
666	939,15	1004,79	726,04	63,941	435,93	8,0682	28147,4	20339,0	226,017
667	940,15	1005,94	726,90	64,206	434,59	8,0694	28179,8	20363,0	226,052
668	941,15	1007,10	727,76	64,472	433,25	8,0706	28212,2	20387,1	226,086
669	942,15	1008,25	728,62	64,740	431,92	8,0719	28244,5	20411,1	226,121
670	943,15	1009,41	729,48	65,008	430,60	8,0731	28276,9	20435,2	226,155
671	944,15	1010,56	730,34	65,276	429,28	8,0743	28309,3	20459,2	226,189
672	945,15	1011,72	731,20	65,546	427,97	8,0755	28341,7	20483,3	226,224
673	946,15	1012,87	732,06	65,817	426,66	8,0768	28374,1	20507,4	226,258
674	947,15	1014,03	732,92	66,088	425,36	8,0780	28406,5	20531,5	226,292
675	948,15	1015,19	733,78	66,360	424,06	8,0792	28438,9	20555,6	226,326
676	949,15	1016,34	734,64	66,634	422,76	8,0804	28471,3	20579,6	226,360
677	950,15	1017,50	735,50	66,908	421,48	8,0817	28503,7	20603,8	226,395
678	951,15	1018,66	736,36	67,183	420,19	8,0829	28536,1	20627,9	226,429
679	952,15	1019,82	737,22	67,459	418,91	8,0841	28568,5	20652,0	226,463
680	953,15	1020,97	738,08	67,736	417,64	8,0853	28601,0	20676,1	226,497
681	954,15	1022,13	738,94	68,013	416,37	8,0865	28633,4	20700,2	226,531
682	955,15	1023,29	739,80	68,292	415,11	8,0877	28665,9	20724,4	226,565
683	956,15	1024,45	740,66	68,571	413,85	8,0889	28698,3	20748,5	226,599
684	957,15	1025,61	741,53	68,852	412,59	8,0902	28730,8	20772,6	226,633
685	958,15	1026,77	742,39	69,133	411,34	8,0914	28763,2	20796,8	226,667
686	959,15	1027,93	743,25	69,415	410,10	8,0926	28795,7	20820,9	226,700
687	960,15	1029,09	744,11	69,699	408,86	8,0938	28828,2	20845,1	226,734
688	961,15	1030,25	744,97	69,983	407,62	8,0950	28860,7	20869,3	226,768
689	962,15	1031,41	745,84	70,268	406,39	8,0962	28893,2	20893,5	226,802
690	963,15	1032,57	746,70	70,553	405,17	8,0974	28925,7	20917,6	226,836
691	964,15	1033,73	747,56	70,840	403,95	8,0986	28958,2	20941,8	226,869
692	965,15	1034,89	748,43	71,128	402,73	8,0998	28990,7	20966,0	226,903
693	966,15	1036,05	749,29	71,416	401,52	8,1010	29023,2	20990,2	226,937
694	967,15	1037,21	750,16	71,706	400,31	8,1022	29055,7	21014,4	226,970
695	968,15	1038,37	751,02	71,997	399,11	8,1034	29088,2	21038,6	227,004
696	969,15	1039,53	751,88	72,288	397,91	8,1046	29120,8	21062,9	227,038
697	970,15	1040,69	752,75	72,580	396,71	8,1058	29153,3	21087,1	227,071
698	971,15	1041,85	753,61	72,874	395,52	8,1070	29185,8	21111,3	227,105
699	972,15	1043,02	754,48	73,168	394,34	8,1082	29218,4	21135,5	227,138
700	973,15	1044,18	755,35	73,463	393,16	8,1094	29251,0	21159,8	227,172
701	974,15	1045,34	756,21	73,759	391,98	8,1106	29283,5	21184,0	227,205
702	975,15	1046,50	757,08	74,056	390,81	8,1118	29316,1	21208,3	227,238
703	976,15	1047,66	757,94	74,354	389,65	8,1130	29348,7	21232,5	227,272
704	977,15	1048,83	758,81	74,653	388,48	8,1142	29381,2	21256,8	227,305
705	978,15	1049,99	759,68	74,953	387,32	8,1154	29413,8	21281,1	227,339
706	979,15	1051,15	760,54	75,254	386,17	8,1165	29446,4	21305,4	227,372
707	980,15	1052,32	761,41	75,555	385,02	8,1177	29479,0	21329,6	227,405
708	981,15	1053,48	762,28	75,858	383,88	8,1189	29511,6	21353,9	227,438
709	982,15	1054,65	763,14	76,162	382,73	8,1201	29544,2	21378,2	227,472
710	983,15	1055,81	764,01	76,467	381,60	8,1213	29576,8	21402,5	227,505
711	984,15	1056,98	764,88	76,772	380,47	8,1225	29609,5	21426,8	227,538
712	985,15	1058,14	765,75	77,079	379,34	8,1237	29642,1	21451,2	227,571
713	986,15	1059,30	766,61	77,386	378,21	8,1248	29674,7	21475,5	227,604
714	987,15	1060,47	767,48	77,695	377,09	8,1260	29707,4	21499,8	227,637
715	988,15	1061,64	768,35	78,004	375,98	8,1272	29740,0	21524,1	227,670
716	989,15	1062,80	769,22	78,315	374,87	8,1284	29772,7	21548,5	227,703
717	990,15	1063,97	770,09	78,626	373,76	8,1296	29805,3	21572,8	227,736
718	991,15	1065,13	770,96	78,939	372,66	8,1307	29838,0	21597,2	227,769
719	992,15	1066,30	771,83	79,252	371,56	8,1319	29870,7	21621,5	227,802

Nitrogen (*Continued*)

t	T	h	u	π_0	θ_0	s^0	H	U	S
720	993,15	1067,47	772,70	79,566	370,46	8,1331	29903,3	21645,9	227,835
721	994,15	1068,63	773,57	79,882	369,37	8,1343	29936,0	21670,2	227,868
722	995,15	1069,80	774,44	80,198	368,28	8,1354	29968,7	21694,6	227,901
723	996,15	1070,97	775,31	80,515	367,20	8,1366	30001,4	21719,0	227,934
724	997,15	1072,13	776,18	80,834	366,12	8,1378	30034,1	21743,4	227,967
725	998,15	1073,30	777,05	81,153	365,05	8,1389	30066,8	21767,8	227,999
726	999,15	1074,47	777,92	81,473	363,98	8,1401	30099,5	21792,2	228,032
727	1000,15	1075,64	778,79	81,795	362,91	8,1413	30132,2	21816,6	228,065
728	1001,15	1076,80	779,66	82,117	361,84	8,1424	30164,9	21841,0	228,098
729	1002,15	1077,97	780,53	82,440	360,79	8,1436	30197,7	21865,4	228,130
730	1003,15	1079,14	781,40	82,765	359,73	8,1448	30230,4	21889,8	228,163
731	1004,15	1080,31	782,28	83,090	358,68	8,1459	30263,1	21914,2	228,195
732	1005,15	1081,48	783,15	83,416	357,63	8,1471	30295,9	21938,6	228,228
733	1006,15	1082,65	784,02	83,744	356,59	8,1483	30328,6	21963,1	228,261
734	1007,15	1083,82	784,89	84,072	355,55	8,1494	30361,4	21987,5	228,293
735	1008,15	1084,99	785,77	84,402	354,51	8,1506	30394,1	22012,0	228,326
736	1009,15	1086,16	786,64	84,732	353,48	8,1517	30426,9	22036,4	228,358
737	1010,15	1087,33	787,51	85,064	352,45	8,1529	30459,7	22060,9	228,391
738	1011,15	1088,50	788,39	85,396	351,43	8,1541	30492,5	22085,4	228,423
739	1012,15	1089,67	789,26	85,729	350,41	8,1552	30525,3	22109,8	228,455
740	1013,15	1090,84	790,13	86,064	349,39	8,1564	30558,0	22134,3	228,488
741	1014,15	1092,01	791,01	86,400	348,38	8,1575	30590,8	22158,8	228,520
742	1015,15	1093,18	791,88	86,736	347,37	8,1587	30623,6	22183,3	228,553
743	1016,15	1094,35	792,75	87,074	346,36	8,1598	30656,4	22207,8	228,585
744	1017,15	1095,52	793,63	87,412	345,36	8,1610	30689,3	22232,3	228,617
745	1018,15	1096,69	794,50	87,752	344,36	8,1621	30722,1	22256,8	228,649
746	1019,15	1097,86	795,38	88,093	343,36	8,1633	30754,9	22281,3	228,682
747	1020,15	1099,04	796,25	88,435	342,37	8,1644	30787,7	22305,8	228,714
748	1021,15	1100,21	797,13	88,778	341,38	8,1656	30820,6	22330,3	228,746
749	1022,15	1101,38	798,01	89,122	340,40	8,1667	30853,4	22354,8	228,778
750	1023,15	1102,55	798,88	89,466	339,42	8,1679	30886,3	22379,4	228,810
751	1024,15	1103,73	799,76	89,812	338,44	8,1690	30919,1	22403,9	228,842
752	1025,15	1104,90	800,63	90,160	337,47	8,1702	30952,0	22428,5	228,874
753	1026,15	1106,07	801,51	90,508	336,50	8,1713	30984,8	22453,0	228,906
754	1027,15	1107,25	802,39	90,857	335,53	8,1725	31017,7	22477,6	228,938
755	1028,15	1108,42	803,26	91,207	334,57	8,1736	31050,6	22502,1	228,970
756	1029,15	1109,59	804,14	91,558	333,61	8,1747	31083,5	22526,7	229,002
757	1030,15	1110,77	805,02	91,911	332,65	8,1759	31116,3	22551,3	229,034
758	1031,15	1111,94	805,89	92,264	331,70	8,1770	31149,2	22575,8	229,066
759	1032,15	1113,12	806,77	92,619	330,75	8,1782	31182,1	22600,4	229,098
760	1033,15	1114,29	807,65	92,975	329,80	8,1793	31215,0	22625,0	229,130
761	1034,15	1115,46	808,53	93,331	328,86	8,1804	31247,9	22649,6	229,162
762	1035,15	1116,64	809,41	93,689	327,92	8,1816	31280,9	22674,2	229,194
763	1036,15	1117,81	810,28	94,048	326,99	8,1827	31313,8	22698,8	229,225
764	1037,15	1118,99	811,16	94,408	326,05	8,1838	31346,7	22723,4	229,257
765	1038,15	1120,17	812,04	94,769	325,13	8,1850	31379,6	22748,0	229,289
766	1039,15	1121,34	812,92	95,131	324,20	8,1861	31412,6	22772,7	229,321
767	1040,15	1122,52	813,80	95,494	323,28	8,1872	31445,5	22797,3	229,352
768	1041,15	1123,69	814,68	95,859	322,36	8,1884	31478,5	22821,9	229,384
769	1042,15	1124,87	815,56	96,224	321,44	8,1895	31511,4	22846,6	229,416
770	1043,15	1126,05	816,44	96,591	320,53	8,1906	31544,4	22871,2	229,447
771	1044,15	1127,22	817,32	96,958	319,62	8,1918	31577,3	22895,8	229,479
772	1045,15	1128,40	818,20	97,327	318,71	8,1929	31610,3	22920,5	229,510
773	1046,15	1129,58	819,08	97,697	317,81	8,1940	31643,3	22945,2	229,542
774	1047,15	1130,75	819,96	98,068	316,91	8,1951	31676,3	22969,8	229,573

t	T	h	u	π_0	θ_0	s^0	H	U	S
775	1048,15	1131,93	820,84	98,440	316,02	8,1963	31709,3	22994,5	229,605
776	1049,15	1133,11	821,72	98,813	315,12	8,1974	31742,2	23019,2	229,636
777	1050,15	1134,29	822,60	99,187	314,23	8,1985	31775,2	23043,9	229,668
778	1051,15	1135,47	823,48	99,563	313,35	8,1996	31808,2	23068,5	229,699
779	1052,15	1136,64	824,36	99,939	312,46	8,2007	31841,2	23093,2	229,731
780	1053,15	1137,82	825,25	100,31	311,58	8,2019	31874,3	23117,9	229,762
781	1054,15	1139,00	826,13	100,69	310,70	8,2030	31907,3	23142,6	229,793
782	1055,15	1140,18	827,01	101,07	309,83	8,2041	31940,3	23167,4	229,825
783	1056,15	1141,36	827,89	101,45	308,96	8,2052	31973,3	23192,1	229,856
784	1057,15	1142,54	828,77	101,83	308,09	8,2063	32006,4	23216,8	229,887
785	1058,15	1143,72	829,66	102,22	307,23	8,2074	32039,4	23241,5	229,918
786	1059,15	1144,90	830,54	102,60	306,36	8,2086	32072,5	23266,2	229,950
787	1060,15	1146,08	831,42	102,99	305,50	8,2097	32105,5	23291,0	229,981
788	1061,15	1147,26	832,31	103,37	304,65	8,2108	32138,6	23315,7	230,012
789	1062,15	1148,44	833,19	103,76	303,80	8,2119	32171,6	23340,5	230,043
790	1063,15	1149,62	834,07	104,15	302,95	8,2130	32204,7	23365,2	230,074
791	1064,15	1150,80	834,96	104,54	302,10	8,2141	32237,8	23390,0	230,105
792	1065,15	1151,98	835,84	104,93	301,26	8,2152	32270,8	23414,7	230,136
793	1066,15	1153,16	836,73	105,33	300,41	8,2163	32303,9	23439,5	230,167
794	1067,15	1154,34	837,61	105,72	299,58	8,2174	32337,0	23464,3	230,198
795	1068,15	1155,52	838,49	106,11	298,74	8,2185	32370,1	23489,1	230,229
796	1069,15	1156,70	839,38	106,51	297,91	8,2197	32403,2	23513,9	230,260
797	1070,15	1157,89	840,26	106,91	297,08	8,2208	32436,3	23538,6	230,291
798	1071,15	1159,07	841,15	107,31	296,25	8,2219	32469,2	23563,4	230,322
799	1072,15	1160,25	842,03	107,71	295,43	8,2230	32502,5	23588,2	230,353
800	1073,15	1161,43	842,92	108,11	294,61	8,2241	32535,7	23613,0	230,384
801	1074,15	1162,61	843,81	108,51	293,79	8,2252	32568,8	23637,9	230,415
802	1075,15	1163,80	844,69	108,91	292,98	8,2263	32601,9	23662,7	230,446
803	1076,15	1164,98	845,58	109,32	292,16	8,2274	32635,0	23687,5	230,477
804	1077,15	1166,16	846,46	109,72	291,36	8,2285	32668,2	23712,3	230,507
805	1078,15	1167,35	847,35	110,13	290,55	8,2296	32701,3	23737,2	230,538
806	1079,15	1168,53	848,24	110,54	289,75	8,2307	32734,5	23762,0	230,569
807	1080,15	1169,71	849,12	110,95	288,94	8,2318	32767,6	23786,8	230,600
808	1081,15	1170,90	850,01	111,36	288,15	8,2329	32800,8	23811,7	230,630
809	1082,15	1172,08	850,90	111,77	287,35	8,2340	32834,0	23836,5	230,661
810	1083,15	1173,27	851,79	112,18	286,56	8,2350	32867,1	23861,4	230,692
811	1084,15	1174,45	852,67	112,59	285,77	8,2361	32900,3	23886,3	230,722
812	1085,15	1175,63	853,56	113,01	284,98	8,2372	32933,5	23911,1	230,753
813	1086,15	1176,82	854,45	113,42	284,20	8,2383	32966,7	23936,0	230,783
814	1087,15	1178,00	855,34	113,84	283,42	8,2394	32999,9	23960,9	230,814
815	1088,15	1179,19	856,22	114,26	282,64	8,2405	33033,1	23985,8	230,844
816	1089,15	1180,37	857,11	114,68	281,86	8,2416	33066,3	24010,7	230,875
817	1090,15	1181,56	858,00	115,10	281,09	8,2427	33099,5	24035,5	230,905
818	1091,15	1182,75	858,89	115,52	280,32	8,2438	33132,7	24060,4	230,936
819	1092,15	1183,93	859,78	115,95	279,55	8,2449	33165,9	24085,4	230,966
820	1093,15	1185,12	860,67	116,37	278,78	8,2459	33199,2	24110,3	230,997
821	1094,15	1186,30	861,56	116,80	278,02	8,2470	33232,4	24135,2	231,027
822	1095,15	1187,49	862,45	117,23	277,26	8,2481	33265,6	24160,1	231,057
823	1096,15	1188,68	863,34	117,65	276,50	8,2492	33298,9	24185,0	231,088
824	1097,15	1189,86	864,23	118,08	275,75	8,2503	33332,1	24210,0	231,118
825	1098,15	1191,05	865,12	118,52	275,00	8,2514	33365,4	24234,9	231,148
826	1099,15	1192,24	866,01	118,95	274,25	8,2524	33398,6	24259,8	231,179
827	1100,15	1193,42	866,90	119,38	273,50	8,2535	33431,9	24284,8	231,209
828	1101,15	1194,61	867,79	119,82	272,75	8,2546	33465,1	24309,7	231,239
829	1102,15	1195,80	868,68	120,25	272,01	8,2557	33498,4	24334,7	231,269

t	T	h	u	π_0	θ_0	s^0	H	U	S
830	1103,15	1196,99	869,57	120,69	271,27	8,2567	33531,7	24359,6	231,300
831	1104,15	1198,18	870,46	121,13	270,54	8,2578	33565,0	24384,6	231,330
832	1105,15	1199,36	871,35	121,57	269,80	8,2589	33598,3	24409,6	231,360
833	1106,15	1200,55	872,25	122,01	269,07	8,2600	33631,5	24434,6	231,390
834	1107,15	1201,74	873,14	122,45	268,34	8,2610	33664,8	24459,5	231,420
835	1108,15	1202,93	874,03	122,89	267,61	8,2621	33698,1	24484,5	231,450
836	1109,15	1204,12	874,92	123,34	266,89	8,2632	33731,4	24509,5	231,480
837	1110,15	1205,31	875,81	123,79	266,17	8,2643	33764,8	24534,5	231,510
838	1111,15	1206,50	876,71	124,23	265,45	8,2653	33798,1	24559,5	231,540
839	1112,15	1207,69	877,60	124,68	264,73	8,2664	33831,4	24584,5	231,570
840	1113,15	1208,88	878,49	125,13	264,01	8,2675	33864,7	24609,5	231,600
841	1114,15	1210,07	879,38	125,58	263,30	8,2685	33898,1	24634,6	231,630
842	1115,15	1211,26	880,28	126,04	262,59	8,2696	33931,4	24659,6	231,660
843	1116,15	1212,45	881,17	126,49	261,88	8,2707	33964,7	24684,6	231,690
844	1117,15	1213,64	882,06	126,94	261,18	8,2717	33998,1	24709,6	231,720
845	1118,15	1214,83	882,96	127,40	260,48	8,2728	34031,4	24734,7	231,749
846	1119,15	1216,02	883,85	127,86	259,78	8,2739	34064,8	24759,7	231,779
847	1120,15	1217,21	884,75	128,32	259,08	8,2749	34098,1	24784,8	231,809
848	1121,15	1218,40	885,64	128,78	258,38	8,2760	34131,5	24809,8	231,839
849	1122,15	1219,59	886,54	129,24	257,69	8,2771	34164,9	24834,9	231,869
850	1123,15	1220,78	887,43	129,70	257,00	8,2781	34198,3	24859,9	231,898
851	1124,15	1221,97	888,32	130,17	256,31	8,2792	34231,6	24885,0	231,928
852	1125,15	1223,17	889,22	130,63	255,62	8,2802	34265,0	24910,1	231,958
853	1126,15	1224,36	890,11	131,10	254,94	8,2813	34298,4	24935,1	231,987
854	1127,15	1225,55	891,01	131,57	254,26	8,2824	34331,8	24960,2	232,017
855	1128,15	1226,74	891,91	132,04	253,58	8,2834	34365,2	24985,3	232,047
856	1129,15	1227,93	892,80	132,51	252,90	8,2845	34398,6	25010,4	232,076
857	1130,15	1229,13	893,70	132,98	252,23	8,2855	34432,0	25035,5	232,106
858	1131,15	1230,32	894,59	133,45	251,55	8,2866	34465,4	25060,6	232,135
859	1132,15	1231,51	895,49	133,93	250,88	8,2876	34498,9	25085,7	232,165
860	1133,15	1232,71	896,39	134,41	250,21	8,2887	34532,3	25110,8	232,194
861	1134,15	1233,90	897,28	134,88	249,55	8,2897	34565,7	25135,9	232,224
862	1135,15	1235,09	898,18	135,36	248,89	8,2908	34599,1	25161,0	232,253
863	1136,15	1236,29	899,08	135,84	248,22	8,2918	34632,6	25186,2	232,283
864	1137,15	1237,48	899,97	136,32	247,56	8,2929	34666,0	25211,3	232,312
865	1138,15	1238,67	900,87	136,81	246,91	8,2939	34699,5	25236,4	232,342
866	1139,15	1239,87	901,77	137,29	246,25	8,2950	34732,9	25261,6	232,371
867	1140,15	1241,06	902,67	137,78	245,60	8,2960	34766,4	25286,7	232,400
868	1141,15	1242,26	903,56	138,26	244,95	8,2971	34799,9	25311,9	232,430
869	1142,15	1243,45	904,46	138,75	244,30	8,2981	34833,3	25337,0	232,459
870	1143,15	1244,65	905,36	139,24	243,66	8,2992	34866,8	25362,2	232,488
871	1144,15	1245,84	906,26	139,73	243,01	8,3002	34900,3	25387,3	232,518
872	1145,15	1247,04	907,16	140,23	242,37	8,3013	34933,8	25412,5	232,547
873	1146,15	1248,23	908,05	140,72	241,73	8,3023	34967,2	25437,7	232,576
874	1147,15	1249,43	908,95	141,21	241,09	8,3034	35000,7	25462,9	232,605
875	1148,15	1250,62	909,85	141,71	240,46	8,3044	35034,2	25488,0	232,634
876	1149,15	1251,82	910,75	142,21	239,82	8,3054	35067,7	25513,2	232,664
877	1150,15	1253,02	911,65	142,71	239,19	8,3065	35101,2	25538,4	232,693
878	1151,15	1254,21	912,55	143,21	238,56	8,3075	35134,8	25563,6	232,722
879	1152,15	1255,41	913,45	143,71	237,94	8,3086	35168,3	25588,8	232,751
880	1153,15	1256,61	914,35	144,21	237,31	8,3096	35201,8	25614,0	232,780
881	1154,15	1257,80	915,25	144,72	236,69	8,3106	35235,3	25639,2	232,809
882	1155,15	1259,00	916,15	145,23	236,07	8,3117	35268,9	25664,5	232,838
883	1156,15	1260,20	917,05	145,73	235,45	8,3127	35302,4	25689,7	232,867
884	1157,15	1261,39	917,95	146,24	234,83	8,3137	35335,9	25714,9	232,896

Nitrogen (*Continued*)

t	T	h	u	π_0	θ_0	s^0	H	U	S
885	1158,15	1262,59	918,85	146,75	234,22	8,3148	35369,5	25740,1	232,925
886	1159,15	1263,79	919,75	147,27	233,60	8,3158	35403,0	25765,4	232,954
887	1160,15	1264,99	920,65	147,78	232,99	8,3168	35436,6	25790,6	232,983
888	1161,15	1266,18	921,55	148,29	232,38	8,3179	35470,1	25815,9	233,012
889	1162,15	1267,38	922,46	148,81	231,78	8,3189	35503,7	25841,1	233,041
890	1163,15	1268,58	923,36	149,33	231,17	8,3199	35537,3	25866,4	233,070
891	1164,15	1269,78	924,26	149,85	230,57	8,3210	35570,8	25891,6	233,099
892	1165,15	1270,98	925,16	150,37	229,97	8,3220	35604,4	25916,9	233,127
893	1166,15	1272,18	926,06	150,89	229,37	8,3230	35638,0	25942,1	233,156
894	1167,15	1273,38	926,96	151,41	228,77	8,3241	35671,6	25967,4	233,185
895	1168,15	1274,57	927,87	151,94	228,18	8,3251	35705,2	25992,7	233,214
896	1169,15	1275,77	928,77	152,46	227,59	8,3261	35738,8	26018,0	233,243
897	1170,15	1276,97	929,67	152,99	227,00	8,3271	35772,4	26043,3	233,271
898	1171,15	1278,17	930,57	153,52	226,41	8,3282	35806,0	26068,6	233,300
899	1172,15	1279,37	931,48	154,05	225,82	8,3292	35839,6	26093,9	233,329
900	1173,15	1280,57	932,38	154,58	225,23	8,3302	35873,2	26119,2	233,357
901	1174,15	1281,77	933,28	155,12	224,65	8,3312	35906,8	26144,5	233,386
902	1175,15	1282,97	934,19	155,65	224,07	8,3322	35940,4	26169,8	233,415
903	1176,15	1284,17	935,09	156,19	223,49	8,3333	35974,1	26195,1	233,443
904	1177,15	1285,37	935,99	156,73	222,91	8,3343	36007,7	26220,4	233,472
905	1178,15	1286,58	936,90	157,27	222,34	8,3353	36041,3	26245,7	233,500
906	1179,15	1287,78	937,80	157,81	221,76	8,3363	36075,0	26271,1	233,529
907	1180,15	1288,98	938,71	158,35	221,19	8,3373	36108,6	26296,4	233,557
908	1181,15	1290,18	939,61	158,89	220,62	8,3384	36142,3	26321,7	233,586
909	1182,15	1291,38	940,52	159,44	220,05	8,3394	36175,9	26347,1	233,614
910	1183,15	1292,58	941,42	159,98	219,48	8,3404	36209,6	26372,4	233,643
911	1184,15	1293,78	942,33	160,53	218,92	8,3414	36243,3	26397,8	233,671
912	1185,15	1294,99	943,23	161,08	218,36	8,3424	36276,9	26423,1	233,700
913	1186,15	1296,19	944,14	161,63	217,80	8,3434	36310,6	26448,5	233,728
914	1187,15	1297,39	945,04	162,19	217,24	8,3445	36344,3	26473,8	233,757
915	1188,15	1298,59	945,95	162,74	216,68	8,3455	36378,0	26499,2	233,785
916	1189,15	1299,79	946,85	163,30	216,12	8,3465	36411,7	26524,6	233,813
917	1190,15	1301,00	947,76	163,85	215,57	8,3475	36445,4	26550,0	233,842
918	1191,15	1302,20	948,67	164,41	215,02	8,3485	36479,1	26575,3	233,870
919	1192,15	1303,40	949,57	164,97	214,47	8,3495	36512,8	26600,7	233,898
920	1193,15	1304,61	950,48	165,53	213,92	8,3505	36546,5	26626,1	233,926
921	1194,15	1305,81	951,38	166,10	213,37	8,3515	36580,2	26651,5	233,955
922	1195,15	1307,01	952,29	166,66	212,83	8,3525	36613,9	26676,9	233,983
923	1196,15	1308,22	953,20	167,23	212,28	8,3535	36647,6	26702,3	234,011
924	1197,15	1309,42	954,11	167,80	211,74	8,3545	36681,3	26727,7	234,039
925	1198,15	1310,62	955,01	168,37	211,20	8,3556	36715,1	26753,1	234,067
926	1199,15	1311,83	955,92	168,94	210,67	8,3566	36748,8	26778,6	234,096
927	1200,15	1313,03	956,83	169,51	210,13	8,3576	36782,5	26804,0	234,124
928	1201,15	1314,24	957,74	170,08	209,59	8,3586	36816,3	26829,4	234,152
929	1202,15	1315,44	958,64	170,66	209,06	8,3596	36850,0	26854,8	234,180
930	1203,15	1316,65	959,55	171,23	208,53	8,3606	36883,8	26880,3	234,208
931	1204,15	1317,85	960,46	171,81	208,00	8,3616	36917,5	26905,7	234,236
932	1205,15	1319,06	961,37	172,39	207,47	8,3626	36951,3	26931,2	234,264
933	1206,15	1320,26	962,28	172,98	206,95	8,3636	36985,0	26956,6	234,292
934	1207,15	1321,47	963,18	173,56	206,42	8,3646	37018,8	26982,1	234,320
935	1208,15	1322,67	964,09	174,14	205,90	8,3656	37052,6	27007,5	234,348
936	1209,15	1323,88	965,00	174,73	205,38	8,3666	37086,4	27033,0	234,376
937	1210,15	1325,09	965,91	175,32	204,86	8,3676	37120,1	27058,5	234,404
938	1211,15	1326,29	966,82	175,91	204,34	8,3686	37153,9	27083,9	234,432
939	1212,15	1327,50	967,73	176,50	203,83	8,3696	37187,7	27109,4	234,460

Nitrogen (*Continued*)

t	T	h	u	π_0	θ_0	s^0	H	U	S
940	1213,15	1328,70	968,64	177,09	203,31	8,3705	37221,5	27134,9	234,487
941	1214,15	1329,91	969,55	177,68	202,80	8,3715	37255,3	27160,4	234,515
942	1215,15	1331,12	970,46	178,28	202,29	8,3725	37289,1	27185,9	234,543
943	1216,15	1332,32	971,37	178,88	201,78	8,3735	37322,9	27211,4	234,571
944	1217,15	1333,53	972,28	179,48	201,27	8,3745	37356,7	27236,9	234,599
945	1218,15	1334,74	973,19	180,08	200,76	8,3755	37390,6	27262,4	234,627
946	1219,15	1335,95	974,10	180,68	200,26	8,3765	37424,4	27287,9	234,654
947	1220,15	1337,15	975,01	181,28	199,76	8,3775	37458,2	27313,4	234,682
948	1221,15	1338,36	975,92	181,89	199,25	8,3785	37492,0	27338,9	234,710
949	1222,15	1339,57	976,83	182,50	198,75	8,3795	37525,9	27364,4	234,737
950	1223,15	1340,78	977,74	183,10	198,26	8,3805	37559,7	27389,9	234,765
951	1224,15	1341,98	978,66	183,71	197,76	8,3814	37593,5	27415,5	234,793
952	1225,15	1343,19	979,57	184,33	197,26	8,3824	37627,4	27441,0	234,820
953	1226,15	1344,40	980,48	184,94	196,77	8,3834	37661,2	27466,5	234,848
954	1227,15	1345,61	981,39	185,55	196,28	8,3844	37695,1	27492,1	234,876
955	1228,15	1346,82	982,30	186,17	195,79	8,3854	37729,0	27517,6	234,903
956	1229,15	1348,03	983,21	186,79	195,30	8,3864	37762,8	27543,2	234,931
957	1230,15	1349,24	984,13	187,41	194,81	8,3874	37796,7	27568,7	234,958
958	1231,15	1350,45	985,04	188,03	194,32	8,3883	37830,6	27594,3	234,986
959	1232,15	1351,65	985,95	188,65	193,84	8,3893	37864,4	27619,8	235,013
960	1233,15	1352,86	986,86	189,28	193,36	8,3903	37898,3	27645,4	235,041
961	1234,15	1354,07	987,78	189,90	192,88	8,3913	37932,2	27671,0	235,068
962	1235,15	1355,28	988,69	190,53	192,40	8,3923	37966,1	27696,6	235,096
963	1236,15	1356,49	989,60	191,16	191,92	8,3932	38000,0	27722,1	235,123
964	1237,15	1357,70	990,52	191,79	191,44	8,3942	38033,9	27747,7	235,151
965	1238,15	1358,91	991,43	192,43	190,96	8,3952	38067,8	27773,3	235,178
966	1239,15	1360,12	992,34	193,06	190,49	8,3962	38101,7	27798,9	235,205
967	1240,15	1361,33	993,26	193,70	190,02	8,3971	38135,6	27824,5	235,233
968	1241,15	1362,55	994,17	194,33	189,55	8,3981	38169,5	27850,1	235,260
969	1242,15	1363,76	995,08	194,97	189,08	8,3991	38203,5	27875,7	235,287
970	1243,15	1364,97	996,00	195,62	188,61	8,4001	38237,4	27901,3	235,315
971	1244,15	1366,18	996,91	196,26	188,14	8,4010	38271,3	27926,9	235,342
972	1245,15	1367,39	997,83	196,90	187,68	8,4020	38305,2	27952,6	235,369
973	1246,15	1368,60	998,74	197,55	187,21	8,4030	38339,2	27978,2	235,396
974	1247,15	1369,81	999,66	198,20	186,75	8,4040	38373,1	28003,8	235,424
975	1248,15	1371,02	1000,57	198,85	186,29	8,4049	38407,1	28029,4	235,451
976	1249,15	1372,24	1001,49	199,50	185,83	8,4059	38441,0	28055,1	235,478
977	1250,15	1373,45	1002,40	200,15	185,37	8,4069	38475,0	28080,7	235,505
978	1251,15	1374,66	1003,32	200,81	184,92	8,4078	38508,9	28106,4	235,532
979	1252,15	1375,87	1004,23	201,46	184,46	8,4088	38542,9	28132,0	235,560
980	1253,15	1377,09	1005,15	202,12	184,01	8,4098	38576,9	28157,7	235,587
981	1254,15	1378,30	1006,07	202,78	183,56	8,4108	38610,8	28183,3	235,614
982	1255,15	1379,51	1006,98	203,44	183,10	8,4117	38644,8	28209,0	235,641
983	1256,15	1380,72	1007,90	204,11	182,65	8,4127	38678,8	28234,6	235,668
984	1257,15	1381,94	1008,81	204,77	182,21	8,4136	38712,8	28260,3	235,695
985	1258,15	1383,15	1009,73	205,44	181,76	8,4146	38746,7	28286,0	235,722
986	1259,15	1384,36	1010,65	206,11	181,31	8,4156	38780,7	28311,6	235,749
987	1260,15	1385,58	1011,56	206,78	180,87	8,4165	38814,7	28337,3	235,776
988	1261,15	1386,79	1012,48	207,45	180,43	8,4175	38848,7	28363,0	235,803
989	1262,15	1388,00	1013,40	208,12	179,99	8,4185	38882,7	28388,7	235,830
990	1263,15	1389,22	1014,31	208,80	179,55	8,4194	38916,7	28414,4	235,857
991	1264,15	1390,43	1015,23	209,47	179,11	8,4204	38950,8	28440,1	235,884
992	1265,15	1391,65	1016,15	210,15	178,67	8,4213	38984,8	28465,8	235,911
993	1266,15	1392,86	1017,07	210,83	178,23	8,4223	39018,8	28491,5	235,937
994	1267,15	1394,08	1017,98	211,51	177,80	8,4233	39052,8	28517,2	235,964

Nitrogen (*Continued*)

t	T	h	u	π_0	θ_0	s^0	H	U	S
995	1268,15	1395,29	1018,90	212,20	177,37	8,4242	39086,8	28542,9	235,991
996	1269,15	1396,51	1019,82	212,88	176,93	8,4252	39120,9	28568,6	236,018
997	1270,15	1397,72	1020,74	213,57	176,50	8,4261	39154,9	28594,4	236,045
998	1271,15	1398,94	1021,66	214,26	176,07	8,4271	39189,0	28620,1	236,072
999	1272,15	1400,15	1022,58	214,95	175,65	8,4281	39223,0	28645,8	236,098
1000	1273,15	1401,37	1023,49	215,65	175,22	8,4290	39257,0	28671,6	236,125
1001	1274,15	1402,58	1024,41	216,34	174,79	8,4300	39291,1	28697,3	236,152
1002	1275,15	1403,80	1025,33	217,04	174,37	8,4309	39325,2	28723,0	236,179
1003	1276,15	1405,01	1026,25	217,73	173,95	8,4319	39359,2	28748,8	236,205
1004	1277,15	1406,23	1027,17	218,43	173,53	8,4328	39393,3	28774,5	236,232
1005	1278,15	1407,45	1028,09	219,14	173,11	8,4338	39427,4	28800,3	236,259
1006	1279,15	1408,66	1029,01	219,84	172,69	8,4347	39461,4	28826,0	236,285
1007	1280,15	1409,88	1029,93	220,54	172,27	8,4357	39495,5	28851,8	236,312
1008	1281,15	1411,10	1030,85	221,25	171,85	8,4366	39529,6	28877,6	236,339
1009	1282,15	1412,31	1031,77	221,96	171,44	8,4376	39563,7	28903,4	236,365
1010	1283,15	1413,53	1032,69	222,67	171,02	8,4385	39597,8	28929,1	236,392
1011	1284,15	1414,75	1033,61	223,38	170,61	8,4395	39631,9	28954,9	236,418
1012	1285,15	1415,96	1034,53	224,10	170,20	8,4404	39666,0	28980,7	236,445
1013	1286,15	1417,18	1035,45	224,81	169,79	8,4414	39700,1	29006,5	236,471
1014	1287,15	1418,40	1036,37	225,53	169,38	8,4423	39734,2	29032,3	236,498
1015	1288,15	1419,62	1037,29	226,25	168,97	8,4433	39768,3	29058,1	236,524
1016	1289,15	1420,83	1038,21	226,97	168,57	8,4442	39802,4	29083,9	236,551
1017	1290,15	1422,05	1039,13	227,70	168,16	8,4451	39836,5	29109,7	236,577
1018	1291,15	1423,27	1040,05	228,42	167,76	8,4461	39870,6	29135,5	236,604
1019	1292,15	1424,49	1040,98	229,15	167,36	8,4470	39904,7	29161,3	236,630
1020	1293,15	1425,71	1041,90	229,88	166,95	8,4480	39938,9	29187,1	236,657
1021	1294,15	1426,92	1042,82	230,61	166,55	8,4489	39973,0	29212,9	236,683
1022	1295,15	1428,14	1043,74	231,34	166,15	8,4499	40007,1	29238,7	236,709
1023	1296,15	1429,36	1044,66	232,07	165,76	8,4508	40041,3	29264,6	236,736
1024	1297,15	1430,58	1045,59	232,81	165,36	8,4517	40075,4	29290,4	236,762
1025	1298,15	1431,80	1046,51	233,55	164,96	8,4527	40109,6	29316,2	236,788
1026	1299,15	1433,02	1047,43	234,29	164,57	8,4536	40143,7	29342,1	236,815
1027	1300,15	1434,24	1048,35	235,03	164,18	8,4546	40177,9	29367,9	236,841
1028	1301,15	1435,46	1049,27	235,77	163,78	8,4555	40212,1	29393,8	236,867
1029	1302,15	1436,68	1050,20	236,52	163,39	8,4564	40246,2	29419,6	236,893
1030	1303,15	1437,90	1051,12	237,27	163,00	8,4574	40280,4	29445,5	236,920
1031	1304,15	1439,12	1052,04	238,02	162,62	8,4583	40314,6	29471,3	236,946
1032	1305,15	1440,34	1052,97	238,77	162,23	8,4592	40348,7	29497,2	236,972
1033	1306,15	1441,56	1053,89	239,52	161,84	8,4602	40382,9	29523,1	236,998
1034	1307,15	1442,78	1054,81	240,28	161,46	8,4611	40417,1	29548,9	237,024
1035	1308,15	1444,00	1055,74	241,03	161,07	8,4620	40451,3	29574,8	237,051
1036	1309,15	1445,22	1056,66	241,79	160,69	8,4630	40485,5	29600,7	237,077
1037	1310,15	1446,44	1057,59	242,55	160,31	8,4639	40519,7	29626,6	237,103
1038	1311,15	1447,66	1058,51	243,31	159,93	8,4648	40553,9	29652,4	237,129
1039	1312,15	1448,88	1059,43	244,08	159,55	8,4658	40588,1	29678,3	237,155
1040	1313,15	1450,10	1060,36	244,84	159,17	8,4667	40622,3	29704,2	237,181
1041	1314,15	1451,32	1061,28	245,61	158,80	8,4676	40656,5	29730,1	237,207
1042	1315,15	1452,54	1062,21	246,38	158,42	8,4686	40690,7	29756,0	237,233
1043	1316,15	1453,77	1063,13	247,15	158,04	8,4695	40724,9	29781,9	237,259
1044	1317,15	1454,99	1064,06	247,93	157,67	8,4704	40759,2	29807,8	237,285
1045	1318,15	1456,21	1064,98	248,70	157,30	8,4713	40793,4	29833,8	237,311
1046	1319,15	1457,43	1065,91	249,48	156,93	8,4723	40827,6	29859,7	237,337
1047	1320,15	1458,65	1066,83	250,26	156,56	8,4732	40861,9	29885,6	237,363
1048	1321,15	1459,88	1067,76	251,04	156,19	8,4741	40896,1	29911,5	237,389
1049	1322,15	1461,10	1068,68	251,83	155,82	8,4750	40930,3	29937,4	237,415

Nitrogen (*Continued*)

t	T	h	u	π_0	θ_0	s^0	H	U	S
1050	1323,15	1462,32	1069,61	252.61	155,45	8,4760	40964,6	29963,4	237,441
1051	1324,15	1463,54	1070,53	253,40	155,09	8,4769	40998,8	29989,3	237,467
1052	1325,15	1464,77	1071,46	254,19	154,72	8,4778	41033,1	30015,3	237,492
1053	1326,15	1465,99	1072,39	254,98	154,36	8,4787	41067,4	30041,2	237,518
1054	1327,15	1467,21	1073,31	255,77	154,00	8,4797	41101,6	30067,1	237,544
1055	1328,15	1468,44	1074,24	256,57	153,63	8,4806	41135,9	30093,1	237,570
1056	1329,15	1469,66	1075,17	257,37	153,27	8,4815	41170,2	30119,1	237,596
1057	1330,15	1470,88	1076,09	258,17	152,91	8,4824	41204,4	30145,0	237,621
1058	1331,15	1472,11	1077,02	258,97	152,55	8,4833	41238,7	30171,0	237,647
1059	1332,15	1473,33	1077,95	259,77	152,20	8,4843	41273,0	30196,9	237,673
1060	1333,15	1474,55	1078,87	260,57	151,84	8,4852	41307,3	30222,9	237,699
1061	1334,15	1475,78	1079,80	261,38	151,49	8,4861	41341,6	30248,9	237,724
1062	1335,15	1477,00	1080,73	262,19	151,13	8,4870	41375,9	30274,9	237,750
1063	1336,15	1478,23	1081,66	263,00	150,78	8,4879	41410,1	30300,8	237.776
1064	1337,15	1479,45	1082,58	263,81	150,43	8,4888	41444,4	30326,8	237,801
1065	1338,15	1480,68	1083,51	264,63	150,08	8,4898	41478,8	30352,8	237,827
1066	1339,15	1481,90	1084,44	265,45	149,73	8,4907	41513,1	30378,8	237,853
1067	1340,15	1483,12	1085,37	266,27	149,38	8,4916	41547,4	30404,8	237,878
1068	134i,15	1484,35	1086,29	267,09	149,03	8,4925	41581,7	30430,8	237,904
1069	1342,15	1485,57	1087,22	267,91	148,68	8,4934	41616,0	30456,8	237,929
1070	1343,15	1486,80	1088,15	268,73	148,34	8,4943	41650,3	30482,8	237,955
1071	1344,15	1488,03	1089,08	269,56	147,99	8,4952	41684,7	30508,8	237,981
1072	1345,15	1489,25	1090,01	270,39	147,65	8,4962	41719,0	30534,9	238,006
1073	1346,15	1490,48	1090,94	271,22	147,30	8,4971	41753,3	30560,9	238,032
1074	1347,15	1491,70	1091,87	272,05	146,96	8,4980	41787,7	30586,9	238,057
1075	1348,15	1492,93	1092,80	272,89	146,62	8,4989	41822,0	30612,9	238,083
1076	1349,15	1494,15	1093,73	273,73	146,28	8,4998	41856,3	30639,0	238,108
1077	1350,15	1495,38	1094,65	274,57	145,94	8,5007	41890,7	30665,0	238,134
1078	1351,15	1496,61	1095,58	275,41	145,60	8,5016	41925,0	30691,0	238,159
1079	1352,15	1497,83	1096,51	276,25	145,27	8,5025	41959,4	30717,1	238,184
1080	1353,15	1499,06	1097,44	277,10	144,93	8,5034	41993,8	30743,1	238,210
1081	1354,15	1500,29	1098,37	277,94	144,59	8,5043	42028,1	30769,2	238,235
1082	1355,15	1501,51	1099,30	278,79	144,26	8,5052	42062,5	30795,2	238,261
1083	1356,15	1502,74	1100,23	279,64	143,93	8,5061	42096,9	30821,8	238,286
1084	1357,15	1503,97	1101,16	280,50	143,60	8,5070	42131,2	30847,3	238,311
1085	1358,15	1505,19	1102,09	281,35	143,26	8,5079	42165,6	30873,4	238,337
1086	1359,15	1506,42	1103,02	282,21	142,93	8,5089	42200,0	30899,5	238,362
1087	1360,15	1507,65	1103,96	283,07	142,60	8,5098	42234,4	30925,5	238,387
1088	1361,15	1508,88	1104,89	283,93	142,28	8,5107	42268,8	30951,6	238,412
1089	1362,15	1510,11	1105,82	284,80	141,95	8,5116	42303,2	30977,7	238,438
1090	1363,15	1511,33	1106,75	285,66	141,62	8,5125	42337,6	31003,8	238,463
1091	1364,15	1512,56	1107,68	286,53	141,30	8,5134	42372,0	31029,9	238,488
1092	1365,15	1513,79	1108,61	287,40	140,97	8,5143	42406,4	31056,0	238,513
1093	1366,15	1515,02	1109,54	288,27	140,65	8,5152	42440,8	31082,1	238,539
1094	1367,15	1516,25	1110,47	289,15	140,33	8,5161	42475,2	31108,2	238,564
1095	1368,15	1517,47	1111,41	290,02	140,00	8,5170	42509,6	31134,3	238,589
1096	1369,15	1518,70	1112,34	290,90	139,68	8,5179	42544,0	31160,4	238,614
1097	1370,15	1519,93	1113,27	291,78	139,36	8,5188	42578,5	31186,5	238.639
1098	1371,15	1521,16	1114,20	292,67	139,04	8,5196	42612,9	31212,6	238,664
1099	1372,15	1522,39	1115,13	293,55	138,73	8,5205	42647,3	31238,7	238,689
1100	1373,15	1523,62	1116,07	294,44	138,41	8,5214	42681,7	31264,8	238,714
1101	1374,15	1524,85	1117,00	295,33	138,09	8,5223	42716,2	31290,9	238,740
1102	1375,14	1526,08	1117,93	296,22	137,78	8,5232	42750,6	31317,1	238,765
1103	1376,15	1527,31	1118,86	297,11	137,46	8,5241	42785,1	31343,2	238,790
1104	1377,15	1528,54	1119,80	298,01	137,15	8,5250	42819,5	31369,3	238,815

Nitrogen (*Continued*)

t	T	h	u	π_0	θ_0	s^0	ll	U	S
1105	1378,15	1529,77	1120,73	298,91	136,84	8,5259	42854,0	31395,5	238,840
1106	1379,15	1531,00	1121,66	299,81	136,53	8,5268	42888,4	31421,6	238,865
1107	1380,15	1532,23	1122,60	300,71	136,22	8,5277	42922,9	31447,8	238,890
1108	1381,15	1533,46	1123,53	301,61	135,91	8,5286	42957,4	31473,9	238,915
1109	1382 15	1534,69	1124,46	302,52	135,60	8,5295	42991,8	31500,1	238,940
1110	1383,15	1535,92	1125,40	303,43	135,29	8,5304	43026,3	31526,2	238,964
1111	1384,15	1537,15	1126,33	304,34	134,98	8,5313	43060,8	31552,4	238,989
1112	1385,15	1538,38	1127,27	305,25	134,67	8,5321	43095,2	31578,5	239,014
1113	1386,15	1539,61	1128,20	306,16	134,37	8,5330	43129,7	31604,7	239,039
1114	1387,15	1540,84	1129,13	307,08	134,06	8,5339	43164,2	31630,9	239,064
1115	1388,15	1542,07	1130,07	308,00	133,76	8,5348	43198,7	31657,0	239,089
1116	1389,15	1543,30	1131,00	308,92	133,46	8,5357	43233,2	31683,2	239.114
1117	1390,15	1544,54	1131,94	309,85	133,16	8,5366	43267,7	31709,4	239,139
1118	1391,15	1545,77	1132,87	310,77	132,85	8,5375	43302,2	31735,6	239,163
1119	1392,15	1547,00	1133,81	311,70	132,55	8,5383	43336,7	31761,8	239,188
1120	1393,15	1548,23	1134,74	312,63	132,25	8,5392	43371,2	31788,0	239,213
1121	1394,15	1549,46	1135,68	313,56	131,96	8,5401	43405,7	31814,2	239,238
1122	1395,15	1550,69	1136,61	314,50	131,66	8,5410	43440,2	31840,4	239,262
1123	1396,15	1551,93	1137,55	315,43	131,36	8,5419	43474,7	31866,6	239,287
1124	1397,15	1553,16	1138,48	316,37	131,06	8,5428	43509,2	31892,8	239,312
1125	1398,15	1554,39	1139,42	317,31	130,77	8,5436	43543,8	31919,0	239,337
1126	1399,15	1555,62	1140,35	318,26	130,47	8,5445	43578,3	31945,2	239,361
1127	1400,15	1556,86	1141,29	319,20	130,18	8,5454	43612,8	31971,4	239,386
1128	1401,15	1558,09	1142,23	320,15	129,89	8,5463	43647,4	31997,6	239,411
1129	1402,15	1559,32	1143,16	321,10	129,60	8,5472	43681,9	32023,8	239,435
1130	1403,15	1560,55	1144,10	322,05	129,31	8,5480	43716,4	32050,1	239,460
1131	1404,15	1561,79	1145,03	323,01	129,02	8,5489	43751,0	32076,3	239,485
1132	1405,15	1563,02	1145,97	323,97	128,73	8,5498	43785,5	32102,5	239,509
1133	1406,15	1564,25	1146,91	324,93	128,44	8,5507	43820,1	32128,8	239,534
1134	1407,15	1565,49	1147,84	325,89	128,15	8,5516	43854,6	32155,0	239,558
1135	1408,15	1566,72	1148,78	326,85	127,86	8,5524	43889,2	32181,3	239,583
1136	1409,15	1567,96	1149,72	327,82	127,58	8,5533	43923,8	32207,5	239,607
1137	1410,15	1569,19	1150,66	328,78	127,29	8,5542	43958,3	32233,8	239,632
1138	1411,15	1570,42	1151,59	329,76	127,01	8,5551	43992,9	32260,0	239,656
1139	1412,15	1571,66	1152,53	330,73	126,72	8,5559	44027,5	32286,3	239,681
1140	1413,15	1572,89	1153,47	331,70	126,44	8,5568	44062,0	32312,5	239,705
1141	1414,15	1574,13	1154,40	332,68	126,16	8,5577	44096,6	32338,8	239,730
1142	1415,15	1575,36	1155,34	333,66	125,88	8,5586	44131,2	32365,1	239,754
1143	1416,15	1576,60	1156,28	334,64	125,59	8,5594	44165,8	32391,3	239,779
1144	1417,15	1577,83	1157,22	335,63	125,31	8,5603	44200,4	32417,6	239,803
1145	1418,15	1579,06	1158,16	336,61	125,04	8,5612	44235,0	32443,9	239,827
1146	1419,15	1580,30	1159,09	337,60	124,76	8,5620	44269,6	32470,2	239,852
1147	1420,15	1581,54	1160,03	338,59	124,48	8,5629	44304,2	32496,5	239,876
1148	1421,15	1582,77	1160,97	339,59	124,20	8,5638	44338,8	32522,8	239,901
1149	1422,15	1584,01	1161,91	340,58	123,93	8,5646	44373,4	32549,0	239,925
1150	1423,15	1585,24	1162,85	341,58	123,65	8,5655	44408,0	32575,3	239,949
1151	1424,15	1586,48	1163,79	342,58	123,38	8,5664	44442,6	32601,6	239,974
1152	1425,15	1587,71	1164,73	343,58	123,10	8,5673	44477,2	32627,9	239,998
1153	1426,15	1588,95	1165,67	344,59	122,83	8,5681	44511,8	32654,3	240,022
1154	1427,15	1590,18	1166,60	345,60	122,56	8,5690	44546,5	32680,6	240,046
1155	1428,15	1591,42	1167,54	346,60	122,29	8,5699	44581,1	32706,9	240,071
1156	1429,15	1592,66	1168,48	347,62	122,02	8,5707	44615,7	32733,2	240,095
1157	1430,15	1593,89	1169,42	348,63	121,75	8,5716	44650,4	32759,5	240,119
1158	1431,15	1595,13	1170,36	349,65	121,48	8,5724	44685,0	32785,8	240,143
1159	1432,15	1596,37	1171,30	350,67	121,21	8,5733	44719,6	32812,2	240,168

Nitrogen (*Continued*)

t	T	h	u	π_0	θ_0	s^0	H	U	S
1160	1433,15	1597,60	1172,24	351,69	120,94	8,5742	44754,3	32838,5	240,192
1161	1434,15	1598,84	1173,18	352,71	120,67	8,5750	44788,9	32864,8	240,216
1162	1435,15	1600,08	1174,12	353,74	120,41	8,5759	44823,6	32891,2	240,240
1163	1436,15	1601,31	1175,06	354,77	120,14	8,5768	44858,2	32917,5	240,264
1164	1437,15	1602,55	1176,00	355,80	119,88	8,5776	44892,9	32943,8	240,288
1165	1438,15	1603,79	1176,94	356,83	119,61	8,5785	44927,6	32970,2	240,312
1166	1439,15	1605,03	1177,88	357,87	119,35	8,5793	44962,2	32996,5	240,337
1167	1440,15	1606,26	1178,83	358,90	119,09	8,5802	44996,9	33022,9	240,361
1168	1441,15	1607,50	1179,77	359,94	118,83	8,5811	45031,6	33049,3	240,385
1169	1442,15	1608,74	1180,71	360,99	118,57	8,5819	45066,2	33075,6	240,409
1170	1443,15	1609,98	1181,65	362,03	118,31	8,5828	45100,9	33102,0	240,433
1171	1444,15	1611,21	1182,59	363,08	118,05	8,5836	45135,6	33128,3	240,457
1172	1445,15	1612,45	1183,53	364,13	117,79	8,5845	45170,3	33154,7	240,481
1173	1446,15	1613,69	1184,47	365,18	117,53	8,5853	45205,0	33181,1	240,505
1174	1447,15	1614,93	1185,41	366,24	117,27	8,5862	45239,7	33207,5	240,529
1175	1448,15	1616,17	1186,36	367,29	117,01	8,5871	45274,4	33233,9	240,553
1176	1449,15	1617,41	1187,30	368,35	116,76	8,5879	45309,1	33260,2	240,577
1177	1450,15	1618,65	1188,24	369,42	116,50	8,5888	45343,8	33286,6	240,601
1178	1451,15	1619,88	1189,18	370,48	116,25	8,5896	45378,5	33313,0	240,625
1179	1452,15	1621,12	1190,12	371,55	115,99	8,5905	45413,2	33339,4	240,648
1180	1453,15	1622,36	1191,07	372,62	115,74	8,5913	45447,9	33365,8	240,672
1181	1454,15	1623,60	1192,01	373,69	115,49	8,5922	45482,6	33392,2	240,696
1182	1455,15	1624,84	1192,95	374,76	115,24	8,5930	45517,3	33418,6	240,720
1183	1456,15	1626,08	1193,89	375,84	114,99	8,5939	45552,0	33445,0	240,744
1184	1457,15	1627,32	1194,84	376,92	114,74	8,5947	45586,8	33471,4	240,768
1185	1458,15	1628,56	1195,78	378,00	114,49	8,5956	45621,5	33497,8	240,792
1186	1459,15	1629,80	1196,72	379,09	114,24	8,5964	45656,2	33524,3	240,815
1187	1460,15	1631,04	1197,67	380,17	113,99	8,5973	45691,0	33550,7	240,839
1188	1461,15	1632,28	1198,61	381,26	113,74	8,5981	45725,7	33577,1	240,863
1189	1462,15	1633,52	1199,55	382,35	113,49	8,5990	45760,4	33603,5	240,887
1190	1463,15	1634,76	1200,50	383,45	113,25	8,5998	45795,2	33630,0	240,911
1191	1464,15	1636,00	1201,44	384,54	113,00	8,6007	45829,9	33656,4	240,934
1192	1465,15	1637,24	1202,38	385,64	112,76	8,6015	45864,7	33682,8	240,958
1193	1466,15	1638,48	1203,33	386,74	112,51	8,6024	45899,4	33709,3	240,982
1194	1467,15	1639,72	1204,27	387,85	112,27	8,6032	45934,2	33735,7	241,005
1195	1468,15	1640,96	1205,21	388,95	112,02	8,6041	45969,0	33762,2	241,029
1196	1469,15	1642,20	1206,16	390,06	111,78	8,6049	46003,7	33788,6	241,053
1197	1470,15	1643,45	1207,10	391,17	111,54	8,6058	46038,5	33815,1	241,076
1198	1471,15	1644,69	1208,05	392,29	111,30	8,6066	46073,3	33841,5	241,100
1199	1472,15	1645,93	1208,99	393,40	111,06	8,6074	46108,0	33868,0	241,124
1200	1473,15	1647,17	1209,94	394,52	110,82	8,6083	46142,8	33894,4	241,147
1201	1474,15	1648,41	1210,88	395,64	110,58	8,6091	46177,6	33920,9	241,171
1202	1475,15	1649,65	1211,83	396,77	110,34	8,6100	46212,4	33947,4	241,195
1203	1476,15	1650,89	1212,77	397,90	110,10	8,6108	46247,2	33973,8	241,218
1204	1477,15	1652,14	1213,72	399,02	109,87	8,6117	46282,0	34000,3	241,242
1205	1478,15	1653,38	1214,66	400,16	109,63	8,6125	46316,7	34026,8	241,265
1206	1479,15	1654,62	1215,61	401,29	109,39	8,6133	46351,5	34053,3	241,289
1207	1480,15	1655,86	1216,55	402,43	109,16	8,6142	46386,3	34079,8	241,312
1208	1481,15	1657,11	1217,50	403,57	108,92	8,6150	46421,1	34106,3	241,336
1209	1482,15	1658,35	1218,44	404,71	108,69	8,6159	46456,0	34132,8	241,359
1210	1483,15	1659,59	1219,39	405,85	108,46	8,6167	46490,8	34159,2	241,383
1211	1484,15	1660,83	1220,34	407,00	108,22	8,6175	46525,6	34185,7	241,406
1212	1485,15	1662,08	1221,28	408,15	107,99	8,6184	46560,4	34212,2	241,430
1213	1486,15	1663,32	1222,23	409,30	107,76	8,6192	46595,2	34238,7	241,453
1214	1487,15	1664,56	1223,17	410,46	107,53	8,6200	46630,0	34265,3	241,477

Nitrogen (*Continued*)

t	T	h	u	π_0	θ_0	s^0	H	U	S
1215	1488,15	1665,80	1224,12	411,61	107,30	8,6209	46664,9	34291,8	241,500
1216	1489,15	1667,05	1225,07	412,77	107,07	8,6217	46699,7	34318,3	241,523
1217	1490,15	1668,29	1226,01	413,94	106,84	8,6225	46734,5	34344,8	241,547
1218	1491,15	1669,53	1226,96	415,10	106,61	8,6234	46769,4	34371,3	241,570
1219	1492,15	1670,78	1227,91	416,27	106,38	8,6242	46804,2	34397,8	241,593
1220	1493,15	1672,02	1228,85	417,44	106,16	8,6250	46839,0	34424,4	241,617
1221	1494,15	1673,27	1229,80	418,61	105,93	8,6259	46873,9	33450,9	241,640
1222	1495,15	1674,51	1230,75	419,79	105,71	8,6267	46908,7	34477,4	241,663
1223	1496,15	1675,75	1231,70	420,97	105,48	8,6275	46943,6	34504,0	241,687
1224	1497,15	1677,00	1232,64	422,15	105,25	8,6284	46978,4	34530,5	241,710
1225	1498,15	1678,24	1233,59	423,33	105,03	8,6292	47013,3	34557,0	241,733
1226	1499,15	1679,49	1234,54	424,52	104,81	8,6300	47048,1	34583,6	241,757
1227	1500,15	1680,73	1235,49	425,70	104,58	8,6309	47083,0	34610,1	241,780
1228	1501,15	1681,98	1236,43	426,90	104,36	8,6317	47117,9	34636,7	241,803
1229	1502,15	1683,22	1237,38	428,09	104,14	8,6325	47152,7	34663,2	241,826
1230	1503,15	1684,47	1238,33	429,29	103,92	8,6333	47187,6	34689,8	241,849
1231	1504,15	1685,71	1239,28	430,48	103,70	8,6342	47222,5	34716,4	241,873
1232	1505,15	1686,96	1240,23	431,69	103,48	8,6350	47257,4	34742,9	241,896
1233	1506,15	1688,20	1241,17	432,89	103,26	8,6358	47292,2	34769,5	241,919
1234	1507,15	1689,45	1242,12	434,10	103,04	8,6367	47327,1	34796,1	241,942
1235	1508,15	1690,69	1243,07	435,31	102,82	8,6375	47362,0	34822,6	241,965
1236	1509,15	1691,94	1244,02	436,52	102,60	8,6383	47396,9	34849,2	241,988
1237	1510,15	1693,18	1244,97	437,74	102,39	8,6391	47431,8	34875,8	242,012
1238	1511,15	1694,43	1245,92	438,95	102,17	8,6400	47466,7	34902,4	242,035
1239	1512,15	1695,67	1246,87	440,17	101,96	8,6408	47501,6	34929,0	242,058
1240	1513,15	1696,92	1247,82	441,40	101,74	8,6416	47536,5	34955,5	242,081
1241	1514,15	1698,17	1248,76	442,62	101,53	8,6424	47571,4	34982,1	242,104
1242	1515,15	1699,41	1249,71	443,85	101,31	8,6433	47606,3	35008,7	242,127
1243	1516,15	1700,66	1250,66	445,08	101,10	8,6441	47641,2	35035,3	242,150
1244	1517,15	1701,90	1251,61	446,32	100,88	8,6449	47676,1	35061,9	242,173
1245	1518,15	1703,15	1252,56	447,55	100,67	8,6457	47711,1	35088,5	242,196
1246	1519,15	1704,40	1253,51	448,79	100,46	8,6465	47746,0	35115,1	242,219
1247	1520,15	1705,64	1254,46	450,04	100,25	8,6474	47780,9	35141,8	242,242
1248	1521,15	1706,89	1255,41	451,28	100,04	8,6482	47815,8	35168,4	242,265
1249	1522,15	1708,14	1256,36	452,53	99,832	8,6490	47850,8	35195,0	242,288
1250	1523,15	1709,39	1257,31	453,78	99,622	8,6498	47885,7	35221,6	242,311
1251	1524,15	1710,63	1258,26	455,03	99,413	8,6506	47920,6	35248,2	242,334
1252	1525,15	1711,88	1259,21	456,29	99,205	8,6515	47955,6	35274,8	242,357
1253	1526,15	1713,13	1260,16	457,55	98,997	8,6523	47990,5	35301,5	242,380
1254	1527,15	1714,37	1261,11	458,81	98,789	8,6531	48025,5	35328,1	242,402
1255	1528,15	1715,62	1262,07	460,07	98,582	8,6539	48060,4	35354,7	242,425
1256	1529,15	1716,87	1263,02	461,34	98,376	8,6547	48095,4	35381,4	242,448
1257	1530,15	1718,12	1263,97	462,61	98,170	8,6555	48130,3	35408,0	242,471
1258	1531,15	1719,37	1264,92	463,88	97,965	8,6564	48165,3	35434,7	242,494
1259	1532,15	1720,61	1265,87	465,16	97,760	8,6572	48200,2	35461,3	242,517
1260	1533,15	1721,86	1266,82	466,43	97,556	8,6580	48235,2	35488,0	242,539
1261	1534,15	1723,11	1267,77	467,72	97,352	8,6588	48270,2	35514,6	242,562
1262	1535,15	1724,36	1268,72	469,00	97,149	8,6596	48305,1	35541,3	242,585
1263	1536,15	1725,61	1269,68	470,29	96,945	8,6604	48340,1	35567,9	242,608
1264	1537,15	1726,85	1270,63	471,57	96,744	8,6612	48375,1	35594,6	242,631
1265	1538,15	1728,10	1271,58	472,87	96,543	8,6620	48410,0	35621,2	242,653
1266	1539,15	1729,35	1272,53	474,16	96,341	8,6629	48445,0	35647,9	242,676
1267	1540,15	1730,60	1273,48	475,46	96,141	8,6637	48480,0	35674,6	242,699
1268	1541,15	1731,85	1274,43	476,76	95,941	8,6645	48515,0	35701,2	242,722
1269	1542,15	1733,10	1275,39	478,06	95,742	8,6653	48550,0	35727,9	242,744

Nitrogen (*Continued*)

t	T	h	u	π_0	θ_0	s^0	H	U	S
1270	1543,15	1734,35	1276,34	479,37	95,543	8,6661	48585,0	35754,6	242,767
1271	1544,15	1735,60	1277,29	480,68	95,344	8,6669	48620,0	35781,3	242,790
1272	1545,15	1736,85	1278,24	481,99	95,146	8,6677	48655,0	35808,0	242,812
1273	1546,15	1738,10	1279,20	483,30	94,949	8,6685	48690,0	35834,7	242,835
1274	1547,15	1739,35	1280,15	484,62	94,752	8,6693	48725,0	35861,3	242,858
1275	1548,15	1740,60	1281,10	485,94	94,556	8,6701	48760,0	35888,0	242,880
1276	1549,15	1741,84	1282,06	487,26	94,360	8,6709	48795,0	35914,7	242,903
1277	1550,15	1743,09	1283,01	488,59	94,165	8,6718	48830,0	35941,4	242,925
1278	1551,15	1744,34	1283,96	489,92	93,970	8,6726	48865,0	35968,1	242,948
1279	1552,15	1745,59	1284,91	491,25	93,775	8,6734	48900,0	35994,8	242,970
1280	1553,15	1746,84	1285,87	492,59	93,582	8,6742	48935,1	36021,5	242,993
1281	1554,15	1748,10	1286,82	493,92	93,388	8,6750	48970,1	36048,3	243,016
1282	1555,15	1749,35	1287,78	495,26	93,196	8,6758	49005,1	36075,0	243,038
1283	1556,15	1750,60	1288,73	496,61	93,003	8,6766	49040,1	36101,7	243,061
1284	1557,15	1751,85	1289,68	497,95	92,811	8,6774	49075,2	36128,4	243,083
1285	1558,15	1753,10	1290,64	499,30	92,620	8,6782	49110,2	36155,1	243,106
1286	1559,15	1754,35	1291,59	500,65	92,429	8,6790	49145,3	36181,8	243,128
1287	1560,15	1755,60	1292,54	502,01	92,239	8,6798	49180,3	36208,6	243,151
1288	1561,15	1756,85	1293,50	503,37	92,049	8,6806	49215,3	36235,3	243,173
1289	1562,15	1758,10	1294,45	504,73	91,860	8,6814	49250,4	36262,0	243,195
1290	1563,15	1759,35	1295,41	506,09	91,671	8,6822	49285,4	36288,8	243,218
1291	1564,15	1760,60	1296,36	507,46	91,483	8,6830	49320,5	36315,5	243,240
1292	1565,15	1761,85	1297,32	508,83	91,295	8,6838	49355,5	36342,2	243,263
1293	1566,15	1763,11	1298,27	510,20	91,107	8,6846	49390,6	36369,0	243,285
1294	1567,15	1764,36	1299,23	511,57	90,921	8,6854	49425,7	36395,7	243,307
1295	1568,15	1765,61	1300,18	512,95	90,734	8,6862	49460,7	36422,5	243,330
1296	1569,15	1766,86	1301,14	514,33	90,548	8,6870	49495,8	36449,2	243,352
1297	1570,15	1768,11	1302,09	515,72	90,363	8,6878	49530,9	36476,0	243,375
1298	1571,15	1769,37	1303,05	517,10	90,178	8,6886	49565,9	36502,8	243,397
1299	1572,15	1770,62	1304,00	518,49	89,993	8,6894	49601,0	36529,5	243,419
1300	1573,15	1771,87	1304,96	519,89	89,809	8,6902	49636,1	36556,3	243,442
1301	1574,15	1773,12	1305,91	521,28	89,626	8,6910	49671,2	36583,0	243,464
1302	1575,15	1774,37	1306,87	522,68	89,443	8,6918	49706,3	36609,8	243,486
1303	1576,15	1775,63	1307,82	524,08	89,260	8,6926	49741,3	36636,6	243,508
1304	1577,15	1776,88	1308,78	525,49	89,078	8,6934	49776,4	36663,4	243,531
1305	1578,15	1778,13	1309,74	526,89	88,896	8,6942	49811,5	36690,1	243,553
1306	1579,15	1779,38	1310,69	528,31	88,715	8,6949	49846,6	36716,9	243,575
1307	1580,15	1780,64	1311,65	529,72	88,534	8,6957	49881,7	36743,7	243,597
1308	1581,15	1781,89	1312,60	531,14	88,354	8,6965	49916,8	36770,5	243,620
1309	1582,15	1783,14	1313,56	532,56	88,174	8,6973	49951,9	36797,3	243,642
1310	1583,15	1784,40	1314,52	533,98	87,995	8,6981	49987,0	36824,1	243,664
1311	1584,15	1785,65	1315,47	535,40	87,816	8,6989	50022,1	36850,9	243,686
1312	1585,15	1786,90	1316,43	536,83	87,638	8,6997	50057,3	36877,7	243,708
1313	1586,15	1788,16	1317,39	538,26	87,460	8,7005	50092,4	36904,5	243,730
1314	1587,15	1789,41	1318,34	539,70	87,282	8,7013	50127,5	36931,3	243,752
1315	1588,15	1790,66	1319,30	541,14	87,105	8,7021	50162,6	36958,1	243,775
1316	1589,15	1791,92	1320,26	542,58	86,929	8,7029	50197,7	36984,9	243,797
1317	1590,15	1793,17	1321,21	544,02	86,752	8,7036	50232,8	37011,7	243,819
1318	1591,15	1794,43	1322,17	545,47	86,577	8,7044	50268,0	37038,5	243,841
1319	1592,15	1795,68	1323,13	546,92	86,401	8,7052	50303,1	37065,3	243,863
1320	1593,15	1796,93	1324,09	548,37	86,227	8,7060	50338,2	37092,1	243,885
1321	1594,15	1798,19	1325,04	549,83	86,052	8,7068	50373,4	37119,0	243,907
1322	1595,15	1799,44	1326,00	551,29	85,878	8,7076	50408,5	37145,8	243,929
1323	1596,15	1800,70	1326,96	552,75	85,705	8,7084	50443,7	37172,6	243,951
1324	1597,15	1801,95	1327,92	554,21	85,532	8,7092	50478,8	37199,4	243,973

t	T	h	u	$\pi_\bullet$	$\theta_\bullet$	$s^\bullet$	H	U	S
1325	1598,15	1803,21	1328,87	555,68	85,359	8,7099	50513,9	37226,3	243,995
1326	1599,15	1804,46	1329,83	557,15	85,187	8,7107	50549,1	37253,1	244,017
1327	1600,15	1805,72	1330,79	558,63	85,015	8,7115	50584,2	37279,9	244,039
1328	1601,15	1806,97	1331,75	560,11	84,844	8,7123	50619,4	37306,8	244,061
1329	1602,15	1808,23	1332,71	561,59	84,673	8,7131	50654,6	37333,6	244,083
1330	1603,15	1809,48	1333,66	563,07	84,503	8,7139	50689,7	37360,5	244,105
1331	1604,15	1810,74	1334,62	564,56	84,333	8,7146	50724,9	37387,3	244,127
1332	1605,15	1811,99	1335,58	566,05	84,163	8,7154	50760,0	37414,2	244,149
1333	1606,15	1813,25	1336,54	567,54	83,994	8,7162	50795,2	37441,0	244,171
1334	1607,15	1814,50	1337,50	569,04	83,825	8,7170	50830,4	37467,9	244,193
1335	1608,15	1815,76	1338,46	570,54	83,657	8,7178	50865,6	37494,7	244,214
1336	1609,15	1817,01	1339,42	572,04	83,489	8,7186	50900,7	37521,6	244,236
1337	1610,15	1818,27	1340,38	573,54	83,322	8,7193	50935,9	37548,5	244,258
1338	1611,15	1819,53	1341,33	575,05	83,155	8,7201	50971,1	37575,3	244,280
1339	1612,15	1820,78	1342,29	576,56	82,988	8,7209	51006,3	37602,2	244,302
1340	1613,15	1822,04	1343,25	578,08	82,822	8,7217	51041,5	37629,1	244,324
1341	1614,15	1823,29	1344,21	579,60	82,656	8,7225	51076,7	37656,0	244,345
1342	1615,15	1824,55	1345,17	581,12	82,491	8,7232	51111,8	37682,8	244,367
1343	1616,15	1825,81	1346,13	582,64	82,326	8,7240	51147,0	37709,7	244,389
1344	1617,15	1827,06	1347,09	584,17	82,162	8,7248	51182,2	37736,6	244,411
1345	1618,15	1828,32	1348,05	585,70	81,998	8,7256	51217,4	37763,5	244,433
1346	1619,15	1829,58	1349,01	587,24	81,834	8,7263	51252,6	37790,4	244,454
1347	1620,15	1830,83	1349,97	588,77	81,671	8,7271	51287,8	37817,3	244,476
1348	1621,15	1832,09	1350,93	590,31	81,508	8,7279	51323,1	37844,1	244,498
1349	1622,15	1833,35	1351,89	591,86	81,345	8,7287	51358,3	37871,0	244,520
1350	1623,15	1834,60	1352,85	593,40	81,183	8,7294	51393,5	37897,9	244,541
1351	1624,15	1835,86	1353,81	594,95	81,022	8,7302	51428,7	37924,8	244,563
1352	1625,15	1837,12	1354,77	596,51	80,861	8,7310	51463,9	37951,7	244,585
1353	1626,15	1838,37	1355,73	598,06	80,700	8,7318	51499,1	37978,6	244,606
1354	1627,15	1839,63	1356,69	599,62	80,539	8,7325	51534,3	38005,6	244,628
1355	1628,15	1840,89	1357,65	601,19	80,379	8,7333	51569,6	38032,5	244,650
1356	1629,15	1842,15	1358,61	602,75	80,220	8,7341	51604,8	38059,4	244,671
1357	1630,15	1843,40	1359,57	604,32	80,061	8,7348	51640,0	38086,3	244,693
1358	1631,15	1844,66	1360,54	605,89	79,902	8,7356	51675,3	38113,2	244,714
1359	1632,15	1845,92	1361,50	607,47	79,743	8,7364	51710,5	38140,1	244,736
1360	1633,15	1847,18	1362,46	609,05	79,585	8,7372	51745,7	38167,1	244,758
1361	1634,15	1848,44	1363,42	610,63	79,428	8,7379	51781,0	38194,0	244,779
1362	1635,15	1849,69	1364,38	612,22	79,271	8,7387	51816,2	38220,9	244,801
1363	1636,15	1850,95	1365,34	613,81	79,114	8,7395	51851,5	38247,8	244,822
1364	1637,15	1852,21	1366,30	615,40	78,957	8,7402	51886,7	38274,8	244,844
1365	1638,15	1853,47	1367,26	616,99	78,801	8,7410	51922,0	38301,7	244,865
1366	1639,15	1854,73	1368,23	618,59	78,646	8,7418	51957,2	38328,7	244,887
1367	1640,15	1855,99	1369,19	620,19	78,490	8,7425	51992,5	38355,6	244,908
1368	1641,15	1857,24	1370,15	621,80	78,335	8,7433	52027,7	38382,5	244,930
1369	1642,15	1858,50	1371,11	623,41	78,181	8,7441	52063,0	38409,5	244,951
1370	1643,15	1859,76	1372,07	625,02	78,027	8,7448	52098,3	38436,4	244,973
1371	1644,15	1861,02	1373,04	626,63	77,873	8,7456	52133,5	38463,4	244,994
1372	1645,15	1862,28	1374,00	628,25	77,720	8,7464	52168,8	38490,3	245,016
1373	1646,15	1863,54	1374,96	629,87	77,567	8,7471	52204,1	38517,3	245,037
1374	1647,15	1864,80	1375,92	631,50	77,414	8,7479	52239,3	38544,3	245,059
1375	1648,15	1866,06	1376,88	633,13	77,262	8,7487	52274,6	38571,2	245,080
1376	1649,15	1867,32	1377,85	634,76	77,110	8,7494	52309,9	38598,2	245,101
1377	1650,15	1868,58	1378,81	636,39	76,959	8,7502	52345,2	38625,1	245,123
1378	1651,15	1869,84	1379,77	638,03	76,808	8,7510	52380,5	38652,1	245,144
1379	1652,15	1871,10	1380,74	639,67	76,657	8,7517	52415,7	38679,1	245,165

Nitrogen (*Continued*)

t	T	h	u	π_0	θ_0	s^0	H	U	S
1380	1653,15	1872,35	1381,70	641,32	76,507	8,7525	52451,0	38706,1	245,187
1381	1654,15	1873,61	1382,66	642,97	76,357	8,7532	52486,3	38733,0	245,208
1382	1655,15	1874,87	1383,62	644,62	76,207	8,7540	52521,6	38760,0	245,229
1383	1656,15	1876,13	1384,59	646,27	76,058	8,7548	52556,9	38787,0	245,251
1384	1657,15	1877,39	1385,55	647,93	75,909	8,7555	52592,2	38814,0	245,272
1385	1658,15	1878,65	1386,51	649,59	75,760	8,7563	52627,5	38841,0	245,293
1386	1659,15	1879,91	1387,48	651,26	75,612	8,7570	52662,8	38867,9	245,315
1387	1660,15	1881,17	1388,44	652,93	75,465	8,7578	52698,1	38894,9	245,336
1388	1661,15	1882,44	1389,40	654,60	75,317	8,7586	52733,4	38921,9	245,357
1389	1662,15	1883,70	1390,37	656,27	75,170	8,7593	52768,7	38948,9	245,378
1390	1663,15	1884,96	1391,33	657,95	75,024	8,7601	52804,0	38975,9	245,400
1391	1664,15	1886,22	1392,30	659,63	74,877	8,7608	52839,4	39002,9	245,421
1392	1665,15	1887,48	1393,26	661,32	74,731	8,7616	52874,7	39029,9	245,442
1393	1666,15	1888,74	1394,22	663,01	74,586	8,7624	52910,0	39056,9	245,463
1394	1667,15	1890,00	1395,19	664,70	74,440	8,7631	52945,3	39083,9	245,485
1395	1668,15	1891,26	1396,15	666,39	74,296	8,7639	52980,6	39110,9	245,506
1396	1669,15	1892,52	1397,12	668,09	74,151	8,7646	53016,0	39138,0	245,527
1397	1670,15	1893,78	1398,08	669,80	74,007	8,7654	53051,3	39165,0	245,548
1398	1671,15	1895,04	1399,04	671,50	73,863	8,7661	53086,6	39192,0	245,569
1399	1672,15	1896,31	1400,01	673,21	73,720	8,7669	53122,0	39219,0	245,590
1400	1673,15	1897,57	1400,97	674,92	73,577	8,7676	53157,3	39246,0	245,611
1401	1674,15	1898,83	1401,94	676,64	73,434	8,7684	53192,6	39273,1	245,633
1402	1675,15	1900,09	1402,90	678,36	73,291	8,7691	53228,0	39300,1	245,654
1403	1676,15	1901,35	1403,87	680,08	73,149	8,7699	53263,3	39327,1	245,675
1404	1677,15	1902,61	1404,83	681,81	73,008	8,7707	53298,7	39354,1	245,696
1405	1678,15	1903,87	1405,80	683,54	72,866	8,7714	53334,0	39381,2	245,717
1406	1679,15	1905,14	1406,76	685,27	72,725	8,7722	53369,4	39408,2	245,738
1407	1680,15	1906,40	1407,73	687,01	72,585	8,7729	53404,7	39435,3	245,759
1408	1681,15	1907,66	1408,69	688,75	72,444	8,7737	53440,1	39462,3	245,780
1409	1682,15	1908,92	1409,66	690,49	72,304	8,7744	53475,4	39489,3	245,801
1410	1683,15	1910,19	1410,62	692,24	72,165	8,7752	53510,8	39516,4	245,822
1411	1684,15	1911,45	1411,59	693,99	72,025	8,7759	53546,1	39543,4	245,843
1412	1685,15	1912,71	1412,56	695,75	71,886	8,7767	53581,5	39570,5	245,864
1413	1686,15	1913,97	1413,52	697,50	71,748	8,7774	53616,9	39597,5	245,885
1414	1687,15	1915,23	1414,49	699,27	71,610	8,7782	53652,2	39624,6	245,906
1415	1688,15	1916,50	1415,45	701,03	71,472	8,7789	53687,6	39651,6	245,927
1416	1689,15	1917,76	1416,42	702,80	71,334	8,7797	53723,0	39678,7	245,948
1417	1690,15	1919,02	1417,38	704,57	71,197	8,7804	53758,4	39705,8	245,969
1418	1691,15	1920,29	1418,35	706,35	71,060	8,7811	53793,7	39732,8	245,990
1419	1692,15	1921,55	1419,32	708,13	70,923	8,7819	53829,1	39759,9	246,011
1420	1693,15	1922,81	1420,28	709,91	70,787	8,7826	53864,5	39787,0	246,032
1421	1694,15	1924,08	1421,25	711,69	70,651	8,7834	53899,9	39814,0	246,053
1422	1695,15	1925,34	1422,22	713,48	70,515	8,7841	53935,3	39841,1	246,073
1423	1696,15	1926,60	1423,18	715,28	70,380	8,7849	53970,7	39868,2	246,094
1424	1697,15	1927,87	1424,15	717,07	70,245	8,7856	54006,1	39895,3	246,115
1425	1698,15	1929,13	1425,12	718,88	70,110	8,7864	54041,5	39922,3	246,136
1426	1699,15	1930,39	1426,08	720,68	69,976	8,7871	54076,9	39949,4	246,157
1427	1700,15	1931,66	1427,05	722,49	69,842	8,7879	54112,3	39976,5	246,178
1428	1701,15	1932,92	1428,02	724,30	69,708	8,7886	54147,7	40003,6	246,198
1429	1702,15	1934,18	1428,98	726,11	69,575	8,7893	54183,1	40030,7	246,219
1430	1703,15	1935,45	1429,95	727,93	69,442	8,7901	54218,5	40057,8	246,240
1431	1704,15	1936,71	1430,92	729,75	69,309	8,7908	54253,9	40084,9	246,261
1432	1705,15	1937,98	1431,89	731,58	69,177	8,7916	54289,3	40112,0	246,282
1433	1706,15	1939,24	1432,85	733,41	69,045	8,7923	54324,7	40139,1	246,302
1434	1707,15	1940,50	1433,82	735,24	68,913	8,7930	54360,1	40166,2	246,323
1435	1708,15	1941,77	1434,79	737,08	68,782	8,7938	54395,5	40193,3	246,344
1436	1709,15	1943,03	1435,75	738,92	68,651	8,7945	54431,0	40220,4	246,365
1437	1710,15	1944,30	1436,72	740,76	68,520	8,7953	54466,4	40247,5	246,385
1438	1711,15	1945,56	1437,69	742,61	68,389	8,7960	54501,8	40274,6	246,406
1439	1712,15	1946,83	1438,66	744,46	68,259	8,7967	54537,2	40301,7	246,427

t	T	h	u	π_0	θ_0	s^b	H	U	S
1440	1713,15	1948,09	1439,63	746,31	68,129	8,7975	54572,7	40328,8	246,447
1441	1714,15	1949,36	1440,59	748,17	68,000	8,7982	54608,1	40355,9	246,468
1442	1715,15	1950,62	1441,56	750,03	67,871	8,7990	54643,5	40383,1	246,489
1443	1716,15	1951,89	1442,53	751,90	67,742	8,7997	54679,0	40410,2	246,509
1444	1717,15	1953,15	1443,50	753,77	67,613	8,8004	54714,4	40437,3	246,530
1445	1718,15	1954,42	1444,47	755,64	67,485	8,8012	54749,8	40464,4	246,551
1446	1719,15	1955,68	1445,44	757,52	67,357	8,8019	54785,3	40491,6	246,571
1447	1720,15	1956,95	1446,40	759,40	67,229	8,8026	54820,7	40518,7	246,592
1448	1721,15	1958,21	1447,37	761,28	67,102	8,8034	54856,2	40545,8	246,613
1449	1722,15	1959,48	1448,34	763,17	66,975	8,8041	54891,6	40573,0	246,633
1450	1723,15	1960,74	1449,31	765,06	66,848	8,8048	54927,1	40600,1	246,654
1451	1724,15	1962,01	1450,28	766,96	66,721	8,8056	54962,5	40627,2	246,674
1452	1725,15	1963,27	1451,25	768,85	66,595	8,8063	54998,0	40654,4	246,695
1453	1726,15	1964,54	1452,22	770,76	66,469	8,8070	55033,4	40681,5	246,715
1454	1727,15	1965,81	1453,19	772,66	66,344	8,8078	55068,9	40708,7	246,736
1455	1728,15	1967,07	1454,15	774,57	66,218	8,8085	55104,4	40735,8	246,756
1456	1729,15	1968,34	1455,12	776,49	66,093	8,8092	55139,8	40763,0	246,777
1457	1730,15	1969,60	1456,09	778,40	65,969	8,8100	55175,3	40790,1	246,797
1458	1731,15	1970,87	1457,06	780,33	65,844	8,8107	55210,8	40817,3	246,818
1459	1732,15	1972,14	1458,03	782,25	65,720	8,8114	55246,2	40844,4	246,838
1460	1733,15	1973,40	1459,00	784,18	65,596	8,8122	55281,7	40871,6	246,859
1461	1734,15	1974,67	1459,97	786,11	65,473	8,8129	55317,2	40898,8	246,879
1462	1735,15	1975,94	1460,94	788,05	65,350	8,8136	55352,7	40925,9	246,900
1463	1736,15	1977,20	1461,91	789,99	65,227	8,8144	55388,1	40953,1	246,920
1464	1737,15	1978,47	1462,88	791,93	65,104	8,8151	55423,6	40980,2	246,941
1465	1738,15	1979,74	1463,85	793,88	64,982	8,8158	55459,1	41007,4	246,961
1466	1739,15	1981,00	1464,82	795,83	64,860	8,8166	55494,6	41034,6	246,982
1467	1740,15	1982,27	1465,79	797,79	64,738	8,8173	55530,1	41061,8	247,002
1468	1741,15	1983,54	1466,76	799,74	64,617	8,8180	55565,6	41089,0	247,022
1469	1742,15	1984,80	1467,73	801,71	64,496	8,8187	55601,1	41116,1	247,043
1470	1743,15	1986,07	1468,70	803,67	64,375	8,8195	55636,6	41143,3	247,063
1471	1744,15	1987,34	1469,67	805,64	64,254	8,8202	55672,1	41170,5	247,083
1472	1745,15	1988,60	1470,64	807,62	64,134	8,8209	55707,6	41197,7	247,104
1473	1746,15	1989,87	1471,61	809,60	64,014	8,8216	55743,1	41224,9	247,124
1474	1747,15	1991,14	1472,58	811,58	63,894	8,8224	55778,6	41252,1	247,144
1475	1748,15	1992,41	1473,55	813,56	63,774	8,8231	55814,1	41279,3	247,165
1476	1749,15	1993,67	1474,52	815,55	63,655	8,8238	55849,6	41306,5	247,185
1477	1750,15	1994,94	1475,50	817,55	63,536	8,8245	55885,1	41333,7	247,205
1478	1751,15	1996,21	1476,47	819,54	63,418	8,8253	55920,6	41360,9	247,226
1479	1752,15	1997,48	1477,44	821,54	63,299	8,8260	55956,2	41388,1	247,246
1480	1753,15	1998,75	1478,41	823,55	63,181	8,8267	55991,7	41415,3	247,266
1481	1754,15	2000,01	1479,38	825,56	63,064	8,8274	56027,2	41442,5	247,286
1482	1755,15	2001,28	1480,35	827,57	62,946	8,8282	56062,7	41469,7	247,307
1483	1756,15	2002,55	1481,32	829,59	62,829	8,8289	56098,2	41496,9	247,327
1484	1757,15	2003,82	1482,29	831,61	62,712	8,8296	56133,8	41524,1	247,347
1485	1758,15	2005,09	1483,27	833,63	62,595	8,8303	56169,3	41551,3	247,367
1486	1759,15	2006,36	1484,24	835,66	62,479	8,8310	56204,8	41578,5	247,388
1487	1760,15	2007,62	1485,21	837,69	62,363	8,8318	56240,4	41605,8	247,408
1488	1761,15	2008,89	1486,18	839,73	62,247	8,8325	56275,9	41633,0	247,428
1489	1762,15	2010,16	1487,15	841,77	62,131	8,8332	56311,4	41660,2	247,448
1490	1763,15	2011,43	1488,12	843,81	62,016	8,8339	56347,0	41687,4	247,468
1491	1764,15	2012,70	1489,10	845,86	61,901	8,8346	56382,5	41714,7	247,488
1492	1765,15	2013,97	1490,07	847,91	61,786	8,8354	56418,1	41741,9	247,509
1493	1766,15	2015,24	1491,04	849,97	61,672	8,8361	56453,6	41769,1	247,529
1494	1767,15	2016,51	1492,01	852,03	61,557	8,8368	56489,2	41796,4	247,549
1495	1768,15	2017,77	1492,99	854,09	61,443	8,8375	56524,7	41823,6	247,569
1496	1769,15	2019,04	1493,96	856,16	61,330	8,8382	56560,3	41850,8	247,589
1497	1770,15	2020,31	1494,93	858,23	61,216	8,8390	56595,8	41878,1	247,609
1498	1771,15	2021,58	1495,90	860,31	61,103	8,8397	56631,4	41905,3	247,629
1499	1772,15	2022,85	1496,88	862,39	60,990	8,8404	56667,0	41932,6	247,649
1500	1773,15	2024,12	1497,85	864,47	60,877	8,8411	56702,5	41959,8	247,669

Table III.5 Atmospheric nitrogen ($\mu = 28.15$)

t	T	c_p	μc_p	c_v	μc_v	$k = \dfrac{c_p}{c_v}$
−50	223,15	1,0306	29,001	0,7349	20,687	1,402
−25	248,15	1,0305	29,008	0,7351	20,694	1,402
0	273,15	1,0305	29,008	0,7351	20,694	1,402
25	298,15	1,0307	29,014	0,7354	20,700	1,402
50	323,15	1,0311	29,029	0,7359	20,715	1,401
75	348,15	1,0322	29,055	0,7368	20,741	1,401
100	373,15	1,0335	29,093	0,7382	20,779	1,400
125	398,15	1,0353	29,144	0,7400	20,830	1,399
150	423,15	1,0376	29,208	0,7422	20,894	1,398
175	448,15	1,0404	29,286	0,7450	20,972	1,396
200	473,15	1,0436	29,376	0,7482	21,062	1,395
250	523,15	1,0513	29,593	0,7559	21,279	1,391
300	573,15	1,0605	29,852	0,7651	21,538	1,386
350	623,15	1,0709	30,145	0,7755	21,831	1,381
400	673,15	1,0822	30,463	0,7868	22,149	1,375
450	723,15	1,0940	30,796	0,7986	22,482	1,370
500	773,15	1,1061	31,136	0,8107	22,822	1,364
550	823.15	1,1181	31,474	0,8227	23,160	1,359
600	873,15	1,1298	31,805	0,8345	23,491	1,354
650	923,15	1,1412	32,124	0,8458	23,810	1,349
700	973,15	1,1520	32,428	0,8566	24,114	1,345
750	1023,15	1,1621	32,714	0,8668	24,400	1,341
800	1073,15	1,1717	32,984	0,8764	24,670	1,337
850	1123,15	1,1807	33,236	0,8853	24,922	1,334
900	1173,15	1,1891	33,472	0,8937	25,158	1,331
950	1223,15	1,1969	33,693	0,9016	25,379	1,328
1000	1273,15	1,2043	33,901	0,9090	25,587	1,325
1050	1323,15	1,2113	34,097	0,9159	25,783	1,322
1100	1373,15	1,2179	34,283	0,9225	25,969	1,320
1150	1423,15	1,2241	34,458	0,9287	26,144	1,318
1200	1473,15	1,2299	34,622	0,9346	26,308	1,316
1250	1523,15	1,2354	34,776	0,9400	26,462	1,314
1300	1573,15	1,2405	34,919	0,9451	26,605	1,312
1350	1623,15	1,2452	35,051	0,9498	26,737	1,311
1400	1673,15	1,2495	35,174	0,9542	26,860	1,310
1450	1723,15	1,2536	35,290	0,9583	26,976	1,308
1500	1773,15	1,2578	35,406	0,9624	27,092	1,307

Table III.6 Atmospheric nitrogen ($\mu = 28.15$)

t	T	h	u	π_0	θ_0	s^0	H	U	S
−50	223,15	229,85	163,94	0,3477	18951	6,4890	6470,4	4615,0	182,665
−49	224,15	230,88	164,68	0,3532	18741	6,4936	6499,4	4635,7	182,795
−48	225,15	231,91	165,41	0,3587	18535	6,4982	6528,4	4656,4	182,924
−47	226,15	232,94	166,15	0,3643	18331	6,5027	6557,4	4677,1	183,052
−46	227,15	233,97	166,88	0,3700	18131	6,5073	6586,4	4697,8	183,180
−45	228,15	235,01	167,62	0,3757	17934	6,5118	6615,4	4718,5	183,308
−44	229,15	236,04	168,35	0,3815	17740	6,5163	6644,4	4739,2	183,435
−43	230,15	237,07	169,09	0,3873	17548	6,5208	6673,4	4759,9	183,561
−42	231,15	238,10	169,82	0,3932	17360	6,5253	6702,4	4780,6	183,687
−41	232,15	239,13	170,56	0,3992	17175	6,5297	6731,4	4801,3	183,812
−40	233,15	240,16	171,29	0,4052	16992	6,5342	6760,4	4821,9	183,937
−39	234,15	241,19	172,03	0,4113	16812	6,5386	6789,5	4842,6	184,061
−38	235,15	242,22	172,76	0,4175	16634	6,5430	6818,5	4863,3	184,184
−37	236,15	243,25	173,50	0,4237	16460	6,5473	6847,5	4884,0	184,308
−36	237,15	244,28	174,24	0,4300	16287	6,5517	6876,5	4904,7	184,430
−35	238,15	245,31	174,97	0,4364	16118	6,5560	6905,5	4925,4	184,552
−34	239,15	246,34	175,71	0,4428	15950	6,5603	6934,5	4946,1	184,674
−33	240,15	247,37	176,44	0,4493	15786	6,5646	6963,5	4966,8	184,795
−32	241,15	248,40	177,18	0,4558	15623	6,5689	6992,5	4987,5	184,915
−31	242,15	249,43	177,91	0,4625	15463	6,5732	7021,5	5008,2	185,035
−30	243,15	250,46	178,65	0,4692	15305	6,5774	7050,5	5028,9	185,155
−29	244,15	251,49	179,38	0,4759	15150	6,5817	7079,5	5049,6	185,274
−28	245,15	252,52	180,12	0,4828	14996	6,5859	7108,5	5070,3	185,393
−27	246,15	253,55	180,85	0,4897	14845	6,5901	7137,6	5091,0	185,511
−26	247,15	254,58	181,59	0 4967	14696	6,5943	7166,6	5111,7	185,628
−25	248,15	255,62	182,32	0,5037	14549	6,5984	7195,6	5132,4	185,745
−24	249,15	256,65	183,06	0,5108	14404	6,6026	7224,6	5153,0	185,862
−23	250,15	257,68	183,79	0,5180	14261	6,6067	7253,6	5173,7	185,978
−22	251,15	258,71	184,53	0,5253	14120	6,6108	7282,6	5194,4	186,094
−21	252,15	259,74	185,26	0,5326	13982	6,6149	7311,6	5215,1	186,209
−20	253,15	260,77	186,00	0,5400	13844	6,6190	7340,6	5235,8	186,324
−19	254,15	261,80	186,73	0,5475	13709	6,6230	7369,6	5256,5	186,438
−18	255,15	262,83	187,47	0,5550	13576	6,6271	7398,6	5277,2	186,552
−17	256,15	263,86	188,20	0,5627	13444	6,6311	7427,6	5297,9	186,666
−16	257,15	264,89	188,94	0,5704	13315	6,6351	7456,6	5318,6	186,779
−15	258,15	265,92	189,67	0,5781	13187	6,6391	7485,6	5339,3	186,891
−14	259,15	266,95	190,41	0,5860	13060	6,6431	7514,7	5360,0	187,003
−13	260,15	267,98	191,14	0,5939	12936	6,6471	7543,7	5380,7	187,115
−12	261,15	269,01	191,88	0,6019	12813	6,6510	7572,7	5401,4	187,227
−11	262,15	270,04	192,61	0,6100	12692	6,6550	7601,7	5422,0	187,337
−10	263,15	271,07	193,35	0,6182	12572	6,6589	7630,7	5442,7	187,448
−9	264,15	272,10	194,08	0,6264	12454	6,6628	7659,7	5463,4	187,558
−8	265,15	273,13	194,82	0,6347	12337	6,6667	7688,7	5484,1	187,667
−7	266,15	274,16	195,55	0,6431	12222	6,6706	7717,7	5504,8	187,777
−6	267,15	275,19	196,29	0,6516	12109	6,6744	7746,7	5525,5	187,885
−5	268,15	276,22	197,02	0,6601	11997	6,6783	7775,7	5546,2	187,994
−4	269,15	277,25	197,76	0,6688	11886	6,6821	7804,7	5566,9	188,102
−3	270,15	278,29	198,49	0,6775	11777	6,6859	7833,7	5587,6	188,209
−2	271,15	279,32	199,23	0,6863	11669	6,6898	7862,7	5608,3	188,317
−1	272,15	280,35	199,96	0,6951	11562	6,6935	7891,7	5629,0	188,423
0	273,15	281,38	200,70	0,7041	11457	6,6973	7920,8	5649,7	188,530
1	274,15	282,41	201,43	0,7131	11354	6,7011	7949,8	5670,4	188,636
2	275,15	283,44	202,17	0,7222	11251	6,7048	7978,8	5691,1	188,741
3	276,15	284,47	202,90	0,7314	11150	6,7086	8007,8	5711,8	188,847
4	277,15	285,50	203,64	0,7407	11050	6,7123	8036,8	5732,4	188,951

t	T	h	u	π_0	θ_0	s^0	H	U	S
5	278,15	286,53	204,37	0,7501	10952	6,7160	8065,8	5753,1	189,056
6	279,15	287,56	205,11	0,7595	10854	6,7197	8094,8	5773,8	189,160
7	280,15	288,59	205,84	0,7691	10758	6,7234	8123,8	5794,5	189,264
8	281,15	289,62	206,58	0,7787	10663	6,7271	8152,8	5815,2	189,367
9	282,15	290,65	207,31	0,7884	10569	6,7307	8181,8	5835,9	189,470
10	283,15	291,68	208,05	0,7982	10477	6,7344	8210,8	5856,6	189,573
11	284,15	292,71	208,79	0,8081	10385	6,7380	8239,8	5877,3	189,675
12	285,15	293,74	209,52	0,8180	10295	6,7416	8268,9	5898,0	189,777
13	286,15	294,77	210,26	0,8281	10205	6,7452	8297,9	5918,7	189,878
14	287,15	295,80	210,99	0,8382	10117	6,7488	8326,9	5939,4	189,980
15	288,15	296,83	211,73	0,8485	10030	6,7524	8355,9	5960,1	190,080
16	289,15	297,86	212,46	0,8588	9944,2	6,7560	8384,9	5980,8	190,181
17	290,15	298,90	213,20	0,8692	9859,1	6,7595	8413,9	6001,5	190,281
18	291,15	299,93	213,93	0,8797	9775,0	6,7631	8442,9	6022,2	190,381
19	292,15	300,96	214,67	0,8903	9691,9	6,7666	8471,9	6042,9	190,480
20	293,15	301,99	215,40	0,9010	9609,8	6,7701	8500,9	6063,6	190,580
21	294,15	303,02	216,14	0,9117	9528,7	6,7737	8530,0	6084,3	190,678
22	295,15	304,05	216,87	0,9226	9448,5	6,7772	8559,0	6105,0	190,777
23	296,15	305,08	217,61	0,9335	9369,3	6,7806	8588,0	6125,7	190,875
24	297,15	306,11	218,34	0,9446	9291,0	6,7841	8617,0	6146,4	190,973
25	298,15	307,14	219,08	0,9557	9213,6	6,7876	8646,0	6167,1	191,070
26	299,15	308,17	219,81	0,9670	9137,1	6,7910	8675,0	6187,8	191,167
27	300,15	309,20	220,55	0,9783	9061,5	6,7945	8704,0	6208,5	191,264
28	301,15	310,23	221,28	0,9897	8986,8	6,7979	8733,1	6229,2	191,361
29	302,15	311,26	222,02	1,0012	8912,9	6,8013	8762,1	6249,9	191,457
30	303,15	312,29	222,76	1,0128	8839,9	6,8047	8791,1	6270,6	191,553
31	304,15	313,32	223,49	1,0245	8767,7	6,8081	8820,1	6291,3	191,648
32	305,15	314,36	224,23	1,0364	8696,4	6,8115	8849,1	6312,0	191,744
33	306,15	315,39	224,96	1,0483	8625,8	6,8149	8878,1	6332,7	191,839
34	307,15	316,42	225,70	1,0603	8556,0	6,8182	8907,2	6353,4	191,933
35	308,15	317,45	226,43	1,0724	8487,1	6,8216	8936,2	6374,1	192,028
36	309,15	318,48	227,17	1,0845	8418,9	6,8249	8965,2	6394,8	192,122
37	310,15	319,51	227,90	1,0968	8351,4	6,8283	8994,2	6415,5	192,215
38	311,15	320,54	228,64	1,1092	8284,7	6,8316	9023,2	6436,2	192,309
39	312,15	321,57	229,38	1,1217	8218,8	6,8349	9052,3	6456,9	192,402
40	313,15	322,60	230,11	1,1343	8153,6	6,8382	9081,3	6477,6	192,495
41	314,15	323,63	230,85	1,1470	8089,1	6,8415	9110,3	6498,3	192,587
42	315,15	324,66	231,58	1,1598	8025,3	6,8447	9139,3	6519,0	192,679
43	316,15	325,70	232,32	1,1727	7962,2	6,8480	9168,3	6539,7	192,771
44	317,15	326,73	233,05	1,1857	7899,8	6,8513	9197,4	6560,5	192,863
45	318,15	327,76	233,79	1,1988	7838,1	6,8545	9226,4	6581,2	192,954
46	319,15	328,79	234,52	1,2120	7777,1	6,8577	9255,4	6601,9	193,045
47	320,15	329,82	235,26	1,2253	7716,7	6,8610	9284,4	6622,6	193,136
48	321,15	330,85	236,00	1,2387	7657,0	6,8642	9313,5	6643,3	193,227
49	322,15	331,88	236,73	1,2523	7597,9	6,8674	9342,5	6664,0	193,317
50	323,15	332,91	237,47	1,2659	7539,5	6,8706	9371,5	6684,7	193,407
51	324,15	333,95	238,20	1,2796	7481,7	6,8738	9400,6	6705,4	193,497
52	325,15	334,98	238,94	1,2935	7424,5	6,8769	9429,6	6726,2	193,586
53	326,15	336,01	239,68	1,3074	7367,9	6,8801	9458,6	6746,9	193,675
54	327,15	337,04	240,41	1,3215	7311,9	6,8833	9487,7	6767,6	193,764
55	328,15	338,07	241,15	1,3356	7256,5	6,8864	9516,7	6788,3	193,853
56	329,15	339,10	241,88	1,3499	7201,7	6,8896	9545,7	6809,0	193,941
57	330,15	340,13	242,62	1,3643	7147,4	6,8927	9574,8	6829,8	194,029
58	331,15	341,16	243,36	1,3787	7093,8	6,8958	9603,8	6850,5	194,117
59	332,15	342,20	244,09	1,3933	7040,7	6,8989	9632,8	6871,2	194,205

t	T	h	u	π_0	θ_0	s^0	H	U	S
60	333,15	343,23	244,83	1,4080	6988,1	6,9020	9661,9	6891,9	194,292
61	334,15	344,26	245,56	1,4229	6936,1	6,9051	9690,9	6912,6	194,379
62	335,15	345,29	246,30	1,4378	6884,6	6,9082	9719,9	6933,4	194,466
63	336,15	346,32	247,04	1,4528	6833,7	6,9113	9749,0	6954,1	194,552
64	337,15	347,35	247,77	1,4680	6783,2	6,9143	9778,0	6974,8	194,638
65	338,15	348,39	248,51	1,4833	6733,3	6,9174	9807,1	6995,6	194,724
66	339,15	349,42	249,25	1,4986	6684,0	6,9204	9836,1	7016,3	194,810
67	340,15	350,45	249,98	1,5141	6635,1	6,9235	9865,2	7037,0	194,896
68	341,15	351,48	250,72	1,5297	6586,7	6,9265	9894,2	7057,7	194,981
69	342,15	352,51	251,46	1,5455	6538,8	6,9295	9923,3	7078,5	195,066
70	343,15	353,55	252,19	1,5613	6491,4	6,9325	9952,3	7099,2	195,151
71	344,15	354,58	252,93	1,5773	6444,4	6,9355	9981,4	7119,9	195,235
72	345,15	355,61	253,67	1,5933	6398,0	6,9385	10010,4	7140,7	195,320
73	346,15	356,64	254,40	1,6095	6352,0	6,9415	10039,5	7161,4	195,404
74	347,15	357,67	255,14	1,6258	6306,4	6,9445	10068,5	7182,2	195,487
75	348,15	358,71	255,88	1,6422	6261,3	6,9475	10097,6	7202,9	195,571
76	349,15	359,74	256,61	1,6588	6216,7	6,9504	10126,6	7223,6	195,654
77	350,15	360,77	257,35	1,6755	6172,5	6,9534	10155,7	7244,4	195,738
78	351,15	361,80	258,09	1,6922	6128,7	6,9563	10184,7	7265,1	195,820
79	352,15	362,83	258,82	1,7091	6085,4	6,9593	10213,8	7285,9	195,903
80	353,15	363,87	259,56	1,7262	6042,5	6,9622	10242,9	7306,6	195,985
81	354,15	364,90	260,30	1,7433	6000,0	6,9651	10271,9	7327,4	196,068
82	355,15	365,93	261,03	1,7606	5957,9	6,9680	10301,0	7348,1	196,150
83	356,15	366,96	261,77	1,7780	5916,3	6,9709	10330,0	7368,9	196,231
84	357,15	368,00	262,51	1,7955	5875,0	6,9738	10359,1	7389,6	196,313
85	358,15	369,03	263,25	1,8131	5834,1	6,9767	10388,2	7410,4	196,394
86	359,15	370,06	263,98	1,8309	5793,7	6,9796	10417,3	7431,1	196,475
87	360,15	371,09	264,72	1,8488	5753,6	6,9824	10446,3	7451,9	196,556
88	361,15	372,13	265,46	1,8668	5713,9	6,9853	10475,4	7472,6	196,637
89	362,15	373,16	266,20	1,8849	5674,6	6,9882	10504,5	7493,4	196,717
90	363,15	374,19	266,93	1,9032	5635,6	6,9910	10533,5	7514,2	196,797
91	364,15	375,23	267,67	1,9216	5597,1	6,9939	10562,6	7534,9	196,877
92	365,15	376,26	268,41	1,9401	5558,9	6,9967	10591,7	7555,7	196,957
93	366,15	377,29	269,15	1,9587	5521,0	6,9995	10620,8	7576,5	197,036
94	367,15	378,33	269,88	1,9775	5483,5	7,0023	10649,9	7597,2	197,116
95	368,15	379,36	270,62	1,9964	5446,4	7,0051	10679,0	7618,0	197,195
96	369,15	380,39	271,36	2,0155	5409,6	7,0079	10708,0	7638,8	197,274
97	370,15	381,43	272,10	2,0346	5373,2	7,0107	10737,1	7659,5	197,352
98	371,15	382,46	272,84	2,0539	5337,1	7,0135	10766,2	7680,3	197,431
99	372,15	383,49	273,57	2,0734	5301,3	7,0163	10795,3	7701,1	197,509
100	373,15	384,53	274,31	2,0929	5265,9	7,0191	10824,4	7721,9	197,587
101	374,15	385,56	275,05	2,1126	5230,8	7,0218	10853,5	7742,7	197,665
102	375,15	386,59	275,79	2,1324	5196,0	7,0246	10882,6	7763,4	197,743
103	376,15	387,63	276,53	2,1524	5161,5	7,0274	10911,7	7784,2	198,820
104	377,15	388,66	277,26	2,1725	5127,4	7,0301	10940,8	7805,0	197,897
105	378,15	389,69	278,00	2,1927	5093,6	7,0328	10969,9	7825,8	197,975
106	379,15	390,73	278,74	2,2131	5060,0	7,0356	10999,0	7846,6	198,051
107	380,15	391,76	279,48	2,2336	5026,8	7,0383	11028,1	7867,4	198,128
108	381,15	392,80	280,22	2,2542	4993,9	7,0410	11057,2	7888,2	198,205
109	382,15	393,83	280,96	2,2750	4961,3	7,0437	11086,3	7909,0	198,281
110	383,15	394,86	281,70	2,2959	4929,0	7,0464	11115,4	7929,8	198,357
111	384,15	395,90	282,44	2,3170	4896,9	7,0491	11144,5	7950,6	198,433
112	385,15	396,93	283,17	2,3381	4865,2	7,0518	11173,6	7971,4	198,508
113	386,15	397,97	283,91	2,3595	4833,7	7,0545	11202,8	7992,2	198,584
114	387,15	399,00	284,65	2,3809	4802,5	7,0572	11231,9	8013,0	198,659

Atmospheric nitrogen (*Continued*)

t	T	h	u	π_0	θ_0	s^0	H	U	S
115	388,15	400,04	285,39	2,4026	4771,6	7,0598	11261,0	8033,8	198,734
116	389,15	401,07	286,13	2,4243	4741,0	7,0625	11290,1	8054,6	198,809
117	390,15	402,11	286,87	2,4462	4710,6	7,0652	11319,3	8075,4	198,884
118	391,15	403,14	287,61	2,4682	4680,6	7,0678	11348,4	8096,2	198,959
119	392,15	404,17	288,35	2,4904	4650,7	7,0704	11377,5	8117,0	199,033
120	393,15	405,21	289,09	2,5127	4621,2	7,0731	11406,6	8137,8	199,107
121	394,15	406,24	289,83	2,5352	4591,9	7,0757	11435,8	8158,7	199,181
122	395,15	407,28	290,57	2,5578	4562,8	7,0783	11464,9	8179,5	199,255
123	396,15	408,31	291,31	2,5806	4534,0	7,0809	11494,1	8200,3	199,329
124	397,15	409,35	292,05	2,6035	4505,5	7,0836	11523,2	8221,1	199,402
125	398,15	410,39	292,79	2,6265	4477,2	7,0862	11552,3	8242,0	199,475
126	399,15	411,42	293,53	2,6497	4449,1	7,0888	11581,5	8262,8	199,549
127	400,15	412,46	294,27	2,6731	4421,3	7,0913	11610,6	8283,6	199,621
128	401,15	413,49	295,01	2,6966	4393,8	7,0939	11639,8	8304,5	199,694
129	402,15	414,53	295,75	2,7202	4366,4	7,0965	11668,9	8325,3	199,767
130	403,15	415,56	296,49	2,7440	4339,3	7,0991	11698,1	8346,1	199,839
131	404,15	416,60	297,23	2,7679	4312,5	7,1017	11727,2	8367,0	199,911
132	405,15	417,63	297,97	2,7920	4285,8	7,1042	11756,4	8387,8	199,984
133	406,15	418,67	298,71	2,8163	4259,4	7,1068	11785,6	8408,7	200,055
134	407,15	419,71	299,45	2,8407	4233,2	7,1093	11814,7	8429,5	200,127
135	408,15	420,74	300,19	2,8652	4207,3	7,1119	11843,9	8450,4	200,199
136	409,15	421,78	300,93	2,8899	4181,5	7,1144	11873,1	8471,2	200,270
137	410,15	422,81	301,67	2,9148	4156,0	7,1169	11902,2	8492,1	200,341
138	411,15	423,85	302,41	2,9398	4130,7	7,1194	11931,4	8512,9	200,412
139	412,15	424,89	303,15	2,9650	4105,6	7,1220	11960,6	8533,8	200,483
140	413,15	425,92	303,90	2,9903	4080,7	7,1245	11989,8	8554,7	200,554
141	414,15	426,96	304,64	3,0158	4056,0	7,1270	12019,0	8575,5	200,624
142	415,15	428,00	305,38	3,0414	4031,5	7,1295	12048,1	8596,4	200,695
143	416,15	429,03	306,12	3,0672	4007,2	7,1320	12077,3	8617,3	200,765
144	417,15	430,07	306,86	3,0932	3983,2	7,1345	12106,5	8638,2	200,835
145	418,15	431,11	307,60	3,1193	3959,3	7,1369	12135,7	8659,0	200,905
146	419,15	432,15	308,35	3,1456	3935,6	7,1394	12164,9	8679,9	200,975
147	420,15	433,18	309,09	3,1720	3912,1	7,1419	12194,1	8700,8	201,044
148	421,15	434,22	309,83	3,1986	3888,8	7,1444	12223,3	8721,7	201,114
149	422,15	435,26	310,57	3,2254	3865,7	7,1468	12252,5	8742,6	201,183
150	423,15	436,30	311,31	3,2523	3842,8	7,1493	12281,7	8763,5	201,252
151	424,15	437,33	312,06	3,2794	3820,1	7,1517	12310,9	8784,4	201,321
152	425,15	438,37	312,80	3,3066	3797,5	7,1542	12340,1	8805,3	201,390
153	426,15	439,41	313,54	3,3340	3775,2	7,1566	12369,4	8826,2	201,458
154	427,15	440,45	314,28	3,3616	3753,0	7,1590	12398,6	8847,1	201,527
155	428,15	441,48	315,03	3,3893	3731,0	7,1615	12427,8	8868,0	201,595
156	429,15	442,52	315,77	3,4172	3709,2	7,1639	12457,0	8888,9	201,663
157	430,15	443,56	316,51	3,4453	3687,5	7,1663	12486,2	8909,8	201,732
158	431,15	444,60	317,25	3,4735	3666,0	7,1687	12515,5	8930,7	201,799
159	432,15	445,64	318,00	3,5020	3644,7	7,1711	12544,7	8951,6	201,867
160	433,15	446,68	318,74	3,5305	3623,6	7,1735	12573,9	8972,6	201,935
161	434,15	447,72	319,48	3,5593	3602,6	7,1759	12603,2	8993,5	202,002
162	435,15	448,75	320,23	3,5882	3581,8	7,1783	12632,4	9014,4	202,069
163	436,15	449,79	320,97	3,6173	3561,2	7,1807	12661,7	9035,3	202,137
164	437,15	450,83	321,72	3,6466	3540,7	7,1831	12690,9	9056,3	202,203
165	438,15	451,87	322,46	3,6760	3520,4	7,1854	12720,2	9077,2	202,270
166	439,15	452,91	323,20	3,7056	3500,2	7,1878	12749,4	9098,2	202,337
167	440,15	453,95	323,95	3,7354	3480,2	7,1902	12778,7	9119,1	202,404
168	441,15	454,99	324,69	3,7653	3460,4	7,1925	12807,9	9140,0	202,470
169	442,15	456,03	325,44	3,7954	3440,7	7,1949	12837,2	9161,0	202,536

t	T	h	u	π_0	θ_0	s^0	H	U	S
170	443,15	457,07	326,18	3,8258	3421,2	7,1972	12866,5	9181,9	202,602
171	444,15	458,11	326,92	3,8562	3401,8	7,1996	12895,8	9202,9	202,668
172	445,15	459,15	327,67	3,8869	3382,6	7,2019	12925,0	9223,9	202,734
173	446,15	460,19	328,41	3,9177	3363,5	7,2043	12954,3	9244,8	202,800
174	447,15	461,23	329,16	3,9487	3344,5	7,2066	12983,6	9265,8	202,865
175	448,15	462,27	329,90	3,9799	3325,8	7,2089	13012,9	9286,8	202,931
176	449,15	463,31	330,65	4,0113	3307,1	7,2112	13042,2	9307,7	202,996
177	450,15	464,35	331,39	4,0428	3288,6	7,2135	13071,4	9328,7	203,061
178	451,15	465,39	332,14	4,0746	3270,2	7,2159	13100,7	9349,7	203,126
179	452,15	466,43	332,88	4,1065	3252,0	7,2182	13130,0	9370,7	203,191
180	453,15	467,47	333,63	4,1386	3233,9	7,2205	13159,3	9391,7	203,256
181	454,15	468,51	334,37	4,1709	3216,0	7,2228	13188,6	9412,7	203,320
182	455,15	469,55	335,12	4,2033	3198,2	7,2250	13217,9	9433,6	203,385
183	456,15	470,60	335,87	4,2360	3180,5	7,2273	13247,3	9454,6	203,449
184	457,15	471,64	336,61	4,2688	3163,0	7,2296	13276,6	9475,6	203,513
185	458,15	472,68	337,36	4,3018	3145,5	7,2319	13305,9	9496,6	203,578
186	459,15	473,72	338,10	4,3350	3128,3	7,2342	13335,2	9517,7	203,641
187	460,15	474,76	338,85	4,3684	3111,1	7,2364	13364,5	9538,7	203,705
188	461,15	475,80	339,60	4,4020	3094,1	7,2387	13393,9	9559,7	203,769
189	462,15	476,85	340,34	4,4358	3077,2	7,2409	13423,2	9580,7	203,832
190	463,15	477,89	341,09	4,4697	3060,4	7,2432	13452,5	9601,7	203,896
191	464,15	478,93	341,84	4,5039	3043,8	7,2454	13481,9	9622,7	203,959
192	465,15	479,97	342,59	4,5382	3027,2	7,2477	13511,2	9643,8	204,022
193	466,15	481,01	343,33	4,5728	3010,8	7,2499	13540,6	9664,8	204,085
194	467,15	482,06	344,08	4,6075	2994,6	7,2522	13569,9	9685,8	204,148
195	468,15	483,10	344,83	4,6424	2978,4	7,2544	13599,3	9706,9	204,211
196	469,15	484,14	345,57	4,6775	2962,4	7,2566	13628,6	9727,9	204,274
197	470,15	485,19	346,32	4,7128	2946,4	7,2588	13658,0	9749,0	204,336
198	471,15	486,23	347,07	4,7483	2930,6	7,2610	13687,4	9770,0	204,399
199	472,15	487,27	347,82	4,7840	2914,9	7,2633	13716,7	9796,1	204,461
200	473,15	488,32	348,57	4,8199	2899,4	7,2655	13746,1	9812,1	204,523
201	474,15	489,36	349,31	4,8560	2883,9	7,2677	13775,5	9833,2	204,585
202	475,15	490,40	350,06	4,8923	2868,5	7,2699	13804,9	9854,3	204,647
203	476,15	491,45	350,81	4,9288	2853,3	7,2721	13834,3	9875,3	204,709
204	477,15	492,49	351,56	4,9655	2838,2	7,2743	13863,6	9896,4	204,770
205	478,15	493,54	352,31	5,0024	2823,1	7,2764	13893,0	9917,5	204,832
206	479,15	494,58	353,06	5,0394	2808,2	7,2786	13922,4	9938,6	204,893
207	480,15	495,62	353,81	5,0767	2793,4	7,2808	13951,8	9959,7	204,955
208	481,15	496,67	354,56	5,1142	2778,7	7,2830	13981,2	9980,8	205,016
209	482,15	497,71	355,31	5,1519	2764,1	7,2851	14010,7	10001,9	205,077
210	483,15	498,76	356,06	5,1898	2749,6	7,2873	14040,1	10023,0	205,138
211	484,15	499,80	356,80	5,2279	2735,2	7,2895	14069,5	10044,1	205,199
212	485,15	500,85	357,55	5,2662	2720,9	7,2916	14098,9	10065,2	205,259
213	486,15	501,89	358,30	5,3048	2706,7	7,2938	14128,3	10086,3	205,320
214	487,15	502,94	359,05	5,3435	2692,6	7,2959	14157,8	10107,4	205,380
215	488,15	503,99	359,81	5,3824	2678,7	7,2981	14187,2	10128,5	205,441
216	489,15	505,03	360,56	5,4216	2664,8	7,3002	14216,6	10149,6	205,501
217	490,15	506,08	361,31	5,4609	2651,0	7,3023	14246,1	10170,8	205,561
218	491,15	507,12	362,06	5,5005	2637,3	7,3045	14275,5	10191,9	205,621
219	492,15	508,17	362,81	5,5402	2623,7	7,3066	14305,0	10213,0	205,681
220	493,15	509,22	363,56	5,5802	2610,2	7,3087	14334,4	10234,2	205,741
221	494,15	510,26	364,31	5,6204	2596,8	7,3109	14363,9	10255,3	205,800
222	495,15	511,31	365,06	5,6608	2583,4	7,3130	14393,3	10276,5	205,860
223	496,15	512,36	365,81	5,7014	2570,2	7,3151	14422,8	10297,6	205,919
224	497,15	513,40	366,56	5,7423	2557,1	7,3172	14452,3	10318,8	205,979

Atmospheric nitrogen (*Continued*)

t	T	h	u	π_0	θ_0	s^0	H	U	S
225	498,15	514,45	367,32	5,7833	2544,0	7,3193	14481,8	10339,9	206,038
226	499,15	515,50	368,07	5,8246	2531,1	7,3214	14511,2	10361,1	206,097
227	500,15	516,54	368,82	5,8661	2518,2	7,3235	14540,7	10382,3	206,156
228	501,15	517,59	369,57	5,9078	2505,4	7,3256	14570,2	10403,5	206,215
229	502,15	518,64	370,32	5,9497	2492,7	7,3277	14599,7	10424,6	206,274
230	503,15	519,69	371,08	5,9919	2480,1	7,3298	14629,2	10445,8	206,333
231	504,15	520,74	371,83	6,0342	2467,6	7,3318	14658,7	10467,0	206,391
232	505,15	521,78	372,58	6,0768	2455,2	7,3339	14688,2	10488,2	296,450
233	506,15	522,83	373,34	6,i196	2442,8	7,3360	14717,7	10509,4	206,508
234	507,15	523,88	374,09	6,1627	2430,6	7,3381	14747,2	10530,6	206,566
235	508,15	524,93	374,84	6,2059	2418,4	7,3401	14776,8	10551,8	206,624
236	509,15	525,98	375,60	6,2494	2406,3	7,3422	14806,3	10573,0	206,682
237	510,15	527,03	376,35	6,2931	2394,3	7,3442	14835,8	10594,2	206,740
238	511,15	528,08	377,10	6,3370	2382,3	7,3463	14865,4	10615,4	206,798
239	512,15	529,13	377,86	6,3812	2370,5	7,3483	14894,9	10636,7	206,856
240	513,15	530,18	378,61	6,4256	2358,7	7,3504	14924,4	10657,9	206,914
241	514,15	531,23	379,37	6,4702	2347,0	7,3524	14954,0	10679,1	206,971
242	515,15	532,27	380,12	6,5150	2335,4	7,3545	14983,5	10700,4	207,029
243	516,15	533,32	380,87	6,5601	2323,8	7,3565	15013,1	10721,6	207,086
244	517,15	534,38	381,63	6,6054	2312,4	7,3585	15042,7	10742,9	207,143
245	518,15	535,43	382,38	6,6509	2301,0	7,3606	15072,2	10764,1	207,200
246	519,15	536,48	383,14	6,6967	2289,7	7,3626	15101,8	10785,4	207,257
247	520,15	537,53	383,89	6,7427	2278,4	7,3646	15131,4	10806,6	207,314
248	521,15	538,58	384,65	6,7889	2267,3	7,3666	15161,0	10827,9	207,371
249	522,1ɔ	539,63	385,41	6,8354	2256,2	7,3687	15190,5	10849,2	207,428
250	523,15	540,68	386,16	6,8821	2245,2	7,3707	15220,1	10870,4	207,484
251	524,15	541,73	386,92	6,9290	2234,2	7,3727	15249,7	10891,7	207,541
252	525,15	542,78	387,67	6,9762	2223,3	7,3747	15279,3	10913,0	207,597
253	526,15	543,83	388,43	7,0236	2212,5	7,3767	15308,9	10934,3	207,654
254	527,15	544,89	389,19	7,0713	2201,8	7,3787	15338,5	10955,6	207,710
255	528,15	545,94	389,94	7,1192	2191,1	7,3807	15368,2	10976,9	207,766
256	529,15	546,99	390,70	7,1673	2180,5	7,3827	15397,8	10998,2	207,822
257	530,15	548,04	391,46	7,2157	2170,0	7,3846	15427,4	11019,5	207,878
258	531,15	549,10	392,21	7,2643	2159,6	7,3866	15457,0	11040,8	207,934
259	532,15	550,15	392,97	7,3132	2149,2	7,3886	15486,7	11062,1	207,989
260	533,15	551,20	393,73	7,3623	2138,8	7,3906	15516,3	11083,5	208,045
261	534,15	552,25	394,49	7,4116	2128,6	7,3926	15545,9	11104,8	208,101
262	535,15	553,31	395,24	7,4612	2118,4	7,3945	15575,6	11126,1	208,156
263	536,15	554,36	396,00	7,5111	2108,3	7,3965	15605,2	11147,5	208,211
264	537,15	555,41	396,76	7,5612	2098,2	7,3985	15634,9	11168,8	208,267
265	538,15	556,47	397,52	7,6115	2088,2	7,4004	15664,6	11190,2	208,322
266	539,15	557,52	398,28	7,6621	2078,3	7,4024	15694,2	11211,5	208,377
267	540,15	558,58	399,04	7,7129	2068,4	7,4043	15723,9	11232,9	208,432
268	541,15	559,63	399,80	7,7640	2058,6	7,4063	15753,6	11254,2	208,487
269	542,15	560,68	400,55	7,8154	2048,8	7,4082	15783,3	11275,6	208,542
270	543,15	561,74	401,31	7,8670	2039,2	7,4102	15813,0	11297,0	208,596
271	544,15	562,79	402,07	7,9188	2029,5	7,4121	15842,7	11318,4	208,651
272	545,15	563,85	402,83	7,9709	2020,0	7,4140	15872,4	11339,8	208,705
273	546,15	564,90	403,59	8,0233	2010,5	7,4160	15902,1	11361,1	208,760
274	547,15	565,96	404,35	8,0759	2001,0	7,4179	15931,8	11382,5	208,814
275	548,15	567,02	405,11	8,1288	1991,7	7,4198	15961,5	11403,9	208,869
276	549,15	568,07	405,87	8,1819	1982,3	7,4218	15991,2	11425,3	208,923
277	550,15	569,13	406,63	8,2353	1973,1	7,4237	16020,9	11446,8	208,977
278	551,15	570,18	407,40	8,2890	1963,9	7,4256	16050,7	11468,2	209,031
279	552,15	571,24	408,16	8,3429	195ʔ4,7	7,4275	16080,4	11489,6	209,085

Atmospheric nitrogen (*Continued*)

t	T	h	u	π_0	θ_0	s^0	H	U	S
280	553,15	572,30	408,92	8,3971	1945,6	7,4294	16110,1	11511,0	209,138
281	554,15	573,35	409,68	8,4515	1936,6	7,4313	16139,9	11532,5	209,192
282	555,15	574,41	410,44	8,5062	1927,6	7,4332	16169,6	11553,9	209,246
283	556,15	575,47	411,20	8,5612	1918,7	7,4351	16199,4	11575,3	209,299
284	557,15	576,52	411,96	8,6164	1909,8	7,4370	16229,2	11596,8	209,353
285	558,15	577,58	412,73	8,6719	1901,0	7,4389	16258,9	11618,2	209,406
286	559,15	578,64	413,49	8,7277	1892,2	7,4408	16288,7	11639,7	209,460
287	560,15	579,70	414,25	8,7837	1883,5	7,4427	16318,5	11661,2	209,513
288	561,15	580,76	415,01	8,8400	1874,8	7,4446	16348,3	11682,6	209,566
289	562,15	581,81	415,78	8,8966	1866,2	7,4465	16378,0	11704,1	209,619
290	563,15	582,87	416,54	8,9534	1857,7	7,4484	16407,8	11725,6	209,672
291	564,15	583,93	417,30	9,0105	1849,2	7,4503	16437,6	11747,1	209,725
292	565,15	584,99	418,07	9,0679	1840,8	7,4521	16467,4	11768,6	209,778
293	566,15	586,05	418,83	9,1256	1832,4	7,4540	16497,3	11790,1	209,830
294	567,15	587,11	419,59	9,1835	1824,0	7,4559	16527,1	11811,6	209,883
295	568,15	588,17	420,36	9,2417	1815,7	7,4577	16556,9	11833,1	209,935
296	569,15	589,23	421,12	9,3002	1807,5	7,4596	16586,7	11854,6	209,988
297	570,15	590,29	421,89	9,3590	1799,3	7,4615	16616,6	11876,1	210,040
298	571,15	591,35	422,65	9,4180	1791,1	7,4633	16646,4	11897,6	210,092
299	572,15	592,41	423,42	9,4773	1783,0	7,4652	16676,2	11919,1	210,145
300	573,15	593,47	424,18	9,5369	1775,0	7,4670	16706,1	11940,7	210,197
301	574,15	594,53	424,95	9,5968	1767,0	7,4689	16735,9	11962,2	210,249
302	575,15	595,59	425,71	9,6570	1759,1	7,4707	16765,8	11983,8	210,301
303	576,15	596,65	426,48	9,7174	1751,1	7,4726	16795,7	12005,3	210,353
304	577,15	597,71	427,24	9,7782	1743,3	7,4744	16825,5	12026,9	210,405
305	578,15	598,77	428,01	9,8392	1735,5	7,4762	16855,4	12048,4	210,456
306	579,15	599,83	428,77	9,9005	1727,7	7,4781	16885,3	12070,0	210,508
307	580,15	600,89	429,54	9,9621	1720,0	7,4799	16915,2	12091,6	210,559
308	581,15	601,96	430,31	10,023	1712,3	7,4817	16945,1	12113,2	210,611
309	582,15	603,02	431,07	10,086	1704,7	7,4836	16975,0	12134,7	210,662
310	583,15	604,08	431,84	10,148	1697,1	7,4854	17004,9	12156,3	210,714
311	584,15	605,14	432,61	10,211	1689,6	7,4872	17034,8	12177,9	210,765
312	585,15	606,21	433,38	10,274	1682,1	7,4890	17064,7	12199,5	210,816
313	586,15	607,27	434,14	10,337	1674,6	7,4908	17094,6	12221,1	210,867
314	587,15	608,33	434,91	10,401	1667,2	7,4927	17124,6	12242,8	210,918
315	588,15	609,40	435,68	10,465	1659,9	7,4945	17154,5	12264,4	210,969
316	589,15	610,46	436,45	10,529	1652,6	7,4963	17184,4	12286,0	211,020
317	590,15	611,52	437,22	10,594	1645,3	7,4981	17214,4	12307,6	211,071
318	591,15	612,59	437,98	10,658	1638,1	7,4999	17244,3	12329,3	211,121
319	592,15	613,65	438,75	10,723	1630,9	7,5017	17274,3	12350,9	211,172
320	593,15	614,72	439,52	10,789	1623,7	7,5035	17304,3	12372,6	211,223
321	594,15	615,78	440,29	10,855	1616,6	7,5053	17334,2	12394,2	211,273
322	595,15	616,85	441,06	10,921	1609,5	7,5071	17364,2	12415,9	211,324
323	596,15	617,91	441,83	10,987	1602,5	7,5088	17394,2	12437,5	211,374
324	597,15	618,98	442,60	11,053	1595,5	7,5106	17424,2	12459,2	211,424
325	598,15	620,04	443,37	11,120	1588,6	7,5124	17454,2	12480,9	211,474
326	599,15	621,11	444,14	11,188	1581,7	7,5142	17484,2	12502,6	211,524
327	600,15	622,17	444,91	11,255	1574,8	7,5160	17514,2	12524,3	211,574
328	601,15	623,24	445,68	11,323	1568,0	7,5177	17544,2	12546,0	211,624
329	602,15	624,30	446,45	11,391	1561,2	7,5195	17574,2	12567,7	211,674
330	603,15	625,37	447,22	11,460	1554,4	7,5213	17604,2	12589,4	211,724
331	604,15	626,44	448,00	11,528	1547,7	7,5231	17634,2	12611,1	211,774
332	605,15	627,51	448,77	11,597	1541,1	7,5248	17664,3	12632,8	211,824
333	606,15	628,57	449,54	11,667	1534,4	7,5266	17694,3	12654,5	211,873
334	607,15	629,64	450,31	11,737	1527,8	7,5283	17724,4	12676,3	211,923

t	T	h	u	π_0	θ_0	s^0	H	U	S
335	608,15	630,71	451,08	11,807	1521,3	7,5301	17754,4	12698,0	211,972
336	609,15	631,77	451,86	11,877	1514,8	7,5318	17784,5	12719,7	212,022
337	610,15	632,84	452,63	11,948	1508,3	7,5336	17814,5	12741,5	212,071
338	611,15	633,91	453,40	12,019	150!,8	7,5353	17844,6	12763,2	212,120
339	612,15	634,98	454,17	12,090	1495,4	7,5371	17874,7	12785,0	212,169
340	613,15	636,05	454,95	12,161	1489,0	7,5388	17904,8	12806,8	212,218
341	614,15	637,12	455,72	12,233	1482,7	7,5406	17934,8	12828,5	212,267
342	615,15	638,19	456,49	12,306	1476,4	7,5423	17964,9	12850,3	212,316
343	616,15	639,26	457,27	12,378	1470,1	7,5441	17995,0	12872,1	212,365
344	617,15	640,32	458,04	12,451	1463,9	7,5458	18025,1	12893,9	212,414
345	618,15	641,39	458,82	12,524	1457,7	7,5475	18055,2	12915,7	212,463
346	619,15	642,46	459,59	12,598	1451,5	7,5493	18085,4	12937,5	212,511
347	620,15	643,53	460,37	12,672	1445,4	7,5510	18115,5	12959,3	212,560
348	621,15	644,60	461,14	12,746	1439,3	7,5527	18145,6	12981,1	212,609
349	622,15	645,68	461,92	12,821	1433,2	7,5544	18175,8	13002,9	212,657
350	623,15	646,75	462,69	12,895	1427,2	7,5561	18205,9	13024,8	212,706
351	624,15	647,82	463,47	12,971	J421,2	7,5579	18236,0	13046,6	212,754
352	625,15	648,89	464,24	13,046	ı415,2	7,5596	18266,2	13068,4	212,802
353	626,15	649,96	465,02	13,122	1409,3	7,5613	18296,4	13090,3	212,850
354	627,15	651,03	465,80	13,198	1403,4	7,5630	18326,5	13112,1	212,898
355	628,15	652,10	466,57	13,275	1397,5	7,5647	18356,7	13134,0	212,947
356	629,15	653,18	467,35	13,352	1391,7	7,5664	18386,9	13155,9	212,995
357	630,15	654,25	468,13	13,429	1385,9	7,5681	18417,1	13177,7	213,043
358	631,15	655,32	468,90	13,506	1380,1	7,5698	18447,3	13199,6	213,090
359	632,15	656,39	469,68	13,584	1374,4	7,5715	18477,5	13221,5	213,138
360	633,15	657,47	470,46	13,662	1368,7	7,5732	18507,7	13243,4	213,186
361	634,15	658,54	471,24	13,741	1363,0	7,5749	18537,9	13265,3	2.3,234
362	635,15	659,61	472,01	13,820	1357,4	7,5766	18568,1	13287,2	213,281
363	636,15	660,69	472,79	13,899	1351,7	7,5783	18598,3	13309,1	213,329
364	637,15	661,76	473,57	13,979	1346,2	7,5800	18628,5	13331,0	213,376
365	638,15	662,83	474,35	14,059	1340,6	7,5817	18658,8	13352,9	213,424
366	639,15	663,91	475,13	14,139	1335,1	7,5833	18689,0	13374,9	213,471
367	640,15	664,98	475,91	14,220	1329,6	7,5850	18719,3	13396,8	213,518
368	641,15	666,06	476,69	14,301	1324,1	7,5867	18749,5	13418,7	213,566
369	642,15	667,13	477,47	14,382	1318,7	7,5884	18779,8	13440,7	213,613
370	643,15	668,21	478,25	14,464	1313,3	7,5900	18810,0	13462,6	213,660
371	644,15	669,28	479,03	14,546	1307,9	7,5917	18840,3	13484,6	213,707
372	645,15	670,36	479,81	14,628	1302,5	7,5934	18870,6	13506,6	213,754
373	646,15	671,43	480,59	14,711	1297,2	7,5951	18900,9	13528,5	213,801
374	647,15	672,51	481,37	14,794	1291,9	7,5967	18931,2	13550,5	213,848
375	648,15	673,59	482,15	14,878	1286,7	7,5984	18961,5	13572,5	213,894
376	649,15	674,66	482,93	14,961	1281,4	7,6000	18991,8	13594,5	213,941
377	650,15	675,74	483,71	15,046	1276,2	7,6017	19022,1	13616,5	213,988
378	651,15	676,82	484,49	15,130	1271,0	7,6034	19052,4	13638,5	214,034
379	652,15	677,89	485,27	15,215	1265,9	7,6050	19082,7	13660,5	214,081
380	653,15	678,97	486,06	15,300	1260,8	7,6067	19113,1	13682,5	214,127
381	654,15	680,05	486,84	15,386	1255,7	7,6083	19143,4	13704,5	214,174
382	655,15	681,13	487,62	15,472	1250,6	7,6099	19173,7	13726,6	214,220
383	656,15	682,21	488,40	15,558	1245,5	7,6116	19204,1	13748,6	214,266
384	657,15	683,28	489,19	15,645	1240,5	7,6132	19234,4	13770,6	214,313
385	658,15	684,36	489,97	15,732	1235,5	7,6149	19264,8	13792,7	214,359
386	659,15	685,44	490,75	15,820	1230,6	7,6165	19295,2	13814,7	214,405
387	660,15	686,52	491,54	15,908	1225,6	7,6181	19325,6	13836,8	214,451
388	661,15	687,60	492,32	15,996	1220,7	7,6198	19355,9	13858,9	214,497
389	662,15	688,68	493,11	16,084	1215,8	7,6214	19386,3	13880,9	214,543

t	T	h	u	π_0	θ_0	s^0	H	U	S
390	663,15	689,76	493,89	16,173	1211,0	7,6230	19416,7	13903,0	214,589
391	664,15	690,84	494,68	16,263	1206,1	7,6247	19447,1	13925,1	214,635
392	665,15	691,92	495,46	16,353	1201,3	7,6263	19477,5	13947,2	214,680
393	666,15	693,00	496,25	16,443	1196,5	7,6279	19507,9	13969,3	214,726
394	667,15	694,08	497,03	16,533	1191,8	7,6295	19538,4	13991,4	214,772
395	668,15	695,16	497,82	16,624	1187,0	7,6312	19568,8	14013,5	214,817
396	669,15	696,24	498,60	16,715	1182,3	7,6328	19599,2	14035,6	214,863
397	670,15	697,32	499,39	16,807	1177,6	7,6344	19629,7	14057,8	214,908
398	671,15	698,41	500,17	16,899	1173,0	7,6360	19660,1	14079,9	214,954
399	672,15	699,49	500,96	16,991	1168,3	7,6376	19690,6	14102,0	214,999
400	673,15	700,57	501,75	17,084	1163,7	7,6392	19721,0	14124,2	215,044
401	674,15	701,65	502,53	17,177	1159,1	7,6408	19751,5	14146,3	215,089
402	675,15	702,73	503,32	17,271	1154,5	7,6424	19782,0	14168,5	215,135
403	676,15	703,82	504,11	17,365	1150,0	7,6440	19812,4	14190,7	215,180
404	677,15	704,90	504,90	17,459	1145,5	7,6456	19842,9	14212,8	215,225
405	678,15	705,98	505,68	17,554	1141,0	7,6472	19873,4	14235,0	215,270
406	679,15	707,07	506,47	17,649	1136,5	7,6488	19903,9	14257,2	215,315
407	680,15	708,15	507,26	17,745	1132,0	7,6504	19934,4	14279,4	215,360
408	681,15	709,23	508,05	17,841	1127,6	7,6520	19964,9	14301,6	215,404
409	682,15	710,32	508,84	17,937	1123,2	7,6536	19995,5	14323,8	215,449
410	683,15	711,40	509,63	18,034	1118,8	7,6552	20026,0	14346,0	215,494
411	684,15	712,49	510,42	18,131	1114,4	7,6568	20056,5	14368,2	215,539
412	685,15	713,57	511,21	18,228	1110,1	7,6584	20087,1	14390,4	215,583
413	686,15	714,66	512,00	18,326	1105,8	7,6600	20117,6	14412,7	215,628
414	687,15	715,74	512,79	18,425	1101,5	7,6615	20148,2	14434,9	215,672
415	688,15	716,83	513,58	18,523	1097,2	7,6631	20178,7	14457,2	215,717
416	689,15	717,91	514,37	18,622	1092,9	7,6647	20209,3	14479,4	215,761
417	690,15	719,00	515,16	18,722	1088,7	7,6663	20239,9	14501,7	215,805
418	691,15	720,09	515,95	18,822	1084,5	7,6678	20270,4	14523,9	215,850
419	692,15	721,17	516,74	18,922	1080,3	7,6694	20301,0	14546,2	215,894
420	693,15	722,26	517,53	19,023	1076,1	7,6710	20331,6	14568,5	215,938
421	694,15	723,35	518,32	19,124	1072,0	7,6725	20362,2	14590,8	215,982
422	695,15	724,43	519,11	19,226	1067,9	7,6741	20392,8	14613,0	216,026
423	696,15	725,52	519,91	19,328	1063,7	7,6757	20423,4	14635,3	216,070
424	697,15	726,61	520,70	19,430	1059,7	7,6772	20454,0	14657,6	216,114
425	698,15	727,70	521,49	19,533	1055,6	7,6788	20484,7	14680,0	216,158
426	699,15	728,78	522,28	19,637	1051,5	7,6803	20515,3	14702,3	216,202
427	700,15	729,87	523,08	19,740	1047,5	7,6819	20545,9	14724,6	216,246
428	701,15	730,96	523,87	19,844	1043,5	7,6835	20576,6	14746,9	216,289
429	702,15	732,05	524,66	19,949	1039,5	7,6850	20607,2	14769,3	216,333
430	703,15	733,14	525,46	20,054	1035,5	7,6866	20637,9	14791,6	216,377
431	704,15	734,23	526,25	20,159	1031,6	7,6881	20668,6	14814,0	216,420
432	705,15	735,32	527,05	20,265	1027,7	7,6897	20699,2	14836,3	216,464
433	706,15	736,41	527,84	20,371	1023,8	7,6912	20729,9	14858,7	216,507
434	707,15	737,50	528,63	20,478	1019,9	7,6927	20760,6	14881,1	216,551
435	708,15	738,59	529,43	20,585	1016,0	7,6943	20791,3	14903,4	216,594
436	709,15	739,68	530,22	20,693	1012,1	7,6958	20822,0	14925,8	216,637
437	710,15	740,77	531,02	20,801	1008,3	7,6974	20852,7	14948,2	216,681
438	711,15	741,86	531,82	20,909	1004,5	7,6989	20883,4	14970,6	216,724
439	712,15	742,95	532,61	21,018	1000,7	7,7004	20914,1	14993,0	216,767
440	713,15	744,04	533,41	21,127	996,97	7,7020	20944,8	15015,4	216,810
441	714,15	745,14	534,20	21,237	993,21	7,7035	20975,6	15037,8	216,853
442	715,15	746,23	535,00	21,347	989,47	7,7050	21006,3	15060,3	216,896
443	716,15	747,32	535,80	21,458	985,74	7,7065	21037,1	15082,7	216,939
444	717,15	748,41	536,59	21,569	982,04	7,7081	21067,8	15105,1	216,982

t	T	h	u	π_0	θ_0	s^v	H	U	S
445	718,15	749,51	537,39	21,680	978,35	7,7096	21098,6	15127,6	217,025
446	719,15	750,60	538,19	21,792	974,68	7,7111	21129,3	15150,0	217,068
447	720,15	751,69	538,99	21,904	971,03	7,7126	21160,1	15172,5	217,111
448	721,15	752,78	539,79	22,017	967,40	7,7141	21190,9	15195,0	217,153
449	722,15	753,88	540,58	22,130	963,78	7,7157	21221,7	15217,4	217,196
450	723,15	754,97	541,38	22,244	960,18	7,7172	21252,5	15239,9	217,239
451	724,15	756,07	542,18	22,358	956,60	7,7187	21283,3	15262,4	217,281
452	725,15	757,16	542,98	22,473	953,03	7,7202	21314,1	15284,9	217,324
453	726,15	758,26	543,78	22,588	949,49	7,7217	21344,9	15307,4	217,366
454	727,15	759,35	544,58	22,704	945,96	7,7232	21375,7	15329,9	217,408
455	728,15	760,45	545,38	22,820	942,44	7,7247	21406,5	15352,4	217,451
456	729,15	761,54	546,18	22,936	938,95	7,7262	21437,4	15374,9	217,493
457	730,15	762,64	546,98	23,053	935,47	7,7277	21468,2	15397,4	217,535
458	731,15	763,73	547,78	23,170	932,00	7,7292	21499,1	15420,0	217,578
459	732,15	764,83	548,58	23,288	928,56	7,7307	21529,9	15442,5	217,620
460	733,15	765,92	549,38	23,406	925,13	7,7322	21560,8	15465,1	217,662
461	734,15	767,02	550,18	23,525	921,71	7,7337	21591,6	15487,6	217,704
462	735,15	768,12	550,98	23,644	918,31	7,7352	21622,5	15510,2	217,746
463	736,15	769,21	551,78	23,764	914,93	7,7367	21653,4	15532,7	217,788
464	737,15	770,31	552,59	23,884	911,57	7,7382	21684,3	15555,3	217,830
465	738,15	771,41	553,39	24,005	908,22	7,7397	21715,2	15577,9	217,872
466	739,15	772,51	554,19	24,126	904,88	7,7412	21746,1	15600,5	217,914
467	740,15	773,61	554,99	24,247	901,56	7,7426	21777,0	15623,1	217,955
468	741,15	774,70	555,80	24,369	898,26	7,7441	21807,9	15645,7	217,997
469	742,15	775,80	556,60	24,492	894,97	7,7456	21838,8	15668,3	218,039
470	743,15	776,90	557,40	24,615	891,70	7,7471	21869,7	15690,9	218,081
471	744,15	778,00	558,21	24,738	888,45	7,7486	21900,7	15713,5	218,122
472	745,15	779,10	559,01	24,862	885,20	7,7500	21931,6	15736,1	218,164
473	746,15	780,20	559,81	24,987	881,98	7,7515	21962,6	15758,8	218,205
474	747,15	781,30	560,62	25,112	878,77	7,7530	21993,5	15781,4	218,247
475	748,15	782,40	561,42	25,237	875,57	7,7545	22024,5	15804,1	218,288
476	749,15	783,50	562,23	25,363	872,39	7,7559	22055,5	15826,7	218,329
477	750,15	784,60	563,03	25,489	869,23	7,7574	22086,4	15849,4	218,371
478	751,15	785,70	563,84	25,616	866,07	7,7589	22117,4	15872,0	218,412
479	752,15	786,80	564,64	25,743	862,94	7,7603	22148,4	15894,7	218,453
480	753,15	787,90	565,45	25,871	859,82	7,7618	22179,4	15917,4	218,494
481	754,15	789,00	566,26	26,000	856,71	7,7633	22210,4	15940,1	218,536
482	755,15	790,10	567,06	26,128	853,61	7,7647	22241,4	15962,8	218,577
483	756,15	791,21	567,87	26,258	850,54	7,7662	22272,4	15985,5	218,618
484	757,15	792,31	568,67	26,388	847,47	7,7676	22303,5	16008,2	218,659
485	758,15	793,41	569,48	26,518	844,42	7,7691	22334,5	16030,9	218,700
486	759,15	794,51	570,29	26,649	841,38	7,7705	22365,5	16053,6	218,741
487	760,15	795,62	571,10	26,780	838,36	7,7720	22396,6	16076,4	218,781
488	761,15	796,72	571,90	26,912	835,35	7,7734	22427,6	16099,1	218,822
489	762,15	797,82	572,71	27,044	832,36	7,7749	22458,7	16121,8	218,863
490	763,15	798,92	573,52	27,177	829,38	7,7763	22489,7	16144,6	218,904
491	764,15	800,03	574,33	27,310	826,41	7,7778	22520,8	16167,4	218,944
492	765,15	801,13	575,14	27,444	823,46	7,7792	22551,9	16190,1	218,985
493	766,15	802,24	575,95	27,578	820,52	7,7807	22583,0	16212,9	219,026
494	767,15	803,34	576,76	27,713	817,59	7,7821	22614,1	16235,7	219,066
495	768,15	804,45	577,56	27,849	814,68	7,7835	22645,2	16258,4	219,107
496	769,15	805,55	578,37	27,985	811,78	7,7850	22676,3	16281,2	219,147
497	770,15	806,66	579,18	28,121	808,89	7,7864	22707,4	16304,0	219,188
498	771,15	807,76	579,99	28,258	806,02	7,7879	22738,5	16326,8	219,228
499	772,15	808,87	580,80	28,395	803,15	7,7893	22769,6	16349,7	219,268

Atmospheric nitrogen (*Continued*)

t	T	h	u	π_0	θ_0	s^0	H	U	S
500	773,15	809,97	581,62	28,533	800,31	7,7907	22800,8	16372,5	219,309
501	774,15	811,08	582,43	28,672	797,47	7,7921	22881,9	16395,3	219,379
502	775,15	812,19	583,24	28,811	794,65	7,7936	22863,0	16418,1	219,389
503	776,15	813,29	584,05	28,950	791,84	7,7950	22894,2	16441,0	219,429
504	777,15	814,40	584,86	29,090	789,04	7,7964	22925,4	16463,8	219,469
505	778,15	815,51	585,67	29,231	786,26	7,7978	22956,5	16486,7	219,509
506	779,15	816,61	586,48	29,372	783,49	7,7993	22987,7	16509,5	219,550
507	780,15	817,72	587,30	29,513	780,73	7,8007	23018,9	16532,4	219,589
508	781,15	818,83	588,11	29,656	777,98	7,8021	23050,1	16555,3	219,629
509	782,15	819,94	588,92	29,798	775,25	7,8035	23081,3	16578,1	219,669
510	783,15	821,05	589,73	29,941	772,53	7,8049	23112,5	16601,0	219,709
511	784,15	822,15	590,55	30,085	769,82	7,8064	23143,7	16623,9	219,749
512	785,15	823,26	591,36	30,230	767,12	7,8078	23174,9	16646,8	219,789
513	786,15	824,37	592,17	30,374	764,44	7,8092	23206,1	16669,7	219,829
514	787,15	825,48	592,99	30,520	761,76	7,8106	23237,3	16692,6	219,868
515	788,15	826,59	593,80	30,666	759,10	7,8120	23268,6	16715,6	219,908
516	789,15	827,70	594,62	30,812	756,45	7,8134	23299,8	16738,5	219,948
517	790,15	828,81	595,43	30,959	753,81	7,8148	23331,0	16761,4	219,987
518	791,15	829,92	596,25	31,107	751,19	7,8162	23362,3	16784,4	220,027
519	792,15	831,03	597,06	31,255	748,57	7,8176	23393,6	16807,3	220,066
520	793,15	832,14	597,88	31,403	745,97	7,8190	23424,8	16830,3	220,106
521	794,15	833,25	598,69	31,553	743,38	7,8204	23456,1	16853,2	220,145
522	795,15	834,37	599,51	31,702	740,80	7,8218	23487,4	16876,2	220,184
523	796,15	835,48	600,33	31,853	738,23	7,8232	23518,7	16899,2	220,224
524	797,15	836,59	601,14	32,003	735,68	7,8246	23550,0	16922,1	220,263
525	798,15	837,70	601,96	32,155	733,13	7,8260	23581,3	16945,1	220,302
526	799,15	838,81	602,77	32,307	730,60	7,8274	23612,6	16968,1	220,341
527	800,15	839,93	603,59	32,459	728,07	7,8288	23643,9	16991,1	220,381
528	801,15	841,04	604,41	32,612	725,56	7,8302	23675,2	17014,1	220,420
529	802,15	842,15	605,23	32,766	723,06	7,8316	23706,5	17037,1	220,459
530	803,15	843,26	606,04	32,920	720,57	7,8330	23737,9	17060,2	220,498
531	804,15	844,38	606,86	33,075	718,09	7,8343	23769,2	17083,2	220,537
532	805,15	845,49	607,68	33,230	715,62	7,8357	23800,6	17106,2	220,576
533	806,15	846,60	608,50	33,386	713,16	7,8371	23831,9	17129,3	220,615
534	807,15	847,72	609,32	33,543	710,72	7,8385	23863,3	17152,3	220,654
535	808,15	848,83	610,14	33,700	708,28	7,8399	23894,7	17175,4	220,692
536	809,15	849,95	610,96	33,858	705,86	7,8413	23926,0	17198,3	220,731
537	810,15	851,06	611,78	34,016	703,44	7,8426	23957,4	17221,5	220,770
538	811,15	852,18	612,60	34,175	701,04	7,8440	23988,8	17244,6	220,809
539	812,15	853,29	613,42	34,334	698,64	7,8454	24020,2	17267,7	220,847
540	813,15	854,41	614,24	34,494	696,26	7,8467	24051,6	17290,7	220,886
541	814,15	855,52	615,06	34,654	693,89	7,8481	24083,0	17313,8	220,925
542	815,15	856,64	615,88	34,816	691,52	7,8495	24114,4	17336,9	220,963
543	816,15	857,76	616,70	34,977	689,17	7,8509	24145,9	17360,1	221,002
544	817,15	858,87	617,52	35,140	686,83	7,8522	24177,3	17383,2	221,040
545	818,15	859,99	618,34	35,302	684,50	7,8536	24208,7	17406,3	221,079
546	819,15	861,11	619,16	35,466	682,18	7,8550	24240,2	17429,4	221,117
547	820,15	862,22	619,98	35,630	679,86	7,8563	24271,6	17452,6	221,155
548	821,15	863,34	620,81	35,795	677,56	7,8577	24303,1	17475,7	221,194
549	822,15	864,46	621,63	35,960	675,27	7,8590	24334,5	17498,8	221,232
550	823,15	865,58	622,45	36,126	672,99	7,8604	24366,0	17522,0	221,270
551	824,15	866,70	623,27	36,292	670,72	7,8618	24397,5	17545,2	221,309
552	825,15	867,81	624,10	36,459	668,45	7,8631	24429,0	17568,3	221,347
553	826,15	868,93	624,92	36,627	666,20	7,8645	24460,5	17591,5	221,385
554	827,15	870,05	625,74	36,795	663,96	7,8658	24492,0	17614,7	221,423

t	T	h	u	π_0	θ_0	s^0	H	U	S
555	828,15	871,17	626,57	36,964	661,72	7,8672	24523,5	17637,9	221,461
556	829,15	872,29	627,39	37,133	659,50	7,8685	24555,0	17661,1	221,499
557	830,15	873,41	628,22	37,303	657,28	7,8699	24586,5	17684,3	221,537
558	831,15	874,53	629,04	37,474	655,08	7,8712	24618,0	17707,5	221,575
559	832,15	875,65	629,87	37,645	652,88	7,8726	24649,5	17730,7	221,613
560	833,15	876,77	630,69	37,817	650,70	7,8739	24681,1	17753,9	221,651
561	834,15	877,89	631,52	37,990	648,52	7,8753	24712,6	17777,2	221,689
562	835,15	879,01	632,34	38,163	646,35	7,8766	24744,2	17800,4	221,726
563	836,15	880,13	633,17	38,337	644,19	7,8779	24775,7	17823,6	221,764
564	837,15	881,25	633,99	38,511	642,04	7,8793	24807,3	17846,9	221,802
565	838,15	882,38	634,82	38,686	639,90	7,8806	24838,9	17870,2	221,840
566	839,15	883,50	635,65	38,861	637,77	7,8820	24870,5	17893,4	221,877
567	840,15	884,62	636,47	39,038	635,65	7,8833	24902,0	17916,7	221,915
568	841,15	885,74	637,30	39,214	633,54	7,8846	24933,6	17940,0	221,952
569	842,15	886,86	638,13	39,392	631,43	7,8860	24965,2	17963,2	221,990
570	843,15	887,99	638,95	39,570	629,34	7,8873	24996,8	17986,5	222,027
571	844,15	889,11	639,78	39,749	627,25	7,8886	25028,4	18009,8	222,065
572	845,15	890,23	640,61	39,928	625,17	7,8900	25060,1	18033,1	222,102
573	846,15	891,36	641,44	40,108	623,10	7,8913	25091,7	18056,4	222,140
574	847,15	892,48	642,27	40,289	621,04	7,8926	25123,3	18079,8	222,177
575	848,15	893,60	643,09	40,470	618,99	7,8939	25154,9	18103,1	222,214
576	849,15	894,73	643,92	40,652	616,95	7,8953	25186,6	18126,4	222,252
577	850,15	895,85	644,75	40,834	614,91	7,8966	25218,2	18149,7	222,289
578	851,15	896,98	645,58	41,018	612,89	7,8979	25249,9	18173,1	222,326
579	852,15	898,10	646,41	41,201	610,87	7,8992	25281,6	18196,4	222,363
580	853,15	899,23	647,24	41,386	608,86	7,9006	25313,2	18219,8	222,401
581	854,15	900,35	648,07	41,571	606,86	7,9019	25344,9	18243,2	222,438
582	855,15	901,48	648,90	41,757	604,87	7,9032	25376,6	18266,5	222,475
583	856,15	902,60	649,73	41,943	602,88	7,9045	25408,3	18289,9	222,512
584	857,15	903,73	650,56	42,130	600,91	7,9058	25440,0	18313,3	222,549
585	858,15	904,86	651,39	42,318	598,94	7,9071	25471,7	18336,7	222,586
586	859,15	905,98	652,22	42,506	596,98	7,9084	25503,4	18360,1	222,623
587	860,15	907,11	653,05	42,696	595,03	7,9098	25535,1	18383,5	222,660
588	861,15	908,24	653,89	42,885	593,08	7,9111	25566,8	18406,9	222,696
589	862,15	909,36	654,72	43,076	591,15	7,9124	25598,6	18430,3	222,733
590	863,15	910,49	655,55	43,267	589,22	7,9137	25630,3	18453,6	222,770
591	864,15	911,62	656,38	43,458	587,30	7,9150	25662,0	18477,1	222,807
592	865,15	912,75	658,21	43,651	585,39	7,9163	25693,8	18500,6	222,843
593	866,15	913,87	658,05	43,844	583,49	7,9176	25725,6	18524,0	222,880
594	867,15	915,00	658,88	44,037	581,59	7,9189	25757,3	18547,5	222,917
595	868,15	916,13	659,71	44,232	579,70	7,9202	25789,1	18570,9	222,953
596	869,15	917,26	660,55	44,427	577,82	7,9215	25820,9	18594,4	222,990
597	870,15	918,39	661,38	44,622	575,95	7,9228	25852,6	18617,9	223,027
598	871,15	919,52	662,21	44,819	574,09	7,9241	25884,4	18641,3	223,063
599	872,15	920,65	663,05	45,016	572,23	7,9254	25916,2	18664,8	223,100
600	873,15	921,78	663,88	45,214	570,38	7,9267	25948,0	18688,3	223,136
601	874,15	922,91	664,72	45,412	568,54	7,9280	25979,8	18711,8	223,172
602	875,15	924,04	665,55	45,511	566,70	7,9293	26011,7	18735,3	223,209
603	876,15	925,17	666,39	45,811	564,88	7,9306	26043,5	18758,8	223,245
604	877,15	926,30	667,22	46,011	563,06	7,9318	26075,3	18782,3	223,281
605	878,15	927,43	668,06	46,213	561,25	7,9331	26107,1	18805,8	223,318
606	879,15	928,56	668,89	46,414	559,44	7,9344	26139,0	18829,4	223,354
607	880,15	929,69	669,73	46,617	557,64	7,9357	26170,8	18852,9	223,390
608	881,15	930,82	670,57	46,820	555,85	7,9370	26202,7	18876,4	223,426
609	882,15	931,95	671,40	47,024	554,07	7,9383	26234,5	18900,0	223,462

Atmospheric nitrogen (*Continued*)

t	T	h	u	π_0	θ_0	s^0	H	U	S
610	883,15	933,09	672,24	47,229	552,30	7,9396	26266,4	18923,5	223,499
611	884,15	934,22	673,08	47,434	550,53	7,9408	26298,3	18947,1	223,535
612	885,15	935,35	673,91	47,640	548,77	7,9421	26330,2	18970,7	223,571
613	886,15	936,48	674,75	47,847	547,01	7,9434	26362,0	18994,2	223,607
614	887,15	937,62	675,59	48,054	545,27	7,9447	26393,9	19017,8	223,643
615	888,15	938,75	676,43	48,263	543,53	7,9460	26425,8	19041,4	223,679
616	889,15	939,88	677,26	48,471	541,79	7,9472	26457,7	19065,0	223,714
617	890,15	941,02	678,10	48,681	540,07	7,9485	26489,6	19088,6	223,750
618	891,15	942,15	678,94	48,891	538,35	7,9498	26521,6	19112,2	223,786
619	892,15	943,29	679,78	49,102	536,64	7,9510	26553,5	19135,8	223,822
620	893,15	944,42	680,62	40,314	534,93	7,9523	26585,4	19159,4	223,858
621	894,15	945,55	681,46	49,526	533,23	7,9536	26617,4	19183,0	223,893
622	895,15	946,69	682,30	49,740	531,54	7,9549	26649,3	19206,7	223,929
623	896,15	947,82	683,14	49,953	529,86	7,9561	26681,3	19230,3	223,965
624	897,15	948,96	683,98	50,168	528,18	7,9574	26713,2	19253,9	224,001
625	898,15	950,09	684,82	50,383	526,51	7,9587	26745,2	19277,6	224,036
626	899,15	951,23	685,66	50,599	524,85	7,9599	26777,1	19301,2	224,072
627	900,15	952,37	686,50	50,816	523,19	7,9612	26809,1	19324,9	224,107
628	901,15	953,50	687,34	51,034	521,54	7,9624	26841,1	19348,6	224,143
629	902,15	954,64	688,18	51,252	519,89	7,9637	26873,1	19372,2	224,178
630	903,15	955,78	689,02	51,471	518,26	7,9650	26905,1	19395,9	224,214
631	904,15	956,91	689,86	51,691	516,62	7,9662	26937,1	19419,6	224,249
632	905,15	958,05	690,70	51,911	515,00	7,9675	26969,1	19443,3	224,284
633	906,15	959,19	691,55	52,132	513,38	7,9687	27001,1	19467,0	224,320
634	907,15	960,32	692,39	52,354	511,77	7,9700	27033,1	19490,7	224,355
635	908,15	961,46	693,23	52,577	510,16	7,9712	27065,1	19514,4	224,390
636	909,15	962,60	694,07	52,800	508,56	7,9725	27097,2	19538,1	224,426
637	910,15	963,74	694,92	53,024	506,97	7,9737	27129,2	19561,9	224,461
638	911,15	964,88	695,76	53,249	505,38	7,9750	27161,3	19585,6	224,496
639	912,15	966,01	696,60	53,475	503,80	7,9762	27193,3	19609,3	224,531
640	913,15	967,15	697,44	53,701	502,23	7,9775	27225,4	19633,1	224,566
641	914,15	968,29	698,29	53,928	500,66	7,9787	27257,4	19656,8	224,601
642	915,15	969,43	699,13	54,156	499,10	7,9800	27289,5	19680,6	224,637
643	916,15	970,57	699,98	54,385	497,55	7,9812	27321,6	19704,3	224,672
644	917,15	971,71	700,82	54,614	496,00	7,9825	27353,7	19728,1	224,707
645	918,15	972,85	701,67	54,845	494,45	7,9837	27385,8	19751,9	224,742
646	919,15	973,99	702,51	55,075	492,92	7,9850	27417,9	19775,7	224,776
647	920,15	975,13	703,36	55,307	491,39	7,9862	27450,0	19799,5	224,811
648	921,15	976,27	704,20	55,540	489,86	7,9874	27482,1	19823,2	224,846
649	922,15	977,41	705,05	55,773	488,34	7,9887	27514,2	19847,0	224,881
650	923,15	978,55	705,89	56,007	486,83	7,9899	27546,3	19870,9	224,916
651	924,15	979,70	706,74	56,242	485,32	7,9911	27578,4	19894,7	224,951
652	925,15	980,84	707,58	56,477	483,82	7,9924	27610,6	19918,5	224,985
653	926,15	981,98	708,43	56,714	482,32	7,9936	27642,7	19942,3	225,020
654	927,15	983,12	709,28	56,951	480,83	7,9948	27674,8	19966,1	225,055
655	928,15	984,26	710,12	57,189	479,35	7,9961	27707,0	19990,0	225,089
656	929,15	985,41	710,97	57,427	477,87	7,9973	27739,2	20013,8	225,124
657	930,15	986,55	711,82	57,667	476,40	7,9985	27771,3	20037,7	225,159
658	931,15	987,69	712,67	57,907	474,93	7,9998	27803,5	20061,5	225,193
659	932,15	988,83	713,51	58,148	473,47	8,0010	27835,7	20085,4	225,228
660	933,15	989,98	714,36	58,390	472,02	8,0022	27867,9	20109,3	225,262
661	934,15	991,12	715,21	58,633	470,57	8,0034	27900,0	20133,1	225,297
662	935,15	992,26	716,06	58,876	469,12	8,0047	27932,2	20157,0	225,331
663	936,15	993,41	716,91	59,120	467,69	8,0059	27964,4	20180,9	225,366
664	937,15	994,55	717,75	59,365	466,25	8,0071	27996,6	20204,8	225,400

t	T	h	u	π_0	θ_0	s^0	H	U	S
665	938,15	995,70	718,60	59,611	464,83	8,0083	28028,9	20228,7	225,434
666	939,15	996,84	719,45	59,858	463,40	8,0095	28061,1	20252,6	225,469
667	940,15	997,99	720,30	60,105	461,99	8,0108	28093,3	20276,5	225,503
668	941,15	999,13	721,15	60,353	460,58	8,0120	28125,5	20300,4	225,537
669	942,15	1000,28	722,00	60,602	459,17	8,0132	28157,8	20324,4	225,572
670	943,15	1001,42	722,85	60,852	457,77	8,0144	28190,0	20348,3	225,606
671	944,15	1002,57	723,70	61,103	456,38	8,0156	28222,3	20372,2	225,640
672	945,15	1003,71	724,55	61,354	454,99	8,0168	28254,5	20396,2	225,674
673	946,15	1004,86	725,40	61,607	453,61	8,0181	28286,8	20420,1	225,708
674	947,15	1006,01	726,25	61,860	452,23	8,0193	28319,1	20444,1	225,742
675	948,15	1007,15	727,11	62,114	450,85	8,0205	28351,3	20468,0	225,776
676	949,15	1008,30	727,96	62,368	449,49	8,0217	28383,6	20492,0	225,810
677	950,15	1009,45	728,81	62,624	448,12	8,0229	28415,9	20516,0	225,844
678	951,15	1010,59	729,66	62,880	446,77	8,0241	28448,2	20539,9	225,878
679	952,15	1011,74	730,51	63,137	445,41	8,0253	28480,5	20563,9	225,912
680	953,15	1012,89	731,36	63,395	444,07	8,0265	28512,8	20587,9	225,946
681	954,15	1014,04	732,22	63,654	442,72	8,0277	28545,1	20611,9	225,980
682	955,15	1015,18	733,07	63,914	441,39	8,0289	28577,4	20635,9	226,014
683	956,15	1016,33	733,92	64,175	440,06	8,0301	28609,7	20659,9	226,048
684	957,15	1017,48	734,78	64,436	438,73	8,0313	28642,1	20683,9	226,082
685	958,15	1018,63	735,63	64,698	437,41	8,0325	28674,4	20708,0	226,115
686	959,15	1019,78	736,48	64,961	436,09	8,0337	28706,8	20732,0	226,149
687	960,15	1020,93	737,34	65,225	434,78	8,0349	28739,1	20756,0	226,183
688	961,15	1022,08	738,19	65,490	433,47	8,0361	28771,5	20780,1	226,216
689	962,15	1023,23	739,04	65,755	432,17	8,0373	28803,8	20804,1	226,250
690	963,15	1024,38	739,90	66,022	430,87	8,0385	28836,2	20828,2	226,284
691	964,15	1025,53	740,75	66,289	429,58	8,0397	28868,5	20852,2	226,317
692	965,15	1026,68	741,61	66,557	428,30	8,0409	28900,9	20876,3	226,351
693	966,15	1027,83	742,46	66,826	427,01	8,0421	28933,3	20900,3	226,384
694	967,15	1028,98	743,32	67,096	425,74	8,0433	28965,7	20924,4	226,418
695	968,15	1030,13	744,17	67,367	424,46	8,0445	28998,1	20948,5	226,451
696	969,15	1031,28	745,03	67,638	423,20	8,0456	29030,5	20972,6	226,485
697	970,15	1032,43	745,89	67,911	421,93	8,0468	29062,9	20996,7	226,518
698	971,15	1033,58	746,74	68,184	420,68	8,0480	29095,3	21020,8	226,552
699	972,15	1034,73	747,60	68,458	419,42	8,0492	29127,7	21044,9	226,585
700	973,15	1035,88	748,45	68,733	418,17	8,0504	29160,2	21069,0	226,618
701	974,15	1037,04	749,31	69,009	416,93	8,0516	29192,6	21093,1	226,652
702	975,15	1038,19	750,17	69,286	415,69	8,0528	29225,0	21117,2	226,685
703	976,15	1039,34	751,03	69,564	414,46	8,0539	29257,5	21141,4	226,718
704	977,15	1040,49	751,88	69,842	413,23	8,0551	29289,9	21165,5	226,751
705	978,15	1041,65	752,74	70,122	412,00	8,0563	29322,4	21189,6	226,785
706	979,15	1042,80	753,60	70,402	410,78	8,0575	29354,8	21213,8	226,818
707	980,15	1043,95	754,46	70,683	409,56	8,0586	29387,3	21237,9	226,851
708	981,15	1045,11	755,31	70,965	408,35	8,0598	29419,8	21262,1	226,884
709	982,15	1046,26	756,17	71,248	407,15	8,0610	29452,2	21286,2	226,917
710	983,15	1047,41	757,03	71,532	405,94	8,0622	29484,7	21310,4	226,950
711	984,15	1048,57	757,89	71,817	404,74	8,0633	29517,2	21334,6	226,983
712	985,15	1049,72	758,75	72,102	403,55	8,0645	29549,7	21358,8	227,016
713	986,15	1050,88	759,61	72,389	402,36	8,0657	29582,2	21383,0	227,049
714	987,15	1052,03	760,47	72,676	401,18	8,0669	29614,7	21407,1	227,082
715	988,15	1053,19	761,33	72,965	400,00	8,0680	29647,2	21431,3	227,115
716	989,15	1054,34	762,19	73,254	398,82	8,0692	29679,7	21455,6	227,148
717	990,15	1055,50	763,05	73,544	397,65	8,0704	29712,3	21479,8	227,181
718	991,15	1056,65	763,91	73,835	396,48	8,0715	29744,8	21504,0	227,214
719	992,15	1057,81	764,77	74,127	395,32	8,0727	29777,3	21528,2	227,246

Atmospheric nitrogen (*Continued*)

t	T	h	u	π_0	θ_0	s^0	H	U	S
720	993,15	1058,97	765,63	74,420	394,16	8,0739	29809,9	21552,4	227,279
721	994,15	1060,12	766,49	74,713	393,00	8,0750	29842,4	21576,7	227,312
722	995,15	1061,28	767,35	75,008	391,85	8,0762	29875,0	21600,9	227,345
723	996,15	1062,43	768,21	75,304	390,71	8,0773	29907,5	21625,1	227,377
724	997,15	1063,59	769,07	75,600	389,57	8,0785	29940,1	21649,4	227,410
725	998,15	1064,75	769,93	75,898	388,43	8,0797	29972,7	21673,6	227,443
726	999,15	1065,91	770,80	76,196	387,30	8,0808	30005,3	21697,9	227,475
727	1000,15	1067,06	771,66	76,495	386,17	8,0820	30037,8	21722,2	227,508
728	1001,15	1068,22	772,52	76,796	385,04	8,0831	30070,4	21746,4	227,540
729	1002,15	1069,38	773,38	77,097	383,92	8,0843	30103,0	21770,7	227,573
730	1003,15	1070,54	774,25	77,399	382,80	8,0855	30135,6	21795,0	227,606
731	1004,15	1071,70	775,11	77,702	381,69	8,0866	30168,2	21819,3	227,638
732	1005,15	1072,85	775,97	78,006	380,58	8,0878	30200,8	21843,6	227,670
733	1006,15	1074,01	776,83	78,311	379,48	8,0889	30233,4	21867,9	227,703
734	1007,15	1075,17	777,70	78,616	378,38	8,0901	30266,1	21892,2	227,735
735	1008,15	1076,33	778,56	78,923	377,28	8,0912	30298,7	21916,5	227,768
736	1009,15	1077,49	779,43	79,231	376,19	8,0924	30331,3	21940,8	227,800
737	1010,15	1078,65	780,29	79,540	375,10	8,0935	30364,0	21965,2	227,832
738	1011,15	1079,81	781,15	79,849	374,02	8,0947	30396,6	21989,5	227,865
739	1012,15	1080,97	782,02	80,160	372,94	8,0958	30429,3	22013,8	227,897
740	1013,15	1082,13	782,88	80,471	371,86	8,0970	30461,9	22038,2	227,929
741	1014,15	1083,29	783,75	80,784	370,79	8,0981	30494,6	22062,5	227,961
742	1015,15	1084,45	784,61	81,097	369,72	8,0992	30527,2	22086,9	227,994
743	1016,15	1085,61	785,48	81,411	368,65	8,1004	30559,9	22111,2	228,026
744	1017,15	1086,77	786,34	81,727	367,59	8,1015	30592,6	22135,6	228,058
745	1018,15	1087,93	787,21	82,043	366,53	8,1027	30625,3	22160,0	228,090
746	1019,15	1089,09	788,08	82,360	365,48	8,1038	30658,0	22184,3	228,122
747	1020,15	1090,25	788,94	82,679	364,43	8,1049	30690,7	22208,7	228,154
748	1021,15	1091,42	789,81	82,998	363,39	8,1061	30723,4	22233,1	228,186
749	1022,15	1092,58	790,67	83,318	362,34	8,1072	30756,1	22257,5	228,218
750	1023,15	1093,74	791,54	83,639	361,31	8,1084	30788,8	22281,9	228,250
751	1024,15	1094,90	792,41	83,961	360,27	8,1095	30821,5	22306,3	228,282
752	1025,15	1096,06	793,28	84,284	359,24	8,1106	30854,2	22330,7	228,314
753	1026,15	1097,23	794,14	84,609	358,21	8,1118	30886,9	22355,1	228,346
754	1027,15	1098,39	795,01	84,934	357,19	8,1129	30919,7	22379,5	228,378
755	1028,15	1099,55	795,88	85,260	356,17	8,1140	30952,4	22404,0	228,410
756	1029,15	1100,72	796,75	85,587	355,15	8,1152	30985,2	22428,4	228,442
757	1030,15	1101,88	797,61	85,915	354,14	8,1163	31017,9	22452,8	228,473
758	1031,15	1103,04	798,48	86,244	353,13	8,1174	31050,7	22477,3	228,505
759	1032,15	1104,21	799,35	86,574	352,13	8,1185	31083,4	22501,7	228,537
760	1033,15	1105,37	800,22	86,905	351,13	8,1197	31116,2	22526,2	228,569
761	1034,15	1106,54	801,09	87,237	350,13	8,1208	31149,0	22550,6	228,600
762	1035,15	1107,70	801,96	87,570	349,13	8,1219	31181,8	22575,1	228,632
763	1036,15	1108,86	802,83	87,904	348,14	8,1230	31214,5	22599,6	228,664
764	1037,15	1110,03	803,70	88,239	347,16	8,1242	31247,3	22624,0	228,695
765	1038,15	1111,19	804,57	88,575	346,17	8,1253	31280,1	22648,5	228,727
766	1039,15	1112,36	805,44	88,912	345,19	8,1264	31312,9	22673,0	228,759
767	1040,15	1113,52	806,31	89,250	344,22	8,1275	31345,7	22697,5	228,790
768	1041,15	1114,69	807,18	89,589	343,24	8,1287	31378,5	22722,0	228,822
769	1042,15	1115,86	808,05	89,929	342,27	8,1298	31411,3	22746,5	228,853
770	1043,15	1117,02	808,92	90,271	341,31	8,1309	31444,2	22771,0	228,885
771	1044,15	1118,19	809,79	90,613	340,34	8,1320	31477,0	22795,5	228,916
772	1045,15	1119,35	810,66	90,956	339,38	8,1331	31509,8	22820,0	228,948
773	1046,15	1120,52	811,53	91,300	338,43	8,1342	31542,7	22844,5	228,979
774	1047,15	1121,69	812,40	91,645	337,48	8,1354	31575,5	22869,1	229,010

t	T	h	u	π_0	θ_0	s^0	H	U	S
775	1048,15	1122,85	813,27	91,992	336,53	8,1365	31608,4	22893,6	229,042
776	1049,15	1124,02	814,14	92,339	335,58	8,1376	31641,2	22918,2	229,073
777	1050,15	1125,19	815,02	92,687	334,64	8,1387	31674,1	22942,7	229,104
778	1051,15	1126,36	815,89	93,037	333,70	8,1398	31706,9	22967,2	229,136
779	1052,15	1127,52	816,76	93,387	332,76	8,1409	31739,8	22991,8	229,167
780	1053,15	1128,69	817,63	93,738	331,83	8,1420	31772,7	23016,4	229,198
781	1054,15	1129,86	818,51	94,091	330,90	8,1431	31805,6	23040,9	229,229
782	1055,15	1131,03	819,38	94,444	329,98	8,1442	31838,4	23065,5	229,260
783	1056,15	1132,20	820,25	94,799	329,05	8,1453	31871,3	23090,1	229,292
784	1057,15	1133,37	821,12	95,155	328,13	8,1465	31904,2	23114,7	229,323
785	1058,15	1134,53	822,00	95,511	327,22	8,1476	31937,1	23139,2	229,354
786	1059,15	1135,70	822,87	95,869	326,30	8,1487	31970,0	23163,8	229,385
787	1060,15	1136,87	823,75	96,228	325,39	8,1498	32003,0	23188,4	229,416
788	1061,15	1138,04	824,62	96,588	324,49	8,1509	32035,9	23213,0	229,447
789	1062,15	1139,21	825,49	96,949	323,58	8,1520	32068,8	23237,7	229,478
790	1063,15	1140,38	826,37	97,311	322,68	8,1531	32101,7	23262,3	229,509
791	1064,15	1141,55	827,24	97,674	321,79	8,1542	32134,7	23286,9	229,540
792	1065,15	1142,72	828,12	98,038	320,89	8,1553	32167,6	23311,5	229,571
793	1066,15	1143,89	828,99	98,403	320,00	8,1564	32200,5	23336,1	229,602
794	1067,15	1145,06	829,87	98.769	319,12	8,1575	32233,5	23360,8	229,633
795	1068,15	1146,23	830,74	99,137	318,23	8,1586	32266,5	23385,4	229,664
796	1069,15	1147,40	831,62	99,505	317,35	8,1597	32299,4	23410,1	229,694
797	1070,15	1148,57	832,49	99,875	316,47	8,1608	32332,4	23434,7	229,725
798	1071,15	1149,75	833,37	100,24	315,60	8,1618	32365,3	23459,4	229,756
799	1072,15	1150,92	834,25	100,61	314,72	8,1629	32398,3	23484,0	229,787
800	1073,15	1152,09	835,12	100,99	313,85	8,1640	32431,3	23508,7	229,818
801	1074,15	1153,26	836,00	101,36	312,99	8,1651	32464,3	23533,4	229,848
802	1075,15	1154,43	836,88	101,73	312,12	8,1662	32497,3	23558,0	229,879
803	1076,15	1155,60	837,75	102,11	311,26	8,1673	32530,3	23582,7	229,910
804	1077,15	1156,78	838,63	102,49	310,41	8,1684	32563,3	23607,4	229,940
805	1078,15	1157,95	839,51	102,87	309,55	8,1695	32596,3	23632,1	229,971
806	1079,15	1159,12	840,38	103,24	308,70	8,1706	32629,3	23656,8	230,002
807	1080,15	1160,30	841,26	103,63	307,85	8,1717	32662,3	23681,5	230,032
808	1081,15	1161,47	842,14	104,01	307,01	8,1727	32695,3	23706,2	230,063
809	1082,15	1162,64	843,02	104,39	306,16	8,1738	32728,4	23730,9	230,093
810	1083,15	1163,82	843,90	104,77	305,33	8,1749	32761,4	23755,6	230,124
811	1084,15	1164,99	844,77	105,16	304,49	8,1760	32794,4	23780,4	230,154
812	1085,15	1166,16	845,65	105,54	303,65	8,1771	32827,5	23805,1	230,185
813	1086,15	1167,34	846,53	105,93	302,82	8,1782	32860,5	23829,8	230,215
814	1087,15	1168,51	847,41	106,32	302,00	8,1792	32893,6	23854,6	230,246
815	1088,15	1169,69	848,29	106,71	301,17	8,1803	32926,6	23879,3	230,276
816	1089,15	1170,86	849,17	107,10	300,35	8,1814	32959,7	23904,1	230,306
817	1090,15	1172,03	850,05	107,49	299,53	8,1825	32992,8	23928,8	230,337
818	1091,15	1173,21	850,93	107,88	298,71	8,1836	33025,8	23953,6	230,367
819	1092,15	1174,38	851,81	108,28	297,90	8,1846	33058,9	23978,3	230,397
820	1093,15	1175,56	852,69	108,67	297,09	8,1857	33092,0	24003,1	230,428
821	1094,15	1176,74	853,57	109,07	296,28	8,1868	33125,1	24027,9	230,458
822	1095,15	1177,91	854,45	109,47	295,47	8,1879	33158,2	24052,7	230,488
823	1096,15	1179,09	855,33	109,87	294,67	8,1889	33191,3	24077,5	230,518
824	1097,15	1180,26	856,41	110,26	293,87	8,1900	33224,4	24102,2	230,548
825	1098,15	1181,44	857,09	110,67	293,07	8,1911	33257,5	24127,0	230,579
826	1099,15	1182,62	857,97	111,07	292,28	8,1921	33290,6	24151,8	230,609
827	1100,15	1183,79	858,85	111,47	291,49	8,1932	33323,7	24176,6	230,639
828	1101,15	1184,97	859,73	111,87	290,70	8,1943	33356,9	24201,5	230,669
829	1102,15	1186,15	860,61	112,28	289,91	8,1953	33390,0	24226,3	230,699

t	T	h	u	π_0	θ_0	s^0	H	U	S
830	1103,15	1187,32	861,50	112,69	289,13	8,1964	33423,1	24251,1	230,729
831	1104,15	1188,50	862,38	113,09	288,35	8,1975	33456,3	24275,9	230,759
832	1105,15	1189,68	863,26	113,50	287,57	8,1985	33489,4	24300,7	230,789
833	1106,15	1190,85	864,14	113,91	286,79	8,1996	33522,6	24325,6	230,819
834	1107,15	1192,03	865,02	114,32	286,02	8,2007	33555,7	24350,4	230,849
835	1108,15	1193,21	865,91	114,74	285,25	8,2017	33588,9	24375,3	230,879
836	1109,15	1194,39	866,79	115,15	284,48	8,2028	33622,0	24400,1	230,909
837	1110,15	1195,57	867,67	115,57	283,71	8,2039	33655,2	24425,0	230,939
838	1111,15	1196,75	868,56	115,98	282,95	8,2049	33688,4	24449,8	230,969
839	1112,15	1197,92	869,44	116,40	282,19	8,2060	33721,6	24474,7	230,999
840	1113,15	1199,10	870,32	116,82	281,43	8,2070	33754,7	24499,6	231,028
841	1114,15	1200,28	871,21	117,24	280,68	8,2081	33787,9	24524,4	231,058
842	1115,15	1201,46	872,09	117,66	279,93	8,2092	33821,1	24549,3	231,088
843	1116,15	1202,64	872,97	118,08	279,18	8,2102	33854,3	24574,2	231,118
844	1117,15	1203,82	873,86	118,50	278,43	8,2113	33887,5	24599,1	231,147
845	1118,15	1205,00	874,74	118,93	277,68	8,2123	33920,7	24624,0	231,177
846	1119,15	1206,18	875,63	119,35	276,94	8,2134	33954,0	24648,9	231,207
847	1120,15	1207,36	876,51	119,78	276,20	8,2144	33987,2	24673,8	231,236
848	1121,15	1208,54	877,40	120,21	275,46	8,2155	34020,4	24698,7	231,266
849	1122,15	1209,72	878,28	120,64	274,73	8,2165	34053,6	24723,6	231,296
850	1123,15	1210,90	879,17	121,07	274,00	8,2176	34086,9	24748,5	231,325
851	1124,15	1212,08	880,05	121,50	273,27	8,2186	34120,1	24773,5	231,355
852	1125,15	1213,26	880,94	121,93	272,54	8,2197	34153,3	24798,4	231,385
853	1126,15	1214,44	881,82	122,36	271,81	8,2207	34186,6	24823,3	231,414
854	1127,15	1215,63	882,71	122,80	271,09	8,2218	34219,8	24848,3	231,444
855	1128,15	1216,81	883,60	123,24	270,37	8,2228	34253,1	24873,2	231,473
856	1129,15	1217,99	884,48	123,67	269,65	8,2239	34286,4	24898,1	231,503
857	1130,15	1219,17	885,37	124,11	268,94	8,2249	34319,6	24923,1	231,532
858	1131,15	1220,35	886,25	124,55	268,22	8,2260	34352,9	24948,1	231,561
859	1132,15	1221,53	887,14	124,99	267,51	8,2270	34386,2	24973,0	231,591
860	1133,15	1222,72	888,03	125,44	266,81	8,2281	34419,5	24998,0	231,620
861	1134,15	1223,90	888,92	125,88	266,10	8,2291	34452,7	25023,0	231,650
862	1135,15	1225,08	889,80	126,32	265,40	8,2302	34486,0	25047,9	231,679
863	1136,15	1226,26	890,69	126,77	264,69	8,2312	34519,3	25072,9	231,708
864	1137,15	1227,45	891,58	127,22	264,00	8,2322	34552,6	25097,9	231,738
865	1138,15	1228,63	892,47	127,67	263,30	8,2333	34585,9	25122,9	231,767
866	1139,15	1229,81	893,35	128,12	262,60	8,2343	34619,3	25147,9	231,796
867	1140,15	1231,00	894,24	128,57	261,91	8,2354	34652,6	25172,9	231,825
868	1141,15	1232,18	895,13	129,02	261,22	8,2364	34685,9	25197,9	231,854
869	1142,15	1233,36	896,02	129,47	260,54	8,2374	34719,2	25222,9	231,884
870	1143,15	1234,55	896,91	129,93	259,85	8,2385	34752,5	25247,9	231,913
871	1144,15	1235,73	897,80	130,39	259,17	8,2395	34785,9	25272,9	231,942
872	1145,15	1236,92	898,68	130,84	258,49	8,2405	34819,2	25298,0	231,971
873	1146,15	1238,10	899,57	131,30	257,81	8,2416	34852,6	25323,0	232,000
874	1147,15	1239,29	900,46	131,76	257,13	8,2426	34885,9	25348,0	232,029
875	1148,15	1240,47	901,35	132,22	256,46	8,2436	34919,3	25373,1	232,058
876	1149,15	1241,66	902,24	132,69	255,79	8,2447	34952,6	25398,1	232,087
877	1150,15	1242,84	903,13	133,15	255,12	8,2457	34986,0	25423,2	232,116
878	1151,15	1244,03	904,02	133,62	254,45	8,2467	35019,4	25448,2	232,145
879	1152,15	1245,21	904,91	134,08	253,79	8,2478	35052,7	25473,3	232,174
880	1153,15	1246,40	905,80	134,55	253,12	8,2488	35086,1	25498,3	232,203
881	1154,15	1247,58	906,69	135,02	252,46	8,2498	35119,5	25523,4	232,232
882	1155,15	1248,77	907,58	135,49	251,80	8,2508	35152,9	25548,5	232,261
883	1156,15	1249,96	908,47	135,96	251,15	8,2519	35186,3	25573,6	232,290
884	1157,15	1251,14	909,37	136,43	250,49	8,2529	35219,7	25598,6	332,319

Atmospheric nitrogen (*Continued*)

t	T	h	u	π_0	θ_0	s^0	H	U	S
885	1158,15	1252,33	910,26	136,91	249,84	8,2539	35253,1	25623,7	232,348
886	1159,15	1253,52	911,15	137,38	249,19	8,2549	35286,5	25648,8	232,377
887	1160,15	1254,70	912,04	137,86	248,54	8,2560	35319,9	25673,9	232,405
888	1161,15	1255,89	912,93	138,34	247,90	8,2570	35353,3	25699,0	232,434
889	1162,15	1257,08	913,82	138,82	247,25	8,2580	35386,7	25724,1	232,463
890	1163,15	1258,26	914,71	139,30	246,61	8,2590	35420,1	25749,2	232,492
891	1164,15	1259,45	915,61	139,78	245,97	8,2601	35453,6	25774,3	232,521
892	1165,15	1260,64	916,50	140,26	245,34	8,2611	35487,0	25799,5	232,549
893	1166,15	1261,83	917,39	140,75	244,70	8,2621	35520,4	25824,6	232.578
894	1167,15	1263,01	918,28	141,24	244,07	8,2631	35553,9	25849,7	232,607
895	1168,15	1264,20	919,18	141,72	243,44	8,2641	35587,3	25874,8	232,635
896	1169,15	1265,39	920,07	142,21	242,81	8,2651	35620,8	25900,0	232,664
897	1170,15	1266,58	920,96	142,70	242,18	8,2662	35654,2	25925,1	232,692
898	1171,15	1267,77	921,86	143,19	241,56	8,2672	35687,7	25950,3	232,721
899	1172,15	1268,96	922,75	143,69	240,93	8,2682	35721,1	25975,4	232,750
900	1173,15	1270,15	923,64	144,18	240,31	8,2692	35754,6	26000,6	232,778
901	1174,15	1271,34	924,54	144,68	239,69	8,2702	35788,1	26025,7	232,807
902	1175,15	1272,52	925,43	145,17	239,03	8,2712	35821,6	26050,9	232,835
903	1176,15	1273,71	926,33	145,67	238,46	8,2722	35855,1	26076,1	232,864
904	1177,15	1274,90	927,22	146,17	237,85	8,2733	35888,5	26101,2	232,892
905	1178,15	1276,09	928,11	146,67	237,24	8,2743	35922,0	26126,4	232,921
906	1179,15	1277,28	929,01	147,17	236,63	8,2753	35955,5	26151,6	232,949
907	1180,15	1278,47	929,90	147,68	236,02	8,2763	35989,0	26176,8	232,977
908	1181,15	1279,66	930,80	148,18	235,42	8,2773	36022,5	26202,0	233,006
909	1182,15	1280,85	931,69	148,69	234,81	8.2783	36056,0	26227,2	233,034
910	1183,15	1282,04	932,59	149,20	234,21	8,2793	36089,6	26252,4	233,062
911	1184,15	1283,24	933,48	149,70	233,61	8,2803	36123,1	26277,6	233,091
912	1185,15	1284,43	934,38	150,22	233,02	8,2813	36156,6	26302,8	233,119
913	1186,15	1285,62	935,28	150,73	232,42	8,2823	36190,1	26328,0	233,147
914	1187,15	1286,81	936,17	151,24	231,83	8,2833	36223,7	26353,2	233,176
915	1188,15	1288,00	937,07	151,75	231,24	8,2843	36257,2	26378,4	233,204
916	1189,15	1289,19	937,96	152,27	230,65	8,2853	36290,7	26403,7	233,232
917	1190,15	1290,38	938,86	152,79	230,06	8,2863	36324,3	26428,9	233,260
918	1191,15	1291,58	939,76	153,31	229,47	8,2873	36357,8	26454,1	233,288
919	1192,15	1292,77	940,65	153,83	228,89	8,2883	36391,4	26479,4	233,317
920	1193,15	1293,96	941,55	154,35	228,31	8,2893	36425,0	26504,6	233,345
921	1194,15	1295,15	942,45	154,87	227,73	8,2903	36458,5	26529,9	233,373
922	1195,15	1296,34	943,34	155,40	227,15	8,2913	36492,1	26555,1	233,401
923	1196,15	1297,54	944,24	155,92	226,57	8,2923	36525,7	26580,4	233,429
924	1197,15	1298,73	945,14	156,45	226,00	8,2933	36559,2	26605,6	233,457
925	1198,15	1299,92	946,04	156,98	225,43	8,2943	36592,8	26630,9	233,485
926	1199,15	1301,12	946,93	157,51	224,86	8,2953	36626,4	26656,2	233,513
927	1200,15	1302,31	947,83	158,04	224,29	8,2963	36660,0	26681,5	233,541
928	1201,15	1303,50	948,73	158,57	223,72	8,2973	36693,6	26706,7	233,569
929	1202,15	1304,70	949,63	159,11	223,15	8,2983	36727,2	26732,0	233,597
930	1203,15	1305,89	950,53	159,64	222,59	8,2993	36760,8	26757,3	233,625
931	1204,15	1307,08	951,42	160,18	222,03	8,3003	36794,4	26782,6	233,653
932	1205,15	1308,28	952,32	160,72	221,47	8,3013	36828,0	26807,9	233,681
933	1206,15	1309,47	953,22	161,26	220,91	8,3023	36861,6	26833,2	233,709
934	1207,15	1310,67	954,12	161,80	220,35	8,3033	36895,3	26858,5	233,737
935	1208,15	1311,86	955,02	162,34	219,80	8,3042	36928,9	26883,8	233,764
936	1209,15	1313,06	955,92	162,88	219,25	8,3052	36962,5	26909,1	233,792
937	1210,15	1314,25	956,82	163,43	218,69	8,3062	36996,2	26934,5	233,820
938	1211,15	1315,45	957,72	163,98	218,14	8,3072	37029,8	26959,8	233,848
939	1212,15	1316,64	958,62	164,53	217,60	8,3082	37063,4	26985,1	233,876

Atmospheric nitrogen (*Continued*)

t	T	h	u	π_0	θ_0	s^0	H	U	S
940	1213,15	1317,84	959,52	165,08	217,05	8,3092	37097,1	27010,5	233,903
941	1214,15	1319,03	960,42	165,63	216,51	8,3102	37130,7	27035,8	233,931
942	1215,15	1320,23	961,32	166,18	215,96	8,3111	37164,4	27061,1	233,959
943	1216,15	1321,42	962,22	166,73	215,42	8,3121	37198,1	27086,5	233,986
944	1217,15	1322,62	963,12	167,29	214,88	8,3131	37231,7	27111,8	234,014
945	1218,15	1323,81	964,02	167,85	214,35	8,3141	37265,4	27137,2	234,042
946	1219,15	1325,01	964,92	168,41	213,81	8,3151	37299,1	27162,5	234,069
947	1220,15	1326,21	965,82	168,97	213,28	8,3161	37332,7	27187,9	234,097
948	1221,15	1327,40	966,72	169,53	212,74	8,3170	37366,4	27213,3	234,125
949	1222,15	1328,60	967,63	170,09	212,21	8,3180	37400,1	27238,6	234,152
950	1223,15	1329,80	968,53	170,66	211,68	8,3190	37433,8	27264,0	234,180
951	1224,15	1330,99	969,43	171,22	211,16	8,3200	37467,5	27289,4	234,207
952	1225,15	1332,19	970,33	171,79	210,63	8,3210	37501,2	27314,8	234,235
953	1226,15	1333,39	971,23	172,36	210,11	8,3219	37534,9	27340,2	234,262
954	1227,15	1334,59	972,13	172,93	209,58	8,3229	37568,6	27365,6	234,290
955	1228,15	1335,78	973,04	173,50	209,06	8,3239	37602,3	27391,0	234,317
956	1229,15	1336,98	973,94	174,08	208,54	8,3249	37636,0	27416,4	234,345
957	1230,15	1338,18	973,84	174,65	208,03	8,3258	37669,8	27441,8	234,372
958	1231,15	1339,38	974,74	175,23	207,51	8,3268	37703,5	27467,2	234,400
959	1232,15	1340,58	976,65	175,81	207,00	8,3278	37737,2	27492,6	234,427
960	1233,15	1341,77	977,55	176,38	206,48	8,3287	37770,9	27518,0	234,454
961	1234,15	1342,97	978,45	176,97	205,97	8,3297	37804,7	27543,4	234,482
962	1235,15	1344,17	979,36	177,55	205,46	8,3307	37838,4	27568,9	234,509
963	1236,15	1345,37	980,26	178,13	204,95	8,3317	37872,2	27594,3	234,536
964	1237,15	1346,57	981,16	178,72	204,45	8,3326	37905,9	27619,7	234,564
965	1238,15	1347,77	982,07	179,31	203,94	8,3336	37939,7	27645,2	234,591
966	1239,15	1348,97	982,97	179,89	203,44	8,3346	37973,4	27670,6	234,618
967	1240,15	1350,17	983,87	180,48	202,94	8,3355	38007,2	27696,1	234,645
968	1241,15	1351,37	984,78	181,08	202,44	8,3365	38041,0	27721,5	234,673
969	1242,15	1352,57	985,68	181,67	201,94	8,3375	38074,7	27747,0	234,700
970	1243,15	1353,77	986,59	182,26	201,44	8,3384	38108,5	27772,5	234,727
971	1244,15	1354,97	987,49	182,86	200,95	8,3394	38142,3	27797,9	234,754
972	1245,15	1356,17	988,40	183,46	200,45	8,3404	38176,1	27823,4	234,781
973	1246,15	1357,37	989,30	184,06	199,96	8,3413	38209,9	27848,9	234,808
974	1247,15	1358,57	990,21	184,66	199,47	8,3423	38243,7	27874,3	234,835
975	1248,15	1359,77	991,11	185,26	198,98	8,3433	38277,5	27899,8	234,863
976	1249,15	1360,97	992,02	185,87	198,49	8,3442	38311,3	27925,3	234,890
977	1250,15	1362,17	992,92	186,47	198,01	8,3452	38345,1	27950,8	234,917
978	1251,15	1363,37	993,83	187,08	197,52	8,3461	38378,9	27976,3	234,944
979	1252,15	1364,57	994,73	187,69	197,04	8,3471	39412,7	28001,8	234,971
980	1253,15	1365,77	995,64	188,30	196,56	8,3481	38446,5	28027,3	234,998
981	1254,15	1366,97	996,55	188,91	196,08	8,3490	38480,3	28052,8	235,025
982	1255,15	1368,18	997,45	189,52	195,60	8,3500	38514,1	28078,3	235,052
983	1256,15	1369,38	998,36	190,14	195,12	8,3509	38548,0	28103,8	235,079
984	1257,15	1370,58	999,27	190,76	194,64	8,3519	38581,8	28129,3	235,106
985	1258,15	1371,78	1000,17	191,37	194,17	8,3528	38615,6	28154,9	235,132
986	1259,15	1372,98	1001,08	191,99	193,70	8,3538	38649,5	28180,4	235,159
987	1260,15	1374,19	1001,99	192,62	193,22	8,3547	38683,3	28205,9	235,186
988	1261,15	1375,39	1002,89	193,24	192,75	8,3557	38717,2	28231,5	235,213
989	1262,15	1376,59	1003,80	193,86	192,28	8,3567	38751,0	28257,0	235,240
990	1263,15	1377,79	1004,71	194,49	191,82	8,3576	38784,9	28282,5	235,267
991	1264,15	1379,00	1005,62	195,12	191,35	8,3586	38818,8	28308,1	235,293
992	1265,15	1380,20	1006,52	195,75	190,89	8,3595	38852,6	28333,6	235,320
993	1266,15	1381,40	1007,43	196,38	190,42	8,3605	38886,5	28359,2	235,347
994	1267,15	1382,61	1008,34	197,01	189,96	8,3614	38920,4	28384,8	235,374

113

t	T	h	u	π_0	θ_0	s^0	H	U	S
995	1268,15	1383,81	1009,25	197,65	189,50	8,3624	38954,2	28410,3	235,400
996	1269,15	1385,01	1010,16	198,28	189,04	8,3633	38988,1	28435,9	235,427
997	1270,15	1386,22	1011,06	198,92	188,59	8,3643	39022,0	28461,5	235,454
998	1271,15	1387,42	1011,97	199,56	188,13	8,3652	39055,9	28487,0	235,481
999	1272,15	1388,63	1012,88	200,20	187,68	8,3662	39089,8	28512,6	235,507
1000	1273,15	1389,83	1013,79	200,84	187,22	8,3671	39123,7	28538,2	235,534
1001	1274,15	1391,03	1014,70	201,49	186,77	8,3680	39157,6	28563,8	235,560
1002	1275,15	1392,24	1015,61	202,13	186,32	8,3690	39191,5	28589,4	235,587
1003	1276,15	1393,44	1016,52	202,78	185,87	8,3699	39225,4	28615,0	235,614
1004	1277,15	1394,65	1017,43	203,43	185,42	8,3709	39259,3	28640,6	235,640
1005	1278,15	1395,85	1018,34	204,08	184,98	8,3718	39293,3	28666,2	235,667
1006	1279,15	1397,06	1019,25	204,73	184,53	8,3728	39327,2	28691,8	235,693
1007	1280,15	1398,26	1020,16	205,38	184,09	8,3737	39361,1	28717,4	235,720
1008	1281,15	1399,47	1021,07	206,04	183,65	8,3746	39395,0	28743,0	235,746
1009	1282,15	1400,67	1021,98	206,70	183,20	8,3756	39429,0	28768,7	235,773
1010	1283,15	1401,88	1022,89	207,36	182,76	8,3765	39462,9	28794,3	235,799
1011	1284,15	1403,09	1023,80	208,02	182,33	8,3775	39496,9	28819,9	235,826
1012	1285,15	1404,29	1024,71	208,68	181,89	8,3784	39530,8	28845,5	235,852
1013	1286,15	1405,50	1025,62	209,34	181,45	8,3793	39564,8	28871,2	235,879
1014	1287,15	1406,70	1026,53	210,01	181,02	8,3803	39598,7	28896,8	235,905
1015	1288,15	1407,91	1027,44	210,68	180,59	8,3812	39632,7	28922,5	235,931
1016	1289,15	1409,12	1028,35	211,34	180,15	8,3822	39666,6	28948,1	235,958
1017	1290,15	1410,32	1029,26	212,01	179,72	8,3831	39700,6	28973,8	235,984
1018	1291,15	1411,53	1030,17	212,69	179,29	8,3840	39734,6	28999,4	236,010
1019	1292,15	1412,74	1031,09	213,36	178,87	8,3850	39768,5	29025,1	236,037
1020	1293,15	1413,94	1032,00	214,04	178,44	8,3859	39802,5	29050,7	236,063
1021	1294,15	1415,15	1032,91	214,71	178,01	8,3868	39836,5	29076,4	236,089
1022	1295,15	1416,36	1033,82	215,39	177,59	8,3878	39870,5	29102,1	236,115
1023	1296,15	1417,57	1034,73	216,07	177,17	8,3887	39904,5	29127,8	236,142
1024	1297,15	1418,77	1035,65	216,76	176,75	8,3896	39938,5	29153,4	236,168
1025	1298,15	1419,98	1036,56	217,44	176,33	8,3906	39972,5	29179,1	236,194
1026	1299,15	1421,19	1037,47	218,13	175,91	8,3915	40006,5	29204,8	236,220
1027	1300,15	1422,40	1038,38	218,81	175,49	8,3924	40040,5	29230,5	236,246
1028	1301,15	1423,61	1039,30	219,50	175,07	8,3933	40074,5	29256,2	236,273
1029	1302,15	1424,81	1040,21	220,19	174,66	8,3943	40108,5	29281,9	236,299
1030	1303,15	1426,02	1041,12	220,89	174,24	8,3952	40142,5	29307,6	236,325
1031	1304,15	1427,23	1042,04	221,58	173,83	8,3961	40176,6	29333,3	236,351
1032	1305,15	1428,44	1042,95	222,28	173,42	8,3971	40210,6	29359,0	236,377
1033	1306,15	1429,65	1043,86	222,97	173,01	8,3980	40244,6	29384,7	236,403
1034	1307,15	1430,86	1044,78	223,67	172,60	8,3989	40278,6	29410,5	236,429
1035	1308,15	1432,07	1045,69	224,38	172,19	8,3998	40312,7	29436,2	236,455
1036	1309,15	1433,28	1046,60	225,08	171,79	8,4008	40346,7	29461,9	236,481
1037	1310,15	1434,49	1047,52	225,78	171,38	8,4017	40380,8	29487,6	236,507
1038	1311,15	1435,70	1048,43	226,49	170,98	8,4026	40414,8	29513,4	236,533
1039	1312,15	1436,90	1049,35	227,20	170,57	8,4035	40448,9	29539,1	236,559
1040	1313,15	1438,11	1050,26	227,91	170,17	8,4044	40482,9	29564,9	236,585
1041	1314,15	1439,32	1051,18	228,62	169,77	8,4054	40517,0	29590,6	236,611
1042	1315,15	1440,53	1052,09	229,33	169,37	8,4063	40551,1	29616,4	236,637
1043	1316,15	1441,75	1053,01	230,05	168,97	8,4072	40585,1	29642,1	236,663
1044	1317,15	1442,96	1053,92	230,77	168,57	8,4081	40619,2	29667,9	236,689
1045	1318,15	1444,17	1054,84	231,49	168,18	8,4090	40653,3	29693,6	236,715
1046	1319,15	1445,38	1055,75	232,21	167,78	8,4100	40687,4	29719,4	236,740
1047	1320,15	1446,59	1056,67	232,93	167,39	8,4109	40721,4	29745,2	236,766
1048	1321,15	1447,80	1057,58	233,65	167,00	8,4118	40755,5	29770,9	236,792
1049	1322,15	1449,01	1058,50	234,38	166,61	8,4127	40789,6	29796,7	236,818

Atmospheric nitrogen (*Continued*)

t	T	h	u	π_0	θ_0	s^0	H	U	S
1050	1323,15	1450,22	1059,41	235,11	166,22	8,4136	40823,7	29822,5	236,844
1051	1324,15	1451,43	1060,33	235,84	165,83	8,4145	40857,8	29848,3	236,869
1052	1325,15	1452,64	1061,25	236,57	165,44	8,4155	40891,9	29874,1	236,895
1053	1326,15	1453,86	1062,16	237,30	165,05	8,4164	40926,0	29899,9	236,921
1054	1327,15	1455,07	1063,08	238,04	164,67	8,4173	40960,1	29925,7	236,947
1055	1328,15	1456,28	1064,00	238,77	164,28	8,4182	40994,2	29951,5	236,972
1056	1329,15	1457,49	1064,91	239,51	163,90	8,4191	41028,4	29977,3	236,998
1057	1330,15	1458,70	1065,83	240,25	163,52	8,4200	41062,5	30003,1	237,024
1058	1331,15	1459,92	1066,75	240,99	163,14	8,4209	41096,6	30028,9	237,049
1059	1332,15	1461,13	1067,66	241,74	162,76	8,4218	41130,7	30054,7	237,075
1060	1333,15	1462,34	1068,58	242,48	162,38	8,4228	41164,9	30080,5	237,100
1061	1334,15	1463,55	1069,50	243,23	162,00	8,4237	41199,0	30106,3	237,126
1062	1335,15	1464,77	1070,41	243,98	161,62	8,4246	41233,2	30132,2	237,152
1063	1336,15	1465,98	1071,33	244,73	161,25	8,4255	41267,3	30158,0	237,177
1064	1337,15	1467,19	1072,25	245,49	160,87	8,4264	41301,4	30183,8	237,203
1065	1338,15	1468,40	1073,17	246,24	160,50	8,4273	41335,6	30209,7	237,228
1066	1339,15	1469,62	1074,09	247,00	160,13	8,4282	41369,8	30235,5	237,254
1067	1340,15	1470,83	1075,00	247,76	159,76	8,4291	41403,9	30261,4	237,279
1068	1341,15	1472,05	1075,92	248,52	159,39	8,4300	41438,1	30287,2	237,305
1069	1342,15	1473,26	1076,84	249,28	159,02	8,4309	41472,2	30313,1	237,330
1070	1343,15	1474,47	1077,76	250,04	158,65	8,4318	41506,4	30338,9	237,356
1071	1344,15	1475,69	1078,68	250,81	158,28	8,4327	41540,6	30364,8	237,381
1072	1345,15	1476,90	1079,60	251,58	157,92	8,4336	41574,8	30390,6	237,407
1073	1346,15	1478,12	1080,52	252,35	157,55	8,4345	41609,0	30416,5	237,432
1074	1347,15	1479,33	1081,43	253,12	157,19	8,4354	41643,1	30442,4	237,457
1075	1348,15	1480,54	1082,35	253,89	156,83	8,4363	41677,3	30468,3	237,483
1076	1349,15	1481,76	1083,27	254,67	156,46	8,4372	41711,5	30494,1	237,508
1077	1350,15	1482,97	1084,19	255,45	156,10	8,4381	41745,7	30520,0	237,533
1078	1351,15	1484,19	1085,11	256,22	155,74	8,4390	41779,9	30545,9	237,559
1079	1352,15	1485,40	1086,03	257,01	155,39	8,4399	41814,1	30571,8	237,584
1080	1353,15	1486,62	1086,95	257,79	155,03	8,4408	41848,3	30597,7	237,609
1081	1354,15	1487,83	1087,87	258,57	154,67	8,4417	41882,5	30623,6	237,635
1082	1355,15	1489,05	1088,79	259,36	154,32	8,4426	41916,8	30649,5	237,660
1083	1356,15	1490,27	1089,71	260,15	153,96	8,4435	41951,0	30675,4	237,685
1084	1357,15	1491,48	1090,63	260,94	153,61	8,4444	41985,3	30701,3	237,710
1085	1358,15	1492,70	1091,55	261,73	153,26	8,4453	42019,4	30727,2	237,735
1086	1359,15	1493,91	1092,47	262,53	152,91	8,4462	42053,7	30753,1	237,761
1087	1360,15	1495,13	1093,39	263,32	152,56	8,4471	42087,9	30779,0	237,786
1088	1361,15	1496,35	1094,31	264,12	152,21	8,4480	42122,1	30805,0	237,811
1089	1362,15	1497,56	1095,24	264,92	151,86	8,4489	42156,4	30830,9	237,836
1090	1363,15	1498,78	1096,16	265,72	151,51	8,4498	42190,6	30856,8	237,861
1091	1364,15	1499,99	1097,08	266,53	151,17	8,4507	42224,9	30882,8	237,886
1092	1365,15	1501,21	1098,00	267,33	150,82	8,4516	42259,1	30908,7	237,912
1093	1366,15	1502,43	1098,92	268,14	150,48	8,4525	42293,4	30934,6	237,937
1094	1367,15	1503,65	1099,84	268,95	150,13	8,4533	42327,6	30960,6	237,962
1095	1368,15	1504,86	1100,76	269,76	149,79	8,4542	42361,9	30986,5	237,987
1096	1369,15	1506,08	1101,69	270,57	149,45	8,4551	42396,2	31012,5	238,012
1097	1370,15	1507,30	1102,61	271,39	149,11	8,4560	42430,4	31038,4	238,037
1098	1371,15	1508,51	1103,53	272,21	148,77	8,4569	42464,7	31064,4	238,062
1099	1372,15	1509,73	1104,45	273,03	148,43	8,4578	42499,0	31090,4	238,087
1100	1373,15	1510,95	1105,38	273,85	148,09	8,4587	42533,3	31116,3	238,112
1101	1374,15	1512,17	1106,30	274,67	147,76	8,4596	42567,5	31142,3	238,137
1102	1375,15	1513,39	1107,22	275,50	147,42	8,4604	42601,8	31168,3	238,162
1103	1376,15	1514,60	1108,14	276,32	147,09	8,4613	42636,1	31194,2	238,187
1104	1377,15	1515,82	1109,07	277,15	146,75	8,4622	42670,4	31220,2	238,211

Atmospheric nitrogen (*Continued*)

t	T	h	u	π_0	θ_0	s^0	H	U	S
1105	1378,15	1517,04	1109,99	277,98	146,42	8,4631	42704,7	31246,2	238,236
1106	1379,15	1518,26	1110,91	278,82	146,09	8,4640	42739,0	31272,2	238,261
1107	1380,15	1519,48	1111,84	279,65	145,76	8,4649	42773,3	31298,2	238,286
1108	1381,15	1520,70	1112,76	280,49	145,43	8,4658	42807,6	31324,2	238,311
1109	1382,15	1521,92	1113,68	281,33	145,10	8,4666	42841,9	31350,2	238,336
1110	1383,15	1523,14	1114,61	282,17	144,77	8,4675	42876,3	31376,2	238,361
1111	1384,15	1524,35	1115,53	283,01	144,45	8,4684	42910,6	31402,2	238,385
1112	1385,15	1525,57	1116,45	283,86	144,12	8,4693	42944,9	31428,2	238,410
1113	1386,15	1526,79	1117,38	284,70	143,80	8,4702	42979,2	31454,2	238,435
1114	1387,15	1528,01	1118,30	285,55	143,47	8,4710	43013,6	31480,2	238,460
1115	1388,15	1529,23	1119,23	286,40	143,15	8,4719	43047,9	31506,2	238,485
1116	1389,15	1530,45	1120,15	287,26	142,83	8,4728	43082,2	31532,3	238,509
1117	1390,15	1531,67	1121,08	288,11	142,50	8,4737	43116,6	31558,3	238,534
1118	1391,15	1532,89	1122,00	288,97	142,18	8,4746	43150,9	31584,3	238,559
1119	1392,15	1534,11	1122,93	289,83	141,86	8,4754	43185,3	31610,4	238,583
1120	1393,15	1535,33	1123,85	290,69	141,55	8,4763	43219,6	31636,4	238,608
1121	1394,15	1536,55	1124,78	291,55	141,23	8,4772	43254,0	31662,4	238,633
1122	1395,15	1537,77	1125,70	292,42	140,91	8,4781	43288,3	31688,5	238,657
1123	1396,15	1538,99	1126,63	293,29	140,60	8,4789	43322,7	31714,5	238,682
1124	1397,15	1540,22	1127,55	294,16	140,28	8,4798	43357,1	31740,6	238,707
1125	1398,15	1541,44	1128,48	295,03	139,97	8,4807	43391,4	31766,6	238,731
1126	1399,15	1542,66	1129,40	295,90	139,65	8,4816	43425,8	31792,7	238,756
1127	1400,15	1543,88	1130,33	296,78	139,34	8,4824	43460,2	31818,8	238,780
1128	1401,15	1545,10	1131,25	297,65	139,03	8,4833	43494,6	31844,8	238,805
1129	1402,15	1546,32	1132,18	298,53	138,72	8,4842	43528,9	31870,9	238,829
1130	1403,15	1547,54	1133,11	299,41	138,41	8,4850	43563,3	31897,0	238,854
1131	1404,15	1548,76	1134,03	300,30	138,10	8,4859	43597,7	31923,0	238,878
1132	1405,15	1549,99	1134,96	301,18	137,79	8,4868	43632,1	31949,1	238,903
1133	1406,15	1551,21	1135,89	302,07	137,48	8,4876	43666,5	31975,2	238,927
1134	1407,15	1552,43	1136,81	302,96	137,18	8,4885	43700,9	32001,3	238,952
1135	1408,15	1553,65	1137,74	303,85	136,87	8,4894	43735,3	32027,4	238,976
1136	1409,15	1554,87	1138,67	304,75	136,57	8,4903	43769,7	32053,5	239,001
1137	1410,15	1556,10	1139,59	305,64	136,26	8,4911	43804,1	32079,6	239,025
1138	1411,15	1557,32	1140,52	306,54	135,96	8,4920	43838,6	32105,7	239,049
1139	1412,15	1558,54	1141,45	307,44	135,66	8,4929	43873,0	32131,8	239,074
1140	1413,15	1559,77	1142,38	308,34	135,36	8,4937	43907,4	32157,9	239,098
1141	1414,15	1560,99	1143,30	309,25	135,06	8,4946	43941,8	32184,0	239,123
1142	1415,15	1562,21	1144,23	310,15	134,76	8,4954	43976,2	32210,1	239,147
1143	1416,15	1563,43	1145,16	311,06	134,46	8,4963	44010,7	32236,2	239,171
1144	1417,15	1564,66	1146,09	311,97	134,16	8,4972	44045,1	32262,3	239,195
1145	1418,15	1565,88	1147,01	312,89	133,86	8,4980	44079,6	32288,5	239,220
1146	1419,15	1567,10	1147,94	313,80	133,57	8,4989	44114,0	32314,6	239,244
1147	1420,15	1568,33	1148,87	314,72	133,27	8,4998	44148,4	32340,7	239,268
1148	1421,15	1569,55	1149,80	315,64	132,98	8,5006	44182,9	32366,9	239,293
1149	1422,15	1570,78	1150,73	316,56	132,68	8,5015	44217,3	32393,0	239,317
1150	1423,15	1572,00	1151,66	317,48	132,39	8,5023	44251,8	32419,1	239,341
1151	1424,15	1573,22	1152,59	318,41	132,10	8,5032	44286,3	32445,3	239,365
1152	1425,15	1574,45	1153,51	319,34	131,81	8,5041	44320,7	32471,4	239,389
1153	1426,15	1575,67	1154,44	320,27	131,52	8,5049	44355,2	32497,6	239,414
1154	1427,15	1576,90	1155,37	321,20	131,23	8,5058	44389,7	32523,7	239,438
1155	1428,15	1578,12	1156,30	322,13	130,94	8,5066	44424,1	32549,9	239,462
1156	1429,15	1579,35	1157,23	323,07	130,65	8,5075	44458,6	32576,1	239,486
1157	1430,15	1580,57	1158,16	324,01	130,36	8,5084	44493,1	32602,2	239,510
1158	1431,15	1581,80	1159,09	324,95	130,08	8,5092	44527,6	32628,4	239,534
1159	1432,15	1583,02	1160,02	325,89	129,79	8,5101	44562,0	32654,6	239,558

t	T	h	u	π_0	θ_0	s^0	H	U	S
1160	1433,15	1584,25	1160,95	326,84	129,51	8,5109	45596,5	32680,7	239,582
1161	1434,15	1585,47	1161,88	327,78	129,22	8,5118	44631,0	32706,9	239,606
1162	1435,15	1586,70	1162,81	328,73	128,94	8,5126	44665,5	32733,1	239,631
1163	1436,15	1587,92	1163,74	329,68	128,66	8,5135	44700,0	32759,3	239,655
1164	1437,15	1589,15	1164,67	330,64	128,37	8,5143	44734,5	32785,5	239,679
1165	1438,15	1590,37	1165,60	331,59	128,09	8,5152	44769,0	32811,7	239,703
1166	1439,15	1591,60	1166,53	332,55	127,81	8,5160	44803,5	32837,9	239,727
1167	1440,15	1592,83	1167,46	333,51	127,53	8,5169	44838,1	32864,1	239,751
1168	1441,15	1594,05	1168,39	334,47	127,26	8,5177	44872,6	32890,3	239,774
1169	1442,15	1595,28	1169,32	335,44	126,98	8,5186	44907,1	32916,5	239,798
1170	1443,15	1596,50	1170,25	336,41	126,70	8,5194	44941,6	32942,7	239,822
1171	1444,15	1597,73	1171,19	337,37	126,42	8,5203	44976,1	32968,9	239,846
1172	1445,15	1598,96	1172,12	338,35	126,15	8,5211	45010,7	32995,1	239,870
1173	1446,15	1600,18	1173,05	339,32	125,87	8,5220	45045,2	33021,3	239,894
1174	1447,15	1601,41	1173,98	340,29	125,60	8,5228	45079,7	33047,5	239,918
1175	1448,15	1602,64	1174,91	341,27	125,33	8,5237	45114,3	33073,8	239,942
1176	1449,15	1603,87	1175,84	342,25	125,05	8,5245	45148,8	33100,0	239,966
1177	1450,15	1605,09	1176,78	343,24	124,78	8,5254	45183,4	33126,2	239,989
1178	1451,15	1606,32	1177,71	344,22	124,51	8,5262	45217,9	33152,5	240,013
1179	1452,15	1607,55	1178,64	345,21	124,24	8,5271	45252,5	33178,7	240,037
1180	1453,15	1608,78	1179,57	346,20	123,97	8,5279	45287,0	33204,9	240,061
1181	1454,15	1610,00	1180,50	347,19	123,70	8,5288	45321,6	33231,2	240,085
1182	1455,15	1611,23	1181,44	348,18	123,43	8,5296	45356,1	33257,4	240,108
1183	1456,15	1612,46	1182,37	349,18	123,17	8,5304	45390,7	33283,7	240,132
1184	1457,15	1613,69	1183,30	350,17	122,90	8,5313	45425,3	33309,9	240,156
1185	1458,15	1614,91	1184,23	351,17	122,63	8,5321	45459,9	33336,2	240,180
1186	1459,15	1616,14	1185,17	352,18	122,37	8,5330	45494,4	33362,5	240,203
1187	1460,15	1617,37	1186,10	353,18	122,10	8,5338	45529,0	33388,7	240,227
1188	1461,15	1618,60	1187,03	354,19	121,84	8,5347	45563,6	33415,0	240,251
1189	1462,15	1619,83	1187,97	355,20	121,58	8,5355	45598,2	33441,3	240,274
1190	1463,15	1621,06	1188,90	356,21	121,31	8,5363	45632,8	33467,5	240,298
1191	1464,15	1622,29	1189,83	357,22	121,05	8,5372	45667,4	33493,8	240,322
1192	1465,15	1623,51	1190,77	358,24	120,79	8,5380	45701,9	33520,1	240,345
1193	1466,15	1624,74	1191,70	359,26	120,53	8,5389	45736,5	33546,4	240,369
1194	1467,15	1625,97	1192,63	360,28	120,27	8,5397	45771,1	33572,7	240,392
1195	1468,15	1627,20	1193,57	361,30	120,01	8,5405	45805,8	33598,9	240,416
1196	1469,15	1628,43	1194,50	362,33	119,75	8,5414	45840,4	33625,2	240,440
1197	1470,15	1629,66	1195,44	363,36	119,50	8,5422	45875,0	33651,5	240,463
1198	1471,15	1630,89	1196,37	364,39	119,24	8,5430	45909,6	33677,8	240,487
1199	1472,15	1632,12	1197,31	365,42	118,98	8,5439	45944,2	33704,1	240,510
1200	1473,15	1633,35	1198,24	366,45	118,73	8,5447	45978,8	33730,4	240,534
1201	1474,15	1634,58	1199,17	367,49	118,47	8,5455	46013,4	33756,8	240,557
1202	1475,15	1635,81	1200,11	368,53	118,22	8,5464	46048,1	33783,1	240,581
1203	1476,15	1637,04	1201,04	369,57	117,97	8,5472	46082,7	33809,4	240,604
1204	1477,15	1638,27	1201,98	370,62	117,71	8,5480	46117,3	33835,7	240,628
1205	1478,15	1639,50	1202,91	371,66	117,46	8,5489	46152,0	33862,0	240,651
1206	1479,15	1640,73	1203,85	372,71	117,21	8,5497	46186,6	33888,3	240,674
1207	1480,15	1641,96	1204,78	373,76	116,96	8,5505	46221,3	33914,7	240,698
1208	1481,15	1643,19	1205,72	374,82	116,71	8,5514	46255,9	33941,0	240,721
1209	1482,15	1644,42	1206,66	375,87	116,46	8,5522	46290,5	33967,3	240,745
1210	1483,15	1645,66	1207,59	376,93	116,21	8,5530	46325,2	33993,7	240,768
1211	1484,15	1646,89	1208,53	377,99	115,97	8,5539	46359,9	34020,0	240,791
1212	1485,15	1648,12	1209,46	379,05	115,72	8,5547	46394,5	34046,4	240,815
1213	1486,15	1649,35	1210,40	380,12	115,47	8,5555	46429,2	34072,7	240,838
1214	1487,15	1650,58	1211,33	381,18	115,23	8,5564	46463,8	34099,1	240,861

Atmospheric nitrogen (*Continued*)

t	T	h	u	π_0	θ_0	s^0	H	U	S
1215	1488,15	1651,81	1212,27	382,25	114,98	8,5572	46498,5	34125,4	240,885
1216	1489,15	1653,04	1213,21	383,33	114,74	8,5580	46533,2	34151,8	240,908
1217	1490,15	1654,28	1214,14	384,40	114,49	8,5588	46567,8	34178,1	240,931
1218	1491,15	1655,51	1215,08	385,48	114,25	8,5597	46602,5	34204,5	240,955
1219	1492,15	1656,74	1216,02	386,56	114,01	8,5605	46637,2	34230,9	240,978
1220	1493,15	1657,97	1216,95	387,64	113,76	8,5613	46671,9	34257,2	241,001
1221	1494,15	1659,20	1217,89	388,72	113,52	8,5621	46706,6	34283,6	241,024
1222	1495,15	1660,44	1218,83	389,81	113,28	8,5630	46741,3	34310,0	241,047
1223	1496,15	1661,67	1219,76	390,90	113,04	8,5638	46776,0	34336,4	241,071
1224	1497,15	1662,90	1220,70	391,99	112,80	8,5646	46810,7	34362,7	241,094
1225	1498,15	1664,13	1221,64	393,09	112,56	8,5654	46845,4	34389,1	241,117
1226	1499,15	1665,37	1222,58	394,18	112,33	8,5663	46880,1	34415,5	241,140
1227	1500,15	1666,60	1223,51	395,28	112,09	8,5671	46914,8	34441,9	241,163
1228	1501,15	1667,83	1224,45	396,38	111,85	8,5679	46949,5	34468,3	241,186
1229	1502,15	1669,06	1225,39	397,48	111,61	8,5687	46984,2	34494,7	241,210
1230	1503,15	1670,30	1226,33	398,59	111,38	8,5695	47018,9	34521,1	241,233
1231	1504,15	1671,53	1227,26	399,70	111,14	8,5704	47053,6	34547,5	241,256
1232	1505,15	1672,76	1228,20	400,81	110,91	8,5712	47088,3	34573,9	241,279
1233	1506,15	1674,00	1229,14	401,92	110,68	8,5720	47123,1	34600,3	241,302
1234	1507,15	1675,23	1230,08	403,04	110,44	8,5728	47157,8	34626,7	241,325
1235	1508,15	1676,47	1231,02	404,16	110,21	8,5736	47192,5	34653,1	241,348
1236	1509,15	1677,70	1231,96	405,28	109,98	8,5745	47227,2	34679,5	241,371
1237	1510,15	1678,93	1232,89	406,40	109,75	8,5753	47262,0	34706,0	241,394
1238	1511,15	1680,17	1233,83	407,53	109,52	8,5761	47296,7	34732,4	241,417
1239	1512,15	1681,40	1234,77	408,65	109,29	8,5769	47331,5	34758,8	241,440
1240	1513,15	1682,64	1235,71	409,78	109,06	8,5777	47366,2	34785,2	241,463
1241	1514,15	1683,87	1236,65	410,92	108,83	8,5785	47400,9	34811,7	241,486
1242	1515,15	1685,10	1237,59	412,05	108,60	8,5794	47435,7	34838,1	241,509
1243	1516,15	1686,34	1238,53	413,19	108,37	8,5802	47470,4	34864,6	241,532
1244	1517,15	1687,57	1239,47	414,33	108,15	8,5810	47505,2	34891,0	241,555
1245	1518,15	1688,81	1240,41	415,47	107,92	8,5818	47540,0	34917,4	241,578
1246	1519,15	1690,04	1241,35	416,62	107,69	8,5826	47574,7	34943,9	241,600
1247	1520,15	1691,28	1242,29	417,77	107,47	8,5834	47609,5	34970,3	241,623
1248	1521,15	1692,51	1243,23	418,92	107,24	8,5842	47644,3	34996,8	241,646
1249	1522,15	1693,75	1244,17	420,07	107,02	8,5850	47679,0	35023,3	241,669
1250	1523,15	1694,98	1245,11	421,23	106,80	8,5859	47713,8	35049,7	241,692
1251	1524,15	1696,22	1246,05	422,38	106,57	8,5867	47748,6	35076,2	241,715
1252	1525,15	1697,46	1246,99	423,55	106,35	8,5875	47783,4	35102,6	241,738
1253	1526,15	1698,69	1247,93	424,71	106,13	8,5883	47818,1	35129,1	241,760
1254	1527,15	1699,93	1248,87	425,87	105,91	8,5891	47852,9	35155,6	241,783
1255	1528,15	1701,16	1249,81	427,04	105,69	8,5899	47887,7	35182,1	241,806
1256	1529,15	1702,40	1250,75	428,21	105,47	8,5907	47922,5	35208,5	241,829
1257	1530,15	1703,63	1251,69	429,39	105,25	8,5915	47957,3	35235,0	241,851
1258	1531,15	1704,87	1252,63	430,56	105,03	8,5923	47992,1	35261,5	241,874
1259	1532,15	1706,11	1253,57	431,74	104,81	8,5931	48026,9	35288,0	241,897
1260	1533,15	1707,34	1254,51	432,92	104,59	8,5939	48061,7	35314,5	241,920
1261	1534,15	1708,58	1255,45	434,10	104,38	8,5948	48096,5	35341,0	241,942
1262	1535,15	1709,82	1256,39	435,29	104,16	8,5956	48131,3	35367,5	241,965
1263	1536,15	1711,05	1257,33	436,48	103,94	8,5964	48166,1	35394,0	241,988
1264	1537,15	1712,29	1258,28	437,67	103,73	8,5972	48201,0	35420,5	242,010
1265	1538,15	1713,53	1259,22	438,86	103,51	8,5980	48235,8	35447,0	242,033
1266	1539,15	1714,76	1260,16	440,06	103,30	8,5988	48270,6	35473,5	242,056
1267	1540,15	1716,00	1261,10	441,26	103,09	8,5996	48305,4	35500,0	242,078
1268	1541,15	1717,24	1262,04	442,46	102,87	8,6004	48340,2	35526,5	242,101
1269	1542,15	1718,48	1262,98	443,66	102,66	8,6012	48375,1	35553,0	242,123

t	T	h	u	π_0	θ_0	s^0	H	U	S
1270	1543,15	1719,71	1263,93	444,87	102,45	8,6020	48409,9	35579,5	242,146
1271	1544,15	1720,95	1264,87	446,08	102,24	8,6028	48444,7	35606,1	242,168
1272	1545,15	1722,19	1265,81	447,29	102,03	8,6036	48479,6	35632,6	242,191
1273	1546,15	1723,43	1266,75	448,50	101,81	8,6044	48514,4	35659,1	242,214
1274	1547,15	1724,66	1267,70	449,72	101,61	8,6052	48549,3	35685,6	242,236
1275	1548,15	1725,90	1268,64	450,94	101,40	8,6060	48584,1	35712,2	242,259
1276	1549,15	1727,14	1269,58	452,16	101,19	8,6068	48619,0	35738,7	242,281
1277	1550,15	1728,38	1270,52	453,39	100,98	8,6076	48653,8	35765,2	242,304
1278	1551,15	1729,62	1271,47	454,62	100,77	8,6084	48688,7	35791,8	242,326
1279	1552,15	1730,85	1272,41	455,85	100,56	8,6092	48723,5	35818,3	242,349
1280	1553,15	1732,09	1273,35	457,08	100,36	8,6100	48758,4	35844,9	242,371
1281	1554,15	1733,33	1274,30	458,31	100,15	8,6108	48793,3	35871,4	242,393
1282	1555,15	1734,57	1275,24	459,55	99,950	8,6116	48828,1	35898,0	242,416
1283	1556,15	1735,81	1276,18	460,79	99,745	8,6124	48863,0	35924,5	242,438
1284	1557,15	1737,05	1277,13	462,04	99,541	8,6132	48897,9	35951,1	242,461
1285	1558,15	1738,29	1278,07	463,28	99,337	8,6140	48932,7	35977,6	242,483
1286	1559,15	1739,52	1279,01	464,53	99,133	8,6148	48967,6	36004,2	242,505
1287	1560,15	1740,76	1279,96	465,78	98,931	8,6156	49002,5	36030,8	242,528
1288	1561,15	1742,00	1280,90	467,04	98,728	8,6163	49037,4	36057,3	242,550
1289	1562,15	1743,24	1281,84	468,29	98,526	8,6171	49072,3	36083,9	242,573
1290	1563,15	1744,48	1282,79	469,55	98,325	8,6179	49107,2	36110,5	242,595
1291	1564,15	1745,72	1283,73	470,81	98,124	8,6187	49142,1	36137,1	242,617
1292	1565,15	1746,96	1284,68	472,08	97,924	8,6195	49177,0	36163,7	242,639
1293	1566,15	1748,20	1285,62	473,34	97,724	8,6203	49211,9	36190,2	242,662
1294	1567,15	1749,44	1286,57	474,61	97,525	8,6211	49246,8	36216,8	242,684
1295	1568,15	1750,68	1287,51	475,89	97,326	8,6219	49281,7	36243,4	242,706
1296	1569,15	1751,92	1288,45	477,16	97,128	8,6227	49316,6	36270,0	242,729
1297	1570,15	1753,16	1289,40	478,44	96,930	8,6235	49351,5	36296,6	242,751
1298	1571,15	1754,40	1290,34	479,72	96,733	8,6243	49386,4	36323,2	242,773
1299	1572,15	1755,64	1291,29	481,01	96,536	8,6251	49421,3	36349,8	242,795
1300	1573,15	1756,88	1292,23	482,29	96,340	8,6258	49456,2	36376,4	242,817
1301	1574,15	1758,12	1293,18	483,58	96,144	8,6266	49491,1	36403,0	242,840
1302	1575,15	1759,36	1294,12	484,87	95,949	8,6274	49526,1	36429,6	242,862
1303	1576,15	1760,60	1295,07	486,17	95,754	8,6282	49561,0	36456,2	242,884
1304	1577,15	1761,84	1296,02	487,46	95,560	8,6290	49595,9	36482,8	242,906
1305	1578,15	1763,09	1296,96	488,76	95,367	8,6298	49630,8	36509,5	242,928
1306	1579,15	1764,33	1297,91	490,07	95,173	8,6306	49665,8	36536,1	242,950
1307	1580,15	1765,57	1298,85	491,37	94,981	8,6314	49700,7	36562,7	242,973
1308	1581,15	1766,81	1299,80	492,68	94,788	8,6321	49735,7	36589,3	242,995
1309	1582,15	1768,05	1300,74	493,99	94,597	8,6329	49770,6	36616,0	243,017
1310	1583,15	1769,29	1301,69	495,30	94,405	8,6337	49805,5	36642,6	243,039
1311	1584,15	1770,53	1302,64	496,62	94,215	8,6345	49840,5	36669,2	243,061
1312	1585,15	1771,77	1303,58	497,94	94,024	8,6353	49875,4	36695,9	243,083
1313	1586,15	1773,02	1304,53	499,26	93,835	8,6361	49910,4	36722,5	243,105
1314	1587,15	1774,26	1305,48	500,59	93,645	8,6368	49945,3	36749,1	243,127
1315	1588,15	1775,50	1306,42	501,91	93,456	8,6376	49980,3	36775,8	243,149
1316	1589,15	1776,74	1307,37	503,24	93,268	8,6384	50015,3	36802,4	243,171
1317	1590,15	1777,98	1308,32	504,58	93,080	8,6392	50050,2	36829,1	243,193
1318	1591,15	1779,23	1309,26	505,91	92,893	8,6400	50085,2	36855,7	243,215
1319	1592,15	1780,47	1310,21	507,25	92,706	8,6407	50120,2	36882,4	243,237
1320	1593,15	1781,71	1311,16	508,59	92,519	8,6415	50155,1	36909,0	243,259
1321	1594,15	1782,95	1312,10	509,94	92,333	8,6423	50190,1	36935,7	243,281
1322	1595,15	1784,19	1313,05	511,28	92,148	8,6431	50225,1	36962,4	243,303
1323	1596,15	1785,44	1314,00	512,63	91,963	8,6439	50260,1	36989,0	243,325
1324	1597,15	1786,68	1314,94	513,99	91,778	8,6446	50295,1	37015,7	243,347

t	T	h	u	π_0	θ_0	s^0	H	U	S
1325	1598,15	1787,92	1315,89	515,34	91,594	8,6454	50330,0	37042,4	243,369
1326	1599,15	1789,17	1316,84	516,70	91,411	8,6462	50365,0	37069,0	243,390
1327	1600,15	1790,41	1317,79	518,06	91,228	8,6470	50400,0	37095,7	243,412
1328	1601,15	1791,65	1318,73	519,43	91,045	8,6477	50435,0	37122,4	243,434
1329	1602,15	1792,90	1319,68	520,79	90,863	8,6485	50470,0	37149,1	243,456
1330	1603,15	1794,14	1320,63	522,16	90,681	8,6493	50505,0	37175,8	243,478
1331	1604,15	1795,38	1321,58	523,53	90,500	8,6501	50540,0	37202,4	243,500
1332	1605,15	1796,63	1322,53	524,91	90,319	8,6509	50575,0	37229,1	243,521
1333	1606,15	1797,87	1323,47	526,29	90,138	8,6516	50610,0	37255,8	243,543
1334	1607,15	1799,11	1324,42	527,67	89,958	8,6524	50645,0	37282,5	243,565
1335	1608,15	1800,36	1325,37	529,05	89,779	8,6532	50680,0	37309,2	243,587
1336	1609,15	1801,60	1326,32	530,44	89,600	8,6539	50715,0	37335,9	243,609
1337	1610,15	1802,84	1327,27	531,83	89,421	8,6547	50750,1	37362,6	243,630
1338	1611,15	1804,09	1328,22	533,22	89,243	8,6555	50785,1	37389,3	243,652
1339	1612,15	1805,33	1329,17	534,62	89,065	8,6563	50820,1	37416,0	243,674
1340	1613,15	1806,58	1330,11	536,02	88,888	8,6570	50855,1	37442,7	243,696
1341	1614,15	1807,82	1331,06	537,42	88,711	8,6578	50890,2	37469,4	243,717
1342	1615,15	1809,07	1332,01	538,82	88,535	8,6586	50925,2	37496,2	243,739
1343	1616,15	1810,31	1332,96	540,23	88,359	8,6593	50960,2	37522,9	243,761
1344	1617,15	1811,55	1333,91	541,64	88,184	8,6601	50995,2	37549,6	243,782
1345	1618,15	1812,80	1334,86	543,05	88,009	8,6609	51030,3	37576,3	243,804
1346	1619,15	1814,04	1335,81	544,47	87,834	8,6617	51065,3	37603,9	243,826
1347	1620,15	1815,29	1336,76	545,89	87,660	8,6624	51100,4	37629,8	243,847
1348	1621,15	1816,53	1337,71	547,31	87,486	8,6632	51135,4	37656,5	243,869
1349	1622,15	1817,78	1338,66	548,73	87,313	8,6640	51170,5	37683,2	243,890
1350	1623,15	1819,02	1339,61	550,16	87,140	8,6647	51205,5	37710,0	243,912
1351	1624,15	1820,27	1340,56	551,59	86,968	8,6655	51240,6	37736,7	243,934
1352	1625,15	1821,51	1341,51	553,02	86,796	8,6663	51275,6	37763,5	243,955
1353	1626,15	1822,76	1342,46	554,46	86,624	8,6670	51310,7	37790,2	243,977
1354	1627,15	1824,00	1343,41	555,90	86,453	8,6678	51345,7	37816,9	243,998
1355	1628,15	1825,25	1344,36	557,34	86,282	8,6686	51380,8	37843,7	244,020
1356	1629,15	1826,50	1345,31	558,79	86,112	8,6693	51415,9	37870,4	244,041
1357	1630,15	1827,74	1346,26	560,23	85,942	8,6701	51450,9	37897,2	244,063
1358	1631,15	1828,99	1347,21	561,68	85,773	8,6709	51486,0	37923,9	244,084
1359	1632,15	1830,23	1348,16	563,14	85,604	8,6716	51521,1	37950,7	244,106
1360	1633,15	1831,48	1349,11	564,60	85,435	8,6724	51556,1	37977,5	244,127
1361	1634,15	1832,73	1350,06	566,06	85,267	8,6731	51591,2	38004,2	244,149
1362	1635,15	1833,97	1351,01	567,52	85,099	8,6739	51626,3	38031,0	244,170
1363	1636,15	1835,22	1351,96	568,98	84,932	8,6747	51661,4	38057,8	244,192
1364	1637,15	1836,46	1352,91	570,45	84,765	8,6754	51696,5	38084,5	244,213
1365	1638,15	1837,71	1353,87	571,92	84,598	8,6762	51731,6	38111,3	244,235
1366	1639,15	1838,96	1354,82	573,40	84,432	8,6769	51766,6	38138,1	244,256
1367	1640,15	1840,20	1355,77	574,88	84,266	8,6777	51801,7	38164,9	244,278
1368	1641,15	1841,45	1356,72	576,36	84,101	8,6785	51836,8	38191,6	244,299
1369	1642,15	1842,70	1357,67	577,84	83,936	8,6792	51871,9	38218,4	244,320
1370	1643,15	1843,94	1358,62	579,33	83,772	8,6800	51907,0	38245,2	244,342
1371	1644,15	1845,19	1359,57	580,82	83,608	8,6807	51942,1	38272,0	244,363
1372	1645,15	1846,44	1360,53	582,31	83,444	8,6815	51977,2	38298,8	244,384
1373	1646,15	1847,69	1361,48	583,81	83,281	8,6823	52012,3	38325,6	244,406
1374	1647,15	1848,93	1362,43	585,31	83,118	8,6830	52047,5	38352,4	244,427
1375	1648,15	1850,18	1363,38	586,81	82,956	8,6838	52082,6	38379,2	244,448
1376	1649,15	1851,43	1364,33	588,32	82,794	8,6845	52117,7	38406,0	244,470
1377	1650,15	1852,67	1365,29	589,82	82,632	8,6853	52152,8	38432,8	244,491
1378	1651,15	1853,92	1366,24	591,33	82,471	8,6860	52187,9	38459,6	244,512
1379	1652,15	1855,17	1367,19	592,85	82,310	8,6868	52223,0	38486,4	244,533

Atmospheric nitrogen (*Continued*)

t	T	h	u	π_0	θ_0	s^0	H	U	S
1380	1653,15	1856,42	1368,14	594,37	82,149	8,6876	52258,2	38513,2	244,555
1381	1654,15	1857,67	1369,09	595,89	81,989	8,6883	52293,3	38540,0	244,576
1382	1655,15	1858,91	1370,05	597,41	81,830	8,6891	52328,4	38566,8	244,597
1383	1656,15	1860,16	1371,00	598,94	81,671	8,6898	52363,6	38593,6	244,618
1384	1657,15	1861,41	1371,95	600,47	81,512	8,6906	52398,7	38620,5	244,640
1385	1658,15	1862,66	1372,91	602,00	81,353	8,6913	52433,8	38647,3	244,661
1386	1659,15	1863,91	1373,86	603,54	81,195	8,6921	52469,0	38674,1	244,682
1387	1660,15	1865,15	1374,81	605,07	81,037	8,6928	52504,1	38700,9	244,703
1388	1661,15	1866,40	1375,76	606,62	80,880	8,6936	52539,2	38727,8	244,724
1389	1662,15	1867,65	1376,72	608,16	80,723	8,6943	52574,4	38754,6	244,745
1390	1663,15	1868,90	1377,67	609,71	80,567	8,6951	52609,5	38781,4	244,767
1391	1664,15	1870,15	1378,62	611,26	80,410	8,6958	52644,7	38808,3	244,788
1392	1665,15	1871,40	1379,58	612,82	80,255	8,6966	52679,8	38835,1	244,809
1393	1666,15	1872,65	1380,53	614,37	80,099	8,6973	52715,0	38861,9	244,830
1394	1667,15	1873,90	1381,48	615,93	79,944	8,6981	52750,2	38888,8	244,851
1395	1668,15	1875,14	1382,44	617,50	79,790	8,6988	52785,3	38915,6	244,872
1396	1669,15	1876,39	1383,39	619,07	79,635	8,6996	52820,5	38942,5	244,893
1397	1670,15	1877,64	1384,35	620,64	79,481	8,7003	52855,6	38969,3	244,914
1398	1671,15	1878,89	1385,30	622,21	79,328	8,7011	52890,8	38996,2	244,935
1399	1672,15	1880,14	1386,25	623,78	79,175	8,7018	52926,0	39023,0	244,956
1400	1673,15	1881,39	1387,21	625,36	79,022	8,7026	52961,2	39049,9	244,977
1401	1674,15	1882,64	1388,16	626,95	78,870	8,7033	52996,3	39076,8	244,998
1402	1675,15	1883,89	1389,12	628,53	78,718	8,7041	53031,5	39103,6	245,019
1403	1676,15	1885,14	1390,07	630,12	78,566	8,7048	53066,7	39130,5	245,040
1404	1677,15	1886,39	1391,03	631,71	78,415	8,7056	53101,9	39157,4	245,061
1405	1678,15	1887,64	1391,98	633,31	78,264	8,7063	53137,1	39184,2	245,082
1406	1679,15	1888,89	1392,93	634,91	78,113	8,7070	53172,2	39211,1	245,103
1407	1680,15	1890,14	1393,89	636,51	77,963	8,7078	53207,4	39238,0	245,124
1408	1681,15	1891,39	1394,84	638,12	77,813	8,7085	53242,6	39264,9	245,145
1409	1682,15	1892,64	1395,80	639,72	77,664	8,7093	53277,8	39291,7	245,166
1410	1683,15	1893,89	1396,75	641,34	77,515	8,7100	53313,0	39318,6	245,187
1411	1684,15	1895,14	1397,71	642,95	77,366	8,7108	53348,2	39345,5	245,208
1412	1685,15	1896,39	1398,66	644,57	77,218	8,7115	53383,4	39372,4	245,229
1413	1686,15	1897,64	1399,62	646,19	77,070	8,7122	53418,6	39399,3	245,250
1414	1687,15	1898,89	1400,57	647,81	76,922	8,7130	53453,8	39426,2	245,271
1415	1688,15	1900,14	1401,53	649,44	76,775	8,7137	53489,0	39453,1	245,291
1416	1689,15	1901,39	1402,49	651,07	76,628	8,7145	53524,2	39480,0	245,312
1417	1690,15	1902,64	1403,44	652,71	76,481	8,7152	53559,5	39506,9	245,333
1418	1691,15	1903,90	1404,40	654,34	76,335	8,7160	53594,7	39533,8	245,354
1419	1692,15	1905,15	1405,35	655,98	76,189	8,7167	53629,9	39560,7	245,375
1420	1693,15	1906,40	1406,31	657,63	76,044	8,7174	53665,1	39587,6	245,396
1421	1694,15	1907,65	1407,26	659,27	75,898	8,7182	53700,3	39614,5	245,416
1422	1695,15	1908,90	1408,22	660,92	75,754	8,7189	53735,6	39641,4	245,437
1423	1696,15	1910,15	1409,18	662,58	75,609	8,7196	53770,8	39668,3	245,458
1424	1697,15	1911,40	1410,13	664,23	75,465	8,7204	53806,0	39695,2	245,479
1425	1698,15	1912,66	1411,09	665,89	75,321	8,7211	53841,2	39722,1	245,499
1426	1699,15	1913,91	1412,04	667,56	75,178	8,7219	53876,5	39749,0	245,520
1427	1700,15	1915,16	1413,00	669,22	75,035	8,7226	53911,7	39776,0	245,541
1428	1701,15	1916,41	1413,96	670,89	74,892	8,7233	53946,9	39802,9	245,562
1429	1702,15	1917,66	1414,91	672,57	74,750	8,7241	53982,2	39829,8	245,582
1430	1703,15	1918,91	1415,87	674,24	74,608	8,7248	54017,4	39856,7	245,603
1431	1704,15	1920,17	1416,83	675,92	74,466	8,7255	54052,7	39883,7	245,624
1432	1705,15	1921,42	1417,78	677,61	74,324	8,7263	54087,9	39910,6	245,644
1433	1706,15	1922,67	1418,74	679,29	74,183	8,7270	54123,2	39937,5	245,665
1434	1707,15	1923,92	1419,70	680,98	74,043	8,7277	54158,4	39964,5	245,686
1435	1708,15	1925,18	1420,65	682,68	73,902	8,7285	54193,7	39991,4	245,706
1436	1709,15	1926,43	1421,61	684,37	73,762	8,7292	54228,9	40018,4	245,727
1437	1710,15	1927,68	1422,57	686,07	73,623	8,7299	54264,2	40045,3	245,748
1438	1711,15	1928,93	1423,53	687,77	73,483	8,7307	54299,5	40072,3	245,769
1439	1712,15	1930,19	1424,48	689,48	73,344	8,7314	54334,7	40099,2	245,789

Atmospheric nitrogen (*Continued*)

t	T	h	u	π_0	θ_v	s^v	H	U	S
1440	1713,15	1931,44	1425,44	691,19	73,206	8,7321	54370,0	40126,2	245,809
1441	1714,15	1932,69	1426,40	692,90	73,067	8,7329	54405,3	40153,1	245,830
1442	1715,15	1933,94	1427,36	694,62	72,929	8,7336	54440,5	40180,1	245,851
1443	1716,15	1935,20	1428,31	696,34	72,792	8,7343	54475,8	40207,0	245,871
1444	1717,15	1936,45	1429,27	698,06	72,654	8,7351	54511,1	40234,0	245,892
1445	1718,15	1937,70	1430,23	699,79	72,517	8,7358	54546,3	40260,9	245,912
1446	1719,15	1938,96	1431,19	701,52	72,380	8,7365	54581,6	40287,9	245,933
1447	1720,15	1940,21	1432,14	703,25	72,244	8,7372	54616,9	40314,9	245,953
1448	1721,15	1941,46	1433,10	704,99	72,108	8,7380	54652,2	40341,8	245,974
1449	1722,15	1942,72	1434,06	706,73	71,972	8,7387	54687,5	40368,8	245,994
1450	1723,15	1943,97	1435,02	708,47	71,837	8,7394	54722,8	40395,8	246,015
1451	1724,15	1945,22	1435,98	710,22	71,702	8,7402	54758,1	40422,8	246,035
1452	1725,15	1946,48	1436,94	711,97	71,567	8,7409	54793,4	40449,7	246,056
1453	1726,15	1947,73	1437,89	713,72	71,432	8,7416	54828,6	40476,7	246,076
1454	1727,15	1948,99	1438,85	715,48	71,298	8,7423	54863,9	40503,7	246,097
1455	1728,15	1950,24	1439,81	717,24	71,165	8,7431	54899,2	40530,7	246,117
1456	1729,15	1951,49	1440,77	719,00	71,031	8,7438	54934,6	40557,7	246,138
1457	1730,15	1952,75	1441,73	720,77	70,898	8,7445	54969,9	40584,7	246,158
1458	1731,15	1954,00	1442,69	722,54	70,765	8,7452	55005,2	40611,7	246,178
1459	1732,15	1955,26	1443,65	724,32	70,632	8,7460	55040,5	40638,7	246,199
1460	1733,15	1956,51	1444,61	726,09	70,500	8,7467	55075,8	40665,7	246,219
1461	1734,15	1957,77	1445,57	727,88	70,368	8,7474	55111,1	40692,7	246,239
1462	1735,15	1959,02	1446,52	729,66	70,237	8,7481	55146,4	40719,7	246,260
1463	1736,15	1960,27	1447,48	731,45	70,105	8,7489	55181,7	40746,7	246,280
1464	1737,15	1961,53	1448,44	733,24	69,974	8,7496	55217,1	40773,7	246,301
1465	1738,15	1962,78	1449,40	735,03	69,844	8,7503	55252,4	40800,7	246,321
1466	1739,15	1964,04	1450,36	736,83	69,713	8,7510	55287,7	40827,7	246,341
1467	1740,15	1965,29	1451,32	738,63	69,583	8,7517	55323,0	40854,7	246,361
1468	1741,15	1966,55	1452,28	740,44	69,453	8,7525	55358,4	40881,7	246,382
1469	1742,15	1967,80	1453,24	742,25	69,324	8,7532	55393,7	40908,7	246,402
1470	1743,15	1969,06	1454,20	744,06	69,195	8,7539	55429,0	40935,8	246,422
1471	1744,15	1970,31	1455,16	745,88	69,066	8,7546	55464,4	40962,8	246,443
1472	1745,15	1971,57	1456,12	747,70	68,937	8,7553	55499,7	40989,8	246,463
1473	1746,15	1972,83	1457,08	749,52	68,809	7,7561	55535,0	41016,8	246,483
1474	1747,15	1974,08	1458,04	751,35	68,681	8,7568	55570,4	41043,9	246,503
1475	1748,15	1975,34	1459,00	753,18	68,553	8,7575	55605,7	41070,9	246,524
1476	1749,15	1976,59	1459,96	755,01	68,426	8,7582	55641,1	41097,9	246,544
1477	1750,15	1977,85	1460,92	756,85	68,299	8,7589	55676,4	41125,0	246,564
1478	1751,15	1979,10	1461,88	758,69	68,172	8,7597	55711,8	41152,0	246,584
1479	1752,15	1980,36	1462,84	760,53	68,046	8,7604	55747,1	41179,1	246,604
1480	1753,15	1981,62	1463,80	762,38	67,920	8,7611	55782,5	41206,1	246,625
1481	1754,15	1982,87	1464,77	764,23	67,794	8,7618	55817,9	41233,1	246,645
1482	1755,15	1984,13	1465,73	766,08	67,668	8,7625	55853,2	41260,2	246,665
1483	1756,15	1985,39	1466,69	767,94	67,543	8,7632	55888,6	41287,2	246,685
1484	1757,15	1986,64	1467,65	769,80	67,418	8,7639	55924,0	41314,3	246,705
1485	1758,15	1987,90	1468,61	771,67	67,293	8,7647	55959,3	41341,3	246,725
1486	1759,15	1989,15	1469,57	773,54	67,169	8,7654	55994,7	41368,4	246,745
1487	1760,15	1990,41	1470,53	775,41	67,045	8,7661	56030,1	41395,5	246,765
1488	1761,15	1991,67	1471,49	777,29	66,921	8,7668	56065,5	41422,5	246,786
1489	1762,15	1992,92	1472,45	779,17	66,797	8,7675	56100,8	41449,6	246,806
1490	1763,15	1994,18	1473,42	781,05	66,674	8,7682	56136,2	41476,7	246,826
1491	1764,15	1995,44	1474,38	782,94	66,551	8,7689	56171,6	41503,7	246,846
1492	1765,15	1996,70	1475,34	784,83	66,428	8,7697	56207,0	41530,8	246,866
1493	1766,15	1997,95	1476,30	786,72	66,306	8,7704	56242,4	41557,9	246,886
1494	1767,15	1999,21	1477,26	788,62	66,184	8,7711	56277,8	41585,0	246,906
1495	1768,15	2000,47	1478,22	790,52	66,062	8,7718	56313,2	41612,0	246,926
1496	1769,15	2001,72	1479,19	792,43	65,941	8,7725	56348,5	41639,1	246,946
1497	1770,15	2002,98	1480,15	794,34	65,819	8,7732	56383,9	41666,2	246,966
1498	1771,15	2004,24	1481,11	796,25	65,698	8,7739	56419,3	41693,3	246,986
1499	1772,15	2005,50	1482,07	798,16	65,578	8,7746	56454,8	41720,4	247,006
1500	1773,15	2006,76	1483,04	800,08	65,457	8,7753	56490,2	41747,5	247,026

Table III.7 Oxygen ($\mu = 31.9988$)

t	T	c_p	μc_p	c_v	μc_v	$k = \dfrac{c_p}{c_v}$
−50	223,15	0,9107	29,142	0,6509	20,828	1,399
−25	248,15	0,9112	29,157	0,6513	20,843	1,399
0	273,15	0,9132	29,224	0,6534	20,910	1,398
25	298,15	0,9167	29,334	0,6569	21,020	1,396
50	323,15	0,9213	29,482	0,6615	21,168	1,393
75	348,15	0,9268	29,659	0,6670	21,345	1,390
100	373,15	0,9331	29,860	0,6733	21,546	1,386
125	398,15	0,9400	30,081	0,6802	21,767	1,382
150	423,15	0,9473	30,315	0,6875	22,001	1,378
175	448,15	0,9550	30,559	0,6952	22,245	1,374
200	473,15	0,9628	30,810	0,7030	22,496	1,370
250	523,15	0,9787	31,318	0,7189	23,004	1,361
300	573,15	0,9944	31,820	0,7346	23,506	1,354
350	623,15	1,0094	32,302	0,7496	23,988	1,347
400	673,15	1,0236	32,754	0,7638	24,440	1,340
450	723,15	1,0366	33,172	0,7768	24,858	1,334
500	773,15	1,0485	33,553	0,7887	25,239	1,329
550	823,15	1,0593	33,898	0,7995	25,584	1,325
600	873,15	1,0691	34,210	0,8092	25,896	1,321
650	923,15	1,0779	34,492	0,8181	26,178	1,318
700	973,15	1,0858	34,747	0,8260	26,433	1,314
750	1023,15	1,0931	34,979	0,8333	26,665	1,312
800	1073,15	1,0998	35,192	0,8399	26,878	1,309
850	1123,15	1,1059	35,390	0,8461	27,076	1,307
900	1173,15	1,1118	35,576	0,8519	27,262	1,305
950	1223,15	1,1172	35,751	0,8574	27,437	1,303
1000	1273,15	1,1224	35,916	0,8626	27,602	1,301
1050	1323,15	1,1272	36,072	0,8674	27,758	1,300
1100	1373,15	1,1319	36,220	0,8721	27,906	1,298
1150	1423,15	1,1362	36,359	0,8764	28,045	1,296
1200	1473,15	1,1403	36,489	0,8805	28,175	1,295
1250	1523,15	1,1441	36,610	0,8842	28,296	1,294
1300	1573,15	1,1477	36,725	0,8878	28,411	1,293
1350	1623,15	1,1512	36,837	0,8913	28,523	1,292
1400	1673,15	1,1547	36,951	0,8949	28,637	1,290
1450	1723,15	1,1586	37,075	0,8988	28,761	1,289
1500	1773,15	1,1632	37,223	0,9034	28,909	1,288

Table III.8 Oxygen ($\mu = 31.9988$)

t	T	h	u	π_0	θ_0	s^0	H	U	S
—50	223,15	202,19	144,21	1,8586	3119,5	6,1440	6469,8	4614,5	196,600
—49	224,15	203,10	144,86	1,8880	3084,8	6,1481	6499,0	4635,3	196,730
—48	225,15	204,01	145,51	1,9176	3050,6	6,1521	6528,1	4656,1	196,860
—47	226,15	204,92	146,16	1,9477	3016,9	6,1561	6557,3	4677,0	196,989
—46	227,15	205,83	146,81	1,9780	2983,8	6,1602	6586,4	4697,8	197,118
—45	228,15	206,74	147,46	2,0087	2951,1	6,1642	6615,5	4718,6	197,246
—44	229,15	207,65	148,11	2,0397	2919,0	6,1681	6641,7	4739,4	197,373
—43	230,15	208,56	148,76	2,0711	2887,3	6,1721	6673,8	4760,3	197,500
—42	231,15	209,48	149,41	2,1028	2856,1	6,1761	6703,0	4781,1	197,626
—41	232,15	210,39	150,07	2,1349	2825,4	6,1800	6732,1	4801,9	197,752
—40	233,15	211,30	150,72	2,1673	2795,1	6,1839	6761,2	4822,7	197,877
—39	234,15	212,21	151,37	2,2000	2765,3	6,1878	6790,4	4843,6	198,002
—38	235,15	213,12	152,02	2,2331	2736,0	6,1917	6819,5	4864,4	198,126
—37	236,15	214,03	152,67	2,2666	2707,0	6,1955	6848,7	4885,2	198,250
—36	237,15	214,94	153,32	2,3004	2678,5	6,1994	6877,8	4906,1	198,373
—35	238,15	215,85	153,97	2,3346	2650,4	6,2032	6907,0	4926,9	198,496
—34	239,15	216,76	154,62	2,3692	2622,8	6,2070	6936,1	4947,7	198,618
—33	240,15	217,67	155,27	2,4041	2595,5	6,2108	6965,2	4968,5	198,740
—32	241,15	218,58	155,92	2,4393	2568,6	6,2146	6994,4	4989,4	198,861
—31	242,15	219,49	156,57	2,4750	2542,1	6,2184	7023,5	5010,2	198,981
—30	243,15	220,40	157,23	2,5110	2516,0	6,2222	7052,7	5031,0	199,101
—29	244,15	221,32	157,88	2,5474	2490,2	6,2259	7081,8	5051,9	199,221
—28	245,15	222,23	158,53	2,5842	2464,9	6,2296	7111,0	5072,7	199,340
—27	246,15	223,14	159,18	2,6213	2439,8	6,2333	7140,1	5093,5	199,459
—26	247,15	224,05	159,83	2,6589	2415,2	6,2370	7169,3	5114,4	199,577
—25	248,15	224,96	160,48	2,6968	2390,8	6,2407	7198,5	5135,2	199,695
—24	249,15	225,87	161,13	2,7351	2366,9	6,2444	7227,6	5156,1	199,812
—23	250,15	226,78	161,78	2,7738	2343,2	6,2480	7256,8	5176,9	199,929
—22	251,15	227,69	162,44	2,8129	2319,9	6,2516	7285,9	5197,8	200,045
—21	252,15	228,61	163,09	2,8523	2296,9	6,2553	7315,1	5218,6	200,161
—20	253,15	229,52	163,74	2,8922	2274,2	6,2589	7344,3	5239,5	200,277
—19	254,15	230,43	164,39	2,9325	2251,8	6,2625	7373,4	5260,3	200,392
—18	255,15	231,34	165,04	2,9732	2229,8	6,2661	7402,6	5281,2	200,506
—17	256,15	232,25	165,69	3,0143	2208,0	6,2696	7431,8	5302,0	200,620
—16	257,15	233,16	166,35	3,0558	2186,5	6,2732	7460,9	5322,9	200,734
—15	258,15	234,07	167,00	3,0977	2165,3	6,2767	7490,1	5343,8	200,847
—14	259,15	234,99	167,65	3,1400	2144,4	6,2802	7519,3	5364,0	200,960
—13	260,15	235,90	168,30	3,1827	2123,8	6,2837	7548,5	5385,5	201,072
—12	261,15	236,81	168,95	3,2259	2103,4	6,2872	7577,7	5406,4	201,184
—11	262,15	237,72	169,61	3,2694	2083,3	6,2907	7606,8	5427,2	201,296
—10	263,15	238,64	170,26	3,3134	2063,5	6,2942	7636,0	5448,1	201,407
—9	264,15	239,55	170,91	3,3578	2044,0	6,2977	7665,2	5469,0	201,518
—8	265,15	240,46	171,56	3,4027	2024,6	6,3011	7694,4	5489,9	201,628
—7	266,15	241,37	172,22	3,4480	2005,6	6,3045	7723,6	5510,7	201,738
—6	267,15	242,28	172,87	3,4937	1986,8	6,3080	7752,8	5531,6	201,847
—5	268,15	243,20	173,52	3,5398	1968,2	6,3114	7782,0	5552,5	201,957
—4	269,15	244,11	174,18	3,5864	1949,9	6,3148	7811,2	5573,4	202,065
—3	270,15	245,02	174,83	3,6335	1931,8	6,3182	7840,4	5594,3	202,174
—2	271,15	245,94	175,48	3,6810	1913,9	6,3215	7869,7	5615,2	202,282
—1	272,15	246,85	176,14	3,7289	1896,3	6,3249	7898,9	5636,1	202,389
0	273,15	247,76	176,79	3,7773	1878,9	6,3282	7928,1	5657,0	202,496
1	274,15	248,68	177,44	3,8261	1861,7	6,3316	7957,3	5677,9	202,603
2	275,15	249,59	178,10	3,8754	1844,7	6,3349	7986,6	5698,8	202,710
3	276,15	250,50	178,75	3,9251	1828,0	6,3382	8015,8	5719,8	202,816
4	277,15	251,42	179,40	3,9753	1811,4	6,3415	8045,0	5740,7	202,921

Oxygen (*Continued*)

t	T	h	u	π_0	θ_0	s^0	H	U	S
5	278,15	252,33	180,06	4,0260	1795,1	6,3448	8074,3	5761,6	203,027
6	279,15	253,24	180,71	4,0772	1778,9	6,3481	8103,5	5782,5	203,132
7	280,15	254,16	181,37	4,1288	1763,0	6,3514	8132,8	5803,5	203,236
8	281,15	255,07	182,02	4,1809	1747,2	6,3546	8162,0	5824,4	203,340
9	282,15	255,99	182,67	4,2334	1731,7	6,3579	8191,3	5845,4	203,444
10	283,15	256,90	183,33	4,2865	1716,3	6,3611	8220,5	5866,3	203,548
11	284,15	257,82	183,98	4,3400	1701,1	6,3643	8249,8	5887,3	203,651
12	285,15	258,73	184,64	4,3940	1686,1	6,3675	8279,1	5908,2	203,754
13	286,15	259,65	185,29	4,4485	1671,3	6,3707	8308,3	5929,2	203,856
14	287,15	260,56	185,95	4,5035	1656,7	6,3739	8337,6	5950,1	203,958
15	288,15	261,48	186,60	4,5589	1642,2	6,3771	8366,9	5971,1	204,060
16	289,15	262,39	187,26	4,6149	1627,9	6,3803	8396,2	5992,1	204,162
17	290,15	263,31	187,91	4,6714	1613,8	6,3835	8425,5	6013,1	204,263
18	291,15	264,22	188,57	4,7284	1599,9	6,3866	8454,8	6034,0	204,364
19	292,15	265,14	189,23	4,7859	1586,1	6,3897	8484,1	6055,0	204,464
20	293,15	266,05	189,88	4,8438	1572,5	6,3929	8513,4	6076,0	204,564
21	294,15	266,97	190,54	4,9024	1559,0	6,3960	8542,7	6097,0	204,664
22	295,15	267,89	191,20	4,9624	1545,7	6,3991	8572,0	6118,0	204,763
23	296,15	268,80	191,85	5,0209	1532,5	6,4022	8601,3	6139,0	204,863
24	297,15	269,72	192,51	5,0810	1519,5	6,4053	8630,7	6160,0	204,962
25	298,15	270,63	193,17	5,1415	1506,7	6,4084	8660,0	6181,1	205,060
26	299,15	271,55	193,82	5,2026	1494,0	6,4114	8689,3	6202,1	205,158
27	300,15	272,47	194,48	5,2643	1481,4	6,4145	8718,7	6223,1	205,256
28	301,15	273,39	195,14	5,3265	1469,0	6,4175	8748,0	6244,1	205,354
29	302,15	274,30	195,79	5,3892	1456,7	6,4206	8777,4	6265,2	205,451
30	303,15	275,22	196,45	5,4524	1444,6	6,4236	8806,7	6286,2	205,548
31	304,15	276,14	197,11	5,5162	1432,6	6,4266	8836,1	6307,3	205,645
32	305,15	277,06	197,77	5,5805	1420,8	6,4297	8865,5	6328,3	205,741
33	306,15	277,97	198,43	5,6454	1409,0	6,4327	8894,8	6349,4	205,837
34	307,15	278,89	199,08	5,7108	1397,4	6,4357	8924,2	6370,4	205,933
35	308,15	279,81	199,74	5,7768	1386,0	6,4386	8953,6	6391,5	206,029
36	309,15	280,73	200,40	5,8434	1374,6	6,4416	8983,0	6412,6	206,124
37	310,15	281,65	201,06	5,9105	1363,4	6,4446	9012,4	6433,7	206,219
38	311,15	282,57	201,72	5,9781	1352,3	6,4475	9041,8	6454,8	206,314
39	312,15	283,49	202,38	6,0464	1341,4	6,4505	9071,2	6475,9	206,408
40	313,15	284,41	203,04	6,1152	1330,5	6,4534	9100,6	6497,0	206,502
41	314,15	285,32	203,70	6,1846	1319,8	6,4564	9130,0	6518,1	206,596
42	315,15	286,24	204,36	6,2545	1309,2	6,4593	9159,5	6539,2	206,689
43	316,15	287,16	205,02	6,3251	1298,7	6,4622	9188,9	6560,3	206,783
44	317,15	288,08	205,68	6,3962	1288,3	6,4651	9218,3	6581,4	206,876
45	318,15	289,00	206,34	6,4679	1278,0	6,4680	9247,8	6602,6	206,968
46	319,15	289,92	207,00	6,5402	1267,9	6,4709	9277,2	6623,7	207,061
47	320,15	290,85	207,66	6,6131	1257,8	6,4738	9306,7	6644,8	207,153
48	321,15	291,77	208,32	6,6866	1247,9	6,4766	9336,2	6666,0	207,245
49	322,15	292,69	208,98	6,7607	1238,1	6,4795	9365,6	6687,2	207,336
50	323,15	293,61	209,64	6,8354	1228,3	6,4824	9395,1	6708,3	207,428
51	324,15	294,53	210,30	6,9107	1218,7	6,4852	9424,6	6729,5	207,519
52	325,15	295,45	210,97	6,9866	1209,2	6,4880	9454,1	6750,7	207,610
53	326,15	296,37	211,63	7,0632	1199,8	6,4909	9483,6	6771,9	207,700
54	327,15	297,30	212,29	7,1403	1190,4	6,4937	9513,1	6793,0	207,791
55	328,15	298,22	212,95	7,2181	1181,2	6,4965	9542,6	6814,2	207,881
56	329,15	299,14	213,62	7,2965	1172,1	6,4993	9572,1	6835,4	207,970
57	330,15	300,06	214,28	7,3755	1163,0	6,5021	9601,7	6856,7	208,060
58	331,15	300,99	214,94	7,4552	1154,1	6,5049	9631,2	6877,9	208,149
59	332,15	301,91	215,60	7,5355	1145,3	6,5077	9660,7	6899,1	208,238

Oxygen (*Continued*)

t	T	h	u	π_0	θ_0	s^0	H	U	S
60	333,15	302,83	216,27	7,6164	1136,5	6,5105	9690,3	6920,3	208,327
61	334,15	303,76	216,93	7,6980	1127,8	6,5132	9719,8	6941,6	208,416
62	335,15	304,68	217,60	7,7802	1119,2	6,5160	9749,4	6962,8	208,504
63	336,15	305,60	218,26	7,8631	1110,8	6,5188	9779,0	6984,1	208,592
64	337,15	306,53	218,92	7,9466	1102,3	6,5215	9808,5	7005,3	208,680
65	338,15	307,45	219,59	8,0307	1094,0	6,5242	9838,1	7026,6	208,768
66	339,15	308,38	220,25	8,1156	1085,8	6,5270	9867,7	7047,9	208,855
67	340,15	309,30	220,92	8,2011	1077,6	6,5297	9897,3	7069,1	208,942
68	341,15	310,23	221,58	8,2873	1069,6	6,5324	9926,9	7090,4	209,029
69	342,15	311,15	222,25	8,3741	1061,6	6,5351	9956,5	7111,7	209,116
70	343,15	312,08	222,92	8,4616	1053,7	6,5378	9986,1	7133,0	209,202
71	344,15	313,00	223,58	8,5498	1045,8	6,5405	10015,7	7154,3	209,288
72	345,15	313,93	224,25	8,6387	1038,1	6,5432	10045,4	7175,7	209,374
73	346,15	314,86	224,91	8,7282	1030,4	6,5459	10075,0	7197,0	209,460
74	347,15	315,78	225,58	8,8185	1022,8	6,5485	10104,7	7218,3	209,546
75	348,15	316,71	226,25	8,9094	1015,3	6,5512	10134,3	7239,7	209,631
76	349,15	317,64	226,92	9,0010	1007,8	6,5539	10164,0	7261,0	209,716
77	350,15	318,56	227,58	9,0934	1000,5	6,5565	10193,7	7282,4	209,801
78	351,15	319,49	228,25	9,1864	993,21	6,5592	10223,3	7303,7	209,885
79	352,15	320,42	228,92	9,2802	985,97	6,5618	10253,0	7325,1	209,970
80	353,15	321,35	229,59	9,3746	978,81	6,5644	10282,7	7346,5	210,054
81	354,15	322,28	230,25	9,4698	971,72	6,5671	10312,4	7367,9	210,138
82	355,15	323,20	230,92	9,5657	964,69	6,5697	10342,1	7389,3	210,222
83	356,15	324,13	231,59	9,6623	957,73	6,5723	10371,8	7410,7	210,305
84	357,15	325,06	232,26	9,7597	950,84	6,5749	10401,6	7432,1	210,389
85	358,15	325,99	232,93	9,8577	944,02	6,5775	10431,3	7453,5	210,472
86	359,15	326,92	233,60	9,9565	937,26	6,5801	10461,0	7474,9	210,555
87	360,15	327,85	234,27	10,056	930,57	6,5827	10490,8	7496,4	210,638
88	361,15	328,78	234,94	10,156	923,94	6,5852	10520,5	7517,8	210,720
89	362,15	329,71	235,61	10,257	917,37	6,5878	10550,3	7539,2	210,802
90	363,15	330,64	236,28	10,359	910,86	6,5904	10580,1	7560,7	210,884
91	364,15	331,57	236,95	10,461	904,42	6,5929	10609,9	7582,2	210,966
92	365,15	332,50	237,62	10,565	898,04	6,5955	10639,7	7603,6	211,048
93	366,15	333,43	238,29	10,669	891,71	6,5980	10669,5	7625,1	211,130
94	367,15	334,36	238,97	10,773	885,45	6,6006	10699,3	7646,6	211,211
95	368,15	335,30	239,64	10,879	879,24	6,6031	10729,1	7668,1	211,292
96	369,15	336,23	240,31	10,985	873,09	6,6056	10758,9	7689,6	211,373
97	370,15	337,16	240,98	11,093	867,00	6,6082	10788,7	7711,1	211,454
98	371,15	338,09	241,66	11,201	860,97	6,6107	10818,6	7732,7	211,534
99	372,15	339,03	242,33	11,309	854,99	6,6132	10848,4	7754,2	211,614
100	373,15	339,96	243,00	11,419	849,06	6,6157	10878,3	7775,7	211,694
101	374,15	340,89	243,67	11,529	843,19	6,6182	10908,1	7797,3	211,774
102	375,15	341,83	244,35	11,640	837,38	6,6207	10938,0	7818,9	211,854
103	376,15	342,76	245,02	11,752	831,62	6,6232	10967,9	7840,4	211,934
104	377,15	343,69	245,70	11,865	825,91	6,6257	10997,8	7862,0	212,013
105	378,15	344,63	246,37	11,978	820,25	6,6281	11027,7	7883,6	212,092
106	379,15	345,56	247,05	12,093	814,64	6,6306	11057,6	7905,2	212,171
107	380,15	346,50	247,72	12,208	809,09	6,6331	11087,5	7926,8	212,250
108	381,15	347,43	248,40	12,324	803,58	6,6355	11117,4	7948,4	212,329
109	382,15	348,37	249,07	12,441	798,12	6,6380	11147,4	7970,0	212,407
110	383,15	349,30	249,75	12,558	792,72	6,6404	11177,3	7991,6	212,485
111	384,15	350,24	250,42	12,677	787,36	6,6429	11207,3	8013,3	212,563
112	385,15	351,18	251,10	12,796	782,05	6,6453	11237,2	8034,9	212,641
113	386,15	352,11	251,78	12,916	776,78	6,6477	11267,2	8056,6	212,719
114	387,15	353,05	252,45	13,037	771,57	6,6501	11297,2	8078,2	212,796

126

Oxygen (*Continued*)

t	r	h	u	π_0	θ_0	s^0	H	U	S
115	388,15	353,99	253,13	13,159	766,40	6,6525	11327,1	8099,9	212,874
116	389,15	354,92	253,81	13,282	761,27	6,6550	11357,1	8121,6	212,951
117	390,15	355,86	254,49	13,405	756,19	6,6574	11387,1	8143,3	213,028
118	391,15	356,80	255,17	13,530	751,16	6,6598	11417,2	8165,0	213,105
119	392,15	357,74	255,84	13,655	746,17	6,6622	11447,2	8186,7	213,181
120	393,15	358,68	256,52	13,781	741,22	6,6646	11477,0	8208,4	213,258
121	394,15	359,62	257,20	13,908	736,32	6,6669	11507,3	8230,1	213,334
122	395,15	360,55	257,88	14,036	731,45	6,6693	11537,3	8251,9	213,410
123	396,15	361,49	258,56	14,165	726,64	6,6717	11567,4	8273,6	213,486
124	397,15	362,43	259,24	14,295	721,86	6,6741	11597,4	8295,4	213,562
125	398,15	363,37	259,92	14,426	717,12	6,6764	11627,5	8317,1	213,638
126	399,15	364,31	260,60	14,557	712,43	6,6788	11657,6	8338,9	213,713
127	400,15	365,25	261,28	14,690	707,77	6,6811	11687,7	8360,7	213,789
128	401,15	366,19	261,96	14,823	703,16	6,6835	11717,8	8382,5	213,864
129	402,15	367,14	262,64	14,957	698,58	6,6858	11747,9	8404,3	213,939
130	403,15	368,08	263,32	15,092	694,05	6,6882	11778,0	8426,1	214,014
131	404,15	369,02	264,01	15,228	689,55	6,6905	11808,1	8447,9	214,088
132	405,15	369,96	264,69	15,366	685,09	6,6928	11838,3	8469,7	214,163
133	406,15	370,90	265,37	15,504	680,67	6,6952	11868,4	8491,5	214,237
134	407,15	371,85	266,05	15,642	676,29	6,6975	11898,6	8513,4	214,311
135	408,15	372,79	266,74	15,782	671,94	6,6998	11928,8	8535,2	214,385
136	409,15	373,73	267,42	15,923	667,63	6,7021	11958,9	8557,1	214,459
137	410,15	374,67	268,10	16,065	663,36	6,7044	11989,1	8579,0	214,533
138	411,15	375,62	268,79	16,208	659,12	6,7067	12019,3	8600,9	214,606
139	412,15	376,56	269,47	16,351	654,92	6,7090	12049,5	8622,7	214,680
140	413,15	377,51	270,16	16,496	650,75	6,7113	12079,7	8644,6	214,753
141	414,15	378,45	270,84	16,641	646,62	6,7136	12110,0	8666,6	214,826
142	415,15	379,40	271,53	16,788	642,52	6,7158	12140,2	8688,5	214,899
143	416,15	380,34	272,21	16,936	638,46	6,7181	12170,4	8710,4	214,972
144	417,15	381,29	272,90	17,084	634,42	6,7204	12200,7	8732,3	215,044
145	418,15	382,23	273,58	17,234	630,43	6,7226	12231,0	8754,3	215,117
146	419,15	383,18	274,27	17,384	626,46	6,7249	12261,2	8776,3	215,189
147	420,15	384,12	274,95	17,536	622,53	6,7272	12291,5	8798,2	215,261
148	421,15	385,07	275,64	17,688	618,63	6,7294	12321,8	8820,2	215,333
149	422,15	386,02	276,33	17,842	614,76	6,7317	12352,1	8842,2	215,405
150	423,15	386,97	277,02	17,996	610,93	6,7339	12382,4	8864,2	215,477
151	424,15	387,91	277,70	18,152	607,12	6,7361	12412,7	8886,2	215,548
152	425,15	388,86	278,39	18,309	603,35	6,7384	12443,1	8908,2	215,620
153	426,15	389,81	279,08	18,466	599,61	6,7406	12473,4	8930,2	215,691
154	427,15	390,76	279,77	18,625	595,89	6,7428	12503,8	8952,3	215,762
155	428,15	391,71	280,46	18,785	592,21	6,7450	12534,1	8974,3	215,833
156	429,15	392,65	281,15	18,945	588,56	6,7473	12564,5	8996,4	215,904
157	430,15	393,60	281,84	19,107	584,93	6,7495	15594,9	9018,4	215,975
158	431,15	394,55	282,53	19,270	581,34	6,7517	12625,2	9040,5	216,045
159	432,15	395,50	283,22	19,434	577,77	6,7539	12655,6	9062,6	216,116
160	433,15	396,45	283,91	19,599	574,23	6,7561	12686,1	9084,7	216,186
161	434,15	397,40	284,60	19,765	570,73	6,7583	12716,5	9106,8	216,256
162	435,15	398,36	285,29	19,932	567,24	6,7604	12746,9	9128,9	216,326
163	436,15	399,31	285,98	20,100	563,79	6,7626	12777,3	9151,0	216,396
164	437,15	400,26	286,67	20,270	560,36	6,7648	12807,8	9173,1	216,466
165	438,15	401,21	287,36	20,440	556,96	6,7670	12838,2	9195,3	216,535
166	439,15	402,16	288,06	20,611	553,59	6,7691	12868,7	9217,4	216,605
167	440,15	403,11	288,75	20,784	550,25	6,7713	12899,2	9239,6	216,674
168	441,15	404,07	289,44	20,958	546,93	6,7735	12929,7	9261,8	216,743
169	442,15	405,02	290,13	21,132	543,63	6,7756	12960,2	9283,9	216,812

Oxygen (*Continued*)

t	T	h	u	π_0	θ_0	s^0	H	U	S
170	443,15	405,97	290,83	21,308	540,36	6,7778	12990,7	9306,1	216,881
171	444,15	406,93	291,52	21,485	537,12	6,7799	13021,2	9328,3	216,950
172	445,15	407,88	292,22	21,663	533,91	6,7821	13051,7	9350,5	217,019
173	446,15	408,84	292,91	21,843	530,71	6,7842	13082,2	9372,8	217,087
174	447,15	409,79	293,60	22,023	527,55	6,7864	13112,8	9395,0	217,155
175	448,15	410,74	294,30	22,205	524,40	6,7885	13143,3	9417,2	217,224
176	449,15	411,70	294,99	22,387	521,28	6,7906	13173,9	9439,5	217,292
177	450,15	412,66	295,69	22,571	518,19	6,7927	13204,5	9461,7	217,360
178	451,15	413,61	296,39	22,756	515,12	6,7949	13235,1	9484,0	217,428
179	452,15	414,57	297,08	22,942	512,07	6,7970	13265,6	9506,3	217,495
180	453,15	415,52	297,78	23,130	509,05	6,7991	13296,3	9528,6	217,563
181	454,15	416,48	298,48	23,318	506,05	6,8012	13326,9	9550,9	217,631
182	455,15	417,44	299,17	23,508	503,07	6,8033	13357,5	9573,2	217,698
183	456,15	418,39	299,87	23,699	500,11	6,8054	13388,1	9595,5	217,765
184	457,15	419,35	300,57	23,891	497,18	6,8075	13418,8	9617,8	217,832
185	458,15	420,31	301,27	24,084	494,27	6,8096	13449,4	9640,2	217,899
186	459,15	421,27	301,97	24,278	491,38	6,8117	13480,1	9662,5	217,966
187	460,15	422,23	302,66	24,474	488,52	6,8138	13510,8	9684,9	218,033
188	461,15	423,19	303,36	24,671	485,67	6,8159	13541,4	9707,3	218,099
189	462,15	424,15	304,06	24,869	482,85	6,8179	13572,1	9729,6	218,166
190	463,15	425,10	304,76	25,068	480,05	6,8200	13602,8	9752,0	218,232
191	464,15	426,06	305,46	25,269	477,26	6,8221	13633,6	9774,4	218,298
192	465,15	427,02	306,16	25,471	474,50	6,8242	13664,3	9796,8	218,365
193	466,15	427,99	306,86	25,674	471,76	6,8262	13695,0	9819,3	218,431
194	467,15	428,95	307,56	25,878	469,04	6,8283	13725,8	9841,7	218,497
195	468,15	429,91	308,27	26,083	466,34	6,8303	13756,5	9864,1	218,562
196	469,15	430,87	308,97	26,290	463,67	6,8324	13787,3	9886,6	218,628
197	470,15	431,83	309,67	26,498	461,01	6,8344	13818,1	9909,0	218,693
198	471,15	432,79	310,37	26,707	458,37	6,8365	13848,8	9931,5	218,759
199	472,15	433,75	311,07	26,918	455,75	6,8385	13879,6	9954,0	218,824
200	473,15	434,72	311,78	27,130	453,15	6,8405	13910,4	9976,5	218,889
201	474,15	435,68	312,48	27,343	450,56	6,8426	13941,2	9999,0	218,954
202	475,15	436,64	313,18	27,557	448,00	6,8446	13972,1	10021,5	219,019
203	476,15	437,61	313,89	27,773	445,46	6,8466	14002,9	10044,0	219,084
204	477,15	438,57	314,59	27,990	442,93	6,8487	14033,8	10066,5	219,149
205	478,15	439,54	315,30	28,208	440,43	6,8507	14064,6	10089,1	219,213
206	479,15	440,50	316,00	28,428	437,94	6,8527	14095,5	10111,6	219,278
207	480,15	441,47	316,71	28,649	435,47	6,8547	14126,4	10134,2	219,342
208	481,15	442,43	317,41	28,871	433,01	6,8567	14157,2	10156,8	219,407
209	482,15	443,40	318,12	29,095	430,58	6,8587	14188,1	10179,3	219,471
210	483,15	444,36	318,82	29,320	428,16	6,8607	14219,0	10201,9	219,535
211	484,15	445,33	319,53	29,546	425,76	6,8627	14250,0	10224,5	219,599
212	485,15	446,29	320,24	29,774	423,38	6,8647	14280,9	10247,1	219,662
213	486,15	447,26	320,94	30,003	421,01	6,8667	14311,8	10269,8	219,726
214	487,15	448,23	321,65	30,233	418,67	6,8687	14342,8	10292,4	219,790
215	488,15	449,20	322,36	30,465	416,33	6,8707	14373,7	10315,0	219,853
216	489,15	450,16	323,07	30,698	414,02	6,8727	14404,7	10337,7	219,917
217	490,15	451,13	323,77	30,932	411,72	6,8746	14435,7	10360,4	219,980
218	491,15	452,10	324,48	31,168	409,44	6,8766	14466,7	10383,0	220,043
219	492,15	453,07	325,19	31,405	407,17	6,8786	14497,7	10405,7	220,106
220	493,15	454,04	325,90	31,644	404,92	6,8805	14528,7	10428,4	220,169
221	494,15	455,01	326,61	31,884	402,69	6,8825	14559,7	10451,1	220,232
222	495,15	455,98	327,32	32,126	400,47	6,8845	14590,7	10473,8	220,295
223	496,15	456,95	328,03	32,368	398,27	6,8864	14621,7	10496,6	220,357
224	497,15	457,92	328,74	32,613	396,08	6,8884	14652,8	10519,3	220,420

Oxygen (*Continued*)

t	T	h	u	π_0	θ_0	s^0	H	U	S
225	498,15	458,89	329,45	32,858	393,91	6,8903	14683,9	10542,0	220,482
226	499,15	459,86	330,16	33,106	391,76	6,8923	14714,9	10564,8	220,544
227	500,15	460,83	330,87	33,354	389,61	6,8942	14746,0	10587,5	220,607
228	501,15	461,80	331,59	33,604	387,49	6,8962	14777,1	10610,3	220,669
229	502,15	462,77	332,30	33,856	385,38	6,8981	14808,2	10633,1	220,731
230	503,15	463,75	333,01	34,109	383,28	6,9000	14839,3	10655,9	220,793
231	504,15	464.72	333,72	34,363	381,20	6,9020	14870,4	10678,7	220,854
232	505,15	465,69	334,44	34,619	379,13	6,9039	14901,6	10701,5	220,916
233	506,15	466,66	335,14	34,876	377,08	6,9058	14932,7	10724,4	220,978
234	507,15	467,64	335,86	35,135	375,04	6,9077	14963,8	10747,2	221,039
235	508,15	468,61	336,58	35,396	373,02	6,9097	14995,0	10770,0	221,101
236	509,15	469,59	337,29	35,658	371,01	6,9116	15026,2	10792,9	221,162
237	510,15	470,56	338,01	35,921	369,01	6,9135	15057,4	10815,8	221,223
238	511,15	471,53	338,72	36,186	367,03	6,9154	15088,5	10838,6	221,284
239	512,15	472,51	339,44	36,452	365,06	6,9173	15119,7	10861,5	221,345
240	513,15	473.49	340,15	36,720	363,10	6,9192	15151,0	10884,4	221,406
241	514,15	474.46	340,87	36,989	361,16	6,9211	15182,2	10907,3	221,467
242	515,15	475,44	341,58	37,260	359,23	6,9230	15213,4	10930,2	221,527
243	516,15	476,41	342,30	37,533	357,31	6,9249	15244,7	10953,2	221,588
244	517,15	477,39	343,02	37,807	355,41	6,9268	15275,9	10976,1	221,649
245	518,15	478,37	343,73	38,083	353,52	6,9287	15307,2	10999,1	221,709
246	519,15	479,34	344,45	38,360	351,64	6,9305	15338,4	11022,0	221,769
247	520,15	480,32	345,17	38,639	349,78	6,9324	15369,7	11045,0	221,829
248	521,15	481,30	345,89	38,919	347,93	6,9343	15401,0	11068,0	221,890
249	522,15	482,28	346,61	39,201	346,09	6,9362	15432,3	11091,0	221,950
250	523,15	483,26	347,32	39,484	344,26	6,9381	15463,6	11113,9	222,009
251	524,15	484,24	348,04	39,769	342,45	6,9399	15495,0	11137,0	222,069
252	525,15	485,21	348,76	40,056	340,64	6,9418	15526,3	11160,0	222,129
253	526,15	486,19	349,48	40,344	338,85	6,9437	15557,6	11183,0	222,189
254	527,15	487,17	350,20	40,634	337,08	6,9455	15589,0	11206,0	222,248
255	528,15	488,15	350,92	40,926	335,31	6,9474	15620,4	11229,1	222,308
256	529,15	489,13	351,64	41,219	333,55	6,9492	15651,7	11252,2	222,367
257	530,15	490,12	352,36	41,514	331,81	6,9511	15683,1	11275,2	222,426
258	531,15	491,10	353,09	41,810	330,08	6,9529	15714,5	11298,3	222,485
259	532,15	492,08	353,81	42,108	328,36	6,9548	15745,9	11321,4	222,5₋4
260	533,15	493,06	354,53	42,408	326,65	6,9566	15777,3	11344,5	222,603
261	534,15	494,04	355,25	42,710	324,96	6,9585	15808,7	11367,6	222,662
262	535,15	495,02	355,97	43,013	323,27	6,9603	15840,2	11390,7	222,721
263	536,15	496,01	356,70	43,318	321,59	6,9621	15871,6	11413,9	222,780
264	537,15	496,99	357,42	43,624	319,93	6,9640	15903,1	11437,0	222,838
265	538,15	497,97	358,14	43,932	318,28	6,9658	15934,5	11460,2	222,897
266	539,15	498,96	358,87	44,242	316,64	6,9676	15966,0	11483,3	222,955
267	540,15	499,94	359,59	44,554	315,01	6,9694	15997,5	11506,5	223,014
268	541,15	500,93	360,32	44,867	313,39	6,9713	16029,0	11529,7	223,072
269	542,15	501,91	361,04	45,182	311,78	6,9731	16060,5	11552,9	223,130
270	543,15	502,89	361,77	45,498	310,18	6,9749	16092,0	11576,1	223,188
271	544,15	503,88	362,49	45,817	308,59	6,9767	16123,6	11599,3	223,246
272	545,15	504,87	363,22	46,137	307,01	6,9785	16155,1	11622,5	223,304
273	546,15	505,85	363,94	46,459	305,44	6,9803	16186,6	11645,7	223,362
274	547,15	506,84	364,67	46,783	303,88	6,9821	16218,2	11669,0	223,420
275	548,15	507,82	365,40	47,108	302,34	6,9839	16249,8	11692,2	223,477
276	549,15	508,81	366,12	47,435	300,80	6,9857	16281,3	11715,5	223,535
277	550,15	509,80	366,85	47,764	299,27	6,9875	16312,9	11738,7	223,592
278	551,15	510,79	367,58	48,095	297,75	6,9893	16344,5	11762,0	223,650
279	552,15	511,77	368,31	48,428	296,24	6,9911	16376,1	11785,3	223,707

Oxygen (*Continued*)

t	T	h	u	π_0	θ_0	s^0	H	U	S
280	553,15	512,76	369,03	48,762	294,74	6,9929	16407,7	11808,6	223,764
281	554,15	513,75	369,76	49,098	293,26	6,9947	16439,4	11831,9	223,821
282	555,15	514,74	370,49	49,436	291,78	6,9965	16471,0	11855,3	223,878
283	556,15	515,73	371,22	49,776	290,31	6,9982	16502,6	11878,6	223,935
284	557,15	516,72	371,95	50,118	288,85	7,0000	16534,3	11901,9	223,992
285	558,15	517,71	372,68	50,461	287,40	7,0018	16566,0	11925,3	224,049
286	559,15	518,70	373,41	50,806	285,95	7,0036	16597,6	11948,6	224,106
287	560,15	519,69	374,14	51,154	284,52	7,0053	16629,3	11972,0	224,162
288	561,15	520,68	374,87	51,503	283,10	7,0071	16661,0	11995,4	224,219
289	562,15	521,67	375,60	51,854	281,68	7,0089	16692,7	12018,8	224,275
290	563,15	522,66	376,33	52,206	280,28	7,0106	16724,5	12042,2	224,332
291	564,15	523,65	377,06	52,561	278,88	7,0124	16756,2	12065,6	224,388
292	565,15	524,64	377,80	52,917	277,49	7,0141	16787,9	12089,0	224,444
293	566,15	525,63	378,53	53,276	276,11	7,0159	16819,7	12112,5	224,500
294	567,15	526,63	379,26	53,636	274,74	7,0176	16851,4	12135,9	224,556
295	568,15	527,62	379,99	53,998	273,38	7,0194	16883,2	12159,4	224,612
296	569,15	528,61	380,73	54,363	272,03	7,0211	16915,0	12182,8	224,668
297	570,15	529,61	381,46	54,729	270,68	7,0229	16946,7	12206,3	224,724
298	571,15	530,60	382,19	55,097	269,35	7,0246	16978,5	12229,8	224,780
299	572,15	531,59	382,93	55,467	268,02	7,0264	17010,3	12253,3	224,835
300	573,15	532,59	383,66	55,838	266,70	7,0281	17042,2	12276,8	224,891
301	574,15	533,58	384,40	56,212	265,39	7,0298	17074,0	12300,3	224,946
302	575,15	534,58	385,13	56,588	264,08	7,0316	17105,8	12323,8	225,002
303	576,15	535,57	385,87	56,966	262,79	7,0333	17137,7	12347,3	225,057
304	577,15	536,57	386,60	57,346	261,50	7,0350	17169,5	12370,9	225,112
305	578,15	537,56	387,34	57,727	260,22	7,0367	17201,4	12394,4	225,167
306	579,15	538,56	388,08	58,111	258,95	7,0385	17233,3	12418,0	225,223
307	580,15	539,56	388,81	58,497	257,69	7,0402	17265,1	12441,5	225,278
308	586,15	540,55	389,55	58,884	256,43	7,0419	17297,0	12465,1	225,332
309	582,15	541,55	390,29	59,274	255,18	7,0436	17328,9	12488,7	225,387
310	583,15	542,55	391,02	59,666	253,94	7,0453	17360,9	12512,3	225,442
311	584,15	543,54	391,76	60,060	252,71	7,0470	17392,8	12535,9	225,497
312	585,15	544,54	392,50	60,456	251,49	7,0487	17424,7	12559,5	225,551
313	586,15	545,54	393,24	60,854	250,27	7,0504	17456,7	12583,2	225,606
314	587,15	546,54	393,98	61,254	249,06	7,0522	17488,6	12606,8	225,660
315	588,15	547,54	394,72	61,656	247,86	7,0539	17520,6	12630,5	225,715
316	589,15	548,54	395,46	62,060	246,66	7,0555	17552,5	12654,1	225,769
317	590,15	549,54	396,20	62,466	245,47	7,0572	17584,5	12677,8	225,823
318	591,15	550,54	396,94	62,874	244,29	7,0589	17616,5	12701,5	225,878
319	592,15	551,54	397,68	63,284	243,12	7,0606	17648,5	12725,1	225,932
320	593,15	552,54	398,42	63,697	241,95	7,0623	17680,5	12748,8	225,986
321	594,15	553,54	399,16	64,111	240,79	7,0640	17712,6	12772,5	226,040
322	595,15	554,54	399,90	64,528	239,64	7,0657	17744,6	12796,3	226,093
323	596,15	555,54	400,64	64,947	238,50	7,0674	17776,6	12820,0	226,147
324	597,15	556,54	401,38	65,368	237,36	7,0690	17808,7	12843,7	226,201
325	598,15	557,54	402,12	65,791	236,23	7,0707	17840,7	12867,5	226,255
326	599,15	558,55	402,87	66,216	235,10	7,0724	17872,8	12891,2	226,308
327	600,15	559,55	403,61	66,644	233,98	7,0741	17904,9	12915,0	226,362
328	601,15	560,55	404,35	67,073	232,87	7,0757	17937,0	12938,8	226,415
329	602,15	561,55	405,09	67,505	231,77	7,0774	17969,1	12962,5	226,468
330	603,15	562,56	405,84	67,939	230,67	7,0791	18001,2	12986,3	226,522
331	604,15	563,56	406,58	68,375	229,58	7,0807	18033,3	13010,1	226,575
332	605,15	564,57	407,33	68,814	228,49	7,0824	18065,4	13034,0	226,628
333	606,15	565,57	408,07	69,254	227,42	7,0840	18097,6	13057,8	226,681
334	607,15	566,57	408,82	69,697	226,34	7,0857	18129,7	13081,6	226,734

Oxygen (*Continued*)

t	T	h	u	π_u	θ_0	s^0	H	U	S
335	608,15	567,58	409,56	70,142	225,28	7,0874	18161,9	13105,5	226,787
336	609,15	568,58	410,31	70,589	224,22	7,0890	18194,0	13129,3	226,840
337	610,15	569,59	411,05	71,039	223,17	7,0907	18226,2	13153,2	226,893
338	611,15	570,60	411,80	71,490	222,12	7,0923	18258,4	13177,0	226,945
339	612,15	571,60	412,54	71,944	221,08	7,0940	18290,6	13200,9	226,998
340	613,15	572,61	413,29	72,401	220,04	7,0956	18322,8	13224,8	227,051
341	614,15	573,61	414,04	72,859	219,02	7,0972	18355,0	13248,7	227,103
342	615,15	574,62	414,78	73,320	217,99	7,0989	18387,2	13272,6	227,155
343	616,15	575,63	415,53	73,783	216,98	7,1005	18419,4	13296,5	227,208
344	617,15	576,64	416,28	74,249	215,97	7,1021	18451,7	13320,4	227,260
345	618,15	577,64	417,03	74,716	214,96	7,1038	18483,9	13344,4	227,312
346	619,15	578,65	417,78	75,186	213,96	7,1054	18516,2	13368,3	287,364
347	620,15	579,66	418,52	75,659	212,97	7,1070	18548,5	13392,3	227,417
348	621,15	580,67	419,27	76,134	211,98	7,1087	18580,7	13416,2	227,469
349	622,15	581,68	420,02	76,611	211,00	7,1103	18613,0	13440,2	227,520
350	623,15	582,69	420,77	77,090	210,03	7,1119	18645,3	13464,2	227,572
351	624,15	583,70	421,52	77,572	209,06	7,1135	18677,6	13488,2	227,624
352	625,15	584,71	422,27	78,056	208,10	7,1151	18710,0	13512,2	227,676
353	626,15	585,72	423,02	78,543	207,14	7,1168	18742,3	13536,2	227,728
354	627,15	586,73	423,77	79,032	206,18	7,1184	18774,6	13560,2	227,779
355	628,15	587,74	424,52	79,523	205,24	7,1200	18807,0	13584,3	227,831
356	629,15	588,75	425,28	80,017	204,30	7,1216	18839,3	13608,3	227,882
357	630,15	589,76	426,03	80,513	203,36	7,1232	18871,7	13632,3	227,934
358	631,15	590,77	426,78	81,012	202,43	7,1248	18904,0	13656,4	227,985
359	632,15	591,79	427,53	81,513	201,50	7,1264	18936,4	13680,5	228,036
360	633,15	592,80	428,28	82,016	200,58	7,1280	18968,8	13704,5	228,087
361	634,15	593,81	429,04	82,522	199,67	7,1296	19001,2	13723,6	228,138
362	635,15	594,82	429,79	83,030	198,76	7,1312	19033,6	13752,7	228,190
363	636,15	595,84	430,54	83,541	197,85	7,1328	19066,0	13776,8	228,241
364	637,15	596,85	431,30	84,055	196,95	7,1344	19098,5	13800,9	228,291
365	638,15	597,86	432,05	84,570	196,06	7,1360	19130,9	13825,1	228,342
366	639,15	598,88	432,80	85,089	195,17	7,1376	19163,3	13849,2	228,393
367	640,15	599,89	433,56	85,610	194,29	7,1391	19195,8	13873,3	228,444
368	641,15	600,91	434,31	86,133	193,41	7,1407	19228,3	13897,5	228,495
369	642,15	601,92	435,07	86,659	192,53	7,1423	19260,7	13921,6	228,545
370	643,15	602,94	435,82	87,187	191,67	7,1439	19293,2	13945,8	228,596
371	644,15	603,95	436,58	87,718	190,80	7,1455	19325,7	13970,0	228,646
372	645,15	604,97	437,33	88,252	189,94	7,1470	19358,2	13994,2	228,697
373	646,15	605,98	438,09	88,788	189,09	7,1486	19390,7	14018,4	228,747
374	647,15	607,00	438,85	89,326	188,24	7,1502	19423,2	14042,6	228,797
375	648,15	608,02	439,60	89,868	187,39	7,1518	19455,8	14066,8	228,847
376	649,15	609,03	440,36	90,411	186,55	7,1533	19488,3	14091,0	228,898
377	650,15	610,05	441,12	90,958	185,72	7,1549	19520,9	14115,2	228,948
378	651,15	611,07	441,88	91,507	184,89	7,1564	19553,4	14139,5	228,998
379	652,15	612,08	442,63	92,059	184,06	7,1580	19586,0	14163,7	229,048
380	653,15	613,10	443,39	92,613	183,24	7,1596	19618,5	14188,0	229,098
381	654,15	614,12	444,15	93,170	182,43	7,1611	19651,1	14212,3	229,147
382	655,15	615,14	444,91	93,729	181,61	7,1627	19683,7	14236,5	229,197
383	656,15	616,16	445,67	94,291	180,81	7,1642	19716,3	14260,8	229,247
384	657,15	617,18	446,43	94,856	180,00	7,1658	19748,9	14285,1	229,297
385	658,15	618,20	447,19	95,424	179,21	7,1673	19781,5	14309,4	229,346
386	659,15	619,22	447,95	95,994	178,41	7,1689	19814,2	14333,7	229,396
387	660,15	620,24	448,71	96,567	177,62	7,1704	19846,8	14358,0	229,445
388	661,15	621,26	449,47	97,142	176,84	7,1720	19879,4	14382,4	229,495
389	662,15	622,28	450,23	97,721	176,06	7,1735	19912,1	14406,7	229,544

t	T	h	u	π_0	θ_0	s^0	H	U	S
390	663,15	623,30	450,99	98,302	175,28	7,1751	19944,8	14431,1	229,593
391	664,15	624,32	451,75	98,886	174,51	7,1766	19977,4	14455,4	229,643
392	665,15	625,34	452,5!	99,472	173,74	7,1781	20010,1	14479,8	229,692
393	666,15	626,36	453,27	100,06	172,98	7,1797	20042,8	14504,2	229,741
394	667,15	627,38	454,03	100,65	172,22	7,1812	20075,5	14528,5	229,790
395	668,15	628,40	454,80	101,24	171,46	7,1827	20108,2	14552,9	229,839
396	669,15	629,43	455,56	101,84	170,71	7,1843	20140,9	14577,3	229,888
397	670,15	630,45	456,32	102,44	169,97	7,1858	20173,6	14601,7	229,937
398	671,15	631,47	457,08	103,04	169,22	7,1873	20206,4	14626,2	229,985
399	672,15	632,50	457,85	103,65	168,48	7,1888	20239,1	14650,6	230,034
400	673,15	633,52	458,61	104,26	167,75	7,1904	20271,9	14675,0	230,083
401	674,15	634,54	459,38	104,87	167,02	7,1919	20304,6	14699,5	230,131
402	675,15	635,57	460,14	105,49	166,29	7,1934	20337,4	14723,9	230,180
403	676,15	636,59	460,90	106,10	165,57	7,1949	20370,2	14748,4	230,229
404	677,15	637,62	461,67	106,72	164,85	7,1964	20403,0	14772,8	230,277
405	678,15	638,64	462,43	107,35	164,14	7,1979	20435,7	14797,3	230,325
406	679,15	639,67	463,20	107,97	163,43	7,1994	20468,5	14821,8	230,374
407	680,15	640,69	463,96	108,60	162,72	7,2010	20501,4	14846,3	230,422
408	681,15	641,72	464,73	109,23	162,02	7,2025	20534,2	14870,8	230,470
409	682,15	642,74	465,50	109,87	161,32	7,2040	20567,0	14895,3	230,518
410	683,15	643,77	466,26	110,50	160,62	7,2055	20599,8	14919,8	230,566
411	684,15	644,80	467,03	111,14	159,93	7,2070	20632,7	14944,4	230,615
412	685,15	645,82	467,80	111,79	159,24	7,2085	20665,5	14968,9	230,663
413	686,15	646,85	468,56	112,43	158,56	7,2100	20698,4	14993,5	230,710
414	687,15	647,88	469,33	113,08	157,88	7,2115	20731,3	15018,0	230,758
415	688,15	648,90	470,10	113,74	157,20	7,2130	20764,1	15042,6	230,806
416	689,15	649,93	470,87	114,39	156,53	7,2145	20797,0	15067,2	230,854
417	690,15	650,96	471,63	115,05	155,86	7,2159	20829,9	15091,7	230,902
418	691,15	651,99	472,40	115,71	155,19	7,2174	20862,8	15116,3	230,949
419	692,15	653,02	473,17	116,37	154,53	7,2189	20895,7	15140,9	230,997
420	693,15	654,05	473,94	117,04	153,87	7,2204	20928,7	15165,5	231,044
421	694,15	655,07	474,71	117,71	153,21	7,2219	20961,6	15190,2	231,092
422	695,15	656,10	475,48	118,39	152,56	7,2234	20994,5	15214,8	231,139
423	696,15	657,13	476,25	119,06	151,91	7,2249	21027,5	15239,4	231,187
424	697,15	658,16	477,02	119,74	151,27	7,2263	21060,4	15264,0	231,234
425	698,15	659,19	477,79	120,42	150,63	7,2278	21093,4	15288,7	231,281
426	699,15	660,22	478,56	121,11	149,99	7,2293	21126,4	15313,4	231,328
427	700,15	661,25	479,33	121,80	149,35	7,2308	21159,4	15338,0	231,375
428	701,15	662,29	480,10	122,49	148,72	7,2322	21192,3	15362,7	231,423
429	702,15	663,32	480,87	123,18	148,10	7,2337	21225,3	15387,4	231,470
430	703,15	664,35	481,65	123,88	147,47	7,2352	21258,3	15412,1	231,517
431	704,15	665,37	482,42	124,58	146,85	7,2366	21291,4	15436,8	231,563
432	705,15	666,41	483,19	125,29	146,23	7,2381	21324,4	15461,5	231,610
433	706,15	667,44	483,96	125,99	145,62	7,2396	21357,4	15486,2	231,657
434	707,15	668,48	484,73	126,70	145,01	7,2410	21390,4	15510,9	231,704
435	708,15	669,51	485,51	127,42	144,40	7,2425	21423,5	15535,6	231,751
436	709,15	670,54	486,28	128,14	143,79	7,2439	21456,5	15560,4	231,797
437	710,15	671,58	487,05	128,86	143,!9	7,2454	21489,6	15585,1	231,844
438	711,15	672,61	487,83	129,58	142,59	7,2468	21522,7	15609,9	231,890
439	712,15	673,64	488,60	130,30	142,00	7,2483	21555,8	15634,6	231,937
440	713,15	674,68	489,38	131,03	141,40	7,2497	21588,8	15659,4	231,983
441	714,15	675,71	490,15	131,77	140,81	7,2512	21621,9	15684,2	232,030
442	715,15	676,75	490,92	132,50	140,23	7,2526	21655,0	15709,0	232,076
443	716,15	677,78	491,70	133,24	139,64	7,2541	21688,2	15733,8	232,122
444	717,15	678,82	492,47	133,99	139,06	7,2555	21721,3	15758,6	232,168

Oxygen (*Continued*)

τ	T	h	u	π₀	θ₀	s⁰	H	U	S
445	718,15	679,85	493,25	134,73	138,49	7,2570	21754,4	15783,4	232,215
446	719,15	680,89	494,03	135,48	137,91	7,2584	21787,5	15808,2	232,261
447	720,15	681,92	494,80	136,23	137,34	7,2599	21820,7	15833,1	232,307
448	721,15	682,96	495,58	136,99	136,77	7,2613	21853,8	15857,9	232,353
449	722,15	683,99	496,35	137,75	136,21	7,2627	21887,0	15882,7	232,399
450	723,15	685,03	497,13	138,51	135,65	7,2642	21920,2	15907,6	232,445
451	724,15	686,07	497,91	139,28	135,09	7,2656	21953,3	15932,5	232,490
452	725,15	687,10	498,69	140,05	134,53	7,2670	21986,5	15957,3	232,536
453	726,15	688,14	499,46	140,82	133,98	7,2685	22019,7	15982,2	232,582
454	727,15	689,18	500,24	141,60	133,43	7,2699	22052,9	16007,1	232,628
455	728,15	690,22	501,02	142,37	132,88	7,2713	22086,1	16032,0	232,673
456	729,15	691,26	501,80	143,16	132,33	7,2727	22119,3	16056,9	232,719
457	730,15	692,29	502,57	143,94	131,79	7,2742	22152,6	16081,8	232,764
458	731,15	693,33	503,35	144,73	131,25	7,2756	22185,8	16106,7	232,810
459	732,15	694,37	504,13	145,54	130,72	7,2770	22219,0	16131,6	232,855
460	733,15	695,41	504,91	146,32	130,18	7,2784	22252,3	16156,6	232,901
461	734,15	696,45	505,69	147,12	129,65	7,2798	22285,5	16181,5	232,946
462	735,15	697,49	506,47	147,93	129,12	7,2813	22318,8	16206,5	232,991
463	736,15	698,53	507,25	148,73	128,60	7,2827	22352,1	16231,4	233,037
464	737,15	699,57	508,03	149,54	128,07	7,2841	22385,3	16256,4	233,082
465	738,15	700,61	508,81	150,36	127,55	7,2855	22418,6	16281,3	233,127
466	739,15	701,65	509,59	151,17	127,03	7,2869	22451,9	16306,3	233,172
467	740,15	702,69	510,37	151,99	126,52	7,2883	22485,2	16331,3	233,217
468	741,15	703,73	511,15	152,82	126,01	7,2897	22518,5	16356,3	233,262
469	742,15	704,77	511,94	153,65	125,50	7,2911	22551,8	16381,3	233,307
470	743,15	705,81	512,72	154,48	124,99	7,2925	22585,2	16406,3	233,352
471	744,15	706,85	513,50	155,31	124,48	7,2939	22618,5	16431,3	233,397
472	745,15	707,90	514,28	156,15	123,98	7,2953	22651,8	16456,4	233,441
473	746,15	708,94	515,06	156,99	123,48	7,2967	22685,2	16481,4	233,486
474	747,15	709,98	515,85	157,84	122,99	7,2981	22718,5	16506,4	233,531
475	748,15	711,02	516,63	158,69	122,49	7,2995	22751,9	16531,5	233,575
476	749,15	712,07	517,41	159,54	122,00	7,3009	22785,3	16556,5	233,620
477	750,15	713,11	518,19	160,40	121,51	7,3023	22818,7	16581,6	233,664
478	751,15	714,15	518,98	161,26	121,02	7,3037	22852,0	16606,7	233,709
479	752,15	715,20	519,76	162,12	120,54	7,3051	22885,4	16631,8	233,753
480	753,15	716,24	520,55	162,99	120,06	7,3065	22918,8	16656,8	233,798
481	754,15	717,28	521,33	163,86	119,58	7,3078	22952,2	16681,9	233,842
482	755,15	718,33	522,11	164,74	119,10	7,3092	22985,7	16707,0	233,886
483	756,15	719,37	522,90	165,62	118,62	7,3106	23019,1	16732,1	233,931
484	757,15	720,42	523,68	166,50	118,15	7,3120	23052,5	16757,3	233,975
485	758,15	721,46	524,47	167,39	117,68	7,3134	23086,0	16782,4	234,019
486	759,15	722,51	525,25	168,28	117,21	7,3147	23119,4	16807,5	234,063
487	760,15	723,55	526,04	169,17	116,75	7,3161	23152,9	16832,7	234,107
488	761,15	724,60	526,83	170,07	116,28	7,3175	23186,3	16857,8	234,151
489	762,15	725,65	527,61	170,97	115,82	7,3189	23219,8	16883,0	234,195
490	763,15	726,69	528,40	171,87	115,36	7,3202	23253,3	16908,1	234,239
491	764,15	727,74	529,19	172,78	114,91	7,3216	23286,7	16933,3	234,283
492	765,15	728,78	529,97	173,69	114,45	7,3230	23320,2	16958,5	234,326
493	766,15	729,83	530,76	174,61	114,00	7,3243	23353,7	16983,6	234,370
494	767,15	730,88	531,55	175,53	113,55	7,3257	23387,2	17008,8	234,414
495	768,15	731,93	532,33	176,46	113,10	7,3271	23420,8	17034,0	234,458
496	769,15	732,97	533,12	177,38	112,66	7,3284	23454,3	17059,2	234,501
497	770,15	734,02	533,91	178,31	112,22	7,3298	23487,8	17084,5	234,545
498	771,15	735,07	534,70	179,25	111,77	7,3312	23521,3	17109,7	234,588
499	772,15	736,12	535,49	180,19	111,34	7,3325	23554,9	17134,9	234,632

t	T	h	u	π_0	θ_0	s^0	H	U	S
500	773,15	737,17	536,27	181,13	110,90	7,3339	23588,4	17160,1	234,675
501	774,15	738,21	537,06	182,08	110,47	7,3352	23622,0	17185,4	234,719
502	775,15	739,26	537,85	183,03	110,03	7,3366	23655,5	17210,6	234,762
503	776,15	740,31	538,64	183,99	109,60	7,3379	23689,1	17235,9	234,805
504	777,15	741,36	539,43	184,95	109,18	7,3393	23722,7	17261,2	234,848
505	778,15	742,41	540,22	185,91	108,75	7,3406	23756,3	17286,4	234,892
506	779,15	743,46	541,01	186,88	108,33	7,3420	23789,9	17311,7	234,935
507	780,15	744,51	541,80	187,85	107,90	7,3433	23823,5	17337,0	234,978
508	781,15	745,56	542,59	188,82	107,48	7,3447	23857,1	17362,3	235,021
509	782,15	746,61	543,38	189,80	107,07	7,3460	23890,7	17387,6	235,064
510	783,15	747,66	544,17	190,79	106,65	7,3474	23924,3	17412,9	235,107
511	784,15	748,71	544,96	191,77	106,24	7,3487	23957,9	17438,2	235,150
512	785,15	749,77	545,76	192,76	105,83	7,3500	23991,6	17463,5	235,193
513	786,15	750,82	546,55	193,76	105,42	7,3514	24025,2	17488,9	235,235
514	787,15	751,87	547,34	194,76	105,01	7,3527	24058,9	17514,2	235,278
515	788,15	752,92	548,13	195,76	104,60	7,3541	24092,5	17539,5	235,321
516	789,15	753,97	548,92	196,77	104,20	7,3554	24126,2	17564,9	235,364
517	790,15	755,02	549,72	197,78	103,80	7,3567	24159,9	17590,2	235,406
518	791,15	756,08	550,51	198,80	103,40	7,3581	24193,5	17615,6	235,449
519	792,15	757,13	551,30	199,82	103,00	7,3594	24227,2	17641,0	235,491
520	793,15	758,18	552,09	200,84	102,60	7,3607	24260,9	17666,3	235,534
521	794,15	759,24	552,89	201,87	102,21	7,3620	24294,6	17691,7	235,576
522	795,15	760,29	553,68	202,90	101,82	7,3634	24328,3	17717,1	235,619
523	796,15	761,34	554,47	203,94	101,43	7,3647	24362,0	17742,5	235,661
524	797,15	762,40	555,27	204,98	101,04	7,3660	24395,8	17767,9	235,703
525	798,15	763,45	556,06	206,03	100,65	7,3673	24429,5	17793,3	235,746
526	799,15	764,50	556,86	207,08	100,27	7,3687	24463,2	17818,8	235,788
527	800,15	765,56	557,65	208,13	99,890	7,3700	24497,0	17844,2	235,830
528	801,15	766,61	558,45	209,19	99,510	7,3713	24530,7	17869,6	235,872
529	802,15	767,67	559,24	210,25	99,130	7,3726	24564,5	17895,1	235,914
530	803,15	768,72	560,04	211,32	98,753	7,3739	24598,2	17920,5	235,957
531	804,15	769,78	560,83	212,39	98,378	7,3752	24632,0	17946,0	235,999
532	805,15	770,83	561,63	213,46	98,004	7,3765	24665,8	17971,4	236,041
533	806,15	771,89	562,42	214,54	97,632	7,3779	24699,5	17996,9	236,082
534	807,15	772,95	563,22	215,62	97,262	7,3792	24733,3	18022,4	236,124
535	808,15	774,00	564,02	216,71	96,894	7,3805	24767,1	18047,8	236,166
536	809,15	775,06	564,81	217,80	96,527	7,3819	24800,9	18073,3	236,208
537	810,15	776,11	565,61	218,90	96,162	7,3831	24834,7	18098,8	236,250
538	811,15	777,17	566,41	220,00	95,799	7,3844	24868,6	18124,3	236,291
539	812,15	778,23	567,20	221,11	95,438	7,3857	24902,4	18149,8	236,333
540	813,15	779,29	568,00	222,22	95,078	7,3870	24936,2	18175,3	236,375
541	814,15	780,34	568,80	223,33	94,720	7,3883	24970,0	18200,9	236,416
542	815,15	781,40	569,60	224,45	94,364	7,3896	25003,9	18226,4	236,458
543	816,15	782,46	570,39	225,57	94,009	7,3909	25037,7	18251,9	236,499
544	817,15	783,52	571,19	226,70	93,656	7,3922	25071,6	18277,5	236,541
545	818,15	784,57	571,99	227,83	93,305	7,3935	25105,4	18303,0	236,582
546	819,15	785,63	572,79	228,97	92,955	7,3948	25139,3	18328,6	236,624
547	820,15	786,69	573,59	230,11	92,607	7,3961	25173,2	18354,1	236,665
548	821,15	787,75	574,39	231,26	92,261	7,3973	25207,1	18379,7	236,706
549	822,15	788,81	575,19	232,41	91,916	7,3986	25241,0	18405,3	236,747
550	823,15	789,87	575,99	233,56	91,573	7,3999	25274,9	18430,9	236,789
551	824,15	790,93	576,79	234,72	91,231	7,4012	25308,8	18456,4	236,830
552	825,15	791,99	577,59	235,88	90,891	7,4025	25342,7	18482,0	236,871
553	826,15	793,05	578,39	237,05	90,553	7,4038	25376,6	18507,6	236,912
554	827,15	794,11	579,19	238,22	90,216	7,4051	25410,5	18533,2	236,953

Oxygen (*Continued*)

t	T	h	u	π_0	θ_0	s^v	H	U	S
555	828,15	795,17	579,99	239,40	89,881	7,4063	25444,4	18558,9	236,994
556	829,15	796,23	580,78	240,58	89,548	7,4076	25478,4	18584,5	237,035
557	830,15	797,29	581,59	241,77	89,215	7,4089	25512,3	18610,1	237,076
558	831,15	798,35	582,39	242,96	88,885	7,4102	25546,3	18635,7	237,117
559	832,15	799,41	583,19	244,16	88,556	7,4115	25580,2	18661,4	237,158
560	833,15	800,47	583,99	245,36	88,229	7,4127	25614,2	18687,0	237,198
561	834,15	801,53	584,79	246,56	87,903	7,4140	25648,1	18712,7	237,239
562	835,15	802,60	585,59	247,77	87,578	7,4153	25682,1	18738,3	237,280
563	836,15	803,66	586,40	248,99	87,255	7,4165	25716,1	18764,0	237,321
564	837,15	804,72	587,20	250,21	86,934	7,4178	25750,1	18789,7	237,361
565	838,15	805,78	588,00	251,43	86,614	7,4191	25784,1	18815,3	237,402
566	839,15	806,84	588,80	252,66	86,295	7,4203	25818,1	18841,0	237,442
567	840,15	807,91	589,61	253,89	85,979	7,4216	25852,1	18866,7	237,483
568	841,15	808,97	590,41	255,13	85,663	7,4229	25886,1	18892,4	237,523
569	842,15	810,03	591,21	256,38	85,349	7,4241	25920,1	18918,1	237,564
570	843,15	811,10	592,02	257,62	85,036	7,4254	25954,1	18943,8	237,604
571	844,15	812,16	592,82	258,88	84,725	7,4267	25988,2	18969,5	237,644
572	845,15	813,22	593,62	260,14	84,415	7,4279	26022,2	18995,3	237,685
573	846,15	814,29	594,43	261,40	84,107	7,4292	26056,2	19021,0	237,725
574	847,15	815,35	595,23	262,67	83,800	7,4304	26090,3	19046,7	237,765
575	848,15	816,42	596,04	263,94	83,495	7,4317	26124,3	19072,5	237,805
576	849,15	817,48	596,84	265,21	83,190	7,4329	26158,3	19098,2	237,845
577	850,15	818,55	597,65	266,50	82,888	7,4342	26192,5	19124,0	237,886
578	851,15	819,61	598,45	267,78	82,586	7,4355	26226,5	19149,7	237,926
579	852,15	820,68	599,26	269,08	82,286	7,4367	26260,6	19175,5	237,966
580	853,15	821,74	600,06	270,37	81,988	7,4380	26294,7	19201,3	238,006
581	854,15	822,81	600,87	271,67	81,691	7,4392	26328,8	19227,0	238,046
582	855,15	823,87	601,67	272,98	81,395	7,4404	26362,9	19252,8	238,085
583	856,15	824,94	602,48	274,29	81,100	7,4417	26397,0	19278,6	238,125
584	857,15	826,00	603,29	275,61	80,807	7,4429	26431,1	19304,4	238,165
585	858,15	827,07	604,09	276,93	80,515	7,4442	26465,2	19330,2	238,205
586	859,15	828,14	604,90	278,26	80,225	7,4454	26499,4	19356,0	238,245
587	860,15	829,20	605,71	279,59	79,935	7,4467	26533,5	19381,8	238,284
588	861,15	830,27	606,51	280,93	79,647	7,4479	26567,6	19407,7	238,324
589	862,15	831,34	607,32	282,27	79,361	7,4491	26601,8	19433,5	238,364
590	863,15	832,40	608,13	283,62	79,075	7,4504	26635,9	19459,3	238,403
591	864,15	833,47	608,93	284,97	78,791	7,4516	26670,1	19485,2	238,443
592	865,15	834,54	609,74	286,33	78,509	7,4528	26704,2	19511,0	238,482
593	866,15	835,61	610,55	287,69	78,227	7,4541	26738,4	19536,9	238,522
594	867,15	836,67	611,36	289,06	77,947	7,4553	26772,6	19562,7	238,561
595	868,15	837,74	612,17	290,43	77,668	7,4565	26806,7	19588,6	238,601
596	869,15	838,81	612,97	291,81	77,390	7,4578	26840,9	19614,4	238,640
597	870,15	839,88	613,78	293,19	77,114	7,4590	26875,1	19640,3	238,679
598	871,15	840,95	614,59	294,58	76,839	7,4602	26909,3	19666,2	238,718
599	872,15	842,02	615,40	295,97	76,565	7,4615	26943,5	19692,1	238,758
600	873,15	843,08	616,21	297,37	76,292	7,4627	26977,7	19718,0	238,797
601	874,15	844,15	617,02	298,77	76,020	7,4639	27011,9	19743,9	238,836
602	875,15	845,22	617,83	300,18	75,750	7,4651	27046,1	19769,8	238,875
603	876,15	846,29	618,64	301,60	75,481	7,4664	27080,4	19795,7	238,914
604	877,15	847,36	619,45	303,02	75,213	7,4676	27114,6	19821,6	238,953
605	878,15	848,43	620,26	304,44	74,946	7,4688	27148,8	19847,5	238,992
606	879,15	849,50	621,07	305,87	74,681	7,4700	27183,1	19873,5	239,031
607	880,15	850,57	621,88	307,31	74,417	7,4712	27217,3	19899,4	239,070
608	881,15	851,64	622,69	308,75	74,153	7,4724	27251,6	19925,3	239,109
609	882,15	852,71	623,50	310,20	73,891	7,4737	27285,8	19951,3	239,148

Oxygen (*Continued*)

t	T	h	u	π_0	0_0	s^0	H	U	S
610	883,15	853,78	624,31	311,65	73,631	7,4749	27320,1	19977,2	239,187
611	884,15	854,86	625,12	313,11	73,371	7,4761	27354,4	20003,2	239,226
612	885,15	855,93	625,93	314,57	73,112	7,4773	27388,7	20029,2	239,264
613	886,15	857,00	626,75	316,04	72,855	7,4785	27422,9	20055,1	239,303
614	887,15	858,07	627,56	317,51	72,599	7,4797	27457,2	20081,1	239,342
615	888,15	859,14	628,37	318,99	72,344	7,4809	27491,5	20107,1	239,380
616	889,15	860,21	629,18	320,47	72,090	7,4821	27525,8	20133,1	239,419
617	890,15	861,29	629,99	321,96	71,837	7,4833	27560,1	20159,1	239,458
618	891,15	862,36	630,81	323,46	71,585	7,4845	27594,4	20185,1	239,496
619	892,15	863,43	631,62	324,96	71,334	7,4857	27628,8	20211,1	239,535
620	893,15	864,50	632,43	326,46	71,085	7,4869	27663,1	20237,1	239,573
621	894,15	865,58	633,24	327,98	70,836	7,4881	27697,4	20263,1	239,611
622	895,15	866,65	634,06	329,49	70,589	7,4893	27731,7	20289,1	239,650
623	896,15	867,72	634,87	331,02	70,343	7,4905	27766,1	20315,1	239,688
624	897,15	868,80	635,68	332,55	70,097	7,4917	27800,4	20341,2	239,726
625	898,15	869,87	636,50	334,08	69,853	7,4929	27834,8	20367,2	239,765
626	899,15	870,94	637,31	335,62	69,610	7,4941	27869,1	20393,2	239,803
627	900,15	872,02	638,13	337,16	69,368	7,4953	26903,5	20419,3	239,841
628	901,15	873,09	638,94	338,72	69,127	7,4965	27937,9	20445,3	239,879
629	902,15	874,17	639,76	340,27	68,887	7,4977	27972,2	20471,4	239,917
630	903,15	875,24	640,57	341,83	68,649	7,4989	28006,6	20497,5	239,955
631	904,15	876,31	641,38	343,40	68,411	7,5001	28041,0	20523,5	239,994
632	905,15	877,39	642,20	344,98	68,174	7,5013	28075,4	20549,6	240,032
633	906,15	878,46	643,01	346,56	67,938	7,5025	28109,8	20575,7	240,070
634	907,15	879,54	643,83	348,14	67,704	7,5036	28144,2	20601,8	240,107
635	908,15	880,61	644,65	349,73	67,470	7,5048	28178,6	20627,9	240,145
636	909,15	881,69	645,46	351,33	67,237	7,5060	28213,0	20654,0	240,183
637	910,15	882,77	646,28	352,93	67,006	7,5072	28247,4	20680,1	240,221
638	911,15	883,84	647,09	354,54	66,775	7,5084	28281,9	20706,2	240,259
639	912,15	884,92	647,91	356,15	66,545	7,5096	28316,3	20732,3	240,297
640	913,15	885,99	648,72	357,77	66,317	7,5107	28350,7	20758,4	240,334
641	914,15	887,07	649,54	359,40	66,089	7,5119	28385,2	20784,5	240,372
642	915,15	888,15	650,36	361,03	65,862	7,5131	28419,6	20810,7	240,410
643	916,15	889,22	651,17	362,67	65,637	7,5143	28454,1	20836,8	240,447
644	917,15	890,30	651,99	364,31	65,412	7,5154	28488,5	20863,0	240,485
645	918,15	891,38	652,81	365,96	65,188	7,5166	28523,0	20889,1	240,523
646	919,15	892,45	653,63	367,62	64,965	7,5178	28557,4	20915,3	240,560
647	920,15	893,53	654,44	369,28	64,743	7,5190	28591,9	20941,4	240,598
648	921,15	894,61	655,26	370,94	64,522	7,5201	28626,4	20967,6	240,635
649	922,15	895,69	656,08	372,62	64,302	7,5213	28660,9	20993,7	240,672
650	923,15	896,76	656,90	374,30	64,083	7,5225	28695,4	21019,9	240,710
651	924,15	897,84	657,72	375,98	63,865	7,5236	28729,9	21046,1	240,747
652	925,15	898,92	658,53	377,67	63,648	7,5248	28764,4	21072,3	240,784
653	926,15	900,00	659,35	379,37	63,432	7,5260	28798,9	21098,5	240,822
654	927,15	901,08	660,17	381,07	63,217	7,5271	28833,4	21124,7	240,859
655	928,15	902,16	660,99	382,78	63,002	7,5283	28867,9	21150,9	240,896
656	929,15	903,23	661,81	384,50	62,789	7,5294	28902,4	21177,1	240,933
657	930,15	904,31	662,63	386,22	62,576	7,5306	28936,9	21203,3	240,970
658	931,15	905,39	663,45	387,95	62,364	7,5318	28971,5	21229,5	241,008
659	932,15	906,47	664,27	389,68	62,154	7,5329	29006,0	21255,7	241,045
660	933,15	907,55	665,09	391,42	61,944	7,5341	29040,6	21282,0	241,082
661	934,15	908,63	665,91	393,17	61,735	7,5352	29075,1	21308,2	241,119
662	935,15	909,71	666,73	394,92	61,527	7,5364	29109,7	21334,4	241,156
663	936,15	910,79	667,55	396,68	61,320	7,5376	29144,2	21360,7	241,193
664	937,15	911,87	668,37	398,44	61,113	7,5387	29178,8	21386,9	241,229

t	T	h	u	π_0	θ_0	s^0	H	U	S
665	938,15	912,95	669,19	400,21	60,908	7,5399	29213,3	21413,2	241,266
666	939,15	914,03	670,01	401,99	60,703	7,5410	29247,9	21439,4	241,303
667	940,15	915,11	670,83	403,77	60,499	7,5422	29282,5	21465,7	241,340
668	941,15	916,19	671,65	405,56	60,297	7,5433	29317,1	21492,0	241,377
669	942,15	917,27	672,47	407,36	60,095	7,5445	29351,7	21518,2	241,413
670	943,15	918,36	673,29	409,16	59,893	7,5456	29386,3	21544,5	241,450
671	944,15	919,44	674,11	410,97	59,693	7,5467	29420,9	21570,8	241,487
672	945,15	920,52	674,93	412,78	59,494	7,5479	29455,5	21597,1	241,523
673	946,15	921,60	675,76	414,60	59,295	7,5490	29490,1	21623,4	241,560
674	947,15	922,68	676,58	416,43	59,097	7,5502	29524,7	21649,7	241,597
675	948,15	923,76	677,40	418,26	58,900	7,5513	29559,3	21676,0	241,633
676	949,15	924,84	678,22	420,10	58,704	7,5525	29593,9	21702,3	241,670
677	950,15	925,93	679,04	421,95	58,509	7,5536	29628,6	21728,6	241,706
678	951,15	927,01	679,77	423,80	58,314	7,5547	29663,2	21754,9	241,743
679	952,15	928,09	680,69	425,66	58,121	7,5559	29697,8	21781,3	241,779
680	953,15	929,17	681,51	427,53	57,928	7,5570	29732,5	21807,6	241,815
681	954,15	930,26	682,34	429,40	57,736	7,5581	29767,1	21833,9	241,852
682	955,15	931,34	683,16	431,28	57,544	7,5593	29801,8	21860,3	241,888
683	956,15	932,42	683,98	433,16	57,354	7,5604	29836,4	21886,6	241,924
684	957,15	933,51	684,81	435,06	57,164	7,5615	29871,1	21913,0	241,960
685	958,15	934,59	685,63	436,95	56,975	7,5627	29905,8	21939,3	241,997
686	959,15	935,67	686,45	438,86	56,787	7,5638	29940,5	21965,7	242,033
687	960,15	936,76	687,28	440,77	56,600	7,5649	29975,1	21992,1	242,069
688	961,15	937,84	688,10	442,69	56,413	7,5661	30009,8	22018,4	242,105
689	962,15	938,93	688,93	444,61	56,227	7,5672	30044,5	22044,8	242,141
690	963,15	940,01	689,75	446,55	56,042	7,5683	30079,2	22071,2	242,177
691	964,15	941,09	690,57	448,48	55,858	7,5694	30113,9	22097,6	242,213
692	965,15	942,18	691,40	450,43	55,675	7,5706	30148,6	22124,0	242,249
693	966,15	943,26	692,22	452,38	55,492	7,5717	30183,3	22150,4	242,285
694	967,15	944,35	693,05	454,34	55,310	7,5728	30218,0	22176,8	242,321
695	968,15	945,43	693,87	456,30	55,129	7,5739	30252,8	22203,2	242,357
696	969,15	946,52	694,70	458,28	54,948	7,5751	30287,5	22229,6	242,393
697	970,15	947,60	695,53	460,25	54,768	7,5762	30322,2	22256,0	242,429
698	971,15	948,69	696,35	462,24	54,589	7,5773	30356,9	22282,4	242,464
699	972,15	949,78	697,18	464,23	54,411	7,5784	30391,7	22308,8	242,500
700	973,15	950,86	698,00	466,23	54,234	7,5795	30426,4	22335,3	242,536
701	974,15	951,95	698,83	468,24	54,057	7,5806	30461,2	22361,7	242,572
702	975,15	953,03	699,66	470,25	53,881	7,5818	30495,9	22388,1	242,607
703	976,15	954,12	700,48	472,27	53,705	7,5829	30530,7	22414,6	242,643
704	977,15	955,21	701,31	474,29	53,531	7,5840	30565,4	22441,0	242,678
705	978,15	956,29	702,13	476,33	53,357	7,5851	30600,2	22467,5	242,714
706	979,15	957,38	702,96	478,37	53,184	7,5862	30635,0	22493,9	242,749
707	980,15	958,47	703,79	480,41	53,011	7,5873	30669,8	22520,4	242,785
708	981,15	959,55	704,62	482,47	52,839	7,5884	30704,6	22546,9	242,820
709	982,15	960,64	705,44	484,53	52,668	7,5895	30739,3	22573,3	242,856
710	983,15	961,73	706,27	486,60	52,498	7,5906	30774,1	22599,8	242,891
711	984,15	962,82	707,10	488,67	52,328	7,5917	30808,9	22626,3	242,927
712	985,15	963,90	707,93	490,75	52,159	7,5928	30843,7	22652,8	242,962
713	986,15	964,99	708,75	492,84	51,991	7,5940	30878,5	22679,3	242,997
714	987,15	966,08	709,58	494,94	51,823	7,5951	30913,3	22705,8	243,033
715	988,15	967,17	710,41	497,04	51,656	7,5962	30948,2	22732,3	243,068
716	989,15	968,25	711,24	499,15	51,490	7,5973	30983,0	22758,8	243,103
717	990,15	969,34	712,07	501,27	51,324	7,5984	31017,8	22785,3	243,138
718	991,15	970,43	712,90	503,39	51,159	7,5995	31052,6	22811,8	243,173
719	992,15	971,52	713,72	505,52	50,995	7,6006	31087,5	22838,3	243,209

Oxygen (*Continued*)

t	T	h	u	π_0	θ_0	s^0	li	U	S
720	993,15	972,61	714,55	507,66	50,831	7,6016	31122,3	22864,9	243,244
721	994,15	973,70	715,38	509,81	50,668	7,6027	31157,2	22891,4	243,279
722	995,15	974,79	716,21	511,96	50,506	7,6038	31192,0	22917,9	243,314
723	996,15	975,88	717,04	514,12	50,344	7,6049	31226,9	22944,5	243,349
724	997,15	976,97	717,87	516,29	50,183	7,6060	31261,7	22971,0	243,384
725	998,15	978,05	718,70	518,46	50,023	7,6071	31296,6	22997,6	243,419
726	999,15	979,14	719,53	520,64	49,863	7,6082	31331,4	23024,1	243,454
727	1000,15	980,23	720,36	522,83	49,704	7,6093	31366,3	23050,7	243,489
728	1001,15	981,32	721,19	525,03	49,546	7,6104	31401,2	23077,2	243,523
729	1002,15	982,41	722,02	527,23	49,388	7,6115	31436,1	23103,8	243,558
730	1003,15	983,50	722,85	529,44	49,231	7,6126	31471,0	23130,4	243,593
731	1004,15	984,59	723,68	531,66	49,074	7,6137	31505,9	23156,9	243,628
732	1005,15	985,69	724,51	533,89	48,918	7,6147	31540,8	23183,5	243,662
733	1006,15	986,78	725,34	536,12	48,763	7,6158	31575,7	23210,1	243,697
734	1007,15	987,87	726,17	538,36	48,608	7,6169	31610,6	23236,7	243,732
735	1008,15	988,96	727,01	540,61	48,454	7,6180	31645,5	23263,3	243,766
736	1009,15	990,05	727,84	542,86	48,301	7,6191	31680,4	23289,9	243,801
737	1010,15	991,14	728,67	545,13	48,148	7,6202	31715,3	23316,5	243,836
738	1011,15	992,23	729,50	547,40	47,996	7,6212	31750,2	23343,1	243,870
739	1012,15	993,32	730,33	549,67	47,844	7,6223	31785,1	23369,7	243,905
740	1013,15	994,41	731,16	551,96	47,693	7,6234	31820,1	23396,3	243,939
741	1014,15	995,51	731,99	554,25	47,543	7,6245	31855,0	23423,0	243,974
742	1015,15	996,60	732,83	556,55	47,393	7,6255	31890,0	23449,6	244,008
743	1016,15	997,69	733,66	558,86	47,244	7,6266	31924,9	23476,2	244,043
744	1017,15	998,78	734,49	561,18	47,095	7,6277	31959,8	23502,8	244,077
745	1018,15	999,88	735,32	563,50	46,947	7,6288	31994,8	23529,5	244,111
746	1019,15	1000,97	736,16	565,83	46,799	7,6298	32029,8	23556,1	244,146
747	1020,15	1002,06	736,99	568,17	46,653	7,6309	32064,7	23582,8	244,180
748	1021,15	1003,15	737,82	570,51	46,506	7,6320	32099,7	23609,4	244,214
749	1022,15	1004,25	738,66	572,87	46,361	7,6330	32134,7	23636,1	244,248
750	1023,15	1005,34	739,49	575,23	46,215	7,6341	32169,6	23662,8	244,283
751	1024,15	1006,43	740,32	577,60	46,071	7,6352	32204,6	23689,4	244,317
752	1025,15	1007,53	741,16	579,98	45,927	7,6363	32239,6	23716,1	244,351
753	1026,15	1008,62	741,99	582,36	45,783	7,6373	32274,6	23742,8	244,385
754	1027,15	1009,71	742,82	584,75	45,641	7,6384	32309,6	23769,4	244,419
755	1028,15	1010,81	743,66	587,15	45,498	7,6394	32344,6	23796,1	244,453
756	1029,15	1011,90	744,49	589,56	45,356	7,6405	32379,6	23822,8	244,487
757	1030,15	1012,99	745,33	591,98	45,215	7,6416	32414,6	23849,5	244,521
758	1031,15	1014,09	746,16	594,40	45,075	7,6426	32449,6	23876,2	244,555
759	1032,15	1015,18	746,99	596,83	44,935	7,6437	32484,6	23902,9	244,589
760	1033,15	1016,28	747,83	599,27	44,795	7,6448	32519,6	23929,6	244,623
761	1034,15	1017,37	748,66	601,72	44,656	7,6458	32554,7	23956,3	244,657
762	1035,15	1018,47	749,50	604,17	44,517	7,6469	32589,7	23983,0	244,691
763	1036,15	1019,56	750,33	606,64	44,380	7,6479	32624,7	24009,8	244,725
764	1037,15	1020,66	751,17	609,11	44,242	7,6490	32659,8	24036,5	244,758
765	1038,15	1021,75	752,00	611,59	44,105	7,6500	32694,8	24063,2	244,792
766	1039,15	1022,85	752,84	614,07	43,969	7,6511	32729,9	24089,9	244,826
767	1040,15	1023,94	753,67	616,57	43,833	7,6522	32764,9	24116,7	244,860
768	1041,15	1025,04	754,51	619,07	43,698	7,6532	32800,0	24143,4	244,893
769	1042,15	1026,13	755,35	621,58	43,563	7,6543	32835,0	24170,2	244,927
770	1043,15	1027,23	756,18	624,10	43,429	7,6553	32870,1	24196,9	244,961
771	1044,15	1028,32	757,02	626,63	43,295	7,6564	32905,2	24223,7	244,994
772	1045,15	1029,42	757,85	629,17	43,162	7,6574	32940,2	24250,4	245,028
773	1046,15	1030,52	758,69	631,71	43,029	7,6585	32975,3	24277,2	245,061
774	1047,15	1031,61	759,53	634,26	42,897	7,6595	33010,4	24304,0	245,095

Oxygen (*Continued*)

t	T	h	u	π_θ	θ_θ	s^θ	H	U	S
775	1048,15	1032,71	760,36	636,82	42,766	7,6605	33045,5	24330,7	245,128
776	1049,15	1033,81	761,20	639,39	42,635	7,6616	33080,6	24357,5	245,162
777	1050,15	1034,90	762,04	641,97	42,504	7,6626	33115,7	24384,3	245,195
778	1051,15	1036,00	762,87	644,55	42,374	7,6637	33150,8	24411,1	245,229
779	1052,15	1037,10	763,71	647,14	42,244	7,6647	33185,9	24437,9	245,262
780	1053,15	1038,19	764,55	649,74	42,115	7,6658	33221,0	24464,7	245,295
781	1054,15	1039,29	765,39	652,35	41,986	7,6668	33256,1	24491,4	245,329
782	1055,15	1040,39	766,22	654,97	41,858	7,6678	33291,2	24518,2	245,362
783	1056,15	1041,49	767,06	657,60	41,731	7,6689	33326,3	24545,1	245,395
784	1057,15	1042,58	767,90	660,23	41,603	7,6699	33361,4	24571,9	245,428
785	1058,15	1043,68	768,74	662,87	41,477	7,6710	33396,6	24598,7	245,462
786	1059,15	1044,78	769,58	665,53	41,351	7,6720	33431,7	24625,5	245,495
787	1060,15	1045,88	770,41	668,18	41,225	7,6730	33466,8	24652,3	245,528
788	1061,15	1046,98	771,25	670,85	41,100	7,6741	33502,0	24679,1	245,561
789	1062,15	1048,07	772,09	673,53	40,975	7,6751	33537,1	24706,0	245,594
790	1063,15	1049,17	772,93	676,21	40,851	7,6761	33572,3	24732,8	245,627
791	1064,15	1050,27	773,77	678,91	40,727	7,6772	33607,4	24759,6	245,660
792	1065,15	1051,37	774,61	681,61	40,604	7,6782	33642,6	24786,5	245,693
793	1066,15	1052,47	775,45	684,32	40,481	7,6792	33677,7	24813,3	245,726
794	1067,15	1053,57	776,28	687,04	40,358	7,6803	33712,9	24840,2	245,759
795	1068,15	1054,67	777,12	689,77	40,236	7,6813	33748,1	24867,0	245,792
796	1069,15	1055,77	777,96	692,50	40,115	7,6823	33783,3	24893,9	245,825
797	1070,15	1056,87	778,80	695,25	39,994	7,6834	33818,4	24920,8	245,858
798	1071,15	1057,97	779,64	698,00	39,874	7,6844	33853,6	24947,6	245,891
799	1072,15	1059,06	780,48	700,76	39,753	7,6854	33888,8	24974,5	245,924
800	1073,15	1060,16	781,32	703,53	39,634	7,6864	33924,0	25001,4	245,957
801	1074,15	1061,26	782,16	706,31	39,515	7,6875	33959,2	25028,3	245,989
802	1075,15	1062,36	783,00	709,10	39,396	7,6885	33994,4	25055,1	246,022
803	1076,15	1063,46	783,84	711,90	39,278	7,6895	34029,6	25082,0	246,055
804	1077,15	1064,56	784,68	714,70	39,160	7,6905	34064,8	25108,9	246,088
805	1078,15	1065,67	785,52	717,51	39,042	7,6915	34100,0	25135,8	246,120
806	1079,15	1066,77	786,36	720,34	38,926	7,6926	34135,2	25162,7	246,153
807	1080,15	1067,87	787,21	723,17	38,809	7,6936	34170,4	25189,6	246,186
808	1081,15	1068,97	788,05	726,01	38,693	7,6946	34205,7	25216,5	246,218
809	1082,15	1070,07	788,89	728,86	38,577	7,6956	34240,9	25243,4	246,251
810	1083,15	1071,17	789,73	731,72	38,462	7,6966	34276,1	25270,4	246,283
811	1084,15	1072,27	790,57	734,58	38,347	7,6977	34311,4	25297,3	246,316
812	1085,15	1073,37	791,41	737,46	38,233	7,6987	34346,6	25324,2	246,348
813	1086,15	1074,47	792,25	740,35	38,119	7,6997	34381,8	25351,1	246,381
814	1087,15	1075,57	793,09	743,24	38,006	7,7007	34417,1	25378,1	246,413
815	1088,15	1076,68	793,94	746,14	37,893	7,7017	34452,3	25405,0	246,446
816	1089,15	1077,78	794,78	749,05	37,780	7,7027	34487,6	25431,9	246,478
817	1090,15	1078,88	795,62	751,98	37,668	7,7037	34522,8	25458,9	246,510
818	1091,15	1079,98	796,46	754,91	37,556	7,7047	34558,1	25485,8	246,543
819	1092,15	1081,08	797,30	757,84	37,445	7,7058	34593,4	25512,8	246,575
820	1093,15	1082,19	798,15	760,79	37,334	7,7068	34628,6	25539,8	246,607
821	1094,15	1083,29	798,99	763,75	37,223	7,7078	34663,9	25566,7	246,639
822	1095,15	1084,39	799,83	766,72	37,113	7,7088	34699,2	25593,7	246,672
823	1096,15	1085,49	800,68	769,69	37,003	7,7098	34734,5	25620,6	246,704
824	1097,15	1086,60	801,52	772,68	36,894	7,7108	34769,8	25647,6	246,736
825	1098,15	1087,70	802,36	775,67	36,785	7,7118	34805,1	25674,6	246,768
826	1099,15	1088,80	803,20	778,67	36,677	7,7128	34840,4	25701,6	246,800
827	1100,15	1089,91	804,05	781,68	36,569	7,7138	34875,7	25728,6	246,832
828	1101,15	1091,01	804,89	784,71	36,461	7,7148	34911,0	25755,5	246,865
829	1102,15	1092,11	805,73	787,74	36,354	7,7158	34946,3	25782,5	246,897

t	T	h	u	π_0	θ_0	s^0	H	U	S
830	1103,15	1093,22	806,58	790,78	36,247	7,7168	34981,6	25809,5	246,929
831	1104,15	1094,32	807,42	793,83	36,140	7,7178	35016,9	25836,5	246,961
832	1105,15	1095,42	808,27	796,88	36,034	7,7188	35052,2	25863,5	246,993
833	1106,15	1096,53	809,11	799,95	35,928	7,7198	35087,5	25890,6	247,025
834	1107,15	1097,63	809,95	803,03	35,823	7,7208	35122,9	25917,6	247,056
835	1108,15	1098,73	810,80	806,12	35,718	7,7218	35158,2	25944,6	247,088
836	1109,15	1099,84	811,64	809,21	35,614	7,7228	35193,5	25971,6	247,120
837	1110,15	1100,94	812,49	812,32	35,510	7,7238	35228,8	25998,6	247,152
838	1111,15	1102,05	813,33	815,43	35,406	7,7248	35264,2	26025,7	247,184
839	1112,15	1103,15	814,18	818,56	35,302	7,7258	35299,6	26052,7	247,216
840	1113,15	1104,26	815,02	821,69	35,199	7,7268	35334,8	26079,7	247,247
841	1114,15	1105,36	815,87	824,83	35,097	7,7278	35370,3	26106,8	247,279
842	1115,15	1106,47	816,71	827,99	34,994	7,7288	35405,6	26133,8	247,311
843	1116,15	1107,57	817,56	831,15	34,893	7,7297	35441,0	26160,8	247,343
844	1117,15	1108,68	818,40	834,32	34,791	7,7307	35476,3	26187,9	247,374
845	1118,15	1109,78	819,25	837,50	34,690	7,7317	35511,7	26215,0	247,406
846	1119,15	1110,89	820,09	840,69	34,589	7,7327	35547,1	26242,0	247,438
847	1120,15	1111,99	820,94	843,90	34,489	7,7337	35582,5	26269,1	247,469
848	1121,15	1113,10	821,79	847,11	34,389	7,7347	35617,8	26296,1	247,501
849	1122,15	1114,21	822,63	850,33	34,289	7,7357	35653,2	26323,2	247,532
850	1123,15	1115,31	823,48	853,56	34,190	7,7367	35688,6	26350,3	247,564
851	1124,15	1116,42	824,32	856,80	34,091	7,7376	35724,0	26377,4	247,595
852	1125,15	1117,52	825,17	860,05	33,992	7,7386	35759,4	26404,4	247,627
853	1126,15	1118,63	826,02	863,30	33,894	7,7396	35794,8	26431,5	247,658
854	1127,15	1119,74	826,86	866,57	33,796	7,7406	35830,2	26458,6	247,690
855	1128,15	1120,84	827.71	869,85	33,698	7,7416	35865,6	26485,7	247,721
856	1129,15	1121,95	828,56	873,14	33,601	7,7426	35901,0	26512,8	247,752
857	1130,15	1123,06	829,40	876,44	33,504	7,7435	35936,4	26539,9	247,784
858	1131,15	1124,16	830,25	879,75	33,408	7,7445	35971,9	26567,0	247,815
859	1132,15	1125,27	831,10	883,07	33,312	7,7455	36007,3	26594,1	247,846
860	1133,15	1126,38	831,94	886,39	33,216	7,7465	36042,7	26621,2	247,878
861	1134,15	1127,48	832,79	889,73	33,121	7,7474	36078,1	26648,4	247,909
862	1135,15	1128,59	833,64	893,08	33,026	7,7484	36113,6	26675,5	247,940
863	1136,15	1129,70	834,49	896,44	32,931	7,7494	36149,0	26702,6	247,971
864	1137,15	1130,81	835,34	899,81	32,837	7,7504	36184,5	26729,7	248,003
865	1138,15	1131,91	836,18	903,19	32,742	7,7513	36219,9	26756,9	248,034
866	1139,15	1133,02	837,03	906,57	32,649	7,7523	36255,4	26784,0	248,065
867	1140,15	1134,13	837,88	909,97	32,555	7,7533	36290,8	26811,1	248,096
868	1141,15	1135,24	838,73	913,38	32,462	7,7543	36326,3	26838,3	248,127
869	1142,15	1136,35	839,58	916,80	32,370	7,7552	36361,7	26865,4	248,158
870	1143,15	1137,45	840,42	920,23	32,277	7,7562	36397,2	26892,6	248,189
871	1144,15	1138,56	841,27	923,67	32,185	7,7572	36432,7	26919,7	248,220
872	1145,15	1139,67	842,12	927,12	32,093	7,7581	36468,1	26946,9	248,251
873	1146,15	1140,78	842,97	930,58	32,002	7,7591	36503,6	26974,0	248,282
874	1147,15	1141,89	843,82	934,04	31,911	7,7601	36539,1	27001,2	248,313
875	1148,15	1143,00	844,67	937,52	31,820	7,7610	36574,6	27028,4	248,344
876	1149,15	1144,11	845,52	941,01	31,730	7,7620	36610,1	27055,5	248,375
877	1150,15	1145,22	846,37	944,52	31,640	7,7630	36645,5	27082,7	248,406
878	1151,15	1146,33	847,22	948,03	31,550	7,7639	36681,0	27109,9	248,437
879	1152,15	1147,43	848,07	951,55	31,461	7,7649	36716,5	27137,1	248,467
880	1153,15	1148,54	848,92	955,08	31,372	7,7659	36752,0	27164,3	248,498
881	1154,15	1149,65	849,77	958,62	31,283	7,7668	36787,5	27191,5	248,529
882	1155,15	1150,76	850,61	962,17	31,194	7,7678	36823,0	27218,7	248,560
883	1156,15	1151,87	851,46	965,74	31,106	7,7687	36858,6	27245,9	248,590
884	1157,15	1152,98	852,31	969,31	31,018	7,7697	36894,1	27273,1	248,621

Oxygen (*Continued*)

t	T	h	u	π_0	θ_0	s^0	H	U	S
885	1158,15	1154,09	853,17	972,89	30,931	7,7707	36929,6	27300,3	248,652
886	1159,15	1155,20	854,02	976,49	30,843	7,7716	36965,1	27327,5	248,682
887	1160,15	1156,31	854,87	980,09	30,756	7,7726	37000,6	27354,7	248,713
888	1161,15	1157,42	855,72	983,71	30,670	7,7735	37036,2	27381,9	248,744
889	1162,15	1158,53	856,57	987,33	30,584	7,7745	37071,7	27409,1	248,774
890	1163,15	1159,64	857,42	990,97	30,498	7,7754	37107,2	27436,3	248,805
891	1164,15	1160,76	858,27	994,62	30,412	7,7764	37142,8	27463,6	248,835
892	1165,15	1161,87	859,12	998,27	30,326	7,7774	37178,3	27490,8	248,866
893	1166,15	1162,98	859,97	1001,9	30,241	7,7783	37213,9	27518,0	248,896
894	1167,15	1164,09	860,82	1005,6	30,157	7,7793	37249,4	27545,3	248,927
895	1168,15	1165,20	861,67	1009,3	30,072	7,7802	37285,0	27572,5	248,957
896	1169,15	1166,31	862,52	1013,0	29,988	7,7812	37320,6	27599,8	248,988
897	1170,15	1167,42	863,38	1016,7	29,904	7,7821	37356,1	27627,0	249,018
898	1171,15	1168,53	864,23	1020,4	29,820	7,7831	37391,7	27654,3	249,049
899	1172,15	1169,65	865,08	1024,1	29,737	7,7840	37427,3	27681,5	249,079
900	1173,15	1170,76	865,93	1027,9	29,654	7,7850	37462,8	27708,8	249,109
901	1174,15	1171,87	866,78	1031,6	29,571	7,7859	37498,4	27736,0	249,140
902	1175,15	1172,98	867,64	1035,4	29,489	7,7869	37534,0	27763,3	249,170
903	1176,15	1174,09	868,49	1039,2	29,407	7,7878	37569,6	27790,6	249,200
904	1177,15	1175,21	869,34	1043,0	29,325	7,7887	37605,2	27817,9	249,230
905	1178,15	1176,32	870,19	1046,8	29,243	7,7897	37640,8	27845,1	249,261
906	1179,15	1177,43	871,05	1050,6	29,162	7,7906	37676,3	27872,4	249,291
907	1180,15	1178,54	871,90	1054,4	29,081	7,7916	37711,9	27899,7	249,321
908	1181,15	1179,66	872,75	1058,2	29,000	7,7925	37747,5	27927,0	249,351
909	1182,15	1180,77	873,60	1062,1	28,920	7,7935	37783,2	27954,3	249,381
910	1183,15	1181,88	874,46	1065,9	28,839	7,7944	37818,8	27981,6	249,411
911	1184,15	1182,99	875,31	1069,8	28,760	7,7953	37854,4	28008,9	249,441
912	1185,15	1184,11	876,16	1073,7	28,680	7,7963	37890,0	28036,2	249,472
913	1186,15	1185,22	877,02	1077,5	28,601	7,7972	37925,6	28063,5	249,502
914	1187,15	1186,33	877,87	1081,4	28,522	7,7982	37961,2	28090,8	249,532
915	1188,15	1187,45	878,72	1085,3	28,443	7,7991	37996,9	28118,1	249,562
916	1189,15	1188,56	879,58	1089,3	28,364	7,8000	38032,5	28145,4	249,592
917	1190,15	1189,67	880,43	1093,2	28,286	7,8010	38068,1	28172,7	249,622
918	1191,15	1190,79	881,29	1097,1	28,208	7,8019	38103,8	28200,1	249,651
919	1192,15	1191,90	882,14	1101,1	28,130	7,8028	38139,4	28227,4	249,681
920	1193,15	1193,02	882,99	1105,1	28,053	7,8038	38175,1	28254,7	249,711
921	1194,15	1194,13	883,85	1109,0	27,976	7,8047	38210,7	28282,1	249,741
922	1195,15	1195,24	884,70	1113,0	27,899	7,8056	38246,4	28309,4	249,771
923	1196,15	1196,36	885,56	1117,0	27,822	7,8066	38282,0	28336,7	249,801
924	1197,15	1197,47	886,41	1121,0	27,746	7,8075	38317,7	28364,1	249,831
925	1198,15	1198,59	887,27	1125,1	27,670	7,8084	38353,3	28391,4	249,860
926	1199,15	1199,70	888,12	1129,1	27,594	7,8094	38389,0	28418,8	249,890
927	1200,15	1200,82	888,98	1133,1	27,518	7,8103	38424,7	28446,1	249,920
928	1201,15	1201,93	889,83	1137,2	27,443	7,8112	38460,3	28473,5	249,950
929	1202,15	1203,05	890,69	1141,3	27,368	7,8121	38496,0	28500,9	249,979
930	1203,15	1204,16	891,54	1145,3	27,293	7,8131	38531,7	28528,2	250,009
931	1204,15	1205,28	892,40	1149,4	27,219	7,8140	38567,3	28555,6	250,039
932	1205,15	1206,39	893,25	1153,5	27,144	7,8149	38603,1	28583,0	250,068
933	1206,15	1207,51	894,11	1157,7	27,070	7,8158	38638,8	28610,3	250,098
934	1207,15	1208,62	894,96	1161,8	26,997	7,8168	38674,5	28637,7	250,127
935	1208,15	1209,74	895,82	1165,9	26,923	7,8177	38710,2	28665,1	250,157
936	1209,15	1210,85	896,67	1170,1	26,850	7,8186	38745,9	28692,5	250,186
937	1210,15	1211,97	897,53	1174,2	26,777	7,8195	38781,6	28719,9	250,216
938	1211,15	1213,09	898,39	1178,4	26,704	7,8205	38817,3	28747,3	250,246
939	1212,15	1214,20	899,24	1182,6	26,631	7,8214	38853,0	28774,7	250,275

Oxygen (*Continued*)

t	T	h	u	π_0	θ_0	s^0	H	U	S
940	1213,15	1215,32	900,10	1186,8	26,559	7,8223	38888,7	28802,1	250,304
941	1214,15	1216,43	900,95	1191,0	26,487	7,8232	38924,4	28829,5	250,334
942	1215,15	1217,55	901,81	1195,2	26,415	7,8241	38960,1	28856,9	250,363
943	1216,15	1218,67	902,67	1199,4	26,344	7,8251	38995,9	28884,3	250,393
944	1217,15	1219,78	903,52	1203,7	26,273	7,8260	39031,6	28911,7	250,422
945	1218,15	1220,90	904,38	1207,9	26,201	7,8269	39067,3	28939,1	250,451
946	1219,15	1222,02	905,24	1212,2	26,131	7,8278	39103,1	28966,5	250,481
947	1220,15	1223,13	906,10	1216,5	26,060	7,8287	39138,8	28994,0	250,510
948	1221,15	1224,25	906,95	1220,8	25,990	7,8296	39174,5	29021,4	250,539
949	1222,15	1225,37	907,81	1225,1	25,920	7,8306	39210,3	29048,8	250,569
950	1223,15	1226,48	908,67	1229,4	25,850	7,8315	39246,0	29076,3	250,598
951	1224,15	1227,60	909,52	1233,7	25,780	7,8324	39281,8	29103,7	250,627
952	1225,15	1228,72	910,38	1238,1	25,711	7,8333	39317,5	29131,1	250,656
953	1226,15	1229,84	911,24	1242,4	25,642	7,8342	39353,3	29158,6	250,685
954	1227,15	1230,95	912,10	1246,8	25,573	7,8351	39389,1	29186,0	250,715
955	1228,15	1232,07	912,96	1251,2	25,504	7,8360	39424,8	29213,5	250,744
956	1229,15	1233,19	913,81	1255,6	25,436	7,8369	39460,6	29240,9	250,773
957	1230,15	1234,31	914,67	1260,0	25,367	7,8379	39496,4	29268,4	250,802
958	1231,15	1235,43	915,53	1264,4	25,299	7,8388	39532,1	29295,9	250,831
959	1232,15	1236,54	916,39	1268,8	25,232	7,8397	39567,9	29323,3	250,860
960	1233,15	1237,66	917,25	1273,2	25,164	7,8406	39603,7	29350,8	250,889
961	1234,15	1238,78	918,11	1277,7	25,097	7,8415	39639,5	29378,3	250,918
962	1235,15	1239,90	918,96	1282,1	25,030	7,8424	39675,3	29405,7	250,947
963	1236,15	1241,02	919,82	1286,6	24,963	7,8433	39711,1	29433,2	250,976
964	1237,15	1242,14	920,68	1291,1	24,896	7,8442	39746,9	29460,7	251,005
965	1238,15	1243,26	921,54	1295,6	24,830	7,8451	39782,7	29488,2	251,034
966	1239,15	1244,37	922,40	1300,1	24,764	7,8460	39818,5	29515,7	251,063
967	1240,15	1245,49	923,26	1304,6	24,698	7,8469	39854,3	29543,2	251,092
968	1241,15	1246,61	924,12	1309,2	24,632	7,8478	39890,1	29570,7	251,120
969	1242,15	1247,73	924,98	1313,7	24,566	7,8487	39925,9	29598,2	251,149
970	1243,15	1248,85	925,84	1318,3	24,501	7,8496	39961,7	29625,7	251,178
971	1244,15	1249,97	926,70	1322,9	24,436	7,8505	39997,5	29653,2	251,207
972	1245,15	1251,09	927,56	1327,5	24,371	7,8514	40033,4	29680,7	251,236
973	1246,15	1252,21	928,42	1332,1	24,306	7,8523	40069,2	29708,2	251,264
974	1247,15	1253,33	929,28	1336,7	24,242	7,8532	40105,0	29735,7	251,293
975	1248,15	1254,45	930,14	1341,3	24,178	7,8541	40140,9	29763,2	251,322
976	1249,15	1255,57	931,00	1345,9	24,114	7,8550	40176,7	29790,7	251,351
977	1250,15	1256,69	931,86	1350,6	24,050	7,8559	40212,5	29818,3	251,379
978	1251,15	1257,81	932,72	1355,2	23,986	7,8568	40248,4	29845,8	251,408
979	1252,15	1258,93	933,58	1359,9	23,923	7,8577	40284,2	29873,3	251,437
980	1253,15	1260,05	934,44	1364,6	23,860	7,8586	40320,1	29900,9	251,465
981	1254,15	1261,17	935,30	1369,3	23,797	7,8595	40355,9	29928,4	251,494
982	1255,15	1262,29	936,16	1374,0	23,734	7,8604	40391,8	29955,9	251,522
983	1256,15	1263,41	937,02	1378,8	23,672	7,8613	40427,6	29983,5	251,551
984	1257,15	1264,53	937,88	1383,5	23,609	7,8622	40463,5	30011,0	251,580
985	1258,15	1265,65	938,74	1388,2	23,547	7,8630	40499,4	30038,6	251,608
986	1259,15	1266,77	939,60	1393,0	23,485	7,8639	40535,2	30066,1	251,637
987	1260,15	1267,89	940,46	1397,8	23,424	7,8648	40571,1	30093,7	251,665
988	1261,15	1269,02	941,32	1402,6	23,362	7,8657	40607,0	30121,3	251,693
989	1262,15	1270,14	942,19	1407,4	23,301	7,8666	40642,9	30148,8	251,722
990	1263,15	1271,26	943,05	1412,2	23,240	7,8675	40678,7	30176,4	251,750
991	1264,15	1272,38	943,91	1417,0	23,179	7,8684	40714,6	30204,0	251,779
992	1265,15	1273,50	944,77	1421,9	23,118	7,8693	40750,5	30231,5	251,807
993	1266,15	1274,62	945,63	1426,7	23,057	7,8702	40786,4	30259,1	251,835
994	1267,15	1275,74	946,49	1431,6	22,997	7,8710	40822,3	30286,7	251,864

Oxygen (*Continued*)

t	T	h	u	π_0	θ_0	s^0	H	U	S
995	1268,15	1276,87	947,36	1436,5	22,937	7,8719	40858,2	30314,3	251,892
996	1269,15	1277,99	948,22	1441,4	22,877	7,8728	40894,1	30341,9	251,920
997	1270,15	1279,11	949,08	1446,3	22,817	7,8737	40930,0	30369,5	251,949
998	1271,15	1280,23	949,94	1451,2	22,758	7,8746	40965,9	30397,0	251,977
999	1272,15	1281,35	950,81	1456,2	22,699	7,8755	41001,8	30424,6	252,005
1000	1273,15	1282,48	951,67	1461,1	22,639	7,8763	41037,7	30452,2	252,033
1001	1274,15	1283,60	952,53	1466,1	22,581	7,8772	41073,7	30479,8	252,062
1002	1275,15	1284,72	953,39	1471,1	22,522	7,8781	41109,6	30507,5	252,090
1003	1276,15	1285,85	954,26	1476,1	22,463	7,8790	41145,5	30535,1	252,118
1004	1277,15	1286,97	955,12	1481,1	22,405	7,8799	41181,4	30562,7	252,146
1005	1278,15	1288,09	955,98	1486,1	22,347	7,8807	41217,4	30590,3	252,174
1006	1279,15	1289,21	956,85	1491,1	22,289	7,8816	41253,3	30617,9	252,202
1007	1280,15	1290,34	957,71	1496,2	22,231	7,8825	41289,2	30645,5	252,230
1008	1281,15	1291,46	958,57	1501,2	22,173	7,8834	41325,2	30673,2	252,258
1009	1282,15	1292,58	959,44	1506,3	22,116	7,8842	41361,1	30700,8	252,287
1010	1283,15	1293,71	960,30	1511,4	22,059	7,8851	41397,1	30728,4	252,315
1011	1284,15	1294,83	961,16	1516,5	22,002	7,8860	41433,0	30756,1	252,343
1012	1285,15	1295,95	962,03	1521,6	21,945	7,8869	41469,0	30783,7	252,371
1013	1286,15	1297,08	962,89	1526,7	21,888	7,8877	41504,9	30811,3	252,399
1014	1287,15	1298,20	963,75	1531,9	21,832	7,8886	41540,9	30839,0	252,426
1015	1288,15	1299,32	964,62	1537,0	21,775	7,8895	41576,8	30866,6	252,454
1016	1289,15	1300,45	965,48	1542,2	21,719	7,8904	41612,8	30894,3	252,482
1017	1290,15	1301,57	966,35	1547,4	21,663	7,8912	41648,8	30921,9	252,510
1018	1291,15	1302,70	967,21	1552,6	21,607	7,8921	41684,7	30949,6	252,538
1019	1292,15	1303,82	968,08	1557,8	21,552	7,8930	41720,7	30977,2	252,566
1020	1293,15	1304,95	968,94	1563,0	21,496	7,8939	41756,7	31004,9	252,594
1021	1294,15	1306,07	969,80	1568,2	21,441	7,8947	41792,7	31032,6	252,622
1022	1295,15	1307,19	970,67	1573,5	21,386	7,8956	41828,7	31060,2	252,649
1023	1296,15	1308,32	971,53	1578,7	21,331	7,8965	41864,6	31087,9	252,677
1024	1297,15	1309,44	972,40	1584,0	21,277	7,8973	41900,6	31115,6	252,705
1025	1298,15	1310,57	973,26	1589,3	21,222	7,8982	41936,6	31143,3	252,733
1026	1299,15	1311,69	974,13	1594,6	21,168	7,8991	41972,6	31171,0	252,760
1027	1300,15	1312,82	974,99	1599,9	21,114	7,8999	42008,6	31198,6	252,788
1028	1301,15	1313,94	975,86	1605,3	21,060	7,9008	42044,6	31226,3	252,816
1029	1302,15	1315,07	976,72	1610,6	21,006	7,9017	42080,6	31254,0	252,843
1030	1303,15	1316,19	977,59	1616,0	20,952	7,9025	42116,6	31281,7	252,871
1031	1304,15	1317,32	978,46	1621,4	20,899	7,9034	42152,7	31309,4	252,899
1032	1305,15	1318,45	979,32	1626,8	20,845	7,9042	42188,7	31337,1	252,926
1033	1306,15	1319,57	980,19	1632,2	20,792	7,9051	42224,7	31364,8	252,954
1034	1307,15	1320,70	981,05	1637,6	20,739	7,9060	42260,7	31392,5	252,981
1035	1308,15	1321,82	981,92	1643,0	20,687	7,9068	42296,7	31420,2	253,009
1036	1309,15	1322,95	982,79	1648,5	20,634	7,9077	42332,8	31448,0	253,036
1037	1310,15	1324,07	983,65	1653,9	20,581	7,9085	42368,8	31475,7	253,064
1038	1311,15	1325,20	984,52	1659,4	20,529	7,9094	42404,8	31503,4	253,091
1039	1312,15	1326,33	985,38	1664,9	20,477	7,9103	42440,9	31531,1	253,119
1040	1313,15	1327,45	986,25	1670,4	20,425	7,9111	42476,9	31558,8	253,146
1041	1314,15	1328,58	987,12	1675,9	20,373	7,9120	42512,9	31586,6	253,174
1042	1315,15	1329,71	987,98	1681,5	20,322	7,9128	42549,0	31614,3	253,201
1043	1316,15	1330,83	988,85	1687,0	20,270	7,9137	42585,0	31642,0	253,229
1044	1317,15	1331,96	989,72	1692,6	20,219	7,9145	42621,1	31669,8	253,256
1045	1318,15	1333,09	990,58	1698,2	20,168	7,9154	42657,2	31697,5	253,283
1046	1319,15	1334,21	991,45	1703,8	20,117	7,9163	42693,2	31725,3	253,311
1047	1320,15	1335,34	992,32	1709,4	20,066	7,9171	42729,3	31753,0	253,338
1048	1321,15	1336,47	993,19	1715,0	20,015	7,9180	42765,3	31780,8	253,365
1049	1322,15	1337,59	994,05	1720,6	19,965	7,9188	42801,4	31808,5	253,393

Oxygen (*Continued*)

t	T	h	u	π₀	θ₀	s⁰	H	U	S
1050	1323,15	1338,72	994,92	1726,3	19,915	7,9197	42837,5	31836,3	253,420
1051	1324,15	1339,85	995,79	1732,0	19,864	7,9205	42873,6	31864,0	253,447
1052	1325,15	1340,98	996,66	1737,6	19,814	7,9214	42909,6	31891,8	253,474
1053	1326,15	1342,10	997,52	1743,3	19,764	7,9222	42945,7	31919,6	253,502
1054	1327,15	1343,23	998,39	1749,0	19,715	7,9231	42981,8	31947,3	253,529
1055	1328,15	1344,36	999,26	1754,8	19,665	7,9239	43017,9	31975,1	253,556
1056	1329,15	1345,49	1000,13	1760,5	19,616	7,9248	43054,0	32002,9	253,583
1057	1330,15	1346,61	1001,00	1766,3	19,567	7,9256	43090,1	32030,6	253,610
1058	1331,15	1347,74	1001,86	1772,0	19,518	7,9265	43126,2	32058,4	253,637
1059	1332,15	1348,87	1002,73	1777,8	19,469	7,9273	43162,2	32086,2	253,665
1060	1333,15	1350,00	1003,60	1783,6	19,420	7,9282	43198,4	32114,0	253,692
1061	1334,15	1351,13	1004,47	1789,5	19,371	7,9290	43234,5	32141,8	253,719
1062	1335,15	1352,26	1005,34	1795,3	19,323	7,9299	43270,6	32169,6	253,746
1063	1336,15	1353,38	1006,21	1801,1	19,275	7,9307	43306,7	32197,4	253,773
1064	1337,15	1354,51	1007,07	1807,0	19,226	7,9315	43342,8	32225,2	253,800
1065	1338,15	1355,64	1007,94	1812,9	19,178	7,9324	43378,9	32253,0	253,827
1066	1339,15	1356,77	1008,81	1818,8	19,131	7,9332	43415,0	32280,8	253,854
1067	1340,15	1357,90	1009,68	1824,7	19,083	7,9341	43451,1	32308,6	253,881
1068	1341,15	1359,03	1010,55	1830,6	19,035	7,9349	43487,3	32336,4	253,908
1069	1342,15	1360,16	1011,42	1836,5	18,988	7,9358	43523,4	32364,2	253,935
1070	1343,15	1361,29	1012,29	1842,5	18,941	7,9366	43559,5	32392,0	253,962
1071	1344,15	1362,42	1013,16	1848,5	18,893	7,9374	43595,7	32419,8	253,988
1072	1345,15	1363,54	1014,03	1854,4	18,847	7,9383	43631,8	32447,7	254,015
1073	1346,15	1364,67	1014,90	1860,4	18,800	7,9391	43667,9	32475,5	254,042
1074	1347,15	1365,80	1015,77	1866,5	18,753	7,9400	43704,1	32503,3	254,069
1075	1348,15	1366,93	1016,64	1872,5	18,707	7,9408	43740,2	32531,2	254,096
1076	1349,15	1368,06	1017,51	1878,5	18,660	7,9416	43776,4	32559,0	254,123
1077	1350,15	1369,19	1018,38	1884,6	18,614	7,9425	43812,5	32586,8	254,149
1078	1351,15	1370,32	1019,25	1890,7	18,568	7,9433	43848,7	32614,7	254,176
1079	1352,15	1371,45	1020,12	1896,8	18,522	7,9441	43884,8	32642,5	254,203
1080	1353,15	1372,58	1020,99	1902,9	18,476	7,9450	43921,0	32670,4	254,230
1081	1354,15	1373,71	1021,86	1909,0	18,430	7,9458	43957,2	32698,2	254,256
1082	1355,15	1374,84	1022,73	1915,1	18,385	7,9466	43993,3	32726,1	254,283
1083	1356,15	1375,97	1023,60	1921,3	18,340	7,9475	44029,5	32753,9	254,310
1084	1357,15	1377,10	1024,47	1927,5	18,294	7,9483	44065,7	32781,8	254,336
1085	1358,15	1378,23	1025,34	1933,7	18,249	7,9491	44101,8	32809,6	254,363
1086	1359,15	1379,36	1026,21	1939,9	18,204	7,9500	44138,0	32837,5	254,390
1087	1360,15	1380,50	1027,08	1946,1	18,159	7,9508	44174,2	32865,4	254,416
1088	1361,15	1381,63	1027,95	1952,3	18,115	7,9516	44210,4	32893,2	254,443
1089	1362,15	1382,76	1028,82	1958,6	18,070	7,9525	44246,6	32921,1	254,470
1090	1363,15	1383,89	1029,69	1964,8	18,026	7,9533	44282,8	32949,0	254,496
1091	1364,15	1385,02	1030,57	1971,1	17,981	7,9541	44319,0	32976,9	254,523
1092	1365,15	1386,15	1031,44	1977,4	17,937	7,9550	44355,1	33004,7	254,549
1093	1366,15	1387,28	1032,31	1983,7	17,893	7,9558	44391,3	33032,6	254,576
1094	1367,15	1388,41	1033,18	1990,1	17,850	7,9566	44427,5	33060,5	254,602
1095	1368,15	1389,54	1034,05	1996,4	17,806	7,9574	44463,8	33088,4	254,629
1096	1369,15	1390,68	1034,92	2002,8	17,762	7,9583	44500,0	33116,3	254,655
1097	1370,15	1391,81	1035,79	2009,1	17,719	7,9591	44536,2	33144,2	254,682
1098	1371,15	1392,94	1036,67	2015,5	17,675	7,9599	44572,4	33172,1	254,708
1099	1372,15	1394,07	1037,54	2021,9	17,632	7,9607	44608,6	33200,0	254,734
1100	1373,15	1395,20	1038,41	2028,4	17,589	7,9615	44644,8	33227,9	254,761
1101	1374,15	1396,33	1039,28	2034,8	17,546	7,9624	44681,0	33255,8	254,787
1102	1375,15	1397,47	1040,15	2041,3	17,503	7,9632	44717,3	33283,7	254,813
1103	1376,15	1398,60	1041,03	2047,8	17,461	7,9640	44753,5	33311,6	254,840
1104	1377,15	1399,73	1041,90	2054,2	17,418	7,9649	44789,7	33339,5	254,866

Oxygen (*Continued*)

t	T	h	u	π_0	θ_0	s^0	H	U	S
1105	1378,15	1400,86	1042,77	2060,8	17,376	7,9657	44826,0	33367,4	254,892
1106	1379,15	1402,00	1043,64	2067,3	17,334	7,9665	44862,2	33395,4	254,919
1107	1380,15	1403,13	1044,52	2073,8	17,291	7,9673	44898,4	33423,3	254,945
1108	1381,15	1404,26	1045,39	2080,4	17,249	7,9681	44934,7	33451,2	254,971
1109	1382,15	1405,39	1046,26	2087,0	17,208	7,9690	44970,9	33479,1	254,997
1110	1383,15	1406,53	1047,14	2093,5	17,166	7,9698	45007,2	33507,1	255,024
1111	1384,15	1407,66	1048,01	2100,1	17,124	7,9706	45043,4	33535,0	255,050
1112	1385,15	1408,79	1048,88	2106,8	17,083	7,9714	45079,7	33563,0	255,076
1113	1386,15	1409,93	1049,75	2113,4	17,041	7,9722	45115,9	33590,9	255,102
1114	1387,15	1411,06	1050,63	2120,1	17,000	7,9731	45152,2	33618,8	255,128
1115	1388,15	1412,19	1051,50	2126,7	16,959	7,9739	45188,4	33646,8	255,154
1116	1389,15	1413,32	1052,37	2133,4	16,918	7,9747	45224,7	33674,7	255,181
1117	1390,15	1414,46	1053,25	2140,1	16,877	7,9755	45261,0	33702,7	255,207
1118	1391,15	1415,59	1054,12	2146,9	16,836	7,9763	45297,2	33730,6	255,233
1119	1392,15	1416,73	1055,00	2153,6	16,796	7,9771	45333,5	33758,6	255,259
1120	1393,15	1417,86	1055,87	2160,4	16,755	7,9780	45369,8	33786,6	255,285
1121	1394,15	1418,99	1056,74	2167,1	16,715	7,9788	45406,1	33814,5	255,311
1122	1395,15	1420,13	1057,62	2173,9	16,674	7,9796	45442,3	33842,5	255,337
1123	1396,15	1421,26	1058,49	2180,7	16,634	7,9804	45478,6	33870,5	255,363
1124	1397,15	1422,39	1059,37	2187,6	16,594	7,9812	45514,9	33898,4	255,389
1125	1398,15	1423,53	1060,24	2194,4	16,554	7,9820	45551,2	33926,4	255,415
1126	1399,15	1424,66	1061,11	2201,3	16,515	7,9828	45587,5	33954,4	255,441
1127	1400,15	1425,80	1061,99	2208,1	16,475	7,9836	45623,8	33982,4	255,467
1128	1401,15	1426,93	1062,86	2215,0	16,435	7,9844	45660,1	34010,3	255,493
1129	1402,15	1428,07	1063,74	2221,9	16,396	7,9853	45696,4	34038,3	255,519
1130	1403,15	1429,20	1064,61	2228,9	16,357	7,9861	45732,7	34066,3	255,544
1131	1404,15	1430,33	1065,49	2235,8	16,317	7,9869	45769,0	34094,3	255,570
1132	1405,15	1431,47	1066,36	2242,8	16,278	7,9877	45805,3	34122,3	255,596
1133	1406,15	1432,60	1067,24	2249,8	16,239	7,9885	45841,6	34150,3	255,622
1134	1407,15	1433,74	1068,11	2256,8	16,201	7,9893	45877,9	34178,3	255,648
1135	1408,15	1434,87	1068,99	2263,8	16,162	7,9901	45914,2	34206,3	255,674
1136	1409,15	1436,01	1069,86	2270,8	16,123	7,9909	45950,6	34234,3	255,699
1137	1410,15	1437,14	1070,74	2277,8	16,085	7,9917	45986,9	34262,3	255,725
1138	1411,15	1438,28	1071,61	2284,9	16,046	7,9925	46023,2	34290,3	255,751
1139	1412,15	1439,41	1072,49	2292,0	16,008	7,9933	46059,5	34318,3	255,777
1140	1413,15	1440,55	1073,36	2299,1	15,970	7,9941	46095,9	34346,4	255,802
1141	1414,15	1441,69	1074,24	2306,2	15,932	7,9949	46132,2	34374,4	255,828
1142	1415,15	1442,82	1075,12	2313,4	15,894	7,9957	46168,5	34402,4	255,854
1143	1416,15	1443,96	1075,99	2320,5	15,856	7,9965	46204,9	34430,4	255,879
1144	1417,15	1445,09	1076,87	2327,7	15,819	7,9973	46241,2	34458,4	255,905
1145	1418,15	1446,23	1077,74	2334,9	15,781	7,9981	46277,6	34486,5	255,931
1146	1419,15	1447,36	1078,62	2342,1	15,744	7,9989	46313,9	34514,5	255,956
1147	1420,15	1448,50	1079,49	2349,3	15,706	7,9997	46350,3	34542,5	255,982
1148	1421,15	1449,64	1080,37	2356,5	15,669	8,0005	46386,6	34570,6	256,007
1149	1422,15	1450,77	1081,25	2363,8	15,632	8,0013	46423,0	34598,6	256,033
1150	1423,15	1451,91	1082,12	2371,1	15,595	8,0021	46459,3	34626,7	256,059
1151	1424,15	1453,04	1083,00	2378,4	15,558	8,0029	46495,7	34654,7	256,084
1152	1425,15	1454,18	1083,88	2385,7	15,521	8,0037	46532,0	34682,8	256,110
1153	1426,15	1455,32	1084,75	2393,0	15,484	8,0045	46568,4	34710,8	256,135
1154	1427,15	1456,45	1085,63	2400,4	15,448	8,0053	46604,8	34738,9	256,161
1155	1428,15	1457,59	1086,51	2407,7	15,411	8,0061	46641,1	34766,9	256,186
1156	1429,15	1458,73	1087,38	2415,1	15,375	8,0069	46677,5	34795,0	256,212
1157	1430,15	1459,86	1088,26	2422,5	15,339	8,0077	46713,9	34823,0	256,237
1158	1431,15	1461,00	1089,14	2429,9	15,303	8,0085	46750,3	34851,1	256,262
1159	1432,15	1462,14	1090,01	2437,4	15,267	8,0093	46786,7	34879,2	256,288

Oxygen (*Continued*)

t	T	h	u	π_0	θ_0	s^0	H	U	S
1160	1433,15	1463,27	1090,89	2444,8	15,231	8,0101	46823,0	34907,2	256,313
1161	1434,15	1464,41	1091,77	2452,3	15,195	8,0109	46859,4	34935,3	256,339
1162	1435,15	1465,55	1092,65	2459,8	15,159	8,0117	46895,8	34963,4	256,364
1163	1436,15	1466,69	1093,52	2467,3	15,123	8,0125	46932,2	34991,5	256,389
1164	1437,15	1467,82	1094,40	2474,8	15,088	8,0133	46968,8	35019,5	256,415
1165	1438,15	1468,96	1095,28	2482,4	15,053	8,0141	47005,0	35047,6	256,440
1166	1439,15	1470,10	1096,16	2489,9	15,017	8,0148	47041,4	35075,7	256,465
1167	1440,15	1471,24	1097,03	2497,5	14,982	8,0156	47077,8	35103,8	256,491
1168	1441,15	1472,37	1097,91	2505,1	14,947	8,0164	47114,2	35131,9	256,516
1169	1442,15	1473,51	1098,79	2512,8	14,912	8,0172	47150,6	35160,0	256,541
1170	1443,15	1474,65	1099,67	2520,4	14,877	8,0180	47187,0	35188,1	256,566
1171	1444,15	1475,79	1100,55	2528,0	14,842	8,0188	47223,4	35216,2	256,592
1172	1445,15	1476,93	1101,43	2535,7	14,808	8,0196	47259,9	35244,3	256,617
1173	1446,15	1478,06	1102,30	2543,4	14,773	8,0204	47296,3	35272,4	256,642
1174	1447,15	1479,20	1103,18	2551,1	14,739	8,0212	47332,7	35300,5	256,667
1175	1448,15	1480,34	1104,06	2558,9	14,704	8,0219	47369,1	35328,6	256,692
1176	1449,15	1481,48	1104,94	2566,6	14,670	8,0227	47405,5	35356,7	256,718
1177	1450,15	1482,62	1105,82	2574,4	14,636	8,0235	47442,0	35384,8	256,743
1178	1451,15	1483,76	1106,70	2582,2	14,602	8,0243	47478,4	35412,9	256,768
1179	1452,15	1484,89	1107,57	2590,0	14,568	8,0251	47514,8	35441,1	256,798
1180	1453,15	1486,03	1108,45	2597,8	14,534	8,0259	47551,3	35469,2	256,818
1181	1454,15	1487,17	1109,33	2605,6	14,500	8,0266	47587,7	35497,3	256,843
1182	1455,15	1488,31	1110,21	2613,5	14,466	8,0274	47624,2	35525,4	256,868
1183	1456,15	1489,45	1111,09	2621,4	14,433	8,0282	47660,6	35553,6	256,893
1184	1457,15	1490,59	1111,97	2629,3	14,399	8,0290	47697,0	35581,7	256,918
1185	1458,15	1491,73	1112,85	2637,2	14,366	8,0298	47733,5	35609,8	256,943
1186	1459,15	1492,87	1113,73	2645,2	14,333	8,0306	47769,9	35638,0	256,968
1187	1460,15	1494,01	1114,61	2653,1	14,299	8,0313	47806,4	35666,1	256,993
1188	1461,15	1495,15	1115,49	2661,1	14,266	8,0321	47842,9	35694,3	257,018
1189	1462,15	1496,28	1116,37	2669,1	14,233	8,0329	47879,3	35722,4	257,043
1190	1463,15	1497,42	1117,25	2677,1	14,200	8,0337	47915,8	35750,5	257,068
1191	1464,15	1498,56	1118,13	2685,1	14,168	8,0345	47952,2	35778,7	257,093
1192	1465,15	1499,70	1119,01	2693,2	14,135	8,0352	47988,7	35806,9	257,118
1193	1466,15	1500,84	1119,89	2701,3	14,102	8,0360	48025,2	35835,0	257,143
1194	1467,15	1501,98	1120,77	2709,3	14,070	8,0368	48061,6	35863,2	257,167
1195	1468,15	1503,12	1121,65	2717,5	14,037	8,0376	48098,1	35891,3	257,192
1196	1469,15	1504,26	1122,53	2725,6	14,005	8,0383	48134,6	35919,5	257,217
1197	1470,15	1505,40	1123,41	2733,7	13,973	8,0391	48171,1	35947,7	257,242
1198	1471,15	1506,54	1124,29	2741,9	13,941	8,0399	48207,6	35975,8	257,267
1199	1472,15	1507,68	1125,17	2750,1	13,908	8,0407	48244,0	36004,0	257,292
1200	1473,15	1508,82	1126,05	2758,3	13,876	8,0414	48280,5	36032,2	257,316
1201	1474,15	1509,96	1126,93	2766,5	13,845	8,0422	48317,0	36060,3	257,341
1202	1475,15	1511,10	1127,81	2774,8	13,813	8,0430	48353,5	36088,5	257,366
1203	1476,15	1512,24	1128,69	2783,0	13,781	8,0438	48390,0	36116,7	257,391
1204	1477,15	1513,39	1129,57	2791,3	13,750	8,0445	48426,5	36144,9	257,415
1205	1478,15	1514,53	1130,45	2799,6	13,718	8,0453	48463,0	36173,1	257,440
1206	1479,15	1515,67	1131,33	2808,0	13,687	8,0461	48499,5	36201,3	257,465
1207	1480,15	1516,81	1132,21	2816,3	13,655	8,0468	48536,0	36229,4	257,489
1208	1481,15	1517,95	1133,09	2824,7	13,624	8,0476	48572,5	36257,6	257,514
1209	1482,15	1519,09	1133,97	2833,1	13,593	8,0484	48609,0	36285,8	257,539
1210	1483,15	1520,23	1134,86	2841,5	13,562	8,0492	48645,5	36314,0	257,563
1211	1484,15	1521,37	1135,74	2849,9	13,531	8,0499	48682,1	36342,2	257,588
1212	1485,15	1522,51	1136,62	2858,3	13,500	8,0507	48718,6	36370,4	257,613
1213	1486,15	1523,65	1137,50	2866,8	13,469	8,0515	48755,1	36398,6	257,637
1214	1487,15	1524,80	1138,38	2875,3	13,438	8,0522	48791,6	36426,8	257,662

Oxygen (*Continued*)

t	T	h	u	π.	θ.	s°	H	U	S
1215	1488,15	1525,94	1139,26	2883,8	13,408	8,0530	48828,1	36455,1	257,686
1216	1489,15	1527,08	1140,14	2892,3	13,377	8,0538	48864,7	36483,3	257,711
1217	1490,15	1528,22	1141,03	2900,8	13,347	8,0545	48901,2	36511,5	257,735
1218	1491,15	1529,36	1141,91	2909,4	13,317	8,0553	48937,7	36539,7	257,760
1219	1492,15	1530,50	1142,79	2918,0	13,286	8,0561	48974,3	36567,9	257,784
1220	1493,15	1531,65	1143,67	2926,6	13,256	8,0568	49010,8	36596,1	257,809
1221	1494,15	1532,79	1144,55	2935,2	13,226	8,0576	49047,3	36624,4	257,833
1222	1495,15	1533,93	1145,44	2943,9	13,196	8,0584	49083,9	36652,6	257,858
1223	1496,15	1535,07	1146,32	2952,5	13,166	8,0591	49120,4	36680,8	257,882
1224	1497,15	1536,21	1147,20	2961,2	13,136	8,0599	49157,0	36709,1	257,907
1225	1498,15	1537,36	1148,08	2969,9	13,106	8,0606	49193,5	36737,3	257,931
1226	1499,15	1538,50	1148,97	2978,6	13,077	8,0614	49230,1	36765,5	257,955
1227	1500,15	1539,64	1149,85	2987,4	13,047	8,0622	49266,6	36793,8	257,980
1228	1501,15	1540,78	1150,73	2996,1	13,018	8,0629	49303,2	36822,0	258,004
1229	1502,15	1541,93	1151,61	3004,9	12,988	8,0637	49339,8	36850,3	258,028
1230	1503,15	1543,07	1152,50	3013,7	12,959	8,0644	49376,3	36878,5	258,053
1231	1504,15	1544,21	1153,38	3022,6	12,930	8,0652	49412,9	36906,8	258,077
1232	1505,15	1545,35	1154,26	3031,4	12,901	8,0660	49449,4	36935,0	258,101
1233	1506,15	1546,50	1155,15	3040,3	12,871	8,0667	49486,0	36963,3	258,126
1234	1507,15	1547,64	1156,03	3019,2	12,842	8,0675	45522,6	36991,5	258,150
1235	1508,15	1548,78	1156,91	3058,1	12,814	8,0682	49559,2	37019,8	258,174
1236	1509,15	1549,92	1157,79	3067,0	12,785	8,0690	49595,7	37048,0	258,198
1237	1510,15	1551,07	1158,68	3076,0	12,756	8,0698	49632,3	37076,3	258,223
1238	1511,15	1552,21	1159,56	3084,9	12,727	8,0705	49668,9	37104,6	258,247
1239	1512,15	1553,35	1160,44	3093,9	12,699	8,0713	49705,5	37132,8	258,271
1240	1513,15	1554,50	1161,33	3102,9	12,670	8,0720	49742,1	37161,1	258,295
1241	1514,15	1555,64	1162,21	3112,0	12,642	8,0728	49778,6	37189,4	258,319
1242	1515,15	1556,78	1163,10	3121,0	12,613	8,0735	49815,2	37217,7	258,344
1243	1516,15	1557,93	1163,98	3130,1	12,585	8,0743	49851,8	37245,9	258,368
1244	1517,15	1559,07	1164,86	3139,2	12,557	8,0750	49888,4	37274,2	258,392
1245	1518,15	1560,22	1165,75	3148,3	12,529	8,0758	49925,0	37302,5	258,416
1246	1519,15	1561,36	1166,63	3157,5	12,501	8,0766	49961,6	37330,8	258,440
1247	1520,15	1562,50	1167,51	3166,6	12,473	8,0773	49998,2	37359,1	258,464
1248	1521,15	1563,65	1168,40	3175,8	12,445	8,0781	50034,8	37387,4	258,488
1249	1522,15	1564,79	1169,28	3185,0	12,417	8,0788	50071,4	37415,7	258,512
1250	1523,15	1565,94	1170,17	3194,2	12,389	8,0796	50108,0	37444,0	258,536
1251	1524,15	1567,08	1171,05	3203,5	12,362	8,0803	50144,7	37472,2	258,560
1252	1525,15	1568,22	1171,94	3212,7	12,334	8,0811	50181,3	37500,5	258,584
1253	1526,15	1569,37	1172,82	3222,0	12,307	8,0818	50217,9	37528,8	258,608
1254	1527,15	1570,51	1173,71	3231,3	12,279	8,0826	50254,5	37557,2	258,632
1255	1528,15	1571,66	1174,59	3240,6	12,252	8,0833	50291,1	37585,5	258,656
1256	1529,15	1572,80	1175,47	3250,0	12,225	8,0841	50327,7	37613,8	258,680
1257	1530,15	1573,95	1176,36	3259,4	12,198	8,0848	50364,4	37642,1	258,704
1258	1531,15	1575,09	1177,24	3268,8	12,170	8,0856	50401,0	37670,4	258,728
1259	1532,15	1576,24	1178,13	3278,2	12,143	8,0863	50437,6	37698,7	258,752
1260	1533,15	1577,38	1179,01	3287,6	12,116	8,0871	50474,3	37727,0	258,776
1261	1534,15	1578,52	1179,90	3297,1	12,090	8,0878	50510,9	37755,3	258,800
1262	1535,15	1579,67	1180,78	3306,6	12,063	8,0885	50547,5	37783,7	258,824
1263	1536,15	1580,81	1181,67	3316,1	12,036	8,0893	50584,2	37812,0	258,848
1264	1537,15	1581,96	1182,55	3325,6	12,009	8,0900	50620,8	37840,3	258,871
1265	1538,15	1583,11	1183,44	3335,1	11,983	8,0908	50657,5	37868,7	258,895
1266	1539,15	1584,25	1184,33	3344,7	11,956	8,0915	50694,1	37897,0	258,919
1267	1540,15	1585,40	1185,21	3354,3	11,930	8,0923	50730,8	37925,3	258,943
1268	1541,15	1586,54	1186,10	3363,9	11,903	8,0930	50767,4	37953,7	258,967
1269	1542,15	1587,69	1186,98	3373,5	11,877	8,0938	50804,1	37982,0	258,990

Oxygen (*Continued*)

t	T	h	u	π_0	θ_0	s^0	H	U	S
1270	1543,15	1588,83	1187,87	3383,2	11,851	8,0945	50840,7	38010,3	259,014
1271	1544,15	1589,98	1188,75	3392,9	11,825	8,0952	50877,4	38038,7	259,038
1272	1545,15	1591,12	1189,64	3402,6	11,799	8,0960	50914,0	38067,0	259,062
1273	1546,15	1592,27	1190,53	3412,3	11,773	8,0967	50950,7	38095,4	259,085
1274	1547,15	1593,41	1191,41	3422,0	11,747	8,0975	50987,4	38123,7	259,109
1275	1548,15	1594,56	1192,30	3431,8	11,721	8,0982	51024,0	38152,1	259,133
1276	1549,15	1595,71	1193,18	3441,6	11,695	8,0989	51060,7	38180,4	259,156
1277	1550,15	1596,85	1194,07	3451,4	11,669	8,0997	51097,4	38208,8	259,180
1278	1551,15	1598,00	1194,96	3461,2	11,644	8,1004	51134,0	38237,2	259,204
1279	1552,15	1599,15	1195,84	3471,1	11,618	8,1012	51170,7	38265,5	259,227
1280	1553,15	1600,29	1196,73	3481,0	11,593	8,1019	51207,4	38293,9	259,251
1281	1554,15	1601,44	1197,62	3490,9	11,567	8,1026	51244,1	38322,2	259,275
1282	1555,15	1602,58	1198,50	3500,8	11,542	8,1034	51280,8	38350,6	259,298
1283	1556,15	1603,73	1199,39	3510,7	11,517	8,1041	51317,5	38379,0	259,322
1284	1557,15	1604,88	1200,28	3520,7	11,491	8,1048	51354,1	38407,4	259,345
1285	1558,15	1606,02	1201,16	3530,7	11,466	8,1056	51390,8	38435,7	259,369
1286	1559,15	1607,17	1202,05	3540,7	11,441	8,1063	51427,5	38464,1	259,393
1287	1560,15	1608,32	1202,94	3550,7	11,416	8,1071	51464,2	38492,5	259,416
1288	1561,15	1609,46	1203,82	3560,8	11,391	8,1078	51500,9	38520,9	259,440
1289	1562,15	1610,61	1204,71	3570,9	11,366	8,1085	51537,6	38549,3	259,463
1290	1563,15	1611,76	1205,60	3581,0	11,342	8,1093	51574,3	38577,6	259,487
1291	1564,15	1612,90	1206,48	3591,1	11,317	8,1100	51611,0	38606,0	259,510
1292	1565,15	1614,05	1207,37	3601,2	11,292	8,1107	51647,7	38634,4	259,533
1293	1566,15	1615,20	1208,26	3611,4	11,268	8,1115	51684,4	38662,8	259,557
1294	1567,15	1616,35	1209,15	3621,6	11,243	8,1122	51721,1	38691,2	259,580
1295	1568,15	1617,49	1210,03	3631,8	11,219	8,1129	51757,9	38719,6	259,604
1296	1569,15	1618,64	1210,92	3642,1	11,194	8,1137	51794,6	38748,0	259,627
1297	1570,15	1619,79	1211,81	3652,3	11,170	8,1144	51831,3	38776,4	259,651
1298	1571,15	1620,94	1212,70	3662,6	11,146	8,1151	51868,0	38804,8	259,674
1299	1572,15	1622,08	1213,58	3672,9	11,121	8,1158	51904,7	38833,2	259,697
1300	1573,15	1623,23	1214,47	3683,2	11,097	8,1166	51941,5	38861,6	259,721
1301	1574,15	1624,38	1215,36	3693,6	11,073	8,1173	51978,2	38890,1	259,744
1302	1575,15	1625,53	1216,25	3704,0	11,049	8,1180	52014,9	38918,5	259,767
1303	1576,15	1626,67	1217,14	3714,4	11,025	8,1188	52051,6	38946,9	259,791
1304	1577,15	1627,82	1218,02	3724,8	11,001	8,1195	52088,4	38975,3	259,814
1305	1578,15	1628,97	1218,91	3735,2	10,977	8,1202	52125,1	39003,7	259,837
1306	1579,15	1630,12	1219,80	3745,7	10,954	8,1209	52161,9	39032,2	259,860
1307	1580,15	1631,27	1220,69	3756,2	10,930	8,1217	52198,6	39060,6	259,884
1308	1581,15	1632,42	1221,58	3766,7	10,906	8,1224	52235,3	39089,0	259,907
1309	1582,15	1633,56	1222,47	3777,3	10,883	8,1231	52272,1	39117,4	259,930
1310	1583,15	1634,71	1223,35	3787,8	10,859	8,1238	52308,8	39145,9	259,953
1311	1584,15	1635,86	1224,24	3798,4	10,836	8,1246	52345,6	39174,3	259,977
1312	1585,15	1637,01	1225,13	3809,0	10,813	8,1253	52382,3	39202,7	260,000
1313	1586,15	1638,16	1226,02	3819,7	10,789	8,1260	52419,1	39231,2	260,023
1314	1587,15	1639,31	1226,91	3830,3	10,766	8,1267	52455,8	39259,6	260,046
1315	1588,15	1640,46	1227,80	3841,0	10,743	8,1275	52492,6	39288,1	260,069
1316	1589,15	1641,60	1228,69	3851,7	10,720	8,1282	52529,4	39316,5	260,092
1317	1590,15	1642,75	1229,58	3862,4	10,697	8,1289	52566,1	39345,0	260,116
1318	1591,15	1643,90	1230,47	3873,2	10,674	8,1296	52602,9	39373,4	260,139
1319	1592,15	1645,05	1231,35	3883,9	10,651	8,1304	52639,6	39401,9	260,162
1320	1593,15	1646,20	1232,24	3894,7	10,628	8,1311	52676,4	39430,3	260,185
1321	1594,15	1647,35	1233,13	3905,6	10,605	8,1318	52713,2	39458,8	260,208
1322	1595,15	1648,50	1234,02	3916,4	10,582	8,1325	52750,0	39487,2	260,231
1323	1596,15	1649,65	1234,91	3927,3	10,560	8,1332	52786,7	39515,7	260,254
1324	1597,15	1650,80	1235,80	3938,2	10,537	8,1340	52823,5	39544,2	260,277

Oxygen (Continued)

t	T	h	u	π_*	θ_0	s^0	H	U	S
1325	1598,15	1651,95	1236,69	3949,1	10,515	8,1347	52860,3	39572,6	260,300
1326	1599,15	1653,10	1237,58	3960,1	10,492	8,1354	52897,1	39601,1	260,323
1327	1600,15	1654,25	1238,47	3971,0	10,470	8,1361	52933,9	39629,6	260,346
1328	1601,15	1655,39	1239,36	3982,0	10,447	8,1368	52970,7	39658,0	260,369
1329	1602,15	1656,54	1240,25	3993,0	10,425	8,1376	53007,4	39686,5	260,392
1330	1603,15	1657,69	1241,14	4004,1	10,403	8,1383	53044,2	39715,0	260,415
1331	1604,15	1658,84	1242,03	4015,1	10,380	8,1390	53081,0	39743,5	260,438
1332	1605,15	1659,99	1242,92	4026,2	10,358	8,1397	53117,8	39771,9	260,461
1333	1606,15	1661,14	1243,81	4037,3	10,336	8,1404	53154,6	39800,4	260,484
1334	1607,15	1662,29	1244,70	4048,5	10,314	8,1411	53191,4	39828,9	260,507
1335	1608,15	1663,44	1245,59	4059,6	10,292	8,1419	53228,2	39857,4	260,530
1336	1609,15	1664,59	1246,48	4070,8	10,270	8,1426	53265,0	39885,9	260,553
1337	1610,15	1665,74	1247,37	4082,0	10,249	8,1433	53301,8	39914,4	260,575
1338	1611,15	1666,90	1248,26	4093,3	10,227	8,1440	53338,6	39942,9	260,598
1339	1612,15	1668,05	1249,15	4104,5	10,205	8,1447	53375,5	39971,4	260,621
1340	1613,15	1669,20	1250,04	4115,8	10,183	8,1454	53412,3	39999,9	260,644
1341	1614,15	1670,35	1250,93	4127,1	10,162	8,1461	53449,1	40028,4	260,667
1342	1615,15	1671,50	1251,82	4138,5	10,140	8,1469	53485,9	40056,9	260,690
1343	1616,15	1672,65	1252,72	4149,8	10,119	8,1476	53522,7	40085,4	260,712
1344	1617,15	1673,80	1253,61	4161,2	10,097	8,1483	53559,5	40113,9	260,735
1345	1618,15	1674,95	1254,50	4172,6	10,076	8,1490	53596,4	40142,4	260,758
1346	1619,15	1676,10	1255,39	4184,0	10,055	8,1497	53633,2	40170,9	260,781
1347	1620,15	1677,25	1256,28	4195,5	10,033	8,1504	53670,0	40199,4	260,803
1348	1621,15	1678,40	1257,17	4207,0	10,012	8,1511	53706,9	40228,0	260,826
1349	1622,15	1679,55	1258,06	4218,5	9,9913	8,1518	53743,7	40256,5	260,849
1350	1623,15	1680,70	1258,95	4230,0	9,9702	8,1525	53780,5	40285,0	260,871
1351	1624,15	1681,86	1259,84	4241,6	9,9492	8,1532	53817,4	40313,5	260,894
1352	1625,15	1683,01	1260,74	4253,2	9,9282	8,1540	53854,2	40342,0	260,917
1353	1626,15	1684,16	1261,63	4264,8	9,9073	8,1547	53891,0	40370,6	260,940
1354	1627,15	1685,31	1262,52	4276,4	9,8864	8,1554	53927,9	40399,1	260,962
1355	1628,15	1686,46	1263,41	4288,1	9,8656	8,1561	53964,7	40427,6	260,985
1356	1629,15	1687,61	1264,30	4299,8	9,8448	8,1568	54001,6	40456,2	261,007
1357	1630,15	1688,76	1265,19	4311,5	9,8241	8,1575	54038,4	40484,7	261,030
1358	1631,15	1689,92	1266,09	4323,2	9,8034	8,1582	54075,3	40513,2	261,053
1359	1632,15	1691,07	1266,98	4335,0	9,7828	8,1589	54112,2	40541,8	261,075
1360	1633,15	1692,22	1267,87	4346,8	9,7623	8,1596	54149,0	40570,3	261,098
1361	1634,15	1693,37	1268,76	4358,6	9,7418	8,1603	54185,9	40598,9	261,120
1362	1635,15	1694,52	1269,65	4370,4	9,7214	8,1610	54222,7	40627,4	261,143
1363	1636,15	1695,68	1270,55	4382,3	9,7010	8,1617	54259,6	40656,0	261,165
1364	1637,15	1696,83	1271,44	4394,2	9,6806	8,1624	54296,5	40684,5	261,188
1365	1638,15	1697,98	1272,33	4406,1	9,6604	8,1631	54333,3	40713,1	261,210
1366	1639,15	1699,13	1273,22	4418,0	9,6401	8,1638	54370,2	40741,6	261,233
1367	1640,15	1700,28	1274,12	4430,0	9,6200	8,1645	54407,1	40770,2	261,255
1368	1641,15	1701,44	1275,01	4442,0	9,5998	8,1652	54444,0	40798,8	261,278
1369	1642,15	1702,59	1275,90	4454,0	9,5798	8,1659	54480,8	40827,3	261,300
1370	1643,15	1703,74	1276,79	4466,0	9,5597	8,1666	54517,7	40855,9	261,323
1371	1644,15	1704,90	1277,69	4478,1	9,5398	8,1673	54554,6	40884,5	261,345
1372	1645,15	1706,05	1278,58	4490,2	9,5199	8,1680	54591,5	40913,0	261,368
1373	1646,15	1707,20	1279,47	4502,3	9,5000	8,1687	54628,4	40941,6	261,390
1374	1647,15	1708,35	1280,37	4514,5	9,4802	8,1694	54665,3	40970,2	261,413
1375	1648,15	1709,51	1281,26	4526,6	9,4604	8,1701	54702,2	40998,8	261,435
1376	1649,15	1710,66	1282,15	4538,8	9,4407	8,1708	54739,0	41027,3	261,457
1377	1650,15	1711,81	1283,05	4551,1	9,4211	8,1715	54775,9	41055,9	261,480
1378	1651,15	1712,97	1283,94	4563,3	9,4015	8,1722	54812,8	41084,5	261,502
1379	1652,15	1714,12	1284,83	4575,6	9,3819	8,1729	54849,7	41113,1	261,524

Oxygen (*Continued*)

t	T	h	u	π_0	θ_0	s^0	H	U	S
1380	1653,15	1715,27	1285,73	4587,9	9,3624	8,1736	54886,6	41141,7	261,547
1381	1654,15	1716,43	1286,62	4600,2	9,3430	8,1743	54923,6	41170,3	261,569
1382	1655,15	1717,58	1287,51	4612,6	9,3236	8,1750	54960,5	41198,9	261,591
1383	1656,15	1718,73	1288,41	4625,0	9,3042	8,1757	54997,4	41227,5	261,614
1384	1657,15	1719,89	1289,30	4637,4	9,2849	8,1764	55034,3	41256,1	261,636
1385	1658,15	1721,04	1290,19	4649,8	9,2657	8,1771	55071,2	41284,7	261,658
1386	1659,15	1722,19	1291,09	4662,3	9,2465	8,1778	55108,1	41313,3	261,680
1387	1660,15	1723,35	1291,98	4674,8	9,2274	8,1785	55145,0	41341,9	261,703
1388	1661,15	1724,50	1292,88	4687,3	9,2083	8,1792	55182,0	41370,5	261,725
1389	1662,15	1725,65	1293,77	4699,8	9,1892	8,1799	55218,9	41399,1	261,747
1390	1663,15	1726,81	1294,66	4712,4	9,1702	8,1806	55255,8	41427,7	261,769
1391	1664,15	1727,96	1295,56	4725,0	9,1513	8,1813	55292,7	41456,3	261,792
1392	1665,15	1729,12	1296,45	4737,6	9,1324	8,1820	55329,7	41484,9	261,814
1393	1666,15	1730,27	1297,35	4750,3	9,1135	8,1827	55366,6	41513,5	261,836
1394	1667,15	1731,43	1298,24	4763,0	9,0947	8,1834	55403,5	41542,2	261,858
1395	1668,15	1732,58	1299,14	4775,7	9,0759	8,1841	55440,5	41570,8	261,880
1396	1669,15	1733,73	1300,03	4788,4	9,0572	8,1848	55477,4	41599,4	261,902
1397	1670,15	1734,89	1300,93	4801,2	9,0386	8,1854	55514,4	41628,0	261,924
1398	1671,15	1736,04	1301,82	4814,0	9,0200	8,1861	55551,3	41656,7	261,947
1399	1672,15	1737,20	1302,71	4826,8	9,0014	8,1868	55588,2	41685,3	261,969
1400	1673,15	1738,35	1303,61	4839,6	8,9829	8,1875	55625,2	41713,9	261,991
1401	1674,15	1739,51	1304,50	4852,5	8,9644	8,1882	55662,1	41742,6	262,013
1402	1675,15	1740,66	1305,40	4865,4	8,9460	8,1889	55699,1	41771,2	262,035
1403	1676,15	1741,82	1306,29	4878,3	8,9276	8,1896	55736,1	41799,9	262,057
1404	1677,15	1742,97	1307,19	4891,2	8,9093	8,1903	55773,0	41828,5	262,079
1405	1678,15	1744,13	1308,09	4904,2	8,8910	8,1910	55810,0	41857,2	262,101
1406	1679,15	1745,28	1308,98	4917,2	8,8728	8,1917	55846,9	41885,8	262,123
1407	1680,15	1746,44	1309,88	4930,3	8,8546	8,1923	55883,9	41914,5	262,145
1408	1681,15	1747,59	1310,77	4943,3	8,8365	8,1930	55920,9	41943,1	262,167
1409	1682,15	1748,75	1311,67	4956,4	8,8184	8,1937	55957,9	41971,8	262,189
1410	1683,15	1749,90	1312,56	4969,5	8,8003	8,1944	55994,8	42000,4	262,211
1411	1684,15	1751,06	1313,46	4982,7	8,7823	8,1951	56031,8	42029,1	262,233
1412	1685,15	1752,22	1314,35	4995,8	8,7644	8,1958	56068,8	42057,7	262,255
1413	1686,15	1753,37	1315,25	5009,0	8,7465	8,1965	56105,8	42086,4	262,277
1414	1687,15	1754,53	1316,15	5022,3	8,7286	8,1971	56142,7	42115,1	262,299
1415	1688,15	1755,68	1317,04	5035,5	8,7108	8,1978	56179,7	42143,8	262,321
1416	1689,15	1756,84	1317,94	5048,8	8,6930	8,1985	56216,7	42172,4	262,343
1417	1690,15	1757,99	1318,83	5062,1	8,6753	8,1992	56253,7	42201,1	262,365
1418	1691,15	1759,15	1319,73	5075,5	8,6576	8,1999	56290,7	42229,8	262,386
1419	1692,15	1760,31	1320,63	5088,8	8,6400	8,2006	56327,7	42258,5	262,408
1420	1693,15	1761,46	1321,52	5102,2	8,6224	8,2012	56364,7	42287,1	262,430
1421	1694,15	1762,62	1322,42	5115,6	8,6048	8,2019	56401,7	42315,8	262,452
1422	1695,15	1763,78	1323,32	5129,1	8,5873	8,2026	56438,7	42344,5	262,474
1423	1696,15	1764,93	1324,21	5142,6	8,5699	8,2033	56475,7	42373,2	262,496
1424	1697,15	1766,09	1325,11	5156,1	8,5525	8,2040	56512,7	42401,9	262,517
1425	1698,15	1767,24	1326,01	5169,6	8,5351	8,2047	56549,7	42430,6	262,539
1426	1699,15	1768,40	1326,90	5183,2	8,5178	8,2053	56586,7	42459,3	262,561
1427	1700,15	1769,56	1327,80	5196,8	8,5005	8,2060	56623,7	42488,0	262,583
1428	1701,15	1770,72	1328,70	5210,4	8,4832	8,2067	56660,8	42516,7	262,605
1429	1702,15	1771,87	1329,59	5224,1	8,4660	8,2074	56697,8	42545,4	262,626
1430	1703,15	1773,03	1330,49	5237,7	8,4489	8,2081	56734,8	42574,1	262,648
1431	1704,15	1774,19	1331,39	5251,5	8,4318	8,2087	56771,8	42602,8	262,670
1432	1705,15	1775,34	1332,29	5265,2	8,4147	8,2094	56808,9	42631,5	262,692
1433	1706,15	1776,50	1333,18	5279,0	8,3977	8,2101	56845,9	42660,3	262,713
1434	1707,15	1777,66	1334,08	5292,8	8,3807	8,2108	56882,9	42689,0	262,735
1435	1708,15	1778,82	1334,98	5306,6	8,3638	8,2115	56919,9	42717,7	262,757
1436	1709,15	1779,97	1335,88	5320,4	8,3469	8,2121	56957,0	42746,4	262,778
1437	1710,15	1781,13	1336,77	5334,3	8,3300	8,2128	56994,0	42775,1	262,800
1438	1711,15	1782,29	1337,67	5348,2	8,3132	8,2135	57031,1	42803,9	262,822
1439	1712,15	1783,45	1338,57	5362,2	8,2965	8,2142	57068,1	42832,6	262,843

Oxygen (*Continued*)

t	T	h	u	π₀	θ₀	s°	H	U	S
1440	1713,15	1784,60	1339,47	5376,1	8,2797	8,2148	57105,2	42861,3	262,865
1441	1714,15	1785,76	1340,36	5390,1	8,2631	8,2155	57142,2	42890,1	262,887
1442	1715,15	1786,92	1341,26	5404,2	8,2464	8,2162	57179,3	42918,8	262,908
1443	1716,15	1788,08	1342,16	5418,2	8,2298	8,2169	57216,3	42947,5	262,930
1444	1717,15	1789,24	1343,06	5432,3	8,2133	8,2175	57253,4	42976,3	262,951
1445	1718,15	1790,39	1343,96	5446,4	8,1968	8,2182	57290,4	43005,0	262,973
1446	1719,15	1791,55	1344,86	5460,6	8,1803	8,2189	57327,5	43033,8	262,994
1447	1720,15	1792,71	1345,75	5474,7	8,1638	8,2196	57364,6	43062,5	263,016
1448	1721,15	1793,87	1346,65	5488,9	8,1475	8,2202	57401,6	43091,3	263,038
1449	1722,15	1795,03	1347,55	5503,2	8,1311	8,2209	57438,7	43120,0	263,059
1450	1723,15	1796,19	1348,45	5517,4	8,1148	8,2216	57475,8	43148,8	263,081
1451	1724,15	1797,34	1349,35	5531,7	8,0985	8,2223	57512,9	43177,6	263,102
1452	1725,15	1798,50	1350,25	5546,0	8,0823	8,2229	57549,9	43206,3	263,124
1453	1726,15	1799,66	1351,15	5560,4	8,0661	8,2236	57587,0	43235,1	263,145
1454	1727,15	1800,82	1352,05	5574,8	8,0500	8,2243	57624,1	43263,9	263,167
1455	1728,15	1801,98	1352,95	5589,2	8,0338	8,2249	57661,2	43292,6	263,188
1456	1729,15	1803,14	1353,85	5603,6	8,0178	8,2256	57698,3	43321,4	263,210
1457	1730,15	1804,30	1354,74	5618,1	8,0017	8,2263	57735,4	43350,2	263,231
1458	1731,15	1805,46	1355,64	5632,6	7,9858	8,2269	57772,5	43379,0	263,252
1459	1732,15	1806,62	1356,54	5647,1	7,9698	8,2276	57809,6	43407,8	263,274
1460	1733,15	1807,78	1357,44	5661,7	7,9539	8,2283	57846,7	43436,5	263,295
1461	1734,15	1808,94	1358,34	5676,3	7,9380	8,2290	57883,8	43465,3	263,317
1462	1735,15	1810,10	1359,24	5690,9	7,9222	8,2296	57920,9	43494,1	263,338
1463	1736,15	1811,26	1360,14	5705,6	7,9064	8,2303	57958,0	43522,9	263,359
1464	1737,15	1812,41	1361,04	5720,3	7,8907	8,2310	57995,1	43551,7	263,381
1465	1738,15	1813,57	1361,94	5735,0	7,8749	8,2316	58032,2	43580,5	263,402
1466	1739,15	1814,73	1362,84	5749,7	7,8593	8,2323	58069,3	43609,3	263,424
1467	1740,15	1815,89	1363,74	5764,5	7,8436	8,2330	58106,5	43638,1	263,445
1468	1741,15	1817,05	1364,64	5779,3	7,8280	8,2336	58143,6	43666,9	263,466
1469	1742,15	1818,22	1365,54	5794,1	7,8125	8,2343	58180,7	43695,8	263,487
1470	1743,15	1819,38	1366,44	5809,0	7,7969	8,2350	58217,8	43724,6	263,509
1471	1744,15	1820,54	1367,34	5823,9	7,7815	8,2356	58255,0	43753,4	263,530
1472	1745,15	1821,70	1368,25	5838,8	7,7660	8,2363	58292,1	43782,2	263,551
1473	1746,15	1822,86	1369,15	5853,8	7,7506	8,2370	58329,2	43811,0	263,573
1474	1747,15	1824,02	1370,05	5868,8	7,7352	8,2376	58366,4	43839,9	263,594
1475	1748,15	1825,18	1370,95	5883,8	7,7199	8,2383	58403,5	43868,7	263,615
1476	1749,15	1826,34	1371,85	5898,9	7,7046	8,2389	58440,7	43897,5	263,636
1477	1750,15	1827,50	1372,75	5913,9	7,6894	8,2396	58477,8	43926,4	263,658
1478	1751,15	1828,66	1373,65	5929,1	7,6741	8,2403	58515,0	43955,2	263,679
1479	1752,15	1829,82	1374,55	5944,2	7,6590	8,2409	58552,1	43984,0	263,700
1480	1753,15	1830,98	1375,45	5959,4	7,6438	8,2416	58589,3	44012,9	263,721
1481	1754,15	1832,15	1376,36	5974,6	7,6287	8,2423	58626,4	44041,7	263,742
1482	1755,15	1833,31	1377,26	5989,8	7,6136	8,2429	58663,6	44070,6	263,764
1483	1756,15	1834,47	1378,16	6005,1	7,5986	8,2436	58700,8	44099,4	263,785
1484	1757,15	1835,63	1379,06	6020,4	7,5836	8,2442	58738,0	44128,3	263,806
1485	1758,15	1836,79	1379,96	6035,7	7,5686	8,2449	58775,1	44157,1	263,827
1486	1759,15	1837,95	1380,86	6051,1	7,5537	8,2456	58812,3	44186,0	263,848
1487	1760,15	1839,12	1381,77	6066,5	7,5388	8,2462	58849,5	44214,9	263,869
1488	1761,15	1840,28	1382,67	6081,9	7,5240	8,2469	58886,7	44243,7	263,891
1489	1762,15	1841,44	1383,57	6097,4	7,5092	8,2475	58923,9	44272,6	263,912
1490	1763,15	1842,60	1384,47	6112,9	7,4944	8,2482	58961,0	44301,5	263,933
1491	1764,15	1843,76	1385,38	6128,4	7,4796	8,2489	58998,2	44330,4	263,954
1492	1765,15	1844,93	1386,28	6144,0	7,4649	8,2495	59035,4	44359,2	263,975
1493	1766,15	1846,09	1387,18	6159,5	7,4502	8,2502	59072,6	44388,1	263,996
1494	1767,15	1847,25	1388,08	6175,2	7,4356	8,2508	59109,8	44417,0	264,017
1495	1768,15	1848,41	1388,99	6190,8	7,4210	8,2515	59147,0	44445,9	264,038
1496	1769,15	1849,58	1389,89	6206,5	7,4064	8,2522	59184,2	44474,8	264,059
1497	1770,15	1850,74	1390,79	6222,2	7,3919	8,2528	59221,5	44503,7	264,080
1498	1771,15	1851,90	1391,70	6238,0	7,3774	8,2535	59258,7	44532,6	264,101
1499	1772,15	1853,07	1392,60	6253,7	7,3629	8,2541	59295,9	44561,5	264,122
1500	1773,15	1854,23	1393,50	6269,6	7,3485	8,2548	59333,1	44590,4	264,143

Table III.9 Carbon dioxide ($\mu = 44.009$)

t	T	c_p	μc_p	c_v	μc_v	$k = \dfrac{c_p}{c_v}$
—50	223,15	0,7605	33,470	0,5716	25,156	1,330
—25	248,15	0,7899	34,765	0,6010	26,451	1,314
0	273,15	0,8178	35,989	0,6288	27,675	1,301
25	298,15	0,8441	37,148	0,6552	28,834	1,288
50	328,15	0,8690	38,246	0,6801	29,932	1,278
75	348,15	0,8927	39,288	0,7038	30,974	1,268
100	373,15	0,9152	40,278	0,7263	31,964	1,260
125	398,15	0,9366	41,218	0,7476	32,904	1,253
150	423,15	0,9569	42,113	0,7680	33,799	1,246
175	448,15	0,9762	42,965	0,7873	34,651	1,240
200	473,15	0,9947	43,776	0,8058	35,462	1,234
250	523,15	1,0290	45,285	0,8401	36,971	1,225
300	573,15	1,0602	46,657	0,8712	38,343	1,217
350	623,15	1,0885	47,906	0,8996	39,592	1,210
400	673,15	1,1143	49,042	0,9254	40,728	1,204
450	723,15	1,1379	50,078	0,9490	41,764	1,199
500	773,15	1,1593	51,022	0,9704	42,708	1,195
550	823,15	1,1789	51,882	0,9900	43,568	1,191
600	873,15	1,1967	52,667	1,0078	44,353	1,187
650	923,15	1,2130	53,384	1,0241	45,070	1,184
700	973,15	1,2279	54,040	1,0390	45,726	1,182
750	1023,15	1,2416	54,643	1,0527	46,329	1,179
800	1073,15	1,2542	55,197	1,0653	46,883	1,177
850	1123,15	1,2658	55,708	1,0769	47,394	1,175
900	1173,15	1,2766	56,181	1,0876	47,867	1,174
950	1223,15	1,2865	56,620	1,0976	48,306	1,172
1000	1273,15	1,2958	57,028	1,1069	48,714	1,171
1050	1323,15	1,3044	57,407	1,1155	49,093	1,169
1100	1373,15	1,3124	57,760	1,1235	49,446	1,168
1150	1423,15	1,3199	58,087	1,1310	49,773	1,167
1200	1473,15	1,3267	58,390	1,1378	50,076	1,166
1260	1523,15	1,3331	58,668	1,1442	50,354	1,165
1300	1573,15	1,3389	58,925	1,1500	50,611	1,164
1350	1623,15	1,3443	59,163	1,1554	50,849	1,164
1400	1673,15	1,3494	59,385	1,1604	51,071	1,163
1450	1723,15	1,3542	59,600	1,1653	51,286	1,162
1500	1773,15	1,3592	59,818	1,1703	51,504	1,161

Table III.10 Carbon dioxide (μ = 44.009)

t	T	h	u	π_0	θ_0	s^0	H	U	S
—50	223,15	152,61	110,45	0,04221	99876	4,6222	6716,0	4860,7	203,420
—49	224,15	153,37	111,02	0,04298	98532	4,6256	6749,5	4885,9	203,570
—48	225,15	154,13	111,59	0,04376	97210	4,6290	6783,1	4911,1	203,719
—47	226,15	154,89	112,17	0,04455	95909	4,6324	6816,7	4936,4	203,868
—46	227,15	155,66	112,74	0,04535	94627	4,6358	6850,3	4961,7	204,016
—45	228,15	156,42	113,32	0,04617	93366	4,6392	6884,1	4987,1	204,164
—44	229,15	157,19	113,90	0,04699	92125	4,6425	6917,8	5012,6	204,312
—43	230,15	157,96	114,48	0,04783	90903	4,6459	6951,6	5038,1	204,459
—42	231,15	158,73	115,06	0,04868	89700	4,6492	6985,5	5063,6	204,606
—41	232,15	159,50	115,64	0,04955	88515	4,6525	7019,4	5089,2	204,753
—40	233,15	160,27	116,22	0,05043	87349	4,6558	7053,4	5114,9	204,899
—39	234,15	161,04	116,81	0,05132	86201	4,6591	7087,4	5140,6	205,044
—38	235,15	161,82	117,39	0,05222	85070	4,6624	7121,5	5166,3	205,189
—37	236,15	162,59	117,98	0,05314	83957	4,6657	7155,6	5192,2	205,334
—36	237,15	163,37	118,57	0,05407	82860	4,6690	7189,8	5218,0	205,479
—35	238,15	164,15	119,16	0,05502	81781	4,6723	7224,0	5243,9	205,523
—34	239,15	164,93	119,75	0,05597	80717	4,6756	7258,3	5269,9	205,766
—33	240,15	165,71	120,34	0,05695	79670	4,6788	7292,6	5295,9	205,910
—32	241,15	166,49	120,93	0,05793	78639	4,6821	7327,0	5322,0	206,053
—31	242,15	167,27	121,52	0,05894	77623	4,6853	7361,4	5348,1	206,195
—30	243,15	168,05	122,12	0,05995	76623	4,6885	7395,9	5374,3	206,337
—29	244,15	168,84	122,71	0,06098	75637	4,6917	7430,5	5400,5	206,479
—28	245,15	169,63	123,31	0,06203	74667	4,6950	7465,0	5426,8	206,620
—27	246,15	170,41	123,91	0,06309	73711	4,6982	7499,7	5453,1	206,761
—26	247,15	171,20	124,51	0,06417	72769	4,7014	7534,4	5479,5	206,902
—25	248,15	171,99	125,11	0,06526	71841	4,7045	7569,1	5505,9	207,042
—24	249,15	172,78	125,71	0,06636	70927	4,7077	7603,9	5532,4	207,182
—23	250,15	173,57	126,31	0,06749	70026	4,7109	7638,7	5558,9	207,322
—22	251,15	174,37	126,92	0,06863	69139	4,7141	7673,6	5585,5	207,461
—21	252,15	175,16	127,52	0,06978	68264	4,7172	7708,6	5612,1	207,600
—20	253,15	175,95	128,13	0,07096	67403	4,7204	7743,6	5638,8	207,738
—19	254,15	176,75	128,73	0,07214	66554	4,7235	7778,6	5665,5	207,876
—18	255,15	177,55	129,34	0,07335	65718	4,7266	7813,7	5692,3	208,014
—17	256,15	178,35	129,95	0,07457	64894	4,7297	7848,8	5719,1	208,152
—16	257,15	179,15	130,56	0,07581	64082	4,7329	7884,0	5746,0	208,289
—15	258,15	179,95	131,18	0,07707	63281	4,7360	7919,3	5772,9	208,425
—14	259,15	180,75	131,79	0,07834	62493	4,7391	7954,5	5799,9	208,562
—13	260,15	181,55	132,40	0,07964	61715	4,7422	7989,9	5826,9	208,698
—12	261,15	182,35	133,02	0,08095	60949	4,7452	8025,3	5854,0	208,834
—11	262,15	183,16	133,63	0,08228	60194	4,7483	8060,7	5881,1	208,969
—10	263,15	183,97	134,25	0,08363	59450	4,7514	8096,2	5908,2	209,104
—9	264,15	184,77	134,87	0,08499	59716	4,7545	8131,7	5935,5	209,239
—8	265,15	185,58	135,49	0,08638	58993	4,7575	8167,3	5962,7	209,373
—7	266,15	186,39	136,11	0,08778	57281	4,7606	8202,9	5990,0	209,508
—6	267,15	187,20	136,73	0,08921	56578	4,7636	8238,6	6017,4	209,641
—5	268,15	188,01	137,35	0,09065	55885	4,7666	8274,3	6044,8	209,775
—4	269,15	188,83	137,98	0,09211	55203	4,7697	8310,1	6072,3	209,908
—3	270,15	189,64	138,60	0,09360	54529	4,7727	8345,9	6099,8	210,041
—2	271,15	190,46	139,23	0,09510	53866	4,7757	8381,8	6127,3	210,173
—1	272,15	191,27	139,86	0,09662	53211	4,7787	8417,7	6154,9	210,306
0	273,15	192,09	140,48	0,09817	52566	4,7817	8453,7	6182,6	210,437
1	274,15	192,91	141,11	0,09974	51930	4,7847	8489,7	6210,3	210,569
2	275,15	193,73	141,74	0,1013	51303	4,7877	8525,7	6238,0	210,700
3	276,15	194,55	142,38	0,1029	50685	4,7906	8561,8	6265,8	210,831
4	277,15	195,37	143,01	0,1045	50075	4,7936	8598,0	6293,7	210,962

Carbon dioxide (*Continued*)

t	T	h	u	π_0	θ_0	s^0	H	U	S
5	278,15	196,19	143,64	0,1062	49474	4,7966	8634,2	6321,5	211,092
6	279,15	197,02	144,28	0,1078	48881	4,7995	8670,4	6349,5	211,222
7	280,15	197,84	144,91	0,1095	48296	4,8025	8706,7	6377,5	211,352
8	281,15	198,67	145,55	0,1113	47719	4,8054	8743,1	6405,5	211,482
9	282,15	199,49	146,19	0,1130	47151	4,8084	8779,5	6433,6	211,611
10	283,15	200,32	146,83	0,1148	46590	4,8113	8815,9	6461,7	211,740
11	284,15	201,15	147,47	0,1166	46037	4,8142	8852,9	6489,8	211,868
12	285,15	201,98	148,11	0,1184	45491	4,8171	8888,9	6518,1	211,997
13	286,15	202,81	148,75	0,1202	44953	4,8200	8925,5	6546,3	212,125
14	287,15	203,64	149,39	0,1221	44422	4,8229	8962,1	6574,6	212,253
15	288,15	204,48	150,04	0,1240	43899	4,8258	8998,8	6603,0	212,380
16	289,15	205,31	150,68	0,1259	43382	4,8287	9035,5	6631,4	212,507
17	290,15	206,15	151,33	0,1278	42873	4,8316	9072,3	6659,8	212,634
18	291,15	206,98	151,98	0,1298	42370	4,8345	9109,1	6688,3	212,761
19	292,15	207,82	152,62	0,1318	41875	4,8374	9145,9	6716,9	212,887
20	293,15	208,66	153,27	0,1338	41386	4,8402	9182,8	6745,4	213,013
21	294,15	209,50	153,92	0,1358	40903	4,8431	9219,8	6774,1	213,139
22	295,15	210,34	154,58	0,1379	40427	4,8459	9256,7	6802,7	213,265
23	296,15	211,18	155,23	0,1400	39958	4,8488	9293,8	6831,5	213,390
24	297,15	212,02	155,88	0,1421	39494	4,8516	9330,9	6860,2	213,515
25	298,15	212,87	156,54	0,1442	39037	4,8545	9368,0	6889,0	213,640
26	299,15	213,71	157,19	0,1464	38586	4,8573	9405,2	6917,9	213,764
27	300,15	214,56	157,85	0,1486	38141	4,8601	9442,4	6946,8	213,888
28	301,15	215,40	158,51	0,1509	37702	4,8629	9479,6	6975,7	214,012
29	302,15	216,25	159,17	0,1531	37269	4,8657	9516,9	7004,7	214,136
30	303,15	217,10	159,83	0,1554	36841	4,8685	9554,3	7033,8	214,259
31	304,15	217,95	160,49	0,1577	36419	4,8713	9591,7	7062,8	214,382
32	305,15	218,80	161,15	0,1601	36003	4,8741	9629,1	7092,0	214,505
33	306,15	219,65	161,81	0,1625	35592	4,8769	9666,6	7121,1	214,628
34	307,15	220,50	162,47	0,1649	35187	4,8797	9704,1	7150,3	214,750
35	308,15	221,36	163,14	0,1673	34786	4,8825	9741,7	7179,6	214,872
36	309,15	222,21	163,81	0,1698	34392	4,8852	9779,3	7208,9	214,994
37	310,15	223,07	164,47	0,1723	34002	4,8880	9817,0	7238,3	215,116
38	311,15	223,92	165,14	0,1748	33617	4,8908	9854,7	7267,6	215,237
39	312,15	224,78	165,81	0,1774	33237	4,8935	9892,4	7297,1	215,358
40	313,15	225,64	166,48	0,1800	32863	4,8963	9930,2	7326,6	215,479
41	314,15	226,50	167,15	0,1826	32493	4,8990	9968,0	7356,1	215,600
42	315,15	227,36	167,82	0,1853	32128	4,9017	10005,9	7385,6	215,720
43	316,15	228,22	168,49	0,1880	31767	4,9045	10043,8	7415,2	215,840
44	317,15	229,09	169,17	0,1907	31411	4,9072	10081,8	7444,9	215,960
45	318,15	229,95	169,84	0,1935	31060	4,9099	10119,8	7474,6	216,080
46	319,15	230,81	170,52	0,1963	30713	4,9126	10157,9	7504,3	216,199
47	320,15	231,68	171,19	0,1991	30371	4,9153	10196,0	7534,1	216,319
48	321,15	232,55	171,87	0,2020	30033	4,9180	10234,1	7563,9	216,438
49	322,15	233,41	172,55	0,2049	29700	4,9207	10272,3	7593,8	216,556
50	323,15	234,28	173,23	0,2078	29370	4,9234	10310,5	7623,7	216,675
51	324,15	235,15	173,91	0,2108	29045	4,9261	10348,8	7653,7	216,793
52	325,15	236,02	174,59	0,2138	28724	4,9288	10387,1	7683,7	216,911
53	326,15	236,89	175,28	0,2169	28407	4,9315	10425,4	7713,7	217,029
54	327,15	237,77	175,96	0,2199	28095	4,9341	10463,8	7743,8	217,146
55	328,15	238,64	176,64	0,2231	27786	4,9368	10502,3	7773,9	217,264
56	329,15	239,51	177,33	0,2262	27481	4,9395	10540,8	7804,1	217,381
57	330,15	240,39	178,02	0,2294	27179	4,9421	10579,3	7834,3	217,498
58	331,15	241,27	178,70	0,2327	26882	4,9448	10617,8	7864,5	217,614
59	332,15	242,14	179,39	0,2360	26588	4,9474	10656,4	7894,8	217,731

Carbon dioxide (*Continued*)

t	T	h	u	π₀	θ₀	s°	H	U	S
60	333,15	243,02	180,08	0,2393	26298	4,9500	10695,1	7925,1	217,847
61	334,15	243,90	180,77	0,2426	26012	4,9527	10733,8	7955,5	217,963
62	335,15	244,78	181,46	0,2469	25729	4,9553	10772,5	7985,9	218,078
63	336,15	245,66	182,15	0,2496	25450	4,9579	10811,3	8016,4	218,194
64	337,15	246,54	182,85	0,2530	25174	4,9606	10850,1	8046,9	218,309
65	338,15	247,43	183,54	0,2565	24902	4,9632	10889,0	8077,4	218,424
66	339,15	248,31	184,24	0,2601	24632	4,9658	10927,9	8108,0	218,539
67	340,15	249,19	184,93	0,2637	24367	4,9684	10966,8	8138,6	218,654
68	341,15	250,08	185,63	0,2673	24104	4,9710	11005,8	8169,3	218,768
69	342,15	250,97	186,33	0,2710	23845	4,9736	11044,8	8200,0	218,882
70	343,15	251,85	187,02	0,2748	23589	4,9762	11083,9	8230,8	218,996
71	344,15	252,74	187,72	0,2786	23336	4,9788	11123,0	8261,6	219,110
72	345,15	253,63	188,42	0,2824	23087	4,9813	11162,1	8292,4	219,224
73	346,15	254,52	189,13	0,2863	22840	4,9839	11201,3	8323,3	219,337
74	347,15	255,41	189,83	0,2902	22596	4,9865	11240,5	8354,2	219,450
75	348,15	256,31	190,53	0,2942	22356	4,9891	11279,8	8385,1	219,563
76	349,15	257,20	191,24	0,2982	22118	4,9916	11319,1	8416,1	219,676
77	350,15	258,09	191,94	0,3022	21883	4,9942	11358,4	8447,2	219,789
78	351,15	258,99	192,65	0,3064	21651	4,9967	11397,8	8478,2	219,901
79	352,15	259,88	193,35	0,3105	21422	4,9993	11437,3	8509,3	220,013
80	353,15	260,78	194,06	0,3147	21195	5,0018	11476,7	8540,5	220,125
81	354,15	261,68	194,77	0,3190	20972	5,0044	11516,2	8571,7	220,237
82	355,15	262,58	195,48	0,3233	20751	5,0069	11555,8	8602,9	220,348
83	356,15	263,48	196,19	0,3277	20532	5,0094	11595,4	8634,2	220,460
84	357,15	264,38	196,90	0,3321	20317	5,0119	11635,0	8665,5	220,571
85	358,15	265,28	197,62	0,3365	20104	5,0145	11674,7	8696,9	220,682
86	359,15	266,18	198,33	0,3410	19893	5,0170	11714,4	8728,3	220,792
87	360,15	267,09	199,04	0,3456	19685	5,0195	11754,1	8759,7	220,903
88	361,15	267,99	199,76	0,3502	19479	5,0220	11793,9	8791,2	221,013
89	362,15	268,89	200,47	0,3549	19276	5,0245	11833,8	8822,7	221,123
90	363,15	269,80	201,19	0,3596	19076	5,0270	11873,6	8854,3	221,233
91	364,15	270,71	201,91	0,3644	18877	5,0295	11913,5	8885,8	221,343
92	365,15	271,61	202,63	0,3692	18681	5,0320	11953,5	8917,5	221,453
93	366,15	272,52	203,35	0,3741	18488	5,0345	11993,5	8949,1	221,562
94	367,15	273,43	204,07	0,3791	18297	5,0369	12033,5	8980,9	221,671
95	368,15	274,34	204,79	0,3841	18107	5,0394	12073,6	9012,6	221,780
96	369,15	275,25	205,51	0,3891	17921	5,0419	12113,7	9044,4	221,889
97	370,15	276,17	206,24	0,3942	17736	5,0444	12153,8	9076,2	221,997
98	371,15	277,08	206,96	0,3994	17554	5,0468	12194,0	9108,1	222,106
99	372,15	277,99	207,68	0,4046	17373	5,0493	12234,2	9140,0	222,214
100	373,15	278,91	208,41	0,4099	17195	5,0517	12274,5	9171,9	222,322
101	374,15	279,82	209,14	0,4153	17019	5,0542	12314,8	9203,9	222,430
102	375,15	280,74	209,86	0,4207	16845	5,0566	12355,1	9235,9	222,538
103	376,15	281,66	210,59	0,4262	16673	5,0591	12395,5	9268,0	222,645
104	377,15	282,58	211,32	0,4317	16503	5,0615	12435,9	9300,1	222,752
105	378,15	283,49	212,05	0,4373	16335	5,0640	12476,3	9332,2	222,859
106	379,15	284,41	212,78	0,4430	16169	5,0664	12516,8	9364,4	222,966
107	380,15	285,34	213,52	0,4487	16005	5,0688	12557,3	9396,6	223,073
108	381,15	286,26	214,25	0,4545	15842	5,0712	12597,9	9428,9	223,180
109	382,15	287,18	214,98	0,4603	15682	5,0736	12638,5	9461,2	223,286
110	383,15	288,10	215,72	0,4662	15523	5,0761	12679,2	9493,5	223,392
111	384,15	289,03	216,45	0,4722	15367	5,0785	12719,8	9525,9	223,498
112	385,12	289,95	217,19	0,4783	15212	5,0809	12760,5	9558,3	223,604
113	386,15	290,88	217,93	0,4844	15059	5,0833	12801,3	9590,7	223,710
114	387,15	291,81	218,66	0,4906	14907	5,0857	12842,1	9623,2	223,815

155

Carbon dioxide (*Continued*)

t	T	h	u	π_0	θ_0	s^0	H	U	S
115	388,15	292,73	219,40	0,4968	14758	5,0881	12882,9	9655,7	223,921
116	389,15	293,66	220,14	0,5032	14610	5,0905	12923,8	9688,2	224,026
117	390,15	294,59	220,88	0,5096	14463	5,0928	12964,7	9720,8	224,131
118	391,15	295,52	221,62	0,5160	14319	5,0952	13005,6	9753,5	224,236
119	392,15	296,45	222,37	0,5226	14176	5,0976	13046,6	9786,1	224,340
120	393,15	297,39	223,11	0,5292	14035	5,1000	13087,6	9818,8	224,445
121	394,15	298,32	223,85	0,5358	13895	5,1023	13128,7	9851,6	224,549
122	395,15	299,25	224,60	0,5426	13757	5,1047	13169,8	9884,3	224,653
123	396,15	300,19	225,34	0,5494	13620	5,1071	13210,9	9917,1	224,757
124	397,15	301,12	226,09	0,5563	13485	5,1094	13252,1	9950,0	224,861
125	398,15	302,06	226,84	0,5633	13352	5,1118	13293,3	9982,9	224,964
126	399,15	302,99	227,59	0,5704	13220	5,1141	13334,5	10015,8	225,068
127	400,15	303,93	228,33	0,5775	13089	5,1165	13375,8	10048,8	225,171
128	401,15	304,87	229,08	0,5847	12960	5,1188	13417,1	10081,7	225,274
129	402,15	305,81	229,83	0,5920	12833	5,1212	13458,4	10114,8	225,377
130	403,15	306,75	230,59	0,5993	12707	5,1235	13499,8	10147,8	225,480
131	404,15	307,69	231,34	0,6068	12582	5,1258	13541,2	10181,0	225,583
132	405,15	308,63	232,09	0,6143	12459	5,1282	13582,7	10214,1	225,685
133	406,15	309,58	232,84	0,6219	12337	5,1305	13624,2	10247,3	225,787
134	407,15	310,52	233,60	0,6296	12216	5,1328	13665,7	10280,5	225,889
135	408,15	311,46	234,36	0,6374	12097	5,1351	13707,3	10313,7	225,991
136	409,15	312,41	235,11	0,6452	11979	5,1374	13748,9	10347,0	226,093
137	410,15	313,36	235,87	0,6532	11862	5,1397	13790,5	10380,3	226,195
138	411,15	314,30	236,63	0,6612	11747	5,1420	13832,2	10413,7	226,296
139	412,15	315,25	237,39	0,6693	11633	5,1443	13873,9	10447,1	226,398
140	413,15	316,20	238,14	0,6775	11520	5,1466	13915,6	10480,5	226,499
141	414,15	317,15	238,91	0,6858	11409	5,1489	13957,4	10514,0	226,600
142	415,15	318,10	239,67	0,6941	11298	5,1512	13999,2	10547,5	226,701
143	416,15	319,05	240,43	0,7026	11189	5,1535	14041,1	10581,0	226,801
144	417,15	320,00	241,19	0,7111	11081	5,1558	14082,9	10614,6	226,902
145	418,15	320,95	241,95	0,7198	10974	5,1581	14124,9	10648,2	227,002
146	419,15	321,91	242,72	0,7285	10869	5,1604	14166,8	10681,8	227,102
147	420,15	322,86	243,48	0,7373	10765	5,1626	14208,8	10715,5	227,202
148	421,15	323,82	244,25	0,7462	10661	5,1649	14250,8	10749,2	227,302
149	422,15	324,77	245,02	0,7552	10559	5,1672	14292,9	10783,0	227,402
150	423,15	325,73	245,78	0,7643	10458	5,1694	14335,0	10816,7	227,502
151	424,15	326,69	246,55	0,7735	10358	5,1717	14377,1	10850,6	227,601
152	425,15	327,64	247,32	0,7828	10259	5,1740	14419,2	10884,3	227,700
153	426,15	328,60	248,09	0,7922	10162	5,1762	14461,5	10918,3	227,800
154	427,15	329,56	248,86	0,8017	10065	5,1785	14503,7	10952,2	227,899
155	428,15	330,52	249,63	0,8113	9969,6	5,1807	14546,0	10986,2	227,997
156	429,15	331,48	250,41	0,8210	9875,0	5,1829	14588,3	11020,2	228,096
157	430,15	332,45	251,18	0,8308	9781,4	5,1852	14630,6	11054,2	228,195
158	431,15	333,41	251,95	0,8407	9688,8	5,1874	14673,0	11088,2	228,293
159	432,15	334,37	252,73	0,8507	9597,2	5,1896	14715,4	11122,3	228,391
160	433,15	335,34	253,50	0,8608	9506,6	5,1919	14757,8	11156,5	228,489
161	434,15	336,30	254,28	0,8710	9416,9	5,1941	14800,3	11190,6	228,587
162	435,15	337,27	255,06	0,8813	9328,2	5,1963	14842,8	11224,8	228,685
163	436,15	338,23	255,84	0,8917	9240,5	5,1985	14885,4	11259,0	228,783
164	437,15	339,20	256,61	0,9022	9153,7	5,2008	14928,0	11293,3	228,880
165	438,15	340,17	257,39	0,9128	9067,8	5,2030	14970,6	11327,6	228,978
166	439,15	341,14	258,17	0,9236	8982,9	5,2052	15013,2	11361,9	229,075
167	440,15	342,11	258,95	0,9344	8898,8	5,2074	15055,9	11396,3	229,172
168	441,15	343,08	259,74	0,9454	8815,7	5,2096	15098,6	11430,7	229,269
169	442,15	344,05	260,52	0,9564	8733,4	5,2118	15141,4	11465,1	229,366

Carbon dioxide (*Continued*)

t	T	h	u	π_0	θ_0	s^0	H	U	S
170	443,15	345,02	261,30	0,9676	8652,0	5,2140	15184,1	11499,6	229,462
171	444,15	346,00	262,09	0,9789	8571,4	5,2162	15226,9	11534,1	229,559
172	445,15	346,97	262,87	0,9903	8491,7	5,2184	15269,8	11568,6	229,655
173	446,15	347,94	263,66	1,0019	8412,9	5,2206	15312,7	11603,2	229,751
174	447,15	348,92	264,44	1,0135	8334,9	5,2227	15355,6	11637,8	229,847
175	448,15	349,90	265,23	1,0253	8257,7	5,2249	15398,5	11672,4	229,943
176	449,15	350,87	266,02	1,0371	8181,3	5,2271	15441,5	11707,1	230,039
177	450,15	351,85	266,80	1,0491	8105,7	5,2293	15484,5	11741,8	230,135
178	451,15	352,83	267,59	1,0613	8030,9	5,2314	15527,6	11776,5	230,230
179	452,15	353,81	268,38	1,0735	7956,9	5,2336	15570,7	11811,3	230,326
180	453,15	354,79	269,17	1,0859	7883,7	5,2358	15613,8	11846,1	230,421
181	454,15	355,77	269,97	1,0984	7811,2	5,2379	15656,9	11880,9	230,516
182	455,15	356,75	270,76	1,1110	7739,5	5,2401	15700,1	11915,8	230,611
183	456,15	357,73	271,55	1,1237	7668,5	5,2422	15743,3	11950,7	230,706
184	457,15	358,71	272,34	1,1366	7598,3	5,2444	15786,6	11985,6	230,801
185	458,15	359,70	273,14	1,1496	7528,8	5,2465	15829,8	12020,6	230,895
186	459,15	360,68	273,93	1,1627	7460,0	5,2487	15873,1	12055,6	230,990
187	460,15	361,66	274,73	1,1760	7392,0	5,2508	15916,5	12090,6	231,084
188	461,15	362,65	275,53	1,1894	7324,6	5,2530	15959,9	12125,7	231,178
189	462,15	363,64	276,32	1,2029	7257,9	5,2551	16003,3	12160,8	231,272
190	463,15	364,62	277,12	1,2166	7192,0	5,2572	16046,7	12195,9	231,366
191	464,15	365,61	277,92	1,2304	7126,7	5,2594	16090,2	12231,0	231,460
192	465,15	366,60	278,72	1,2443	7062,1	5,2615	16133,7	12266,2	231,553
193	466,15	367,59	279,52	1,2584	6998,1	5,2636	16177,2	12301,5	231,647
194	467,15	368,58	280,32	1,2726	6934,8	5,2657	16220,8	12336,7	231,740
195	468,15	369,57	281,12	1,2870	6872,1	5,2679	16264,4	12372,0	231,833
196	469,15	370,56	281,93	1,3015	6810,1	5,2700	16308,0	12407,3	231,927
197	470,15	371,55	282,73	1,3161	6748,8	5,2721	16351,7	12442,7	232,020
198	471,15	372,55	283,53	1,3309	6688,0	5,2742	16395,4	12478,1	232,112
199	472,15	373,54	284,34	1,3458	6627,9	5,2763	16439,1	12513,5	232,205
200	473,15	374,53	285,14	1,3609	6568,4	5,2784	16482,9	12548,9	232,298
201	474,15	375,53	285,95	1,3761	6509,5	5,2805	16526,7	12584,4	232,390
202	475,15	376,52	286,76	1,3914	6451,2	5,2826	16570,5	12619,9	232,482
203	476,15	377,52	287,56	1,4070	6393,5	5,2847	16614,3	12655,4	232,575
204	477,15	378,52	288,37	1,4226	6336,3	5,2868	16658,2	12691,0	232,667
205	478,15	379,52	289,18	1,4384	6279,8	5,2889	16702,1	12726,6	232,759
206	479,15	380,52	289,99	1,4544	6223,8	5,2910	16746,1	12762,2	232,851
207	480,15	381,51	290,80	1,1705	6168,4	5,2931	16790,1	12797,9	232,942
208	481,15	382,51	291,61	1,4868	6113,5	5,2951	16834,1	12833,6	233,034
209	482,15	383,51	292,43	1,5033	6059,2	5,2972	16878,1	12869,3	233,125
210	483,15	384,52	293,24	1,5199	6005,5	5,2993	16922,2	12905,1	233,217
211	484,15	385,52	294,05	1,5366	5952,3	5,3014	16966,3	12940,9	233,308
212	485,15	386,52	294,86	1,5536	5899,6	5,3034	17010,4	12976,7	233,399
213	486,15	387,53	295,68	1,5706	5847,4	5,3055	17054,6	13012,6	233,490
214	487,15	388,53	296,49	1,5879	5795,8	5,3076	17098,8	13048,4	233,581
215	488,15	389,53	297,31	1,6053	5744,7	5,3096	17143,0	13084,4	233,671
216	489,15	390,54	298,13	1,6229	5694,1	5,3117	17187,3	13120,3	233,762
217	490,15	391,55	298,95	1,6407	5644,0	5,3137	17231,6	13156,3	233,852
218	491,15	392,55	299,76	1,6586	5594,4	5,3158	17275,9	13192,3	233,943
219	492,15	393,56	300,58	1,6767	5545,3	5,3178	17320,3	13228,3	234,033
220	493,15	394,57	301,40	1,6949	5496,7	5,3199	17364,6	13264,4	234,123
221	494,15	395,58	302,22	1,7134	5448,6	5,3219	17409,0	13300,5	234,213
222	495,15	396,59	303,04	1,7320	5400,9	5,3240	17453,5	13336,6	234,303
223	496,15	397,60	303,86	1,7508	5353,7	5,3260	17498,0	13372,8	234,392
224	497,15	398,61	304,69	1,7698	5307,0	5,3280	17542,5	13409,0	234,482

Carbon dioxide (*Continued*)

t	T	h	u	π_0	θ_0	s^0	H	U	S
225	498,15	399,62	305,51	1,7889	5260,8	5,3301	17587,0	13445,2	234,572
226	499,15	400,64	306,33	1,8082	5215,0	5,3321	17631,6	13481,4	234,661
227	500,15	401,65	307,16	1,8278	5169,6	5,3341	17676,2	13517,7	234,750
228	501,15	402,66	307,98	1,8475	5124,7	5,3362	17720,8	13554,0	234,839
229	502,15	403,68	308,81	1,8673	5080,3	5,3382	17765,4	13590,3	234,928
230	503,15	404,69	309,63	1,8874	5036,2	5,3402	17810,1	13626,7	235,017
231	504,15	405,71	310,46	1,9077	4992,7	5,3422	17854,8	13663,1	235,106
232	505,15	406,73	311,29	1,9281	4949,5	5,3442	17899,6	13699,5	235,195
233	506,15	407,74	512,12	1,9488	4906,8	5,3463	17944,3	13736,0	235,283
234	507,15	408,76	312,95	1,9696	4864,4	5,3483	17989,1	13772,5	235,372
235	508,15	409,78	313,78	1,9906	4822,5	5,3503	18034,0	13809,0	235,460
236	509,15	410,80	314,61	2,0119	4781,0	5,3523	18078,8	13845,6	235,548
237	510,15	411,82	315,44	2,0333	4739,9	5,3543	18123,7	13882,1	235,636
238	511,15	412,84	316,27	2,0549	4699,3	5,3563	18168,7	13918,7	235,724
239	512,15	413,86	317,10	2,0768	4659,0	5,3583	18213,6	13955,4	235,812
240	513,15	414,88	317,94	2,0988	4619,1	5,3603	18258,6	13992,0	235,900
241	514,15	415,91	318,77	2,1210	4579,5	5,3623	18303,6	14028,7	235,987
242	515,15	416,93	319,60	2,1435	4540,4	5,3642	18348,6	14065,5	236,075
243	516,15	417,95	320,44	2,1661	4501,6	5,3662	18393,7	14102,2	236,162
244	517,15	418,98	321,28	2,1890	4463,3	5,3682	18438,8	14139,0	236,250
245	518,15	420,00	322,11	2,2120	4425,2	5,3702	18483,9	14175,8	236,337
246	519,15	421,03	322,95	2,2353	4387,6	5,3722	18529,1	14212,6	236,424
247	520,15	422,06	323,79	2,2588	4350,3	5,3741	18574,3	14249,5	236,511
248	521,15	423,08	324,62	2,2825	4313,4	5,3761	18619,5	14286,4	236,598
249	522,15	424,11	325,46	2,3065	4276,8	5,3781	18664,7	14323,3	236,684
250	523,15	425,14	326,30	2,3306	4240,6	5,3801	18710,0	14360,3	236,771
251	524,15	426,17	327,14	2,3550	4204,7	5,3820	18755,3	14397,3	236,857
252	525,15	427,20	327,99	2,3796	4169,2	5,3840	18800,6	14434,3	236,944
253	526,15	428,23	328,83	2,4044	4134,0	5,3859	18846,0	14471,3	237,030
254	527,15	429,26	329,67	2,4295	4099,2	5,3879	18891,3	14508,4	237,116
255	528,15	430,29	330,51	2,4548	4064,7	5,3899	18936,8	14545,5	237,202
256	529,15	431,33	331,36	2,4803	4030,5	5,3918	18982,2	14582,6	237,288
257	530,15	432,36	332,20	2,5060	3996,6	5,3938	19027,7	14619,8	237,374
258	531,15	433,39	333,04	2,5320	3963,1	5,3957	19073,2	14657,0	237,460
259	532,15	434,43	333,89	2,5582	3929,8	5,3977	19118,7	14694,2	237,546
260	533,15	435,46	334,74	2,5847	3896,9	5,3996	19164,3	14731,4	237,631
261	534,15	436,50	335,58	2,6114	3864,3	5,4015	19209,8	14768,7	237,717
262	535,15	437,53	336,43	2,6383	3832,0	5,4035	19255,5	14806,0	237,802
263	536,15	438,57	337,28	2,6655	3800,1	5,4054	19301,1	14843,3	237,887
264	537,15	439,61	338,13	2,6929	3768,4	5,4074	19346,8	14880,7	237,972
265	538,15	440,65	338,98	2,7206	3737,0	5,4093	19392,5	14918,1	238,057
266	539,15	441,69	339,83	2,7485	3705,9	5,4112	19438,2	14955,5	238,142
267	540,15	442,73	340,68	2,7766	3675,1	5,4131	19483,9	14992,9	238,227
268	541,15	443,77	341,53	2,8051	3644,6	5,4151	19529,7	15030,4	238,311
269	542,15	444,81	342,38	2,8337	3614,4	5,4170	19575,5	15067,9	238,396
270	543,15	445,85	343,23	2,8627	3584,5	5,4189	19621,4	15105,4	238,481
271	544,15	446,89	344,09	2,8919	3554,8	5,4208	19667,2	15142,9	238,565
272	545,15	447,93	344,94	2,9213	3525,4	5,4227	19713,1	15180,5	238,649
273	546,15	448,98	345,80	2,9510	3496,3	5,4246	19759,0	15218,1	238,733
274	547,15	450,02	346,65	2,9810	3467,5	5,4266	19805,0	15255,7	238,817
275	548,15	451,07	347,51	3,0113	3438,9	5,4285	19850,9	15293,4	238,901
276	549,15	452,11	348,36	3,0418	3410,7	5,4304	19896,9	15331,1	238,985
277	550,15	453,16	349,22	3,0726	3382,6	5,4323	19943,0	15368,8	239,069
278	551,15	454,20	350,08	3,1037	3354,9	5,4342	19989,0	15406,5	239,153
279	552,15	455,25	350,94	3,1350	3327,3	5,4361	20035,1	15444,3	239,236

158

Carbon dioxide (*Continued*)

t	T	h	u	π_0	θ_0	s^0	H	U	S
280	553,15	456,30	351,79	3,1666	3300,1	5,4380	20081,2	15482,1	239,319
281	554,15	457,35	352,65	3,1985	3273,1	5,4399	20127,4	15519,9	239,403
282	555,15	458,40	353,51	3,2307	3246,3	5,4418	20173,5	15557,8	239,486
283	556,15	459,45	354,37	3,2632	3219,8	5,4436	20219,7	15595,7	239,569
284	557,15	460,50	355,24	3,2959	3193,6	5,4455	20265,9	15633,6	239,652
285	558,15	461,55	356,10	3,3289	3167,5	5,4474	20312,2	15671,5	239,735
286	559,15	462,60	356,96	3,3623	3141,8	5,4493	20358,5	15709,5	239,818
287	560,15	463,65	357,82	3,3959	3116,2	5,4512	20404,8	15747,4	239,901
288	561,15	464,70	358,69	3,4298	3090,9	5,4531	20451,1	15785,4	239,983
289	562,15	465,76	359,55	3,4640	3065,8	5,4549	20497,4	15823,5	240,066
290	563,15	466,81	360,42	3,4985	3041,0	5,4568	20543,8	15861,6	240,148
291	564,15	467,86	361,28	3,5334	3016,4	5,4587	20590,2	15899,6	240,231
292	565,15	468,92	362,15	3,5685	2992,0	5,4605	20636,6	15937,8	240,313
293	566,15	469,97	363,01	3,6039	2967,8	5,4624	20683,1	15975,9	240,395
294	567,15	471,03	363,88	3,6396	2943,9	5,4643	20729,6	16014,1	240,477
295	568,15	472,09	364,75	3,6757	2920,1	5,4661	20776,1	16052,3	240,559
296	569,15	473,15	365,62	3,7120	2896,6	5,4680	20822,6	16090,5	240,641
297	570,15	474,20	366,49	3,7487	2873,3	5,4698	20869,2	16128,7	240,723
298	571,15	475,26	367,36	3,7857	2850,2	5,4717	20915,8	16167,0	240,804
299	572,15	476,32	368,23	3,8230	2827,3	5,4736	20962,4	16205,3	240,886
300	573,15	477,38	369,10	3,8607	2804,7	5,4754	21009,1	16243,7	240,967
301	574,15	478,44	369,97	3,8986	2782,2	5,4773	21055,7	16282,0	241,049
302	575,15	479,50	370,84	3,9369	2760,0	5,4791	21102,4	16320,4	241,130
303	576,15	480,56	371,71	3,9755	2737,9	5,4809	21149,2	16358,8	241,211
304	577,15	481,63	372,59	4,0145	2716,0	5,4828	21195,9	16397,2	241,292
305	578,15	482,69	373,46	4,0538	2694,4	5,4846	21242,7	16435,7	241,373
306	579,15	483,75	374,34	4,0934	2672,9	5,4865	21289,5	16474,2	241,454
307	580,15	484,82	375,21	4,1334	2651,6	5,4883	21336,3	16512,7	241,535
308	581,15	485,88	376,09	4,1737	2630,5	5,4901	21383,1	16551,2	241,615
309	582,15	486,95	376,96	4,2143	2609,7	5,4920	21430,0	16589,8	241,696
310	583,15	488,01	377,84	4,2553	2588,9	5,4938	21476,9	16628,4	241,776
311	584,15	489,08	378,72	4,2967	2568,4	5,4956	21523,9	16667,0	241,857
312	585,15	490,15	379,60	4,3384	2548,1	5,4974	21570,8	16705,6	241,937
313	586,15	491,21	380,47	4,3805	2527,9	5,4993	21617,8	16744,3	242,017
314	587,15	492,28	381,35	4,4229	2508,0	5,5011	21664,8	16783,0	242,098
315	588,15	493,35	382,23	4,4657	2488,2	5,5029	21711,8	16821,7	242,178
316	589,15	494,42	383,11	4,5088	2468,6	5,5047	21758,9	16860,5	242,257
317	590,15	495,49	383,99	4,5523	2449,1	5,5065	21806,0	16899,2	242,337
318	591,15	496,56	384,88	4,5962	2429,8	5,5084	21853,1	16938,0	242,417
319	592,15	497,63	385,76	4,6404	2410,7	5,5102	21900,2	16976,8	242,497
320	593,15	498,70	386,64	4,6851	2391,8	5,5120	21947,4	17015,7	242,576
321	594,15	499,77	387,52	4,7301	2373,0	5,5138	21994,5	17054,5	242,656
322	595,15	500,85	388,41	4,7754	2354,5	5,5156	22041,9	17093,4	242,735
323	596,15	501,92	389,29	4,8212	2336,0	5,5174	22089,0	17132,4	242,814
324	597,15	502,99	390,18	4,8673	2317,8	5,5192	22136,2	17171,3	242,894
325	598,15	504,07	391,06	4,9139	2299,6	5,5210	22183,5	17210,3	242,973
326	599,15	505,14	391,95	4,9608	2281,7	5,5228	22230,8	17249,3	243,052
327	600,15	506,22	392,84	5,0081	2263,9	5,5246	22278,2	17288,3	243,131
328	601,15	507,29	393,72	5,0559	2246,3	5,5264	22325,5	17327,3	243,210
329	602,15	508,37	394,61	5,1040	2228,8	5,5282	22372,9	17366,4	243,288
330	603,15	509,45	395,50	5,1525	2211,5	5,5299	22420,3	17405,5	243,367
331	604,15	510,53	396,39	5,2014	2194,3	5,5317	22467,8	17444,6	243,446
332	605,15	511,60	397,28	5,2507	2177,3	5,5335	22515,2	17483,7	243,524
333	606,15	512,68	398,17	5,3005	2160,4	5,5353	22562,7	17522,9	243,602
334	607,15	513,76	399,06	5,3507	2143,7	5,5371	22610,2	17562,1	243,681

Carbon dioxide (*Continued*)

t	T	h	u	π_0	θ_0	s^0	H	U	S
335	608,15	514,84	399,95	5,4012	2127,1	5,5388	22657,7	17601,3	243,759
336	609,15	515,92	400,84	5,4522	2110,7	5,5406	22705,3	17640,6	243,837
337	610,15	517,00	401,73	5,5036	2094,4	5,5424	22752,9	17679,8	243,915
338	611,15	518,09	402,62	5,5555	2078,3	5,5442	22800,5	17719,1	243,993
339	612,15	519,17	403,52	5,6078	2062,3	5,5459	22848,1	17758,4	244,071
340	613,15	520,25	404,41	5,6605	2046,4	5,5477	22895,8	17797,8	244,149
341	614,15	521,33	405,31	5,7136	2030,7	5,5495	22943,4	17837,1	244,226
342	615,15	522,42	406,20	5,7672	2015,1	5,5512	22991,1	17876,5	244,304
343	616,15	523,50	407,10	5,8212	1999,6	5,5530	23038,9	17915,9	244,382
344	617,15	524,59	407,99	5,8757	1984,3	5,5548	23086,6	17955,4	244,459
345	618,15	525,67	408,89	5,9306	1969,1	5,5565	23134,4	17994,8	244,536
346	619,15	526,76	409,79	5,9860	1954,1	5,5583	23182,2	18034,3	244,614
347	620,15	527,85	410,68	6,0418	1939,1	5,5600	23230,0	18073,8	244,691
348	621,15	528,93	411,58	6,0981	1924,3	5,5618	23277,8	18113,3	244,768
349	622,15	530,02	412,48	6,1548	1909,7	5,5635	23325,7	18152,9	244,845
350	623,15	531,11	413,38	6,2120	1895,1	5,5653	23373,6	18192,5	244,922
351	624,15	532,20	414,28	6,2697	1880,7	5,5670	23421,5	18232,1	244,999
352	625,15	533,29	415,18	6,3278	1866,4	5,5688	23469,5	18271,7	245,075
353	626,15	534,38	416,08	6,3864	1852,2	5,5705	23517,4	18311,4	245,152
354	627,15	535,47	416,98	6,4455	1838,2	5,5722	23565,4	18351,0	245,229
355	628,15	536,56	417,89	6,5051	1824,2	5,5740	23613,4	18390,7	245,305
356	629,15	537,65	418,79	6,5652	1810,4	5,5757	23661,5	18430,5	245,382
357	630,15	538,74	419,69	6,6257	1796,7	5,5774	23709,5	18470,2	245,458
358	631,15	539,84	420,60	6,6868	1783,2	5,5792	23757,6	18510,0	245,534
359	632,15	540,93	421,50	6,7483	1769,7	5,5809	23805,7	18549,8	245,610
360	633,15	542,02	422,40	6,8103	1756,4	5,5826	23853,8	18589,6	245,686
361	634,15	543,12	423,31	6,8729	1743,1	5,5844	23902,0	18629,4	245,762
362	635,15	544,21	424,21	6,9359	1730,0	5,5861	23950,2	18669,3	245,838
363	636,15	545,31	425,12	6,9994	1717,0	5,5878	23998,4	18709,2	245,914
364	637,15	546,40	426,03	7,0635	1704,1	5,5895	24046,6	18749,1	245,990
365	638,15	547,50	426,94	7,1281	1691,3	5,5913	24094,8	18789,0	246,066
366	639,15	548,59	427,84	7,1932	1678,6	5,5930	24143,1	18829,0	246,141
367	640,15	549,69	428,75	7,2588	1666,1	5,5947	24191,4	18868,9	246,217
368	641,15	550,79	429,66	7,3249	1653,6	5,5964	24239,7	18908,9	246,292
369	642,15	551,89	430,57	7,3916	1641,2	5,5981	24288,1	18949,0	246,367
370	643,15	552,99	431,48	7,4588	1629,0	5,5998	24336,4	18989,0	246,443
371	644,15	554,09	432,39	7,5265	1616,8	5,6015	24384,8	19029,1	246,518
372	645,15	555,19	433,30	7,5948	1604,8	5,6032	24433,2	19069,2	246,593
373	646,15	556,29	434,21	7,6636	1592,8	5,6049	24481,6	19109,3	246,668
374	647,15	557,39	435,13	7,7330	1581,0	5,6066	24530,1	19149,4	246,743
375	648,15	558,49	436,04	7,8029	1569,2	5,6083	24578,6	19189,6	246,818
376	649,15	559,59	436,95	7,8734	1557,6	5,6100	24627,1	19229,8	246,892
377	650,15	560,69	437,86	7,9445	1546,0	5,6117	24675,6	19270,0	246,967
378	651,15	561,80	438,78	8,0161	1534,6	5,6134	24724,1	19310,2	247,042
379	652,15	562,90	439,69	8,0883	1523,2	5,6151	24772,7	19350,5	247,116
380	653,15	564,00	440,61	8,1610	1512,0	5,6168	24821,3	19390,7	247,191
381	654,15	565,11	441,52	8,2343	1500,8	5,6185	24869,9	19431,0	247,265
382	655,15	566,21	442,44	8,3082	1489,7	5,6202	24918,5	19471,3	247,339
383	656,15	567,32	443,36	8,3827	1478,7	5,6219	24967,2	19511,7	247,414
384	657,15	568,43	444,27	8,4578	1467,8	5,6236	25015,9	19552,1	247,488
385	658,15	569,53	445,19	8,5335	1457,0	5,6253	25064,6	19592,4	247,562
386	659,15	570,64	446,11	8,6097	1446,3	5,6269	25113,3	19632,9	247,636
387	660,15	571,75	447,03	8,6866	1435,7	5,6286	25162,0	19673,3	247,710
388	661,15	572,86	447,95	8,7641	1425,2	5,6303	25210,8	19713,7	247,783
389	662,15	573,96	448,87	8,8421	1414,7	5,6320	25259,6	19754,2	247,857

Carbon dioxide (*Continued*)

t	T	h	u	π_0	θ_0	s^0	H	U	S
390	663,15	575,07	449,79	8,9208	1404,4	5,6336	25308,4	19794,7	247,931
391	664,15	576,18	450,71	9,0001	1394,1	5,6353	25357,2	19835,2	248,004
392	665,15	577,29	451,63	9,0801	1383,9	5,6370	25406,1	19875,8	248,078
393	666,15	578,40	452,55	9,1606	1373,8	5,6387	25455,0	19916,3	248,151
394	667,15	579,52	453,47	9,2418	1363,8	5,6403	25503,9	19956,9	248,225
395	668,15	580,63	454,40	9,3236	1353,8	5,6420	25552,8	19997,5	248,298
396	669,15	581,74	455,32	9,4060	1344,0	5,6436	25601,7	20038,2	248,371
397	670,15	582,85	456,24	9,4891	1334,2	5,6453	25650,7	20078,8	248,444
398	671,15	583,96	457,17	9,5729	1324,5	5,6470	25699,7	20119,5	248,517
399	672,15	585,08	458,09	9,6572	1314,9	5,6486	25748,7	20160,2	248,590
400	673,15	586,19	459,02	9,7423	1305,3	5,6503	25797,7	20200,9	248,663
401	674,15	587,31	459,94	9,8280	1295,9	5,6519	25846,8	20241,6	248,736
402	675,15	588,42	460,87	9,9143	1286,5	5,6536	25895,9	20282,4	248,809
403	676,15	589,54	461,80	10,001	1277,2	5,6552	25945,0	20323,2	248,881
404	677,15	590,65	462,72	10,089	1268,0	5,6569	25994,1	20364,0	248,954
405	678,15	591,77	463,65	10,177	1258,8	5,6585	26043,2	20404,8	249,027
406	679,15	592,89	464,58	10,266	1249,7	5,6602	26092,4	20445,7	249,099
407	680,15	594,01	465,51	10,356	1240,7	5,6618	26141,6	20486,5	249,171
408	681,15	595,12	466,44	10,446	1231,8	5,6635	26190,8	20527,4	249,244
409	682,15	596,24	467,37	10,537	1222,9	5,6651	26240,0	20568,3	249,316
410	683,15	597,36	468,30	10,629	1214,1	5,6667	26289,2	20609,3	249,388
411	684,15	598,48	469,23	10,722	1205,4	5,6684	26338,5	20650,2	249,460
412	685,15	599,60	470,16	10,815	1196,8	5,6700	26387,8	20691,2	249,532
413	686,15	600,72	471,09	10,909	1188,2	5,6717	26437,1	20732,2	249,604
414	687,15	601,84	472,02	11,004	1179,7	5,6733	26486,4	20773,2	249,676
415	688,15	602,96	472,95	11,099	1171,3	5,6749	26535,8	20814,2	249,748
416	689,15	604,08	473,89	11,195	1162,9	5,6765	26585,2	20855,3	249,819
417	690,15	605,21	474,82	11,292	1154,6	5,6782	26634,6	20896,4	249,891
418	691,15	606,33	475,75	11,390	1146,3	5,6798	26684,0	20937,5	249,962
419	692,15	607,45	476,69	11,488	1138,2	5,6814	26733,4	20978,6	250,034
420	693,15	608,58	477,62	11,587	1130,1	5,6830	26782,9	21019,7	250,105
421	694,15	609,70	478,56	11,687	1122,0	5,6847	26832,4	21060,9	250,177
422	695,15	610,83	479,49	11,787	1114,1	5,6863	26881,9	21102,1	250,248
423	696,15	611,95	480,43	11,889	1106,2	5,6879	26931,4	21143,3	250,319
424	697,15	613,08	481,37	11,991	1098,3	5,6895	26980,9	21184,5	250,390
425	698,15	614,20	482,31	12,094	1090,5	5,6911	27030,5	21225,8	250,461
426	699,15	615,33	483,24	12,197	1082,8	5,6927	27080,1	21267,0	250,532
427	700,15	616,46	484,18	12,302	1075,2	5,6944	27129,7	21308,3	250,603
428	701,15	617,58	485,12	12,407	1067,6	5,6960	27179,3	21349,6	250,674
429	702,15	618,71	486,06	12,513	1060,0	5,6976	27228,9	21391,0	250,745
430	703,15	619,84	487,00	12,620	1052,5	5,6992	27278,6	21432,3	250,815
431	704,15	620,97	487,94	12,728	1045,1	5,7008	27328,3	21473,7	250,886
432	705,15	622,10	488,88	12,836	1037,3	5,7024	27378,0	21515,1	250,957
433	706,15	623,23	489,82	12,945	1030,5	5,7040	27427,7	21556,5	251,027
434	707,15	624,36	490,76	13,055	1023,2	5,7056	27477,5	21597,9	251,097
435	708,15	625,49	491,70	13,166	1016,0	5,7072	27527,2	21639,4	251,168
436	709,15	626,62	492,65	13,278	1008,9	5,7088	27577,0	21680,8	251,238
437	710,15	627,75	493,59	13,391	1001,8	5,7104	27626,8	21722,3	251,308
438	711,15	628,89	494,53	13,504	994,87	5,7120	27676,6	21763,9	251,378
439	712,15	630,02	495,48	13,618	987,91	5,7136	27726,5	21805,4	251,448
440	713,15	631,15	496,42	13,733	981,01	5,7152	27776,4	21846,9	251,518
441	714,15	632,29	497,36	13,849	974,16	5,7167	27826,3	21888,5	251,588
442	715,15	633,42	498,31	13,966	967,36	5,7183	27876,2	21930,1	251,658
443	716,15	634,55	499,26	14,084	960,62	5,7199	27926,1	21971,7	251,728
444	717,15	635,69	500,20	14,202	953,93	5,7215	27976,0	22013,4	251,798

Carbon dioxide (*Continued*)

t	T	h	u	π_0	θ_0	s^0	H	U	S
445	718,15	636,82	501,15	14,322	947,30	5,7231	28026,0	22055,0	251,867
446	719,15	637,96	502,09	14,442	940,71	5,7247	28076,0	22096,7	251,937
447	720,15	639,10	503,04	14,563	934,18	5,7262	28126,0	22138,4	252,006
448	721,15	640,23	503,99	14,686	927,70	5,7278	28176,0	22180,1	252,076
449	722,15	641,37	504,94	14,809	921,27	5,7294	28226,1	22221,8	252,145
450	723,15	642,51	505,89	14,933	914,89	5,7310	28276,1	22263,6	252,214
451	724,15	643,65	506,84	15,057	908,56	5,7325	28326,2	22305,4	252,283
452	725,15	644,78	507,79	15,183	902,28	5,7341	28376,3	22347,1	252,353
453	726,15	645,92	508,74	15,310	896,05	5,7357	28426,5	22389,0	252,422
454	727,15	647,06	509,69	15,437	889,87	5,7373	28476,6	22430,8	252,491
455	728,15	648,20	510,64	15,566	883,73	5,7388	28526,8	22472,6	252,560
456	729,15	649,34	511,59	15,695	877,64	5,7404	28577,0	22514,5	252,629
457	730,15	650,48	512,54	15,826	871,60	5,7419	28627,2	22556,4	252,697
458	731,15	651,63	513,49	15,957	865,61	5,7435	28677,4	22598,3	252,766
459	732,15	652,77	514,45	16,090	859,67	5,7451	28727,6	22640,2	252,835
460	733,15	653,91	515,40	16,223	853,77	5,7466	28777,9	22682,2	252,903
461	734,15	655,05	516,35	16,357	847,91	5,7482	28828,2	22724,2	252,972
462	735,15	656,20	517,31	16,492	842,10	5,7497	28878,5	22766,2	253,040
463	736,15	657,34	518,26	16,629	836,34	5,7513	28928,8	22808,2	253,109
464	737,15	658,48	519,22	16,766	830,62	5,7528	28979,2	22850,2	253,177
465	738,15	659,63	520,17	16,904	824,95	5,7544	29029,5	22892,2	253,245
466	739,15	660,77	521,13	17,043	819,31	5,7559	29079,9	22934,3	253,314
467	740,15	661,92	522,08	17,184	813,73	5,7575	29130,3	22976,4	253,382
468	741,15	663,06	523,04	17,325	808,18	5,7590	29180,7	23018,5	253,450
469	742,15	664,21	524,00	17,467	802,68	5,7606	29231,1	23060,6	253,518
470	743,15	665,35	524,96	17,611	797,22	5,7621	29281,6	23102,8	253,586
471	744,15	666,50	525,91	17,755	791,80	5,7637	29332,1	23144,9	253,654
472	745,15	667,65	526,87	17,900	786,43	5,7652	29382,6	23187,1	253,721
473	746,15	668,80	527,83	18,047	781,09	5,7668	29433,1	23229,3	253,789
474	747,15	669,95	528,79	18,194	775,80	5,7683	29483,6	23271,5	253,857
475	748,15	671,09	529,75	18,343	770,54	5,7698	29534,2	23313,7	253,924
476	749,15	672,24	530,71	18,492	765,33	5,7714	29584,7	23356,0	253,992
477	750,15	673,39	531,67	18,643	760,16	5,7729	29635,3	23398,3	254,059
478	751,15	674,54	532,63	18,795	755,02	5,7744	29685,9	23440,6	254,127
479	752,15	675,69	533,59	18,948	749,93	5,7760	29736,6	23482,9	254,194
480	753,15	676,84	534,55	19,102	744,87	5,7775	29787,2	23525,2	254,262
481	754,15	677,99	535,52	19,257	739,86	5,7790	29837,9	23567,6	254,329
482	755,15	679,15	536,48	19,413	734,88	5,7805	29888,6	23609,9	254,396
483	756,15	680,30	537,44	19,570	729,94	5,7821	29939,3	23652,3	254,463
484	757,15	681,45	538,41	19,729	725,03	5,7836	29990,0	23694,7	254,530
485	758,15	682,60	539,37	19,888	720,17	5,7851	30040,7	23737,1	254,597
486	759,15	683,76	540,33	20,049	715,34	5,7866	30091,5	23779,6	254,664
487	760,15	684,91	541,30	20,211	710,54	5,7882	30142,2	23822,0	254,731
488	761,15	686,07	542,26	20,374	705,79	5,7897	30193,0	23864,5	254,798
489	762,15	687,22	543,23	20,538	701,07	5,7912	30243,9	23907,0	254,864
490	763,15	688,37	544,20	20,703	696,38	5,7927	30294,7	23949,5	254,931
491	764,15	689,53	545,16	20,870	691,73	5,7942	30345,5	23992,1	254,998
492	765,15	690,69	546,13	21,037	687,12	5,7957	30396,4	24034,6	255,064
493	766,15	691,84	547,10	21,206	682,54	5,7972	30447,3	24077,2	255,130
494	767,15	693,00	548,06	21,376	677,99	5,7987	30498,2	24119,8	255,197
495	768,15	694,16	549,03	21,547	673,48	5,8003	30549,1	24162,4	255,263
496	769,15	695,31	550,00	21,720	669,01	5,8018	30600,0	24205,0	255,329
497	770,15	696,47	550,97	21,894	664,56	5,8033	30651,0	24247,7	255,396
498	771,15	697,63	551,94	22,068	660,15	5,8048	30702,0	24290,3	255,462
499	772,15	698,79	552,91	22,245	655,77	5,8063	30753,0	24333,0	255,528

162

Carbon dioxide (*Continued*)

t	T	h	u	π_0	θ_0	$s^\bullet$	H	U	S
500	773,15	699,95	553,88	22,422	651,43	5,8078	30804,0	24375,7	255,594
501	774,15	701,11	554,85	22,601	647,12	5,8093	30855,0	24418,4	255,660
502	775,15	702,27	555,82	22,780	642,84	5,8108	30906,1	24461,2	255,726
503	776,15	703,43	556,79	22,961	638,59	5,8123	30957,1	24503,9	255,792
504	777,15	704,59	557,76	23,144	634,38	5,8138	31008,2	24546,7	255,857
505	778,15	705,75	558,74	23,327	630,19	5,8152	31059,3	24589,5	255,923
506	779,15	706,91	559,71	23,512	626,04	5,8167	31110,4	24632,3	255,989
507	780,15	708,07	560,68	23,699	621,92	5,8182	31161,6	24675,1	256,054
508	781,15	709,24	561,66	23,886	617,83	5,8197	31212,7	24717,9	256,120
509	782,15	710,40	562,63	24,075	613,76	5,8212	31263,9	24760,8	256,185
510	783,15	711,56	563,60	24,265	609,73	5,8227	31315,1	24803,7	256,251
511	784,15	712,72	564,58	24,457	605,73	5,8242	31366,3	24846,6	256,316
512	785,15	713,89	565,55	24,649	601,76	5,8257	31417,5	24889,5	256,381
513	786,15	715,05	566,53	24,844	597,82	5,8271	31468,8	24932,4	256,447
514	787,15	716,22	567,51	25,039	593,91	5,8286	31520,0	24975,4	256,512
515	788,15	717,38	568,48	25,236	590,02	5,8301	31571,3	25018,3	256,577
516	789,15	718,55	569,46	25,434	586,17	5,8316	31622,6	25061,3	256,642
517	790,15	719,71	570,44	25,634	582,34	5,8331	31673,9	25104,3	256,707
518	791,15	720,88	571,41	25,835	578,54	5,8345	31725,3	25147,3	256,772
519	792,15	722,05	572,39	26,037	574,77	5,8360	31776,6	25190,4	256,837
520	793,15	723,22	573,37	26,241	571,03	5,8375	31828,0	25233,4	256,902
521	794,15	724,38	574,35	26,446	567,32	5,8389	31879,4	25276,5	256,966
522	795,15	725,55	575,33	26,652	563,63	5,8404	31930,8	25319,6	257,031
523	796,15	726,72	576,31	26,860	559,97	5,8419	31982,2	25362,7	257,096
524	797,15	727,89	577,29	27,070	556,33	5,8434	32033,6	25405,8	257,160
525	798,15	729,06	578,27	27,280	552,73	5,8448	32085,1	25448,9	257,225
526	799,15	730,23	579,25	27,493	549,15	5,8463	32136,5	25492,1	257,289
527	800,15	731,40	580,23	27,706	545,59	5,8477	32188,0	25535,3	257,353
528	801,15	732,57	581,21	27,922	542,06	5,8492	32239,5	25578,4	257,418
529	802,15	733,74	582,19	28,138	538,56	5,8507	32291,1	25621,6	257,482
530	803,15	734,91	583,17	28,357	535,08	5,8521	32342,6	25664,9	257,546
531	804,15	736,08	584,16	28,576	531,63	5,8536	32394,1	25708,1	257,610
532	805,15	737,25	585,14	28,797	528,21	5,8550	32445,7	25751,4	257,675
533	806,15	738,42	586,12	29,020	524,80	5,8565	32497,3	25794,6	257,739
534	807,15	739,60	587,11	29,244	521,43	5,8580	32548,9	25837,9	257,803
535	808,15	740,77	588,09	29,470	518,08	5,8594	32600,5	25881,2	257,866
536	809,15	741,94	589,07	29,697	514,75	5,8609	32652,2	25924,6	257,930
537	810,15	743,12	590,06	29,926	511,44	5,8623	32703,8	25967,9	257,994
538	811,15	744,29	591,04	30,156	508,16	5,8638	32755,5	26011,3	258,058
539	812,15	745,47	592,03	30,388	504,91	5,8652	32807,2	26054,7	258,122
540	813,15	746,64	593,02	30,621	501,68	5,8666	32858,9	26098,0	258,185
541	814,15	747,82	594,00	30,856	498,47	5,8681	32910,6	26141,5	258,249
542	815,15	748,99	594,99	31,093	495,28	5,8695	32962,4	26184,9	258,312
543	816,15	750,17	595,98	31,331	492,12	5,8710	33014,1	26228,3	258,376
544	817,15	751,34	596,96	31,571	488,98	5,8724	33065,9	26271,8	258,439
545	818,15	752,52	597,95	31,813	485,86	5,8739	33117,7	26315,3	258,502
546	819,15	753,70	598,94	32,056	482,77	5,8753	34169,5	26358,8	258,566
547	820,15	754,88	599,93	32,300	479,70	5,8767	33221,3	26402,3	258,629
548	821,15	756,05	600,92	32,547	476,64	5,8782	33273,2	26445,8	258,692
549	822,15	757,23	601,91	32,795	473,62	5,8796	33325,0	26489,3	258,755
550	823,15	758,41	602,90	33,044	470,61	5,8810	33376,9	26532,9	258,818
551	824,15	759,59	603,89	33,296	467,62	5,8825	33428,8	26576,5	258,881
552	825,15	760,77	604,88	33,549	464,66	5,8839	33480,7	26620,1	258,944
553	826,15	761,95	605,87	33,803	461,72	5,8853	33532,6	26663,7	259,007
554	827,15	763,13	606,86	34,060	458,80	5,8867	33584,6	26707,3	259,070

Carbon dioxide (*Continued*)

t	T	h	u	π_0	θ_0	s^0	H	U	S
555	828,15	764,31	607,85	34,318	455,90	5,8882	33636,5	26750,9	259,133
556	829,15	765,49	608,84	34,578	453,02	5,8896	33688,5	26794,6	259,195
557	830,15	766,67	609,84	34,839	450,16	5,8910	33740,5	26838,3	259,258
558	831,15	767,85	610,83	35,103	447,32	5,8924	33792,5	26882,0	259,321
559	832,15	769,04	611,82	35,368	444,50	5,8939	33844,5	26925,7	259,383
560	833,15	770,22	612,82	35,635	441,70	5,8953	33896,5	26969,4	259,446
561	834,15	771,40	613,81	35,903	438,92	5,8967	33948,6	27013,1	259,508
562	835,15	772,58	614,80	36,174	436,17	5,8981	34000,7	27056,9	259,571
563	836,15	773,77	615,80	36,446	433,43	5,8995	34052,7	27100,6	259,633
564	837,15	774,95	616,79	36,720	430,71	5,9010	34104,8	27144,4	259,695
565	838,15	776,14	617,79	36,996	428,01	5,9024	34157,0	27188,2	259,757
566	839,15	777,32	618,78	37,273	425,32	5,9038	34209,1	27232,0	259,820
567	840,15	778,51	619,78	37,553	422,66	5,9052	34261,2	27275,9	259,882
568	841,15	779,69	620,78	37,834	420,02	5,9066	34313,4	27319,7	259,944
569	842,15	780,88	621,77	38,117	417,39	5,9080	34365,6	27363,6	260,006
570	843,15	782,06	622,77	38,402	414,79	5,9094	34417,8	27407,5	260,068
571	844,15	783,25	623,77	38,689	412,20	5,9108	34470,0	27451,4	260,130
572	845,15	784,44	624,77	38,978	409,63	5,9122	34522,2	27495,3	260,191
573	846,15	785,62	625,76	39,269	407,08	5,9136	34574,5	27539,2	260,253
574	847,15	786,81	626,76	39,561	404,55	5,9150	34626,7	27583,2	260,315
575	848,15	788,00	627,76	39,856	402,03	5,9164	34679,0	27627,1	260,377
576	849,15	789,19	628,76	40,152	399,53	5,9178	34731,3	27671,1	260,438
577	850,15	790,37	629,76	40,451	397,05	5,9192	34783,6	27715,1	260,500
578	851,15	791,56	630,76	40,751	394,59	5,9206	34835,9	27759,1	260,561
579	852,15	792,75	631,76	41,053	392,14	5,9220	34888,3	27803,1	260,623
580	853,15	793,94	632,76	41,358	389,72	5,9234	34940,6	27847,2	260,684
581	854,15	795,13	633,76	41,664	387,30	5,9248	34993,0	27891,2	260,745
582	855,15	796,32	634,76	41,972	384,91	5,9262	35045,4	27935,3	260,807
583	856,15	797,51	635,76	42,283	382,53	5,9276	35097,8	27979,4	260,868
584	857,15	798,70	636,77	42,595	380,17	5,9290	35150,2	28023,5	260,929
585	858,15	799,90	637,77	42,909	377,83	5,9304	35202,6	28067,6	260,990
586	859,15	801,09	638,77	43,266	375,50	5,9318	35255,1	28111,7	261,051
587	860,15	802,28	639,78	43,544	373,18	5,9332	35307,5	28155,9	261,112
588	861,15	803,47	640,78	43,865	370,89	5,9345	35360,0	28200,0	261,173
589	862,15	804,66	641,78	44,187	368,61	5,9359	35412,5	28244,2	261,234
590	863,15	805,86	642,79	44,512	366,34	5,9373	35465,0	28288,4	261,295
591	864,15	807,05	643,79	44,839	364,10	5,9387	35517,5	28332,6	261,356
592	865,15	808,24	644,80	45,168	361,86	5,9401	35570,1	28376,8	261,417
593	866,15	809,44	645,80	45,499	359,65	5,9415	35622,6	28421,1	261,477
594	867,15	810,63	646,81	45,832	357,44	5,9428	35675,2	28465,3	261,538
595	868,15	811,83	647,81	46,167	355,26	5,9442	35727,8	28509,6	261,599
596	869,15	813,02	648,82	46,504	353,08	5,9456	35780,4	28553,9	261,659
597	870,15	814,22	649,83	46,844	350,93	5,9470	35833,0	28598,2	261,720
598	871,15	815,42	650,83	47,186	348,79	5,9483	35885,6	28642,5	261,780
599	872,15	816,61	651,84	47,530	346,66	5,9497	35938,2	28686,8	261,841
600	873,15	817,81	652,85	47,876	344,55	5,9511	35990,9	28731,2	261,901
601	874,15	819,00	653,86	48,225	342,45	5,9524	36043,6	28775,5	261,961
602	875,15	820,20	654,86	48,575	340,37	5,9538	36096,3	28819,9	262,022
603	876,15	821,40	655,87	48,928	338,30	5,9552	36149,0	28864,3	262,082
604	877,15	822,60	656,88	49,283	336,24	5,9566	36201,7	28908,7	262,142
605	878,15	823,80	657,89	49,641	334,20	5,9579	36254,4	28953,1	262,202
606	879,15	824,99	658,90	50,001	332,18	5,9593	36307,2	28997,6	262,262
607	880,15	826,19	659,91	50,363	330,16	5,9606	36359,9	29042,0	262,322
608	881,15	827,39	660,92	50,727	328,16	5,9620	36412,7	29086,5	262,382
609	882,15	828,59	661,93	51,094	326,18	5,9634	36465,5	29131,0	262,442

Carbon dioxide (*Continued*)

t	T	h	u	π_0	θ_0	s^0	H	U	S
610	883,15	829,79	662,94	51,463	324,21	5,9647	36518,3	29175,4	262,502
611	884,15	830,99	663,95	51,834	322,25	5,9661	36571,1	29220,0	262,561
612	885,15	832,19	664,97	52,208	320,30	5,9674	36624,0	29264,5	262,621
613	886,15	833,39	665,98	52,584	318,37	5,9688	36676,8	29309,0	262,681
614	887,15	834,60	666,99	52,962	316,45	5,9702	36729,7	29353,6	262,740
615	888,15	835,80	668,00	53,343	314,55	5,9715	36782,6	29398,1	262,800
616	889,15	837,00	669,02	53,726	312,66	5,9729	36835,5	29442,7	262,859
617	890,15	838,20	670,03	54,112	310,78	5,9742	36888,4	29487,3	262,919
618	891,15	839,40	671,04	54,500	308,91	5,9756	36941,3	29531,9	262,978
619	892,15	840,61	672,06	54,891	307,06	5,9769	36994,2	29576,5	263,038
620	893,15	841,81	673,07	55,284	305,22	5,9783	37047,2	29621,2	263,097
621	894,15	843,01	674,09	55,679	303,39	5,9796	37100,2	29665,8	263,156
622	895,15	844,22	675,10	56,077	301,57	5,9809	37153,2	29710,5	263,216
623	896,15	845,42	676,12	56,477	299,77	5,9823	37206,1	29755,2	263,275
624	897,15	846,63	677,13	56,881	297,97	5,9836	37259,2	29799,9	263,334
625	898,15	847,83	678,15	57,286	296,19	5,9850	37312,2	29844,6	263,393
626	899,15	849,04	679,16	57,694	294,43	5,9863	37365,2	29889,3	263,452
627	900,15	850,24	680,18	58,105	292,67	5,9877	37418,3	29934,1	263,511
628	901,15	851,45	681,20	58,518	290,93	5,9890	37471,4	29978,8	263,570
629	902,15	852,65	682,21	58,934	289,19	5,9903	37524,4	30023,6	263,629
630	903,15	853,86	683,23	59,353	287,47	5,9917	37577,5	30068,4	263,688
631	904,15	855,07	684,25	59,774	285,77	5,9930	37630,6	30113,2	263,746
632	905,15	856,27	685,27	60,197	284,07	5,9943	37683,8	30158,0	263,805
633	906,15	857,48	686,29	60,624	282,38	5,9957	37736,9	30202,8	263,864
634	907,15	858,69	687,31	61,053	280,71	5,9970	37790,1	30247,6	263,922
635	908,15	859,90	688,33	61,484	279,04	5,9983	37843,2	30292,5	263,981
636	909,15	861,11	689,34	61,919	277,39	5,9997	37896,4	30337,4	264,039
637	910,15	862,31	690,36	62,356	275,75	6,0010	37949,6	30382,3	264,098
638	911,15	863,52	691,38	62,795	274,12	6,0023	38002,8	30427,1	264,156
639	912,15	864,73	692,41	63,238	272,50	6,0037	38056,0	30472,1	264,215
640	913,15	865,94	693,43	63,683	270,89	6,0050	38109,3	30517,0	264,273
641	914,15	867,15	694,45	64,131	269,29	6,0063	38162,5	30561,9	264,331
642	915,15	868,36	695,47	64,582	267,71	6,0076	38215,8	30606,9	264,390
643	916,15	869,57	696,49	65,036	266,13	6,0089	38269,1	30651,8	264,448
644	917,15	870,78	697,51	65,492	264,56	6,0103	38322,4	30696,8	264,506
645	918,15	872,00	698,53	65,951	263,01	6,0116	38375,7	30741,8	264,564
646	919,15	873,21	699,56	66,413	261,46	6,0129	38429,0	30786,8	264,622
647	920,15	874,42	700,58	66,878	259,93	6,0142	38482,3	30831,8	264,680
648	921,15	875,63	701,60	67,346	258,40	6,0155	38535,7	30876,9	264,738
649	922,15	876,84	702,63	67,816	256,89	6,0169	38589,1	30921,9	264,796
650	923,15	878,06	703,65	68,290	255,38	6,0182	38642,4	30967,0	264,854
651	924,15	879,27	704,68	68,766	253,89	6,0195	38695,8	31012,1	264,912
652	925,15	880,48	705,70	69,246	252,40	6,0208	38749,2	31057,2	264,969
653	926,15	881,70	706,72	69,728	250,93	6,0221	38802,6	31102,3	265,027
654	927,15	882,91	707,75	70,213	249,46	6,0234	38856,1	31147,4	265,085
655	928,15	884,13	708,78	70,701	248,01	6,0247	38909,5	31192,5	265,142
656	929,15	885,34	709,80	71,193	246,56	6,0260	38963,0	31237,6	265,200
657	930,15	886,56	710,83	71,687	245,13	6,0273	39016,5	31282,8	265,257
658	931,15	887,77	711,85	72,184	243,70	6,0287	39069,9	31328,0	265,315
659	932,15	888,99	712,88	72,684	242,28	6,0300	39123,4	31373,2	265,372
660	933,15	890,20	713,91	73,188	240,87	6,0313	39177,0	31418,4	265,430
661	934,15	891,42	714,93	73,694	239,48	6,0326	39230,5	31463,6	265,487
662	935,15	892,64	715,96	74,204	238,09	6,0339	39284,0	31508,8	265,544
663	936,15	893,85	716,99	74,716	236,71	6,0352	39337,6	31554,0	265,602
664	937,15	895,07	718,02	75,232	235,33	6,0365	39391,1	31599,3	265,659

Carbon dioxide (*Continued*)

t	T	h	u	π_0	θ_0	s^0	H	U	S
665	938,15	896,29	719,05	75,751	233,97	6,0378	39444,7	31644,6	265,716
666	939,15	897,51	720,08	76,273	232,62	6,0391	39498,3	31689,8	265,773
667	940,15	898,72	721,11	76,798	231,27	6,0404	39551,9	31735,1	265,830
668	941,15	899,94	722,13	77.326	229,94	6,0417	39605,5	31780,4	265,887
669	942,15	901,16	723,16	77,858	228,61	6,0429	39659,2	31825,8	265,944
670	943,15	902,38	724,19	78,392	227,29	6,0442	39712,8	31871,1	266,001
671	944,15	903,60	725,23	78,930	225,98	6,0455	39766,5	31916,4	266,058
672	945,15	904,82	726,26	79,472	224,68	6,0468	39820,2	31961,8	266,115
673	946,15	906,04	727,29	80,016	223,39	6,0481	39873,8	32007,2	266,171
674	947,15	907,26	728,32	80,564	222,10	6,0494	39927,5	32052,5	266,228
675	948,15	908,48	729,35	81,115	220,83	6,0507	39981,3	32097,9	266,285
676	949,15	909,70	730,38	81,669	219,56	6,0520	40035,0	32143,4	266,341
677	950,15	910,92	731,41	82,227	218,30	6,0533	40088,7	32188,8	266,398
678	951,15	912,14	732,45	82,788	217,05	6,0545	40142,5	32234,2	266,455
679	952,15	913,36	733,48	83,353	215,81	6,0558	40196,2	32279,7	266,511
680	953,15	914,59	734,51	83,921	214,57	6,0571	40250,0	32325,1	266,567
681	954,15	915,81	735,55	84,492	213,34	6,0584	40303,8	32370,6	266,624
682	955,15	917,03	736,58	85,067	212,12	6,0597	40357,6	32416,1	266,680
683	956,15	918,25	737,61	85,645	210,91	6,0610	40411,4	32461,6	266,737
684	957,15	919,48	738,65	86,226	209,71	6,0622	40465,3	32507,1	266,793
685	958,15	920,70	739,68	86,811	208,51	6,0635	40519,1	32552,6	266,849
686	959,15	921,92	740,72	87,400	207,33	6,0648	40573,0	32598,2	266,905
687	960,15	923,15	741,75	87,992	206,15	6,0661	40626,8	32643,7	266,961
688	961,15	924,37	742,79	88,588	204,97	6,0673	40680,7	32689,3	267,017
689	962,15	925,60	743,82	89,187	203,81	6,0686	40734,6	32734,9	267,073
690	963,15	926,82	744,86	89,789	202,65	6,0699	40788,5	32780,5	267,129
691	964,15	928,05	745,89	90,396	201,50	6,0712	40842,4	32826,1	267,185
692	965,15	929,27	746,93	91,006	200,36	6,0724	40896,4	32871,7	267,241
693	966,15	930,50	747,97	91,619	199,22	6,0737	40950,3	32917,3	267,297
694	967,15	931,72	749,01	92,236	198,09	6,0750	41004,3	32963,0	267,353
695	968,15	932,95	750,04	92,857	196,97	6,0762	41058,2	33008,6	267,409
696	969,15	934,18	751,08	93,482	195,86	6,0775	41112,2	33054,3	267,465
697	970,15	935,40	752,12	94,110	194,75	6,0788	41166,2	33100,0	267,520
698	971,15	936,63	753,16	94,742	193,65	6,0800	41220,2	33145,7	267,576
699	972,15	937,86	754,20	95,377	192,56	6,0813	41274,2	33191,4	267,631
700	973,15	939,09	755,23	96,017	191,47	6,0826	41328,3	33237,1	267,687
701	974,15	940,31	756,27	96,660	190,39	6,0838	41382,3	33282,8	267,743
702	975,15	941,54	757,31	97,307	189,32	6,0851	41436,4	33328,6	267,798
703	976,15	942,77	758,35	97,958	188,26	6,0863	41490,5	33374,3	267,853
704	977,15	944,00	759,39	98,612	187,20	6,0876	41544,5	33420,1	267,909
705	978,15	945,23	760,43	99,271	186,15	6,0888	41598,6	33465,9	267,964
706	979,15	946,46	761,47	99,933	185,10	6,0901	41652,7	33511,7	268,019
707	980,15	947,69	762,51	100,59	184,07	6,0914	41706,9	33557,5	268,075
708	981,15	948,92	763,56	101,27	183,03	6,0926	41761,0	33603,3	268,130
709	982,15	950,15	764,60	101,94	182,01	6,0939	41815,1	33649,1	268,185
710	983,15	951,38	765,64	102,62	180,99	6,0951	41869,3	33695,0	268,240
711	984,15	952,61	766,68	103,30	179,98	6,0964	41923,5	33740,8	268,295
712	985,15	953,84	767,72	103,99	178,97	6,0976	41977,7	33786,7	268,350
713	986,15	955,07	768,77	104,68	177,97	6,0989	42031,9	33832,6	268,405
714	987,15	956,31	769,81	105,37	176,98	6,1001	42086,1	33878,5	268,460
715	988,15	957,54	770,85	106,07	175,99	6,1014	42140,3	33924,4	268,515
716	989,15	958,77	771,89	106,77	175,01	6,1026	42194,5	33970,3	268,570
717	990,15	960,00	772,94	107,48	174,04	6,1039	42248,8	34016,2	268,625
718	991,15	961,24	773,98	108,19	173,07	6,1051	42303,0	34062,2	268,679
719	992,15	962,47	775,03	108,90	172,11	6,1063	42357,3	34108,1	268,734

166

Carbon dioxide (*Continued*)

t	T	h	u	π_0	θ_0	s^0	H	U	S
720	993,15	963,70	776,07	109,62	171,15	6,1076	42411,6	34154,1	268,789
721	994,15	964,94	777,12	110,34	170,20	6,1088	42465,9	34200,1	268,844
722	995,15	966,17	778,16	111,07	169,26	6,1101	42520,2	34246,1	268,898
723	996,15	967,40	779,21	111,80	168,32	6,1113	42574,5	34292,1	268,953
724	997,15	968,64	780,25	112,54	167,39	6,1125	42628,8	34338,1	269,007
725	998,15	969,87	781,30	113,28	166,46	6,1138	42683,2	34384,1	269,062
726	999,15	971,11	782,34	114,02	165,54	6,1150	42737,5	34430,2	269,116
727	1000,15	972,34	783,39	114,77	164,63	6,1163	42791,9	34476,2	269,170
728	1001,15	973,58	784,44	115,52	163,72	6,1175	42846,2	34522,3	269,225
729	1002,15	974,82	785,48	116,28	162,82	6,1187	42900,6	34568,4	269,279
730	1003,15	976,05	786,53	117,04	161,92	6,1200	42955,0	34614,4	269,333
731	1004,15	977,29	787,58	117,80	161,03	6,1212	43009,5	34660,5	269,388
732	1005,15	978,52	788,63	118,57	160,14	6,1224	43063,9	34706,7	269,442
733	1006,15	979,76	789,67	119,35	159,26	6,1237	43118,3	34752,8	269,496
734	1007,15	981,00	790,72	120,13	158,38	6,1249	43172,8	34798,9	269,550
735	1008,15	982,24	791,77	120,91	157,51	6,1261	43227,2	34845,1	269,604
736	1009,15	983,47	792,82	121,70	156,65	6,1273	43281,7	34891,2	269,658
737	1010,15	984,71	793,87	122,49	155,79	6,1286	43336,2	34937,4	269,712
738	1011,15	985,95	794,92	123,29	154,94	6,1298	43390,7	34983,6	269,766
739	1012,15	987,19	795,97	124,09	154,09	6,1310	43445,2	35029,8	269,820
740	1013,15	988,43	797,02	124,90	153,24	6,1322	43499,7	35076,0	269,874
741	1014,15	989,67	798,07	125,71	152,41	6,1335	43554,2	35122,2	269,927
742	1015,15	990,91	799,12	126,52	151,57	6,1347	43608,8	35168,4	269,981
743	1016,15	992,15	800,17	127,34	150,74	6,1359	43663,3	35214,7	270,035
744	1017,15	993,39	801,22	128,17	149,92	6,1371	43717,9	35260,9	270,089
745	1018,15	994,63	802,27	129,00	149,10	6,1383	43772,5	35307,2	270,142
746	1019,15	995,87	803,32	129,83	148,29	6,1396	43827,1	35353,4	270,196
747	1020,15	997,11	804,37	130,67	147,48	6,1408	43881,7	35399,7	270,249
748	1021,15	998,35	805,43	131,51	146,68	6,1420	43936,3	35446,0	270,303
749	1022,15	999,59	806,48	132,36	145,88	6,1432	43990,9	35492,3	270,356
750	1023,15	1000,83	807,53	133,22	145,09	6,1444	44045,6	35538,7	270,410
751	1024,15	1002,07	808,58	134,07	144,30	6,1456	44100,2	35585,0	270,463
752	1025,15	1003,31	809,64	134,94	143,52	6,1468	44154,9	35631,3	270,516
753	1026,15	1004,56	810,69	135,80	142,74	6,1481	44209,5	35677,7	270,570
754	1027,15	1005,80	811,74	136,68	141,97	6,1493	44264,2	35724,1	270,623
755	1028,15	1007,04	812,80	137,55	141,20	6,1505	44318,9	35770,5	270,676
756	1029,15	1008,29	813,85	138,44	140,44	6,1517	44373,6	35816,8	270,729
757	1030,15	1009,53	814,91	139,33	139,68	6,1529	44428,3	35863,2	270,783
758	1031,15	1010,77	815,96	140,22	138,92	6,1541	44483,1	35909,7	270,836
759	1032,15	1012,02	817,02	141,12	138,17	6,1553	44537,8	35956,1	270,889
760	1033,15	1013,26	818,07	142,02	137,43	6,1565	44592,6	36002,5	270,942
761	1034,15	1014,50	819,13	142,93	136,69	6,1577	44647,3	36049,0	270,995
762	1035,15	1015,75	820,18	143,84	135,95	6,1589	44702,1	36095,4	271,048
763	1036,15	1016,99	821,24	144,76	135,22	6,1601	44756,9	36141,9	271,101
764	1037,15	1018,24	822,30	145,68	134,49	6,1613	44811,7	36188,4	271,153
765	1038,15	1019,48	823,35	146,61	133,77	6,1625	44866,5	36234,9	271,206
766	1039,15	1020,73	824,41	147,54	133,05	6,1637	44921,3	36281,4	271,259
767	1040,15	1021,98	825,47	148,48	132,34	6,1649	44976,1	36327,9	271,312
768	1041,15	1023,22	826,52	149,43	131,63	6,1661	45031,0	36374,4	271,365
769	1042,15	1024,47	827,58	150,38	130,92	6,1673	45085,8	36421,0	271,417
770	1043,15	1025,71	828,64	151,33	130,22	6,1685	45140,7	36467,5	271,470
771	1044,15	1026,96	829,70	152,29	129,52	6,1697	45195,6	36514,1	271,522
772	1045,15	1028,21	830,75	153,26	128,83	6,1709	45250,5	36560,6	271,575
773	1046,15	1029,46	831,81	154,23	128,14	6,1721	45305,3	36607,2	271,627
774	1047,15	1030,70	832,87	155,20	127,46	6,1733	45360,3	36653,8	271,680

Carbon dioxide (*Continued*)

t	T	h	u	π_0	θ_0	s^0	H	U	S
775	1048,15	1031,95	833,93	156,18	126,78	6,1745	45415,2	36700,4	271,732
776	1049,15	1033,20	834,99	157,17	126,10	6,1757	45470,1	36747,0	271,785
777	1050,15	1034,45	836,05	158,16	125,43	6,1769	45525,0	36793,7	271,837
778	1051,15	1035,70	837,11	159,16	124,76	6,1780	45580,0	36840,3	271,889
779	1052,15	1036,95	838,17	160,17	124,10	6,1792	45635,0	36887,0	271,942
780	1053,15	1038,20	839,23	161,18	123,44	6,1804	45689,9	36933,6	271,994
781	1054,15	1039,44	840,29	162,19	122,78	6,1816	45744,9	36980,3	272,046
782	1055,15	1040,69	841,35	163,21	122,13	6,1828	45799,9	37027,0	272,098
783	1056,15	1041,94	842,41	164,24	121,48	6,1840	45854,9	37073,7	272,150
784	1057,15	1043,19	843,47	165,27	120,84	6,1852	45910,0	37120,4	272,202
785	1058,15	1044,44	844,53	166,31	120,20	6,1863	45965,0	37167,1	272,254
786	1059,15	1045,70	845,60	167,35	119,56	6,1875	46020,0	37213,8	272,306
787	1060,15	1046,95	846,66	168,40	118,93	6,1887	46075,1	37260,6	272,358
788	1061,15	1048,20	847,72	169,45	118,30	6,1899	46130,1	37307,3	272,410
789	1062,15	1049,45	848,78	170,51	117,68	6,1911	46185,2	37354,1	272,462
790	1063,15	1050,70	849,84	171,58	117,05	6,1922	46240,3	37400,8	272,514
791	1064,15	1051,95	850,91	172,65	116,44	6,1934	46295,4	37447,6	272,566
792	1065,15	1053,20	851,97	173,73	115,82	6,1946	46350,5	37494,4	272,617
793	1066,15	1054,46	853.03	174,81	115,21	6,1958	46405,6	37541,2	272,669
794	1067,15	1055,71	854,10	175,90	114,61	6,1969	46460,7	37588,0	272,721
795	1068,15	1056,96	855,16	177,00	114,00	6,1981	46515,9	37634,8	272,773
796	1069,15	1058,22	856,23	178,10	113,40	6,1993	46571,0	37681,7	272,824
797	1070,15	1059,47	857,29	179,21	112,81	6,2005	46626,2	37728,5	272,876
798	1071,15	1060,72	858,36	180,32	112,22	6,2016	46681,4	37775,4	272,927
799	1072,15	1061,98	859,42	181,45	111,63	6,2028	46736,5	37822,2	272,979
800	1073,15	1063,23	860,49	182,57	111,04	6,2040	46791,7	37869,1	273,030
801	1074,15	1064,49	861,55	183,70	110,46	6,2051	46846,9	37916,0	273,082
802	1075,15	1065,74	862,62	184,84	109,88	6,2063	46902,1	37962,9	273,133
803	1076,15	1066,99	863,68	185,99	109,31	6,2075	46957,4	38009,8	273,184
804	1077,15	1068,25	864,75	187,14	108,74	6,2086	47012,6	38056,7	273,236
805	1078,15	1069,50	865,82	188,30	108,17	6,2098	47067,8	38103,7	273,287
806	1079,15	1070,76	866,88	189,46	107,60	6,2110	47123,1	38150,6	273,338
807	1080,15	1072,02	867,95	190,63	107,04	6,2121	47178,4	38197,5	273,389
808	1081,15	1073,27	869,02	191,81	106,48	6,2133	47233,6	38244,5	273,440
809	1082,15	1074,53	870,08	192,99	105,93	6,2144	47288,9	38291,5	273,492
810	1083,15	1075,78	871,15	194,18	105,38	6,2156	47344,2	38338,5	273,543
811	1084,15	1077,04	872,22	195,37	104,83	6,2168	47399,5	38385,5	273,594
812	1085,15	1078,30	873,29	196,58	104,28	6,2179	47454,8	38432,5	273,645
813	1086,15	1079,56	874,35	197,78	103,74	6,2191	47510,2	38479,5	273,696
814	1087,15	1080,81	875,42	199,00	103,20	6,2202	47565,5	38526,5	273,747
815	1088,15	1082,07	876,49	200,22	102,67	6,2214	47620,9	38573,5	273,797
816	1089,15	1083,33	877,56	201,45	102,14	6,2226	47676 2	38620,6	273,848
817	1090,15	1084,59	878,63	202,68	101,61	6,2237	47731,6	38667,6	273,899
818	1091,15	1085,85	879,70	203,93	101,08	6,2249	47787,0	38714,7	273,950
819	1092,15	1087,10	880,77	205,17	100,56	6,2260	47842,4	38761,8	274,001
820	1093,15	1088,36	881,84	206,43	100,04	6,2272	47897,8	38808,9	274,051
821	1094,15	1089,62	882,91	207,69	99,526	6,2283	47953,2	38856,0	274,102
822	1095,15	1090,88	883,98	208,96	99,012	6,2295	48008,6	38903,1	274,153
823	1096,15	1092,14	885,05	210,24	98,501	6,2306	48064,0	38950,2	274,203
824	1097,15	1093,40	886,12	211,52	97,993	6,2318	48119,5	38997,3	274,254
825	1098,15	1094,66	887,19	212,81	97,488	6,2329	48174,9	39044,4	274,304
826	1099,15	1095,92	888,26	214,10	96,986	6,2341	48230,4	39091,6	274,355
827	1100,15	1097,18	889,34	215,41	96,488	6,2352	48285,9	39138,8	274,405
828	1101,15	1098,44	890,41	216,72	95,991	6,2364	48341,3	39185,9	274,456
829	1102,15	1099,70	891,48	218,03	95,498	6,2375	48396,8	39233,1	274,506

Carbon dioxide (*Continued*)

t	T	h	u	π_0	θ_0	s^0	H	U	S
830	1103,15	1100,96	892,55	219,36	95,008	6,2386	48452,3	39280,3	274,556
831	1104,15	1102,23	893,62	220,69	94,521	6,2398	48507,8	39327,5	274,607
832	1105,15	1103,49	894,70	222,03	94,036	6,2409	48563,4	39374,7	274,657
833	1106,15	1104,75	895,77	223,37	93,554	6,2421	48618,9	39421,9	274,707
834	1107,15	1106,01	896,84	224,72	93,075	6,2432	48674,4	39469,1	274,757
835	1108,15	1107,27	897,92	226,08	92,599	6,2443	48730,0	39516,4	274,807
836	1109,15	1108,54	898,99	227,45	92,126	6,2455	48785,6	39563,6	274,858
837	1110,15	1109,80	900,06	228,83	91,655	6,2466	48841,1	39610,9	274,908
838	1111,15	1111,06	901,14	230,21	91,187	6,2478	48896,7	39658,2	274,958
839	1112,15	1112,32	902,21	231,60	90,722	6,2489	48952,3	39705,4	275,008
840	1113,15	1113,59	903,29	232,99	90,259	6,2500	49007,9	39752,7	275,058
841	1114,15	1114,85	904,36	234,40	89,799	6,2512	49063,5	39800,0	275,108
842	1115,15	1116,12	905,44	235,81	89,342	6,2523	49119,2	39847,3	275,158
843	1116,15	1117,38	906,51	237,23	88,887	6,2534	49174,8	39894,7	275,207
844	1117,15	1118,64	907,59	238,65	88,435	6,2546	49230,4	39942,0	275,257
845	1118,15	1119,91	908,66	240,08	87,986	6,2557	49286,1	39989,3	275,307
846	1119,15	1121,17	909,74	241,53	87,539	6,2568	49341,7	40036,7	275,357
847	1120,15	1122,44	910,81	242,97	87,095	6,2580	49397,4	40084,0	275,407
848	1121,15	1123,70	911,89	244,43	86,653	6,2591	49453,1	40131,4	275,456
849	1122,15	1124,97	912,97	245,90	86,214	6,2602	49508,8	40178,8	275,506
850	1123,15	1126,24	914,04	247,37	85,778	6,2613	49564,5	40226,2	275,555
851	1124,15	1127,50	915,12	248,85	85,344	6,2625	49620,2	40273,6	275,605
852	1125,15	1128,77	916,20	250,33	84,912	6,2636	49675,9	40321,0	275,655
853	1126,15	1130,03	917,28	251,83	84,483	6,2647	49731,7	40368,4	275,704
854	1127,15	1131,30	918,35	253,33	84,056	6,2658	49787,4	40415,8	275,754
855	1128,15	1132,57	919,43	254,84	83,632	6,2670	49843,2	40463,3	275,803
856	1129,15	1133,83	920,51	256,36	83,210	6,2681	49898,9	40510,7	275,852
857	1130,15	1135,10	921,59	257,89	82,791	6,2692	49954,7	40558,2	275,902
858	1131,15	1136,37	922,67	259,42	82,374	6,2703	50010,5	40605,6	275,951
859	1132,15	1137,64	923,75	260,97	81,959	6,2715	50066,3	40653,1	276,000
860	1133,15	1138,90	924,82	262,52	81,547	6,2726	50122,1	40700,6	276,050
861	1134,15	1140,17	925,90	264,08	81,137	6,2737	50177,9	40748,1	276,099
862	1135,15	1141,44	926,98	265,64	80,730	6,2748	50233,7	40795,6	276,148
863	1136,15	1142,71	928,06	267,22	80,324	6,2759	50289,5	40843,1	276,197
864	1137,15	1143,98	929,14	268,80	79,921	6,2770	50345,4	40890,6	276,246
865	1138,15	1145,25	930,22	270,39	79,521	6,2782	50401,2	40938,2	276,296
866	1139,15	1146,52	931,30	271,99	79,122	6,2793	50457,1	40985,7	276,345
867	1140,15	1147,79	932,38	273,60	78,726	6,2804	50512,9	41033,3	276,394
868	1141,15	1149,06	933,46	275,22	78,333	6,2815	50568,8	41080,8	276,443
869	1142,15	1150,33	934,55	276,85	77,941	6,2826	50624,7	41128,4	276,492
870	1143,15	1151,60	935,63	278,48	77,552	6,2837	50680,6	41176,0	276,540
871	1144,15	1152,87	936,71	280,12	77,164	6,2848	50736,5	41223,6	276,589
872	1145,15	1154,14	937,79	281,77	76,779	6,2859	50792,4	41271,2	276,638
873	1146,15	1155,41	938,87	283,43	76,397	6,2871	50848,3	41318,8	276,687
874	1147,15	1156,68	939,95	285,10	76,016	6,2882	50904,3	41366,4	276,736
875	1148,15	1157,95	941,04	286,78	75,637	6,2893	50960,2	41414,0	276,785
876	1149,15	1159,22	942,12	288,46	75,261	6,2904	51016,2	41461,7	276,833
877	1150,15	1160,49	943,20	290,15	74,887	6,2915	51072,1	41509,3	276,882
878	1151,15	1161,76	944,28	291,86	74,515	6,2926	51128,1	41557,0	276,931
879	1152,15	1163,04	945,37	293,57	74,145	6,2937	51184,1	41604,6	276,979
880	1153,15	1164,31	946,45	295,29	73,777	6,2948	51240,1	41652,3	277,028
881	1154,15	1165,58	947,53	297,02	73,411	6,2959	51296,1	41700,0	277,076
882	1155,15	1166,85	948,62	298,76	73,047	6,2970	51352,1	41747,7	277,125
883	1156,15	1168,13	949,70	300,50	72,685	6,2981	51408,1	41795,4	277,173
884	1157,15	1169,40	950,79	302,26	72,326	6,2992	51464,1	41843,1	277,222

169

Carbon dioxide (*Continued*)

t	T	h	u	π_0	θ_0	s^0	H	U	S
885	1158,15	1170,67	951,87	304,02	71,968	6,3003	51520,2	41890,8	277,270
886	1159,15	1171,95	952,95	305,80	71,612	6,3014	51576,2	41938,6	277,318
887	1160,15	1173,22	954,04	307,58	71,258	6,3025	51632,3	41986,3	277,367
888	1161,15	1174,50	955,12	309,37	70,907	6,3036	51688,4	42034,1	277,415
889	1162,15	1175,77	956,21	311,17	70,557	6,3047	51744,4	42081,8	277,463
890	1163,15	1177,04	957,30	312,98	70,209	6,3058	51800,5	42129,6	277,512
891	1164,15	1178,32	958,38	314,80	69,863	6,3069	51856,6	42177,4	277,560
892	1165,15	1179,59	959,47	316,63	69,519	6,3080	51912,7	42225,2	277,608
893	1166,15	1180,87	960,55	318,47	69,177	6,3091	51968,8	42273,0	277,656
894	1167,15	1182,14	961,64	320,32	68,837	6,3102	52024,9	42320,8	277,704
895	1168,15	1183,42	962,73	322,18	68,499	6,3113	52081,1	42368,6	277,752
896	1169,15	1184,69	963,81	324,04	68,163	6,3124	52137,2	42416,4	277,800
897	1170,15	1185,97	964,90	325,92	67,828	6,3134	52193,4	42464,3	277,848
898	1171,15	1187,25	965,99	327,81	67,496	6,3145	52249,5	42512,1	277,896
899	1172,15	1188,52	967,07	329,70	67,165	6,3156	52305,7	42560,0	277,944
900	1173,15	1189,80	968,16	331,61	66,836	6,3167	52361,9	42607,8	277,992
901	1174,15	1191,08	969,25	333,52	66,509	6,3178	52418,0	42655,7	278,040
902	1175,15	1192,35	970,34	335,45	66,184	6,3189	52474,2	42703,6	278,088
903	1176,15	1193,63	971,43	337,38	65,860	6,3200	52530,4	42751,5	278,136
904	1177,15	1194,91	972,51	339,33	65,538	6,3211	52586,7	42799,4	278,183
905	1178,15	1196,18	973,60	341,28	65,219	6,3221	52642,9	42847,3	278,231
906	1179,15	1197,46	974,69	343,24	64,900	6,3232	52699,1	42895,2	278,279
907	1180,15	1198,74	975,78	345,22	64,584	6,3243	52755,4	42943,1	278,327
908	1181,15	1200,02	976,87	347,20	64,269	6,3254	52811,6	42991,0	278,374
909	1182,15	1201,30	977,96	349,19	63,957	6,3265	52867,9	43039,0	278,422
910	1183,15	1202,58	979,05	351,20	63,645	6,3276	52924,1	43086,9	278,469
911	1184,15	1203,85	980,14	353,21	63,336	6,3286	52980,4	43134,9	278,517
912	1185,15	1205,13	981,23	355,24	63,028	6,3297	53036,7	43182,9	278,564
913	1186,15	1206,41	982,32	357,27	62,722	6,3308	53093,0	43230,8	278,612
914	1187,15	1207,69	983,41	359,32	62,418	6,3319	53149,3	43278,8	278,659
915	1188,15	1208,97	984,50	361,37	62,115	6,3330	53205,6	43326,8	278,707
916	1189,15	1210,25	985,59	363,44	61,814	6,3340	53261,9	43374,8	278,754
917	1190,15	1211,53	986,68	365,51	61,515	6,3351	53318,2	43422,8	278,802
918	1191,15	1212,81	987,77	367,60	61,217	6,3362	53374,6	43470,9	278,849
919	1192,15	1214,09	988,86	369,69	60,921	6,3373	53430,9	43518,9	278,896
920	1193,15	1215,37	989,96	371,80	60,627	6,3383	53487,3	43566,9	278,943
921	1194,15	1216,65	991,05	373,92	60,334	6,3394	53543,6	43615,0	278,991
922	1195,15	1217,93	992,14	376,05	60,043	6,3405	53600,0	43663,1	279,038
923	1196,15	1219,21	993,23	378,19	59,753	6,3415	53656,4	43711,1	279,085
924	1197,15	1220,50	994,32	380,34	59,465	6,3426	53712,8	43759,2	279,132
925	1198,15	1221,78	995,42	382,50	59,179	6,3437	53769,2	43807,3	279,179
926	1199,15	1223,06	996,51	384,67	58,894	6,3448	53825,6	43855,4	279,226
927	1200,15	1224,34	997,60	386,85	58,610	6,3458	53882,0	43903,5	279,273
928	1201,15	1225,62	998,70	389,04	58,329	6,3469	53938,4	43951,6	279,320
929	1202,15	1226,91	999,79	391,25	58,048	6,3480	53994,9	43999,7	279,367
930	1203,15	1228,19	1000,88	393,46	57,770	6,3490	54051,3	44047,8	279,414
931	1204,15	1229,47	1001,98	395,69	57,492	6,3501	54107,8	44096,0	279,461
932	1205,15	1230,75	1003,07	397,92	57,217	6,3512	54164,2	44144,1	279,508
933	1206,15	1232,04	1004,16	400,17	56,942	6,3522	54220,7	44192,3	279,555
934	1207,15	1233,32	1005,26	402,43	56,670	6,3533	54277,2	44240,4	279,602
935	1208,15	1234,60	1006,35	404,70	56,399	6,3543	54333,7	44288,6	279,648
936	1209,15	1235,89	1007,45	406,98	56,129	6,3554	54390,2	44336,8	279,695
937	1210,15	1237,17	1008,54	409,28	55,860	6,3565	54446,7	44385,0	279,742
938	1211,15	1238,46	1009,64	411,58	55,594	6,3575	54503,2	44433,2	279,789
939	1212,15	1239,74	1010,73	413,90	55,328	6,3586	54559,7	44481,4	279,835

Carbon dioxide (*Continued*)

t	*T*	*h*	*u*	π_0	θ_0	*s*°	*H*	*U*	*S*
940	1213,15	1241,02	1011,83	416,22	55,064	6,3596	54616,2	44529,6	279,882
941	1214,15	1242,31	1012,93	418,56	54,802	6,3607	54672,8	44577,8	279,928
942	1215,15	1243,59	1014,02	420,91	54,541	6,3618	54729,3	44626,1	279,975
943	1216,15	1244,88	1015,12	423,27	54,281	6,3628	54785,9	44674,3	280,021
944	1217,15	1246,16	1016,21	425,65	54,023	6,3639	54842,4	44722,6	280,068
945	1218,15	1247,45	1017,31	428,03	53,766	6,3649	54899,0	44770,8	280,114
946	1219,15	1248,74	1018,41	430,43	53,510	6,3660	54955,6	44819,1	280,161
947	1220,15	1250,02	1019,50	432,84	53,256	6,3670	55012,2	44867,4	280,207
948	1221,15	1251,31	1020,60	435,26	53,003	6,3681	55068,8	44915,6	280,254
949	1222,15	1252,59	1021,70	437,69	52,752	6,3691	55125,4	44963,9	280,300
950	1223,15	1253,88	1022,80	440,14	52,502	6,3702	55182,0	45012,2	280,346
951	1224,15	1255,17	1023,89	442,59	52,253	6,3713	55238,6	45060,6	280,393
952	1225,15	1256,45	1024,99	445,06	52,006	6,3723	55295,3	45108,9	280,439
953	1226,15	1257,74	1026,09	447,54	51,760	6,3734	55351,9	45157,2	280,485
954	1227,15	1259,03	1027,19	450,03	51,515	6,3744	55408,6	45205,5	280,531
955	1228,15	1260,32	1028,29	452,54	51,272	6,3755	55465,2	45253,9	280,577
956	1229,15	1261,60	1029,39	455,06	51,029	6,3765	55521,9	45302,2	280,623
957	1230,15	1262,89	1030,48	457,59	50,789	6,3775	55578,6	45350,6	280,670
958	1231,15	1264,18	1031,58	460,13	50,549	6,3786	55635,2	45399,0	280,716
959	1232,15	1265,47	1032,68	462,68	50,311	6,3796	55691,9	45447,3	280,762
960	1233,15	1266,76	1033,78	465,25	50,074	6,3807	55748,6	45495,7	280,808
961	1234,15	1268,04	1034,88	467,83	49,838	6,3817	55805,3	45544,1	280,854
962	1235,15	1269,33	1035,98	470,42	49,604	6,3828	55862,1	45592,5	280,899
963	1236,15	1270,62	1037,08	473,02	49,371	6,3838	55918,8	45640,9	280,945
964	1237,15	1271,91	1038,18	475,64	49,139	6,3849	55975,5	45689,3	280,991
965	1238,15	1273,20	1039,28	478,27	48,908	6,3859	56032,3	45737,8	281,037
966	1239,15	1274,49	1040,38	480,91	48,678	6,3869	56089,0	45786,2	281,083
967	1240,15	1275,78	1041,48	483,57	48,450	6,3880	56145,8	45834,6	281,129
968	1241,15	1277,07	1042,58	486,24	48,223	6,3890	56202,5	45883,1	281,174
969	1242,15	1278,36	1043,69	488,92	47,997	6,3901	56259,3	45931,6	281,220
970	1243,15	1279,65	1044,79	491,61	47,773	6,3911	56316,1	45980,0	281,266
971	1244,15	1280,94	1045,89	494,32	47,549	6,3921	56372,9	46028,5	281,312
972	1245,15	1282,23	1046,99	497,04	47,327	6,3932	56429,7	46077,0	281,357
973	1246,15	1283,52	1048,09	499,78	47,106	6,3942	56486,5	46125,5	281,403
974	1247,15	1284,81	1049,19	502,52	46,886	6,3952	56543,3	46174,0	281,448
975	1248,15	1286,10	1050,30	505,28	46,667	6,3963	56600,1	46222,5	281,494
976	1249,15	1287,39	1051,40	508,06	46,450	6,3973	56657,0	46271,0	281,539
977	1250,15	1288,69	1052,50	510,84	46,233	6,3983	56713,8	46319,5	281,585
978	1251,15	1289,98	1053,60	513,64	46,018	6,3994	56770,6	46368,1	281,630
979	1252,15	1291,27	1054,71	516,46	45,804	6,4004	56827,5	46416,6	281,676
980	1253,15	1292,56	1055,81	519,29	45,591	6,4014	56884,4	46465,2	281,721
981	1254,15	1293,85	1056,91	522,13	45,379	6,4025	56941,2	46513,7	281,767
982	1255,15	1295,15	1058,02	524,98	45,168	6,4035	56998,1	46562,3	281,812
983	1256,15	1296,44	1059,12	527,85	44,958	6,4045	57055,0	46610,9	281,857
984	1257,15	1297,73	1060,22	530,73	44,750	6,4056	57111,9	46659,4	281,902
985	1258,15	1299,03	1061,33	533,63	44,542	6,4066	57168,8	46708,0	281,948
986	1259,15	1300,32	1062,43	536,54	44,336	6,4076	57225,7	46756,6	281,993
987	1260,15	1301,61	1063,54	539,46	44,131	6,4086	57282,6	46805,2	282,038
988	1261,15	1302,91	1064,64	542,40	43,927	6,4097	57339,6	46853,8	282,083
989	1262,15	1304,20	1065,75	545,35	43,723	6,4107	57396,5	46902,5	282,128
990	1263,15	1305,49	1066,85	548,32	43,521	6,4117	57453,4	46951,1	282,174
991	1264,15	1306,79	1067,96	551,30	43,320	6,4127	57510,4	46999,7	282,219
992	1265,15	1308,08	1069,06	554,29	43,120	6,4138	57567,4	47048,4	282,264
993	1266,15	1309,38	1070,17	557,30	42,921	6,4148	57624,3	47097,0	282,309
994	1267,15	1310,67	1071,27	560,33	42,724	6,4158	57681,3	47145,7	282,354

Carbon dioxide (*Continued*)

t	T	h	u	π_0	θ_0	s°	H	U	S
995	1268,15	1311,97	1072,38	563,36	42,527	6,4168	57738,3	47194,4	282,399
996	1269,15	1313,26	1073,49	566,42	42,331	6,4179	57795,3	47243,0	282,443
997	1270,15	1314,56	1074,59	569,48	42,136	6,4189	57852,3	47291,7	282,488
998	1271,15	1315,85	1075,70	572,56	41,942	6,4199	57909,3	47340,4	282,533
999	1272,15	1317,15	1076,81	575,66	41,750	6,4209	57966,3	47389,1	282,578
1000	1273,15	1318,44	1077,91	578,77	41,558	6,4219	58023,3	47437,8	282,623
1001	1274,15	1319,74	1079,02	581,90	41,367	6,4230	58080,4	47486,6	282,668
1002	1275,15	1321,03	1080,13	585,04	41,177	6,4240	58137,4	47535,3	282,712
1003	1276,15	1322,33	1081,23	588,19	40,989	6,4250	58194,4	47584,0	282,757
1004	1277,15	1323,63	1082,34	591,36	40,801	6,4260	58251,5	47632,7	282,802
1005	1278,15	1324,92	1083,45	594,55	40,614	6,4270	58308,6	47681,5	282,846
1006	1279,15	1326,22	1084,56	597,75	40,428	6,4280	58365,6	47730,3	282,891
1007	1280,15	1327,52	1085,66	600,96	40,243	6,4290	58422,7	47779,0	282,936
1008	1281,15	1328,81	1086,77	604,19	40,059	6,4301	58479,8	47827,8	282,980
1009	1282,15	1330,11	1087,88	607,44	39,876	6,4311	58536,9	47876,6	283,025
1010	1283,15	1331,41	1088,99	610,70	39,694	6,4321	58594,0	47925,4	283,069
1011	1284,15	1332,71	1090,10	613,98	39,513	6,4331	58651,1	47974,2	283,114
1012	1285,15	1334,00	1091,21	617,27	39,333	6,4341	58708,2	48023,0	283,158
1013	1286,15	1335,30	1092,32	620,58	39,154	6,4351	58765,3	48071,8	283,203
1014	1287,15	1336,60	1093,43	623,90	38,976	6,4361	58822,5	48120,6	283,247
1015	1288,15	1337,90	1094,54	627,24	38,798	6,4371	58879,6	48169,4	283,292
1016	1289,15	1339,20	1095,65	630,59	38,622	6,4381	58936,8	48218,2	283,336
1017	1290,15	1340,50	1096,75	633,96	38,446	6,4391	58993,9	48267,1	283,380
1018	1291,15	1341,80	1097,86	637,35	38,272	6,4401	59051,1	48315,9	283,424
1019	1292,15	1343,09	1098,98	640,75	38,098	6,4412	59108,3	48364,8	283,469
1020	1293,15	1344,39	1100,09	644,17	37,925	6,4422	59165,4	48413,7	283,513
1021	1294,15	1345,69	1101,20	647,60	37,753	6,4432	59222,6	48462,5	283,557
1022	1295,15	1346,99	1102,31	651,05	37,582	6,4442	59279,8	48511,4	283,601
1023	1296,15	1348,29	1103,42	654,52	37,412	6,4452	59337,0	48560,3	283,646
1024	1297,15	1349,59	1104,53	658,00	37,243	6,4462	59394,2	48609,2	283,690
1025	1298,15	1350,89	1105,64	661,50	37,074	6,4472	59451,4	48658,1	283,734
1026	1299,15	1352,19	1106,75	665,02	36,907	6,4482	59508,7	48707,0	283,778
1027	1300,15	1353,49	1107,86	668,55	36,740	6,4492	59565,9	48755,9	283,822
1028	1301,15	1354,79	1108,97	672,10	36,574	6,4502	59623,1	48804,8	283,866
1029	1302,15	1356,10	1110,09	675,66	36,409	6,4512	59680,4	48853,8	283,910
1030	1303,15	1357,40	1111,20	679,24	36,245	6,4522	59737,6	48902,7	283,954
1031	1304,15	1358,70	1112,31	682,84	36,082	6,4532	59794,9	48951,7	283,998
1032	1305,15	1360,00	1113,42	686,46	35,919	6,4542	59852,2	49000,6	284,042
1033	1306,15	1361,30	1114,54	690,09	35,758	6,4552	59909,5	49049,6	284,085
1034	1307,15	1362,60	1115,65	693,73	35,597	6,4562	59966,7	49098,6	284,129
1035	1308,15	1363,90	1116,76	697,40	35,437	6,4572	60024,0	49147,5	284,173
1036	1309,15	1365,21	1117,87	701,08	35,278	6,4582	60081,3	49196,5	284,217
1037	1310,15	1366,51	1118,99	704,78	35,119	6,4591	60138,6	49245,5	284,261
1038	1311,15	1367,81	1120,10	708,50	34,962	6,4601	60196,0	49294,5	284,304
1039	1312,15	1369,11	1121,21	712,23	34,805	6,4611	60253,3	49343,5	284,348
1040	1313,15	1370,42	1122,33	715,98	34,649	6,4621	60310,6	49392,5	284,392
1041	1314,15	1371,72	1123,44	719,75	34,494	6,4631	60367,9	49441,6	284,435
1042	1315,15	1373,02	1124,56	723,54	34,340	6,4641	60425,3	49490,6	284,479
1043	1316,15	1374,32	1125,67	727,34	34,186	6,4651	60482,6	49539,6	284,523
1044	1317,15	1375,63	1126,78	731,16	34,033	6,4661	60540,0	49588,7	284,566
1045	1318,15	1376,93	1127,90	735,00	33,881	6,4671	60597,4	49637,7	284,610
1046	1319,15	1378,23	1129,01	738,86	33,730	6,4681	60654,7	49686,8	284,653
1047	1320,15	1379,54	1130,13	742,73	33,579	6,4691	60712,1	49735,9	284,697
1048	1321,15	1380,84	1131,24	746,62	33,430	6,4700	60769,5	49784,9	284,740
1049	1322,15	1382,15	1132,36	750,53	33,281	6,4710	60826,9	49834,0	284,784

Carbon dioxide (*Continued*)

t	T	h	u	π_0	θ_0	s^0	H	U	S
1050	1323,15	1383,45	1133,47	754,46	33,132	6,4720	60884,3	49883,1	284,827
1051	1324,15	1384,76	1134,59	758,41	32,985	6,4730	60941,7	49932,2	284,870
1052	1325,15	1386,06	1135,71	762,37	32,838	6,4740	60999,1	49981,3	284,914
1053	1326,15	1387,37	1136,82	766,35	32,692	6,4750	61056,6	50030,4	284,957
1054	1327,15	1388,67	1137,94	770,36	32,547	6,4760	61114,0	50079,5	285,000
1055	1328,15	1389,98	1139,05	774,37	32,402	6,4769	61171,4	50128,7	285,044
1056	1329,15	1391,28	1140,17	778,41	32,259	6,4779	61228,9	50177,8	285,087
1057	1330,15	1392,59	1141,29	782,47	32,115	6,4789	61286,3	50226,9	285,130
1058	1331,15	1393,89	1142,40	786,54	31,973	6,4799	61343,8	50276,1	285,173
1059	1332,15	1395,20	1143,52	790,64	31,832	6,4809	61401,3	50325,2	285,216
1060	1333,15	1396,50	1144,64	794,75	31,691	6,4818	61458,7	50374,4	285,260
1061	1334,15	1397,81	1145,76	798,88	31,550	6,4828	61516,2	50423,6	285,303
1062	1335,15	1399,12	1146,87	803,03	31,411	6,4838	61573,7	50472,7	285,346
1063	1336,15	1400,42	1147,99	807,20	31,272	6,4848	61631,2	50521,9	285,389
1064	1337,15	1401,73	1149,11	811,38	31,134	6,4858	61688,7	50571,1	285,432
1065	1338,15	1403,04	1150,23	815,59	30,997	6,4867	61746,2	50620,3	285,475
1066	1339,15	1404,34	1151,34	819,82	30,860	6,4877	61803,8	50669,5	285,518
1067	1340,15	1405,65	1152,46	824,06	30,724	6,4887	61861,3	50718,7	285,561
1068	1341,15	1406,96	1153,58	828,33	30,588	6,4897	61918,8	50767,9	285,604
1069	1342,15	1408,27	1154,70	832,61	30,454	6,4906	61976,4	50817,2	285,646
1070	1343,15	1409,57	1155,82	836,91	30,320	6,4916	62033,9	50866,4	285,689
1071	1344,15	1410,88	1156,94	841,24	30,186	6,4926	62091,5	50915,6	285,732
1072	1345,15	1412,19	1158,06	845,58	30,054	6,4936	62149,0	50964,9	285,775
1073	1346,15	1413,50	1159,18	849,94	29,922	6,4945	62206,6	51014,1	285,818
1074	1347,15	1414,81	1160,29	854,32	29,790	6,4955	62264,2	51063,4	285,861
1075	1348,15	1416,11	1161,41	858,72	29,660	6,4965	62321,7	51112,7	285,903
1076	1349,15	1417,42	1162,53	863,15	29,530	6,4974	62379,3	51162,0	285,946
1077	1350,15	1418,73	1163,65	867,59	29,400	6,4984	62436,9	51211,2	285,989
1078	1351,15	1420,04	1164,77	872,05	29,271	6,4994	62494,5	51260,5	286,031
1079	1352,15	1421,35	1165,89	876,53	29,143	6,5003	62552,2	51309,8	286,074
1080	1353,15	1422,66	1167,01	881,03	29,016	6,5013	62609,8	51359,1	286,116
1081	1354,15	1423,97	1168,13	885,56	28,889	6,5023	62667,4	51408,4	286,159
1082	1355,15	1425,28	1169,26	890,10	28,763	6,5033	62725,0	51457,8	286,202
1083	1356,15	1426,59	1170,38	894,66	28,637	6,5042	62782,7	51507,1	286,244
1084	1357,15	1427,90	1171,50	899,25	28,512	6,5052	62840,3	51556,4	286,287
1085	1358,15	1429,21	1172,62	903,85	28,388	6,5061	62898,0	51605,7	286,329
1086	1359,15	1430,52	1173,74	908,48	28,264	6,5071	62955,6	51655,1	286,372
1087	1360,15	1431,83	1174,86	913,12	28,141	6,5081	63013,3	51704,4	286,414
1088	1361,15	1433,14	1175,98	917,79	28,018	6,5090	63071,0	51753,8	286,456
1089	1362,15	1434,45	1177,10	922,48	27,896	6,5100	63128,6	51803,2	286,499
1090	1363,15	1435,76	1178,23	927,19	27,775	6,5110	63186,3	51852,5	286,541
1091	1364,15	1437,07	1179,35	931,92	27,654	6,5119	63244,0	51901,9	286,583
1092	1365,15	1438,38	1180,47	936,67	27,534	6,5129	63301,7	51951,3	286,626
1093	1366,15	1439,69	1181,59	941,44	27,415	6,5138	63359,4	52000,7	286,668
1094	1367,15	1441,00	1182,72	946,24	27,296	6,5148	63417,2	52050,1	286,710
1095	1368,15	1442,32	1183,84	951,05	27,178	6,5158	63474,9	52099,5	286,752
1096	1369,15	1443,63	1184,96	955,89	27,060	6,5167	63532,6	52148,9	286,794
1097	1370,15	1444,94	1186,08	960,75	26,943	6,5177	63590,3	52198,4	286,837
1098	1371,15	1446,25	1187,21	965,63	26,826	6,5186	63648,1	52247,8	286,879
1099	1372,15	1447,56	1188,33	970,53	26,710	6,5196	63705,8	52297,2	286,921
1100	1373,15	1448,88	1189,45	975,45	26,594	6,5206	63763,6	52346,7	286,963
1101	1374,15	1450,19	1190,58	980,40	26,480	6,5215	63821,4	52396,1	287,005
1102	1375,15	1451,50	1191,70	985,37	26,365	6,5225	63879,1	52445,6	287,047
1103	1376,15	1452,81	1192,82	990,36	26,251	6,5234	63936,9	52495,0	287,089
1104	1377,15	1454,13	1193,95	995,37	26,138	6,5244	63994,7	52544,5	287,131

Carbon dioxide (*Continued*)

t	T	h	u	π_0	θ_0	s^0	H	U	S
1105	1378,15	1455,44	1195,07	1000,4	26,026	6,5253	64052,5	52594,0	287,173
1106	1379,15	1456,75	1196,20	1005,4	25,913	6,5263	64110,3	52644,5	287,215
1107	1380,15	1458,07	1197,32	1010,5	25,802	6,5272	64168,1	52693,9	287,257
1108	1381,15	1459,38	1198,45	1015,6	25,691	6,5282	64225,9	52742,4	287,299
1109	1382,15	1460,69	1199,57	1020,7	25,580	6,5291	64283,7	52791,9	287,340
1110	1383,15	1462,01	1200,70	1025,9	25,471	6,5301	64341,5	52841,5	287,382
1111	1384,15	1463,32	1201,82	1031,0	25,361	6,5310	64399,4	52891,0	287,424
1112	1385,15	1464,64	1202,95	1036,2	25,252	6,5320	64457,2	52940,5	287,466
1113	1386,15	1465,95	1204,07	1041,4	25,144	6,5329	64515,0	52990,0	287,508
1114	1387,15	1467,27	1205,20	1046,7	25,036	6,5339	64572,9	53039,6	287,549
1115	1388,15	1468,58	1206,32	1052,0	24,929	6,5348	64630,7	53089,1	287,591
1116	1389,15	1469,90	1207,45	1057,2	24,822	6,5358	64688,6	53138,6	287,633
1117	1390,15	1471,21	1208,58	1062,5	24,716	6,5367	64746,5	53188,2	287,674
1118	1391,15	1472,53	1209,70	1067,9	24,610	6,5377	64804,4	53237,8	287,716
1119	1392,15	1473,84	1210,83	1073,2	24,505	6,5386	64862,2	53287,3	287,758
1120	1393,15	1475,16	1211,95	1078,6	24,400	6,5396	64920,1	53336,9	287,799
1121	1394,15	1476,47	1213,08	1084,0	24,296	6,5405	64978,0	53386,5	287,841
1122	1395,15	1477,79	1214,21	1089,4	24,192	6,5414	65035,9	53436,1	287,882
1123	1396,15	1479,10	1215,34	1094,9	24,089	6,5424	65093,8	53485,7	287,924
1124	1397,15	1480,42	1216,46	1100,4	23,986	6,5433	65151,8	53535,3	287,965
1125	1398,15	1481,74	1217,59	1105,9	23,884	6,5443	65209,7	53584,9	288,007
1126	1399,15	1483,05	1218,72	1111,4	23,783	6,5452	65267,6	53634,5	288,048
1127	1400,15	1484,37	1219,84	1116,9	23,681	6,5461	65325,5	53684,1	288,089
1128	1401,15	1485,68	1220,97	1122,5	23,581	6,5471	65383,5	53733,8	288,131
1129	1402,15	1487,00	1222,10	1128,1	23,480	6,5480	65441,4	53783,4	288,172
1130	1403,15	1488,32	1223,23	1133,7	23,381	6,5490	65499,4	53833,0	288,213
1131	1404,15	1489,64	1224,36	1139,4	23,281	6,5499	65557,4	53882,7	288,255
1132	1405,15	1490,95	1225,48	1145,0	23,183	6,5508	65615,3	53932,3	288,296
1133	1406,15	1492,27	1226,61	1150,7	23,084	6,5518	65673,3	53982,0	288,337
1134	1407,15	1493,59	1227,74	1156,5	22,987	6,5527	65731,3	54031,7	288,378
1135	1408,15	1494,90	1228,87	1162,2	22,889	6,5537	65789,3	54081,3	288,420
1136	1409,15	1496,22	1230,00	1168,0	22,792	6,5546	65847,3	54131,0	288,461
1137	1410,15	1497,54	1231,13	1173,8	22,696	6,5555	65905,3	54180,7	288,502
1138	1411,15	1498,86	1232,26	1179,6	22,600	6,5565	65963,3	54230,4	288,543
1139	1412,15	1500,18	1233,39	1185,4	22,504	6,5574	66021,3	54280,1	288,584
1140	1413,15	1501,50	1234,52	1191,3	22,409	6,5583	66079,3	54329,8	288,625
1141	1414,15	1502,81	1235,65	1197,2	22,315	6,5593	66137,3	54379,5	288,666
1142	1415,15	1504,13	1236,78	1203,1	22,221	6,5602	66195,4	54429,2	288,707
1143	1416,15	1505,45	1237,90	1209,1	22,127	6,5611	66253,4	54479,0	288,748
1144	1417,15	1506,77	1239,03	1215,0	22,034	6,5621	66311,5	54528,7	288,789
1145	1418,15	1508,09	1240,17	1221,0	21,941	6,5630	66369,5	54578,4	288,830
1146	1419,15	1509,41	1241,30	1227,1	21,849	6,5639	66427,6	54628,2	288,871
1147	1420,15	1510,73	1242,43	1233,1	21,757	6,5648	66485,6	54677,9	288,912
1148	1421,15	1512,05	1243,56	1239,2	21,665	6,5658	66543,7	54727,7	288,953
1149	1422,15	1513,37	1244,69	1245,3	21,574	6,5667	66601,8	54777,4	288,994
1150	1423,15	1514,69	1245,82	1251,4	21,484	6,5676	66659,9	54827,2	289,035
1151	1424,15	1516,01	1246,95	1257,6	21,394	6,5686	66718,0	54877,0	289,075
1152	1425,15	1517,33	1248,08	1263,8	21,304	6,5695	66776,0	54926,8	289,116
1153	1426,15	1518,65	1249,21	1270,0	21,215	6,5704	66834,2	54976,6	289,157
1154	1427,15	1519,97	1250,34	1276,2	21,126	6,5713	66892,3	55026,4	289,198
1155	1428,15	1521,29	1251,47	1282,5	21,037	6,5723	66950,4	55076,2	289,238
1156	1429,15	1522,61	1252,61	1288,8	20,949	6,5732	67008,5	55126,0	289,279
1157	1430,15	1523,93	1253,74	1295,1	20,862	6,5741	67066,6	55175,8	289,320
1158	1431,15	1525,25	1254,87	1301,4	20,774	6,5750	67124,8	55225,6	289,360
1159	1432,15	1526,57	1256,00	1307,8	20,688	6,5760	67182,9	55275,4	289,401

Carbon dioxide (*Continued*)

t	T	h	u	π_0	θ_0	s^0	H	U	S
1160	1433,15	1527,89	1257,13	1314,2	20,601	6,5769	67241,0	55325,2	289,442
1161	1434,15	1529,21	1258,27	1320,6	20,515	6,5778	67299,2	55375,1	289,482
1162	1435,15	1530,54	1259,40	1327,1	20,430	6,5787	67357,4	55424,9	289,523
1163	1436,15	1531,86	1260,53	1333,6	20,345	6,5796	67415,5	55474,8	289,563
1164	1437,15	1533,18	1261,67	1340,1	20,260	6,5806	67473,7	55524,6	289,604
1165	1438,15	1534,50	1262,80	1346,6	20,176	6,5815	67531,9	55574,5	289,644
1166	1439,15	1535,82	1263,93	1353,2	20,092	6,5824	67590,1	55624,4	289,685
1167	1440,15	1537,15	1265,06	1359,8	20,008	6,5833	67648,2	55674,2	289,725
1168	1441,15	1538,47	1266,20	1366,4	19,925	6,5842	67706,4	55724,1	289,765
1169	1442,15	1539,79	1267,33	1373,0	19,842	6,5851	67764,6	55774,0	289,806
1170	1443,15	1541,11	1268,47	1379,7	19,760	6,5861	67822,8	55823,9	289,846
1171	1444,15	1542,44	1269,60	1386,4	19,678	6,5870	67881,1	55873,8	289,886
1172	1445,15	1543,76	1270,73	1393,2	19,596	6,5879	67939,3	55923,7	289,927
1173	1446,15	1545,08	1271,87	1399,9	19,515	6,5888	67997,5	55973,6	289,967
1174	1447,15	1546,41	1273,00	1406,7	19,434	6,5897	68055,7	56023,5	290,007
1175	1448,15	1547,73	1274,14	1413,6	19,354	6,5906	68114,0	56073,5	290,048
1176	1449,15	1549,05	1275,27	1420,4	19,274	6,5916	68172,2	56133,4	290,088
1177	1450,15	1550,38	1276,41	1427,3	19,194	6,5925	68230,5	56173,3	290,128
1178	1451,15	1551,70	1277,45	1434,2	19,115	6,5934	68288,7	56223,3	290,168
1179	1452,15	1553,02	1278,68	1441,1	19,036	6,5943	68347,0	56273,2	290,208
1180	1453,15	1554,35	1279,81	1448,1	18,957	6,5952	68405,3	56323,2	290,248
1181	1454,15	1555,67	1280,95	1455,1	18,879	6,5961	68463,5	56373,1	290,288
1182	1455,15	1557,00	1282,08	1462,1	18,801	6,5970	68521,8	56423,1	290,328
1183	1456,15	1558,32	1283,22	1469,2	18,724	6,5979	68580,1	56473,1	290,369
1184	1457,15	1559,64	1284,35	1476,3	18,647	6,5988	68638,4	56523,0	290,409
1185	1458,15	1560,97	1285,49	1483,4	18,570	6,5998	68696,7	56573,0	290,449
1186	1459,15	1562,29	1286,62	1490,6	18,493	6,6007	68755,0	56623,0	290,488
1187	1460,15	1563,62	1287,76	1497,7	18,417	6,6016	68813,3	56673,0	290,528
1188	1461,15	1564,94	1288,90	1504,9	18,342	6,6025	68871,6	56723,0	290,568
1189	1462,15	1566,27	1290,03	1512,2	18,266	6,6034	68929,9	56773,0	290,608
1190	1463,15	1567,59	1291,17	1519,5	18,191	6,6043	68988,3	56823,0	290,648
1191	1464,15	1568,92	1292,31	1526,8	18,117	6,6052	69046,6	56873,1	290,688
1192	1465,15	1570,25	1293,44	1534,1	18,043	6,6061	69104,9	56923,1	290,728
1193	1466,15	1571,57	1294,58	1541,4	17,969	6,6070	69163,3	56973,1	290,768
1194	1467,15	1572,90	1295,72	1548,8	17,895	6,6079	69221,6	57023,2	290,807
1195	1468,15	1574,22	1296,85	1556,3	17,822	6,6088	69280,0	57073,2	290,847
1196	1469,15	1575,55	1297,99	1563,7	17,749	6,6097	69338,4	57123,2	290,887
1197	1470,15	1576,88	1299,13	1571,2	17,676	6,6106	69396,7	57173,3	290,927
1198	1471,15	1578,20	1300,26	1578,7	17,604	6,6115	69455,1	57223,4	290,966
1199	1472,15	1579,53	1301,40	1586,3	17,532	6,6124	69513,5	57273,4	291,006
1200	1473,15	1580,86	1302,54	1593,9	17,461	6,6133	69571,9	57323,5	291,046
1201	1474,15	1582,18	1303,68	1601,5	17,389	6,6142	69630,3	57373,6	291,085
1202	1475,15	1583,51	1304,82	1609,1	17,319	6,6151	69688,7	57423,7	291,125
1203	1476,15	1584,84	1305,95	1616,8	17,248	6,6160	69747,1	57473,8	291,164
1204	1477,15	1586,16	1307,09	1624,5	17,178	6,6169	69805,5	57523,8	291,204
1205	1478,15	1587,49	1308,23	1632,3	17,108	6,6178	69863,9	57573,9	291,244
1206	1479,15	1588,82	1309,37	1640,0	17,038	6,6187	69922,3	57624,1	291,283
1207	1480,15	1590,15	1310,51	1647,8	16,969	6,6196	69980,7	57674,2	291,323
1208	1481,15	1591,47	1311,65	1655,7	16,900	6,6205	70039,2	57724,3	291,362
1209	1482,15	1592,80	1312,79	1663,6	16,831	6,6214	70097,6	57774,4	291,401
1210	1483,15	1594,13	1313,93	1671,5	16,763	6,6223	70156,1	57824,5	291,441
1211	1484,15	1595,46	1315,06	1679,4	16,695	6,6232	70214,5	57874,7	291,480
1212	1485,15	1596,79	1316,20	1687,4	16,627	6,6241	70273,0	57924,8	291,520
1213	1486,15	1598,11	1317,34	1695,5	16,560	6,6250	70331,4	57975,0	291,559
1214	1487,15	1599,44	1318,48	1703,4	16,493	6,6259	70389,9	58025,1	291,598

Carbon dioxide (*Continued*)

t	T	h	u	π_0	θ_0	$s^\bullet$	H	U	S
1215	1488,15	1600,77	1319,62	1711,5	16,426	6,6268	70448,4	58075,3	291,638
1216	1489,15	1602,10	1320,76	1719,6	16,360	6,6277	70506,8	58125,4	291,677
1217	1490,15	1603,43	1321,90	1727,7	16,294	6,6286	70565,3	58175,6	291,716
1218	1491,15	1604,76	1323,04	1735,9	16,228	6,6294	70623,8	58225,8	291,755
1219	1492,15	1606,09	1324,18	1744,1	16,162	6,6303	70682,3	58276,0	291,795
1220	1493,15	1607,42	1325,32	1752,3	16,097	6,6312	70740,8	58326,2	291,834
1221	1494,15	1608,75	1326,46	1760,6	16,032	6,6321	70799,3	58376,3	291,873
1222	1495,15	1610,08	1327,60	1768,9	15,968	6,6330	70857,8	58426,5	291,912
1223	1496,15	1611,41	1328,75	1777,3	15,903	6,6339	70916,3	58476,7	291,951
1224	1497,15	1612,74	1329,89	1785,7	15,839	6,6348	70974,9	58527,0	291,990
1225	1498,15	1614,07	1331,03	1794,1	15,775	6,6357	71033,4	58577,2	292,029
1226	1499,15	1615,40	1332,17	1802,5	15,712	6,6366	71091,9	58627,4	292,068
1227	1500,15	1616,73	1333,31	1811,0	15,649	6,6374	71150,5	58677,6	292,108
1228	1501,15	1618,06	1334,45	1819,5	15,586	6,6383	71209,0	58727,8	292,147
1229	1502,15	1619,39	1335,59	1828,1	15,523	6,6392	71267,6	58778,1	292,186
1230	1503,15	1620,72	1336,73	1836,7	15,461	6,6401	71326,1	58828,3	292,225
1231	1504,15	1622,05	1337,88	1845,3	15,399	6,6410	71384,7	58878,6	292,263
1232	1505,15	1623,38	1339,02	1853,9	15,337	6,6419	71443,3	58928,8	292,302
1233	1506,15	1624,71	1340,16	1862,6	15,276	6,6428	71501,8	58979,1	292,341
1234	1507,15	1626,04	1341,30	1871,4	15,215	6,6436	71560,4	59029,3	292,380
1235	1508,15	1627,37	1342,44	1880,1	15,154	6,6445	71619,0	59079,6	292,419
1236	1509,15	1628,70	1343,59	1888,9	15,093	6,6454	71677,6	59129,9	292,458
1237	1510,15	1630,03	1344,73	1897,8	15,033	6,6463	71736,2	59180,2	292,497
1238	1511,15	1631,37	1345,87	1906,7	14,973	6,6472	71794,8	59230,5	292,535
1239	1512,15	1632,70	1347,01	1915,6	14,913	6,6481	71853,4	59280,8	292,574
1240	1513,15	1634,03	1348,16	1924,5	14,853	6,6489	71912,0	59331,1	292,613
1241	1514,15	1635,36	1349,30	1933,5	14,794	6,6498	71970,6	59381,4	292,652
1242	1515,15	1636,69	1350,44	1942,5	14,735	6,6507	72029,2	59431,7	292,690
1243	1516,15	1638,03	1351,59	1951,6	14,676	6,6516	72087,9	59482,0	292,729
1244	1517,15	1639,36	1352,73	1960,7	14,618	6,6525	72146,5	59532,3	292,768
1245	1518,15	1640,69	1353,87	1969,8	14,560	6,6533	72205,1	59582,6	292,806
1246	1519,15	1642,02	1355,02	1979,0	14,502	6,6542	72263,8	59632,9	292,845
1247	1520,15	1643,36	1356,16	1988,2	14,444	6,6551	72322,4	59683,3	292,884
1248	1521,15	1644,69	1357,30	1997,4	14,387	6,6560	72381,1	59733,6	292,922
1249	1522,15	1646,02	1358,45	2006,7	14,330	6,6568	72439,7	59784,0	292,961
1250	1523,15	1647,35	1359,59	2016,0	14,273	6,6577	72498,4	59834,3	292,999
1251	1524,15	1648,69	1360,74	2025,4	14,216	6,6586	72557,1	59884,7	293,038
1252	1525,15	1650,02	1361,88	2034,8	14,160	6,6595	72615,8	59935,0	293,076
1253	1526,15	1651,35	1363,03	2044,2	14,104	6,6603	72674,4	59985,4	293,115
1254	1527,15	1652,69	1364,17	2053,7	14,048	6,6612	72733,1	60035,8	293,153
1255	1528,15	1654,02	1365,32	2063,2	13,992	6,6621	72791,8	60086,2	293,192
1256	1529,15	1655,36	1366,46	2072,8	13,937	6,6630	72850,5	60136,5	293,230
1257	1530,15	1656,69	1367,61	2082,4	13,882	6,6638	72909,2	60186,9	293,268
1258	1531,15	1658,02	1368,75	2092,0	13,827	6,6647	72967,9	60237,3	293,307
1259	1532,15	1659,36	1369,90	2101,6	13,772	6,6656	73026,6	60287,7	293,345
1260	1533,15	1660,69	1371,04	2111,4	13,718	6,6664	73085,4	60338,1	293,383
1261	1534,15	1662,03	1372,19	2121,1	13,664	6,6673	73144,1	60388,5	293,422
1262	1535,15	1663,36	1373,33	2130,9	13,610	6,6682	73202,8	60439,0	293,460
1263	1536,15	1664,69	1374,48	2140,7	13,556	6,6690	73261,6	60489,4	293,498
1264	1537,15	1666,03	1375,62	2150,6	13,503	6,6699	73320,3	60539,8	293,536
1265	1538,15	1667,36	1376,77	2160,5	13,450	6,6708	73379,0	60590,2	293,575
1266	1539,15	1668,70	1377,92	2170,4	13,397	6,6717	73437,8	60640,7	293,613
1267	1540,15	1670,03	1379,06	2180,4	13,344	6,6725	73496,5	60691,1	293,651
1268	1541,15	1671,37	1380,21	2190,4	13,292	6,6734	73555,3	60741,6	293,689
1269	1542,15	1672,70	1381,35	2200,5	13,239	6,6743	73614,1	60792,0	293,727

Carbon dioxide (Continued)

t	T	h	u	π_e	θ_e	s^0	H	U	S
1270	1543,15	1674,04	1382,50	2210,6	13,187	6,6751	73672,8	60842,5	293,765
1271	1544,15	1675,38	1383,65	2220,8	13,136	6,6760	73731,6	60892,9	293,803
1272	1545,15	1676,71	1384,79	2230,9	13,084	6,6768	73790,4	60943,4	293,841
1273	1546,15	1678,05	1385,94	2241,2	13,033	6,6777	73849,2	60993,9	293,879
1274	1547,15	1679,38	1387,09	2251,4	12,982	6,6786	73908,0	61044,3	293,917
1275	1548,15	1680,72	1388,23	2261,8	12,931	6,6794	73966,8	61094,8	293,955
1276	1549,15	1682,06	1389,38	2272,1	12,880	6,6803	74025,6	61145,3	293,993
1277	1550,15	1683,39	1390,53	2282,5	12,830	6,6812	74084,4	61195,8	294,031
1278	1551,15	1684,73	1391,68	2292,9	12,780	6,6820	74143,2	61246,3	294,069
1279	1552,15	1686,06	1392,82	2303,4	12,730	6,6829	74202,0	61296,8	294,107
1280	1553,15	1687,40	1393,97	2313,9	12,680	6,6837	74260,8	61347,3	294,145
1281	1554,15	1688,74	1395,12	2324,5	12,631	6,6846	74319,7	61397,8	294,183
1282	1555,15	1690,07	1396,27	2335,1	12,581	6,6855	74378,5	61448,3	294,221
1283	1556,15	1691,41	1397,42	2345,7	12,532	6,6863	74437,3	61498,9	294,259
1284	1557,15	1692,75	1398,56	2356,4	12,484	6,6872	74496,2	61549,4	294,296
1285	1558,15	1694,09	1399,71	2367,2	12,435	6,6880	74555,0	61599,9	294,334
1286	1559,15	1695,42	1400,86	2377,9	12,387	6,6889	74613,9	61650,5	294,372
1287	1560,15	1696,76	1402,01	2388,8	12,338	6,6898	74672,7	61701,0	294,410
1288	1561,15	1698,10	1403,16	2399,6	12,290	6,6906	74731,6	61751,6	294,447
1289	1562,15	1699,44	1404,31	2410,5	12,243	6,6915	74790,5	61802,1	294,485
1290	1563,15	1700,77	1405,46	2421,5	12,195	6,6923	74849,3	61852,7	294,523
1291	1564,15	1702,11	1406,60	2432,5	12,148	6,6932	74908,2	61903,2	294,560
1292	1565,15	1703,45	1407,75	2443,5	12,101	6,6940	74967,1	61953,8	294,598
1293	1566,15	1704,79	1408,90	2454,6	12,054	6,6949	75026,0	62004,4	294,636
1294	1567,15	1706,13	1410,05	2465,7	12,007	6,6958	75084,9	62055,0	294,673
1295	1568,15	1707,46	1411,20	2476,9	11,960	6,6966	75143,8	62105,5	294,711
1296	1569,15	1708,80	1412,35	2488,1	11,914	6,6975	75202,7	62156,1	294,748
1297	1570,15	1710,14	1413,50	2499,3	11,868	6,6983	75261,6	62206,7	294,786
1298	1571,15	1711,48	1414,65	2510,6	11,822	6,6992	75320,5	62257,3	294,823
1299	1572,15	1712,82	1415,80	2522,0	11,776	6,7000	75379,4	62307,9	294,861
1300	1573,15	1714,16	1416,95	2533,4	11,731	6,7009	75438,3	62358,5	294,898
1301	1574,15	1715,50	1418,10	2544,8	11,686	6,7017	75497,3	62409,1	294,936
1302	1575,15	1716,84	1419,25	2556,3	11,641	6,7026	75556,2	62459,8	294,973
1303	1576,15	1718,17	1420,40	2567,8	11,596	6,7034	75615,1	62510,4	295,011
1304	1577,15	1719,51	1421,55	2579,4	11,551	6,7043	75674,1	62561,0	295,048
1305	1578,15	1720,85	1422,70	2591,0	11,506	6,7051	75733,0	62611,6	295,085
1306	1579,15	1722,19	1423,85	2602,7	11,462	6,7060	75792,0	62662,3	295,123
1307	1580,15	1723,53	1425,00	2614,4	11,418	6,7068	75850,9	62712,9	295,160
1308	1581,15	1724,87	1426,15	2626,1	11,374	6,7077	75909,9	62763,6	295,197
1309	1582,15	1726,21	1427,30	2637,9	11,330	6,7085	75968,9	62814,2	295,235
1310	1583,15	1727,55	1428,46	2649,8	11,287	6,7094	76027,8	62864,9	295,272
1311	1584,15	1728,89	1429,61	2661,7	11,244	6,7102	76086,8	62915,5	295,309
1312	1585,15	1730,23	1430,76	2673,6	11,200	6,7110	76145,8	62966,2	295,346
1313	1586,15	1731,57	1431,91	2685,6	11,157	6,7119	76204,8	63016,9	295,384
1314	1587,15	1732,91	1433,06	2697,7	11,115	6,7127	76263,8	63067,6	295,421
1315	1588,15	1734,25	1434,21	2709,7	11,072	6,7136	76322,8	63118,2	295,458
1316	1589,15	1735,59	1435,36	2721,9	11,030	6,7144	76381,8	63168,9	295,495
1317	1590,15	1736,94	1436,52	2734,0	10,987	6,7153	76440,8	63219,6	295,532
1318	1591,15	1738,28	1437,67	2746,3	10,945	6,7161	76499,8	63270,3	295,569
1319	1592,15	1739,62	1438,82	2758,5	10,903	6,7170	76558,8	63321,0	295,606
1320	1593,15	1740,96	1439,97	2770,9	10,862	6,7178	76617,8	63371,7	295,643
1321	1594,15	1742,30	1441,12	2783,2	10,820	6,7186	76676,9	63422,4	295,680
1322	1595,15	1743,64	1442,28	2795,7	10,779	6,7195	76735,9	63473,2	295,717
1323	1596,15	1744,98	1443,43	2808,1	10,738	6,7203	76794,9	63523,9	295,754
1324	1597,15	1746,32	1444,58	2820,6	10,697	6,7212	76854,0	63574,6	295,791

Carbon dioxide (*Continued*)

t	T	h	u	π_0	$\dot\theta_0$	s^0	H	U	S
1325	1598,15	1747,67	1445,73	2833,2	10,656	6,7220	76913,0	63625,3	295,828
1326	1599,15	1749,01	1446,89	2845,8	10,616	6,7228	76972,1	63676,1	295,865
1327	1600,15	1750,35	1448,04	2858,5	10,575	6,7237	77031,1	63726,8	295,902
1328	1601,15	1751,69	1449,19	2871,2	10,535	6,7245	77090,2	63777,5	295,939
1329	1602,15	1753,03	1450,35	2884,0	10,495	6,7254	77149,2	63828,3	295,976
1330	1603,15	1754,38	1451,50	2896,8	10,455	6,7262	77208,3	63879,0	296,013
1331	1604,15	1755,72	1452,65	2909,6	10,415	6,7270	77267,4	63929,8	296,050
1332	1605,15	1757,06	1453,81	2922,6	10,376	6,7279	77326,4	63980,6	296,087
1333	1606,15	1758,40	1454,96	2935,5	10,336	6,7287	77385,5	64031,3	296,123
1334	1607,15	1759,74	1456,11	2948,5	10,297	6,7295	77444,6	64082,1	296,160
1335	1608,15	1761,09	1457,27	2961,6	10,258	6,7304	77503,7	64132,9	296,197
1336	1609,15	1762,43	1458,42	2974,7	10,219	6,7312	77562,8	64183,7	296,234
1337	1610,15	1763,77	1459,58	2987,9	10,180	6,7320	77621,9	64234,4	296,270
1338	1611,15	1765,12	1460,73	3001,1	10,142	6,7329	77681,0	64285,2	296,307
1339	1612,15	1766,46	1461,88	3014,4	10,104	6,7337	77740,1	64336,0	296,344
1340	1613,15	1767,80	1463,04	3027,7	10,065	6,7345	77799,2	64386,8	296,380
1341	1614,15	1769,15	1464,19	3041,0	10,027	6,7354	77858,3	64437,6	296,417
1342	1615,15	1770,49	1465,35	3054,5	9,9898	6,7362	77917,5	64488,4	296,454
1343	1616,15	1771,83	1466,50	3067,9	9,9521	6,7370	77976,6	64539,3	296,490
1344	1617,15	1773,18	1467,66	3081,5	9,9145	6,7379	78035,7	64590,1	296,527
1345	1618,15	1774,52	1468,81	3095,1	9,8772	6,7387	78094,9	64640,9	296,563
1346	1619,15	1775,86	1469,97	3108,7	9,8399	6,7395	78154,0	64691,7	296,600
1347	1620,15	1777,21	1471,12	3122,4	9,8028	6,7404	78213,2	64742,6	296,636
1348	1621,15	1778,55	1472,28	3136,1	9,7659	6,7412	78272,3	64793,4	296,673
1349	1622,15	1779,90	1473,43	3149,9	9,7292	6,7420	78331,5	64844,2	296,709
1350	1623,15	1781,24	1474,59	3163,7	9,6926	6,7428	78390,6	64895,1	296,746
1351	1624,15	1782,59	1475,74	3177,6	9,6561	6,7437	78449,8	64945,9	296,782
1352	1625,15	1783,93	1476,90	3191,6	9,6199	6,7445	78509,0	64996,8	296,819
1353	1626,15	1785,27	1478,05	3205,6	9,5837	6,7453	78568,1	65047,7	296,855
1354	1627,15	1786,62	1479,21	3219,6	9,5477	6,7462	78627,3	65098,5	296,891
1355	1628,15	1787,96	1480,37	3233,8	9,5119	6,7470	78686,5	65149,4	296,928
1356	1629,15	1789,31	1481,52	3247,9	9,4763	6,7478	78745,7	65200,3	296,964
1357	1630,15	1790,65	1482,68	3262,1	9,4407	6,7486	78804,9	65251,1	297,000
1358	1631,15	1792,00	1483,83	3276,4	9,4054	6,7495	78864,1	65302,0	297,037
1359	1632,15	1793,34	1484,99	3290,7	9,3702	6,7503	78923,3	65352,9	297,073
1360	1633,15	1794,69	1486,15	3305,1	9,3351	6,7511	78982,5	65403,8	297,109
1361	1634,15	1796,03	1487,30	3319,6	9,3002	6,7519	79041,7	65454,7	297,146
1362	1635,15	1797,38	1488,46	3334,1	9,2654	6,7528	79100,9	65505,6	297,182
1363	1636,15	1798,73	1489,62	3348,6	9,2308	6,7536	79160,1	65556,5	297,218
1364	1637,15	1800,07	1490,77	3363,2	9,1963	6,7544	79219,3	65607,4	297,254
1365	1638,15	1801,42	1491,93	3377,9	9,1620	6,7552	79278,6	65658,3	297,290
1366	1639,15	1802,76	1493,09	3392,6	9,1278	6,7560	79337,8	65709,2	297,327
1367	1640,15	1804,11	1494,24	3407,4	9,0938	6,7569	79397,0	65760,2	297,363
1368	1641,15	1805,46	1495,40	3422,2	9,0599	6,7577	79456,3	65811,1	297,399
1369	1642,15	1806,80	1496,56	3437,1	9,0262	6,7585	79515,5	65862,0	297,435
1370	1643,15	1808,15	1497,72	3452,0	8,9926	6,7593	79574,8	65913,0	297,471
1371	1644,15	1809,49	1498,87	3467,0	8,9591	6,7601	79634,0	65963,9	297,507
1372	1645,15	1810,84	1500,03	3482,1	8,9258	6,7610	79693,3	66014,9	297,543
1373	1646,15	1812,19	1501,19	3497,2	8,8926	6,7618	79752,6	66065,8	297,579
1374	1647,15	1813,53	1502,35	3512,4	8,8596	6,7626	79811,8	66116,8	297,615
1375	1648,15	1814,88	1503,50	3527,6	8,8267	6,7634	79871,1	66167,7	297,651
1376	1649,15	1816,23	1504,66	3542,9	8,7939	6,7642	79930,4	66218,7	297,687
1377	1650,15	1817,58	1505,82	3558,2	8,7613	6,7650	79989,7	66269,6	297,723
1378	1651,15	1818,92	1506,98	3573,7	8,7288	6,7659	80049,0	66320,6	297,759
1379	1652,15	1820,27	1508,14	3589,1	8,6965	6,7667	80108,2	66371,6	297,795

Carbon dioxide (*Continued*)

t	T	h	u	π_0	θ_0	s°	H	U	S
1380	1653,15	1821,62	1509,30	3604,6	8,6643	6,7675	80167,5	66422,6	297,831
1381	1654,15	1822,96	1510,45	3620,2	8,6322	6,7683	80226,8	66473,6	297,866
1382	1655,15	1824,31	1511,61	3635,9	8,6003	6,7691	80286,1	66524,6	297,902
1383	1656,15	1825,66	1512,77	3651,6	8,5685	6,7699	80345,5	66575,5	297,938
1384	1657,15	1827,01	1513,93	3667,3	8,5368	6,7707	80404,8	66626,5	297,974
1385	1658,15	1828,36	1515,09	3683,1	8,5053	6,7716	80464,1	66677,5	298,010
1386	1659,15	1829,70	1516,25	3699,0	8,4739	6,7724	80523,4	66728,6	298,045
1387	1660,15	1831,05	1517,41	3714,9	8,4426	6,7732	80582,7	66779,6	298,081
1388	1661,15	1832,40	1518,57	3730,9	8,4115	6,7740	80642,1	66830,6	298,117
1389	1662,15	1833,75	1519,73	3747,0	8,3805	6,7748	80701,4	66881,6	298,153
1 390	1663,15	1835,10	1520,89	3763,1	8,3496	6,7756	80760,7	66932,6	298,188
1391	1664,15	1836,44	1522,04	3779,3	8,3188	6,7764	80820,1	66983,7	298,224
1392	1665,15	1837,79	1523,20	3795,5	8,2882	6,7772	80879,4	67034,7	298,260
1393	1666,15	1839,14	1524,36	3811,8	8,2577	6,7781	80938,8	67085,7	298,295
1394	1667,15	1840,49	1525,52	3828,2	8,2274	6,7789	80998,1	67136,8	298,331
1395	1668,15	1841,84	1526,68	3844,6	8,1971	6,7797	81057,5	67187,8	298,366
1396	1669,15	1843,19	1527,84	3861,1	8,1670	6,7805	81116,9	67238,9	298,402
1397	1670,15	1844,54	1529,00	3877,7	8,1371	6,7813	81176,2	67289,9	298,438
1398	1671,15	1845,89	1530,16	3894,3	8,1072	6,7821	81235,6	67341,0	298,473
1399	1672,15	1847,24	1531,32	3910,9	8,0775	6,7829	81295,0	67392,1	298,509
1400	1673,15	1848,58	1532,48	3927,7	8,0479	6,7837	81354,4	67443,1	298,544
1401	1674,15	1849,93	1533,65	3944,5	8,0184	6,7845	81413,8	67494,2	298,580
1402	1675,15	1851,28	1534,81	3961,3	7,9890	6,7853	81473,2	67545,3	298,615
1403	1676,15	1852,63	1535,97	3978,3	7,9598	6,7861	81532,5	67596,4	298,651
1404	1677,15	1853,98	1537,13	3995,3	7,9307	6,7869	81592,0	67647,4	298,686
1405	1678,15	1855,33	1538,29	4012,3	7,9017	6,7877	81651,4	67698,5	298,721
1406	1679,15	1856,68	1539,45	4029,4	7,8728	6,7885	81710,8	67749,6	298,757
1407	1680,15	1858,03	1540,61	4046,6	7,8440	6,7893	81770,2	67800,7	298,792
1408	1681,15	1859,38	1541,77	4063,8	7,8154	6,7901	81829,6	67851,8	298,828
1409	1682,15	1860,73	1542,93	4081,2	7,7869	6,7909	81889,0	67902,9	298,863
1410	1683,15	1862,08	1544,09	4098,5	7,7585	6,7918	81948,4	67954,0	298,898
1411	1684,15	1863,43	1545,26	4116,0	7,7302	6,7926	82007,9	68005,2	298,933
1412	1685,15	1864,78	1546,42	4133,5	7,7020	6,7934	82067,3	68056,3	298,969
1413	1686,15	1866,14	1547,58	4151,0	7,6740	6,7942	82126,7	68107,4	299,004
1414	1687,15	1867,49	1548,74	4168,7	7,6461	6,7950	82186,2	68158,5	299,039
1415	1688,15	1868,84	1549,90	4186,4	7,6183	6,7958	82245,6	68209,7	299,074
1416	1689,15	1870,19	1551,06	4204,1	7,5906	6,7966	82305,1	68260,8	299,110
1417	1690,15	1871,54	1552,23	4222,0	7,5630	6,7974	82364,5	68311,9	299,145
1418	1691,15	1872,89	1553,39	4239,9	7,5355	6,7982	82424,0	68363,1	299,180
1419	1692,15	1874,24	1554,55	4257,8	7,5082	6,7990	82483,5	68414,2	299,215
1420	1693,15	1875,59	1555,71	4275,9	7,4809	6,7998	82542,9	68465,4	299,250
1421	1694,15	1876,94	1556,88	4294,0	7,4538	6,8006	82602,4	68516,6	299,285
1422	1695,15	1878,30	1558,04	4312,1	7,4268	6,8013	82661,9	68567,7	299,321
1423	1696,15	1879,65	1559,20	4330,3	7,3999	6,8021	82721,4	68618,9	299,356
1424	1697,15	1881,00	1560,36	4348,6	7,3731	6,8029	82780,9	68670,1	299,391
1425	1698,15	1882,35	1561,53	4367,0	7,3464	6,8037	82840,4	68721,2	299,426
1426	1699,15	1883,70	1562,69	4385,5	7,3198	6,8045	82899,9	68772,4	299,461
1427	1700,15	1885,05	1563,85	4404,0	7,2933	6,8053	82959,4	68823,6	299,496
1428	1701,15	1886,41	1565,02	4422,5	7,2670	6,8061	83018,9	68874,8	299,531
1429	1702,15	1887,76	1566,18	4441,2	7,2407	6,8069	83078,4	68926,0	299,566
1430	1703,15	1889,11	1567,34	4459,9	7,2146	6,8077	83137,9	68977,2	299,601
1431	1704,15	1890,46	1568,51	4478,7	7,1886	6,8085	83197,4	69028,4	299,636
1432	1705,15	1891,82	1569,67	4497,5	7,1626	6,8093	83256,9	69079,5	299,671
1433	1706,15	1893,17	1570,83	4516,4	7,1368	6,8101	83316,4	69130,8	299,705
1434	1707,15	1894,52	1572,00	4535,4	7,1111	6,8109	83376,0	69182,0	299,740
1435	1708,15	1895,87	1573,16	4554,5	7,0855	6,8117	83435,5	69233,2	299,775
1436	1709,15	1897,23	1574,32	4573,6	7,0600	6,8125	83495,0	69284,5	299,810
1437	1710,15	1898,58	1575,49	4592,8	7,0346	6,8133	83554,6	69335,7	299,845
1438	1711,15	1899,93	1576,65	4612,1	7,0093	6,8141	83614,1	69386,9	299,880
1439	1712,15	1901,29	1577,82	4631,4	6,9841	6,8148	83673,7	69438,2	299,914

Carbon dioxide (*Continued*)

t	T	h	u	π_0	θ_0	s°	H	U	S
1440	1713,15	1902,64	1578,98	4650,8	6,9590	6,8156	83733,2	69489,4	299,949
1441	1714,15	1903,99	1580,15	4670,3	6,9341	6,8164	83792,8	69540,6	299,984
1442	1715,15	1905,35	1581,31	4689,8	6,9092	6,8172	83852,4	69591,9	300,019
1443	1716,15	1906,70	1582,48	4709,5	6,8844	6,8180	83911,9	69643,1	300,053
1444	1717,15	1908,05	1583,64	4729,2	6,8597	6,8188	83971,5	69694,4	300,088
1445	1718,15	1909,41	1584,80	4748,9	6,8351	6,8196	84031,1	69745,7	300,123
1446	1719,15	1910,76	1585,97	4768,8	6,8107	6,8204	84090,7	69796,9	300,158
1447	1720,15	1912,11	1587,13	4788,7	6,7863	6,8212	84150,2	69848,2	300,192
1448	1721,15	1913,47	1588,30	4808,7	6,7620	6,8219	84209,8	69899,5	300,227
1449	1722,15	1914,82	1589,46	4828,7	6,7378	6,8227	84269,4	69950,8	300,261
1450	1723,15	1916,18	1590,63	4848,9	6,7137	6,8235	84329,0	70002,0	300,296
1451	1724,15	1917,53	1591,80	4869,1	6,6897	6,8243	84388,6	70053,3	300,331
1452	1725,15	1918,89	1592,96	4889,4	6,6659	6,8251	84448,2	70104,6	300,365
1453	1726,15	1920,24	1594,13	4909,7	6,6421	6,8259	84507,8	70155,9	300,400
1454	1727,15	1921,59	1595,29	4930,2	6,6184	6,8267	84567,5	70207,2	300,434
1455	1728,15	1922,95	1596,46	4950,7	6,5948	6,8274	84627,1	70258,5	300,469
1456	1729,15	1924,30	1597,62	4971,2	6,5713	6,8282	84686,7	70309,8	300,503
1457	1730,15	1925,66	1598,79	4991,9	6,5479	6,8290	84746,3	70361,1	300,538
1458	1731,15	1927,01	1599,96	5012,6	6,5246	6,8298	84806,0	70412,5	300,572
1459	1732,15	1928,37	1601,12	5033,4	6,5013	6,8306	84865,6	70463,8	300,607
1460	1733,15	1929,72	1602,29	5054,3	6,4782	6,8314	84925,2	70515,1	300,641
1461	1734,15	1931,08	1603,45	5075,3	6,4552	6,8321	84984,9	70566,4	300,675
1462	1735,15	1932,43	1604,62	5096,3	6,4323	6,8329	85044,5	70617,8	300,710
1463	1736 15	1933,79	1605,79	5117,4	6,4094	6,8337	85104,2	70669,1	300,744
1464	1737,15	1935,15	1606,95	5138,6	6,3867	6,8345	85163,8	70720,5	300,779
1465	1738,15	1936,50	1608,12	5159,9	6,3640	6,8353	85223,5	70771,8	300,813
1466	1739,15	1937,86	1609,29	5181,2	6,3414	6,8360	85283,2	70823,2	300,847
1467	1740,15	1939,21	1610,46	5202,6	6,3190	6,8368	85342,8	70874,5	300,881
1468	1741,15	1940,57	1611,62	5224,1	6,2966	6,8376	85402,5	70925,9	300,916
1469	1742,15	1941,93	1612,79	5245,7	6,2743	6,8384	85462,2	70977,2	300,950
1470	1743,15	1943,28	1613,96	5267,4	6,2521	6,8392	85521,9	71028,6	300,984
1471	1744,15	1944,64	1615,12	5289,1	6,2300	6,8399	85581,6	71080,0	301,018
1472	1745,15	1945,99	1616,29	5310,9	6,2079	6,8407	85641,3	71131,4	301,053
1473	1746,15	1947,35	1617,46	5332,8	6,1860	6,8415	85701,0	71182,7	301,087
1474	1747,15	1948,71	1618,63	5354,8	6,1641	6,8423	85760,7	71234,1	301,121
1475	1748,15	1950,06	1619,79	5376,8	6,1424	6,8430	85820,4	71285,5	301,155
1476	1749,15	1951,42	1620,96	5398,9	6,1207	6,8438	85880,1	71336,9	301,189
1477	1750,15	1952,78	1622,13	5421,1	6.0991	6,8446	85939,8	71388,3	301,224
1478	1751,15	1954,13	1623,30	5443,4	6,0776	6,8454	85999,5	71439,7	301,258
1479	1752,15	1955,49	1624,47	5465,8	6,0562	6,8461	86059,2	71491,1	301,292
1480	1753,15	1956,85	1625,63	5488,2	6,0349	6,8469	86119,0	71542,5	301,326
1481	1754,15	1958,21	1626,80	5510,8	6,0136	6,8477	86178,7	71594,0	301,360
1482	1755,15	1959,56	1627,97	5533,4	5,9925	6,8485	86238,4	71645,4	301,394
1483	1756,15	1960,92	1629,14	5556,1	5,9714	6,8492	86298,2	71696,8	301,428
1484	1757,15	1962,28	1630,31	5578,9	5,9504	6,8500	86357,9	71748,2	301,462
1485	1758,15	1963,64	1631,48	5601,7	5,9295	6,8508	86417,7	71799,7	301,496
1486	1759,15	1964,99	1632,65	5624,6	5,9087	6,8516	86477,4	71851,1	301,530
1487	1760,15	1966,35	1633,81	5647,7	5,8879	6,8523	86537,2	71902,6	301,564
1488	1761,15	1967,71	1634,98	5670,8	5,8673	6,8531	86596,9	71954,0	301,598
1489	1762,15	1969,07	1636,15	5694,0	5,8467	6,8539	86656,7	72005,5	301,632
1490	1763,15	1970,43	1637,32	5717,2	5,8262	6,8546	86716,5	72056,9	301,666
1491	1764,15	1971,78	1638,49	5740,6	5,8058	6,8554	86776,2	72108,4	301,700
1492	1765,15	1973,14	1639,66	5764,0	5,7855	6,8562	86836,0	72159,8	301,733
1493	1766,15	1974,50	1640,83	5787,5	5,7652	6,8569	86895,8	72211,3	301,767
1494	1767,15	1975,86	1642,00	5811,1	5,7451	6,8577	86955,6	72262,8	301,801
1495	1768,15	1977,22	1643,17	5834,8	5,7250	6,8585	87015,4	72314,3	301,835
1496	1769,15	1978,58	1644,34	5858,6	5,7050	6,8593	87075,2	72365,8	301,869
1497	1770,15	1979,94	1645,51	5882,5	5,6850	6,8600	87135,0	72417,2	301,903
1498	1771,15	1981,29	1646,68	5906,4	5,6652	6,8608	87194,8	72468,7	301,936
1499	1772,15	1982,65	1647,85	5930,5	5,6454	6,8616	87254,6	72520,2	301,970
1500	1773,15	1984,01	1649,02	5954,6	5,6257	6,8623	87314,4	72571,7	302,004

Table III.11 Steam ($\mu = 18.016$)

t	T	c_p	μc_p	c_v	μc_v	$k\dfrac{c_p}{c_v}$
0	273,15	1,8597	33,504	1,3982	25,190	1,330
25	298,15	1,8644	33,590	1,4030	25,276	1,329
50	323,15	1,8714	33,715	1,4099	25,401	1,327
75	348,15	1,8800	33,870	1,4185	25,556	1,325
100	373,15	1,8900	34,051	1,4286	25,737	1,323
125	398,15	1,9012	34,252	1,4397	25,938	1,321
150	423,15	1,9134	34,471	1,4519	26,157	1,318
175	448,15	1,9263	34,704	1,4648	26,390	1,315
200	473,15	1,9399	34,949	1,4784	26,635	1,312
250	523,15	1,9688	35,470	1,5073	27,156	1,306
300	573,15	1,9994	36,022	1,5380	27,708	1,300
350	623,15	2,0315	36,599	1,5700	28,285	1,294
400	673,15	2,0646	37,195	1,6031	28,881	1,288
450	723,15	2,0984	37,805	1,6369	29,491	1,282
500	773,15	2,1329	38,426	1,6714	30,112	1,276
550	823,15	2,1677	39,054	1,7063	30,740	1,270
600	873,15	2,2030	39,689	1,7415	31,375	1,265
650	923,15	2,2383	40,326	1,7769	32,012	1,260
700	973,15	2,2738	40,964	1,8123	32,650	1,255
750	1023,15	2,3091	41,600	1,8476	33,286	1,250
800	1073,15	2,3441	42,232	1,8827	33,918	1,245
850	1123,15	2,3788	42,857	1,9174	34,543	1,241
900	1173,15	2,4130	43,472	1,9515	35,158	1,236
950	1223,15	2,4466	44,077	1,9851	35,763	1,232
1000	1273,15	2,4793	44,667	2,0178	36,353	1,229
1050	1323,15	2,5112	45,241	2,0497	36,927	1,225
1100	1373,15	2,5420	45,797	2,0805	37,483	1,222
1150	1423,15	2,5718	46,333	2,1103	38,019	1,219
1200	1473,15	2,6004	46,849	2,1389	38,535	1,216
1250	1523,15	2,6279	47,344	2,1664	39,030	1,213
1300	1573,15	2,6541	47,817	2,1927	39,503	1,210
1350	1623,15	2,6793	48,270	2,2178	39,956	1,208
1400	1673,15	2,7034	48,704	2,2419	40,390	1,206
1450	1723,15	2,7267	49,124	2,2652	40,810	1,204
1500	1773,15	2,7494	49,534	2,2880	41,220	1,202

Table III.12 Steam ($\mu = 18.016$)

t	T	h	u	π_0	θ_0	s^0	H	U	S
−50	223,15	409,89	306,90	0,2239	45988	9,9359	7384,5	5529,2	179,005
−49	224,15	411,75	308,30	0,2280	45369	9,9442	7418,0	5554,3	179,155
−48	225,15	413,61	309,70	0,2321	44762	9,9525	7451,5	5579,5	179,304
−47	226,15	415,46	311,10	0,2363	44165	9,9607	7485,0	5604,7	179,452
−46	227,15	417,32	312,49	0,2405	43579	9,9689	7518,5	5629,9	179,600
−45	228,15	419,18	313,89	0,2448	43003	9,9771	7552,0	5655,0	179,747
−44	229,15	421,04	315,29	0,2491	42438	9,9852	7585,5	5680,2	179,893
−43	230,15	422,90	316,68	0,2536	41882	9,9933	7618,9	5705,4	180,039
−42	231,15	424,76	318,08	0,2580	41336	10,0013	7652,4	5730,5	180,184
−41	232,15	426,62	319,48	0,2625	40799	10,0094	7685,9	5755,7	180,329
−40	233,15	428,47	320,87	0,2671	40272	10,0174	7719,4	5780,9	180,473
−39	234,15	430,33	322,27	0,2718	39754	10,0253	7752,8	5806,0	180,616
−38	235,15	432,19	323,67	0,2765	39244	10,0332	7786,3	5831,2	180,759
−37	236,15	434,05	325,06	0,2812	38744	10,0411	7819,8	5856,3	180,901
−36	237,15	435,91	326,46	0,2861	38252	10,0490	7853,3	5881,5	181,042
−35	238,15	437,76	327,86	0,2910	37768	10,0568	7886,7	5906,7	181,183
−34	239,15	439,62	329,25	0,2959	37292	10,0646	7920,2	5931,8	181,323
−33	240,15	441,48	330,65	0,3009	36824	10,0723	7953,7	5957,0	181,463
−32	241,15	443,34	332,04	0,3060	36364	10,0800	7987,1	5982,1	181,602
−31	242,15	445,19	333,44	0,3111	35912	10,0877	8020,6	6007,3	181,740
−30	243,15	447,05	334,84	0,3163	35467	10,0954	8054,1	6032,4	181,878
−29	244,15	448,91	336,23	0,3216	35029	10,1030	8087,5	6057,6	182,016
−28	245,15	450,77	337,63	0,3269	34599	10,1106	8121,0	6082,7	182,152
−27	246,15	452,62	339,03	0,3324	34175	10,1182	8154,5	6107,9	182,289
−26	247,15	454,48	340,42	0,3378	33759	10,1257	8187,9	6133,0	182,424
−25	248,15	456,34	341,82	0,3434	33349	10,1332	8221,4	6158,2	182,559
−24	249,15	458,20	343,21	0,3490	32945	10,1407	8254,9	6183,3	182,694
−23	250,15	460,05	344,61	0,3546	32549	10,1481	8288,3	6208,5	182,828
−22	251,15	461,91	346,01	0,3604	32158	10,1555	8321,8	6233,6	182,962
−21	252,15	463,77	347,40	0,3662	31774	10,1629	8355,3	6258,8	183,095
−20	253,15	465,63	348,80	0,3721	31396	10,1702	8388,7	6284,0	183,227
−19	254,15	467,49	350,19	0,3780	31023	10,1776	8422,2	6309,1	183,359
−18	255,15	469,34	351,59	0,3840	30657	10,1849	8455,7	6334,3	183,491
−17	256,15	471,20	352,99	0,3901	30296	10,1921	8489,2	6359,4	183,621
−16	257,15	473,06	354,38	0,3963	29941	10,1994	8522,6	6384,6	183,752
−15	258,15	474,92	355,78	0,4026	29591	10,2066	8556,1	6409,7	183,882
−14	259,15	476,77	357,18	0,4089	29247	10,2138	8589,6	6434,9	184,011
−13	260,15	478,63	358,57	0,4153	28908	10,2209	8623,1	6460,1	184,140
−12	261,15	480,49	359,97	0,4217	28575	10,2281	8656,5	6485,2	184,269
−11	262,15	482,35	361,37	0,4283	28246	10,2352	8690,0	6510,4	184,397
−10	263,15	484,21	362,76	0,4349	27922	10,2422	8723,5	6535,6	184,524
−9	264,15	486,07	364,16	0,4416	27604	10,2493	8757,0	6560,7	184,651
−8	265,15	487,93	365,56	0,4483	27290	10,2563	8790,5	6585,9	184,778
−7	266,15	489,78	366,96	0,4552	26980	10,2633	8823,9	6611,1	184,904
−6	267,15	491,64	368,35	0,4621	26676	10,2703	8857,4	6636,2	185,029
−5	268,15	493,50	369,75	0,4691	26376	10,2772	8890,9	6661,4	185,154
−4	269,15	495,36	371,15	0,4762	26080	10,2841	8924,4	6686,6	185,279
−3	270,15	497,22	372,55	0,4834	25789	10,2910	8957,9	6711,8	185,403
−2	271,15	499,08	373,94	0,4906	25502	10,2979	8991,4	6737,0	185,527
−1	272,15	500,94	375,34	0,4980	25219	10,3047	9024,9	6762,1	185,650
0	273,15	502,80	376,74	0,5054	24940	10,3116	9058,4	6787,3	185,773
1	274,15	504,66	378,14	0,5129	24666	10,3184	9091,9	6812,5	185,896
2	275,15	506,52	379,54	0,5205	24395	10,3251	9125,4	6837,7	186,018
3	276,15	508,38	380,93	0,5281	24128	10,3319	9158,9	6862,9	186,139
4	277,15	510,24	382,33	0,5359	23865	10,3386	9192,4	6888,1	186,260

Steam (*Continued*)

t	T	h	u	π_0	θ_0	$s^\bullet$	H	U	S
5	278,15	512,10	383,73	0,5437	23606	10,3453	9226,0	6913,3	186,381
6	279,15	513,96	385,13	0,5516	23351	10,3520	9259,5	6938,5	186,501
7	280,15	515,82	386,53	0,5597	23099	10,3586	9293,0	6963,7	186,621
8	281,15	517,68	387,93	0,5678	22851	10,3653	9326,5	6988,9	186,741
9	282,15	519,54	389,33	0,5759	22606	10,3719	9360,0	7014,1	186,860
10	283,15	521,40	390,73	0,5842	22365	10,3785	9393,6	7039,3	186,978
11	284,15	523,26	392,13	0,5926	22127	10,3850	9427,1	7064,6	187,096
12	285,15	525,12	393,53	0,6010	21893	10,3916	9460,6	7089,8	187,214
13	286,15	526,99	394,93	0,6096	21661	10,3981	9494,2	7115,0	187,332
14	287,15	528,85	396,33	0,6182	21433	10,4046	9527,7	7140,2	187,449
15	288,15	530,71	397,73	0,6270	21208	10,4110	9561,3	7165,5	187,565
16	289,15	532,57	399,13	0,6358	20986	10,4175	9594,8	7190,7	187,682
17	290,15	534,44	400,53	0,6447	20768	10,4239	9628,4	7216,0	187,797
18	291,15	536,30	401,93	0,6537	20552	10,4303	9661,9	7241,2	187,913
19	292,15	538,16	403,33	0,6628	20339	10,4367	9695,5	7266,4	188,028
20	293,15	540,02	404,73	0,6721	20129	10,4431	9729,1	7291,7	188,143
21	294,15	541,89	406,14	0,6814	19922	10,4494	9762,6	7317,0	188,257
22	295,15	543,75	407,54	0,6908	19717	10,4558	9796,2	7342,2	188,371
23	296,15	545,61	408,94	0,7003	19516	10,4621	9829,8	7367,5	188,484
24	297,15	547,48	410,34	0,7099	19317	10,4683	9863,4	7392,7	188,598
25	298,15	549,34	411,75	0,7196	19121	10,4746	9897,0	7418,0	188,711
26	299,15	551,21	413,15	0,7294	18927	10,4809	9930,6	7443,3	188,823
27	300,15	553,07	414,55	0,7393	18736	10,4871	9964,1	7468,6	188,935
28	301,15	554,94	415,96	0,7493	18547	10,4933	9997,7	7493,9	189,047
29	302,15	556,80	417,36	0,7594	18361	10,4995	10031,3	7519,2	189,158
30	303,15	558,67	418,76	0,7696	18177	10,5056	10065,0	7544,4	189,269
31	304,15	560,53	420,17	0,7799	17996	10,5118	10098,6	7569,7	189,380
32	305,15	562,40	421,57	0,7903	17817	10,5179	10132,2	7595,0	189,490
33	306,15	564,27	422,98	0,8009	17641	10,5240	10165,8	7620,4	189,600
34	307,15	566,13	424,38	0,8115	17467	10,5301	10199,4	7645,7	189,710
35	308,15	568,00	425,79	0,8222	17294	10,5362	10233,1	7671,0	189,819
36	309,15	569,87	427,19	0,8331	17125	10,5422	10266,7	7696,3	189,928
37	310,15	571,73	428,60	0,8440	16957	10,5482	10300,3	7721,6	190,037
38	311,15	573,60	430,00	0,8551	16792	10,5542	10334,0	7747,0	190,145
39	312,15	575,47	431,41	0,8663	16628	10,5602	10367,6	7772,3	190,253
40	313,15	577,34	432,82	0,8776	16467	10,5662	10401,3	7797,6	190,361
41	314,15	579,21	434,22	0,8890	16308	10,5722	10435,0	7823,0	190,468
42	315,15	581,07	435,63	0,9005	16150	10,5781	10468,6	7848,3	190,575
43	316,15	582,94	437,04	0,9121	15995	10,5840	10502,3	7873,7	190,682
44	317,15	584,81	438,45	0,9239	15842	10,5899	10536,0	7899,1	190,788
45	318,15	586,68	439,86	0,9357	15690	10,5958	10569,7	7924,4	190,894
46	319,15	588,55	441,26	0,9477	15541	10,6017	10603,4	7949,8	191,000
47	320,15	590,42	442,67	0,9598	15393	10,6075	10637,1	7975,2	191,105
48	321,15	592,29	444,08	0,9720	15247	10,6134	10670,8	8000,6	191,211
49	322,15	594,16	445,49	0,9843	15103	10,6192	10704,5	8026,0	191,315
50	323,15	596,03	446,90	0,9968	14961	10,6250	10738,2	8051,4	191,420
51	324,15	597,91	448,31	1,0093	14820	10,6308	10771,9	8076,8	191,524
52	325,15	599,78	449,72	1,0220	14681	10,6365	10805,6	8102,2	191,628
53	326,15	601,65	451,13	1,0348	14544	10,6423	10839,3	8127,6	191,731
54	327,15	603,52	452,54	1,0478	14409	10,6480	10873,1	8153,0	191,835
55	328,15	605,40	453,95	1,0608	14275	10,6537	10906,8	8178,4	191,938
56	329,15	607,27	455,37	1,0740	14142	10,6594	10940,6	8203,9	192,040
57	330,15	609,14	456,78	1,0873	14012	10,6651	10974,3	8229,3	192,143
58	331,15	611,02	458,19	1,1008	13883	10,6708	11008,1	8254,7	192,245
59	332,15	612,89	459,60	1,1143	13755	10,6764	11041,8	8280,2	192,347

Steam (*Continued*)

t	T	h	u	π_0	θ_0	s^0	H	U	S
60	333,15	614,76	461,02	1,1280	13629	10,6821	11075,6	8305,6	192,448
61	334,15	616,64	462,43	1,1418	13505	10,6877	11109,4	8331,1	192,549
62	335,15	618,51	463,84	1,1558	13382	10,6933	11143,1	8356,6	192,650
63	336,15	620,39	465,26	1,1698	13260	10,6989	11176,9	8382,0	192,751
64	337,15	622,27	466,67	1,1841	13140	10,7045	11210,7	8407,5	192,851
65	338,15	624,14	468,08	1,1984	13021	10,7100	11244,5	8433,0	192,951
66	339,15	626,02	469,50	1,2129	12904	10,7155	11278,3	8458,5	193,051
67	340,15	627,89	470,92	1,2275	12788	10,7211	11312,2	8484,0	193,151
68	341,15	629,77	472,33	1,2422	12673	10,7266	11346,0	8509,5	193,250
69	342,15	631,65	473,75	1,2571	12560	10,7321	11379,8	8535,0	193,349
70	343,15	633,53	475,16	1,2721	12448	10,7376	11413,6	8560,5	193,448
71	344,15	635,41	476,58	1,2873	12337	10,7430	11447,5	8586,1	193,546
72	345,15	637,28	478,00	1,3026	12228	10,7485	11481,3	8611,6	193,645
73	346,15	639,16	479,41	1,3180	12119	10,7539	11515,2	8637,1	193,743
74	347,15	641,04	480,83	1,3336	12012	10,7593	11549,0	8662,7	193,840
75	348,15	642,92	482,25	1,3493	11907	10,7647	11582,9	8688,2	193,938
76	349,15	644,80	483,67	1,3652	11802	10,7701	11616,8	8713,8	194,035
77	350,15	646,68	485,09	1,3812	11699	10,7755	11650,6	8739,4	194,132
78	351,15	648,56	486,51	1,3973	11597	10,7809	11684,5	8764,9	194,228
79	352,15	650,45	487,93	1,4136	11496	10,7862	11718,4	8790,5	194,325
80	353,15	652,33	489,35	1,4301	11396	10,7916	11752,3	8816,1	194,421
81	354,15	654,21	490,77	1,4466	11297	10,7969	11786,2	8841,7	194,517
82	355,15	656,09	492,19	1,4634	11199	10,8022	11820,1	8867,3	194,612
83	356,15	657,97	493,61	1,4803	11103	10,8075	11854,1	8892,9	194,708
84	357,15	659,86	495,03	1,4973	11007	10,8128	11888,0	8918,5	194,803
85	358,15	661,74	496,45	1,5145	10913	10,8180	11921,9	8944,1	194,898
86	359,15	663,63	497,88	1,5318	10820	10,8233	11955,9	8969,8	194,992
87	360,15	665,51	499,30	1,5493	10727	10,8285	11989,8	8995,4	195,087
88	361,15	667,39	500,72	1,5670	10636	10,8338	12023,8	9021,0	195,181
89	362,15	669,28	502,15	1,5848	10545	10,8390	12057,7	9046,7	195,275
90	363,15	671,17	503,57	1,6027	10456	10,8442	12091,7	9072,3	195,368
91	364,15	673,05	505,00	1,6208	10368	10,8494	12125,7	9098,0	195,462
92	365,15	674,94	506,42	1,6391	10280	10,8545	12159,7	9123,7	195,555
93	366,15	676,82	507,85	1,6575	10194	10,8597	12193,7	9149,4	195,648
94	367,15	678,71	509,27	1,6761	10108	10,8648	12227,7	9175,0	195,741
95	368,15	680,60	510,70	1,6949	10024	10,8700	12261,7	9200,7	195,833
96	369,15	682,49	512,12	1,7138	9940,4	10,8751	12295,7	9226,4	195,926
97	370,15	684,38	513,55	1,7329	9857,6	10,8802	12329,7	9252,1	196,018
98	371,15	686,27	514,98	1,7521	9775,7	10,8853	12363,8	9277,9	196,109
99	372,15	688,15	516,41	1,7715	9694,7	10,8904	12397,8	9303,6	196,201
100	373,15	690,04	517,84	1,7911	9614,5	10,8954	12431,8	9329,3	196,292
101	374,15	691,93	519,26	1,8108	9535,2	10,9005	12465,9	9355,1	196,384
102	375,15	693,83	520,69	1,8307	9456,7	10,9056	12500,0	9380,8	196,474
103	376,15	695,72	522,12	1,8508	9379,0	10,9106	12534,0	9406,6	196,565
104	377,15	697,61	523,55	1,8711	9302,2	10,9156	12568,1	9432,3	196,656
105	378,15	699,50	524,98	1,8915	9226,1	10,9206	12602,2	9458,1	196,746
106	379,15	701,39	526,41	1,9121	9150,9	10,9256	12636,3	9483,9	196,836
107	380,15	703,28	527,84	1,9329	9076,5	10,9306	12670,4	9509,7	196,926
108	381,15	705,18	529,28	1,9538	9002,8	10,9356	12704,5	9535,4	197,015
109	382,15	707,07	530,71	1,9749	8929,9	10,9105	12738,6	9561,3	197,105
110	383,15	708,97	532,14	1,9962	8857,7	10,9455	12772,7	9587,1	197,194
111	384,15	710,86	533,57	2,0177	8786,3	10,9504	12806,9	9612,9	197,283
112	385,15	712,76	535,01	2,0393	8715,6	10,9554	12841,0	9638,7	197,372
113	386,15	714,65	536,44	2,0612	8645,7	10,9603	12875,1	9664,5	197,460
114	387,15	716,55	537,88	2,0832	8576,5	10,9652	12909,3	9690,4	197,549

Steam (*Continued*)

t	T	h	u	π_o	θ_o	s^o	H	U	S
115	388,15	718,44	539,31	2,1054	8508,0	10,9701	12943,5	9716,2	197,637
116	389,15	720,34	540,75	2,1278	8440,2	10,9749	12977,6	9742,1	197,725
117	390,15	722,24	542,18	2,1504	8373,0	10,9798	13011,8	9768,0	197,812
118	391,15	724,13	543,62	2,1731	8306,6	10,9847	13046,0	9793,8	197,900
119	392,15	726,03	545,06	2,1960	8240,8	10,9895	13080,2	9819,7	197,987
120	393,15	727,93	546,49	2,2192	8175,7	10,9944	13114,4	9845,6	198,074
121	394,15	729,83	547,93	2,2425	8111,3	10,9992	13148,6	9871,5	198,161
122	395,15	731,73	549,37	2,2660	8047,5	11,0040	13182,9	9897,4	198,248
123	396,15	733,63	550,81	2,2897	7984,4	11,0088	13217,1	9923,3	198,334
124	397,15	735,53	552,25	2,3136	7921,8	11,0136	13251,3	9949,3	198,421
125	398,15	737,43	553,68	2,3377	7859,9	11,0184	13285,6	9975,2	198,507
126	399,15	739,33	555,12	2,3620	7798,7	11,0231	13319,8	10001,1	198,593
127	400,15	741,23	556,57	2,3865	7738,0	11,0279	13354,1	10027,1	198,679
128	401,15	743,14	558,01	2,4111	7677,9	11,0326	13388,4	10053,0	198,764
129	402,15	745,04	559,45	2,4360	7618,5	11,0374	13422,6	10079,0	198,849
130	403,15	746,94	560,89	2,4611	7559,6	11,0421	13456,9	10105,0	198,935
131	404,15	748,85	562,33	2,4864	7501,3	11,0468	13491,2	10131,0	199,020
132	405,15	750,75	563,77	2,5119	7443,6	11,0515	13525,5	10157,0	199,104
133	406,15	752,66	565,22	2,5376	7386,4	11,0562	13559,9	10183,0	199,189
134	407,15	754,56	566,66	2,5635	7329,8	11,0609	13594,2	10209,0	199,273
135	408,15	756,47	568,11	2,5896	7273,7	11,0656	13628,5	10235,0	199,358
136	409,15	758,37	569,55	2,6159	7218,2	11,0702	13662,9	10261,0	199,442
137	410,15	760,28	570,99	2,6424	7163,3	11,0749	13697,2	10287,0	199,525
138	411,15	762,19	572,44	2,6691	7108,8	11,0795	13731,6	10313,1	199,609
139	412,15	764,09	573,89	2,6960	7054,9	11,0842	13765,9	10339,1	199,693
140	413,15	766,00	575,33	2,7232	7001,5	11,0888	13800,3	10365,2	199,776
141	414,15	767,91	576,78	2,7505	6948,7	11,0934	13834,7	10391,3	199,859
142	415,15	769,82	578,23	2,7781	6896,3	11,0980	13869,1	10417,4	199,942
143	416,15	771,73	579,68	2,8059	6844,4	11,1026	13903,5	10443,4	200,025
144	417,15	773,64	581,12	2,8339	6793,1	11,1072	13937,9	10469,5	200,107
145	418,15	775,55	582,57	2,8622	6742,2	11,1118	13972,3	10495,6	200,190
146	419,15	777,46	584,02	2,8906	6691,8	11,1163	14006,7	10521,8	200,272
147	420,15	779,37	585,47	2,9193	6641,9	11,1209	14041,2	10547,9	200,354
148	421,15	781,29	586,92	2,9482	6592,4	11,1254	14075,6	10574,0	200,436
149	422,15	783,20	588,37	2,9773	6543,5	11,1300	14110,1	10600,2	200,518
150	423,15	785,11	589,83	3,0066	6494,9	11,1345	14144,6	10626,3	200,599
151	424,15	787,02	591,28	3,0362	6446,9	11,1390	14179,0	10652,5	200,681
152	425,15	788,94	592,73	3,0660	6399,3	11,1435	14213,5	10678,6	200,762
153	426,15	790,85	594,18	3,0960	6352,1	11,1480	14248,0	10704,8	200,843
154	427,15	792,77	595,64	3,1263	6305,4	11,1525	14282,5	10731,0	200,924
155	428,15	794,68	597,09	3,1568	6259,1	11,1570	14317,0	10757,2	201,004
156	429,15	796,60	598,55	3,1875	6213,2	11,1615	14351,5	10783,4	201,085
157	430,15	798,52	600,00	3,2185	6167,8	11,1659	14386,1	10809,6	201,165
158	431,15	800,43	601,46	3,2497	6122,8	11,1704	14420,6	10835,9	201,245
159	432,15	802,35	602,91	3,2811	6078,2	11,1748	14455,2	10862,1	201,326
160	433,15	804,27	604,37	3,3128	6034,0	11,1793	14489,7	10888,3	201,405
161	434,15	806,19	605,83	3,3447	5990,3	11,1837	14524,3	10914,6	201,485
162	435,15	808,11	607,28	3,3769	5946,9	11,1881	14558,9	10940,8	201,565
163	436,15	810,03	608,74	3,4093	5903,9	11,1925	14593,4	10967,1	201,644
164	437,15	811,95	610,20	3,4419	5861,3	11,1969	14628,0	10993,4	201,723
165	438,15	813,87	611,66	3,4748	5819,1	11,2013	14662,6	11019,7	201,802
166	439,15	815,79	613,12	3,5079	5777,3	11,2057	14697,3	11046,0	201,881
167	440,15	817,71	614,58	3,5413	5735,9	11,2100	14731,9	11072,3	201,960
168	441,15	819,63	616,04	3,5749	5694,8	11,2144	14766,5	11098,6	202,039
169	442,15	821,56	617,50	3,6088	5654,2	11,2188	14801,1	11124,9	202,117

185

Steam (*Continued*)

t	T	h	u	π_0	θ_0	s^0	H	U	S
170	443,15	823,48	618,96	3,6430	5613,8	11,2231	14835,8	11151,3	202,195
171	444,15	825,40	620,43	3,6774	5573,9	11,2274	14870,5	11177,6	202,273
172	445,15	827,33	621,89	3,7120	5534,3	11,2318	14905,1	11204,0	202,351
173	446,15	829,25	623,35	3,7469	5495,1	11,2361	14939,8	11230,3	202,429
174	447,15	831,18	624,82	3,7821	5456,2	11,2404	14974,5	11256,7	202,507
175	448,15	833,10	626,28	3,8175	5417,6	11,2447	15009,2	11283,1	202,584
176	449,15	835,03	627,75	3,8532	5379,4	11,2490	15043,9	11309,5	202,662
177	450,15	836,96	629,21	3,8891	5341,6	11,2533	15078,6	11335,9	202,739
178	451,15	838,88	630,68	3,9253	5304,1	11,2576	15113,3	11362,3	202,816
179	452,15	840,81	632,14	3,9618	5266,9	11,2618	15148,1	11388,7	202,893
180	453,15	842,74	633,61	3,9986	5230,0	11,2661	15182,8	11415,2	202,970
181	454,15	844,67	635,08	4,0356	5193,5	11,2703	15217,6	11441,6	203,046
182	455,15	846,60	636,55	4,0729	5157,2	11,2746	15252,4	11468,0	203,123
183	456,15	848,53	638,02	4,1104	5121,4	11,2788	15287,1	11494,5	203,199
184	457,15	850,46	639,49	4,1483	5085,8	11,2830	15321,9	11521,0	203,275
185	458,15	852,39	640,96	4,1864	5050,5	11,2873	15356,7	11547,5	203,351
186	459,15	854,32	642,43	4,2248	5015,5	11,2915	15391,5	11574,0	203,427
187	460,15	856,26	643,90	4,2634	4980,9	11,2957	15426,3	11600,5	203,503
188	461,15	858,19	645,37	4,3024	4946,5	11,2999	15461,2	11627,0	203,579
189	462,15	860,12	646,84	4,3416	4912,5	11,3041	15496,0	11653,5	203,654
190	463,15	862,06	648,31	4,3811	4878,7	11,3082	15530,8	11680,0	203,729
191	464,15	863,99	649,79	4,4209	4845,2	11,3124	15565,7	11706,5	203,804
192	465,15	865,93	651,26	4,4610	4812,0	11,3166	15600,5	11733,1	203,880
193	466,15	867,86	652,73	4,5013	4779,1	11,3207	15635,4	11759,7	203,954
194	467,15	869,80	654,21	4,5420	4746,5	11,3249	15670,3	11786,2	204,029
195	468,15	871,74	655,68	4,5829	4714,2	11,3290	15705,2	11812,8	204,104
196	469,15	873,67	657,16	4,6242	4682,1	11,3332	15740,1	11839,4	204,178
197	470,15	875,61	658,64	4,6657	4650,3	11,3373	15775,0	11866,0	204,253
198	471,15	877,55	660,11	4,7075	4618,8	11,3414	15809,9	11892,6	204,327
199	472,15	879,49	661,59	4,7496	4587,6	11,3455	15844,9	11919,2	204,401
200	473,15	881,43	663,07	4,7921	4556,6	11,3496	15879,8	11945,9	204,475
201	474,15	883,37	664,55	4,8348	4525,9	11,3537	15914,8	11972,5	204,549
202	475,15	885,31	666,03	4,8778	4495,4	11,3578	15949,7	11999,1	204,622
203	476,15	887,25	667,51	4,9211	4465,2	11,3619	15984,7	12025,8	204,696
204	477,15	889,19	668,99	4,9648	4435,3	11,3660	16019,7	12052,5	204,769
205	478,15	891,13	670,47	5,0087	4405,6	11,3700	16054,7	12079,1	204,842
206	479,15	893,08	671,95	5,0530	4376,1	11,3741	16089,7	12105,8	204,916
207	480,15	895,02	673,43	5,0975	4346,9	11,3781	16124,7	12132,5	204,989
208	481,15	896,97	674,91	5,1424	4318,0	11,3822	16159,7	12159,2	205,061
209	482,15	898,91	676,40	5,1876	4289,3	11,3862	16194,8	12186,0	205,134
210	483,15	900,86	677,88	5,2331	4260,8	11,3903	16229,8	12212,7	205,207
211	484,15	902,80	679,36	5,2789	4232,5	11,3943	16264,9	12239,4	205,279
212	485,15	904,75	680,85	5,3250	4204,5	11,3983	16299,9	12266,2	205,352
213	486,15	906,69	682,34	5,3715	4176,8	11,4023	16335,0	12292,9	205,424
214	487,15	908,64	683,82	5,4183	4149,2	11,4063	16370,1	12319,7	205,496
215	488,15	910,59	685,31	5,4654	4121,9	11,4103	16405,2	12346,5	205,568
216	489,15	912,54	686,79	5,5128	4094,8	11,4143	16440,3	12373,3	205,640
217	490,15	914,49	688,28	5,5606	4067,9	11,4183	16475,4	12400,1	205,711
218	491,15	916,44	689,77	5,6086	4041,3	11,4222	16510,5	12426,9	205,783
219	492,15	918,39	691,26	5,6571	4014,9	11,4262	16545,7	12453,7	205,855
220	493,15	920,34	692,75	5,7058	3988,7	11,4302	16580,8	12480,6	205,926
221	494,15	922,29	694,24	5,7549	3962,6	11,4341	16616,0	12507,4	205,997
222	495,15	924,24	695,73	5,8043	3936,9	11,4381	16651,1	12534,3	206,068
223	496,15	926,19	697,22	5,8541	3911,3	11,4420	16686,3	12561,1	206,139
224	497,15	928,15	698,71	5,9042	3885,9	11,4459	16721,5	12588,0	206,210

Steam (*Continued*)

t	T	h	u	π_0	θ_0	$s^\bullet$	H	U	S
225	498,15	930,10	700,20	5,9546	3860,7	11,4499	16756,7	12614,9	206,281
226	499,15	932,06	701,70	6,0054	3835,8	11,4538	16791,9	12641,8	206,351
227	500,15	934,01	703,19	6,0565	3811,0	11,4577	16827,1	12668,7	206,422
228	501,15	935,97	704,68	6,1080	3786,4	11,4616	16862,4	12695,6	206,492
229	502,15	937,92	706,18	6,1598	3762,1	11,4655	16897,6	12722,5	206,562
230	503,15	939,88	707,67	6,2120	3737,9	11,4694	16932,9	12749,5	206,633
231	504,15	941,84	709,17	6,2645	3713,9	11,4733	16968,1	12776,4	206,703
232	505,15	943,79	710,67	6,3174	3690,2	11,4772	17003,4	12803,4	206,772
233	506,15	945,75	712,16	6,3707	3666,6	11,4810	17038,7	12830,3	206,842
234	507,15	947,71	713,66	6,4243	3643,1	11,4849	17074,0	12857,3	206,912
235	508,15	949,67	715,16	6,4782	3619,9	11,4888	17109,3	12884,3	206,981
236	509,15	951,63	716,66	6,5326	3596,9	11,4926	17144,6	12911,3	207,051
237	510,15	953,59	718,16	6,5872	3574,0	11,4965	17179,9	12938,3	207,120
238	511,15	955,55	719,66	6,6423	3551,4	11,5003	17215,2	12965,3	207,189
239	512,15	957,52	721,16	6,6977	3528,9	11,5041	17250,6	12992,4	207,258
240	513,15	959,48	722,66	6,7535	3506,5	11,5080	17285,9	13019,4	207,327
241	514,15	961,44	724,16	6,8097	3484,4	11,5118	17321,3	13046,5	207,396
242	515,15	963,40	725,66	6,8662	3462,4	11,5156	17356,7	13073,5	207,465
243	516,15	965,37	727,16	6,9231	3440,6	11,5194	17392,1	13100,6	207,534
244	517,15	967,33	728,67	6,9804	3419,0	11,5232	17427,5	13127,7	207,602
245	518,15	969,30	730,17	7,0381	3397,6	11,5270	17462,9	13154,8	207,671
246	519,15	971,27	731,68	7,0961	3376,3	11,5308	17498,3	13181,9	207,739
247	520,15	973,23	733,18	7,1545	3355,1	11,5346	17533,7	13209,0	207,807
248	521,15	975,20	734,69	7,2134	3334,2	11,5384	17569,2	13236,1	207,875
249	522,15	977,17	736,19	7,2726	3313,4	11,5421	17604,6	13263,3	207,943
250	523,15	979,14	737,70	7,3322	3292,7	11,5459	17640,1	13290,4	208,011
251	524,15	981,10	739,21	7,3921	3272,3	11,5497	17675,6	13317,6	208,079
252	525,15	983,07	740,72	7,4525	3251,9	11,5534	17711,1	13344,8	208,146
253	526,15	985,04	742,23	7,5133	3231,8	11,5572	17746,6	13371,9	208,214
254	527,15	987,02	743,73	7,5745	3211,8	11,5609	17782,1	13399,1	208,281
255	528,15	988,99	745,24	7,6360	3191,9	11,5646	17817,6	13426,3	208,349
256	529,15	990,96	746,76	7,6980	3172,2	11,5684	17853,1	13453,5	208,416
257	530,15	992,93	748,27	7,7604	3152,7	11,5721	17888,7	13480,8	208,483
258	531,15	994,90	749,78	7,8232	3133,3	11,5758	17924,2	13508,0	208,550
259	532,15	996,88	751,29	7,8864	3114,0	11,5795	17959,8	13535,3	208,617
260	533,15	998,85	752,80	7,9500	3094,9	11,5832	17995,3	13562,5	208,684
261	534,15	1000,83	754,32	8,0140	3075,9	11,5869	18030,9	13589,8	208,750
262	535,15	1002,80	755,83	8,0784	3057,1	11,5906	18066,5	13617,1	208,817
263	536,15	1004,78	757,35	8,1433	3038,4	11,5943	18102,1	13644,3	208,883
264	537,15	1006,76	758,86	8,2085	3019,9	11,5980	18137,7	13671,7	208,950
265	538,15	1008,73	760,38	8,2742	3001,5	11,6017	18173,4	13699,0	209,016
266	539,15	1010,71	761,89	8,3403	2983,3	11,6054	18209,0	13726,3	209,082
267	540,15	1012,69	763,41	8,4068	2965,1	11,6090	18244,6	13753,6	209,148
268	541,15	1014,67	764,93	8,4738	2947,2	11,6127	18280,3	13781,0	209,214
269	542,15	1016,65	766,45	8,5412	2929,3	11,6163	18316,0	13808,3	209,280
270	543,15	1018,63	767,97	8,6090	2911,6	11,6200	18351,7	13835,7	209,346
271	544,15	1020,61	769,49	8,6772	2894,0	11,6236	18387,4	13863,1	209,411
272	545,15	1022,59	771,01	8,7459	2876,6	11,6273	18423,1	13890,5	209,477
273	546,15	1024,58	772,53	8,8150	2859,2	11,6309	18458,8	13917,9	209,542
274	547,15	1026,56	774,05	8,8846	2842,0	11,6345	18494,5	13945,3	209,608
275	548,15	1028,54	775,57	8,9546	2825,0	11,6382	18530,2	13972,7	209,673
276	549,15	1030,53	777,09	9,0251	2808,0	11,6418	18566,0	14000,1	209,738
277	550,15	1032,51	778,62	9,0959	2791,2	11,6454	18601,7	14027,6	209,803
278	551,15	1034,50	780,14	9,1673	2774,5	11,6490	18637,5	14055,0	209,868
279	552,15	1036,48	781,67	9,2391	2758,0	11,6526	18673,3	14082,5	209,933

Steam (*Continued*)

t	T	h	u	π_0	θ_0	$s^\bullet$	H	U	S
280	553,15	1038,47	783,19	9,3113	2741,5	11,6562	18709,1	14110,0	209,998
281	554,15	1040,46	784,72	9,3840	2725,2	11,6598	18744,9	14137,5	210,062
282	555,15	1042,45	786,24	9,4572	2709,0	11,6634	18780,7	14165,0	210,127
283	556,15	1044,43	787,77	9,5308	2692,9	11,6669	18816,5	14192,5	210,192
284	557,15	1046,42	789,30	9,6049	2677,0	11,6705	18852,4	14220,0	210,256
285	558,15	1048,41	790,83	9,6795	2661,1	11,6741	18888,2	14247,5	210,320
286	559,15	1050,40	792,35	9,7545	2645,4	11,6776	18924,1	14275,1	210,384
287	560,15	1052,39	793,88	9,8300	2629,8	11,6812	18959,9	14302,6	210,448
288	561,15	1054,39	795,41	9,9059	2614,2	11,6847	18995,8	14330,2	210,512
289	562,15	1056,38	796,95	9,9823	2598,9	11,6883	19031,7	14357,8	210,576
290	563,15	1058,37	798,48	10,059	2583,6	11,6918	19067,6	14385,4	210,640
291	564,15	1060,36	800,01	10,136	2568,4	11,6954	19103,5	14413,0	210,704
292	565,15	1062,36	801,54	10,214	2553,3	11,6989	19139,5	14440,6	210,768
293	566,15	1064,35	803,07	10,292	2538,4	11,7024	19175,4	14468,2	210,831
294	567,15	1066,35	804,61	10,371	2523,5	11,7060	19211,3	14495,8	210,894
295	568,15	1068,34	806,14	10,451	2508,8	11,7095	19247,3	14523,5	210,958
296	569,15	1070,34	807,68	10,530	2494,2	11,7130	19283,3	14551,1	211,021
297	570,15	1072,34	809,21	10,611	2479,6	11,7165	19319,3	14578,8	211,084
298	571,15	1074,34	810,75	10,692	2465,2	11,7200	19355,2	14606,5	211,147
299	572,15	1076,34	812,29	10,773	2450,9	11,7235	19391,3	14634,2	211,210
300	573,15	1078,33	813,82	10,855	2436,7	11,7270	19427,3	14661,9	211,273
301	574,15	1080,33	815,36	10,937	2422,6	11,7305	19463,3	14689,6	211,336
302	575,15	1082,33	816,90	11,020	2408,5	11,7339	19499,3	14717,3	211,399
303	576,15	1084,34	818,44	11,103	2394,6	11,7374	19535,4	14745,0	211,461
304	577,15	1086,34	819,98	11,187	2380,8	11,7409	19571,4	14772,8	211,524
305	578,15	1088,34	821,52	11,271	2367,1	11,7444	19607,5	14800,5	211,586
306	579,15	1090,34	823,06	11,356	2353,5	11,7478	19643,6	14828,3	211,649
307	580,15	1092,35	824,61	11,441	2339,9	11,7513	19679,7	14856,1	211,711
308	581,15	1094,35	826,15	11,527	2326,5	11,7547	19715,8	14883,9	211,773
309	582,15	1096,35	827,69	11,614	2313,2	11,7582	19751,9	14911,7	211,835
310	583,15	1098,36	829,24	11,701	2299,9	11,7616	19788,1	14939,5	211,897
311	584,15	1100,37	830,78	11,788	2286,8	11,7651	19824,2	14967,3	211,959
312	585,15	1102,37	832,33	11,876	2273,7	11,7685	19860,3	14995,2	212,021
313	586,15	1104,38	833,87	11,965	2260,7	11,7719	19896,5	15023,0	212,083
314	587,15	1106,39	835,42	12,054	2247,9	11,7753	19932,7	15050,9	212,144
315	588,15	1108,40	836,96	12,143	2235,1	11,7788	19968,9	15078,8	212,206
316	589,15	1110,41	838,51	12,234	2222,4	11,7822	20005,1	15106,6	212,267
317	590,15	1112,42	840,06	12,324	2209,8	11,7856	20041,3	15134,5	212,329
318	591,15	1114,43	841,61	12,415	2197,3	11,7890	20077,5	15162,4	212,390
319	592,15	1116,44	843,16	12,507	2184,8	11,7924	20113,7	15190,4	212,451
320	593,15	1118,45	844,71	12,600	2172,5	11,7958	20150,0	15218,3	212,513
321	594,15	1120,46	846,26	12,692	2160,2	11,7992	20186,2	15246,2	212,574
322	595,15	1122,47	847,81	12,786	2148,0	11,8025	20222,5	15274,2	212,635
323	596,15	1124,49	849,36	12,880	2135,9	11,8059	20258,8	15302,1	212,696
324	597,15	1126,50	850,92	12,975	2123,9	11,8093	20295,1	15330,1	212,756
325	598,15	1128,52	852,47	13,070	2112,0	11,8127	20331,4	15358,1	212,817
326	599,15	1130,53	854,02	13,165	2100,1	11,8160	20367,7	15386,1	212,878
327	600,15	1132,55	855,58	13,262	2088,4	11,8194	20404,0	15414,1	212,938
328	601,15	1134,57	857,13	13,358	2076,7	11,8228	20440,3	15442,1	212,999
329	602,15	1136,58	858,69	13,456	2065,1	11,8261	20476,7	15470,2	213,059
330	603,15	1138,60	860,25	13,554	2053,6	11,8295	20513,1	15498,2	213,120
331	604,15	1140,62	861,80	13,652	2042,1	11,8328	20549,4	15526,3	213,180
332	605,15	1142,64	863,36	13,752	2030,7	11,8361	20585,8	15554,3	213,240
333	606,15	1144,66	864,92	13,851	2019,4	11,8395	20622,2	15582,4	213,300
334	607,15	1146,68	866,48	13,952	2008,2	11,8428	20658,6	15610,5	213,360

t	T	h	u	π_0	θ_0	s^0	H	U	S
335	608,15	1148,70	868,04	14,053	1997,1	11,8461	20695,0	15638,6	213,420
336	609,15	1150,72	869,60	14,154	1986,0	11,8495	20731,5	15666,7	213,480
337	610,15	1152,75	871,16	14,256	1975,0	11,8528	20767,9	15694,9	213,540
338	611,15	1154,77	872,72	14,359	1964,1	11,8561	20804,4	15723,0	213,599
339	612,15	1156,79	874,29	14,462	1953,3	11,8594	20840,8	15751,1	213,659
340	613,15	1158,82	875,85	14,566	1942,5	11,8627	20877,3	15779,3	213,719
341	614,15	1160,84	877,41	14,671	1931,8	11,8660	20913,8	15807,5	213,778
342	615,15	1162,87	878,98	14,776	1921,2	11,8693	20950,3	15835,7	213,837
343	616,15	1164,90	880,54	14,882	1910,6	11,8726	20986,8	15863,9	213,897
344	617,15	1166,92	882,11	14,988	1900,2	11,8759	21023,3	15892,1	213,956
345	618,15	1168,95	883,68	15,095	1889,7	11,8792	21059,8	15920,3	214,015
346	619,15	1170,98	885,24	15,203	1879,4	11,8824	21096,4	15948,5	214,074
347	620,15	1173,01	886,81	15,311	1869,1	11,8857	21132,9	15976,8	214,133
348	621,15	1175,04	888,38	15,420	1858,9	11,8890	21169,5	16005,0	214,192
349	622,15	1177,07	889,95	15,530	1848,8	11,8923	21206,1	16033,3	214,251
350	623,15	1179,10	891,52	15,640	1838,7	11,8955	21242,7	16061,6	214,310
351	624,15	1181,13	893,09	15,750	1828,7	11,8988	21279,3	16089,9	214,368
352	625,15	1183,17	894,66	15,862	1818,8	11,9020	21315,9	16118,2	214,427
353	626,15	1185,20	896,23	15,974	1808,9	11,9053	21352,5	16146,5	214,486
354	627,15	1187,23	897,80	16,087	1799,1	11,9085	21389,2	16174,8	214,544
355	628,15	1189,27	899,37	16,200	1789,3	11,9118	21425,8	16203,1	214,602
356	629,15	1191,30	900,95	16,314	1779,7	11,9150	21462,5	16231,5	214,661
357	630,15	1193,34	902,52	16,429	1770,0	11,9182	21499,2	16259,8	214,719
358	631,15	1195,37	904,10	16,544	1760,5	11,9215	21535,9	16288,2	214,777
359	632,15	1197,41	905,67	16,660	1751,0	11,9247	21572,6	16316,6	214,835
360	633,15	1199,45	907,25	16,777	1741,6	11,9279	21609,3	16345,0	214,893
361	634,15	1201,49	908,83	16,894	1732,2	11,9311	21646,0	16373,4	214,951
362	635,15	1203,53	910,40	17,012	1722,9	11,9343	21682,7	16401,8	215,009
363	636,15	1205,57	911,98	17,131	1713,7	11,9376	21719,5	16430,3	215,067
364	637,15	1207,61	913,56	17,250	1704,5	11,9408	21756,2	16458,7	215,125
365	638,15	1209,65	915,14	17,370	1695,4	11,9440	21793,0	16487,2	215,182
366	639,15	1211,69	916,72	17,491	1686,3	11,9472	21829,8	16515,6	215,240
367	640,15	1213,73	918,30	17,613	1677,3	11,9503	21866,6	16544,1	215,297
368	641,15	1215,77	919,88	17,735	1668,3	11,9535	21903,4	16572,6	215,355
369	642,15	1217,82	921,46	17,858	1659,4	11,9567	21940,2	16601,1	215,412
370	643,15	1219,86	923,05	17,981	1650,6	11,9599	21977,0	16629,6	215,470
371	644,15	1221,91	924,63	18,105	1641,8	11,9631	22013,9	16658,1	215,527
372	645,15	1223,95	926,21	18,230	1633,1	11,9663	22050,7	16686,7	215,584
373	646,15	1226,00	927,80	18,356	1624,5	11,9694	22087,6	16715,2	215,641
374	647,15	1228,04	929,38	18,482	1615,8	11,9726	22124,5	16743,8	215,698
375	648,15	1230,09	930,97	18,609	1607,3	11,9757	22161,3	16772,4	215,755
376	649,15	1232,14	932,56	18,737	1598,8	11,9789	22198,2	16800,9	215,812
377	650,15	1234,19	934,14	18,865	1590,4	11,9821	22235,2	16829,5	215,869
378	651,15	1236,24	935,73	18,995	1582,0	11,9852	22272,1	16858,2	215,926
379	652,15	1238,29	937,32	19,125	1573,6	11,9884	22309,0	16886,8	215,982
380	653,15	1240,34	938,91	19,255	1565,4	11,9915	22346,0	16915,4	216,039
381	654,15	1242,39	940,50	19,387	1557,1	11,9946	22382,9	16944,1	216,095
382	655,15	1244,44	942,09	19,519	1548,9	11,9978	22419,9	16972,7	216,152
383	656,15	1246,50	943,68	19,652	1540,8	12,0009	22456,9	17001,4	216,208
384	657,15	1248,55	945,27	19,785	1532,7	12,0040	22493,9	17030,1	216,265
385	658,15	1250,60	946,87	19,920	1524,7	12,0072	22530,9	17058,8	216,321
386	659,15	1252,66	948,46	20,055	1516,8	12,0103	22567,9	17087,5	216,377
387	660,15	1254,71	950,05	20,191	1508,8	12,0134	22604,9	17116,2	216,433
388	661,15	1256,77	951,65	20,327	1501,0	12,0165	22642,0	17144,9	216,489
389	662,15	1258,83	953,24	20,465	1493,1	12,0196	22679,0	17173,6	216,545

t	T	h	u	π_0	θ_0	s^0	H	U	S
390	663,15	1260,88	954,84	20,603	1485,4	12,0227	22716,1	17202,4	216,601
391	664,15	1262,94	956,44	20,742	1477,6	12,0258	22753,2	17231,2	216,657
392	665,15	1265,00	958,03	20,881	1470,0	12,0289	22790,3	17259,9	216,713
393	666,15	1267,06	959,63	21,022	1462,3	12,0320	22827,4	17288,7	216,769
394	667,15	1269,12	961,23	21,163	1454,8	12,0351	22864,5	17317,5	216,824
395	668,15	1271,18	962,83	21,305	1447,2	12,0382	22901,6	17346,3	216,880
396	669,15	1273,24	964,43	21,448	1439,7	12,0413	22938,8	17375,2	216,935
397	670,15	1275,31	966,03	21,592	1432,3	12,0443	22975,9	17404,0	216,991
398	671,15	1277,37	967,63	21,736	1424,9	12,0474	23013,1	17432,9	217,046
399	672,15	1279,43	969,23	21,881	1417,6	12,0505	23050,3	17461,7	217,102
400	673,15	1281,50	970,84	22,027	1410,3	12,0536	23087,4	17490,6	217,157
401	674,15	1283,56	972,44	22,174	1403,0	12,0566	23124,6	17519,5	217,212
402	675,15	1285,63	974,04	22,321	1395,8	12,0597	23161,9	17548,4	217,267
403	676,15	1287,69	975,65	22,470	1388,6	12,0627	23199,1	17577,3	217,322
404	677,15	1289,76	977,25	22,619	1381,5	12,0658	23236,3	17606,2	217,377
405	678,15	1291,83	978,86	22,769	1374,4	12,0689	23273,6	17635,1	217,432
406	679,15	1293,90	980,47	22,920	1367,4	12,0719	23310,8	17664,1	217,487
407	680,15	1295,96	982,07	23,072	1360,4	12,0749	23348,1	17693,1	217,542
408	681,15	1298,03	983,68	23,224	1353,5	12,0780	23385,4	17722,0	217,597
409	682,15	1300,10	985,29	23,378	1346,6	12,0810	23422,7	17751,0	217,652
410	683,15	1302,18	986,90	23,532	1339,7	12,0841	23460,0	17780,0	217,706
411	684,15	1304,25	988,51	23,687	1332,9	12,0871	23497,3	17809,0	217,761
412	685,15	1306,32	990,12	23,843	1326,1	12,0901	23534,6	17838,0	217,815
413	686,15	1308,39	991,73	23,999	1319,4	12,0931	23572,0	17867,1	217,870
414	687,15	1310,47	993,34	24,157	1312,7	12,0962	23609,3	17896,1	217,924
415	688,15	1312,54	994,96	24,315	1306,0	12,0992	23646,7	17925,2	217,979
416	689,15	1314,61	996,57	24,475	1299,4	12,1022	23684,1	17954,2	218,033
417	690,15	1316,69	998,18	24,635	1292,8	12,1052	23721,5	17983,3	218,087
418	691,15	1318,77	999,80	24,796	1286,3	12,1082	23758,9	18012,4	218,141
419	692,15	1320,84	1001,42	24,958	1279,8	12,1112	23796,3	18041,5	218,195
420	693,15	1322,92	1003,03	25,120	1273,4	12,1142	23833,7	18070,6	218,249
421	694,15	1325 00	1004,65	25,284	1266,9	12,1172	23871,2	18099,7	218,303
422	695,15	1327,08	1006,27	25,448	1260,6	12,1202	23908,6	18128,9	218,357
423	696,15	1329,16	1007,88	25,614	1254,2	12,1232	23946,1	18158,0	218,411
424	697,15	1331,24	1009,50	25,780	1247,9	12,1262	23983,6	18187,2	218,465
425	698,15	1333,32	1011,12	25,947	1241,7	12,1292	24021,1	18216,4	218,519
426	699,15	1335,40	1012,74	26,115	1235,4	12,1321	24058,6	18245,6	218,572
427	700,15	1337,48	1014,36	26,284	1229,3	12,1351	24096,1	18274,8	218,626
428	701,15	1339,57	1015,98	26,454	1223,1	12,1381	24133,6	18304,0	218,680
429	702,15	1341,65	1017,61	26,625	1217,0	12,1410	24171,2	18333,2	218,733
430	703,15	1343,73	1019,23	26,797	1210,9	12,1440	24208,7	18362,4	218,787
431	704,15	1345,82	1020,85	26,969	1204,9	12,1470	24246,3	18391,7	218,840
432	705,15	1347,91	1022,48	27,143	1198,9	12,1499	24283,9	18421,0	218,893
433	706,15	1349,99	1024,10	27,317	1192,9	12,1529	24321,5	18450,2	218,947
434	707,15	1352,08	1025,73	27,493	1187,0	12,1558	24359,1	18479,5	219,000
435	708,15	1354,17	1027,35	27,669	1181,1	12,1588	24396,7	18508,8	219,053
436	709,15	1356,26	1028,98	27,846	1175,2	12,1617	24434,3	18538,1	219,106
437	710,15	1358,34	1030,61	28,024	1169,4	12,1647	24471,9	18567,5	219,159
438	711,15	1360,43	1032,24	28,204	1163,6	12,1676	24509,6	18596,8	219,212
439	712,15	1362,52	1033,87	28,384	1157,8	12,1706	24547,2	18626,1	219,265
440	713,15	1364,62	1035,50	28,565	1152,1	12,1735	24584,9	18655,5	219,318
441	714,15	1366,71	1037,13	28,747	1146,4	12,1764	24622,6	18684,9	219,371
442	715,15	1368,80	1038,76	28,930	1140,8	12,1794	24660,3	18714,3	219,423
443	716,15	1370,89	1040,39	29,114	1135,2	12,1823	24698,0	18743,7	219,476
444	717,15	1372,99	1042,02	29,299	1129,6	12,1852	24735,7	18773,1	219,529

Steam (*Continued*)

t	T	h	u	π_0	θ_0	s^o	H	U	S
445	718,15	1375,08	1043,65	29,484	1124,0	12,1881	24773,5	18802,5	219,581
446	719,15	1377,18	1045,29	29,671	1118,5	12,1910	24811,2	18831,9	219,634
447	720,15	1379,27	1046,92	29,859	1113,0	12,1940	24849,0	18861,4	219,686
448	721 15	1381,37	1048,56	30,048	1107,5	12,1969	24886,8	18890,8	219,739
449	722,15	1383,47	1050,19	30,238	1102,1	12,1998	24924,5	18920,3	219,791
450	723,15	1385,57	1051,83	30,429	1096,7	12,2027	24962,3	18949,8	219,843
451	724,15	1387,66	1053,47	30,620	1091,3	12,2056	25000,2	18979,3	219,896
452	725,15	1389,76	1055,11	30,813	1086,0	12,2085	25038,0	19008,8	219,948
453	726,15	1391,86	1056,74	31,007	1080,7	12,2114	25075,8	19038,3	220,000
454	727,15	1393,96	1058,38	31,202	1075,4	12,2143	25113,7	19067,8	220,052
455	728,15	1396,07	1060,02	31,398	1070,2	12,2171	25151,5	19097,4	220,104
456	729,15	1398,17	1061,66	31,595	1065,0	12,2200	25189,4	19126,9	220,156
457	730,15	1400,27	1063,31	31,793	1059,8	12,2229	25227,3	19156,5	220,208
458	731,15	1402,37	1064,95	31,992	1054,7	12,2258	25265,2	19186,1	220,260
459	732,15	1404,48	1066,59	32,192	1049,6	12,2287	25303,1	19215,7	220,312
460	733,15	1406,58	1068,23	32,393	1044,5	12,2315	25341,0	19245,3	220,363
461	734,15	1408,69	1069,88	32,595	1039,4	12,2344	25378,9	19274,9	220,415
462	735,15	1410,79	1071,52	32,798	1034,4	12,2373	25416,9	19304,5	220,467
463	736,15	1412,90	1073,17	33,002	1029,4	12,2401	25454,8	19334,2	220,518
464	737,15	1415,01	1074,81	33,207	1024,4	12,2430	25492,8	19363,8	220,570
465	738,15	1417,12	1076,46	33,413	1019,5	12,2459	25530,8	19393,5	220,621
466	739,15	1419,23	1078,11	33,621	1014,5	12,2487	25568,8	19423,2	220,673
467	740,15	1421,34	1079,76	33,829	1009,7	12,2516	25606,8	19452,9	220,724
468	741,15	1423,45	1081,41	34,039	1004,8	12,2544	25644,8	19482,6	220,776
469	742,15	1425,56	1083,05	34,249	1000,0	12,2573	25682,9	19512,3	220,827
470	743,15	1427,67	1084,70	34,461	995,20	12,2601	25720,9	19542,0	220,878
471	744,15	1429,78	1086,36	34,674	990,43	12,2629	25759,0	19571,8	220,929
472	745,15	1431,90	1088,01	34,888	985,68	12,2658	25797,0	19601,5	220,980
473	746,15	1434,01	1089,66	35,103	980,96	12,2686	25835,1	19631,3	221,031
474	747,15	1436,12	1091,31	35,319	976,26	12,2715	25873,2	19661,1	221,082
475	748,15	1438,24	1092,97	35,536	971,60	12,2743	25911,3	19690,9	221,133
476	749,15	1440,35	1094,62	35,754	966,95	12,2771	25949,4	19720,7	221,184
477	750,15	1442,47	1096,28	35,974	962,34	12,2799	25987,6	19750,5	221,235
478	751,15	1444,59	1097,93	36,194	957,75	12,2828	26025,7	19780,3	221,286
479	752,15	1446,71	1099,59	36,416	953,19	12,2856	26063,9	19810,2	221,337
480	753,15	1448,82	1101,25	36,639	948,65	12,2884	26102,0	19840,0	221,388
481	754,15	1450,94	1102,90	36,863	944,14	12,2912	26140,2	19869,9	221,438
482	755,15	1453,06	1104,56	37,088	939,66	12,2940	26178,4	19899,8	221,489
483	756,15	1455,18	1106,22	37,314	935,20	12,2968	26216,6	19929,7	221,539
484	757,15	1457,31	1107,88	37,541	930,76	12,2996	26254,8	19959,6	221,590
485	758,15	1459,43	1109,54	37,770	926,35	12,3024	26293,1	19989,5	221,640
486	759,15	1461,55	1111,20	37,999	921,97	12,3052	26331,3	20019,4	221,691
487	760,15	1463,67	1112,86	38,230	917,61	12,3080	26369,6	20049,4	221,741
488	761,15	1465,80	1114,53	38,462	913,27	12,3108	26407,8	20079,3	221,791
489	762,15	1467,92	1116,19	38,696	908,96	12,3136	26446,1	20109,3	221,842
490	763,15	1470,05	1117,85	38,930	904,67	12,3164	26484,4	20139,3	221,892
491	764,15	1472,18	1119,52	39,166	900,41	12,3192	26522,7	20169,3	221,942
492	765,15	1474,30	1121,18	39,402	896,17	12,3219	26561,0	20199,3	221,992
493	766,15	1476,43	1122,85	39,640	891,95	12,3247	26599,4	20229,3	222,042
494	767,15	1478,56	1124,52	39,880	887,76	12,3275	26637,7	20259,3	222,092
495	768,15	1480,69	1126,19	40,120	883,59	12,3303	26676,1	20289,4	222,142
496	769,15	1482,82	1127,85	40,361	879,45	12,3330	26714,4	20319,4	222,192
497	770,15	1484,95	1129,52	40,604	875,32	12,3358	26752,8	20349,5	222,242
498	771,15	1487,08	1131,19	40,848	871,22	12,3386	26791,2	20379,6	222,292
499	772,15	1489,21	1132,86	41,094	867,15	12,3413	26829,6	20409,6	222,342

Steam (*Continued*)

t	T	h	u	π_0	θ_0	s^θ	H	U	S
500	773,15	1491,34	1134,53	41,340	863,09	12,3441	26868,0	20439,7	222,391
501	774,15	1493,48	1136,20	41,588	859,06	12,3469	26906,5	20469,9	222,441
502	775,15	1495,61	1137,88	41,837	855,05	12,3496	26944,9	20500,0	222,491
503	776,15	1497,74	1139,55	42,087	851,07	12,3524	26983,4	20530,1	222,540
504	777,15	1499,88	1141,22	42,339	847,10	12,3551	27021,8	20560,3	222,590
505	778,15	1502,02	1142,90	42,591	843,16	12,3579	27060,3	20590,5	222,639
506	779,15	1504,15	1144,57	42,845	839,24	12,3606	27098,8	20620,6	222,689
507	780,15	1506,29	1146,25	43,101	835,34	12,3633	27137,3	20650,8	222,738
508	781,15	1508,43	1147,93	43,357	831,46	12,3661	27175,8	20681,0	222,787
509	782,15	1510,57	1149,60	43,615	827,60	12,3688	27214,4	20711,3	222,837
510	783,15	1512,71	1151,28	43,874	823,77	12,3716	27252,9	20741,5	222,886
511	784,15	1514,85	1152,96	44,134	819,95	12,3743	27291,5	20771,7	222,935
512	785,15	1516,99	1154,64	44,396	816,16	12,3770	27330,0	20802,0	222,984
513	786,15	1519,13	1156,32	44,659	812,38	12,3797	27368,6	20832,2	223,033
514	787,15	1521,27	1158,00	44,923	808,63	12,3825	27407,2	20862,5	223,082
515	788,15	1523,41	1159,68	45,189	804,90	12,3852	27445,8	20892,8	223,131
516	789,15	1525,56	1161,36	45,456	801,19	12,3879	27484,4	20923,1	223,180
517	790,15	1527,70	1163,05	45,724	797,50	12,3906	27523,1	20953,4	223,229
518	791,15	1529,85	1164,73	45,994	793,83	12,3933	27561,7	20983,8	223,278
519	792,15	1531,99	1166,41	46,265	790,18	12,3960	27600,4	21014,1	223,327
520	793,15	1534,14	1168,10	46,537	786,54	12,3987	27639,0	21044,5	223,376
521	794,15	1536,29	1169,78	46,811	782,93	12,4015	27677,7	21074,8	223,425
522	795,15	1538,43	1171,47	47,086	779,34	12,4042	27716,4	21105,2	223,473
523	796,15	1540,58	1173,16	47,362	775,77	12,4069	27755,1	21135,6	223,522
524	797,15	1542,73	1174,85	47,640	772,22	12,4096	27793,8	21166,0	223,571
525	798,15	1544,88	1176,53	47,919	768,68	12,4122	27832,6	21196,4	223,619
526	799,15	1547,03	1178,22	48,199	765,17	12,4149	27871,3	21226,9	223,668
527	800,15	1549,18	1179,91	48,481	761,67	12,4176	27910,1	21257,3	223,716
528	801,15	1551,33	1181,60	48,764	758,19	12,4203	27948,9	21287,8	223,764
529	802,15	1553,49	1183,29	49,049	754,74	12,4230	27987,6	21318,2	223,813
530	803,15	1555,64	1184,99	49,334	751,30	12,4257	28026,4	21348,7	223,861
531	804,15	1557,80	1186,68	49,622	747,88	12,4284	28065,2	21379,2	223,909
532	805,15	1559,95	1188,37	49,911	744,47	12,4310	28104,1	21409,7	223,958
533	806,15	1562,11	1190,07	50,201	741,09	12,4337	28142,9	21440,2	224,006
534	807,15	1564,26	1191,76	50,492	737,72	12,4364	28181,7	21470,8	224,054
535	808,15	1566,42	1193,46	50,785	734,38	12,4391	28220,6	21501,3	224,102
536	809,15	1568,58	1195,15	51,080	731,05	12,4417	28259,5	21531,9	224,150
537	810,15	1570,73	1196,85	51,376	727,73	12,4444	28298,3	21562,4	224,198
538	811,15	1572,89	1198,55	51,673	724,44	12,4471	28337,2	21593,0	224,246
539	812,15	1575,05	1200,24	51,972	721,16	12,4497	28376,2	21623,6	224,294
540	813,15	1577,21	1201,94	52,272	717,90	12,4524	28415,1	21654,2	224,342
541	814,15	1579,37	1203,64	52,574	714,66	12,4550	28454,0	21684,8	224,390
542	815,15	1581,54	1205,34	52,877	711,44	12,4577	28493,0	21715,5	224,438
543	816,15	1583,70	1207,04	53,182	708,23	12,4603	28531,9	21746,1	224,486
544	817,15	1585,86	1208,75	53,488	705,04	12,4630	28570,9	21776,8	224,533
545	818,15	1588,03	1210,45	53,795	701,86	12,4656	28609,9	21807,4	224,581
546	819,15	1590,19	1212,15	54,105	698,71	12,4683	28648,9	21838,1	224,629
547	820,15	1592,36	1213,86	54,415	695,57	12,4709	28687,9	21868,8	224,676
548	821,15	1594,52	1215,56	54,727	692,44	12,4736	28726,9	21899,5	224,724
549	822,15	1596,69	1217,26	55,041	689,34	12,4762	28765,9	21930,2	224,771
550	823,15	1598,86	1218,97	55,356	686,25	12,4788	28805,0	21961,0	224,819
551	824,15	1601,02	1220,68	55,673	683,17	12,4815	28844,0	21991,7	224,866
552	825,15	1603,19	1222,38	55,991	680,12	12,4841	28883,1	22022,5	224,913
553	826,15	1605,36	1224,09	56,310	677,07	12,4867	28922,2	22053,3	224,961
554	827,15	1607,53	1225,80	56,632	674,05	12,4894	28961,3	22084,0	225,008

Steam (*Continued*)

t	T	h	u	π_0	ϑ_0	s^0	H	U	S
555	828,15	1609,70	1227,51	56,955	671,04	12,4920	29000,4	22114,8	225,055
556	829,15	1611,87	1229,22	57,279	668,04	12,4946	29039,5	22145,6	225,103
557	830,15	1614,05	1230,93	57,605	665,07	12,4972	29078,7	22176,5	225,150
558	831,15	1616,22	1232,64	57,932	662,10	12,4998	29117,8	22207,3	225,197
559	832,15	1618,39	1234,36	58,261	659,16	12,5024	29157,0	22238,1	225,244
560	833,15	1620,57	1236,07	58,592	656,22	12,5051	29196,2	22269,0	225,291
561	834,15	1622,74	1237,78	58,924	653,31	12,5077	29235,3	22299,9	225,338
562	835,15	1624,92	1239,50	59,258	650,41	12,5103	29274,5	22330,8	225,385
563	836,15	1627,10	1241,21	59,593	647,52	12,5129	29313,8	22361,7	225,432
564	837,15	1629,27	1242,93	59,930	644,65	12,5155	29353,0	22392,6	225,479
565	838,15	1631,45	1244,64	60,269	641,79	12,5181	29392,2	22423,5	225,526
566	839,15	1633,63	1246,36	60,609	638,95	12,5207	29431,5	22454,4	225,572
567	840,15	1635,81	1248,08	60,951	636,13	12,5233	29470,7	22485,4	225,619
568	841,15	1637,99	1249,80	61,294	633,31	12,5259	29510,0	22516,3	225,666
569	842,15	1640,17	1251,52	61,639	630,52	12,5285	29549,3	22547,3	225,713
570	843,15	1642,35	1253,24	61,986	627,73	12,5310	29588,6	22578,3	225,759
571	844,15	1644,53	1254,96	62,335	624,97	12,5336	29627,9	22609,3	225,806
572	845,15	1646,72	1256,68	62,685	622,21	12,5362	29667,2	22640,3	225,852
573	846,15	1648,90	1258,40	63,036	619,47	12,5388	29706,6	22671,3	225,899
574	847,15	1651,08	1260,12	63,390	616,75	12,5414	29745,9	22702,4	225,945
575	848,15	1653,27	1261,85	63,745	614,04	12,5440	29785,3	22733,4	225,992
576	849,15	1655,45	1263,57	64,101	611,34	12,5465	29824,7	22764,5	226,038
577	850,15	1657,64	1265,30	64,460	608,66	12,5491	29864,1	22795,6	226,085
578	851,15	1659,83	1267,02	64,820	605,99	12,5517	29903,5	22826,7	226,131
579	852,15	1662,02	1268,75	65,182	603,33	12,5542	29942,9	22857,8	226,177
580	853,15	1664,20	1270,47	65,545	600,69	12,5568	29982,3	22888,9	226,223
581	854,15	1666,39	1272,20	65,911	598,06	12,5594	30021,7	22920,0	226,270
582	855,15	1668,58	1273,93	66,278	595,45	12,5619	30061,2	22951,1	226,316
583	856,15	1670,77	1275,66	66,646	592,84	12,5645	30100,7	22982,3	226,362
584	857,15	1672,97	1277,39	67,017	590,26	12,5671	30140,1	23013,4	226,408
585	858,15	1675,16	1279,12	67,389	587,68	12,5696	30179,6	23044,6	226,454
586	859,15	1677,35	1280,85	67,763	585,12	12,5722	30219,1	23075,8	226,500
587	860,15	1679,54	1282,58	68,138	582,57	12,5747	30258,7	23107,0	226,546
588	861,15	1681,74	1284,32	68,516	580,03	12,5773	30298,2	23138,2	226,592
589	862,15	1683,93	1286,05	68,895	577,51	12,5798	30337,7	23169,5	226,638
590	863,15	1686,13	1287,78	69,276	575,00	12,5824	30377,3	23200,7	226,684
591	864,15	1688,32	1289,52	69,659	572,50	12,5849	30416,9	23232,0	226,730
592	865,15	1690,52	1291,25	70,043	570,02	12,5874	30456,4	23263,2	226,775
593	866,15	1692,72	1292,99	70,430	567,55	12,5900	30496,0	23294,5	226,821
594	867,15	1694,92	1294,73	70,818	565,09	12,5925	30535,6	23325,8	226,867
595	868,15	1697,12	1296,46	71,208	562,64	12,5950	30575,3	23357,1	226,912
596	869,15	1699,32	1298,20	71,600	560,21	12,5976	30614,9	23388,4	226,958
597	870,15	1701,52	1299,94	71,994	557,79	12,6001	30654,5	23419,7	227,004
598	871,15	1703,72	1301,68	72,389	555,38	12,6026	30694,2	23451,1	227,049
599	872,15	1705,92	1303,42	72,786	552,98	12,6052	30733,9	23482,4	227,095
600	873,15	1708,12	1305,16	73,185	550,59	12,6077	30773,5	23513,8	227,140
601	874,15	1710,33	1306,90	73,587	548,22	12,6102	30813,2	23543,2	227,186
602	875,15	1712,53	1308,65	73,989	545,86	12,6127	30852,9	23576,6	227,231
603	876,15	1714,74	1310,39	74,394	543,51	12,6153	30892,7	23608,0	227,276
604	877,15	1716,94	1312,13	74,801	541,17	12,6178	30932,4	23639,4	227,322
605	878,15	1719,15	1313,88	75,209	538,84	12,6203	30972,1	23670,8	227,367
606	879,15	1721,35	1315,62	75,620	536,53	12,6228	31011,9	23702,3	227,412
607	880,15	1723,56	1317,37	76,032	534,23	12,6253	31051,7	23733,7	227,457
608	881,15	1725,77	1319,12	76,446	531,94	12,6278	31091,5	23765,2	227,503
609	882,15	1727,98	1320,86	76,863	529,66	12,6303	31131,2	23796,7	227,548

Steam (*Continued*)

t	T	h	u	π_0	θ_0	s^0	H	U	S
610	883,15	1730,19	1322,61	77,281	527,39	12,6328	31171,1	23828,2	227,593
611	884,15	1732,40	1324,36	77,701	525,13	12,6353	31210,9	23859,7	227,638
612	885,15	1734,61	1326,11	78,123	522,89	12,6378	31250,7	23891,2	227,683
613	886,15	1736,82	1327,86	78,546	520,65	12,6403	31290,6	23922,7	227,728
614	887,15	1739,03	1329,61	78,972	518,43	12,6428	31330,4	23954,3	227,773
615	888,15	1741,25	1331,36	79,400	516,22	12,6453	31370,3	23985,9	227,818
616	889,15	1743,46	1333,12	79,830	514,01	12,6478	31410,2	24017,4	227,863
617	890,15	1745,68	1334,87	80,262	511,82	12,6503	31450,1	24049,0	227,908
618	891,15	1747,89	1336,62	80,695	509,65	12,6528	31490,0	24080,6	227,952
619	892,15	1750,11	1338,38	81,131	507,48	12,6553	31529,9	24112,2	227,997
620	893,15	1752,32	1340,13	81,569	505,32	12,6577	31569,9	24143,8	228,042
621	894,15	1754,54	1341,89	82,009	503,17	12,6602	31609,8	24175,5	228,087
622	895,15	1756,76	1343,65	82,450	501,04	12,6627	31649,8	24207,1	228,131
623	896,15	1758,98	1345,40	82,894	498,91	12,6652	31689,7	24238,8	228,176
624	897,15	1761,20	1347,16	83,340	496,80	12,6677	31729,7	24270,5	228,220
625	898,15	1763,42	1348,92	83,788	494,69	12,6701	31769,7	24302,1	228,265
626	899,15	1765,64	1350,68	84,238	492,60	12,6726	31809,7	24333,8	228,310
627	900,15	1767,86	1352,44	84,690	490,51	12,6751	31849,8	24365,6	228,354
628	901,15	1770,08	1354,20	85,144	488,44	12,6775	31889,8	24397,3	228,398
629	902,15	1772,31	1355,96	85,600	486,38	12,6800	31929,9	24429,0	228,443
630	903,15	1774,53	1357,72	86,058	484,32	12,6825	31969,9	24460,8	228,487
631	904,15	1776,75	1359,49	86,518	482,28	12,6849	32010,0	24492,5	228,532
632	905,15	1778,98	1361,25	86,981	480,25	12,6874	32050,1	24524,3	228,576
633	906,15	1781,21	1363,02	87,445	478,22	12,6898	32090,2	24556,1	228,620
634	907,15	1783,43	1364,78	87,912	476,21	12,6923	32130,3	24587,9	228,664
635	908,15	1785,66	1366,55	88,381	474,21	12,6948	32170,4	24619,7	228,709
636	909,15	1787,89	1368,31	88,851	472,21	12,6972	32210,6	24651,5	228,753
637	910,15	1790,12	1370,08	89,324	470,23	12,6997	32250,7	24683,4	228,797
638	911,15	1792,35	1371,85	89,800	468,25	12,7021	32290,9	24715,2	228,841
639	912,15	1794,58	1373,62	90,277	466,29	12,7046	32331,1	24747,1	228,885
640	913,15	1796,81	1375,39	90,756	464,34	12,7070	32371,3	24779,0	228,929
641	914,15	1799,04	1377,16	91,238	462,39	12,7094	32411,5	24810,9	228,973
642	915,15	1801,27	1378,93	91,722	460,45	12,7119	32451,7	24842,8	229,017
643	916,15	1803,50	1380,70	92,208	458,53	12,7143	32491,9	24874,7	229,061
644	917,15	1805,74	1382,47	92,696	456,61	12,7168	32532,2	24906,6	229,105
645	918,15	1807,97	1384,24	93,186	454,70	12,7192	32572,4	24938,5	229,149
646	919,15	1810,21	1386,02	93,679	452,81	12,7216	32612,7	24970,5	229,193
647	920,15	1812,44	1387,79	94,174	450,92	12,7241	32653,0	25002,5	229,237
648	921,15	1814,68	1389,57	94,671	449,04	12,7265	32693,3	25034,4	229,280
649	922,15	1816,92	1391,34	95,170	447,17	12,7289	32733,6	25066,4	229,324
650	923,15	1819,15	1393,12	95,671	445,30	12,7313	32773,9	25098,4	229,368
651	924,15	1821,39	1394,90	96,175	443,45	12,7338	32814,2	25130,5	229,411
652	925,15	1823,63	1396,67	96,681	441,61	12,7362	32854,6	25162,5	229,455
653	926,15	1825,87	1398,45	97,189	439,77	12,7386	32894,9	25194,5	229,499
654	927,15	1828,11	1400,23	97,700	437,95	12,7410	32935,3	25226,6	229,542
655	928,15	1830,36	1402,01	98,213	436,13	12,7434	32975,7	25258,7	229,586
656	929,15	1832,60	1403,79	98,728	434,32	12,7459	33016,1	25290,7	229,629
657	930,15	1834,84	1405,57	99,246	432,52	12,7483	33056,5	25322,8	229,673
658	931,15	1837,08	1407,36	99,765	430,73	12,7507	33096,9	25354,9	229,716
659	932,15	1839,33	1409,14	100,28	428,95	12,7531	33137,3	25387,1	229,760
660	933,15	1841,57	1410,92	100,81	427,17	12,7555	33177,8	25419,2	229,803
661	934,15	1843,82	1412,71	101,33	425,41	12,7579	33218,2	25451,3	229,846
662	935,15	1846,07	1414,49	101,86	423,65	12,7603	33258,7	25483,5	229,890
663	936,15	1848,31	1416,28	102,40	421,90	12,7627	33299,2	25515,7	229,933
664	937,15	1850,56	1418,06	102,93	420,16	12,7651	33339,7	25547,9	229,976

Steam (*Continued*)

t	T	h	u	π_0	θ_0	s^0	H	U	S
665	938,15	1852,81	1419,85	103,47	418,43	12,7675	33380,2	25580,1	230,019
666	939,15	1855,06	1421,64	104,00	416,71	12,7699	33420,7	25612,3	230,062
667	940,15	1857,31	1423,43	104,54	414,99	12,7723	33461,3	25644,5	230,106
668	941,15	1859,56	1425,22	105,09	413,29	12,7747	33501,8	25676,7	230,149
669	942,15	1861,81	1427,01	105,63	411,59	12,7771	33542,4	25709,0	230,192
670	943,15	1864,06	1428,80	106,18	409,90	12,7795	33583,0	25741,2	230,235
671	944,15	1866,32	1430,59	106,73	408,21	12,7819	33623,6	25773,5	230,278
672	945,15	1868,57	1432,38	107,29	406,54	12,7842	33664,2	25805,8	230,321
673	946,15	1870,82	1434,17	107,84	404,87	12,7866	33704,8	25838,1	230,364
674	947,15	1873,08	1435,97	108,40	403,21	12,7890	33745,4	25870,4	230,407
675	948,15	1875,33	1437,76	108,96	401,56	12,7914	33786,0	25902,7	230,450
676	949,15	1877,59	1439,56	109,52	399,92	12,7938	33826,7	25935,1	230,492
677	950,15	1879,85	1441,35	110,09	398,29	12,7961	33867,3	25967,4	230,535
678	951,15	1882,11	1443,15	110,66	396,66	12,7985	33908,0	25999,8	230,578
679	952,15	1884,36	1444,95	111,23	395,04	12,8009	33948,7	26032,1	230,621
680	953,15	1886,62	1446,74	111,80	393,43	12,8033	33989,4	26064,5	230,663
681	954,15	1888,88	1448,54	112,38	391,82	12,8056	34030,1	26096,9	230,706
682	955,15	1891,14	1450,34	112,95	390,23	12,8080	34070,9	26129,3	230,749
683	956,15	1893,41	1452,14	113,54	388,64	12,8104	34111,6	26161,8	230,791
684	957,15	1895,67	1453,94	114,12	387,05	12,8127	34152,4	26194,2	230,834
685	958,15	1897,93	1455,74	114,70	385,48	12,8151	34193,1	26226,7	230,877
686	959,15	1900,19	1457,55	115,29	383,91	12,8174	34233,9	26259,1	230,919
687	960,15	1902,46	1459,35	115,88	382,35	12,8198	34274,7	26291,6	230,962
688	961,15	1904,72	1461,15	116,48	380,80	12,8222	34315,5	26324,1	231,004
689	962,15	1906,99	1462,96	117,07	379,26	12,8245	34356,3	26356,6	231,047
690	963,15	1909,26	1464,76	117,67	377,72	12,8269	34397,1	26389,1	231,089
691	964,15	1911,52	1466,57	118,27	376,19	12,8292	34438,0	26421,6	231,131
692	965,15	1913,79	1468,37	118,88	374,67	12,8316	34478,8	26454,2	231,174
693	966,15	1916,06	1470,18	119,48	373,15	12,8339	34519,7	26486,7	231,216
694	967,15	1918,33	1471,99	120,09	371,64	12,8363	34560,6	26519,3	231,258
695	968,15	1920,60	1473,79	120,71	370,14	12,8386	34601,5	26551,9	231,301
696	969,15	1922,87	1475,60	121,32	368,64	12,8410	34642,4	26584,5	231,343
697	970,15	1925,14	1477,41	121,94	367,16	12,8433	34683,3	26617,1	231,385
698	971,15	1927,41	1479,22	122,56	365,67	12,8456	34724,2	26649,7	231,427
699	972,15	1929,68	1481,04	123,18	364,20	12,8480	34765,2	26682,3	231,469
700	973,15	1931,96	1482,85	123,81	362,73	12,8503	34806,1	26715,0	231,511
701	974,15	1934,23	1484,66	124,43	361,27	12,8527	34847,1	26747,6	231,554
702	975,15	1936,51	1486,47	125,07	359,82	12,8550	34888,1	26780,3	231,596
703	976,15	1938,78	1488,29	125,70	358,37	12,8573	34929,1	26813,0	231,638
704	977,15	1941,06	1490,10	126,34	356,93	12,8597	34970,1	26845,7	231,680
705	978,15	1943,33	1491,92	126,97	355,50	12,8620	35011,1	26878,4	231,722
706	979,15	1945,61	1493,73	127,62	354,07	12,8643	35052,2	26911,1	231,763
707	980,15	1947,89	1495,55	128,26	352,65	12,8666	35093,2	26943,8	231,805
708	981,15	1950,17	1497,37	128,91	351,24	12,8690	35134,3	26976,6	231,847
709	982,15	1952,45	1499,19	129,56	349,83	12,8713	35175,3	27009,3	231,889
710	983,15	1954,73	1501,01	130,21	348,43	12,8736	35216,4	27042,1	231,931
711	984,15	1957,01	1502,82	130,87	347,04	12,8759	35257,5	27074,9	231,973
712	985,15	1959,29	1504,65	131,53	345,65	12,8782	35298,6	27107,7	232,014
713	986,15	1961,58	1506,47	132,19	344,27	12,8806	35339,8	27140,5	232,056
714	987,15	1963,86	1508,29	132,85	342,90	12,8829	35380,9	27173,3	232,098
715	988,15	1966,14	1510,11	133,52	341,53	12,8852	35422,0	27206,2	232,140
716	989,15	1968,43	1511,93	134,19	340,17	12,8875	35463,2	27239,0	232,181
717	990,15	1970,71	1513,76	134,86	338,81	12,8898	35504,4	27271,9	232,223
718	991,15	1973,00	1515,58	135,54	337,46	12,8921	35545,6	27304,7	232,264
719	992,15	1975,29	1517,41	136,22	336,12	12,8944	35586,8	27337,6	232,306

Steam (*Continued*)

t	T	h	u	π_0	θ_0	s^0	H	U	S
720	993,15	1977,57	1519,23	136,90	334,78	12,8967	35628,0	27370,5	232,347
721	994,15	1979,86	1521,06	137,59	333,45	12,8990	35669,2	27403,4	232,389
722	995,15	1982,15	1522,89	138,27	332,12	12,9013	35710,4	27436,3	232,430
723	996,15	1984,44	1524,72	138,96	330,80	12,9036	35751,7	27469,3	232,472
724	997,15	1986,73	1526,54	139,66	329,49	12,9059	35792,9	27502,2	232,513
725	998,15	1989,02	1528,37	140,35	328,19	12,9082	35834,2	27535,2	232,555
726	999,15	1991,31	1530,20	141,05	326,88	12,9105	35875,5	27568,2	232,596
727	1000,15	1993,61	1532,04	141,76	325,59	12,9128	35916,8	27601,2	232,637
728	1001,15	1995,90	1533,87	142,46	324,30	12,9151	35958,1	27634,2	232,679
729	1002,15	1998,19	1535,70	143,17	323,02	12,9174	35999,5	27667,2	232,720
730	1003,15	2000,49	1537,53	143,88	321,74	12,9197	36040,8	27700,2	232,761
731	1004,15	2002,78	1539,37	144,60	320,47	12,9220	36082,1	27733,2	232,802
732	1005,15	2005,08	1541,20	145,32	319,20	12,9243	36123,5	27766,3	232,843
733	1006,15	2007,38	1543,04	146,04	317,94	12,9265	36164,9	27799,3	232,885
734	1007,15	2009,67	1544,87	146,76	316,69	12,9288	36206,3	27832,4	232,926
735	1008,15	2011,97	1546,71	147,49	315,44	12,9311	36247,7	27865,5	232,967
736	1009,15	2014,27	1548,55	148,22	314,20	12,9334	36289,1	27898,6	233,008
737	1010,15	2016,57	1550,38	148,95	312,96	12,9357	36330,5	27931,7	233,049
738	1011,15	2018,87	1552,22	149,69	311,73	12,9379	36372,0	27964,9	233,090
739	1012,15	2021,17	1554,06	150,43	310,51	12,9402	36413,4	27998,0	233,131
740	1013,15	2023,47	1555,90	151,17	309,28	12,9425	36454,9	28031,1	233,172
741	1014,15	2025,78	1557,74	151,92	308,07	12,9448	36496,4	28064,3	233,213
742	1015,15	2028,08	1559,59	152,67	306,86	12,9470	36537,9	28097,5	233,254
743	1016,15	2030,38	1561,43	153,42	305,66	12,9493	36579,4	28130,7	233,294
744	1017,15	2032,69	1563,27	154,17	304,46	12,9516	36620,9	28163,9	233,335
745	1018,15	2034,99	1565,11	154,93	303,27	12,9538	36662,4	28197,1	233,376
746	1019,15	2037,30	1566,96	155,69	302,08	12,9561	36704,0	28230,3	233,417
747	1020,15	2039,60	1568,80	156,46	300,90	12,9583	36745,5	28263,6	233,458
748	1021,15	2041,91	1570,65	157,23	299,72	12,9606	36787,1	28296,8	233,498
749	1022,15	2044,22	1572,50	158,00	298,55	12,9629	36828,7	28330,1	233,539
750	1023,15	2046,53	1574,34	158,77	297,38	12,9651	36870,3	28363,4	233,580
751	1024,15	2048,84	1576,19	159,55	296,22	12,9674	36911,9	28396,7	233,620
752	1025,15	2051,15	1578,04	160,33	295,07	12,9696	36953,5	28430,0	233,661
753	1026,15	2053,46	1579,89	161,12	293,91	12,9719	36995,1	28463,3	233,702
754	1027,15	2055,77	1581,74	161,91	292,77	12,9741	37036,8	28496,6	233,742
755	1028,15	2058,08	1583,59	162,70	291,63	12,9764	37078,4	28530,0	233,783
756	1029,15	2060,40	1585,44	163,49	290,49	12,9786	37120,1	28563,3	233,823
757	1030,15	2062,71	1587,29	164,29	289,36	12,9809	37161,8	28596,7	233,864
758	1031,15	2065,02	1589,15	165,09	288,24	12,9831	37203,5	28630,1	233,904
759	1032,15	2067,34	1591,00	165,90	287,12	12,9854	37245,2	28663,4	233,945
760	1033,15	2069,65	1592,85	166,70	286,00	12,9876	37286,9	28696,9	233,985
761	1034,15	2071,97	1594,71	167,51	284,89	12,9899	37328,6	28730,3	234,025
762	1035,15	2074,29	1596,56	168,33	283,79	12,9921	37370,4	28763,7	234,066
763	1036,15	2076,61	1598,42	169,15	282,69	12,9943	37412,1	28797,1	234,106
764	1037,15	2078,92	1600,28	169,97	281,59	12,9966	37453,9	28830,6	234,146
765	1038,15	2081,24	1602,14	170,79	280,50	12,9988	37495,7	28864,1	234,187
766	1039,15	2083,56	1603,99	171,62	279,42	13,0010	37537,5	28897,6	234,227
767	1040,15	2085,88	1605,85	172,46	278,34	13,0033	37579,3	28931,0	234,267
768	1041,15	2088,21	1607,71	173,29	277,26	13,0055	37621,1	28964,6	234,307
769	1042,15	2090,53	1609,57	174,13	276,19	13,0077	37662,9	28998,1	234,347
770	1043,15	2092,85	1611,43	174,97	275,12	13,0100	37704,8	29031,6	234,387
771	1044,15	2095,17	1613,30	175,82	274,06	13,0122	37746,6	29065,2	234,428
772	1045,15	2097,50	1615,16	176,67	273,01	13,0144	37788,5	29098,7	234,468
773	1046,15	2099,82	1617,02	177,52	271,95	13,0166	37830,4	29132,3	234,508
774	1047,15	2102,15	1618,89	178,38	270,91	13,0189	37872,3	29165,9	234,548

t	T	h	u	π_0	0_0	s^0	H	U	S
775	1048,15	2104,47	1620,75	179,24	269,86	13,0211	37914,2	29199,5	234,588
776	1049,15	2106,80	1622,62	180,10	268,83	13,0233	37956,1	29233,1	234,628
777	1050,15	2109,13	1624,48	180,97	267,79	13,0255	37998,1	29266,7	234,668
778	1051,15	2111,46	1626,35	181,84	266,76	13,0277	38040,0	29300,3	234,708
779	1052,15	2113,79	1628,22	182,72	265,74	13,0299	38082,0	29334,0	234,747
780	1053,15	2116,12	1630,09	183,59	264,72	13,0322	38124,0	29367,6	234,787
781	1054,15	2118,45	1631,96	184,48	263,70	13,0344	38165,9	29401,3	234,827
782	1055,15	2120,78	1633,82	185,36	262,69	13,0366	38207,9	29435,0	234,867
783	1056,15	2123,11	1635,70	186,25	261,68	13,0388	38249,9	29468,7	234,907
784	1057,15	2125,44	1637,57	187,14	260,68	13,0410	38292,0	29502,4	234,947
785	1058,15	2127,78	1639,44	188,04	259,68	13,0432	38334,0	29536,1	234,986
786	1059,15	2130,11	1641,31	188,94	258,69	13,0454	38376,1	29569,8	235,026
787	1060,15	2132,44	1643,18	189,85	257,70	13,0476	38418,1	29603,6	235,066
788	1061,15	2134,78	1645,06	190,75	256,72	13,0498	38460,2	29637,4	235,105
789	1062,15	2137,12	1646,93	191,67	255,74	13,0520	38502,3	29671,1	235,145
790	1063,15	2139,45	1648,81	192,58	254,76	13,0542	38544,4	29704,9	235,185
791	1064,15	2141,79	1650,68	193,50	253,79	13,0564	38586,5	29738,7	235,224
792	1065,15	2144,13	1652,56	194,42	252,82	13,0586	38628,6	29772,5	235,264
793	1066,15	2146,47	1654,44	195,35	251,86	13,0608	38670,8	29806,3	235,303
794	1067,15	2148,81	1656,32	196,28	250,90	13,0630	38712,9	29840,2	235,343
795	1068,15	2151,15	1658,19	197,22	249,94	13,0652	38755,1	29874,0	235,382
796	1069,15	2153,49	1660,07	198,15	248,99	13,0674	38797,2	29907,9	235,422
797	1070,15	2155,83	1661,95	199,10	248,05	13,0696	38839,4	29941,8	235,461
798	1071,15	2158,17	1663,83	200,04	247,11	13,0718	38881,6	29975,6	235,501
799	1072,15	2160,52	1665,72	200,99	246,17	13,0739	38923,8	30009,5	235,540
800	1073,15	2162,86	1667,60	201,95	245,23	13,0761	38966,1	30043,5	235,579
801	1074,15	2165,20	1669,48	202,90	244,30	13,0783	39008,3	30077,4	235,619
802	1075,15	2167,55	1671,37	203,87	243,38	13,0805	39050,6	30111,3	235,658
803	1076,15	2169,89	1673,25	204,83	242,45	13,0827	39092,8	30145,3	235,697
804	1077,15	2172,24	1675,13	205,80	241,54	13,0849	39135,1	30179,2	235,737
805	1078,15	2174,59	1677,02	206,77	240,62	13,0870	39177,4	30213,2	235,776
806	1079,15	2176,94	1678,91	207,75	239,71	13,0892	39219,7	30247,2	235,815
807	1080,15	2179,28	1680,79	208,73	238,81	13,0914	39262,0	30281,2	235,854
808	1081,15	2181,63	1682,68	209,72	237,90	13,0936	39304,3	30315,2	235 894
809	1082,15	2183,98	1684,57	210,71	237,00	13,0957	39346,7	30349,2	235,933
810	1083,15	2186,34	1686,46	211,70	236,11	13,0979	39389,0	30383,3	235,972
811	1084,15	2188,69	1688,35	212,70	235,22	13,1001	39431,4	30417,3	236,011
812	1085,15	2191,04	1690,24	213,70	234,33	13,1022	39473,8	30451,4	236 050
813	1086,15	2193,39	1692,13	214,71	233,45	13,1044	39516,1	30485,4	236,089
814	1087,15	2195,75	1694,02	215,72	232,57	13,1066	39558,5	30519,5	236,128
815	1088,15	2198,10	1695,92	216,73	231,70	13,1087	39601,0	30553,6	236,167
816	1089,15	2200,45	1697,81	217,75	230,82	13,1109	39643,4	30587,7	236,206
817	1090,15	2202,81	1699,70	218,77	229,96	13,1131	39685,8	30621,9	236,245
818	1091,15	2205,17	1701,60	219,80	229,09	13,1152	39728,3	30656,0	236,284
819	1092,15	2207,52	1703,49	220,83	228,23	13,1174	39770,7	30690,2	236,323
820	1093,15	2209,88	1705,39	221,87	227,38	13,1195	39813,2	30724,3	236,362
821	1094,15	2212,24	1707,29	222,90	226,52	13,1217	39855,7	30758,5	236,400
822	1095,15	2214,60	1709,19	223,95	225,67	13,1238	39898,2	30792,7	236,439
823	1096,15	2216,96	1711,08	225,00	224,83	13,1260	39940,7	30826,9	236,478
824	1097,15	2219,32	1712,98	226,05	223,99	13,1282	39983,2	30861,1	236,517
825	1098,15	2221,68	1714,88	227,10	223,15	13,1303	40025,8	30895,3	236,556
826	1099,15	2224,04	1716,78	228,16	222,31	13,1325	40068,3	30929,6	236,594
827	1100,15	2226,40	1718,68	229,23	221,48	13,1346	40110,9	30963,8	236,633
828	1101,15	2228,77	1720,59	230,30	220,65	13,1368	40153,5	30998,1	236,672
829	1102,15	2231,13	1722,49	231,37	219,83	13,1389	40196,1	31032,3	236,710

Steam (*Continued*)

t	T	h	u	π_0	θ_0	s^0	H	U	S
830	1103,15	2233,50	1724,39	232,45	219,01	13,1410	40238,7	31066,6	236,749
831	1104,15	2235,86	1726,29	233,53	218,19	13,1432	40281,3	31100,9	236,788
832	1105,15	2238,23	1728,20	234,62	217,38	13,1453	40323,9	31135,2	236,826
833	1106,15	2240,59	1730,10	235,71	216,57	13,1475	40366,5	31169,6	236,865
834	1107,15	2242,96	1732,01	236,80	215,76	13,1496	40409,2	31203,9	236,903
835	1108,15	2245,33	1733,92	237,90	214,96	13,1517	40451,9	31238,2	236,942
836	1109,15	2247,70	1735,82	239,01	214,16	13,1539	40494,5	31272,6	236,980
837	1110,15	2250,07	1737,73	240,12	213,36	13,1560	40537,2	31307,0	237,019
838	1111,15	2252,44	1739,64	241,23	212,57	13,1582	40579,9	31341,4	237,057
839	1112,15	2254,81	1741,55	242,35	211,78	13,1603	40622,6	31375,8	237,096
840	1113,15	2257,18	1743,46	243,47	210,99	13,1624	40665,4	31410,2	237,134
841	1114,15	2259,55	1745,37	244,59	210,21	13,1645	40708,1	31444,6	237,172
842	1115,15	2261,93	1747,28	245,72	209,43	13,1667	40750,9	31479,0	237,211
843	1116,15	2264,30	1749,19	246,86	208,65	13,1688	40793,6	31513,5	237,249
844	1117,15	2266,67	1751,11	248,00	207,88	13,1709	40836,4	31548,0	237,287
845	1118,15	2269,05	1753,02	249,14	207,11	13,1731	40879,2	31582,4	237,326
846	1119,15	2271,42	1754,94	250,29	206,34	13,1752	40922,0	31616,9	237,364
847	1120,15	2273,80	1756,85	251,45	205,58	13,1773	40964,8	31651,4	237,402
848	1121,15	2276,18	1758,77	252,61	204,82	13,1794	41007,6	31685,9	237,440
849	1122,15	2278,56	1760,68	253,77	204,06	13,1815	41050,5	31720,4	237,479
850	1123,15	2280,93	1762,60	254,94	203,31	13,1837	41093,3	31755,0	237,517
851	1124,15	2283,31	1764,52	256,11	202,56	13,1858	41136,2	31789,5	237,555
852	1125,15	2285,69	1766,43	257,29	201,81	13,1879	41179,1	31824,1	237,593
853	1126,15	2288,07	1768,35	258,47	201,07	13,1900	41221,9	31858,7	237,631
854	1127,15	2290,46	1770,27	259,65	200,33	13,1921	41264,8	31893,3	237,669
855	1128,15	2292,84	1772,19	260,85	199,59	13,1942	41307,8	31927,9	237,707
856	1129,15	2295,22	1774,12	262,04	198,85	13,1963	41350,7	31962,5	237,745
857	1130,15	2297,60	1776,04	263,24	198,12	13,1985	41393,6	31997,1	237,783
858	1131,15	2299,99	1777,96	264,45	197,39	13,2006	41436,6	32031,7	237,821
859	1132,15	2302,37	1779,88	265,66	196,67	13,2027	41479,5	32066,4	237,859
860	1133,15	2304,76	1781,81	266,87	195,95	13,2048	41522,5	32101,0	237,897
861	1134,15	2307,14	1783,73	268,09	195,23	13,2069	41565,5	32135,7	237,935
862	1135,15	2309,53	1785,66	269,32	194,51	13,2090	41608,5	32170,4	237,973
863	1136,15	2311,92	1787,58	270,55	193,80	13,2111	41651,5	32205,1	238,011
864	1137,15	2314,31	1789,51	271,78	193,09	13,2132	41694,5	32239,8	238,049
865	1138,15	2316,69	1791,44	273,02	192,38	13,2153	41737,6	32274,5	238,087
866	1139,15	2319,08	1793,36	274,26	191,67	13,2174	41780,6	32309,2	238,124
867	1140,15	2321,47	1795,29	275,51	190,97	13,2195	41823,7	32344,0	238,162
868	1141,15	2323,86	1797,22	276,77	190,27	13,2216	41866,7	32378,8	238,200
869	1142,15	2326,26	1799,15	278,03	189,58	13,2237	41909,8	32413,5	238,238
870	1143,15	2328,65	1801,08	279,29	188,89	13,2258	41952,9	32448,3	238,275
871	1144,15	2331,04	1803,01	280,56	188,20	13,2279	41996,0	32483,1	238,313
872	1145,15	2333,43	1804,95	281,83	187,51	13,2299	42039,2	32517,9	238,351
873	1146,15	2335,83	1806,88	283,11	186,82	13,2320	42082,3	32552,7	238,388
874	1147,15	2338,22	1808,81	284,40	186,14	13,2341	42125,4	32587,6	238,426
875	1148,15	2340,62	1810,75	285,69	185,46	13,2362	42168,6	32622,4	238,464
876	1149,15	2343,02	1812,68	286,98	184,79	13,2383	42211,8	32657,3	238,501
877	1150,15	2345,41	1814,62	288,28	184,12	13,2404	42255,0	32692,1	238,539
878	1151,15	2347,81	1816,55	289,59	183,45	13,2425	42298,2	32727,0	238,576
879	1152,15	2350,21	1818,49	290,90	182,78	13,2446	42341,4	32761,9	238,614
880	1153,15	2352,61	1820,43	292,21	182,11	13,2466	42384,6	32796,8	238,651
881	1154,15	2355,01	1822,37	293,53	181,45	13,2487	42427,8	32831,7	238,689
882	1155,15	2357,41	1824,30	294,86	180,79	13,2508	42471,1	32866,7	238,726
883	1156,15	2359,81	1826,24	296,19	180,14	13,2529	42514,3	32901,6	238,764
884	1157,15	2362,21	1828,18	297,52	179,48	13,2549	42557,6	32936,6	238,801

Steam (*Continued*)

t	T	h	u	π_0	0_0	s^0	H	U	S
885	1158,15	2364,61	1830,13	298,86	178,83	13,2570	42600,9	32971,5	238,839
886	1159,15	2367,02	1832,07	300,21	178,18	13,2591	42644,2	33006,5	238,876
887	1160,15	2369,42	1834,01	301,56	177,54	13,2612	42687,5	33041,5	238,913
888	1161,15	2371,83	1835,95	302,92	176,90	13,2632	42730,8	33076,5	238,951
889	1162,15	2374,23	1837,90	304,28	176,26	13,2653	42774,1	33111,5	238,988
890	1163,15	2376,64	1839,84	305,65	175,62	13,2674	42817,5	33146,6	239,025
891	1164,15	2379,04	1841,79	307,02	174,98	13,2694	42860,8	33181,6	239,062
892	1165,15	2381,45	1843,73	308,40	174,35	13,2715	42904,2	33216,7	239,100
893	1166,15	2383,86	1845,68	309,78	173,72	13,2736	42947,6	33251,7	239,137
894	1167,15	2386,27	1847,62	311,17	173,09	13,2756	42991,0	33286,8	239,174
895	1168,15	2388,68	1849,57	312,56	172,47	13,2777	43034,4	33321,9	239,211
896	1169,15	2391,09	1851,52	313,96	171,85	13,2798	43077,8	33357,0	239,248
897	1170,15	2393,50	1853,47	315,37	171,23	13,2818	43121,2	33392,1	239,285
898	1171,15	2395,91	1855,42	316,78	170,61	13,2839	43164,7	33427,2	239,323
899	1172,15	2398,32	1857,37	318,20	170,00	13,2860	43208,1	33462,4	239,360
900	1173,15	2400,73	1859,32	319,62	169,39	13,2880	43251,6	33497,5	239,397
901	1174,15	2403,15	1861,27	321,05	168,78	13,2901	43295,1	33532,7	239,434
902	1175,15	2405,56	1863,23	322,48	168,17	13,2921	43338,6	33567,9	239,471
903	1176,15	2407,97	1865,18	323,92	167,56	13,2942	43382,1	33603,1	239,508
904	1177,15	2410,39	1867,13	325,36	166,96	13,2962	43425,6	33638,3	239,545
905	1178,15	2412,81	1869,09	326,81	166,36	13,2983	43469,1	33673,5	239,582
906	1179,15	2415,22	1871,04	328,27	165,77	13,3003	43512,6	33708,7	239,619
907	1180,15	2417,64	1873,00	329,73	165,17	13,3024	43556,2	33743,9	239,656
908	1181,15	2420,06	1874,96	331,19	164,58	13,3044	43599,8	33779,2	239,693
909	1182,15	2422,48	1876,91	332,67	163,99	13,3065	43643,3	33814,5	239,729
910	1183,15	2424,90	1878,87	334,14	163,40	13,3085	43686,9	33849,7	239,766
911	1184,15	2427,32	1880,83	335,63	162,82	13,3106	43730,5	33885,0	239,803
912	1185,15	2429,74	1882,79	337,12	162,24	13,3126	43774,1	33920,3	239,840
913	1186,15	2432,16	1884,75	338,61	161,66	13,3146	43817,8	33955,6	239,877
914	1187,15	2434,58	1886,71	340,11	161,08	13,3167	43861,4	33990,9	239,913
915	1188,15	2437,00	1888,67	341,62	160,50	13,3187	43905,0	34026,3	239,950
916	1189,15	2439,43	1890,63	343,13	159,93	13,3208	43948,7	34061,6	239,987
917	1190,15	2441,85	1892,59	344,65	159,36	13,3228	43992,4	34097,0	240,024
918	1191,15	2444,28	1894,56	346,17	158,79	13,3248	44036,1	34132,4	240,060
919	1192,15	2446,70	1896,52	347,70	158,22	13,3269	44079,8	34167,7	240,097
920	1193,15	2449,13	1898,49	349,24	157,66	13,3289	44123,5	34203,1	240,134
921	1194,15	2451,55	1900,45	350,78	157,10	13,3309	44167,2	34238,5	240,170
922	1195,15	2453,98	1902,42	352,33	156,54	13,3330	44210,9	34271,0	240,207
923	1196,15	2456,41	1904,38	353,88	155,98	13,3350	44254,7	34309,4	240,244
924	1197,15	2458,84	1906,35	355,44	155,43	13,3370	44298,4	34344,8	240,280
925	1198,15	2461,27	1908,32	357,01	154,88	13,3391	44342,2	34380,3	240,317
926	1199,15	2463,70	1910,29	358,58	154,33	13,3411	44386,0	34415,8	240,353
927	1200,15	2466,13	1912,26	360,16	153,78	13,3431	44429,8	34451,2	240,390
928	1201,15	2468,56	1914,23	361,74	153,23	13,3451	44473,6	34486,7	240,426
929	1202,15	2470,99	1916,20	363,33	152,69	13,3472	44517,4	34522,2	240,463
930	1203,15	2473,43	1918,17	364,93	152,15	13,3492	44561,2	34557,8	240,499
931	1204,15	2475,86	1920,14	366,53	151,61	13,3512	44605,1	34593,3	240,535
932	1205,15	2478,29	1922,12	368,14	151,07	13,3532	44648,9	34628,8	240,572
933	1206,15	2480,73	1924,09	369,75	150,54	13,3553	44692,8	34664,4	240,608
934	1207,15	2483,16	1926,06	371,38	150,00	13,3573	44736,7	34699,9	240,645
935	1208,15	2485,60	1928,04	373,00	149,47	13,3593	44780,6	34735,5	240,681
936	1209,15	2488,04	1930,01	374,64	148,94	13,3613	44824,5	34771,1	240,717
937	1210,15	2490,47	1931,99	376,27	148,42	13,3633	44868,4	34806,7	240,754
938	1211,15	2492,91	1933,97	377,92	147,89	13,3653	44912,3	34842,3	240,790
939	1212,15	2495,35	1935,94	379,57	147,37	13,3673	44956,3	34877,9	240,826

Steam (*Continued*)

t	T	h	u	π_0	θ_0	s^θ	H	U	S
940	1213,15	2497,79	1937.92	381,23	146,85	13,3694	45000,2	34913,6	240,862
941	1214,15	2500,23	1939,90	382,90	146,33	13,3714	45044,2	34949,2	240,899
942	1215,15	2502,67	1941,88	384,57	145,82	13,3734	45088,1	34984,9	240,935
943	1216,15	2505,11	1943,86	386,24	145,30	13,3754	45132,1	35020,6	240,971
944	1217,15	2507,56	1945,84	387,93	144,79	13,3774	45176,1	35056,2	241,007
945	1218,15	2510,00	1947,82	389,62	144,28	13,3794	45220,1	35091,9	241,043
946	1219,15	2512,44	1949,80	391,31	143,78	13,3814	45264,2	35127,7	241,079
947	1220,15	2514,89	1951,79	393,02	143,27	13,3834	45308,2	35163,4	241,116
948	1221,15	2517,33	1953,77	394,73	142,77	13,3854	45352,3	35199,1	241,152
949	1222,15	2519,78	1955,75	396,44	142,26	13,3874	45396,3	35234,9	241,188
950	1223,15	2522,22	1957,74	398,16	141,77	13,3894	45440,4	35270,6	241,224
951	1224,15	2524,67	1959,72	399,89	141,27	13,3914	45484,5	35306,4	241,260
952	1225,15	2527,12	1961,71	401,63	140,77	13,3934	45528,6	35342,2	241,296
953	1226,15	2529,57	1963,70	403,37	140,28	13,3954	45572,7	35378,0	241,332
954	1227,15	2532,02	1965,68	405,12	139,79	13,3974	45616,8	35413,8	241,368
955	1228,15	2534,46	1967,67	406,87	139,30	13,3994	45660,9	35449,6	241,404
956	1229,15	2536,91	1969,66	408,64	138,81	13,4014	45705,1	35485,4	241,440
957	1230,15	2539,37	1971,65	410,41	138,32	13,4034	45749,2	35521,2	241,475
958	1231,15	2541,82	1973,64	412,18	137,84	13,4054	45793,4	35557,1	241,511
959	1232,15	2544,27	1975,63	413,96	137,36	13,4074	45837,6	35593,0	241,547
960	1233,15	2546,72	1977,62	415,75	136,88	13,4094	45881,7	35628,8	241,583
961	1234,15	2549,18	1979,61	417,55	136,40	13,4114	45925,9	35664,7	241,619
962	1235,15	2551,63	1981,61	419,35	135,92	13,4133	45970,2	35700,6	241,655
963	1236,15	2554,08	1983,60	421,16	135,45	13,4153	46014,4	35736,5	241,691
964	1237,15	2556,54	1985,59	422,97	134,98	13,4173	46058,6	35772,5	241,726
965	1238,15	2559,00	1987,59	424,80	134,51	13,4193	46102,9	35808,4	241,762
966	1239,15	2561,45	1989,58	426,63	134,04	13,4213	46147,1	35844,3	241,798
967	1240,15	2563,91	1991,58	428,46	133,57	13,4233	46191,4	35880,3	241,833
968	1241,15	2566,37	1993,58	430,31	133,11	13,4252	46235,7	35916,3	241,869
969	1242,15	2568,83	1995,57	432,16	132,64	13,4272	46280,0	35952,2	241,905
970	1243,15	2571,29	1997,57	434,01	132,18	13,4292	46324,3	35988,2	241,941
971	1244,15	2573,75	1999,57	435,88	131,72	13,4312	46368,6	36024,2	241,976
972	1245,15	2576,21	2001,57	437,75	131,26	13,4332	46413,0	36060,3	242,012
973	1246,15	2578,67	2003,57	439,63	130,81	13,4351	46457,3	36096,3	242,047
974	1247,15	2581,13	2005,57	441,51	130,35	13,4371	46501,7	36132,3	242,083
975	1248,15	2583,59	2007,57	443,41	129,90	13,4391	46546,0	36168,4	242,119
976	1249,15	2586,06	2009,57	445,31	129,45	13,4411	46590,4	36204,5	242,154
977	1250,15	2588,52	2011,57	447,21	129,00	13,4430	46634,8	36240,5	242,190
978	1251,15	2590,99	2013,58	449,13	128,56	13,4450	46679,2	36276,6	242,225
979	1252,15	2593,45	2015,58	451,05	128,11	13,4470	46723,6	36312,7	242,261
980	1253,15	2595,92	2017,59	452,97	127,67	13,4489	46768,0	36348,8	242,296
981	1254,15	2598,38	2019,59	454,91	127,23	13,4509	46812,5	36385,0	242,331
982	1255,15	2600,85	2021,60	456,85	126,79	13,4529	46856,9	36421,1	242,367
983	1256,15	2603,32	2023,60	458,80	126,35	13,4548	46901,4	36457,2	242,402
984	1257,15	2605,79	2025,61	460,76	125,91	13,4568	46945,9	36493,4	242,438
985	1258,15	2608,26	2027,62	462,72	125,48	13,4588	46990,4	36529,6	242,473
986	1259,15	2610,73	2029,63	464,70	125,04	13,4607	47034,8	36565,8	242,508
987	1260,15	2613,20	2031,64	466,67	124,61	13,4627	47079,4	36602,0	242,544
988	1261,15	2615,67	2033,65	468,66	124,18	13,4646	47123,9	36638,2	242,579
989	1262,15	2618,14	2035,66	470,65	123,75	13,4666	47168,4	36674,4	242,614
990	1263,15	2620,61	2037,67	472,66	123,33	13,4686	47213,0	36710,6	242,650
991	1264,15	2623,09	2039,68	474,66	122,90	13,4705	47257,5	36746,9	242,685
992	1265,15	2625,56	2041,69	476,68	122,48	13,4725	47302,1	36783,1	242,720
993	1266,15	2628,03	2043,70	478,70	122,06	13,4744	47346,7	36819,4	242,755
994	1267,15	2630,51	2045,72	480,74	121,64	13,4764	47391,3	36855,6	242,791

t	T	h	u	π_0	θ_0	s^0	H	U	S
995	1268,15	2632,98	2047,73	482,77	121,22	13,4783	47435,9	36891,9	242,826
996	1269,15	2635,46	2049,75	484,82	120,80	13,4803	47480,5	36928,2	242,861
997	1270,15	2637,94	2051,76	486,87	120,39	13,4822	47525,1	36964,5	242,896
998	1271,15	2640,42	2053,78	488,94	119,98	13,4842	47569,7	37000,9	242,931
999	1272,15	2642,89	2055,80	491,01	119,56	13,4861	47614,4	37037,2	242,966
1000	1273,15	2645,37	2057,81	493,08	119,15	13,4881	47659,0	37073,6	243,001
1001	1274,15	2647,85	2059,83	495,17	118,75	13,4900	47703,7	37109,9	243,036
1002	1275,15	2650,33	2061,85	497,26	118,34	13,4920	47748,4	37146,3	243,072
1003	1276,15	2652,81	2063,87	499,36	117,93	13,4939	47793,1	37182,7	243,107
1004	1277,15	2655,30	2065,89	501,47	117,53	13,4959	47837,8	37219,1	243,142
1005	1278,15	2657,78	2067,91	503,58	117,13	13,4978	47882,5	37255,5	243,177
1006	1279,15	2660,26	2069,93	505,70	116,73	13,4998	47927,3	37291,9	243,212
1007	1280,15	2662,74	2071,95	507,84	116,33	13,5017	47972,0	37328,3	243,247
1008	1281,15	2665,23	2073,98	509,97	115,93	13,5036	48016,8	37364,7	243,281
1009	1282,15	2667,71	2076,00	512,12	115,54	13,5056	48061,5	37401,2	243,316
1010	1283,15	2670,20	2078,02	514,28	115,14	13,5075	48106,3	37437,7	243,351
1011	1284,15	2672,68	2080,05	516,44	114,75	13,5094	48151,1	37474,1	243,386
1012	1285,15	2675,17	2082,07	518,61	114,36	13,5114	48195,9	37510,6	243,421
1013	1286,15	2677,66	2084,10	520,79	113,97	13,5133	48240,7	37547,1	243,456
1014	1287,15	2680,15	2086,13	522,98	113,58	13,5153	48285,5	37583,6	243,491
1015	1288,15	2682,64	2088,15	525,17	113,19	13,5172	48330,4	37620,2	243,526
1016	1289,15	2685,12	2090,18	527,37	112,81	13,5191	48375,2	37656,7	243,560
1017	1290,15	2687,61	2092,21	529,58	112,42	13,5210	48420,1	37693,2	243,595
1018	1291,15	2690,11	2094,24	531,80	112,04	13,5230	48464,9	37729,8	243,630
1019	1292,15	2692,60	2096,27	534,03	111,66	13,5249	48509,8	37766,4	243,665
1020	1293,15	2695,09	2098,30	536,27	111,28	13,5268	48554,7	37802,9	243,699
1021	1294,15	2697,58	2100,33	538,51	110,90	13,5288	48599,6	37839,5	243,734
1022	1295,15	2700,07	2102,36	540,76	110,53	13,5307	48644,5	37876,1	243,769
1023	1296,15	2702,57	2104,39	543,02	110,15	13,5326	48689,5	37912,7	243,804
1024	1297,15	2705,06	2106,43	545,29	109,78	13,5345	48734,4	37949,4	243,838
1025	1298,15	2707,56	2108,46	547,57	109,41	13,5365	48779,3	37986,0	243,873
1026	1299,15	2710,05	2110,49	549,85	109,03	13,5384	48824,3	38022,6	243,907
1027	1300,15	2712,55	2112,53	552,14	108,67	13,5403	48869,3	38059,3	243,942
1028	1301,15	2715,05	2114,56	554,44	108,30	13,5422	48914,3	38096,0	243,977
1029	1302,15	2717,54	2116,60	556,75	107,93	13,5441	48959,3	38132,7	244,011
1030	1303,15	2720,04	2118,64	559,07	107,57	13,5461	49004,3	38169,4	244,046
1031	1304,15	2722,54	2120,67	561,40	107,20	13,5480	49049,3	38206,1	244,080
1032	1305,15	2725,04	2122,71	563,74	106,84	13,5499	49094,3	38242,8	244,115
1033	1306,15	2727,54	2124,75	566,08	106,48	13,5518	49139,4	38279,5	244,149
1034	1307,15	2730,04	2126,79	568,43	106,12	13,5537	49184,4	38316,2	244,184
1035	1308,15	2732,54	2128,83	570,79	105,76	13,5556	49229,5	38353,0	244,218
1036	1309,15	2735,04	2130,87	573,16	105,40	13,5575	49274,6	38389,8	244,253
1037	1310,15	2737,55	2132,91	575,54	105,05	13,5595	49319,7	38426,5	244,287
1038	1311,15	2740,05	2134,95	577,93	104,70	13,5614	49364,8	38463,3	244,321
1039	1312,15	2742,55	2137,00	580,32	104,34	13,5633	49409,9	38500,1	244,356
1040	1313,15	2745,06	2139,04	582,73	103,99	13,5652	49455,0	38536,9	244,390
1041	1314,15	2747,56	2141,08	585,14	103,64	13,5671	49500,1	38573,7	244,425
1042	1315,15	2750,07	2143,13	587,56	103,29	13,5690	49545,3	38610,6	244,459
1043	1316,15	2752,58	2145,17	589,99	102,95	13,5709	49590,4	38647,4	244,493
1044	1317,15	2755,08	2147,22	592,43	102,60	13,5728	49635,6	38684,3	244,528
1045	1318,15	2757,59	2149,26	594,88	102,25	13,5747	49680,8	38721,1	244,562
1046	1319,15	2760,10	2151,31	597,34	101,91	13,5766	49726,0	38758,0	244,596
1047	1320,15	2762,61	2153,36	599,80	101,57	13,5785	49771,2	38794,9	244,630
1048	1321,15	2765,12	2155,41	602,28	101,23	13,5804	49816,4	38831,8	244,665
1049	1322,15	2767,63	2157,45	604,76	100,89	13,5823	49861,6	38868,7	244,699

Steam (*Continued*)

t	T	h	u	$\pi_\circ$	$\theta_\circ$	s°	H	U	S
1050	1323,15	2770,14	2159,50	607,25	100,55	13,5842	49906,8	38905,6	244,733
1051	1324,15	2772,65	2161,55	609,76	100,21	13,5861	49952,1	38942,6	244,767
1052	1325,15	2775,16	2163,60	612,27	99,883	13,5880	49997,3	38979,5	244,801
1053	1326,15	2777,68	2165,66	614,79	99,549	13,5899	50042,6	39016,5	244,836
1054	1327,15	2780,19	2167,71	617,32	99,216	13,5918	50087,9	39053,4	244,870
1055	1328,15	2782,70	2169,76	619,85	98,884	13,5937	50133,2	39090,4	244,904
1056	1329,15	2785,22	2171,81	622,40	98,553	13,5956	50178,5	39127,4	244,938
1057	1330,15	2787,73	2173,87	624,96	98,224	13,5975	50223,8	39164,4	244,972
1058	1331,15	2790,25	2175,92	627,52	97,896	13,5994	50269,1	39201,4	245,006
1059	1332,15	2792,77	2177,98	630,10	97,569	13,6012	50314,5	39238,4	245,040
1060	1333,15	2795,28	2180,03	632,68	97,244	13,6031	50359,8	39275,5	245,074
1061	1334,15	2797,80	2182,09	635,27	96,919	13,6050	50405,2	39312,5	245,108
1062	1335,15	2800,32	2184,15	637,88	96,596	13,6069	50450,5	39349,6	245,142
1063	1336,15	2802,84	2186,20	640,49	96,274	13,6088	50495,9	39386,6	245,176
1064	1337,15	2805,36	2188,26	643,11	95,954	13,6107	50541,3	39423,7	245,210
1065	1338,15	2807,88	2190,32	645,74	95,634	13,6126	50586,7	39460,8	245,244
1066	1339,15	2810,40	2192,38	648,38	95,316	13,6144	50632,1	39497,9	245,278
1067	1340,15	2812,92	2194,44	651,03	94,999	13,6163	50677,6	39535,0	245,312
1068	1341,15	2815,44	2196,50	653,69	94,683	13,6182	50723,0	39572,1	245,346
1069	1342,15	2817,96	2198,56	656,36	94,368	13,6201	50768,5	39609,3	245,380
1070	1343,15	2820,49	2200,62	659,04	94,055	13,6220	50813,9	39646,4	245,413
1071	1344,15	2823,01	2202,69	661,73	93,743	13,6239	50859,4	39683,6	245,447
1072	1345,15	2825,54	2204,75	664,42	93,432	13,6257	50904,9	39720,7	245,481
1073	1346,15	2828,06	2206,81	667,13	93,122	13,6276	50950,4	39757,9	245,515
1074	1347,15	2830,59	2208,88	669,85	92,813	13,6295	50995,9	39795,1	245,549
1075	1348,15	2833,11	2210,94	672,57	92,505	13,6314	51041,4	39832,3	245,582
1076	1349,15	2835,64	2213,01	675,31	92,199	13,6332	51086,9	39869,5	245,616
1077	1350,15	2838,17	2215,07	678,06	91,893	13,6351	51132,5	39906,8	245,650
1078	1351,15	2840,70	2217,14	680,81	91,589	13,6370	51178,0	39944,0	245,684
1079	1352,15	2843,23	2219,21	683,58	91,286	13,6388	51223,6	39981,2	245,717
1080	1353,15	2845,76	2221,28	686,35	90,984	13,6407	51269,1	40018,5	245,751
1081	1354,15	2848,29	2223,34	689,14	90,683	13,6426	51314,7	40055,8	245,785
1082	1355,15	2850,82	2225,41	691,93	90,384	13,6445	51360,3	40093,0	245,818
1083	1356,15	2853,35	2227,48	694,74	90,085	13,6463	51405,9	40130,3	245,852
1084	1357,15	2855,88	2229,55	697,55	89,788	13,6482	51451,5	40167,6	245,886
1085	1358,15	2858,41	2231,62	700,38	89,492	13,6500	51497,2	40204,9	245,919
1086	1359,15	2860,95	2233,70	703,21	89,196	13,6519	51542,8	40242,3	245,953
1087	1360,15	2863,48	2235,77	706,06	88,902	13,6538	51588,5	40279,6	245,986
1088	1361,15	2866,01	2237,84	708,92	88,609	13,6556	51634,1	40317,0	246,020
1089	1362,15	2868,55	2239,91	711,78	88,317	13,6575	51679,8	40354,3	246,054
1090	1363,15	2871,08	2241,99	714,66	88,027	13,6594	51725,5	40391,7	246,087
1091	1364,15	2873,62	2244,06	717,54	87,737	13,6612	51771,2	40429,1	246,121
1092	1365,15	2876,16	2246,14	720,44	87,448	13,6631	51816,9	40466,4	246,154
1093	1366,15	2878,70	2248,22	723,35	87,161	13,6649	51862,6	40503,8	246,188
1094	1367,15	2881,23	2250,29	726,26	86,874	13,6668	51908,3	40541,3	246,221
1095	1368,15	2883,77	2252,37	729,19	86,589	13,6687	51954,0	40578,7	246,254
1096	1369,15	2886,31	2254,45	732,13	86,304	13,6705	51999,8	40616,1	246,288
1097	1370,15	2888,85	2256,53	735,07	86,021	13,6724	52045,6	40653,6	246,321
1098	1371,15	2891,39	2258,60	738,03	85,739	13,6742	52091,3	40691,0	246,355
1099	1372,15	2893,93	2260,68	741,00	85,458	13,6761	52137,1	40728,5	246,388
1100	1373,15	2896,48	2262,76	743,98	85,177	13,6779	52182,9	40766,0	246,421
1101	1374,15	2899,02	2264,85	746,97	84,898	13,6798	52228,7	40803,5	246,455
1102	1375,15	2901,56	2266,93	749,97	84,620	13,6816	52274,5	40841,0	246,488
1103	1376,15	2904,10	2269,01	752,98	84,343	13,6835	52320,3	40878,5	246,521
1104	1377,15	2906,65	2271,09	756,00	84,067	13,6853	52366,2	40916,0	246,555

Steam (*Continued*)

t	T	h	u	π_0	θ_0	s^0	H	U	S
1105	1378,15	2909,19	2273,17	759,03	83,792	13,6872	52412,0	40953,5	246,588
1106	1379,15	2911,74	2275,26	762,08	83,518	13,6890	52457,9	40991,1	246,621
1107	1380,15	2914,28	2277,34	765,13	83,245	13,6909	52503,7	41028,6	246,654
1108	1381,15	2916,83	2279,43	768,19	82,973	13,6927	52549,6	41066,2	246,688
1109	1382,15	2919,38	2281,51	771,27	82,702	13,6945	52595,5	41103,8	246,721
1110	1383,15	2921,93	2283,60	774,35	82,432	13,6964	52641,4	41141,3	246,754
1111	1384,15	2924,47	2285,69	777,45	82,163	13,6982	52687,3	41178,9	246,787
1112	1385,15	2927,02	2287,77	780,56	81,895	13,7001	52733,3	41216,5	246,820
1113	1386,15	2929,57	2289,86	783,68	81,628	13,7019	52779,2	41254,2	246,854
1114	1387,15	2932,12	2291,95	786,81	81,362	13,7037	52825,1	41291,8	246,887
1115	1388,15	2934,67	2294,04	789,95	81,097	13,7056	52871,1	41329,4	246,920
1116	1389,15	2937,23	2296,13	793,10	80,833	13,7074	52917,1	41367,1	246,953
1117	1390,15	2939,78	2298,22	796,26	80,570	13,7093	52963,0	41404,8	246,986
1118	1391,15	2942,33	2300,31	799,43	80,308	13,7111	53009,0	41442,4	247,019
1119	1392,15	2944,88	2302,40	802,62	80,047	13,7129	53055,0	41480,1	247,052
1120	1393,15	2947,44	2304,50	805,81	79,787	13,7148	53101,0	41517,8	247,085
1121	1394,15	2949,99	2306,59	809,02	79,528	13,7166	53147,0	41555,5	247,118
1122	1395,15	2952,55	2308,68	812,24	79,269	13,7184	53193,1	41593,2	247,151
1123	1396,15	2955,10	2310,78	815,47	79,012	13,7203	53239,1	41631,0	247,184
1124	1397,15	2957,66	2312,87	818,71	78,756	13,7221	53285,2	41668,7	247,217
1125	1398,15	2960,22	2314,97	821,96	78,500	13,7239	53331,2	41706,4	247,250
1126	1399,15	2962,77	2317,06	825,22	78,246	13,7257	53377,3	41744,2	247,283
1127	1400,15	2965,33	2319,16	828,50	77,992	13,7276	53423,4	41782,0	247,316
1128	1401,15	2967,89	2321,26	831,78	77,740	13,7294	53469,5	41819,8	247,349
1129	1402,15	2970,45	2323,35	835,08	77,488	13,7312	53515,6	41857,5	247,382
1130	1403,15	2973,01	2325,45	838,39	77,237	13,7331	53561,7	41895,3	247,415
1131	1404,15	2975,57	2327,55	841,71	76,988	13,7349	53607,8	41933,2	247,448
1132	1405,15	2978,13	2329,65	845,04	76,739	13,7367	53654,0	41971,0	247,480
1133	1406,15	2980,69	2331,75	848,38	76,491	13,7385	53700,1	42008,8	247,513
1134	1407,15	2983,25	2333,85	851,74	76,243	13,7403	53746,3	42046,7	247,546
1135	1408,15	2985,82	2335,95	855,10	75,997	13,7422	53792,5	42084,5	247,579
1136	1409,15	2988,38	2338,05	858,48	75,752	13,7440	53838,6	42122,4	247,612
1137	1410,15	2990,94	2340,16	861,87	75,508	13,7458	53884,8	42160,3	247,644
1138	1411,15	2993,51	2342,26	865,27	75,264	13,7476	53931,0	42198,2	247,677
1139	1412,15	2996,07	2344,36	868,69	75,021	13,7494	53977,2	42236,0	247,710
1140	1413,15	2998,64	2346,47	872,11	74,780	13,7513	54023,5	42274,0	247,743
1141	1414,15	3001,20	2348,57	875,55	74,539	13,7531	54069,7	42311,9	247,775
1142	1415,15	3003,77	2350,68	879,00	74,299	13,7549	54115,9	42349,8	247,808
1143	1416,15	3006,34	2352,78	882,46	74,060	13,7567	54162,2	42387,7	247,841
1144	1417,15	3008,91	2354,89	885,93	73,821	13,7585	54208,5	42425,7	247,873
1145	1418,15	3011,48	2357,00	889,42	73,584	13,7603	54254,7	42463,7	247,906
1146	1419,15	3014,04	2359,10	892,91	73,348	13,7621	54301,0	42501,6	247,939
1147	1420,15	3016,61	2361,21	896,42	73,112	13,7639	54347,3	42539,6	247,971
1148	1421,15	3019,18	2363,32	899,95	72,877	13,7658	54393,6	42577,6	248,004
1149	1422,15	3021,76	2365,43	903,48	72,643	13,7676	54440,0	42615,6	248,036
1150	1423,15	3024,33	2367,54	907,02	72,410	13,7694	54486,3	42653,6	248,069
1151	1424,15	3026,90	2369,65	910,58	72,178	13,7712	54532,6	42691,7	248,101
1152	1425,15	3029,47	2371,76	914,15	71,947	13,7730	54579,0	42729,7	248,134
1153	1426,15	3032,05	2373,88	917,73	71,716	13,7748	54625,3	42767,7	248,167
1154	1427,15	3034,62	2375,99	921,33	71,486	13,7766	54671,7	42805,8	248,199
1155	1428,15	3037,19	2378,10	924,94	71,257	13,7784	54718,1	42843,9	248,232
1156	1429,15	3039,77	2380,21	928,56	71,029	13,7802	54764,5	42881,9	248,264
1157	1430,15	3042,34	2382,33	932,19	70,802	13,7820	54810,9	42920,0	248,296
1158	1431,15	3044,92	2384,44	935,83	70,576	13,7838	54857,3	42958,1	248,329
1159	1432,15	3047,50	2386,56	939,49	70,350	13,7856	54903,7	42996,2	248,361

t	T	h	u	π_0	θ_0	s^0	H	U	S
1160	1433,15	3050,08	2388,67	943,16	70,125	13,7874	54950,2	43034,4	248,394
1161	1434,15	3052,65	2390,79	946,84	69,901	13,7892	54996,6	43072,5	248,426
1162	1435,15	3055,23	2392,91	950,53	69,678	13,7910	55043,1	43110,6	248,459
1163	1436,15	3057,81	2395,03	954,24	69,456	13,7928	55089,5	43148,8	248,491
1164	1437,15	3060,39	2397,14	957,96	69,234	13,7946	55136,0	43186,9	248,523
1165	1438,15	3062,97	2399,26	961,69	69,014	13,7964	55182,5	43225,1	248,556
1166	1439,15	3065,55	2401,38	965,44	68,794	13,7982	55229,0	43263,3	248,588
1167	1440,15	3068,13	2403,50	969,20	68,575	13,8000	55275,5	43301,5	248,620
1168	1441,15	3070,71	2405,62	972,97	68,356	13,8018	55322,0	43339,7	248,652
1169	1442,15	3073,30	2407,74	976,75	68,139	13,8035	55368,5	43377,9	248,685
1170	1443,15	3075,88	2409,86	980,55	67,922	13,8053	55415,1	43416,1	248,717
1171	1444,15	3078,46	2411,99	984,36	67,706	13,8071	55461,6	43454,4	248,749
1172	1445,15	3081,05	2414,11	988,18	67,491	13,8089	55508,2	43492,6	248,781
1173	1446,15	3083,63	2416,23	992,02	67,276	13,8107	55554,7	43530,9	248,814
1174	1447,15	3086,22	2418,36	995,87	67,062	13,8125	55601,3	43569,1	248,846
1175	1448,15	3088,81	2420,48	999,73	66,850	13,8143	55647,9	43607,4	248,878
1176	1449,15	3091,39	2422,61	1003,6	66,637	13,8161	55694,5	43645,7	248,910
1177	1450,15	3093,98	2424,73	1007,4	66,426	13,8179	55741,1	43684,0	248,942
1178	1451,15	3096,57	2426,86	1011,4	66,215	13,8196	55787,7	43722,3	248,975
1179	1452,15	3099,15	2428,99	1015,3	66,006	13,8214	55834,4	43760,6	249,007
1180	1453,15	3101,74	2431,11	1019,2	65,796	13,8232	55881,0	43798,9	249,039
1181	1454,15	3104,33	2433,24	1023,1	65,588	13,8250	55927,7	43837,3	249,071
1182	1455,15	3106,92	2435,37	1027,1	65,380	13,8268	55974,3	43875,6	249,103
1183	1456,15	3109,51	2437,50	1031,1	65,174	13,8285	56021,0	43914,0	249,135
1184	1457,15	3112,11	2439,63	1035,0	64,967	13,8303	56067,7	43952,3	249,167
1185	1458,15	3114,70	2441,76	1039,0	64,762	13,8321	56114,4	43990,7	249,199
1186	1459,15	3117,29	2443,89	1043,0	64,557	13,8339	56161,1	44029,1	249,231
1187	1460,15	3119,88	2446,02	1047,1	64,353	13,8357	56207,8	44067,5	249,263
1188	1461,15	3122,48	2448,15	1051,1	64,150	13,8374	56254,5	44105,9	249,295
1189	1462,15	3125,07	2450,29	1055,2	63,948	13,8392	56301,2	44144,3	249,327
1190	1463,15	3127,66	2452,42	1059,2	63,746	13,8410	56348,0	44182,8	249,359
1191	1464,15	3130,26	2454,55	1063,3	63,545	13,8427	56394,7	44221,2	249,391
1192	1465,15	3132,85	2456,69	1067,4	63,345	13,8445	56441,5	44259,7	249,423
1193	1466,15	3135,45	2458,82	1071,5	63,145	13,8463	56488,3	44298,1	249,455
1194	1467,15	3138,05	2460,96	1075,6	62,946	13,8481	56535,1	44336,6	249,487
1195	1468,15	3140,65	2463,09	1079,7	62,748	13,8498	56581,9	44375,1	249,519
1196	1469,15	3143,24	2465,23	1083,9	62,551	13,8516	56628,7	44413,6	249,550
1197	1470,15	3145,84	2467,37	1088,0	62,354	13,8534	56675,5	44452,1	249,582
1198	1471,15	3148,44	2469,50	1092,2	62,158	13,8551	56722,3	44490,6	249,614
1199	1472,15	3151,04	2471,64	1096,4	61,962	13,8569	56769,1	44529,1	249,646
1200	1473,15	3153,64	2473,78	1100,6	61,768	13,8587	56816,0	44567,6	249,678
1201	1474,15	3156,24	2475,92	1104,8	61,574	13,8604	56862,8	44606,2	249,710
1202	1475,15	3158,84	2478,06	1109,1	61,380	13,8622	56909,7	44644,7	249,741
1203	1476,15	3161,44	2480,20	1113,3	61,188	13,8640	56956,6	44683,3	249,773
1204	1477,15	3164,05	2482,34	1117,6	60,996	13,8657	57003,5	44721,8	249,805
1205	1478,15	3166,65	2484,48	1121,8	60,805	13,8675	57050,4	44760,4	249,837
1206	1479,15	3169,25	2486,62	1126,1	60,614	13,8692	57097,3	44799,0	249,868
1207	1480,15	3171,86	2488,77	1130,4	60,424	13,8710	57144,2	44837,6	249,900
1208	1481,15	3174,46	2490,91	1134,8	60,235	13,8728	57191,1	44876,2	249,932
1209	1482,15	3177,07	2493,05	1139,1	60,046	13,8745	57238,1	44914,8	219,963
1210	1483,15	3179,67	2495,20	1143,4	59,859	13,8763	57285,0	44953,5	249,995
1211	1484,15	3182,28	2497,34	1147,8	59,671	13,8780	57332,0	44992,1	250,027
1212	1485,15	3184,89	2499,49	1152,2	59,485	13,8798	57378,9	45030,8	250,058
1213	1486,15	3187,49	2501,63	1156,6	59,299	13,8815	57425,9	45069,4	250,090
1214	1487,15	3190,10	2503,78	1161,0	59,114	13,8833	57472,9	45108,1	250,122

Steam (*Continued*)

t	T	h	u	π_0	θ_0	s^0	H	U	S
1215	1488,15	3192,71	2505,93	1165,4	58,929	13,8851	57519,9	45146,8	250,153
1216	1489,15	3195,32	2508,07	1169,8	58,745	13,8868	57566,9	45185,5	250,185
1217	1490,15	3197,93	2510,22	1174,3	58,562	13,8886	57613,9	45224,2	250,216
1218	1491,15	3200,54	2512,37	1178,7	58,379	13,8903	57660,9	45262,9	250,248
1219	1492,15	3203,15	2514,52	1183,2	58,197	13,8921	57708,0	45301,6	250,279
1220	1493,15	3205,76	2516,67	1187,7	58,016	13,8938	57755,0	45340,3	250,311
1221	1494,15	3208,37	2518,82	1192,2	57,835	13,8956	57802,1	45379,1	250,342
1222	1495,15	3210,99	2520,97	1196,7	57,655	13,8973	57849,1	45417,8	250,374
1223	1496,15	3213,60	2523,12	1201,3	57,476	13,8991	57896,2	45456,6	250,405
1224	1497,15	3216,21	2525,28	1205,8	57,297	13,9008	57943,3	45495,4	250,437
1225	1498,15	3218,83	2527,43	1210,4	57,119	13,9025	57990,4	45534,2	250,468
1226	1499,15	3221,44	2529,58	1215,0	56,941	13,9043	58037,5	45572,9	250,500
1227	1500,15	3224,06	2531,74	1219,6	56,764	13,9060	58084,6	45611,7	250,531
1228	1501,15	3226,67	2533,89	1224,2	56,588	13,9078	58131,7	45650,6	250,562
1229	1502,15	3229,29	2536,04	1228,8	56,412	13,9095	58178,9	45689,4	250,594
1230	1503,15	3231,91	2538,20	1233,5	56,237	13,9113	58226,0	45728,2	250,625
1231	1504,15	3234,52	2540,36	1238,1	56,063	13,9130	58273,2	45767,0	250,657
1232	1505,15	3237,14	2542,51	1242,8	55,889	13,9147	58320,3	45805,9	250,688
1233	1506,15	3239,76	2544,67	1247,5	55,716	13,9165	58367,5	45844,8	250,719
1234	1507,15	3242,38	2546,83	1252,2	55,543	13,9182	58414,7	45883,6	250,751
1235	1508,15	3245,00	2548,98	1256,9	55,371	13,9200	58461,9	45922,5	250,782
1236	1509,15	3247,62	2551,14	1261,7	55,200	13,9217	58509,1	45961,4	250,813
1237	1510,15	3250,24	2553,30	1266,4	55,029	13,9234	58556,3	46000,3	250,844
1238	1511,15	3252,86	2555,46	1271,2	54,859	13,9252	58603,5	46039,2	250,876
1239	1512,15	3255,48	2557,62	1276,0	54,689	13,9269	58650,8	46078,1	250,907
1240	1513,15	3258,10	2559,78	1280,8	54,520	13,9286	58698,0	46117,0	250,938
1241	1514,15	3260,73	2561,94	1285,6	54,352	13,9304	58745,2	46156,0	250,969
1242	1515,15	3263,35	2564,11	1290,4	54,184	13,9321	58792,5	46194,9	251,001
1243	1516,15	3265,97	2566,27	1295,3	54,017	13,9338	58839,8	46233,9	251,032
1244	1517,15	3268,60	2568,43	1300,2	53,850	13,9356	58887,1	46272,9	251,063
1245	1518,15	3271,22	2570,59	1305,0	53,684	13,9373	58934,4	46311,8	251,094
1246	1519,15	3273,85	2572,76	1309,9	53,519	13,9390	58981,7	46350,8	251,125
1247	1520,15	3276,47	2574,92	1314,8	53,354	13,9407	59029,0	46389,8	251,156
1248	1521,15	3279,10	2577,09	1319,8	53,189	13,9425	59076,3	46428,8	251,188
1249	1522,15	3281,73	2579,25	1324,7	53,026	13,9442	59123,6	46467,8	251,219
1250	1523,15	3284,36	2581,42	1329,7	52,862	13,9459	59171,0	46506,9	251,250
1251	1524,15	3286,98	2583,59	1334,7	52,700	13,9476	59218,3	46545,9	251,281
1252	1525,15	3289,61	2585,75	1339,7	52,538	13,9494	59265,7	46584,9	251,312
1253	1526,15	3292,24	2587,92	1344,7	52,376	13,9511	59313,0	46624,0	251,343
1254	1527,15	3294,87	2590,09	1349,7	52,215	13,9528	59360,4	46663,1	251,374
1255	1528,15	3297,50	2592,26	1354,7	52,055	13,9545	59407,8	46702,1	251,405
1256	1529,15	3300,13	2594,43	1359,8	51,895	13,9563	59455,2	46741,2	251,436
1257	1530,15	3302,76	2596,60	1364,9	51,736	13,9580	59502,6	46780,3	251,467
1258	1531,15	3305,40	2598,77	1370,0	51,577	13,9597	59550,0	46819,4	251,498
1259	1532,15	3308,03	2600,94	1375,1	51,419	13,9614	59597,5	46858,5	251,529
1260	1533,15	3310,66	2603,11	1380,2	51,261	13,9631	59644,9	46897,7	251,560
1261	1534,15	3313,30	2605,28	1385,4	51,104	13,9649	59692,3	46936,8	251,591
1262	1535,15	3315,93	2607,46	1390,5	50,948	13,9666	59739,8	46975,9	251,622
1263	1536,15	3318,56	2609,63	1395,7	50,792	13,9683	59787,3	47015,1	251,653
1264	1537,15	3321,20	2611,80	1400,9	50,636	13,9700	59834,7	47054,2	251,683
1265	1538,15	3323,84	2613,98	1406,1	50,481	13,9717	59882,2	47093,4	251,714
1266	1539,15	3326,47	2616,15	1411,3	50,327	13,9734	59929,7	47132,6	251,745
1267	1540,15	3329,11	2618,33	1416,6	50,173	13,9751	59977,2	47171,8	251,776
1268	1541,15	3331,75	2620,50	1421,9	50,020	13,9769	60024,7	47211,0	251,807
1269	1542,15	3334,38	2622,68	1427,1	49,867	13,9786	60072,2	47250,2	251,838

Steam (*Continued*)

t	T	h	u	π_0	θ_0	s^0	H	U	S
1270	1543,15	3337,02	2624,86	1432,4	49,715	13,9803	60119,8	47289,4	251,869
1271	1544,15	3339,66	2627,03	1437,8	49,563	13,9820	60167,3	47328,6	251,899
1272	1545,15	3342,30	2629,21	1443,1	49,412	13,9837	60214,9	47367,9	251,930
1273	1546,15	3344,94	2631,39	1448,4	49,261	13,9854	60262,4	47407,1	251,961
1274	1547,15	3347,58	2633,57	1453,8	49,111	13,9871	60310,0	47446,4	251,992
1275	1548,15	3350,22	2635,75	1459,2	48,962	13,9888	60357,6	47485,6	252,022
1276	1549,15	3352,86	2637,93	1464,6	48,812	13,9905	60405,2	47524,9	252,053
1277	1550,15	3355,50	2640,11	1470,0	48,664	13,9922	60452,8	47564,2	252,084
1278	1551,15	3358,15	2642,29	1475,5	48,516	13,9939	60500,4	47603,5	252,115
1279	1552,15	3360,79	2644,47	1480,9	48,368	13,9956	60548,0	47642,8	252,145
1280	1553,15	3363,43	2646,65	1486,4	48,221	13,9973	60595,6	47682,1	252,176
1281	1554,15	3366,08	2648,84	1491,9	48,075	13,9990	60643,3	47721,4	252,207
1282	1555,15	3368,72	2651,02	1497,4	47,929	14,0007	60690,9	47760,7	252,237
1283	1556,15	3371,37	2653,20	1502,9	47,783	14,0024	60738,6	47800,1	252,268
1284	1557,15	3374,01	2655,39	1508,5	47,638	14,0041	60786,2	47839,4	252,298
1285	1558,15	3376,66	2657,57	1514,0	47,493	14,0058	60833,9	47878,8	252,329
1286	1559,15	3379,31	2659,76	1519,6	47,349	14,0075	60881,6	47918,2	252,360
1287	1560,15	3381,95	2661,94	1525,2	47,206	14,0092	60929,3	47957,5	252,390
1288	1561,15	3384,60	2664,13	1530,8	47,063	14,0109	60977,0	47996,9	252,421
1289	1562,15	3387,25	2666,31	1536,4	46,920	14,0126	61024,7	48036,3	252,451
1290	1563,15	3389,90	2668,50	1542,1	46,778	14,0143	61072,4	48075,7	252,482
1291	1564,15	3392,55	2670,69	1547,8	46,636	14,0160	61120,1	48115,2	252,512
1292	1565,15	3395,20	2672,88	1553,5	46,495	14,0177	61167,9	48154,6	252,543
1293	1566,15	3397,85	2675,07	1559,2	46,355	14,0194	61215,6	48194,0	252,573
1294	1567,15	3400,50	2677,26	1564,9	46,215	14,0211	61263,4	48233,5	252,604
1295	1568,15	3403,15	2679,45	1570,6	46,075	14,0228	61311,1	48272,9	252,634
1296	1569,15	3405,80	2681,64	1576,4	45,936	14,0245	61358,9	48312,4	252,665
1297	1570,15	3408,45	2683,83	1582,2	45,797	14,0262	61406,7	48351,8	252,695
1298	1571,15	3411,11	2686,02	1588,0	45,659	14,0278	61454,5	48391,3	252,726
1299	1572,15	3413,76	2688,21	1593,8	45,521	14,0295	61502,3	48430,8	252,756
1300	1573,15	3416,41	2690,40	1599,7	45,384	14,0312	61550,1	48470,3	252,787
1301	1574,15	3419,07	2692,60	1605,5	45,247	14,0329	61597,9	48509,8	252,817
1302	1575,15	3421,72	2694,79	1611,4	45,110	14,0346	61645,8	48549,3	252,847
1303	1576,15	3424,38	2696,98	1617,3	44,975	14,0363	61693,6	48588,9	252,878
1304	1577,15	3427,04	2699,18	1623,2	44,839	14,0380	61741,5	48628,4	252,908
1305	1578,15	3429,69	2701,37	1629,1	44,704	14,0397	61789,3	48667,9	252,938
1306	1579,15	3432,35	2703,57	1635,1	44,570	14,0413	61837,2	48707,5	252,969
1307	1580,15	3435,01	2705,77	1641,1	44,436	14,0430	61885,1	48747,1	252,999
1308	1581,15	3437,66	2707,96	1647,0	44,302	14,0447	61933,0	48786,6	253,029
1309	1582,15	3440,32	2710,16	1653,1	44,169	14,0464	61980,9	48826,2	253,060
1310	1583,15	3442,98	2712,36	1659,1	44,036	14,0481	62028,8	48865,8	253,090
1311	1584,15	3445,64	2714,55	1665,1	43,904	14,0497	62076,7	48905,4	253,120
1312	1585,15	3448,30	2716,75	1671,2	43,772	14,0514	62124,6	48945,0	253,150
1313	1586,15	3450,96	2718,95	1677,3	43,641	14,0531	62172,5	48984,6	253,181
1314	1587,15	3453,62	2721,15	1683,4	43,510	14,0548	62220,5	49024,3	253,211
1315	1588,15	3456,29	2723,35	1689,5	43,379	14,0564	62268,4	49063,9	253,241
1316	1589,15	3458,95	2725,55	1695,7	43,249	14,0581	62316,4	49103,6	253,271
1317	1590,15	3461,61	2727,75	1701,8	43,120	14,0598	62364,4	49143,2	253,301
1318	1591,15	3464,27	2729,96	1708,0	42,991	14,0615	62412,3	49182,9	253,331
1319	1592,15	3466,94	2732,16	1714,2	42,862	14,0631	62460,3	49222,5	253,362
1320	1593,15	3469,60	2734,36	1720,4	42,734	14,0648	62508,3	49262,2	253,392
1321	1594,15	3472,27	2736,56	1726,7	42,606	14,0665	62556,3	49301,9	253,422
1322	1595,15	3474,93	2738,77	1733,0	42,479	14,0682	62604,4	49341,6	253,452
1323	1596,15	3477,60	2740,97	1739,2	42,352	14,0698	62652,4	49381,3	253,482
1324	1597,15	3480,26	2743,18	1745,5	42,225	14,0715	62700,4	49421,0	253,512

Steam (*Continued*)

t	*T*	*h*	*u*	π_0	0_0	s^0	*H*	*U*	*S*
1325	1598,15	3482,93	2745,38	1751,9	42,099	14,0732	62748,5	49460,8	253,542
1326	1599,15	3485,60	2747,59	1758,2	41,973	14,0748	62796,5	49500,5	253,572
1327	1600,15	3488 26	2749,79	1764,6	41,848	14,0765	62844,6	49540,3	253,602
1328	1601,15	3490,93	2752,00	1771,0	41,723	14,0782	62892,6	49580,0	253,632
1329	1602,15	3493,60	2754,21	1777,4	41,599	14,0798	62940,7	49619,8	253,662
1330	1603,15	3496,27	2756,41	1783,8	41,475	14,0815	62988,8	49659,6	253,692
1331	1604,15	3498,94	2758,62	1790,2	41,351	14,0832	63036,9	49699,3	253,722
1332	1605,15	3501,61	2760,83	1796,7	41,228	14,0848	63085,0	49739,1	253,752
1333	1606,15	3504,28	2763,04	1803,2	41,106	14,0865	63133,1	49778,9	253,782
1334	1607,15	3506,95	2765,25	1809,7	40,983	14,0882	63181,2	49818,7	253,812
1335	1608,15	3509,62	2767,46	1816,2	40,861	14,0898	63229,4	49858,6	253,842
1336	1609,15	3512,30	2769,67	1822,8	40,740	14,0915	63277,5	49898,4	253,872
1337	1610,15	3514,97	2771,88	1829,3	40,619	14,0931	63325,7	49938,2	253,902
1338	1611,15	3517,64	2774,09	1835,9	40,498	14,0948	63373,8	49978,1	253,932
1339	1612,15	3520,32	2776,31	1842,5	40,378	14,0965	63422,0	50017,9	253,962
1340	1613,15	3522,99	2778,52	1849,2	40,258	14,0981	63470,2	50057,8	253,992
1341	1614,15	3525,66	2780,73	1855,8	40,139	14,0998	63518,4	50097,7	254,022
1342	1615,15	3528,34	2782,95	1862,5	40,019	14,1014	63566,6	50137,5	254,051
1343	1616,15	3531,01	2785,16	1869,2	39,901	14,1031	63614,8	50177,4	254,081
1344	1617,15	3533,69	2787,37	1875,9	39,783	14,1047	63663,0	50217,3	254,111
1345	1618,15	3536,37	2789,59	1882,7	39,665	14,1064	63711,2	50257,2	254,141
1346	1619,15	3539,04	2791,80	1889,4	39,547	14,1081	63759,4	50297,2	254,171
1347	1620,15	3541,72	2794,02	1896,2	39,430	14,1097	63807,7	50337,1	254,200
1348	1621,15	3544,40	2796,24	1903,0	39,313	14,1114	63855,9	50377,0	254,230
1349	1622,15	3547,08	2798,45	1909,8	39,197	14,1130	63904,2	50417,0	254,260
1350	1623,15	3549,76	2800,67	1916,7	39,081	14,1147	63952,4	50456,9	254,290
1351	1624,15	3552,44	2802,89	1923,5	38,966	14,1163	64000,7	50496,9	254,319
1352	1625,15	3555,12	2805,11	1930,4	38,851	14,1180	64049,0	50536,8	254,349
1353	1626,15	3557,80	2807,33	1937,3	38,736	14,1196	64097,3	50576,8	254,379
1354	1627,15	3560,48	2809,55	1944,3	38,622	14,1213	64145,6	50616,8	254,409
1355	1628,15	3563,16	2811,77	1951,2	38,508	14,1229	64193,9	50656,8	254,438
1356	1629,15	3565,84	2813,99	1958,2	38,394	14,1246	64242,2	50696,8	254,468
1357	1630,15	3568,53	2816,21	1965,2	38,281	14,1262	64290,6	50736,8	254,498
1358	1631,15	3571,21	2818,43	1972,2	38,168	14,1278	64338,9	50776,9	254,527
1359	1632,15	3573,89	2820,65	1979,2	38,055	14,1295	64387,2	50816,9	254,557
1360	1633,15	3576,58	2822,88	1986,3	37,943	14,1311	64435,6	50856,9	254,586
1361	1634,15	3579,26	2825,10	1993,4	37,832	14,1328	64484,0	50897,0	254,616
1362	1635,15	3581,95	2827,32	2000,5	37,720	14,1344	64532,3	50937,0	254,646
1363	1636,15	3584,63	2829,55	2007,6	37,609	14,1361	64580,7	50977,1	254,675
1364	1637,15	3587,32	2831,77	2014,8	37,499	14,1377	64629,1	51017,2	254,705
1365	1638,15	3590,00	2834,00	2022,0	37,389	14,1393	64677,5	51057,3	254,734
1366	1639,15	3592,69	2836,22	2029,2	37,279	14,1410	64725,9	51097,4	254,764
1367	1640,15	3595,38	2838,45	2036,4	37,169	14,1426	64774,3	51137,5	254,793
1368	1641,15	3598,07	2840,67	2043,6	37,060	14,1443	64822,8	51177,6	254,823
1369	1642,15	3600,75	2842,90	2050,9	36,951	14,1459	64871,2	51217,7	254,852
1370	1643,15	3603,44	2845,13	2058,2	36,843	14,1475	64919,6	51257,8	254,882
1371	1644,15	3606,13	2847,36	2065,5	36,735	14,1492	64968,1	51298,0	254,911
1372	1645,15	3608,82	2849,58	2072,8	36,627	14,1508	65016,6	51338,1	254,941
1373	1646,15	3611,51	2851,81	2080,2	36,520	14,1524	65065,0	51378,3	254,970
1374	1647,15	3614,20	2854,04	2087,5	36,413	14,1541	65113,5	51418,4	255,000
1375	1648,15	3616,90	2856,27	2094,9	36,306	14,1557	65162,0	51458,6	255,029
1376	1649,15	3619,59	2858,50	2102,4	36,200	14,1573	65210,5	51498,8	255,059
1377	1650,15	3622,28	2860,73	2109,8	36,094	14,1590	65259,0	51539,0	255,088
1378	1651,15	3624,97	2862,96	2117,3	35,989	14,1606	65307,5	51579,2	255,117
1379	1652,15	3627,67	2865,20	2124,8	35,883	14,1622	65356,0	51619,4	255,147

Steam (*Continued*)

t	T	h	u	π_0	θ_0	s^0	H	U	S
1380	1653,15	3630,36	2867,43	2132,3	35,778	14,1639	65404,5	51659,6	255,176
1381	1654,15	3633,05	2869,66	2139,8	35,674	14,1655	65453,1	51699,8	255,205
1382	1655,15	3635,75	2871,89	2147,4	35,570	14,1671	65501,6	51740,0	255,235
1383	1656,15	3638,44	2874,13	2155,0	35,466	14,1687	65550,2	51780,3	255,264
1384	1657,15	3641,14	2876,36	2162,6	35,362	14,1704	65598,8	51820,5	255,293
1385	1658,15	3643,83	2878,60	2170,2	35,259	14,1720	65647,3	51860,8	255,323
1386	1659,15	3646,53	2880,83	2177,9	35,157	14,1736	65695,9	51901,1	255,352
1387	1660,15	3649,23	2883,07	2185,6	35,054	14,1753	65744,5	51941,3	255,381
1388	1661,15	3651,93	2885,30	2193,3	34,952	14,1769	65793,1	51981,6	255,411
1389	1662,15	3654,62	2887,54	2201,0	34,850	14,1785	65841,7	52021,9	255,440
1390	1663,15	3657,32	2889,78	2208,8	34,749	14,1801	65890,3	52062,2	255,469
1391	1664,15	3660,02	2892,01	2216,5	34,648	14,1817	65939,0	52102,5	255,498
1392	1665,15	3662,72	2894,25	2224,3	34,547	14,1834	65987,6	52142,8	255,528
1393	1666,15	3665,42	2896,49	2232,2	34,446	14,1850	66036,2	52183,2	255,557
1394	1667,15	3668,12	2898,73	2240,0	34,346	14,1866	66084,9	52223,5	255,586
1395	1668,15	3670,82	2900,97	2247,9	34,246	14,1882	66133,5	52263,9	255,615
1396	1669,15	3673,52	2903,21	2255,8	34,147	14,1898	66182,2	52304,2	255,644
1397	1670,15	3676,23	2905,45	2263,7	34,048	14,1915	66230,9	52344,6	255,673
1398	1671,15	3678,93	2907,69	2271,7	33,949	14,1931	66279,6	52384,9	255,703
1399	1672,15	3681,63	2909,93	2279,6	33,851	14,1947	66328,3	52425,3	255,732
1400	1673,15	3684,33	2912,17	2287,6	33,752	14,1963	66377,0	52465,7	255,761
1401	1674,15	3687,04	2914,42	2295,7	33,655	14,1979	66425,7	52506,1	255,790
1402	1675,15	3689,74	2916,66	2303,7	33,557	14,1995	66474,4	52546,5	255,819
1403	1676,15	3692,45	2918,90	2311,8	33,460	14,2012	66523,1	52586,9	255,848
1404	1677,15	3695,15	2921,14	2319,9	33,363	14,2028	66571,9	52627,3	255,877
1405	1678,15	3697,86	2923,39	2328,0	33,266	14,2044	66620,6	52667,8	255,906
1406	1679,15	3700,56	2925,63	2336,1	33,170	14,2060	66669,4	52708,2	255,935
1407	1680,15	3703,27	2927,88	2344,3	33,074	14,2076	66718,1	52748,7	255,964
1408	1681,15	3705,98	2930,12	2352,5	32,979	14,2092	66766,9	52789,1	255,993
1409	1682,15	3708,68	2932,37	2360,7	32,883	14,2108	66815,7	52829,6	256,022
1410	1683,15	3711,39	2934,62	2369,0	32,783	14,2124	66864,5	52870,1	256,051
1411	1684,15	3714,10	2936,86	2377,2	32,694	14,2140	66913,3	52910,5	256,080
1412	1685,15	3716,81	2939,11	2385,5	32,599	14,2157	66962,1	52951,0	256,109
1413	1686,15	3719,52	2941,36	2393,9	32,505	14,2173	67010,9	52991,5	256,138
1414	1687,15	3722,23	2943,61	2402,2	32,412	14,2189	67059,7	53032,0	256,167
1415	1688,15	3724,94	2945,86	2410,6	32,318	14,2205	67108,5	53072,5	256,196
1416	1689,15	3727,65	2948,11	2419,0	32,225	14,2221	67157,4	53113,1	256,225
1417	1690,15	3730,36	2950,36	2427,4	32,132	14,2237	67206,2	53153,6	256,254
1418	1691,15	3733,07	2952,61	2435,9	32,040	14,2253	67255,1	53194,1	256,283
1419	1692,15	3735,79	2954,86	2444,3	31,947	14,2269	67303,9	53234,7	256,312
1420	1693,15	3738,50	2957,11	2452,8	31,855	14,2285	67352,8	53275,3	256,341
1421	1694,15	3741,21	2959,36	2461,4	31,764	14,2301	67401,7	53315,8	256,369
1422	1695,15	3743,93	2961,61	2469,9	31,672	14,2317	67450,6	53356,4	256,398
1423	1696,15	3746,64	2963,86	2478,5	31,581	14,2333	67499,5	53397,0	256,427
1424	1697,15	3749,35	2966,12	2487,1	31,491	14,2349	67548,4	53437,6	256,456
1425	1698,15	3752,07	2968,37	2495,7	31,400	14,2365	67597,3	53478,2	256,485
1426	1699,15	3754,78	2970,62	2504,4	31,310	14,2381	67646,2	53518,8	256,514
1427	1700,15	3757,50	2972,88	2513,1	31,220	14,2397	67695,1	53559,4	256,542
1428	1701,15	3760,22	2975,13	2521,8	31,131	14,2413	67744,1	53600,0	256,571
1429	1702,15	3762,93	2977,39	2530,5	31,041	14,2429	67793,0	53640,6	256,600
1430	1703,15	3765,65	2979,65	2539,3	30,952	14,2445	67842,0	53681,3	256,629
1431	1704,15	3768,37	2981,90	2548,1	30,864	14,2461	67890,9	53721,9	256,657
1432	1705,15	3771,09	2984,16	2556,9	30,775	14,2477	67939,9	53762,6	256,686
1433	1706,15	3773,81	2986,42	2565,8	30,687	14,2493	67988,9	53803,3	256,715
1434	1707,15	3776,53	2988,67	2574,6	30,599	14,2509	68037,9	53843,9	256,743
1435	1708,15	3779,25	2990,93	2583,5	30,512	14,2525	68086,9	53884,6	256,772
1436	1709,15	3781,97	2993,19	2592,5	30,425	14,2540	68135,9	53925,3	256,801
1437	1710,15	3784,69	2995,45	2601,4	30,338	14,2556	68184,9	53966,0	256,830
1438	1711,15	3787,41	2997,71	2610,4	30,251	14,2572	68233,9	54006,7	256,858
1439	1712,15	3790,13	2999,97	2619,4	30,164	14,2588	68283,0	54047,4	256,887

Steam (Continued)

t	T	h	u	π_0	θ_0	s^0	H	U	S
1440	1713,15	3792,85	3002,23	2628,5	30,078	14,2604	68332,0	54088,2	256,915
1441	1714,15	3795,57	3004,49	2637,5	29,992	14,2620	68381,0	54128,9	256,944
1442	1715,15	3798,30	3006,75	2646,6	29,907	14,2636	68430,1	54169,6	256,973
1443	1716,15	3801,02	3009,01	2655,7	29,822	14,2652	68479,2	54210,4	257,001
1444	1717,15	3803,74	3011,28	2664,9	29,736	14,2668	68528,2	54251,1	257,030
1445	1718,15	3806,47	3013,54	2674,0	29,652	14,2683	68577,3	54291,9	257,058
1446	1719,15	3809,19	3015,80	2683,2	29,567	14,2699	68626,4	54332,7	257,087
1447	1720,15	3811,92	3018,07	2692,5	29,483	14,2715	68675,5	54373,5	257,116
1448	1721,15	3814,64	3020,33	2701,7	29,399	14,2731	68724,6	54414,3	257,144
1449	1722,15	3817,37	3022,59	2711,0	29,315	14,2747	68773,7	54455,1	257,173
1450	1723,15	3820,10	3024,86	2720,3	29,232	14,2763	68822,8	54495,9	257,201
1451	1724,15	3822,82	3027,13	2729,7	29,149	14,2778	68872,0	54536,7	257,230
1452	1725,15	3825,55	3029,39	2739,0	29,066	14,2794	68921,1	54577,5	257,258
1453	1726,15	3828,28	3031,66	2748,4	28,983	14,2810	68970,3	54618,3	257,287
1454	1727,15	3831,01	3033,92	2757,9	28,901	14,2826	69019,4	54659,2	257,315
1455	1728,15	3833,74	3036,19	2767,3	28,819	14,2842	69068,6	54700,0	257,344
1456	1729,15	3836,47	3038,46	2776,8	28,737	14,2857	69117,8	54740,9	257,372
1457	1730,15	3839,20	3040,73	2786,3	28,656	14,2873	69166,9	54781,8	257,400
1458	1731,15	3841,93	3043,00	2795,9	28,574	14,2889	69216,1	54822,6	257,429
1459	1732,15	3844,66	3045,27	2805,4	28,493	14,2905	69265,3	54863,5	257,457
1460	1733,15	3847,39	3047,54	2815,0	28,413	14,2921	69314,5	54904,4	257,486
1461	1734,15	3850,12	3049,81	2824,7	28,332	14,2936	69363,7	54945,3	257,514
1462	1735,15	3852,85	3052,08	2834,3	28,252	14,2952	69413,0	54986,2	257,542
1463	1736,15	3855,58	3054,35	2844,0	28,172	14,2968	69462,2	55027,1	257,571
1464	1737,15	3858,32	3056,62	2853,7	28,092	14,2984	69511,4	55068,1	257,599
1465	1738,15	3861,05	3058,89	2863,5	28,013	14,2999	69560,7	55109,0	257,627
1466	1739,15	3863,78	3061,16	2873,2	27,933	14,3015	69609,9	55149,9	257,656
1467	1740,15	3866,52	3063,44	2883,0	27,854	14,3031	69659,2	55190,9	257,684
1468	1741,15	3869,25	3065,71	2892,9	27,776	14,3046	69708,5	55231,8	257,712
1469	1742,15	3871,99	3067,98	2902,7	27,697	14,3062	69757,8	55272,8	257,741
1470	1743,15	3874,72	3070,26	2912,6	27,619	14,3078	69807,0	55313,8	257,769
1471	1744,15	3877,46	3072,53	2922,5	27,541	14,3093	69856,3	55354,8	257,797
1472	1745,15	3880,20	3074,81	2932,5	27,463	14,3109	69905,6	55395,7	257,825
1473	1746,15	3882,93	3077,08	2942,5	27,386	14,3125	69954,9	55436,7	257,854
1474	1747,15	3885,67	3079,36	2952,5	27,309	14,3141	70004,3	55477,7	257,882
1475	1748,15	3888,41	3081,64	2962,5	27,232	14,3156	70053,6	55518,8	257,910
1476	1749,15	3891,15	3083,91	2972,6	27,155	14,3172	70102,9	55559,8	257,938
1477	1750,15	3893,89	3086,19	2982,7	27,079	14,3187	70152,3	55600,8	257,967
1478	1751,15	3896,63	3088,47	2992,8	27,002	14,3203	70201,6	55641,9	257,995
1479	1752,15	3899,37	3090,75	3003,0	26,926	14,3219	70251,0	55682,9	258,023
1480	1753,15	3902,11	3093,03	3013,2	26,851	14,3234	70300,4	55724,0	258,051
1481	1754,15	3904,85	3095,31	3023,4	26,775	14,3250	70349,7	55765,0	258,079
1482	1755,15	3907,59	3097,59	3033,6	26,700	14,3266	70399,1	55806,1	258,107
1483	1756,15	3910,33	3099,87	3043,9	26,625	14,3281	70448,5	55847,2	258,136
1484	1757,15	3913,07	3102,15	3054,2	26,550	14,3297	70497,9	55888,3	258,164
1485	1758,15	3915,82	3104,43	3064,6	26,475	14,3312	70547,3	55929,4	258,192
1486	1759,15	3918,56	3106,71	3074,9	26,401	14,3328	70596,8	55970,5	258,220
1487	1760,15	3921,30	3108,99	3085,4	26,327	14,3344	70646,2	56011,6	258,248
1488	1761,15	3924,05	3111,27	3095,8	26,253	14,3359	70695,6	56052,7	258,276
1489	1762,15	3926,79	3113,56	3106,3	26,180	14,3375	70745,1	56093,8	258,304
1490	1763,15	3929,54	3115,84	3116,8	26,106	14,3390	70794,5	56135,0	258,332
1491	1764,15	3932,28	3118,12	3127,3	26,033	14,3406	70844,0	56176,1	258,360
1492	1765,15	3935,03	3120,41	3137,8	25,960	14,3422	70893,4	56217,3	258,388
1493	1766,15	3937,77	3122,69	3148,4	25,887	14,3437	70942,9	56258,4	258,416
1494	1767,15	3940,52	3124,98	3159,1	25,815	14,3453	70992,4	56299,6	258,444
1495	1768,15	3943,27	3127,26	3169,7	25,743	14,3468	71041,9	56340,8	258,472
1496	1769,15	3946,01	3129,55	3180,4	25,671	14,3484	71091,4	56382,0	258,500
1497	1770,15	3948,76	3131,84	3191,1	25,599	14,3499	71140,9	56423,1	258,528
1498	1771,15	3951,51	3134,12	3201,9	25,528	14,3515	71190,4	56464,3	258,556
1499	1772,15	3954,26	3136,41	3212,7	25,456	14,3530	71239,9	56605,6	258,584
1500	1773,15	3957,01	3138,70	3223,5	25,385	14,3546	71289,5	56546,8	258,612

Table III.13 Carbon monoxide ($\mu = 28.011$)

t	T	c_p	μc_p	c_v	μc_v	$k = \dfrac{c_p}{c_v}$
−50	223,15	1,0393	29 090	0,7417	20,776	1,401
−25	248,15	1,0389	29,099	0,7420	20,785	1,400
0	273,15	1,0389	29,099	0,7421	20,785	1,400
25	298,15	1,0394	29,114	0,7426	20,800	1,400
50	323,15	1,0405	29,145	0,7437	20,831	1,399
75	348,15	1,0422	29,191	0,7453	20,877	1,398
100	373,15	1,0444	29,254	0,7476	20,940	1,397
125	398,15	1,0472	29,333	0,7504	21,019	1,396
150	423,15	1,0506	29,428	0,7538	21,114	1,394
175	448,15	1,0545	29,537	0,7577	21,223	1,392
200	473,15	1,0589	29,659	0,7620	21,345	1,390
250	523,15	1,0689	29,940	0,7721	21,626	1,384
300	573,15	1,0803	30,260	0,7835	21,946	1,379
350	623,15	1,0928	30,608	0,7959	22,294	1,373
400	673,15	1,1057	30,971	0,8089	22,657	1,367
450	723,15	1,1190	31,342	0,8221	23,028	1,361
500	773,15	1,1321	31,710	0,8353	23,396	1,355
550	823,15	1,1449	32,069	0,8481	23,755	1,350
600	873,15	1,1572	32,412	0,8603	24,098	1,345
650	923,15	1,1688	32,738	0,8720	24,424	1,340
700	973,15	1,1797	33,043	0,8829	24,729	1,336
750	1023,15	1,1898	33,327	0,8930	25,013	1,332
800	1073,15	1,1992	33,591	0,9024	25,277	1,329
850	1123,15	1,2080	33,835	0,9111	25,521	1,326
900	1173,15	1,2161	34,062	0,9192	25,748	1,323
950	1223,15	1,2236	34,273	0,9268	25,959	1,320
1000	1273,15	1,2307	34,471	0,9338	26,157	1,318
1050	1323,15	1,2373	34,656	0,9404	26,342	1,316
1100	1373,15	1,2435	34,830	0,9467	26,516	1,314
1150	1423,15	1,2493	34,994	0,9525	26,680	1,312
1200	1473,15	1,2548	35,147	0,9580	26,833	1,310
1250	1523,15	1,2598	35,288	0,9630	26,974	1,308
1300	1573,15	1,2645	35,418	0,9676	27,104	1,307
1350	1623,15	1,2687	35,537	0,9719	27,223	1,305
1400	1673,15	1,2726	35,647	0,9758	27,333	1,304
1450	1723,15	1,2764	35,752	0,9796	27,438	1,303
1500	1773,15	1,2803	35,862	0,9835	27,548	1,302

Table III.14 Carbon monoxide ($\mu = 28.011$)

t	T	h	u	π_0	0_0	s^0	H	U	S
−50	223,15	231,60	165,36	0,7881	8403,6	6,7640	6487,3	4631,9	189,467
−49	224,15	232,64	166,10	0,8006	8310,2	6,7687	6516,4	4652,7	189,598
−48	225,15	233,68	166,85	0,8132	8218,2	6,7733	6545,5	4673,5	189,727
−47	226,15	234,72	167,59	0,8259	8127,6	6,7779	6574,6	4694,3	189,856
−46	227,15	235,75	168,33	0,8387	8038,4	6,7825	6603,7	4715,1	189,985
−45	228,15	236,79	169,07	0,8517	7950,5	6,7871	6632,8	4735,9	190,112
−44	229,15	237,83	169,82	0,8649	7864,0	6,7916	6661,9	4756,7	190,240
−43	230,15	238,87	170,56	0,8782	7778,9	6,7961	6691,1	4777,5	190,367
−42	231,15	239,91	171,30	0,8916	7695,0	6,8006	6720,2	4798,3	190,493
−41	232,15	240,95	172,04	0,9052	7612,4	6,8051	6749,3	4819,1	190,618
−40	233,15	241,99	172,78	0,9189	7531,0	6,8096	6778,4	4839,9	190,743
−39	234,15	243,03	173,53	0,9328	7450,8	6,8140	6807,5	4860,7	190,868
−38	235,15	244,07	174,27	0,9468	7371,8	6,8185	6836,6	4881,5	190,992
−37	236,15	245,11	175,01	0,9609	7294,0	6,8229	6865,7	4902,2	191,116
−36	237,15	246,15	175,75	0,9753	7217,4	6,8273	6894,8	4923,0	191,239
−35	238,15	247,18	176,50	0,9897	7141,8	6,8316	6923,9	4943,8	191,361
−34	239,15	248,22	177,24	1,0044	7067,4	6,8360	6953,0	4964,6	191,483
−33	240,15	249,26	177,98	1,0191	6994,0	6,8403	6982,1	4985,4	191,604
−32	241,15	250,30	178,72	1,0341	6921,7	6,8446	7011,2	5006,2	191,725
−31	242,15	251,34	179,46	1,0492	6850,5	6,8489	7040,3	5027,0	191,846
−30	243,15	252,38	180,21	1,0644	6780,3	6,8532	7069,4	5047,8	191,966
−29	244,15	253,42	180,95	1,0798	6711,1	6,8575	7098,5	5068,5	192,085
−28	245,15	254,46	181,69	1,0954	6642,8	6,8617	7127,6	5089,3	192,204
−27	246,15	255,50	182,43	1,1111	6575,6	6,8660	7156,7	5110,1	192,323
−26	247,15	256,54	183,17	1,1270	6509,3	6,8702	7185,8	5130,9	192,441
−25	248,15	257,57	183,92	1,1430	6443,9	6,8744	7214,9	5151,7	192,558
−24	249,15	258,61	184,66	1,1592	6379,4	6,8785	7244,0	5172,5	192,675
−23	250,15	259,65	185,40	1,1756	6315,9	6,8827	7273,1	5193,3	192,792
−22	251,15	260,69	186,14	1,1921	6253,2	6,8869	7302,2	5214,0	192,908
−21	252,15	261,73	186,88	1,2088	6191,4	6,8910	7331,3	5234,8	193,023
−20	253,15	262,77	187,63	1,2257	6130,4	6,8951	7360,4	5255,6	193,139
−19	254,15	263,81	188,37	1,2427	6070,3	6,8992	7389,5	5276,4	193,253
−18	255,15	264,85	189,11	1,2599	6011,0	6,9033	7418,6	5297,2	193,367
−17	256,15	265,88	189,85	1,2773	5952,5	6,9073	7447,7	5318,0	193,481
−16	257,15	266,92	190,59	1,2948	5894,8	6,9114	7476,8	5338,7	193,595
−15	258,15	267,96	191,34	1,3125	5837,9	6,9154	7505,9	5359,5	193,708
−14	259,15	269,00	192,08	1,3304	5781,8	6,9194	7535,0	5380,3	193,820
−13	260,15	270,04	192,82	1,3484	5726,4	6,9234	7564,1	5401,1	193,932
−12	261,15	271,08	193,56	1,3667	5671,7	6,9274	7593,2	5421,9	194,044
−11	262,15	272,12	194,30	1,3851	5617,8	6,9314	7622,3	5442,7	194,155
−10	263,15	273,16	195,05	1,4036	5564,6	6,9353	7651,4	5463,4	194,266
−9	264,15	274,19	195,79	1,4224	5512,1	6,9393	7680,5	5484,2	194,376
−8	265,15	275,23	196,53	1,4413	5460,2	6,9432	7709,6	5505,0	194,486
−7	266,15	276,27	197,27	1,4605	5409,1	6,9471	7738,7	5525,8	194,596
−6	267,15	277,31	198,01	1,4797	5358,6	6,9510	7767,8	5546,6	194,705
−5	268,15	278,35	198,76	1,4992	5308,8	6,9549	7796,9	5567,4	194,814
−4	269,15	279,39	199,50	1,5189	5259,6	6,9588	7826,0	5588,1	194,922
−3	270,15	280,43	200,24	1,5387	5211,1	6,9626	7855,1	5608,9	195,030
−2	271,15	281,47	200,98	1,5588	5163,2	6,9665	7884,2	5629,7	195,137
−1	272,15	282,51	201,72	1,5790	5115,9	6,9703	7913,3	5650,5	195,244
0	273,15	283,54	202,47	1,5994	5069,2	6,9741	7942,4	5671,3	195,351
1	274,15	284,58	203,21	1,6200	5023,1	6,9779	7971,5	5692,1	195,457
2	275,15	285,62	203,95	1,6407	4977,6	6,9817	8000,6	5712,9	195,563
3	276,15	286,66	204,69	1,6617	4932,7	6,9854	8029,7	5733,6	195,669
4	277,15	287,70	205,43	1,6828	4888,3	6,9892	8058,8	5754,4	195,774

Carbon monoxide (*Continued*)

t	T	h	u	π_0	θ_0	s^0	H	U	S
5	278,15	288,74	206,18	1,7042	4844,5	6,9929	8087,9	5775,2	195,879
6	279,15	289,78	206,92	1,7257	4801,2	6,9967	8117,0	5796,0	195,983
7	280,15	290,82	207,66	1,7475	4758,5	7,0004	8146,1	5816,8	196,087
8	281,15	291,86	208,40	1,7694	4716,3	7,0041	8175,2	5837,6	196,191
9	282,15	292,89	209,14	1,7915	4674,6	7,0078	8204,3	5858,4	196,294
10	283,15	293,93	209,89	1,8139	4633,4	7,0114	8233,4	5879,1	196,397
11	284,15	294,97	210,63	1,8364	4592,7	7,0151	8262,5	5899,9	196,500
12	285,15	296,01	211,37	1,8591	4552,6	7,0188	8291,6	5920,7	196,602
13	286,15	297,05	212,11	1,8820	4512,9	7,0224	8320,7	5941,5	196,704
14	287,15	298,09	212,86	1,9052	4473,7	7,0260	8349,8	5962,3	196,806
15	288,15	299,13	213,60	1,9285	4435,0	7,0296	8378,9	5983,1	196,907
16	289,15	300,17	214,34	1,9520	4396,7	7,0332	8408,0	6003,9	197,008
17	290,15	301,21	215,08	1,9758	4358,9	7,0368	8437,1	6024,7	197,108
18	291,15	302,25	215,83	1,9997	4321,6	7,0404	8466,2	6045,5	197,208
19	292,15	303,29	216,57	2,0238	4284,7	7,0440	8495,3	6066,3	197,308
20	293,15	304,32	217,31	2,0482	4248,2	7,0475	8524,4	6087,1	197,408
21	294,15	305,36	218,05	2,0728	4212,2	7,0510	8553,5	6107,9	197,507
22	295,15	306,40	218,79	2,0975	4176,6	7,0546	8582,7	6128,7	197,606
23	296,15	307,44	219,54	2,1225	4141,4	7,0581	8611,8	6149,5	197,704
24	297,15	308,48	220,28	2,1477	4106,6	7,0616	8640,9	6170,3	197,802
25	298,15	309,52	221,02	2,1732	4072,2	7,0651	8670,0	6191,1	197,900
26	299,15	310,56	221,76	2,1988	4038,3	7,0686	8699,1	6211,9	197,998
27	300,15	311,60	222,51	2,2246	4004,7	7,0720	8728,2	6232,7	198,095
28	301,15	312,64	223,25	2,2507	3971,5	7,0755	8757,3	6253,5	198,192
29	302,15	313,68	223,99	2,2770	3938,7	7,0789	8786,5	6274,3	198,288
30	303,15	314,72	224,74	2,3035	3906,3	7,0824	8815,6	6295,1	198,384
31	304,15	315,76	225,48	2,3302	3874,2	7,0858	8844,7	6315,9	198,480
32	305,15	316,80	226,22	2,3571	3842,5	7,0892	8873,8	6336,7	198,576
33	306,15	317,84	226,96	2,3843	3811,2	7,0926	8902,9	6357,5	198,671
34	307,15	318,88	227,71	2,4117	3780,2	7,0960	8932,1	6378,3	198,766
35	308,15	319,92	228,45	2,4393	3749,6	7,0994	8961,2	6399,1	198,861
36	309,15	320,96	229,19	2,4672	3719,3	7,1027	8990,3	6419,9	198,955
37	310,15	322,00	229,94	2,4952	3689,3	7,1061	9019,4	6440,7	199,049
38	311,15	323,04	230,68	2,5235	3659,7	7,1095	9048,6	6461,5	199,143
39	312,15	324,08	231,42	2,5521	3630,5	7,1128	9077,7	6482,4	199,236
40	313,15	325,12	232,16	2,5808	3601,5	7,1161	9106,8	6503,2	199,329
41	314,15	326,16	232,91	2,6098	3572,9	7,1194	9136,0	6524,0	199,422
42	315,15	327,20	233,65	2,6390	3544,5	7,1227	9165,1	6544,8	199,515
43	316,15	328,24	234,39	2,6685	3516,5	7,1260	9194,2	6565,6	199,607
44	317,15	329,28	235,14	2,6982	3488,8	7,1293	9223,4	6586,4	199,699
45	318,15	330,32	235,88	2,7281	3461,4	7,1326	9252,5	6607,3	199,791
46	319,15	331,36	236,62	2,7583	3434,3	7,1359	9281,6	6628,1	199,882
47	320,15	332,40	237,37	2,7887	3407,5	7,1391	9310,8	6648,9	199,974
48	321,15	333,44	238,11	2,8194	3381,0	7,1424	9339,9	6669,7	200,064
49	322,15	334,48	238,86	2,8502	3354,8	7,1456	9369,1	6690,6	200,155
50	323,15	335,52	239,60	2,8814	3328,8	7,1488	9398,2	6711,4	200,245
51	324,15	336,56	240,34	2,9128	3303,2	7,1520	9427,3	6732,2	200,335
52	325,15	337,60	241,09	2,9444	3277,8	7,1552	9456,5	6753,1	200,425
53	326,15	338,64	241,83	2,9763	3252,6	7,1584	9485,6	6773,9	200,515
54	327,15	339,68	242,57	3,0084	3227,8	7,1616	9514,8	6794,7	200,604
55	328,15	340,72	243,32	3,0407	3203,2	7,1648	9543,9	6815,6	200,693
56	329,15	341,76	244,06	3,0734	3178,9	7,1680	9573,1	6836,5	200,782
57	330,15	342,80	244,81	3,1062	3154,8	7,1711	9602,3	6857,3	200,870
58	331,15	343,84	245,55	3,1393	3131,0	7,1743	9631,4	6878,1	200,958
59	332,15	344,88	246,29	3,1727	3107,4	7,1774	9660,6	6898,9	201,046

Carbon monoxide (*Continued*)

t	T	h	u	π_0	0_0	s^0	H	U	S
60	333,15	345,93	247,04	3,2063	3084,0	7,1805	9689,7	6919,8	201,134
61	334,15	346,97	247,78	3,2402	3060,9	7,1837	9718,9	6940,6	201,221
62	335,15	348,01	248,53	3,2744	3038,1	7,1868	9748,1	6961,5	201,308
63	336,15	349,05	249,27	3,3088	3015,5	7,1899	9777,2	6982,3	201,395
64	337,15	350,09	250,02	3,3434	2993,1	7,1930	9806,4	7003,2	201,482
65	338,15	351,13	250,76	3,3783	2970,9	7,1960	9835,6	7024,0	201,568
66	339,15	352,17	251,50	3,4135	2949,0	7,1991	9864,7	7044,9	201,654
67	340,15	353,21	252,25	3,4490	2927,3	7,2022	9893,9	7065,8	201,740
68	341,15	354,26	252,99	3,4847	2905,8	7,2052	9923,1	7086,6	201,826
69	342,15	355,30	253,74	3,5207	2884,6	7,2083	9952,3	7107,5	201,911
70	343,15	356,34	254,48	3,5569	2863,5	7,2113	9981,4	7128,3	201,997
71	344,15	357,38	255,23	3,5934	2842,7	7,2144	10010,6	7149,2	202,081
72	345,15	358,42	255,97	3,6302	2822,1	7,2174	10039,8	7170,1	202,166
73	346,15	359,47	256,72	3,6673	2801,6	7,2204	10069,0	7191,0	202,251
74	347,15	360,51	257,46	3,7046	2781,4	7,2234	10098,2	7211,8	202,335
75	348,15	361,55	258,21	3,7422	2761,4	7,2264	10127,4	7232,7	202,419
76	349,15	362,59	258,95	3,7801	2741,6	7,2294	10156,6	7253,6	202,502
77	350,15	363,63	259,70	3,8182	2722,0	7,2324	10185,8	7274,5	202,586
78	351,15	364,68	260,45	3,8567	2702,5	7,2353	10215,0	7295,3	202,669
79	352,15	365,72	261,19	3,8954	2683,3	7,2383	10244,1	7316,2	202,752
80	353,15	366,76	261,94	3,9344	2664,3	7,2413	10273,4	7337,1	202,835
81	354,15	367,80	262,68	3,9736	2645,4	7,2442	10302,6	7358,0	202,918
82	355,15	368,85	263,43	4,0132	2626,7	7,2472	10331,8	7378,9	203,000
83	356,15	369,89	264,17	4,0530	2608,2	7,2501	10361,0	7399,8	203,082
84	357,15	370,93	264,92	4,0931	2589,9	7,2530	10390,2	7420,7	203,164
85	358,15	371,98	265,67	4,1336	2571,8	7,2559	10419,4	7441,6	203,246
86	359,15	373,02	266,41	4,1743	2553,8	7,2588	10448,6	7462,5	203,327
87	360,15	374,06	267,16	4,2152	2536,0	7,2617	10477,8	7483,4	203,408
88	361,15	375,10	267,91	4,2565	2518,4	7,2646	10507,0	7504,3	203,489
89	362,15	376,15	268,65	4,2981	2500,9	7,2675	10536,3	7525,2	203,570
90	363,15	377,19	269,40	4,3399	2483,7	7,2704	10565,5	7546,1	203,651
91	364,15	378,23	270,15	4,3821	2466,5	7,2733	10594,7	7567,0	203,731
92	365,15	379,28	270,89	4,4246	2449,6	7,2761	10624,0	7588,0	203,811
93	366,15	380,32	271,64	4,4673	2432,8	7,2790	10653,2	7608,9	203,891
94	367,15	381,37	272,39	4,5104	2416,1	7,2818	10682,4	7629,8	203,971
95	368,15	382,41	273,13	4,5537	2399,7	7,2847	10711,7	7650,7	204,051
96	369,15	383,45	273,88	4,5974	2383,3	7,2875	10740,9	7671,6	204,130
97	370,15	384,50	274,63	4,6413	2367,1	7,2903	10770,2	7692,6	204,209
98	371,15	385,54	275,37	4,6856	2351,1	7,2931	10799,4	7713,5	204,288
99	372,15	386,59	276,12	4,7301	2335,2	7,2959	10828,7	7734,4	204,367
100	373,15	387,63	276,87	4,7750	2319,5	7,2987	10857,9	7755,4	204,445
101	374,15	388,67	277,62	4,8202	2303,9	7,3015	10887,2	7776,3	204,523
102	375,15	389,72	278,36	4,8657	2288,5	7,3043	10916,4	7797,3	204,602
103	376,15	390,76	279,11	4,9115	2273,2	7,3071	10945,7	7818,2	204,679
104	377,15	391,81	279,86	4,9576	2258,0	7,3099	10974,9	7839,2	204,757
105	378,15	392,85	280,61	5,0040	2243,0	7,3127	11004,2	7860,1	204,835
106	379,15	393,90	281,36	5,0508	2228,1	7,3154	11033,5	7881,1	204,912
107	380,15	394,94	282,10	5,0978	2213,4	7,3182	11062,8	7902,0	204,989
108	381,15	395,99	282,85	5,1452	2198,8	7,3209	11092,0	7923,0	205,066
109	382,15	397,03	283,60	5,1929	2184,3	7,3236	11121,3	7944,0	205,143
110	383,15	398,08	284,35	5,2409	2170,0	7,3264	11150,6	7964,9	205,219
111	384,15	399,12	285,10	5,2892	2155,7	7,3291	11179,9	7985,9	205,296
112	385,15	400,17	285,85	5,3379	2141,6	7,3318	11209,2	8006,9	205,372
113	386,15	401,22	286,60	5,3869	2127,7	7,3345	11238,5	8027,8	205,448
114	387,15	402,26	287,35	5,4362	2113,8	7,3372	11267,8	8048,8	205,523

Carbon monoxide (*Continued*)

t	T	h	u	π_0	θ_0	s°	H	U	S
115	388,15	403,31	288,09	5,4859	2100,1	7,3399	11297,1	8069,8	205,599
116	389,15	404,35	288,84	5,5358	2086,5	7,3426	11326,4	8090,8	205,674
117	390,15	405,40	289,59	5,5861	2073,1	7,3453	11355,7	8111,8	205,750
118	391,15	406,45	290,34	5,6368	2059,7	7,3480	11385,0	8132,8	205,825
119	392,15	407,49	291,09	5,6877	2046,5	7,3507	11414,3	8153,8	205,899
120	393,15	408,54	291,84	5,7390	2033,3	7,3533	11443,6	8174,8	205,974
121	394,15	409,59	292,59	5,7907	2020,3	7,3560	11472,9	8195,8	206,049
122	395,15	410,63	293,34	5,8426	2007,4	7,3586	11502,2	8216,8	206,123
123	396,15	411,68	294,09	5,8950	1994,7	7,3613	11531,6	8237,8	206,197
124	397,15	412,73	294,84	5,9476	1982,0	7,3639	11560,9	8258,8	206,271
125	398,15	413,77	295,59	6,0006	1969,4	7,3666	11590,2	8279,8	206,345
126	399,15	414,82	296,34	6,0540	1957,0	7,3692	11619,6	8300,9	206,418
127	400,15	415,87	297,09	6,1077	1944,6	7,3718	11648,9	8321,9	206,492
128	401,15	416,92	297,84	6,1617	1932,4	7,3744	11678,2	8342,9	206,565
129	402,15	417,96	298,59	6,2161	1920,3	7,3770	11707,6	8363,9	206,638
130	403,15	419,01	299,35	6,2708	1908,2	7,3796	11736,9	8385,0	206,711
131	404,15	420,06	300,10	6,3259	1896,3	7,3822	11766,3	8406,0	206,784
132	405,15	421,11	300,85	6,3813	1884,5	7,3848	11795,6	8427,1	206,856
133	406,15	422,16	301,60	6,4371	1872,8	7,3874	11825,0	8448,1	206,929
134	407,15	423,20	302,35	6,4933	1861,1	7,3900	11854,4	8469,1	207,001
135	408,15	424,25	303,10	6,5498	1849,6	7,3926	11883,7	8490,2	207,073
136	409,15	425,30	303,85	6,6066	1838,2	7,3951	11913,1	8511,3	207,145
137	410,15	426,35	304,61	6,6639	1826,9	7,3977	11942,5	8532,3	207,216
138	411,15	427,40	305,36	6,7215	1815,6	7,4002	11971,9	8553,4	207,288
139	412,15	428,45	306,11	6,7794	1804,5	7,4028	12001,2	8574,4	207,359
140	413,15	429,50	306,86	6,8377	1793,4	7,4053	12030,6	8595,5	207,431
141	414,15	430,55	307,61	6,8964	1782,5	7,4079	12060,0	8616,6	207,502
142	415,15	431,59	308,37	6,9555	1771,6	7,4104	12089,4	8637,7	207,572
143	416,15	432,64	309,12	7,0149	1760,8	7,4129	12118,8	8658,8	207,643
144	417,15	433,69	309,87	7,0747	1750,1	7,4154	12148,2	8679,8	207,714
145	418,15	434,74	310,63	7,1348	1739,5	7,4179	12177,6	8700,9	207,784
146	419,15	435,79	311,38	7,1954	1729,0	7,4205	12207,0	8722,0	207,854
147	420,15	436,84	312,13	7,2563	1718,6	7,4230	12236,4	8743,1	207,924
148	421,15	437,89	312,89	7,3176	1708,3	7,4255	12265,9	8764,2	207,994
149	422,15	438,94	313,64	7,3793	1698,0	7,4279	12295,3	8785,3	208,064
150	423,15	439,99	314,39	7,4413	1687,8	7,4304	12324,7	8806,5	208,134
151	424,15	441,05	315,15	7,5037	1677,8	7,4329	12354,1	8827,6	208,203
152	425,15	442,10	315,90	7,5666	1667,8	7,4354	12383,6	8848,7	208,273
153	426,15	443,15	316,65	7,6298	1657,8	7,4379	12413,0	8869,8	208,342
154	427,15	444,20	317,41	7,6933	1648,0	7,4403	12442,4	8890,9	208,411
155	428,15	445,25	318,16	7,7573	1638,2	7,4428	12471,9	8912,1	208,480
156	429,15	446,30	318,92	7,8217	1628,5	7,4452	12501,3	8933,2	208,548
157	430,15	447,35	319,67	7,8864	1618,9	7,4477	12530,8	8954,3	208,617
158	431,15	448,40	320,43	7,9516	1609,4	7,4501	12560,3	8975,5	208,685
159	432,15	449,46	321,18	8,0171	1599,9	7,4526	12589,7	8996,6	208,754
160	433,15	450,51	321,94	8,0831	1590,6	7,4550	12619,2	9017,8	208,822
161	434,15	451,56	322,69	8,1494	1581,2	7,4574	12648,7	9039,0	208,890
162	435,15	452,61	323,45	8,2162	1572,0	7,4598	12678,1	9060,1	208,957
163	436,15	453,67	324,20	8,2833	1562,9	7,4623	12707,6	9081,3	209,025
164	437,15	454,72	324,96	8,3509	1553,8	7,4647	12737,1	9102,5	209,093
165	438,15	455,77	325,72	8,4188	1544,7	7,4671	12766,6	9123,6	209,160
166	439,15	456,82	326,47	8,4872	1535,8	7,4695	12796,1	9144,8	209,227
167	440,15	457,88	327,23	8,5559	1526,9	7,4719	12825,6	9166,0	209,294
168	441,15	458,93	327,98	8,6251	1518,1	7,4743	12855,1	9187,2	209,361
169	442,15	459,98	328,74	8,6947	1509,4	7,4766	12884,6	9208,4	209,428

Carbon monoxide (*Continued*)

t	T	h	u	π_0	θ_0	s^o	II	U	S
170	443,15	461,04	329,50	8,7647	1500,7	7,4790	12914,1	9229,6	209,495
171	444,15	462,09	330,25	8,8351	1492,1	7,4814	12943,6	9250,8	209,561
172	445,15	463,14	331,01	8,9060	1483,6	7,4838	12973,1	9272,0	209,628
173	446,15	464,20	331,77	8,9772	1475,1	7,4861	13002,7	9293,2	209,694
174	447,15	465,25	332,53	9,0489	1466,7	7,4885	13032,2	9314,4	209,760
175	448,15	466,31	333,28	9,1210	1458,4	7,4908	13061,7	9335,6	209,826
176	449,15	467,36	334,04	9,1935	1450,1	7,4932	13091,3	9356,8	209,892
177	450,15	468,42	334,80	9,2664	1441,9	7,4955	13120,8	9378,1	209,958
178	451,15	469,47	335,57	9,3398	1433,7	7,4979	13150,4	9399,3	210,023
179	452,15	470,53	336,32	9,4136	1425,7	7,5002	13179,9	9420,5	210,089
180	453,15	471,58	337,07	9,4878	1417,6	7,5025	13209,5	9441,8	210,154
181	454,15	472,64	337,83	9,5625	1409,7	7,5049	13239,0	9463,0	210,219
182	455,15	473,69	338,59	9,6375	1401,8	7,5072	13268,6	9484,3	210,284
183	456,15	474,75	339,35	9,7131	1393,9	7,5095	13298,2	9505,6	210,349
184	457,15	475,80	340,11	9,7890	1386,1	7,5118	13327,7	9526,8	210,414
185	458,15	476,86	340,87	9,8654	1378,4	7,5141	13357,3	9548,1	210,478
186	459,15	477,92	341,63	9,9423	1370,7	7,5164	13386,9	9569,4	210,543
187	460,15	478,97	342,39	10,019	1363,1	7,5187	13416,5	9590,6	210,607
188	461,15	480,03	343,15	10,097	1355,6	7,5210	13446,1	9611,9	210,672
189	462,15	481,09	343,91	10,175	1348,1	7,5233	13475,7	9633,2	210,736
190	463,15	482,14	344,67	10,254	1340,6	7,5256	13505,3	9654,5	210,800
191	464,15	483,20	345,43	10,333	1333,2	7,5279	13534,9	9675,8	210,864
192	465,15	484,26	346,19	10,412	1325,9	7,5302	13564,5	9697,1	210,927
193	466,15	485,32	346,95	10,492	1318,6	7,5324	13594,2	9718,4	210,991
194	467,15	486,37	347,71	10,573	1311,4	7,5347	13623,8	9739,7	211,054
195	468,15	487,43	348,47	10,653	1304,2	7,5370	13653,4	9761,0	211,118
196	469,15	488,49	349,23	10,735	1297,1	7,5392	13683,1	9782,3	211,181
197	470,15	489,55	349,99	10,817	1290,1	7,5415	13712,7	9803,7	211,244
198	471,15	490,61	350,76	10,899	1283,0	7,5437	13742,3	9825,0	211,307
199	472,15	491,66	351,52	10,982	1276,1	7,5460	13772,0	9846,3	211,370
200	473,15	492,72	352,28	11,065	1269,2	7,5482	13801,6	9867,7	211,433
201	474,15	493,78	353,04	11,148	1262,3	7,5504	13831,3	9889,0	211,495
202	475,15	494,84	353,80	11,233	1255,5	7,5527	13861,0	9910,4	211,558
203	476,15	495,90	354,57	11,317	1248,7	7,5549	13890,7	9931,7	211,620
204	477,15	496,96	355,33	11,402	1242,0	7,5571	13920,3	9953,1	211,682
205	478,15	498,02	356,09	11,488	1235,4	7,5593	13950,0	9974,5	211,745
206	479,15	499,08	356,85	11,574	1228,7	7,5616	13979,7	9995,8	211,807
207	480,15	500,14	357,62	11,660	1222,2	7,5638	14009,4	10017,2	211,869
208	481,15	501,20	358,38	11,747	1215,7	7,5660	14039,1	10038,6	211,930
209	482,15	502,26	359,14	11,835	1209,2	7,5682	14068,8	10060,0	211,992
210	483,15	503,32	359,91	11,923	1202,7	7,5704	14098,5	10081,4	212,054
211	484,15	504,38	360,67	12,011	1196,4	7,5726	14128,2	10102,8	212,115
212	485,15	505,44	361,44	12,100	1190,0	7,5747	14157,9	10124,2	212,176
213	486,15	506,50	362,20	12,189	1183,7	7,5769	14187,7	10145,6	212,237
214	487,15	507,56	362,97	12,279	1177,5	7,5791	14217,4	10167,0	212,299
215	488,15	508,63	363,73	12,370	1171,3	7,5813	14247,1	10188,5	212,360
216	489,15	509,69	364,50	12,461	1165,1	7,5835	14276,9	10209,9	212,420
217	490,15	510,75	365,26	12,552	1159,0	7,5856	14306,6	10231,3	212,481
218	491,15	511,81	366,03	12,644	1152,9	7,5878	14336,4	10252,8	212,542
219	492,15	512,87	366,79	12,736	1146,9	7,5900	14366,1	10274,2	212,602
220	493,15	513,94	367,56	12,829	1140,9	7,5921	14395,9	10295,6	212,663
221	494,15	515,00	368,32	12,923	1134,9	7,5943	14425,7	10317,1	212,723
222	495,15	516,06	369,09	13,016	1129,0	7,5964	14455,4	10338,6	212,783
223	496,15	517,13	369,86	13,111	1123,2	7,5986	14485,2	10360,0	212,843
224	497,15	518,19	370,62	13,206	1117,4	7,6007	14515,0	10381,5	212,903

Carbon monoxide (*Continued*)

t	T	h	u	π_0	θ_0	s^0	H	U	S
225	498,15	519,25	371,39	13,301	1111,6	7,6028	14544,8	10403,0	212,963
226	499,15	520,32	372,16	13,397	1105,8	7,6050	14574,6	10424,5	213,023
227	500,15	521,38	372,92	13,494	1100,1	7,6071	14604,4	10445,9	213,083
228	501,15	522,44	373,69	13,591	1094,4	7,6092	14634,2	10467,4	213,142
229	502,15	523,51	374,46	13,688	1088,8	7,6114	14664,0	10488,9	213,202
230	503,15	524,57	375,23	13,786	1083,2	7,6135	14693,8	10510,4	213,261
231	504,15	525,64	375,99	13,885	1077,7	7,6156	14723,7	10532,0	213,320
232	505,15	526,70	376,76	13,984	1072,2	7,6177	14753,5	10553,5	213,379
233	506,15	527,77	377,53	14,083	1066,7	7,6198	14783,3	10575,0	213,438
234	507,15	528,83	378,30	14,183	1061,3	7,6219	14813,2	10596,5	213,497
235	508,15	529,90	379,07	14,284	1055,9	7,6240	14843,0	10618,1	213,556
236	509,15	530,97	379,84	14,385	1050,5	7,6261	14872,9	10639,6	213,615
237	510,15	532,03	330,61	14,487	1045,2	7,6282	14902,7	10661,1	213,673
238	511,15	533,10	381,37	14,589	1039,9	7,6303	14932,6	10682,7	213,732
239	512,15	534,16	382,14	14,692	1034,6	7,6324	14962,5	10704,3	213,790
240	513,15	535,23	382,91	14,796	1029,4	7,6344	14992,4	10725,8	213,848
241	514,15	536,30	383,68	14,899	1024,2	7,6365	15022,2	10747,4	213,907
242	515,15	537,36	384,45	15,004	1019,1	7,6386	15052,1	10769,0	213,965
243	516,15	538,43	385,23	15,109	1013,9	7,6407	15082,0	10790,5	214,023
244	517,15	539,50	386,00	15,214	1008,9	7,6427	15111,9	10812,1	214,080
245	518,15	540,57	386,77	15,320	1003,8	7,6448	15141,8	10833,7	214,138
246	519,15	541,64	387,54	15,427	998,84	7,6468	15171,7	10855,3	214,196
247	520,15	542,70	388,31	15,534	993,85	7,6489	15201,7	10876,9	214,253
248	521,15	543,77	389,08	15,642	988,91	7,6510	15231,6	10898,5	214,311
249	522,15	544,84	389,85	15,750	983,99	7,6530	15261,5	10920,2	214,368
250	523,15	545,91	390,62	15,859	979,11	7,6551	15291,5	10941,8	214,426
251	524,15	546,98	391,40	15,969	974,25	7,6571	15321,4	10963,4	214,483
252	525,15	548,05	392,17	16,079	969,43	7,6591	15351,4	10985,0	214,540
253	526,15	549,12	392,94	16,189	964,64	7,6612	15381,3	11006,7	214,597
254	527,15	550,19	393,71	16,301	959,89	7,6632	15411,3	11028,3	214,654
255	528,15	551,26	394,49	16,412	955,16	7,6652	15441,2	11050,0	214,711
256	529,15	552,33	395,26	16,525	950,47	7,6672	15471,2	11071,6	214,767
257	530,15	553,40	396,03	16,637	945,80	7,6693	15501,2	11093,3	214,824
258	531,15	554,47	396,81	16,751	941,17	7,6713	15531,2	11115,0	214,880
259	532,15	555,54	397,58	16,865	936,56	7,6733	15561,2	11136,7	214,937
260	533,15	556,61	398,36	16,980	931,99	7,6753	15591,2	11158,3	214,993
261	534,15	557,68	399,13	17,095	927,44	7,6773	15621,2	11180,0	215,049
262	535,15	558,75	399,90	17,211	922,93	7,6793	15651,2	11201,7	215,105
263	536,15	559,82	400,68	17,327	918,44	7,6813	15681,2	11223,4	215,162
264	537,15	560,89	401,45	17,444	913,98	7,6833	15711,2	11245,1	215,217
265	538,15	561,97	402,23	17,562	909,55	7,6853	15741,3	11266,9	215,273
266	539,15	563,04	403,01	17,680	905,15	7,6873	15771,3	11288,6	215,329
267	540,15	564,11	403,78	17,799	900,78	7,6893	15801,3	11310,3	215,385
268	541,15	565,18	404,56	17,918	896,44	7,6913	15881,4	11332,0	215,440
269	542,15	566,26	405,33	18,038	892,12	7,6933	15861,4	11353,8	215,496
270	543,15	567,33	406,11	18,158	887,83	7,6952	15891,5	11375,5	215,551
271	544,15	568,40	406,89	18,280	883,57	7,6972	15921,6	11397,3	215,607
272	545,15	569,48	407,66	18,401	879,33	7,6992	15951,6	11419,0	215,662
273	546,15	570,55	408,44	18,524	875,12	7,7011	15981,7	11440,8	215,717
274	547,15	571,63	409,22	18,647	870,94	7,7031	16011,8	11462,6	215,772
275	548,15	572,70	409,99	18,771	866,79	7,7051	16041,9	11484,4	215,827
276	549,15	573,77	410,77	18,895	862,66	7,7070	16072,0	11506,1	215,882
277	550,15	574,85	411,55	19,020	858,55	7,7090	16102,1	11527,9	215,936
278	551,15	575,92	412,33	19,145	854,47	7,7109	16132,2	11549,7	215,991
279	552,15	577,00	413,11	19,271	850,42	7,7129	16162,3	11571,5	216,046

Carbon monoxide (*Continued*)

t	T	h	u	π_u	θ_e	s^o	H	U	S
280	553,15	578,07	413,89	19,398	846,39	7,7148	16192,5	11593,3	216,100
281	554,15	579,15	414,66	19,525	842,39	7,7168	16222,6	11615,2	216,155
282	555,15	580,23	415,44	19,654	838,42	7,7187	16252,7	11637,0	216,209
283	556,15	581,30	416,22	19,782	834,46	7,7207	16282,9	11658,8	216,263
284	557,15	582,38	417,00	19,911	830,54	7,7226	16313,0	11680,6	216,317
285	558,15	583,46	417,78	20,041	826,63	7,7245	16343,2	11702,5	216,372
286	559,15	584,53	418,56	20,172	822,75	7,7264	16373,3	11724,3	216,426
287	560,15	585,61	419,34	20,303	818,90	7,7284	16403,5	11746,2	216,479
288	561,15	586,69	420,12	20,435	815,07	7,7303	16433,7	11768,1	216,533
289	562,15	587,76	420,90	20,568	811,26	7,7322	16463,9	11789,9	216,587
290	563,15	588,84	421,68	20,701	807,47	7,7341	16494,1	11811,8	216,641
291	564,15	589,92	422,47	20,835	803,71	7,7360	16524,3	11833,7	216,694
292	565,15	591,00	423,25	20,969	799,97	7,7379	16554,5	11855,6	216,748
293	566,15	592,08	424,03	21,104	796,26	7,7399	16584,7	11877,5	216,801
294	567,15	593,16	424,81	21,240	792,56	7,7418	16614,9	11899,4	216,854
295	568,15	594,23	425,59	21,376	788,89	7,7437	16645,1	11921,3	216,908
296	569,15	595,31	426,38	21,514	785,24	7,7456	16675,3	11943,2	216,961
297	570,15	596,39	427,16	21,651	781,62	7,7475	16705,6	11965,1	217,014
298	571,15	597,47	427,94	21,790	778,01	7,7493	16735,8	11987,0	217,067
299	572,15	598,55	428,72	21,929	774,43	7,7512	16766,1	12009,0	217,120
300	573,15	599,63	429,51	22,069	770,87	7,7531	16796,3	12030,9	217,173
301	574,15	600,71	430,29	22,209	767,33	7,7550	16826,6	12052,9	217,225
302	575,15	601,79	431,07	22,350	763,81	7,7569	16856,9	12074,8	217,278
303	576,15	602,88	431,86	22,492	760,32	7,7588	16887,1	12096,8	217,331
304	577,15	603,96	432,64	22,635	756,84	7,7606	16917,4	12118,8	217,383
305	578,15	605,04	433,43	22,778	753,39	7,7625	16947,7	12140,7	217,436
306	579,15	606,12	434,21	22,922	749,95	7,7644	16978,0	12162,7	217,488
307	580,15	607,20	435,00	23,066	746,54	7,7662	17008,3	12184,7	217,540
308	581,15	608,28	435,78	23,212	743,14	7,7681	17038,6	12206,7	217,592
309	582,15	609,37	436,57	23,358	739,77	7,7700	17068,9	12228,7	217,645
310	583,15	610,45	437,35	23,504	736,42	7,7718	17099,3	12250,7	217,697
311	584,15	611,53	438,14	23,652	733,08	7,7737	17129,6	12272,7	217,749
312	585,15	612,61	438,93	23,800	729,77	7,7755	17159,9	12294,8	217,801
313	586,15	613,70	439,71	23,949	726,47	7,7774	17190,3	12316,8	217,852
314	587,15	614,78	440,50	24,098	723,20	7,7792	17220,6	12338,8	217,904
315	588,15	615,86	441,29	24,248	719,94	7,7811	17251,0	12360,9	217,956
316	589,15	616,95	442,07	24,399	716,71	7,7829	17281,4	12382,9	218,007
317	590,15	618,03	442,86	24,551	713,49	7,7848	17311,7	12405,0	218,059
318	591,15	619,12	443,65	24,703	710,29	7,7866	17342,1	12427,0	218,110
319	592,15	620,20	444,44	24,856	707,11	7,7884	17372,5	12449,1	218,162
320	593,15	621,29	445,22	25,010	703,95	7,7903	17402,9	12471,2	218,213
321	594,15	622,37	446,01	25,165	700,81	7,7921	17433,3	12493,3	218,264
322	595,15	623,46	446,80	25,320	697,68	7,7939	17463,7	12515,4	218,315
323	596,15	624,54	447,59	25,476	694,58	7,7957	17494,1	12537,5	218,366
324	597,15	625,63	448,38	25,632	691,49	7,7976	17524,5	12559,6	218,417
325	598,15	626,72	449,17	25,790	688,42	7,7994	17555,0	12581,7	218,468
326	599,15	627,80	449,96	25,948	685,37	7,8012	17585,4	12603,8	218,519
327	600,15	628,89	450,75	26,107	682,33	7,8030	17615,8	12625,9	218,570
328	601,15	629,98	451,54	26,267	679,31	7,8048	17646,3	12648,1	218,621
329	602,15	631,06	452,33	26,427	676,31	7,8066	17676,7	12670,2	218,671
330	603,15	632,15	453,12	26,588	673,33	7,8084	17707,2	12692,4	218,722
331	604,15	633,24	453,91	26,750	670,36	7,8102	17737,7	12714,5	218,772
332	605,15	634,33	454,70	26,913	667,42	7,8120	17768,1	12736,7	218,823
333	606,15	635,42	455,49	27,076	664,48	7,8138	17798,6	12758,8	218,873
334	607,15	636,50	456,29	27,240	661,57	7,8156	17829,1	12781,0	218,923

Carbon monoxide (*Continued*)

t	T	h	u	π_0	θ_0	s^0	H	U	S
335	608,15	637,59	457,08	27,405	658,67	7,8174	17859,6	12803,2	218,973
336	609,15	638,68	457,87	27,571	655,79	7,8192	17890,1	12825,4	219,023
337	610,15	639,77	458,66	27,737	652,92	7,8210	17920,6	12847,6	219,073
338	611,15	640,86	459,46	27,905	650,07	7,8228	17951,2	12869,8	219,123
339	612,15	641,95	460,25	28,073	647,24	7,8245	17981,7	12892,0	219,173
340	613,15	643,04	461,04	28,241	644,42	7,8263	18012,2	12914,2	219,223
341	614,15	644,13	461,83	28,411	641,62	7,8281	18042,8	12936,5	219,273
342	615,15	645,22	462,63	28,581	638,84	7,8299	18073,3	12958,7	219,323
343	616,15	646,31	463,42	28,752	636,07	7,8316	18103,9	12980,9	219,372
344	617,15	647,40	464,22	28,924	633,31	7,8334	18134,4	13003,2	219,422
345	618,15	648,49	465,01	29,097	630,58	7,8352	18165,0	13025,4	219,471
346	619,15	649,59	465,81	29,271	627,85	7,8369	18195,6	13047,7	219,521
347	620,15	650,68	466,60	29,445	625,14	7,8387	18226,1	13070,0	219,570
348	621,15	651,77	467,40	29,620	622,45	7,8405	18256,7	13092,2	219,619
349	622,15	652,86	468,19	29,796	619,77	7,8422	18287,3	13114,5	219,669
350	623,15	653,95	468,99	29,972	617,11	7,8440	18317,9	13136,8	219,718
351	624,15	655,05	469,78	30,150	614,46	7,8457	18348,5	13159,1	219,767
352	625,15	656,14	470,58	30,328	611,83	7,8475	18379,2	13181,4	219,816
353	626,15	657,23	471,38	30,507	609,21	7,8492	18409,8	13203,7	219,865
354	627,15	658,33	472,17	30,687	606,61	7,8510	18440,4	13226,0	219,914
355	628,15	659,42	472,97	30,868	604,02	7,8527	18471,1	13248,4	219,963
356	629,15	660,52	473,77	31,049	601,44	7,8545	18501,7	13270,7	220,011
357	630,15	661,61	474,56	31,232	598,88	7,8562	18532,4	13293,0	220,060
358	631,15	662,70	475,36	31,415	596,33	7,8579	18563,0	13315,4	220,109
359	632,15	663,80	476,16	31,599	593,80	7,8597	18593,7	13337,7	220,157
360	633,15	664,89	476,96	31,784	591,28	7,8614	18624,4	13360,1	220,206
361	634,15	665,99	477,76	31,969	588,78	7,8631	18655,0	13382,5	220,254
362	635,15	667,09	478,56	32,156	586,29	7,8649	18685,7	13404,8	220,302
363	636,15	668,18	479,36	32,343	583,81	7,8666	18716,4	13427,2	220,351
364	637,15	669,28	480,15	32,531	581,34	7,8683	18747,1	13449,6	220,399
365	638,15	670,37	480,95	32,720	578,89	7,8700	18777,8	13472,0	220,447
366	639,15	671,47	481,75	32,910	576,46	7,8717	18808,6	13494,4	220,495
367	640,15	672,57	482,55	33,101	574,03	7,8735	18839,3	13516,8	220,543
368	641,15	673,66	483,35	33,292	571,62	7,8752	18870,0	13539,2	220,591
369	642,15	674,76	484,16	33,485	569,22	7,8769	18900,8	13561,7	220,639
370	643,15	675,86	484,96	33,678	566,84	7,8786	18931,5	13584,1	220,687
371	644,15	676,96	485,76	33,872	564,47	7,8803	18962,3	13606,5	220,735
372	645,15	678,06	486,56	34,067	562,11	7,8820	18993,0	13629,0	220,782
373	646,15	679,15	487,36	34,263	559,76	7,8837	19023,8	13651,4	220,830
374	647,15	680,25	488,16	34,460	557,43	7,8854	19054,6	13673,9	220,878
375	648,15	681,35	488,96	34,657	555,11	7,8871	19085,4	13696,4	220,925
376	649,15	682,45	489,77	34,856	552,80	7,8888	19116,2	13718,9	220,973
377	650,15	683,55	490,57	35,055	550,50	7,8905	19147,0	13741,3	221,020
378	651,15	684,65	491,37	35,255	548,22	7,8922	19177,8	13763,8	221,067
379	652,15	685,75	492,18	35,456	545,94	7,8939	19208,6	13786,3	221,115
380	653,15	686,85	492,98	35,658	543,68	7,8955	19239,4	13808,8	221,162
381	654,15	687,95	493,78	35,861	541,44	7,8972	19270,2	13831,4	221,209
382	655,15	689,05	494,59	36,065	539,20	7,8989	19301,1	13853,9	221,256
383	656,15	690,15	495,39	36,269	536,98	7,9006	19331,9	13876,4	221,303
384	657,15	691,26	496,20	36,475	534,77	7,9023	19362,8	13898,9	221,350
385	658,15	692,36	497,00	36,681	532,57	7,9039	19393,6	13921,5	221,397
386	659,15	693,46	497,81	36,889	530,38	7,9056	19424,5	13944,0	221,444
387	660,15	694,56	498,61	37,097	528,20	7,9073	19455,3	13966,6	221,491
388	661,15	695,66	499,42	37,306	526,04	7,9090	19486,2	13989,2	221,538
389	662,15	696,77	500,22	37,516	523,88	7,9106	19517,1	14011,7	221,584

Carbon monoxide (*Continued*)

t	T	h	u	π_v	0_v	s^o	H	U	S
390	663,15	697,87	501,03	37,727	521,74	7,9123	19548,0	14034,3	221,631
391	664,15	698,97	501,83	37,939	519,61	7,9139	19578,9	14056,9	221,677
392	665,15	700,08	502,64	38,152	517,49	7,9156	19609,8	14079,5	221,724
393	666,15	701,18	503,45	38,365	515,38	7,9173	19640,7	14102,1	221,770
394	667,15	702,28	504,26	38,580	513,28	7,9189	19671,7	14124,7	221,817
395	668,15	703,39	505,06	38,796	511,19	7,9206	19702,6	14147,3	221,863
396	669,15	704,49	505,87	39,012	509,12	7,9222	19733,5	14169,9	221,909
397	670,15	705,60	506,68	39,230	507,05	7,9239	19764,5	14192,6	221,956
398	671,15	706,70	507,49	39,448	505,00	7,9255	19795,4	14215,2	222,002
399	672,15	707,81	508,30	39,667	502,95	7,9272	19826,4	14237,9	222,048
400	673,15	708,91	509,10	39,887	500,92	7,9288	19857,4	14260,5	222,094
401	674,15	710,02	509,91	40,109	498,90	7,9305	19888,3	14283,2	222,140
402	675,15	711,12	510,72	40,331	496,89	7,9321	19919,3	14305,8	222,186
403	676,15	712,23	511,53	40,554	494,88	7,9337	19950,3	14328,5	222,232
404	677,15	713,34	512,34	40,778	492,89	7,9354	19981,3	14351,2	222,277
405	678,15	714,44	513,15	41,003	490,91	7,9370	20012,3	14373,9	222,323
406	679,15	715,55	513,96	41,229	488,94	7,9386	20043,3	14396,6	222,369
407	680,15	716,66	514,77	41,456	486,98	7,9403	20074,3	14419,3	222,415
408	681,15	717,77	515,58	41,684	485,03	7,9419	20105,4	14442,0	222,460
409	682,15	718,87	516,39	41,913	483,09	7,9435	20136,4	14464,7	222,506
410	683,15	719,98	517,21	42,143	481,16	7,9451	20167,4	14487,4	222,551
411	684,15	721,09	518,02	42,373	479,24	7,9468	20198,5	14510,2	222,597
412	685,15	722,20	518,83	42,605	477,33	7,9484	20229,5	14532,9	222,642
413	686,15	723,31	519,64	42,838	475,43	7,9500	20260,6	14555,7	222,687
414	687,15	724,42	520,45	43,072	473,54	7,9516	20291,7	14578,4	222,732
415	688,15	725,53	521,27	43,307	471,65	7,9532	20322,8	14601,2	222,778
416	689,15	726,64	522,08	43,542	469,78	7,9548	20353,8	14624,0	222,823
417	690,15	727,75	522,89	43,779	467,92	7,9564	20384,9	14646,7	222,868
418	691,15	728,86	523,71	44,017	466,07	7,9580	20416,0	14669,5	222,913
419	692,15	729,97	524,52	44,256	464,22	7,9597	20447,1	14692,3	222,958
420	693,15	731,08	525,33	44,495	462,39	7,9613	20478,3	14715,1	223,003
421	694,15	732,19	526,15	44,736	460,56	7,9629	20509,4	14737,9	223,048
422	695,15	733,30	526,96	44,978	458,75	7,9645	20540,5	14760,7	223,093
423	696,15	734,41	527,78	45,221	456,94	7,9661	20571,6	14783,6	223,137
424	697,15	735,53	528,59	45,464	455,14	7,9677	20602,8	14806,4	223,182
425	698,15	736,64	529,41	45,709	453,35	7,9692	20633,9	14829,2	223,227
426	699,15	737,75	530,22	45,955	451,57	7,9708	20665,1	14852,1	223,271
427	700,15	738,86	531,04	46,202	449,80	7,9724	20696,3	14874,9	223,316
428	701,15	739,98	531,86	46,450	448,04	7,9740	20727,4	14897,8	223,360
429	702,15	741,09	532,67	46,699	446,29	7,9756	20758,6	14920,7	223,405
430	703,15	742,20	533,49	46,949	444,55	7,9772	20789,8	14943,5	223,449
431	704,15	743,32	534,31	47,200	442,81	7,9788	20821,0	14966,4	223,493
432	705,15	744,43	535,12	47,452	441,08	7,9804	20852,2	14989,3	223,538
433	706,15	745,54	535,94	47,705	439,37	7,9819	20883,4	15012,2	223,582
434	707,15	746,66	536,76	47,959	437,66	7,9835	20914,7	15035,1	223,626
435	708,15	747,77	537,58	48,214	435,95	7,9851	20945,9	15058,0	223,670
436	709,15	748,89	538,39	48,471	434,26	7,9867	20977,1	15080,9	223,714
437	710,15	750,00	539,21	48,728	432,58	7,9882	21008,4	15103,9	223,758
438	711,15	751,12	540,03	48,986	430,90	7,9898	21039,6	15126,8	223,802
439	712,15	752,24	540,85	49,246	429,23	7,9914	21070,9	15149,8	223,846
440	713,15	753,35	541,67	49,506	427,58	7,9929	21102,1	15172,7	223,890
441	714,15	754,47	542,49	49,768	425,92	7,9945	21133,4	15195,7	223,934
442	715,15	755,58	543,31	50,031	424,28	7,9961	21164,7	15218,6	223,978
443	716,15	756,70	544,13	50,294	422,65	7,9976	21196,0	15241,6	224,021
444	717,15	757,82	544,95	50,559	421,02	7,9992	21227,3	15264,6	224,065

Carbon monoxide (*Continued*)

t	T	h	u	π_0	θ_0	s^0	H	U	S
445	718,15	758,94	545,77	50,825	419,40	8,0007	21258,6	15287,6	224,109
446	719,15	760,05	546,59	51,092	417,79	8,0023	21289,9	15310,6	224,152
447	720,15	761,17	547,41	51,360	416,19	8,0038	21321,2	15333,6	224,196
448	721,15	762,29	548,23	51,630	414,59	8,0054	21352,5	15356,6	224,239
449	722,15	763,41	549,06	51,900	413,00	8,0070	21383,8	15379,6	224,283
450	723,15	764,53	549,88	52,171	411,42	8,0085	21415,2	15402,6	224,326
451	724,15	765,65	550,70	52,444	409,85	8,0100	21446,5	15425,6	224,369
452	725,15	766,77	551,52	52,717	408,29	8,0116	21477,9	15448,7	224,413
453	726,15	767,89	552,34	52,992	406,73	8,0131	21509,2	15471,7	224,456
454	727,15	769,00	553,17	53,268	405,18	8,0147	21540,6	15494,8	224,499
455	728,15	770,13	553,99	53,545	403,64	8,0162	21572,0	15517,8	224,542
456	729,15	771,25	554,81	53,823	402,11	8,0178	21603,4	15540,9	224,585
457	730,15	772,37	555,64	54,102	400,58	8,0193	21634,7	15564,0	224,628
458	731,15	773,49	556,46	54,383	399,06	8,0208	21666,1	15587,1	224,671
459	732,15	774,61	557,29	54,664	397,55	8,0224	21697,5	15610,2	224,714
460	733,15	775,73	558,11	54,947	396,05	8,0239	21729,0	15633,2	224,757
461	734,15	776,85	558,94	55,230	394,55	8,0254	21760,4	15656,4	224,800
462	735,15	777,97	559,76	55,515	393,06	8,0269	21791,8	15679,5	224,843
463	736,15	779,10	560,59	55,801	391,57	8,0285	21823,2	15702,6	2 24,885
464	737,15	780,22	561,41	56,089	390,10	8,0300	21854,7	15725,7	224,928
465	738,15	781,34	562,24	56,377	388,63	8,0315	21886,1	15748,8	224,971
466	739,15	782,46	563,06	56,666	387,17	8,0330	21917,6	15772,0	225,013
467	740,15	783,59	563,89	56,957	385,71	8,0345	21949,0	15795,1	225,056
468	741,15	784,71	564,72	57,249	384,27	8,0361	21980,5	15818,3	225,098
469	742,15	785,83	565,54	57,542	382,83	8,0376	22012,0	15841,5	225,141
470	743,15	786,96	566,37	57,836	381,39	8,0391	22043,5	15864,6	225,183
471	744,15	788,08	567,20	58,131	379,96	8,0406	22075,0	15887,8	225,225
472	745,15	789,21	568,03	58,428	378,54	8,0421	22106,5	15911,0	225,268
473	746,15	790,33	568,85	58,726	377,13	8,0436	22138,0	15934,2	225,310
474	747,15	791,46	569,68	59,025	375,72	8,0451	22169,5	15957,4	225,352
475	748,15	792,58	570,51	59,325	374,32	8,0466	22201,0	15980,6	225,394
476	749,15	793,71	571,34	59,626	372,93	8,0481	22232,6	16003,8	225,436
477	750,15	794,83	572,17	59,928	371,54	8,0496	22264,1	16027,0	225,479
478	751,15	795,96	573,00	60,232	370,16	8,0511	22295,6	16050,3	225,521
479	752,15	797,09	573,83	60,537	368,79	8,0526	22327,2	16073,5	225,563
480	753,15	798,21	574,66	60,843	367,42	8,0541	22358,8	16096,8	225,604
481	754,15	799,34	575,49	61,150	366,06	8,0556	22390,3	16120,0	225,646
482	755,15	800,47	576,32	61,459	364,71	8,0571	22421,9	16143,3	225,688
483	756,15	801,59	577,15	61,768	363,36	8,0586	22453,5	16166,5	225,730
484	757,15	802,72	577,98	62,079	362,02	8,0601	22485,1	16189,8	225,772
485	758,15	803,85	578,81	62,392	360,68	8,0616	22516,7	16213,1	225,813
486	759,15	804,98	579,64	62,705	359,35	8,0631	22548,3	16236,4	225,855
487	760,15	806,11	580,47	63,020	358,03	8,0646	22579,9	16259,7	225,897
488	761,15	807,24	581,31	63,335	356,71	8,0661	22611,5	16283,0	225,938
489	762,15	808,37	582,14	63,653	355,40	8,0675	22643,1	16306,3	225,980
490	763,15	809,49	582,97	63,971	354,10	8,0690	22674,8	16329,6	226,021
491	764,15	810,62	583,80	64,290	352,80	8,0705	22706,4	16352,9	226,063
492	765,15	811,75	584,64	64,611	351,50	8,0720	22738,0	16376,3	226,104
493	766,15	812,88	585,47	64,933	350,22	8,0735	22769,7	16399,6	226,145
494	767,15	814,01	586,30	65,257	348,94	8,0749	22801,4	16423,0	226,187
495	768,15	815,14	587,14	65,581	347,66	8,0764	22833,0	16446,3	226,228
496	769,15	816,28	587,97	65,907	346,39	8,0779	22864,7	16469,7	226,269
497	770,15	817,41	588,81	66,234	345,13	8,0793	22896,4	16493,0	226,310
498	771,15	818,54	589,64	66,563	343,88	8,0808	22928,1	16516,4	226,352
499	772,15	819,67	590,48	66,892	342,62	8,0823	22959,8	16539,8	226,393

Carbon monoxide (*Continued*)

t	T	h	u	π_0	0_0	s^0	H	U	S
500	773,15	820,80	591,31	67,223	341,38	8,0837	22991,5	16563,2	226,434
501	774,15	821,93	592,15	67,556	340,14	8,0852	23023,2	16586,6	226,475
502	775,15	823,07	592,98	67,889	338,91	8,0867	23054,9	16610,0	226,516
503	776,15	824,20	593,82	68,224	337,68	8,0881	23086,6	16633,4	226,556
504	777,15	825,33	594,65	68,560	336,45	8,0896	23118,4	16656,8	226,597
505	778,15	826,47	595,49	68,898	335,24	8,0910	23150,1	16680,3	226,638
506	779,15	827,60	596,33	69,236	334,03	8,0925	23181,9	16703,7	226,679
507	780,15	828,73	597,16	69,576	332,82	8,0940	23213,6	16727,1	226,720
508	781,15	829,87	598,00	69,918	331,62	8,0954	23245,4	16750,6	226,760
509	782,15	831,00	598,84	70,260	330,43	8,0969	23277,2	16774,1	226,801
510	783,15	832,14	599,68	70,604	329,24	8,0983	23308,9	16797,5	226,842
511	784,15	833,27	600,51	70,950	328,05	8,0998	23340,7	16821,0	226,882
512	785,15	834,41	601,35	71,296	326,87	8,1012	23372,5	16844,5	226,923
513	786,15	835,54	602,19	71,644	325,70	8,1026	23404,3	16868,0	226,963
514	787,15	836,68	603,03	71,993	324,53	8,1041	23436,1	16891,4	227,004
515	788,15	837,81	603,87	72,344	323,37	8,1055	23467,9	16914,9	227,044
516	789,15	838,95	604,71	72,696	322,21	8,1070	23499,8	16938,5	227,084
517	790,15	840,08	605,55	73,049	321,06	8,1084	23531,6	16962,0	227,125
518	791,15	841,22	606,39	73,404	319,91	8,1098	23563,4	16985,5	227,165
519	792,15	842,36	607,23	73,760	318,77	8,1113	23595,3	17009,0	227,205
520	793,15	843,49	608,07	74,117	317,64	8,1127	23627,1	17032,6	227,245
521	794,15	844,63	608,91	74,476	316,50	8,1142	23659,0	17056,1	227,285
522	795,15	845,77	609,75	74,836	315,38	8,1156	23690,9	17079,7	227,326
523	796,15	846,91	610,59	75,197	314,26	8,1170	23722,7	17103,2	227,366
524	797,15	848,05	611,43	75,560	313,14	8,1184	23754,6	17126,8	227,406
525	798,15	849,18	612,27	75,924	312,03	8,1199	23786,5	17150,4	227,446
526	799,15	850,32	613,11	76,290	310,92	8,1213	23818,4	17173,9	227,486
527	800,15	851,46	613,96	76,657	309,82	8,1227	23850,3	17197,5	227,525
528	801,15	852,60	614,80	77,025	308,73	8,1241	23882,2	17221,1	227,565
529	802,15	853,74	615,64	77,395	307,64	8,1256	23914,1	17244,7	227,605
530	803,15	854,88	616,48	77,766	306,55	8,1270	23946,0	17268,3	227,645
531	804,15	856,02	617,33	78,138	305,47	8,1284	23978,0	17291,9	227,685
532	805,15	857,16	618,17	78,512	304,39	8,1298	24009,9	17315,6	227,724
533	806,15	858,30	619,01	78,888	303,32	8,1312	24041,9	17339,2	227,764
534	807,15	859,44	619,86	79,264	302,25	8,1326	24073,8	17362,8	227,804
535	808,15	860,58	620,70	79,643	301,19	8,1341	24105,8	17386,5	227,843
536	809,15	861,72	621,55	80,022	300,13	8,1355	24137,7	17410,1	227,883
537	810,15	862,86	622,39	80,403	299,08	8,1369	24169,7	17433,8	227,922
538	811,15	864,01	623,24	80,785	298,03	8,1383	24201,7	17457,4	227,962
539	812,15	865,15	624,08	81,169	296,99	8,1397	24233,7	17481,1	228,001
540	813,15	866,29	624,93	81,555	295,95	8,1411	24265,7	17504,8	228,040
541	814,15	867,43	625,77	81,941	294,91	8,1425	24297,7	17528,5	228,080
542	815,15	868,58	626,62	82,329	293,88	8,1439	24329,7	17552,2	228,119
543	816,15	869,72	627,46	82,719	292,86	8,1453	24361,7	17575,9	228,158
544	817,15	870,86	628,31	83,110	291,84	8,1467	24393,7	17599,6	228,197
545	818,15	872,01	629,16	83,503	290,82	8,1481	24425,7	17623,3	228,237
546	819,15	873,15	630,00	83,897	289,81	8,1495	24457,8	17647,0	228,276
547	820,15	874,29	630,85	84,292	288,80	8,1509	24489,8	17670,8	228,315
548	821,15	875,44	631,70	84,689	287,80	8,1523	24521,9	17694,5	228,354
549	822,15	876,58	632,55	85,087	286,80	8,1537	24553,9	17718,2	228,393
550	823,15	877,73	633,39	85,487	285,81	8,1551	24586,0	17742,0	228,432
551	824,15	878,87	634,24	85,888	284,82	8,1565	24618,1	17765,7	228,471
552	825,15	880,02	635,09	86,291	283,83	8,1579	24650,1	17789,5	228,510
553	826,15	881,16	635,94	86,695	282,85	8,1592	24682,2	17813,3	228,549
554	827,15	882,31	636,79	87,101	281,87	8,1606	24714,3	17837,1	228,587

t	T	h	u	π_0	θ_0	s^0	H	U	S
555	828,15	883,45	637,64	87,508	280,90	8,1620	24746,4	17860,8	228,626
556	829,15	884,60	638,49	87,917	279,93	8,1634	24778,5	17884,6	228,665
557	830,15	885,75	639,34	88,327	278,97	8,1648	24810,6	17908,4	228,704
558	831,15	886,89	640,19	88,739	278,01	8,1662	24842,8	17932,2	228,742
559	832,15	888,04	641,04	89,152	277,05	8,1675	24874,9	17956,1	228,781
560	833,15	889,19	641,89	89,567	276,10	8,1689	24907,0	17979,9	228,820
561	834,15	890,33	642,74	89,984	275,15	8,1703	24939,2	18003,7	228,858
562	835,15	891,48	643,59	90,401	274,21	8,1717	24971,3	18027,5	228,897
563	836,15	892,63	644,44	90,821	273,27	8,1730	25003,5	18051,4	228,935
564	837,15	893,78	645,29	91,242	272,33	8,1744	25035,6	18075,2	228,974
565	838,15	894,93	646,14	91,664	271,40	8,1758	25067,8	18099,1	229,012
566	839,15	896,08	646,99	92,088	270,48	8,1772	25100,0	18122,9	229,050
567	840,15	897,23	647,85	92,514	269,55	8,1785	25132,2	18146,8	229,089
568	841,15	898,37	648,70	92,941	268,63	8,1799	25164,4	18170,7	229,127
569	842,15	899,52	649,55	93,369	267,72	8,1813	25196,6	18194,6	229,165
570	843,15	900,67	650,40	93,799	266,81	8,1826	25228,8	18218,5	229,203
571	844,15	901,82	651,26	94,231	265,90	8,1840	25261,0	18242,4	229,242
572	845,15	902,97	652,11	94,664	265,00	8,1853	25293,2	18266,3	229,280
573	846,15	904,12	652,96	95,099	264,10	8,1867	25325,4	18290,2	229,318
574	847,15	905,27	653,82	95,536	263,20	8,1881	25357,7	18314,1	229,356
575	848,15	906,43	654,67	95,974	262,31	8,1894	25389,9	18338,0	229,394
576	849,15	907,58	655,53	96,413	261,42	8,1908	25422,1	18362,0	229,432
577	850,15	908,73	656,38	96,855	260,54	8,1921	25454,4	18385,9	229,470
578	851,15	909,88	657,24	97,297	259,66	8,1935	25486,6	18409,8	229,508
579	852,15	911,03	658,09	97,742	258,78	8,1948	25518,9	18433,8	229,546
580	853,15	912,18	658,95	98,188	257,91	8,1962	25551,2	18457,8	229,584
581	854,15	913,34	659,80	98,635	257,04	8,1975	25583,5	18481,7	229,621
582	855,15	914,49	660,66	99,084	256,17	8,1989	25615,8	18505,7	229,659
583	856,15	915,64	661,51	99,535	255,31	8,2002	25648,1	18529,7	229,697
584	857,15	916,80	662,37	99,988	254,45	8,2016	25680,4	18553,7	229,735
585	858,15	917,95	663,23	100,44	253,60	8,2029	25712,7	18577,6	229,772
586	859,15	919,10	664,08	100,89	252,74	8,2043	25745,0	18601,6	229,810
587	860,15	920,26	664,94	101,35	251,90	8,2056	25777,3	18625,7	229,848
588	861,15	921,41	665,80	101,81	251,05	8,2070	25809,6	18649,7	229,885
589	862,15	922,56	666,66	102,27	250,21	8,2083	25842,0	18673,7	229,923
590	863,15	923,72	667,51	102,73	249,38	8,2096	25874,3	18697,7	229,950
591	864,15	924,87	668,37	103,20	248,54	8,2110	25906,6	18721,8	229,998
592	865,15	926,03	669,23	103,66	247,71	8,2123	25939,0	18745,8	230,035
593	866,15	927,18	670,09	104,13	246,89	8,2136	25971,4	18769,8	230,072
594	867,15	928,34	670,95	104,60	246,06	8,2150	26003,7	18793,9	230,110
595	868,15	929,50	671,81	105,07	245,24	8,2163	26036,1	18818,0	230,147
596	869,15	930,65	672,67	105,54	244,43	8,2176	26068,5	18842,0	230,184
597	870,15	931,81	673,52	106,01	243,62	8,2190	26100,9	18866,1	230,222
598	871,15	932,96	674,38	106,49	242,81	8,2203	26133,3	18890,2	230,259
599	872,15	934,12	675,24	106,97	242,00	8,2216	26165,7	18914,3	230,296
600	873,15	935,28	676,10	107,45	241,20	8,2230	26198,1	18938,4	230,333
601	874,15	936,44	676,96	107,93	240,40	8,2243	26230,5	18962,5	230,370
602	875,15	937,59	677,83	108,41	239,60	8,2256	26262,9	18986,6	230,407
603	876,15	938,75	678,69	108,89	238,81	8,2269	26295,4	19010,7	230,444
604	877,15	939,91	679,55	109,38	238,02	8,2282	26327,8	19034,8	230,481
605	878,15	941,07	680,41	109,87	237,24	8,2296	26360,2	19058,9	230,518
606	879,15	942,23	681,27	110,35	236,45	8,2309	26392,7	19083,1	230,555
607	880,15	943,38	682,13	110,85	235,68	8,2322	26425,1	19107,2	230,592
608	881,15	944,54	682,99	111,34	234,90	8,2335	26457,6	19131,4	230,629
609	882,15	945,70	683,86	111,83	234,13	8,2348	26490,1	19155,5	230,666

Carbon monoxide (*Continued*)

t	T	h	u	π_0	0_0	s^0	H	U	S
610	883,15	946,86	684,72	112,33	233,36	8,2361	26522,5	19179,7	230,703
611	884,15	948,02	685,58	112,83	232,59	8,2375	26555,0	19203,8	230,739
612	885,15	949,18	686,45	113,33	231,83	8,2388	26587,5	19228,0	230,776
613	886,15	950,34	687,31	113,83	231,07	8,2401	26620,0	19252,2	230,813
614	887,15	951,50	688,17	114,33	230,31	8,2414	26652,5	19276,4	230,849
615	888,15	952,66	·689,04	114,83	229,56	8,2427	26685,0	19300,6	230,886
616	889,15	953,82	689,90	115,34	228,81	8,2440	26717,5	19324,8	230,923
617	890,15	954,98	690,76	115,85	228,06	8,2453	26750,1	19349,0	230,959
618	891,15	956,15	691,63	116,36	227,31	8,2466	26782,6	19373,2	230,996
619	892,15	957,31	692,49	116,87	226,57	8,2479	26815,1	19397,4	231,032
620	893,15	958,47	693,36	117,38	225,83	8,2492	26847,7	19421,7	231,069
621	894,15	959,63	694,22	117,90	225,10	8,2505	26880,2	19445,9	231,105
622	895,15	960,79	695,09	118,42	224,37	8,2518	26912,8	19470,1	231,141
623	896,15	961,96	695,95	118,94	223,64	8,2531	26945,3	19494,4	231,178
624	897,15	963,12	696,82	119,46	222,91	8,2544	26977,9	19518,6	231,214
625	898,15	964,28	697,69	119,98	222,19	8,2557	27010,5	19542,9	231,250
626	899,15	965,44	698,55	120,50	221,47	8,2570	27043,1	19567,2	231,287
627	900,15	966,61	699,42	121,03	220,75	8,2583	27075,6	19591,4	231,323
628	901,15	967,77	700,29	121,56	220,03	8,2596	27108,2	19615,7	231,359
629	902,15	968,93	701,15	122,09	219,32	8,2609	27140,8	19640,0	231,395
630	903,15	970,10	702,02	122,62	218,61	8 2622	27173,4	19664,3	231,431
631	904,15	971,26	702,89	123,15	217,91	8,2634	27206,1	19688,6	231,467
632	905,15	972,43	703,76	123,69	217,20	8,2647	27238,7	19712,9	231,504
633	906,15	973,59	704,62	124,22	216,50	8,2660	27271,3	19737,2	231,540
634	907,15	974,76	705,49	124,76	215,81	8,2673	27303,9	19761,5	231,576
635	908,15	975,92	706,36	125,30	215,11	8,2686	27336,6	19785,8	231,612
636	909,15	977,09	707,23	125,85	214,42	8,2699	27369,2	19810,2	231,647
637	910,15	978,25	708,10	126,39	213,73	8,2712	27401,9	19834,5	231,683
638	911,15	979,42	708,97	126,94	213,05	8,2724	27434,5	19858,9	231,719
639	912,15	980,59	709,84	127,49	212,36	8,2737	27467,2	19883,2	231,755
640	913,15	981,75	710,71	128,04	211,68	8,2750	27499,9	19907,6	231,791
641	914,15	982,92	711,57	128,59	211,00	8,2763	27532,5	19931,9	231,827
642	915,15	984,09	712,44	129,14	210,33	8,2775	27565,2	19956,3	231,862
643	916,15	985,25	713,32	129,70	209,66	8,2788	27597,9	19980,7	231,898
644	917,15	986,42	714,19	130,26	208,99	8,2801	27630,6	20005,1	231,934
645	918,15	987,59	715,06	130,82	208,32	8,2814	27663,3	20029,4	231,969
646	919,15	988,76	715,93	131,38	207,66	8,2826	27696,0	20053,8	232,005
647	920,15	989,92	716,80	131,94	206,99	8,2839	27728,7	20078,2	232,041
648	921,15	991,09	717,67	132,51	206,34	8,2852	27761,5	20102,6	232,076
649	922,15	992,26	718,54	133,07	205,68	8,2864	27794,2	20127,1	232,112
650	923,15	993,43	719,41	133,64	205,03	8,2877	27826,9	20151,5	232,147
651	924,15	994,60	720,29	134,21	204,37	8,2890	27859,7	20175,9	232,182
652	925,15	995,77	721,16	134,79	203,73	8,2902	27892,4	20200,3	232,218
653	926,15	996,94	722,03	135,36	203,08	8,2915	27925,2	20224,8	232,253
654	927,15	998,11	722,90	135,94	202,44	8,2928	27957,9	20249,2	232,289
655	928,15	999,28	723,78	136,52	201,80	8,2940	27990,7	20273,7	232,324
656	929,15	1000,45	724,65	137,10	201,16	8,2953	28023,5	20298,1	232,359
657	930,15	1001,62	725,52	137,68	200,52	8,2965	28056,2	20322,6	232,395
658	931,15	1002,79	726,40	138,26	199,89	8,2978	28089,0	20347,1	232,430
659	932,15	1003,96	727,27	138,85	199,26	8,2991	28121,8	20371,5	232,465
660	933,15	1005,13	728,14	139,44	198,63	8,3003	28154,6	20396,0	232,500
661	934,15	1006,30	729,02	140,03	198,00	8,3016	28187,4	20420,5	232,535
662	935,15	1007,47	729,89	140,62	197,38	8,3028	28220,2	20445,0	232,570
663	936,15	1008,64	730,77	141,22	196,76	8,3041	28253,1	20469,5	232,605
664	937,15	1009,81	731,64	141,81	196,14	8,3053	28285,9	20494,0	232,640

Carbon monoxide (*Continued*)

t	T	h	u	π_0	θ_0	s°	H	U	S
665	938,15	1010,99	732,52	142,41	195,53	8,3066	28318,7	20518,5	232,675
666	939,15	1012,16	733,39	143,01	194,91	8,3078	28351,5	20543,1	232,710
667	940,15	1013,33	734,27	143,61	194,30	8,3091	28384,4	20567,6	232,745
668	941,15	1014,50	735,14	144,22	193,69	8,3103	28417,2	20592,1	232,780
669	942,15	1015,68	736,02	144,82	193,09	8,3116	28450,1	20616,7	232,815
670	943,15	1016,85	736,90	145,43	192,48	8,3128	28482,9	20641,2	232,850
671	944,15	1018,02	737,77	146,04	191,88	8,3141	28515,8	20665,8	232,885
672	945,15	1019,20	738,65	146,66	191,28	8,3153	28548,7	20690,3	232,920
673	946,15	1020,37	739,53	147,27	190,69	8,3165	28581,6	20714,9	232,954
674	947,15	1021,54	740,40	147,89	190,09	8,3178	28614,4	20739,4	232,989
675	948,15	1022,72	741,28	148,51	189,50	8,3190	28647,3	20764,0	233,024
676	949,15	1023,89	742,16	149,13	188,91	8,3203	28680,2	20788,6	233,059
677	950,15	1025,07	743,04	149,75	188,32	8,3215	28713,1	20813,2	233,093
678	951,15	1026,24	743,91	150,37	187,74	8,3227	28746,0	20837,8	233,128
679	952,15	1027,42	744,79	151,00	187,15	8,3240	28779,0	20862,4	233,162
680	953,15	1028,59	745,67	151,63	186,57	8,3252	28811,9	20887,0	233,197
681	954,15	1029,77	746,55	152,26	186,00	8,3264	28844,8	20911,6	233,232
682	955,15	1030,94	747,43	152,89	185,42	8,3277	28877,7	20936,2	233,266
683	956,15	1032,12	748,31	153,53	184,85	8,3289	28910,7	20960,8	233,300
684	957,15	1033,29	749,19	154,17	184,27	8,3301	28943,6	20985,5	233,335
685	958,15	1034,47	750,07	154,81	183,71	8,3313	28976,6	21010,1	233,369
686	959,15	1035,65	750,95	155,45	183,14	8,3326	29009,5	21034,8	233,404
687	960,15	1036,82	751,83	156,09	182,57	8,3338	29042,5	21059,4	233,438
688	961,15	1038,00	752,71	156,74	182,01	8,3350	29075,5	21084,1	233,472
689	962,15	1039,18	753,59	157,38	181,45	8,3362	29108,4	21108,7	233,507
690	963,15	1040,36	754,47	158,03	180,89	8,3375	29141,4	21133,4	233,541
691	964,15	1041,53	755,35	158,69	180,34	8,3387	29174,4	21158,1	233,575
692	965,15	1042,71	756,23	159,34	179,78	8,3399	29207,4	21182,7	233,609
693	966,15	1043,89	757,11	160,00	179,23	8,3411	29240,4	21207,4	233,644
694	967,15	1045,07	757,99	160,66	178,68	8,3424	29273,4	21232,1	233,678
695	968,15	1046,25	758,87	161,32	178,13	8,3436	29306,4	21256,8	233,712
696	969,15	1047,42	759,76	161,98	177,59	8,3448	29339,4	21281,5	233,746
697	970,15	1048,60	760,64	162,64	177,04	8,3460	29372,4	21306,2	233,780
698	971,15	1049,78	761,52	163,31	176,50	8,3472	29405,5	21330,9	233,814
699	972,15	1050,96	762,40	163,98	175,96	8,3484	29438,5	21355,6	233,848
700	973,15	1052,14	763,28	164,65	175,43	8,3496	29471,5	21380,4	233,882
701	974,15	1053,32	764,17	165,32	174,89	8,3509	29504,6	21405,1	233,916
702	975,15	1054,50	765,05	166,00	174,36	8,3521	29537,6	21429,8	233,950
703	976,15	1055,68	765,93	166,68	173,83	8,3533	29570,7	21454,6	233,984
704	977,15	1056,86	766,82	167,36	173,30	8,3545	29603,8	21479,3	234,018
705	978,15	1058,04	767,70	168,04	172,77	8,3557	29636,8	21504,1	234,051
706	979,15	1059,22	768,59	168,72	172,25	8,3569	29669,9	21528,9	234,085
707	980,15	1060,40	769,47	169,41	171,72	8,3581	29703,0	21553,6	234,119
708	981,15	1061,59	770,35	170,10	171,20	8,3593	29736,1	21578,4	234,153
709	982,15	1062,77	771,24	170,79	170,68	8,3605	29769,2	21603,2	234,186
710	983,15	1063,95	772,12	171,48	170,17	8,3617	29802,3	21628,0	234,220
711	984,15	1065,13	773,01	172,18	169,65	8,3629	29835,4	21652,7	234,254
712	985,15	1066,31	773,89	172,88	169,14	8,3641	29868,5	21677,5	234,287
713	986,15	1067,49	774,78	173,58	168,63	8,3653	29901,6	21702,3	234,321
714	987,15	1068,68	775,66	174,28	168,12	8,3665	29934,7	21727,2	234,354
715	988,15	1069,86	776,55	174,98	167,61	8,3677	29967,8	21752,0	234,388
716	989,15	1071,04	777,44	175,69	167,10	8,3689	30001,0	21776,8	234,422
717	990,15	1072,23	778,32	176,40	166,60	8,3701	30034,1	21801,6	234,455
718	991,15	1073,41	779,21	177,11	166,10	8,3713	30067,3	21826,4	234,488
719	992,15	1074,59	780,10	177,83	165,60	8,3725	30100,4	21851,3	234,522

Carbon monoxide (*Continued*)

t	T	h	u	π_0	θ_0	s^0	H	U	S
720	993,15	1075,78	780,98	178,54	165,10	8,3737	30133,6	21876,1	234,555
721	994,15	1076,96	781,87	179,26	164,61	8,3749	30166,7	21901,0	234,589
722	995,15	1078,14	782,76	179,98	164,11	8,3761	30199,9	21925,8	234,622
723	996,15	1079,33	783,64	180,70	163,62	8,3773	30233,1	21950,7	234,655
724	997,15	1080,51	784,53	181,43	163,13	8,3784	30266,3	21975,5	234,689
725	998,15	1081,70	785,42	182,15	162,64	8,3796	30299,4	22000,4	234,722
726	999,15	1082,88	786,31	182,88	162,16	8,3808	30332,6	22025,3	234,755
727	1000,15	1084,07	787,20	183,62	161,67	8,3820	30365,8	22050,2	234,788
728	1001,15	1085,25	788,09	184,35	161,19	8,3832	30399,0	22075,1	234,822
729	1002,15	1086,44	788,97	185,09	160,71	8,3844	30432,2	22099,9	234,855
730	1003,15	1087,62	789,86	185,83	160,23	8,3856	30465,4	22124,8	234,888
731	1004,15	1088,81	790,75	186,57	159,75	8,3867	30498,7	22149,8	234,921
732	1005,15	1090,00	791,64	187,31	159,27	8,3879	30531,9	22174,7	234,954
733	1006,15	1091,18	792,53	188,06	158,80	8,3891	30565,1	22199,6	234,987
734	1007,15	1092,37	793,42	188,81	158,33	8,3903	30598,4	22224,5	235,020
735	1008,15	1093,56	794,31	189,56	157,86	8,3915	30631,6	22249,4	235,053
736	1009,15	1094,74	795,20	190,31	157,39	8,3926	30664,8	22274,4	235,086
737	1010,15	1095,93	796,09	191,06	156,92	8,3938	30698,1	22299,3	235,119
738	1011,15	1097,12	796,98	191,82	156,46	8,3950	30731,4	22324,2	235,152
739	1012,15	1098,30	797,87	192,58	155,99	8,3962	30764,6	22349,2	235,185
740	1013,15	1099,49	798,76	193,34	155,53	8,3973	30797,9	22374,1	235,218
741	1014,15	1100,68	799,65	194,11	155,07	8,3985	30831,2	22399,1	235,250
742	1015,15	1101,87	800,55	194,88	154,61	8,3997	30864,4	22424,1	235,283
743	1016,15	1103,06	801,44	195,65	154,16	8,4008	30897,7	22449,0	235,316
744	1017,15	1104,25	802,33	196,42	153,70	8,4020	30931,0	22474,0	235,349
745	1018,15	1105,43	803,22	197,19	153,25	8,4032	30964,3	22499,0	235,381
746	1019,15	1106,62	804,11	197,97	152,80	8,4043	30997,6	22524,0	235,414
747	1020,15	1107,81	805,00	198,75	152,35	8,4055	31030,9	22549,0	235,447
748	1021,15	1109,00	805,90	199,53	151,90	8,4067	31064,2	22574,0	235,479
749	1022,15	1110,19	806,79	200,32	151,45	8,4078	31097,6	22599,0	235,512
750	1023,15	1111,38	807,68	201,10	151,01	8,4090	31130,9	22624,0	235,545
751	1024,15	1112,57	808,58	201,89	150,57	8,4102	31164,2	22649,0	235,577
752	1025,15	1113,76	809,47	202,68	150,12	8,4113	31197,6	22674,0	235,610
753	1026,15	1114,95	810,36	203,48	149,68	8,4125	31230,9	22699,1	235,642
754	1027,15	1116,14	811,26	204,27	149,25	8,4136	31264,2	22724,1	235,675
755	1028,15	1117,33	812,15	205,07	148,81	8,4148	31297,6	22749,1	235,707
756	1029,15	1118,52	813,04	205,87	148,37	8,4160	31330,9	22774,2	235,740
757	1030,15	1119,71	813,94	206,68	147,94	8,4171	31364,3	22799,2	235,772
758	1031,15	1120,91	814,83	207,48	147,51	8,4183	31397,7	22824,3	235,804
759	1032,15	1122,10	815,73	208,29	147,08	8,4194	31431,1	22849,3	235,837
760	1033,15	1123,29	816,62	209,10	146,65	8,4206	31464,4	22874,4	235,869
761	1034,15	1124,48	817,52	209,92	146,22	8,4217	31497,8	22899,5	235,901
762	1035,15	1125,67	818,41	210,73	145,80	8,4229	31531,2	22924,5	235,934
763	1036,15	1126,86	819,31	211,55	145,37	8,4240	31564,6	22949,6	235,966
764	1037,15	1128,06	820,20	212,37	144,95	8,4252	31598,0	22974,7	235,998
765	1038,15	1129,25	821,10	213,20	144,53	8,4263	31631,4	22999,8	236,030
766	1039,15	1130,44	821,99	214,02	144,11	8,4275	31664,8	23024,9	236,062
767	1040,15	1131,64	822,89	214,85	143,69	8,4286	31698,2	23050,0	236,095
768	1041,15	1132,83	823,79	215,69	143,28	8,4298	31731,7	23075,1	236,127
769	1042,15	1134,02	824,69	216,52	142,86	8,4309	31765,1	23100,2	236,159
770	1043,15	1135,22	825,58	217,36	142,45	8,4321	31798,5	23125,3	236,191
771	1044,15	1136,41	826,48	218,20	142,04	8,4332	31831,9	23150,5	236,223
772	1045,15	1137,60	827,37	219,04	141,62	8,4344	31865,4	23175,6	236,255
773	1046,15	1138,80	828,27	219,88	141,22	8,4355	31898,8	23200,7	236,287
774	1047,15	1139,99	829,17	220,73	140,81	8,4366	31932,3	23225,9	236,319

Carbon monoxide (*Continued*)

t	T	h	u	π_0	θ_0	s^0	H	U	S
775	1048,15	1141,19	830,07	221,58	140,40	8,4378	31965,8	23251,0	236,351
776	1049,15	1142,38	830,96	222,43	140,00	8,4389	31999,2	23276,2	236,383
777	1050,15	1143,58	831,86	223,28	139,60	8,4401	32032,7	23301,3	236,415
778	1051,15	1144,77	832,76	224,14	139,19	8,4412	32066,2	23326,5	236,446
779	1052,15	1145,97	833,66	225,00	138,79	8,4423	32099,6	23351,6	236,478
780	1053,15	1147,16	834,56	225,86	138,40	8,4435	32133,1	23376,8	236,510
781	1054,15	1148,36	835,46	226,73	138,00	8,4446	32166,6	23402,0	236,542
782	1055,15	1149,55	836,36	227,60	137,60	8,4457	32200,1	23427,2	236,574
783	1056,15	1150,75	837,25	228,47	137,21	8,4469	32233,6	23452,3	236,605
784	1057,15	1151,94	838,15	229,34	136,82	8,4480	32267,1	23477,5	236,637
785	1058,15	1153,14	839,05	230,22	136,42	8,4491	32300,6	23502,7	236,669
786	1059,15	1154,34	839,95	231,09	136,03	8,4503	32334,1	23527,9	236,700
787	1060,15	1155,53	840,85	231,98	135,64	8,4514	32367,7	23553,1	236,732
788	1061,15	1156,73	841,75	232,86	135,26	8,4525	32401,2	23578,4	236,764
789	1062,15	1157,93	842,65	233,75	134,87	8,4537	32434,7	23603,6	236,795
790	1063,15	1159,13	843,55	234,63	134,49	8,4548	32468,3	23628,8	236,827
791	1064,15	1160,32	844,45	235,53	134,10	8,4559	32501,8	23654,0	236,858
792	1065,15	1161,52	845,36	236,42	133,72	8,4570	32535,4	23679,3	236,890
793	1066,15	1162,72	846,26	237,32	133,34	8,4582	32568,9	23704,5	236,921
794	1067,15	1163,92	847,16	238,22	132,96	8,4593	32602,5	23729,7	236,953
795	1068,15	1165,11	848,06	239,12	132,59	8,4604	32636,0	23755,0	236,984
796	1069,15	1166,31	848,96	240,02	132,21	8,4615	32669,6	23780,2	237,016
797	1070,15	1167,51	849,86	240,93	131,83	8,4626	32703,2	23805,5	237,047
798	1071,15	1168,71	850,76	241,84	131,46	8,4638	32736,7	23830,8	237,078
799	1072,15	1169,91	851,67	242,76	131,09	8,4649	32770,3	23856,0	237,110
800	1073,15	1171,11	852,57	243,67	130,72	8,4660	32803,9	23881,3	237,141
801	1074,15	1172,31	853,47	244,59	130,35	8,4671	32837,5	23906,6	237,172
802	1075,15	1173,51	854,37	245,51	129,98	8,4682	32871,1	23931,9	237,204
803	1076,15	1174,71	855,28	246,44	129,61	8,4693	32904,7	23957,2	237,235
804	1077,15	1175,91	856,18	247,36	129,25	8,4705	32938,3	23982,4	237,266
805	1078,15	1177,11	857,08	248,29	128,88	8,4716	32971,9	24007,7	237,297
806	1079,15	1178,31	857,99	249,23	128,52	8,4727	33005,5	24033,1	237,328
807	1080,15	1179,51	858,89	250,16	128,16	8,4738	33039,2	24058,4	237,360
808	1081,15	1180,71	859,79	251,10	127,80	8,4749	33072,8	24083,7	237,391
809	1082,15	1181,91	860,70	252,04	127,44	8,4760	33106,4	24109,0	237,422
810	1083,15	1183,11	861,60	252,98	127,08	8,4771	33140,1	24134,3	237,453
811	1084,15	1184,31	862,51	253,93	126,72	8,4782	33173,7	24159,6	237,484
812	1085,15	1185,51	863,41	254,88	126,37	8,4793	33207,4	24185,0	237,515
813	1086,15	1186,71	864,31	255,83	126,01	8,4805	33241,0	24210,3	237,546
814	1087,15	1187,91	865,22	256,79	125,66	8,4816	33274,7	24235,7	237,577
815	1088,15	1189,12	866,12	257,74	125,31	8,4827	33308,3	24261,0	237,608
816	1089,15	1190,32	867,03	258,70	124,96	8,4838	33342,0	24286,4	237,639
817	1090,15	1191,52	867,93	259,67	124,61	8,4849	33375,7	24311,7	237,670
818	1091,15	1192,72	868,84	260,63	124,26	8,4860	33409,4	24337,1	237,701
819	1092,15	1193,93	869,75	261,60	123,91	8,4871	33443,0	24362,5	237,731
820	1093,15	1195,13	870,65	262,57	123,57	8,4882	33476,7	24387,8	237,762
821	1094,15	1196,33	871,56	263,55	123,22	8,4893	33510,4	24413,2	237,793
822	1095,15	1197,53	872,46	264,53	122,88	8,4904	33544,1	24438,6	237,824
823	1096,15	1198,74	873,37	265,51	122,54	8,4915	33577,8	24464,0	237,855
824	1097,15	1199,94	874,28	266,49	122,20	8,4926	33611,5	24489,4	237,885
825	1098,15	1201,14	875,18	267,48	121,86	8,4937	33645,2	24514,8	237,916
826	1099,15	1202,35	876,09	268,47	121,52	8,4948	33679,0	24540,2	237,947
827	1100,15	1203,55	877,00	269,46	121,18	8,4959	33712,7	24565,6	237,977
828	1101,15	1204,76	877,90	270,45	120,85	8,4969	33746,4	24591,0	238,008
829	1102,15	1205,96	878,81	271,45	120,51	8,4980	33780,1	24616,4	238,039

Carbon monoxide (Continued)

t	T	h	u	π₀	θ₀	s°	H	U	S
830	1103,15	1207,16	879,72	272,45	120,18	8,4991	33813,9	24641,8	238,069
831	1104,15	1208,37	880,63	273,46	119,84	8,5002	33847,6	24667,3	238,100
832	1105,15	1209,57	881,54	274,46	119,51	8,5013	33881,4	24692,7	238,130
833	1106,15	1210,78	882,44	275,47	119,18	8,5024	33915,1	24718,1	238,161
834	1107,15	1211,98	883,35	276,48	118,85	8,5035	33948,9	24743,6	238,191
835	1108,15	1213,19	884,26	277,50	118,53	8,5046	33982,6	24769,0	238,222
836	1109,15	1214,39	885,17	278,52	118,20	8,5057	34016,4	24794,5	238,252
837	1110,15	1215,60	886,08	279,54	117,87	8,5068	34050,2	24819,9	238,283
838	1111,15	1216,81	886,99	280,56	117,55	8,5078	34083,9	24845,4	238,313
839	1112,15	1218,01	887,90	281,59	117,23	8,5089	34117,7	24870,9	238,344
840	1113,15	1219,22	888,81	282,62	116,90	8,5100	34151,5	24896,3	238,374
841	1114,15	1220,42	889,71	283,65	116,58	8,5111	34185,3	24921,8	238,404
842	1115,15	1221,63	890,62	284,69	116,26	8,5122	34219,1	24947,3	238,435
843	1116,15	1222,84	891,53	285,73	115,94	8,5133	34252,9	24972,8	238,465
844	1117,15	1224,04	892,44	286,77	115,62	8,5143	34286,7	24998,3	238,495
845	1118,15	1225,25	893,35	287,82	115,31	8,5154	34320,5	25023,8	238,525
846	1119,15	1226,46	894,26	288,87	114,99	8,5165	34354,3	25049,3	238,556
847	1120,15	1227,67	895,18	289,92	114,68	8,5176	34388,1	25074,8	238,586
848	1121,15	1228,87	896,09	290,97	114,36	8,5187	34422 0	25100,3	238,616
849	1122,15	1230,08	897,00	292,03	114,05	8,5197	34455,8	25125,8	238,646
850	1123,15	1231,29	897,91	293,09	113,74	8,5208	34489,6	25151,3	238,676
851	1124,15	1232,50	898,82	294,15	113,43	8,5219	34523,5	25176,8	238,706
852	1125,15	1233,70	899,73	295,22	113,12	8,5230	34557,3	25202,3	238,737
853	1126,15	1234,91	900,64	296,29	112,81	8,5240	34591,2	25227,9	238,767
854	1127,15	1236,12	901,55	297,36	112,51	8,5251	34625,0	25253,4	238,797
855	1128,15	1237,33	902,47	298,44	112,20	8,5262	34658,9	25279,0	238,827
856	1129,15	1238,54	903,38	299,52	111,89	8,5272	34692,7	25304,5	238,857
857	1130,15	1239,75	904,29	300,60	111,59	8,5283	34726,6	25330,1	238,887
858	1131,15	1240,96	905,20	301,68	111,29	8,5294	34760,5	25355,6	238,917
859	1132,15	1242,17	906,11	302,77	110,99	8,5305	34794,3	25381,2	238,947
860	1133,15	1243,38	907,03	303,86	110,68	8,5315	34828,2	25406,7	238,976
861	1134,15	1244,59	907,94	304,96	110,38	8,5326	34862,1	25432,3	239,006
862	1135,15	1245,80	908,85	306,05	110,09	8,5337	34896,0	25457,9	239,036
863	1136,15	1247,01	909,77	307,15	109,79	8,5347	34929,9	25483,5	239,066
864	1137,15	1248,22	910,68	308,26	109,49	8,5358	34963,8	25509,0	239,096
865	1138,15	1249,43	911,59	309,37	109,20	8,5368	34997,7	25534,6	239,126
866	1139,15	1250,64	912,51	310,48	108,90	8,5379	35031,6	25560,2	239,155
867	1140,15	1251,85	913,42	311,59	108,61	8,5390	35065,5	25585,8	239,185
868	1141,15	1253,06	914,33	312,70	108,31	8,5400	35099,4	25611,4	239,215
869	1142,15	1254,27	915,25	313,82	108,02	8,5411	35133,3	25637,0	239,245
870	1143,15	1255,48	916,16	314,95	107,73	8,5422	35167,3	25662,6	239,274
871	1144,15	1256,69	917,08	316,07	107,44	8,5432	35201,2	25688,3	239,304
872	1145,15	1257,90	917,99	317,20	107,15	8,5443	35235,1	25713,9	239,334
873	1146,15	1259,11	918,91	318,33	106,86	8,5453	35269,1	25739,5	239,363
874	1147,15	1260,33	919,82	319,47	106,58	8,5464	35303,0	25765,1	239,393
875	1148,15	1261,54	920,74	320,61	106,29	8,5474	35336,9	25790,8	239,422
876	1149,15	1262,75	921,65	321,75	106,01	8,5485	35370,9	25816,4	239,452
877	1150,15	1263,96	922,57	322,89	105,72	8,5496	35404,9	25842,0	239,482
878	1151,15	1265,18	923,48	324,04	105,44	8,5506	35438,8	25867,7	239,511
879	1152,15	1266,39	924,40	325,19	105,16	8,5517	35472,8	25893,3	239,541
880	1153,15	1267,60	925,31	326,35	104,88	8,5527	35506,8	25919,0	239,570
881	1154,15	1268,81	926,23	327,51	104,60	8,5538	35540,7	25944,7	239,599
882	1155,15	1270,03	927,15	328,67	104,32	8,5548	35574,7	25970,3	239,629
883	1156,15	1271,24	928,06	329,83	104,04	8,5559	35608,7	25996,0	239,658
884	1157,15	1272,45	928,98	331,00	103,76	8,5569	35642,7	26021,7	239,688

Carbon monoxide (*Continued*)

t	T	h	u	π_0	θ_0	s^0	H	U	S
885	1158,15	1273,67	929,90	332,17	103,48	8,5580	35676,7	26047,3	239,717
886	1159,15	1274,88	930,81	333,35	103,21	8,5590	35710,7	26073,0	239,746
887	1160,15	1276,09	931,73	334,52	102,93	8,5601	35744,7	26098,7	239,776
888	1161,15	1277,31	932,65	335,71	102,66	8,5611	35778,7	26124,4	239,805
889	1162,15	1278,52	933,57	336,89	102,39	8,5621	35812,7	26150,1	239,834
890	1163,15	1279,74	934,48	338,08	102,12	8,5632	35846,7	26175,8	239,864
891	1164,15	1280,95	935,40	339,27	101,85	8,5642	35880,7	26201,5	239,893
892	1165,15	1282,17	936,32	340,46	101,58	8,5653	35914,8	26227,2	239,922
893	1166,15	1283,38	937,24	341,66	101,31	8,5663	35948,8	26252,9	239,951
894	1167,15	1284,60	938,15	342,86	101,04	8,5674	35982,8	26278,7	239,980
895	1168,15	1285,81	939,07	344,06	100,77	8,5684	36016,9	26304,4	240,010
896	1169,15	1287,03	939,99	345,27	100,50	8,5694	36050,9	26330,1	240,039
897	1170,15	1288,24	940,91	346,48	100,24	8,5705	36084,9	26355,8	240,068
898	1171,15	1289,46	941,83	347,70	99,978	8,5715	36119,0	26381,6	240,097
899	1172,15	1290,67	942,75	348,91	99,714	8,5726	36153,0	26407,3	240,126
900	1173,15	1291,89	943,67	350,14	99,451	8,5736	36187,1	26433,1	240,155
901	1174,15	1293,11	944,59	351,36	99,189	8,5746	36221,2	26458,8	240,184
902	1175,15	1294,32	945,51	352,59	98,928	8,5757	36255,2	26484,6	240,213
903	1176,15	1295,54	946,43	353,82	98,668	8,5767	36289,3	26510,3	240,242
904	1177,15	1296,75	947,34	355,05	98,408	8,5777	36323,4	26536,1	240,271
905	1178,15	1297,97	948,26	356,29	98,150	8,5788	36357,5	26561,8	240,300
906	1179,15	1299,19	949,18	357,53	97,892	8,5798	36391,6	26587,6	240,329
907	1180,15	1300,40	950,11	358,78	97,635	8,5808	36425,6	26613,4	240,358
908	1181,15	1301,62	951,03	360,03	97,379	8,5819	36459,7	26639,2	240,387
909	1182,15	1302,84	951,95	361,28	97,124	8,5829	36493,8	26665,0	240,415
910	1183,15	1304,06	952,87	362,53	96,869	8,5839	36527,9	26690,7	240,444
911	1184,15	1305,27	953,79	363,79	96,616	8,5850	36562,0	26716,5	240,473
912	1185,15	1306,49	954,71	365,05	96,363	8,5860	36596,2	26742,3	240,502
913	1186,15	1307,71	955,63	366,32	96,111	8,5870	36630,3	26768,1	240,531
914	1187,15	1308,93	956,55	367,59	95,860	8,5880	36664,4	26793,9	240,559
915	1188,15	1310,15	957,47	368,86	95,610	8,5891	36698,5	26819,8	240,588
916	1189,15	1311,37	958,39	370,14	95,360	8,5901	36732,6	26845,6	240,617
917	1190,15	1312,58	959,32	371,42	95,112	8,5911	36766,8	26871,4	240,646
918	1191,15	1313,80	960,24	372,70	94,864	8,5921	36800,9	26897,2	240,674
919	1192,15	1315,02	961,16	373,99	94,617	8,5932	36835,1	26923,0	240,703
920	1193,15	1316,24	962,08	375,28	94,371	8,5942	36869,2	26948,9	240,731
921	1194,15	1317,46	963,00	376,57	94,126	8,5952	36903,4	26974,7	240,760
922	1195,15	1318,68	963,93	377,87	93,881	8,5962	36937,5	27000,5	240,789
923	1196,15	1319,90	964,85	379,17	93,637	8,5972	36971,7	27026,4	240,817
924	1197,15	1321,12	965,77	380,47	93,394	8,5983	37005,8	27052,2	240,846
925	1198,15	1322,34	966,69	381,78	93,152	8,5993	37040,0	27078,1	240,874
926	1199,15	1323,56	967,62	383,09	92,911	8,6003	37074,2	27103,9	240,903
927	1200,15	1324,78	968,54	384,41	92,670	8,6013	37108,3	27129,8	240,931
928	1201,15	1326,00	969,46	385,72	92,430	8,6023	37142,5	27155,7	240,960
929	1202,15	1327,22	970,39	387,05	92,191	8,6033	37176,7	27181,5	240,988
930	1203,15	1328,44	971,31	388,37	91,953	8,6044	37210,9	27207,4	241,017
931	1204,15	1329,66	972,24	389,70	91,716	8,6054	37245,1	27233,3	241,045
932	1205,15	1330,88	973,16	391,03	91,479	8,6064	37279,3	27259,2	241,073
933	1206,15	1332,10	974,08	392,37	91,243	8,6074	37313,5	27285,1	241,102
934	1207,15	1333,32	975,01	393,71	91,008	8,6084	37347,7	27310,9	241,130
935	1208,15	1334,54	975,93	395,05	90,774	8,6094	37381,9	27336,8	241,159
936	1209,15	1335,76	976,86	396,40	90,540	8,6104	37416,1	27362,7	241,187
937	1210,15	1336,99	977,78	397,75	90,307	8,6114	37450,3	27388,6	241,215
938	1211,15	1338,21	978,71	399,11	90,075	8,6125	37484,5	27414,6	241,243
939	1212,15	1339,43	979,63	400,46	89,844	8,6135	37518,8	27440,5	241,272

t	T	h	u	π_0	0_0	s^0	H	U	S
940	1213,15	1340,65	980,56	401,83	89,613	8,6145	37553,0	27466,4	241,300
941	1214,15	1341,87	981,48	403,19	89,383	8,6155	37587,2	27492,3	241,328
942	1215,15	1343,10	982,41	404,56	89,154	8,6165	37621,5	27518,2	241,356
943	1216,15	1344,32	983,33	405,93	88,925	8,6175	37655,7	27544,1	241,384
944	1217,15	1345,54	984,26	407,31	88,698	8,6185	37690,0	27570,1	241,413
945	1218,15	1346,76	985,18	408,69	88,471	8,6195	37724,2	27596,0	241,441
946	1219,15	1347,99	986,11	410,08	88,245	8,6205	37758,5	27622,0	241,469
947	1220,15	1349,21	987,04	411,46	88,019	8,6215	37792,7	27647,9	241,497
948	1221,15	1350,43	987,96	412,85	87,794	8,6225	37827,0	27673,8	241,525
949	1222,15	1351,66	988,89	414,25	87,570	8,6235	37861,3	27699,8	241,553
950	1223,15	1352,88	989,82	415,65	87,347	8,6245	37895,5	27725,8	241,581
951	1224,15	1354,10	990,74	417,05	87,124	8,6255	37929,8	27751,7	241,609
952	1225,15	1355,33	991,67	418,46	86,903	8,6265	37964,1	27777,7	241,637
953	1226,15	1356,55	992,60	419,87	86,681	8,6275	37998,4	27803,7	241,665
954	1227,15	1357,78	993,52	421,28	86,461	8,6285	38032,7	27829,6	241,693
955	1228,15	1359,00	994,45	422,70	86,241	8,6295	38066,9	27855,6	241,721
956	1229,15	1360,22	995,38	424,12	86,022	8,6305	38101,3	27881,6	241,749
957	1230,15	1361,45	996,31	425,55	85,804	8,6315	38135,5	27907,6	241,777
958	1231,15	1362,67	997,24	426,98	85,586	8,6325	38169,8	27933,6	241,805
959	1232,15	1363,90	998,16	428,41	85,369	8,6335	38204,2	27959,5	241,832
960	1233,15	1365,12	999,09	429,85	85,153	8,6345	38238,5	27985,5	241,860
961	1234,15	1366,35	1000,02	431,29	84,937	8,6355	38272,8	28011,5	241,888
962	1235,15	1367,57	1000,95	432,73	84,722	8,6365	38307,1	28037,6	241,916
963	1236,15	1368,80	1001,88	434,18	84,508	8,6375	38341,4	28063,6	241,944
964	1237,15	1370,02	1002,81	435,63	84,295	8,6384	38375,7	28089,6	241,971
965	1238,15	1371,25	1003,73	437,09	84,082	8,6394	38410,1	28115,6	241,999
966	1239,15	1372,48	1004,66	438,55	83,870	8,6404	38444,4	28141,6	242,027
967	1240,15	1373,70	1005,59	440,01	83,658	8,6414	38478,8	28167,6	242,055
968	1241,15	1374,93	1006,52	441,48	83,447	8,6424	38513,1	28193,7	242,082
969	1242,15	1376,15	1007,45	442,95	83,237	8,6434	38547,4	28219,7	242,110
970	1243,15	1377,38	1008,38	444,42	83,028	8,6444	38581,8	28245,7	242,138
971	1244,15	1378,61	1009,31	445,90	82,819	8,6454	38616,2	28271,8	242,165
972	1245,15	1379,83	1010,24	447,38	82,611	8,6463	38650,5	28297,8	242,193
973	1246,15	1381,06	1011,17	448,87	82,403	8,6473	38684,9	28323,9	242,220
974	1247,15	1382,29	1012,10	450,36	82,196	8,6483	38719,2	28349,9	242,248
975	1248,15	1383,51	1013,03	451,86	81,990	8,6493	38753,6	28376,0	242,275
976	1249,15	1384,74	1013,96	453,36	81,784	8,6503	38788,0	28402,0	242,303
977	1250,15	1385,97	1014,89	454,86	81,580	8,6513	38822,4	28428,1	242,331
978	1251,15	1387,20	1015,82	456,37	81,375	8,6522	38856,8	28454,2	242,358
979	1252,15	1388,42	1016,75	457,88	81,172	8,6532	38891,1	28480,3	242,386
980	1253,15	1389,65	1017,68	459,39	80,969	8,6542	38925,5	28506,3	242,413
981	1254,15	1390,88	1018,61	460,91	80,766	8,6552	38959,9	28532,4	242,440
982	1255,15	1392,11	1019,55	462,43	80,565	8,6562	38994,3	28558,5	242,468
983	1256,15	1393,34	1020,48	463,96	80,364	8,6571	39028,7	28584,6	242,495
984	1257,15	1394,56	1021,41	465,49	80,163	8,6581	39063,1	28610,7	242,523
985	1258,15	1395,79	1022,34	467,02	79,963	8,6591	39097,5	28636,8	242,550
986	1259,15	1397,02	1023,27	468,56	79,764	8,6601	39132,0	28662,9	242,577
987	1260,15	1398,25	1024,20	470,10	79,566	8,6610	39166,4	28689,0	242,605
988	1261,15	1399,48	1025,14	471,65	79,368	8,6620	39200,8	28715,1	242,632
989	1262,15	1400,71	1026,07	473,20	79,170	8,6630	39235,2	28741,2	242,659
990	1263,15	1401,94	1027,00	474,75	78,974	8,6640	39269,7	28767,3	242,686
991	1264,15	1403,17	1027,93	476,31	78,777	8,6649	39304,1	28793,4	242,714
992	1265,15	1404,40	1028,87	477,88	78,582	8,6659	39338,5	28819,6	242,741
993	1266,15	1405,63	1029,80	479,44	78,387	8,6669	39373,0	28845,7	242,768
994	1267,15	1406,86	1030,73	481,01	78,193	8,6679	39407,4	28871,8	242,795

Carbon monoxide (*Continued*)

t	T	h	u	π_0	θ_0	s^0	H	U	Δ
995	1268,15	1408,08	1031,66	482,59	77,999	8,6688	39441,9	28897,9	242,823
996	1269,15	1409,31	1032,60	484,17	77,806	8,6698	39476,3	28924,1	242,850
997	1270,15	1410,54	1033,53	485,75	77,614	8,6708	39510,8	28950,2	242,877
998	1271,15	1411,78	1034,46	487,34	77,422	8,6717	39545,2	28976,4	242,904
999	1272,15	1413,01	1035,40	488,93	77,231	8,6727	39579,7	29002,5	242,931
1000	1273,15	1414,24	1036,33	490,52	77,040	8,6737	39614,2	29028,7	242,958
1001	1274,15	1415,47	1037,27	492,12	76,850	8,6746	39648,6	29054,8	242,985
1002	1275,15	1416,70	1038,20	493,73	76,660	8,6756	39683,1	29081,0	243,012
1003	1276,15	1417,93	1039,13	495,33	76,472	8,6766	39717,6	29107,2	243,039
1004	1277,15	1419,16	1040,07	496,95	76,283	8,6775	39752,1	29133,3	243,066
1005	1278,15	1420,39	1041,00	498,56	76,095	8,6785	39786,6	29159,5	243,093
1006	1279,15	1421,62	1041,94	500,18	75,908	8,6795	39821,1	29185,7	243,120
1007	1280,15	1422,85	1042,87	501,81	75,722	8,6804	39855,6	29211,9	243,147
1008	1281,15	1424,09	1043,81	503,43	75,536	8,6814	39890,1	29238,1	243,174
1009	1282,15	1425,32	1044,74	505,07	75,350	8,6823	39924,6	29264,2	243,201
1010	1283,15	1426,55	1045,68	506,70	75,166	8,6833	39959,1	29290,4	243,228
1011	1284,15	1427,78	1046,61	508,34	74,981	8,6843	39993,6	29316,6	243,255
1012	1285,15	1429,01	1047,55	509,99	74,798	8,6852	40028,1	29342,8	243,282
1013	1286,15	1430,25	1048,48	511,64	74,614	8,6862	40062,6	29369,0	243,309
1014	1287,15	1431,48	1049,42	513,29	74,432	8,6871	40097,1	29395,2	243,335
1015	1288,15	1432,71	1050,35	514,95	74,250	8,6881	40131,7	29421,4	243,362
1016	1289,15	1433,94	1051,29	516,61	74,068	8,6891	40166,2	29447,7	243,389
1017	1290,15	1435,18	1052,23	518,28	73,888	8,6900	40200,7	29473,9	243,416
1018	1291,15	1436,41	1053,16	519,95	73,707	8,6910	40235,3	29500,1	243,443
1019	1292,15	1437,64	1054,10	521,63	73,527	8,6919	40269,8	29526,3	243,469
1020	1293,15	1438,88	1055,03	523,31	73,348	8,6929	40304,3	29552,6	243,496
1021	1294,15	1440,11	1055,97	524,99	73,170	8,6938	40338,9	29578,8	243,523
1022	1295,15	1441,34	1056,91	526,68	72,991	8,6948	40373,4	29605,0	243,549
1023	1296,15	1442,58	1057,84	528,37	72,814	8,6957	40408,0	29631,3	243,576
1024	1297,15	1443,81	1058,78	530,07	72,637	8,6967	40442,6	29657,5	243,603
1025	1298,15	1445,04	1059,72	531,77	72,460	8,6976	40477,1	29683,8	243,629
1026	1299,15	1446,28	1060,66	533,47	72,284	8,6986	40511,7	29710,0	243,656
1027	1300,15	1447,51	1061,59	535,18	72,109	8,6995	40546,3	29736,3	243,683
1028	1301,15	1448,75	1062,53	536,89	71,934	8,7005	40580,8	29762,5	243,709
1029	1302,15	1449,98	1063,47	538,61	71,760	8,7014	40615,4	29788,8	243,736
1030	1303,15	1451,22	1064,41	540,34	71,586	8,7024	40650,0	29815,1	243,762
1031	1304,15	1452,45	1065,34	542,06	71,413	8,7033	40684,6	29841,3	243,789
1032	1305,15	1453,68	1066,28	543,79	71,240	8,7043	40719,2	29867,6	243,815
1033	1306,15	1454,92	1067,22	545,53	71,068	8,7052	40753,8	29893,9	243,842
1034	1307,15	1456,15	1068,16	547,27	70,896	8,7062	40788,4	29920,2	243,868
1035	1308,15	1457,39	1069,10	549,01	70,725	8,7071	40823,0	29946,5	243,895
1036	1309,15	1458,63	1070,03	550,76	70,554	8,7081	40857,6	29972,7	243,921
1037	1310,15	1459,86	1070,97	552,52	70,384	8,7090	40892,2	29999,0	243,948
1038	1311,15	1461,10	1071,91	554,27	70,214	8,7099	40926,8	30025,3	243,974
1039	1312,15	1462,33	1072,85	556,04	70,045	8,7109	40961,4	30051,6	244,000
1040	1313,15	1463,57	1073,79	557,80	69,876	8,7118	40996,0	30077,9	244,027
1041	1314,15	1464,80	1074,73	559,57	69,708	8,7128	41030,6	30104,2	244,053
1042	1315,15	1466,04	1075,67	561,35	69,541	8,7137	41065,3	30130,6	244,079
1043	1316,15	1467,28	1076,61	563,13	69,374	8,7146	41099,9	30156,9	244,106
1044	1317,15	1468,51	1077,55	564,91	69,207	8,7156	41134,5	30183,2	244,132
1045	1318,15	1469,75	1078,49	566,70	69,041	8,7165	41169,2	30209,5	244,158
1046	1319,15	1470,99	1079,43	568,49	68,875	8,7175	41203,8	30235,8	244,185
1047	1320,15	1472,22	1080,37	570,29	68,710	8,7184	41238,4	30262,2	244,211
1048	1321,15	1473,46	1081,31	572,09	68,546	8,7193	41273,1	30288,5	244,237
1049	1322,15	1474,70	1082,25	573,90	68,382	8,7203	41307,7	30314,8	244,263

t	T	h	u	π_0	θ_0	s^0	H	U	S
1050	1323,15	1475,93	1083,19	575,71	68,218	8,7212	41342,4	30341,2	244,290
1051	1324,15	1477,17	1084,13	577,53	68,055	8,7221	41377,0	30367,5	244,316
1052	1325,15	1478,41	1085,07	579,35	67,892	8,7231	41411,7	30393,9	244,342
1053	1326,15	1479,65	1086,01	581,17	67,730	8,7240	41446,4	30420,2	244,368
1054	1327,15	1480,88	1086,95	583,00	67,569	8,7249	41481,0	30446,6	244,394
1055	1328,15	1482,12	1087,89	584,84	67,407	8,7259	41515,7	30472,9	244,420
1056	1329,15	1483,36	1088,83	586,68	67,247	8,7268	41550,4	30499,3	244,446
1057	1330,15	1484,60	1089,77	588,52	67,087	8,7277	41585,1	30525,7	244,473
1058	1331,15	1485,84	1090,72	590,37	66,927	8,7287	41619,7	30552,0	244,499
1059	1332,15	1487,07	1091,66	592,22	66,768	8,7296	41654,4	30578,4	244,525
1060	1333,15	1488,31	1092,60	594,08	66,609	8,7305	41689,1	30604,8	244,551
1061	1334,15	1489,55	1093,54	595,94	66,451	8,7315	41723,8	30631,1	244,577
1062	1335,15	1490,79	1094,48	597,80	66,293	8,7324	41758,5	30657,5	244,603
1063	1336,15	1492,03	1095,42	599,68	66,135	8,7333	41793,2	30683,9	244,629
1064	1337,15	1493,27	1096,37	601,55	65,979	8,7342	41827,9	30710,3	244,655
1065	1338,15	1494,51	1097,31	603,43	65,822	8,7352	41862,6	30736,7	244,681
1066	1339,15	1495,75	1098,25	605,32	65,666	8,7361	41897,3	30763,1	244,706
1067	1340,15	1496,99	1099,19	607,21	65,511	8,7370	41932,1	30789,5	244,732
1068	1341,15	1498,22	1100,14	609,10	65,356	8,7379	41966,8	30815,9	244,758
1069	1342,15	1499,46	1101,08	611,00	65,201	8,7389	42001,5	30842,3	244,784
1070	1343,15	1500,70	1102,02	612,90	65,047	8,7398	42036,2	30868,7	244,810
1071	1344,15	1501,94	1102,96	614,81	64,894	8,7407	42070,9	30895,1	244,836
1072	1345,15	1503,18	1103,91	616,72	64,740	8,7416	42105,7	30921,6	244,862
1073	1346,15	1504,42	1104,85	618,64	64,588	8,7425	42140,4	30948,0	244,888
1074	1347,15	1505,66	1105,79	620,56	64,436	8,7435	42175,2	30974,4	244,913
1075	1348,15	1506,90	1106,74	622,49	64,284	8,7444	42209,9	31000,8	244,939
1076	1349,15	1508,14	1107,68	624,42	64,132	8,7453	42244,6	31027,3	244,965
1077	1350,15	1509,39	1108,63	626,36	63,981	8,7462	42279,4	31053,7	244,991
1078	1351,15	1510,63	1109,57	628,30	63,831	8,7471	42314,2	31080,1	245,016
1079	1352,15	1511,87	1110,51	630,25	63,681	8,7481	42348,9	31106,6	245,042
1080	1353,15	1513,11	1111,46	632,20	63,531	8,7490	42383,7	31133,0	245,068
1081	1354,15	1514,35	1112,40	634,15	63,382	8,7499	42418,4	31159,5	245,093
1082	1355,15	1515,59	1113,35	636,11	63,234	8,7508	42453,2	31185,9	245,119
1083	1356,15	1516,83	1114,29	638,08	63,085	8,7517	42488,0	31212,4	245,145
1084	1357,15	1518,07	1115,23	640,05	62,938	8,7526	42522,7	31238,8	245,170
1085	1358,15	1519,31	1116,18	642,03	62,790	8,7536	42557,5	31265,3	245,196
1086	1359,15	1520,56	1117,12	644,01	62,643	8,7545	42592,3	31291,8	245,222
1087	1360,15	1521,80	1118,07	645,99	62,497	8,7554	42627,1	31318,2	245,247
1088	1361,15	1523,04	1119,01	647,98	62,351	8,7563	42661,9	31344,7	245,273
1089	1362,15	1524,28	1119,96	649,97	62,205	8,7572	42696,7	31371,2	245,298
1090	1363,15	1525,52	1120,91	651,97	62,060	8,7581	42731,5	31397,7	245,324
1091	1364,15	1526,77	1121,85	653,98	61,915	8,7590	42766,3	31424,2	245,349
1092	1365,15	1528,01	1122,80	655,99	61,771	8,7599	42801,1	31450,6	245,375
1093	1366,15	1529,25	1123,74	658,00	61,627	8,7609	42835,9	31477,1	245,400
1094	1367,15	1530,49	1124,69	660,02	61,483	8,7618	42870,7	31503,6	245,426
1095	1368,15	1531,74	1125,63	662,04	61,340	8,7627	42905,5	31530,1	245,451
1096	1369,15	1532,98	1126,58	664,07	61,197	8,7636	42940,3	31556,6	245,477
1097	1370,15	1534,22	1127,53	666,10	61,055	8,7645	42975,1	31583,1	245,502
1098	1371,15	1535,47	1128,47	668,14	60,913	8,7654	43009,9	31609,6	245,528
1099	1372,15	1536,71	1129,42	670,19	60,772	8,7663	43044,8	31636,1	245,553
1100	1373,15	1537,95	1130,37	672,23	60,631	8,7672	43079,6	31662,7	245,578
1101	1374,15	1539,20	1131,31	674,29	60,490	8,7681	43114,4	31689,2	245,604
1102	1375,15	1540,44	1132,26	676,35	60,350	8,7690	43149,3	31715,7	245,629
1103	1376,15	1541,68	1133,21	678,41	60,210	8,7699	43184,1	31742,2	245,654
1104	1377,15	1542,93	1134,15	680,48	60,071	8,7708	43218,9	31768,8	245,680

Carbon monoxide (*Continued*)

t	T	h	u	π_0	θ_0	s^0	H	U	S
1105	1378,15	1544,17	1135,10	682,55	59,932	8,7717	43253,8	31795,3	245,705
1106	1379,15	1545,42	1136,05	684,63	59,793	8,7726	43288,6	31821,8	245,730
1107	1380,15	1546,66	1136 99	686,71	59,655	8,7735	43323,5	31848,4	245,755
1108	1381,15	1547,90	1137,94	688,80	59,517	8,7744	43358,3	31874,9	245,781
1109	1382,15	1549,15	1138,89	690,89	59,380	8,7753	43393,2	31901,4	245,806
1110	1383,15	1550,39	1139,84	692,99	59,243	8,7762	43428,1	31928,0	245,831
1111	1384,15	1551,64	1140,79	695,10	59,106	8,7771	43462,9	31954,5	245,856
1112	1385,15	1552,88	1141,73	697,20	58,970	8,7780	43497,8	31981,1	245,882
1113	1386,15	1554,13	1142,68	699,32	58,834	8,7789	43532,7	32007,7	245,907
1114	1387,15	1555,37	1143,63	701,44	58,699	8,7798	43567,5	32034,2	245,932
1115	1388,15	1556,62	1144,58	703,56	58,564	8,7807	43602,4	32060,8	245,957
1116	1389,15	1557,86	1145,53	705,69	58,429	8,7816	43637,3	32087,3	245,982
1117	1390,15	1559,11	1146,48	707,82	58,295	8,7825	43672,2	32113,9	246,007
1118	1391,15	1560,35	1147,42	709,96	58,161	8,7834	43707,1	32140,5	246,032
1119	1392,15	1561,60	1148,37	712,11	58,028	8,7843	43742,0	32167,1	246,057
1120	1393,15	1562,85	1149,32	714,26	57,895	8,7852	43776,9	32193,7	246,082
1121	1394,15	1564,09	1150,27	716,41	57,762	8,7861	43811,8	32220,2	246,107
1122	1395,15	1565,34	1151,22	718,57	57,630	8,7870	43846,7	32246,8	246,132
1123	1396,15	1566,58	1152,17	720,74	57,498	8,7879	43881,6	32273,4	246,157
1124	1397,15	1567,83	1153,12	722,90	57,366	8,7888	43916,5	32300,0	246,182
1125	1398,15	1569,08	1154,07	725,08	57,235	8,7897	43951,4	32326,6	246,207
1126	1399,15	1570,32	1155,02	727,26	57,105	8,7906	43986,3	32353,2	246,232
1127	1400,15	1571,57	1155,97	729,45	56,974	8,7915	44021,2	32379,8	246,257
1128	1401,15	1572,82	1156,92	731,64	56,844	8,7923	44056,2	32406,4	246,282
1129	1402,15	1574,06	1157,87	733,83	56,714	8,7932	44091,1	32433,0	246,307
1130	1403,15	1575,31	1158,82	736,03	56,585	8,7941	44126,0	32459,6	246,332
1131	1404,15	1576,56	1159,77	738,24	56,456	8,7950	44160,9	32486,3	246,357
1132	1405,15	1577,80	1160,72	740,45	56,328	8,7959	44195,9	32512,9	246,382
1133	1406,15	1579,05	1161,67	742,67	56,200	8,7968	44230,8	32539,5	246,407
1134	1407,15	1580,30	1162,62	744,89	56,072	8,7977	44265,8	32566,1	246,432
1135	1408,15	1581,55	1163,57	747,12	55,944	8,7986	44300,7	32592,8	246,456
1136	1409,15	1582,79	1164,52	749,35	55,817	8,7994	44335,6	35619,4	246,481
1137	1410,15	1584,04	1165,47	751,59	55,691	8,8003	44370,6	32646,0	246,506
1138	1411,15	1585,29	1166,42	753,83	55,564	8,8012	44405,6	32672,7	246,531
1139	1412,15	1586,54	1167,37	756,08	55,438	8,8021	44440,5	32699,3	246,556
1140	1413,15	1587,79	1168,33	758,33	55,313	8,8030	44475,5	32726,0	246,580
1141	1414,15	1589,03	1169,28	760,59	55,187	8,8039	44510,4	32752,6	246,605
1142	1415,15	1590,28	1170,23	762,86	55,062	8,8047	44545,4	32779,3	246,630
1143	1416,15	1591,53	1171,18	765,13	54,938	8,8056	44580,4	32805,9	246,654
1144	1417,15	1592,78	1172,13	767,40	54,814	8,8065	44615,3	32832,6	246,679
1145	1418,15	1594,03	1173,08	769,68	54,690	8,8074	44650,3	32859,2	246,704
1146	1419,15	1595,28	1174,04	771,97	54,566	8,8083	44685,3	32885,9	246,728
1147	1420,15	1596,53	1174,99	774,26	54,443	8,8092	44720,3	32912,6	246,753
1148	1421,15	1597,77	1175,94	776,56	54,320	8,8100	44755,3	32939,2	246,778
1149	1422,15	1599,02	1176,89	778,86	54,198	8,8109	44790,3	32965,9	246,802
1150	1423,15	1600,27	1177,84	781,17	54,076	8,8118	44825,3	32992,6	246,827
1151	1424,15	1601,52	1178,80	783,48	53,954	8,8127	44860,2	33019,3	246,852
1152	1425,15	1602,77	1179,75	785,80	53,833	8,8135	44895,2	33046,0	246,876
1153	1426,15	1604,02	1180,70	788,12	53,712	8,8144	44930,2	33072,7	246,901
1154	1427,15	1605,27	1181,66	790,45	53,591	8,8153	44965,3	33099,3	246,925
1155	1428,15	1606,52	1182,61	792,79	53,470	8,8162	45000,3	33126,0	246,950
1156	1429,15	1607,77	1183,56	795,13	53,350	8,8170	45035,3	33152,7	246,974
1157	1430,15	1609,02	1184,51	797,47	53,231	8,8179	45070,3	33179,4	246,999
1158	1431,15	1610,27	1185,47	799,82	53,111	8,8188	45105,3	33206,1	247,023
1159	1432,15	1611,52	1186,42	802,18	52,992	8,8197	45140,3	33232,8	247,048

Carbon monoxide (*Continued*)

t	T	h	u	π_0	θ_0	s^0	H	U	S
1160	1433,15	1612,77	1187,37	804,54	52,874	8,8205	45175,3	33259,6	247,072
1161	1434,15	1614,02	1188,33	806,91	52,755	8,8214	45210,4	33286,3	247,097
1162	1435,15	1615,27	1189,28	809,28	52,637	8,8223	45245,4	33313,0	247,121
1163	1436,15	1616,52	1190,24	811,66	52,520	8,8232	45280,4	33339,7	247,145
1164	1437,15	1617,77	1191,19	814,05	52,402	8,8240	45315,5	33366,4	247,170
1165	1438,15	1619,03	1192,14	816,44	52,285	8,8249	45350,5	33393,1	247,194
1166	1439,15	1620,28	1193,10	818,83	52,168	8,8258	45385,6	33419,9	247,218
1167	1440,15	1621,53	1194,05	821,23	52,052	8,8266	45420,6	33446,6	247,243
1168	1441,15	1622,78	1195,01	823,64	51,936	8,8275	45455,7	33473,3	247,267
1169	1442,15	1624,03	1195,96	826,05	51,820	8,8284	45490,7	33500,1	247,291
1170	1443,15	1625,28	1196,92	828,47	51,705	8,8292	45525,8	33526,8	247,316
1171	1444,15	1626,53	1197,87	830,89	51,590	8,8301	45560,8	33553,6	247,340
1172	1445,15	1627,78	1198,83	833,32	51,475	8,8310	45595,9	33580,3	247,364
1173	1446,15	1629,04	1199,78	835,76	51,361	8,8318	45630,9	33607,1	247,389
1174	1447,15	1630,29	1200,74	838,20	51,247	8,8327	45666,0	33633,8	247,413
1175	1448,15	1631,54	1201,69	840,64	51,133	8,8336	45701,1	33660,6	247,437
1176	1449,15	1632,79	1202,65	843,09	51,019	8,8344	45736,2	33687,3	247,461
1177	1450,15	1634,04	1203,60	845,55	50,906	8,8353	45771,2	33714,1	247,485
1178	1451,15	1635,30	1204,56	848,01	50,793	8,8362	45806,3	33740,9	247,510
1179	1452,15	1636,55	1205,51	850,48	50,681	8,8370	45841,4	33767,6	247,534
1180	1453,15	1637,80	1206,47	852,96	50,569	8,8379	45876,5	33794,4	247,558
1181	1454,15	1639,05	1207,42	855,44	50,457	8,8387	45911,6	33821,2	247,582
1182	1455,15	1640,31	1208,38	857,92	50,345	8,8396	45946,7	33847,9	247,606
1183	1456,15	1641,56	1209,34	860,41	50,234	8,8405	45981,7	33874,7	247,630
1184	1457,15	1642,81	1210,29	862,91	50,123	8,8413	46016,8	33901,5	247,654
1185	1458,15	1644,07	1211,25	865,41	50,012	8,8422	46051,9	33928,3	247,678
1186	1459,15	1645,32	1212,21	867,92	49,902	8,8430	46087,1	33955,1	247,703
1187	1460,15	1646,57	1213,16	870,43	49,792	8,8439	46122,2	33981,9	247,727
1188	1461,15	1647,83	1214,12	872,95	49,682	8,8448	46157,3	34008,7	247,751
1189	1462,15	1649,08	1215,07	875,48	49,573	8,8456	46192,4	34035,5	247,775
1190	1463,15	1650,33	1216,03	878,01	49,463	8,8465	46227,5	34062,3	247,799
1191	1464,15	1651,59	1216,99	880,55	49,355	8,8473	46262,6	34089,1	247,823
1192	1465,15	1652,84	1217,95	883,09	49,246	8,8482	46297,7	34115,9	247,847
1193	1466,15	1654,10	1218,90	885,64	49,138	8,8490	46332,9	34142,7	247,871
1194	1467,15	1655,35	1219,86	888,20	49,030	8,8499	46368,0	34169,5	247,895
1195	1468,15	1656,60	1220,82	890,76	48,922	8,8508	46403,1	34196,3	247,918
1196	1469,15	1657,86	1221,77	893,32	48,815	8,8516	46438,3	34223,1	247,942
1197	1470,15	1659,11	1222,73	895,90	48,708	8,8525	46473,4	34250,0	247,966
1198	1471,15	1660,37	1223,69	898,48	48,601	8,8533	46508,5	34276,8	247,990
1199	1472,15	1661,62	1224,65	901,06	48,495	8,8542	46543,7	34303,6	248,014
1200	1473,15	1662,88	1225,61	903,65	48,389	8,8550	46578,8	34330,4	248,038
1201	1474,15	1664,13	1226,56	906,25	48,283	8,8559	46614,0	34357,3	248,062
1202	1475,15	1665,39	1227,52	908,85	48,177	8,8567	46649,1	34384,1	248,086
1203	1476,15	1666,64	1228,48	911,45	48,072	8,8576	46684,3	34411,0	248,109
1204	1477,15	1667,90	1229,44	914,07	47,967	8,8584	46719,4	34437,8	248,133
1205	1478,15	1669,15	1230,40	916,69	47,862	8,8593	46754,6	34464,6	248,157
1206	1479,15	1670,41	1231,36	919,31	47,758	8,8601	46789,7	34491,5	248,181
1207	1480,15	1671,66	1232,31	921,95	47,654	8,8610	46824,9	34518,3	248,205
1208	1481,15	1672,92	1233,27	924,58	47,550	8,8618	46860,1	34545,2	248,228
1209	1482,15	1674,17	1234,23	927,23	47,446	8,8627	46895,3	34572,1	248,252
1210	1483,15	1675,43	1235,19	929,88	47,343	8,8635	46930,4	34598,9	248,276
1211	1484,15	1676,68	1236,15	932,53	47,240	8,8644	46965,6	34625,8	248,300
1212	1485,15	1677,94	1237,11	935,19	47,137	8,8652	47000,8	34652,6	248,323
1213	1486,15	1679,20	1238,07	937,86	47,035	8,8660	47036,0	34679,5	248,347
1214	1487,15	1680,45	1239,03	940,53	46,933	8,8669	47071,2	34706,4	248,371

Carbon monoxide (*Continued*)

t	T	h	u	π_0	θ_0	S^0	H	U	S
1215	1488,15	1681,71	1239,99	943,21	46,831	8,8677	47106,3	34733,3	248,394
1216	1489,15	1682,97	1240,95	945,90	46,729	8,8686	47141,5	34760,1	248,418
1217	1490,15	1684,22	1241,91	948,59	46,628	8,8694	47176,7	34787,0	248,442
1218	1491,15	1685,48	1242,87	951,29	46,527	8,8703	47211,9	34813,9	248,465
1219	1492,15	1686,73	1243,82	953,99	46,426	8,8711	47247,1	34840,8	248,489
1220	1493,15	1687,99	1244,78	956,70	46,326	8,8720	47282,3	34867,7	248,512
1221	1494,15	1689,25	1245,74	959,42	46,226	8,8728	47317,5	34894,6	248,536
1222	1495,15	1690,51	1246,71	962,14	46,126	8,8736	47352,7	34921,5	248,559
1223	1496,15	1691,76	1247,67	964,87	46,026	8,8745	47388,0	34948,4	248,583
1224	1497,15	1693,02	1248,63	967,60	45,927	8,8753	47423,2	34975,3	248,607
1225	1498,15	1694,28	1249,59	970,34	45,827	8,8762	47458,4	35002,2	248,630
1226	1499,15	1695,53	1250,55	973,09	45,729	8,8770	47493,6	35029,1	248,654
1227	1500,15	1696,79	1251,51	975,84	45,630	8,8778	47528,8	35056,0	248,677
1228	1501,15	1698,05	1252,47	978,60	45,532	8,8787	47564,1	35082,9	248,700
1229	1502,15	1699,31	1253,43	981,37	45,534	8,8795	47599,3	35109,8	248,724
1230	1503,15	1700,56	1254,39	984,14	45,336	8,8803	47634,5	35136,7	248,747
1231	1504,15	1701,82	1255,35	986,92	45,238	8,8812	47669,8	35163,6	248,771
1232	1505,15	1703,08	1256,31	989,70	45,141	8,8820	47705,0	35190,6	248,794
1233	1506,15	1704,34	1257,27	992,49	45,044	8,8829	47740,2	35217,5	248,818
1234	1507,15	1705,60	1258,23	995,29	44,947	8,8837	47775,5	35244,4	248,841
1235	1508,15	1706,86	1259,20	998,09	44,851	8,8845	47810,7	35271,3	248,864
1236	1509,15	1708,11	1260,16	1000,9	44,755	8,8854	47846,0	35298,3	248,888
1237	1510,15	1709,37	1261,12	1003,7	44,659	8,8862	47881,2	35325,2	248,911
1238	1511,15	1710,63	1262,08	1006,5	44,563	8,8870	47916,5	35352,2	248,934
1239	1512,15	1711,89	1263,04	1009,3	44,468	8,8879	47951,7	35379,1	248,958
1240	1513,15	1713,15	1264,00	1012,2	44,372	8,8887	47987,0	35406,0	248,981
1241	1514,15	1714,41	1264,97	1015,0	44,278	8,8895	48022,3	35433,0	249,004
1242	1515,15	1715,67	1265,93	1017,8	44,183	8,8904	48057,5	35459,9	249,028
1243	1516,15	1716,92	1266,89	1020,7	44,088	8,8912	48092,8	35486,9	249,051
1244	1517,15	1718,18	1267,85	1023,5	43,994	8,8920	48128,1	35513,8	249,074
1245	1518,15	1719,44	1268,82	1026,4	43,900	8,8928	48163,3	35540,8	249,097
1246	1519,15	1720,70	1269,78	1029,3	43,807	8,8937	48198,6	35567,8	249,121
1247	1520,15	1721,96	1270,74	1032,2	43,713	8,8945	48233,9	35594,7	249,144
1248	1521,15	1723,22	1271,70	1035,0	43,620	8,8953	48269,2	35621,7	249,167
1249	1522,15	1724,48	1272,67	1037,9	43,527	8,8962	48304,4	35648,7	249,190
1250	1523,15	1725,74	1273,63	1040,8	43,435	8,8970	48339,7	35675,6	249,213
1251	1524,15	1727,00	1274,59	1043,7	43,342	8,8978	48375,0	35702,6	249,237
1252	1525,15	1728,26	1275,56	1046,6	43,250	8,8986	48410,3	35729,6	249,260
1253	1526,15	1729,52	1276,52	1049,6	43,158	8,8995	48445,6	35756,6	249,283
1254	1527,15	1730,78	1277,48	1052,5	43,067	8,9003	48480,9	35783,6	249,306
1255	1528,15	1732,04	1278,45	1055,4	42,975	8,9011	48516,2	35810,5	249,329
1256	1529,15	1733,30	1279,41	1058,4	42,884	8,9019	48551,5	35837,5	249,352
1257	1530,15	1734,56	1280,37	1061,3	42,793	8,9028	48586,8	35864,5	249,375
1258	1531,15	1735,82	1281,34	1064,2	42,703	8,9036	48622,1	35891,5	249,398
1259	1532,15	1737,08	1282,30	1067,2	42,612	8,9044	48657,4	35918,5	249,421
1260	1533,15	1738,34	1283,26	1070,2	42,522	8,9052	48692,8	35945,5	249,444
1261	1534,15	1739,60	1284,23	1073,1	42,432	8,9061	48728,1	35972,5	249,467
1262	1535,15	1740,87	1285,19	1076,1	42,342	8,9069	48763,4	35999,5	249,491
1263	1536,15	1742,13	1286,16	1079,1	42,253	8,9077	48798,7	36026,5	249,514
1264	1537,15	1743,39	1287,12	1082,1	42,164	8,9085	48834,0	36053,5	249,536
1265	1538,15	1744,65	1288,09	1085,1	42,075	8,9093	48869,4	36080,6	249,559
1266	1539,15	1745,91	1289,05	1088,1	41,986	8,9102	48904,7	36107,6	249,582
1267	1540,15	1747,17	1290,01	1091,1	41,897	8,9110	48940,0	36134,6	249,605
1268	1541,15	1748,43	1290,98	1094,1	41,809	8,9118	48975,4	36161,6	249,628
1269	1542,15	1749,69	1291,94	1097,1	41,721	8,9126	49010,7	36188,6	249,651

Carbon monoxide (*Continued*)

t	T	h	u	π_0	$\cdot\,\theta_0$	s^0	H	U	S
1270	1543,15	1750,96	1292,91	1100,1	41,633	8,9134	49046,0	36215,7	249,674
1271	1544,15	1752,22	1293,87	1103,2	41,546	8,9142	49081,4	36242,7	249,697
1272	1545,15	1753,48	1294,84	1106,2	41,458	8,9151	49116,7	36269,7	249,720
1273	1546,15	1754,74	1295,80	1109,3	41,371	8,9159	49152,1	36296,7	249,743
1274	1547,15	1756,00	1296,77	1112,3	41,284	8,9167	49187,4	36323,8	249,766
1275	1548,15	1757,27	1297,73	1115,4	41,198	8,9175	49222,8	36350,8	249,788
1276	1549,15	1758,53	1298,70	1118,4	41,111	8,9183	49258,1	36377,9	249,811
1277	1550,15	1759,79	1299,66	1121,5	41,025	8,9191	49293,5	36404,9	249,834
1278	1551,15	1761,05	1300,63	1124,6	40,939	8,9200	49328,9	36432,0	249,857
1279	1552,15	1762,32	1301,60	1127,7	40,853	8,9208	49364,2	36459,0	249,880
1280	1553,15	1763,58	1302,56	1130,8	40,768	8,9216	49399,6	36486,1	249,903
1281	1554,15	1764,84	1303,53	1133,9	40,682	8,9224	49435,0	36513,1	249,925
1282	1555,15	1766,10	1304,49	1137,0	40,597	8,9232	49470,3	36540,2	249,948
1283	1556,15	1767,37	1305,46	1140,1	40,512	8,9240	49505,7	36567,2	249,971
1284	1557,15	1768,63	1306,43	1143,2	40,428	8,9248	49541,1	36594,3	249,993
1285	1558,15	1769,89	1307,39	1146,3	40,343	8,9256	49576,5	36621,4	250,016
1286	1559,15	1771,16	1308,36	1149,5	40,259	8,9265	49611,8	36648,4	250,039
1287	1560,15	1772,42	1309,32	1152,6	40,175	8,9273	49647,2	36675,5	250,062
1288	1561,15	1773,68	1310,29	1155,8	40,092	8,9281	49682,6	36702,6	250,084
1289	1562,15	1774,95	1311,26	1158,9	40,008	8,9289	49718,0	36729,6	250,107
1290	1563,15	1776,21	1312,22	1162,1	39,925	8,9297	49753,4	36756,7	250,130
1291	1564,15	1777,47	1313,19	1165,3	39,842	8,9305	49788,8	36783,8	250,152
1292	1565,15	1778,74	1314,16	1168,4	39,759	8,9313	49824,2	36810,9	250,175
1293	1566,15	1780,00	1315,12	1171,6	39,676	8,9321	49859,6	36838,0	250,197
1294	1567,15	1781,26	1316,09	1174,8	39,594	8,9329	49895,0	36865,1	250,220
1295	1568,15	1782,53	1317,06	1178,0	39,512	8,9337	49930,4	36892,1	250,243
1296	1569,15	1783,79	1318,03	1181,2	39,430	8,9345	49965,8	36919,2	250,265
1297	1570,15	1785,06	1318,99	1184,4	39,348	8,9353	50001,2	36946,3	250,288
1298	1571,15	1786,32	1319,96	1187,6	39,266	8,9361	50036,6	36973,4	250,310
1299	1572,15	1787,58	1320,93	1190,8	39,185	8,9369	50072,0	37000,5	250,333
1300	1573,15	1788,85	1321,90	1194,1	39,104	8,9378	50107,4	37027,6	250,355
1301	1574,15	1790,11	1322,86	1197,3	39,023	8,9386	50142,9	37054,7	250,378
1302	1575,15	1791,38	1323,83	1200,6	38,942	8,9394	50178,3	37081,8	250,400
1303	1576,15	1792,64	1324,80	1203,8	38,862	8,9402	50213,7	37109,0	250,423
1304	1577,15	1793,91	1325,77	1207,1	38,781	8,9410	50249,1	37136,1	250,445
1305	1578,15	1795,17	1326,74	1210,3	38,701	8,9418	50284,6	37163,2	250,468
1306	1579,15	1796,44	1327,70	1213,6	38,621	8,9426	50320,0	37190,3	250,490
1307	1580,15	1797,70	1328,67	1216,9	38,542	8,9434	50355,4	37217,4	250,513
1308	1581,15	1798,97	1329,64	1220,2	38,462	8,9442	50390,9	37244,5	250,535
1309	1582,15	1800,23	1330,61	1223,5	38,383	8,9450	50426,3	37271,7	250,557
1310	1583,15	1801,50	1331,58	1226,8	38,304	8,9458	50461,7	37298,8	250,580
1311	1584,15	1802,76	1332,55	1230,1	38,225	8,9466	50497,2	37325,9	250,602
1312	1585,15	1804,03	1333,51	1233,4	38,146	8,9474	50532,6	37353,1	250,625
1313	1586,15	1805,29	1334,48	1236,7	38,068	8,9482	50568,1	37380,2	250,647
1314	1587,15	1806,56	1335,45	1240,0	37,990	8,9490	50603,5	37407,3	250,669
1315	1588,15	1807,83	1336,42	1243,4	37,912	8,9498	50639,0	37434,5	250,692
1316	1589,15	1809,09	1337,39	1246,7	37,834	8,9506	50674,5	37461,6	250,714
1317	1590,15	1810,36	1338,36	1250,1	37,756	8,9514	50709,9	37488,8	250,736
1318	1591,15	1811,62	1339,33	1253,4	37,679	8,9521	50745,4	37515,9	250,759
1319	1592,15	1812,89	1340,30	1256,8	37,602	8,9529	50780,8	37543,0	250,781
1320	1593,15	1814,16	1341,27	1260,1	37,525	8,9537	50816,3	37570,2	250,803
1321	1594,15	1815,42	1342,24	1263,5	37,448	8,9545	50851,8	37597,4	250,825
1322	1595,15	1816,69	1343,20	1266,9	37,371	8,9553	50887,2	37624,5	250,848
1323	1596,15	1817,95	1344,17	1270,3	37,295	8,9561	50922,7	37651,7	250,870
1324	1597,15	1819,22	1345,14	1273,7	37,219	8,9569	50958,2	37678,8	250,892

Carbon monoxide (*Continued*)

t	T	h	u	π_0	θ_0	s^0	H	U	S
1325	1598,15	1820,49	1346,11	1277,1	37,142	8,9577	50993,7	37706,0	250,914
1326	1599,15	1821,75	1347,08	1280,5	37,067	8,9585	51029,1	37733,2	250,936
1327	1600,15	1823,02	1348,05	1283,9	36,991	8,9593	51064,6	37760,3	250,959
1328	1601,15	1824,29	1349,02	1287,4	36,916	8,9601	51100,1	37787,5	250,981
1329	1602,15	1825,55	1349,99	1290,8	36,840	8,9609	51135,6	37814,7	251,003
1330	1603,15	1826,82	1350,96	1294,2	36,765	8,9617	51171,1	37841,8	251,025
1331	1604,15	1828,09	1351,93	1297,7	36,690	8,9625	51206,6	37869,0	251,047
1332	1605,15	1829,36	1352,90	1301,2	36,616	8,9632	51242,1	37896,2	251,069
1333	1606,15	1830,62	1353,87	1304,6	36,541	8,9640	51277,6	37923,4	251,091
1334	1607,15	1831,89	1354,85	1308,1	36,467	8,9648	51313,1	37950,6	251,114
1335	1608,15	1833,16	1355,82	1311,6	36,393	8,9656	51348,6	37977,8	251,136
1336	1609,15	1834,42	1356,79	1315,1	36,319	8,9664	51384,1	38004,9	251,158
1337	1610,15	1835,69	1357,76	1318,5	36,245	8,9672	51419,6	38032,1	251,180
1338	1611,15	1836,96	1358,73	1322,0	36,172	8,9680	51455,1	38059,3	251,202
1339	1612,15	1838,23	1359,70	1325,6	36,098	8,9688	51490,6	38086,5	251,224
1340	1613,15	1839,50	1360,67	1329,1	36,025	8,9695	51526,1	38113,7	251,246
1341	1614,15	1840,76	1361,64	1332,6	35,952	8,9703	51561,6	38140,9	251,268
1342	1615,15	1842,03	1362,61	1336,1	35,879	8,9711	51597,1	38168,1	251,290
1343	1616,15	1843,30	1363,58	1339,7	35,807	8,9719	51632,7	38195,3	251,312
1344	1617,15	1844,57	1364,55	1343,2	35,734	8,9727	51668,2	38222,5	251,334
1345	1618,15	1845,84	1365,53	1346,8	35,662	8,9735	51703,7	38249,8	251,356
1346	1619,15	1847,10	1366,50	1350,3	35,590	8,9743	51739,2	38277,0	251,378
1347	1620,15	1848,37	1367,47	1353,9	35,518	8,9750	51774,8	38304,2	251,400
1348	1621,15	1849,64	1368,44	1357,5	35,447	8,9758	51810,3	38331,4	251,422
1349	1622,15	1850,91	1369,41	1361,0	35,375	8,9766	51845,8	38358,6	251,444
1350	1623,15	1852,18	1370,38	1364,6	35,304	8,9774	51881,4	38385,8	251,465
1351	1624,15	1853,45	1371,36	1368,2	35,233	8,9782	51916,9	38413,1	251,487
1352	1625,15	1854,72	1372,33	1371,8	35,162	8,9789	51952,5	38440,3	251,509
1353	1626,15	1855,98	1373,30	1375,5	35,091	8,9797	51988,0	38467,5	251,531
1354	1627,15	1857,25	1374,27	1379,1	35,020	8,9805	52023,5	38494,7	251,553
1355	1628,15	1858,52	1375,24	1382,7	34,950	8,9813	52059,1	38522,0	251,575
1356	1629,15	1859,79	1376,22	1386,3	34,880	8,9821	52094,6	38549,2	251,597
1357	1630,15	1861,06	1377,19	1390,0	34,810	8,9828	52130,2	38576,5	251,618
1358	1631,15	1862,33	1378,16	1393,6	34,740	8,9836	52165,7	38603,7	251,640
1359	1632,15	1863,60	1379,13	1397,3	34,670	8,9844	52201,3	38630,9	251,662
1360	1633,15	1864,87	1380,11	1400,9	34,601	8,9852	52236,9	38658,2	251,684
1361	1634,15	1866,14	1381,08	1404,6	34,531	8,9860	52272,4	38685,4	251,706
1362	1635,15	1867,41	1382,05	1408,3	34,462	8,9867	52308,0	38712,7	251,727
1363	1636,15	1868,68	1383,03	1412,0	34,393	8,9875	52343,5	38739,9	251,749
1364	1637,15	1869,95	1384,00	1415,7	34,324	8,9883	52379,1	38767,2	251,771
1365	1638,15	1871,22	1384,97	1419,4	34,256	8,9891	52414,7	38794,4	251,792
1366	1639,15	1872,49	1385,94	1423,1	34,187	8,9898	52450,3	38821,7	251,814
1367	1640,15	1873,76	1386,92	1426,8	34,119	8,9906	52485,8	38849,0	251,836
1368	1641,15	1875,03	1387,89	1430,5	34,051	8,9914	52521,4	38876,2	251,858
1369	1642,15	1876,30	1388,86	1434,3	33,983	8,9922	52557,0	38903,5	251,879
1370	1643,15	1877,57	1389,84	1438,0	33,915	8,9929	52592,6	38930,7	251,901
1371	1644,15	1878,84	1390,81	1441,8	33,848	8,9937	52628,2	38958,0	251,923
1372	1645,15	1880,11	1391,78	1445,5	33,780	8,9945	52663,7	38985,3	251,944
1373	1646,15	1881,38	1392,76	1449,3	33,713	8,9952	52699,3	39012,6	251,966
1374	1647,15	1882,65	1393,73	1453,1	33,646	8,9960	52734,9	39039,8	251,987
1375	1648,15	1883,92	1394,71	1456,8	33,579	8,9968	52770,5	39067,1	252,009
1376	1649,15	1885,19	1395,68	1460,6	33,512	8,9976	52806,1	39094,4	252,031
1377	1650,15	1886,46	1396,65	1464,4	33,445	8,9983	52841,7	39121,7	252,052
1378	1651,15	1887,73	1397,63	1468,2	33,379	8,9991	52877,3	39149,0	252,074
1379	1652,15	1889,00	1398,60	1472,0	33,313	8,9999	52912,9	39176,2	252,095

Carbon dioxide (*Continued*)

t	T	h	u	π_0	θ_0	s^0	H	U	S
1380	1653,15	1890,28	1399,58	1475,9	33,247	9,0006	52948,5	39203,5	252,117
1381	1654,15	1891,55	1400,55	1479,7	33,181	9,0014	52984,1	39230,8	252,138
1382	1655,15	1892,82	1401,52	1483,5	33,115	9,0022	53019,7	39258,1	252,160
1383	1656,15	1894,09	1402,50	1487,4	33·049	9,0029	53055,3	39285,4	252,181
1384	1657,15	1895,36	1403,47	1491,2	32,984	9,0037	53090,9	39312,7	252,203
1385	1658,15	1896,63	1404,45	1495,1	32,919	9,0045	53126,5	39340,0	252,224
1386	1659,15	1897,90	1405,42	1498,9	32,854	9,0052	53162,2	39367,3	252,246
1387	1660,15	1899,17	1406,40	1502,8	32,789	9,0060	53197,8	39394,6	252,267
1388	1661,15	1900, 5	1407,37	1506,7	32,724	9,0068	53233,4	39421,9	252,289
1389	1662,15	1901,72	1408,35	1510,6	32,659	9,0075	53269,0	39449,2	252,310
1390	1663,15	1902,99	1409,32	1514,5	32,595	9,0083	53304,6	39476,5	252,332
1391	1664,15	1904,26	1410,30	1518,4	32,531	9,0091	53340,3	39503,8	252,353
1392	1665,15	1905,53	1411,27	1522,3	32,466	9,0098	53375,9	39531,2	252,374
1393	1666,15	1906,81	1412,25	1526,2	32,402	9,0106	53411,5	39558,5	252,396
1394	1667,15	1908,08	1413,22	1530,2	32,339	9,0114	53447,2	39585,8	252,417
1395	1668,15	1909,35	1414,20	1534,1	32,275	9,0121	53482,8	39613,1	252,439
1396	1669,15	1910,62	1415,17	1538,0	32,211	9,0129	53518,4	39640,4	252,460
1397	1670,15	1911,89	1416,15	1542,0	32,148	9,0136	53554,1	39667,8	252,481
1398	1671,15	1913,17	1417,13	1546,0	32,085	9,0144	53589,7	39695,1	252,503
1399	1672,15	1914,44	1418,10	1549,9	32,022	9,0152	53625,4	39722,4	252,524
1400	1673,15	1915,71	1419,08	1553,9	31,959	9,0159	53661,0	39749,8	252,545
1401	1674,15	1916,98	1420,05	1557,9	31,896	9,0167	53696,7	39777,1	252,567
1402	1675,15	1918,26	1421,03	1561,9	31,834	9,0175	53732,3	39804,4	252,588
1403	1676,15	1919,53	1422,00	1565,9	31,771	9,0182	53768,0	39831,8	252,609
1404	1677,15	1920,80	1422,98	1569,9	31,709	9,0190	53803,6	39859,1	252,630
1405	1678,15	1922,08	1423,96	1573,9	31,647	9,0197	53839,3	39886,4	252,652
1406	1679,15	1923,35	1424,93	1577,9	31,585	9,0205	53874,9	39913,8	252,673
1407	1680,15	1924,62	1425,91	1582,0	31,523	9,0212	53910,6	39941,1	252,694
1408	1681,15	1925,90	1426,89	1586,0	31,462	9,0220	53946,3	39968,5	252,715
1409	1682,15	1927,17	1427,86	1590,1	31,400	9,0228	53981,9	39995,8	252,737
1410	1683,15	1928,44	1428,84	1594,1	31,339	9,0235	54017,6	40023,2	252,758
1411	1684,15	1929,72	1429,81	1598,2	31,278	9,0243	54053,3	40050,5	252,779
1412	1685,15	1930,99	1430,79	1602,3	31,217	9,0250	54088,9	40077,9	252,800
1413	1686,15	1932,26	1431,77	1606,4	31,156	9,0258	54124,6	40105,3	252,821
1414	1687,15	1933,54	1432,75	1610,4	31,095	9,0265	54160,3	40132,6	252,842
1415	1688,15	1934,81	1433,72	1614,5	31,034	9,0273	54196,0	40160,0	252,864
1416	1689,15	1936,08	1434,70	1618,7	30,974	9,0281	54231,6	40187,4	252,885
1417	1690,15	1937,36	1435,68	1622,8	30,914	9,0288	54267,3	40214,7	252,906
1418	1691,15	1938,63	1436,65	1626,9	30,854	9,0296	54303,0	40242,1	252,927
1419	1692,15	1939,91	1437,63	1631,0	30,794	9,0303	54338,7	40269,5	252,948
1420	1693,15	1941,18	1438,61	1635,2	30,734	9,0311	54374,4	40296,8	252,969
1421	1694,15	1942,45	1439,58	1639,3	30,674	9,0318	54410,1	40324,2	252,990
1422	1695,15	1943,73	1440,56	1643,5	30,614	9,0326	54445,8	40351,6	253,011
1423	1696,15	1945,00	1441,54	1647,6	30,555	9,0333	54481,5	40379,0	253,032
1424	1697,15	1946,28	1442,52	1651,8	30,496	9,0341	54517,2	40406,4	253,053
1425	1698,15	1947,55	1443,49	1656,0	30,437	9,0348	54552,9	40433,7	253,074
1426	1699,15	1948,83	1444,47	1660,2	30,378	9,0356	54588,6	40461,1	253,095
1427	1700,15	1950,10	1445,45	1664,4	30,319	9,0363	54624,3	40488,5	253,116
1428	1701,15	1951,37	1446,43	1668,6	30,260	9,0371	54660,0	40515,9	253,137
1429	1702,15	1952,65	1447,41	1672,8	30,202	9,0378	54695,7	40543,3	253,158
1430	1703,15	1953,92	1448,38	1677,0	30,143	9,0386	54731,4	40570,7	253,179
1431	1704,15	1955,20	1449,36	1681,3	30,085	9,0393	54767,1	40598,1	253,200
1432	1705,15	1956,47	1450,34	1685,5	30,027	9,0401	54802,8	40625,5	253,221
1433	1706,15	1957,75	1451,32	1689,8	29,969	9,0408	54838,5	40652,9	253,242
1434	1707,15	1959,02	1452,30	1694,0	29,911	9,0416	54874,2	40680,3	253,263
1435	1708,15	1960,30	1453,28	1698,3	29,854	9,0423	54910,0	40707,7	253,284
1436	1709,15	1961,58	1454,25	1702,6	29,796	9,0431	54945,7	40735,1	253,305
1437	1710,15	1962,85	1455,23	1706,9	29,739	9,0438	54981,4	40762,5	253,326
1438	1711,15	1964,13	1456,21	1711,1	29,681	9,0445	55017,1	40789,9	253,347
1439	1712,15	1965,40	1457,19	1715,5	29,624	9,0453	55052,9	40817,3	253,368

Carbon dioxide (*Continued*)

t	T	h	u	π_0	θ_0	s^0	H	U	S
1440	1713,15	1966,68	1458,17	1719,8	29,567	9,0460	55088,6	40844,8	253,388
1441	1714,15	1967,95	1459,15	1724,1	29,510	9,0468	55124,3	40872,2	253,409
1442	1715,15	1969,23	1460,13	1728,4	29,454	9,0475	55160,1	40899,6	253,430
1443	1716,15	1970,50	1461,11	1732,7	29,397	9,0483	55195,8	40927,0	253,451
1444	1717,15	1971,78	1462,08	1737,1	29,341	9,0490	55231,5	40954,4	253,472
1445	1718,15	1973,06	1463,06	1741,4	29,284	9,0498	55267,3	40981,9	253,493
1446	1719,15	1974,33	1464,04	1745,8	29,228	9,0505	55303,0	41009,3	253,513
1447	1720,15	1975,61	1465,02	1750,2	29,172	9,0512	55338,8	41036,7	253,534
1448	1721,15	1976,88	1466,00	1754,6	29,116	9,0520	55374,5	41064,2	253,555
1449	1722,15	1978,16	1466,98	1758,9	29,061	9,0527	55410,3	41091,6	253,576
1450	1723,15	1979,44	1467,96	1763,3	29,005	9,0535	55446,0	41119,0	253,596
1451	1724,15	1980,71	1468,94	1767,7	28,949	9,0542	55481,8	41146,5	253,617
1452	1725,15	1981,99	1469,92	1772,2	28,894	9,0549	55517,5	41173,9	253,638
1453	1726,15	1983,27	1470,90	1776,6	28,839	9,0557	55553,3	41201,4	253,659
1454	1727,15	1984,54	1471,88	1781,0	28,784	9,0564	55589,0	41228,8	253,679
1455	1728,15	1985,82	1472,86	1785,4	28,729	9,0572	55624,8	41256,2	253,700
1456	1729,15	1987,10	1473,84	1789,9	28,674	9,0579	55660,6	41283,7	253,721
1457	1730,15	1988,37	1474,82	1794,4	28,619	9,0586	55696,3	41311,1	253,741
1458	1731,15	1989,65	1475,80	1798,8	28,565	9,0594	55732,1	41338,6	253,762
1459	1732,15	1990,93	1476,78	1803,3	28,510	9,0601	55767,9	41366,1	253,783
1460	1733,15	1992,20	1477,76	1807,8	28,456	9,0608	55803,6	41393,5	253,803
1461	1734,15	1993,48	1478,74	1812,3	28,402	9,0616	55839,4	41421,0	253,824
1462	1735,15	1994,76	1479,72	1816,8	28,348	9,0623	55875,2	41448,4	253,845
1463	1736,15	1996,04	1480,70	1821,3	28,294	9,0631	55911,0	41475,9	253,865
1464	1737,15	1997,31	1481,68	1825,8	28,240	9,0638	55946,7	41503,4	253,886
1465	1738,15	1998,59	1482,66	1830,3	28,187	9,0645	55982,5	41530,8	253,907
1466	1739,15	1999,87	1483,64	1834,9	28,133	9,0653	56018,3	41558,3	253,927
1467	1740,15	2001,15	1484,62	1839,4	28,080	9,0660	56054,1	41585,8	253,948
1468	1741,15	2002,42	1485,60	1844,0	28,026	9,0667	56089,9	41613,3	253,968
1469	1742,15	2003,70	1486,59	1848,5	27,973	9,0675	56125,7	41640,7	253,989
1470	1743,15	2004,98	1487,57	1853,1	27,920	9,0682	56161,5	41668,2	254,009
1471	1744,15	2006,26	1488,55	1857,7	27,867	9,0689	56197,3	41695,7	254,030
1472	1745,15	2007,54	1489,53	1862,3	27,815	9,0697	56233,1	41723,2	254,050
1473	1746,15	2008,81	1490,51	1866,9	27,762	9,0704	56268,9	41750,7	254,071
1474	1747,15	2010,09	1491,49	1871,5	27,710	9,0711	56304,7	41778,2	254,091
1475	1748,15	2011,37	1492,47	1876,1	27,657	9,0719	56340,5	41805,6	254,112
1476	1749,15	2012,65	1493,45	1880,7	27,605	9,0726	56376,3	41833,1	254,132
1477	1750,15	2013,93	1494,44	1885,4	27,553	9,0733	56412,1	41860,6	254,153
1478	1751,15	2015,20	1495,42	1890,0	27,501	9,0741	56447,9	41888,1	254,173
1479	1752,15	2016,48	1496,40	1894,7	27,449	9,0748	56483,7	41915,6	254,194
1480	1753,15	2017,76	1497,38	1899,3	27,397	9,0755	56519,5	41943,1	254,214
1481	1754,15	2019,04	1498,36	1904,0	27,346	9,0762	56555,4	41970,6	254,235
1482	1755,15	2020,32	1499,34	1908,7	27,294	9,0770	56591,2	41998,1	254,255
1483	1756,15	2021,60	1500,33	1913,4	27,243	9,0777	56627,0	42025,6	254,275
1484	1757,15	2022,88	1501,31	1918,1	27,191	9,0784	56662,8	42053,2	254,296
1485	1758,15	2024,16	1502,29	1922,8	27,140	9,0792	56698,6	42080,7	254,316
1486	1759,15	2025,44	1503,27	1927,5	27,089	9,0799	56734,5	42108,2	254,337
1487	1760,15	2026,71	1504,26	1932,2	27,038	9,0806	56770,3	42135,7	254,357
1488	1761,15	2027,99	1505,24	1936,9	26,988	9,0813	56806,1	42163,2	254,377
1489	1762,15	2029,27	1506,22	1941,7	26,937	9,0821	56842,0	42190,7	254,398
1490	1763,15	2030,55	1507,20	1946,4	26,886	9,0828	56877,8	42218,3	254,418
1491	1764,15	2031,83	1508,19	1951,2	26,836	9,0835	56913,7	42245,8	254,438
1492	1765,15	2033,11	1509,17	1956,0	26,786	9,0842	56949,5	42273,3	254,459
1493	1766,15	2034,39	1510,15	1960,8	26,735	9,0850	56985,3	42300,8	254,479
1494	1767,15	2035,67	1511,13	1965,6	26,685	9,0857	57021,2	42328,4	254,499
1495	1768,15	2036,95	1512,12	1970,4	26,635	9,0864	57057,0	42355,9	254,519
1496	1769,15	2038,23	1513,10	1975,2	26,586	9,0871	57092,9	42383,4	254,540
1497	1770,15	2039,51	1514,08	1980,0	26,536	9,0879	57128,7	42411,0	254,560
1498	1771,15	2040,79	1515,07	1984,8	26,486	9,0886	57164,6	42438,5	254,580
1499	1772,15	2042,07	1516,05	1989,6	26,437	9,0893	57200,5	42466,1	254,600
1500	1773,15	2043,35	1517,03	1994,5	26,387	9,0900	57236,3	42493,6	254,621

Table III.15 Hydrogen ($\mu = 2.01594$)

t	T	c_p	μc_p	c_v	μc_v	$k = \dfrac{c_p}{c_v}$
—50	223,15	13,826	27,873	9,702	19,559	1,425
—25	248,15	14,032	28,289	9,908	19,975	1,416
0	273,15	14,179	28,584	10,055	20,270	1,410
25	298,15	14,282	28,793	10,158	20,479	1,406
50	323,15	14,355	28,940	10,231	20,626	1,403
75	348,15	14,406	29,042	10,282	20,728	1,401
100	373,15	14,441	29,114	10,318	20,800	1,400
125	398,15	14,466	29,163	10,342	20,849	1,399
150	423,15	14,482	29,196	10,358	20,882	1,398
175	448,15	14,494	29,219	10,370	20,905	1,398
200	473,15	14,502	29,236	10,378	20,922	1,397
250	523,15	14,514	29,261	10,390	20,947	1,397
300	573,15	14,528	29,288	10,404	20,974	1.396
350	623,15	14,547	29,327	10,423	21,013	1,396
400	673,15	14,575	29,383	10,451	21,069	1,395
450	723,15	14,613	29,459	10,489	21,145	1,393
500	773,15	14,660	29,554	10,536	21,240	1,391
550	823,15	14,717	29,669	10,593	21,355	1,389
600	873,15	14,783	29,803	10,659	21,489	1,387
650	923,15	14,857	29,952	10,733	21.638	1,384
700	973,15	14,938	30,116	10,815	21,802	1,381
750	1023,15	15,026	30,293	10,902	21,979	1,378
800	1073,15	15,119	30,480	10,995	22,166	1,375
850	1123,15	15,216	30,675	11,092	22,361	1,372
900	1173,15	15,316	30,877	11,192	22,563	1,368
950	1223,15	15,419	31,085	11,295	22,771	1,365
1000	1273,15	15,524	31,296	11,400	22,982	1,362
1050	1323,15	15,630	31,510	11,506	23,196	1,358
1100	1373,15	15,737	31,726	11,613	23,412	1,355
1150	1423,15	15,844	31,942	11,720	23,628	1,352
1200	1473,15	15,951	32,158	11,827	23,844	1,349
1250	1523,15	16,059	32,375	11,935	24,061	1,346
1300	1573,15	16,166	32,590	12,042	24,276	1,343
1350	1623,15	16,271	32,803	12,147	24,489	1,340
1400	1673,15	16,376	33,014	12,252	24,700	1,337
1450	1723,15	16,479	33,221	12,355	24,907	1,334
1500	1773,15	16,579	33,424	12,455	25,110	1,331

Table III.16 Hydrogen (μ = 2.01594)

t	T	h	u	π_0	θ_0	s^0	H	U	S
−50	223,15	3143,09	2222,74	2,4637	37355	60,6986	6336,3	4480,9	122,365
−49	224,15	3156,92	2232,45	2,5009	36964	60,7605	6364,2	4500,5	122,489
−48	225,15	3170,76	2242,16	2,5386	36578	60,8221	6392,1	4520,1	122,614
−47	226,15	3184,61	2251,89	2,5766	36198	60,8834	6420,0	4539,7	122,737
−46	227,15	3198,47	2261,62	2,6151	35823	60,9446	6447,9	4559,3	122,861
−45	228,15	3212,34	2271,37	2,6540	35453	61,0055	6475,9	4578,9	122,983
−44	229,15	3226,21	2281,12	2,6934	35089	61,0662	6503,9	4598,6	123,106
−43	230,15	3240,10	2290,88	2,7331	34729	61,1267	6531,8	4618,3	123,228
−42	231,15	3254,00	2300,66	2,7734	34374	61,1869	6559,9	4638,0	123,349
−41	232,15	3267,90	2310,44	2,8140	34024	61,2469	6587,9	4657,7	123,470
−40	233,15	3281,81	2320,22	2,8551	33678	61,3067	6615,9	4677,4	123,591
−39	234,15	3295,73	2330,02	2,8967	33338	61,3663	6644,0	4697,2	123,711
−38	235,15	3309,66	2339,83	2,9387	33002	61,4257	6672,1	4716,9	123,830
−37	236,15	3323,60	2349,64	2,9811	32670	61,4848	6700,2	4736,7	123,950
−36	237,15	3337,55	2359,46	3,0240	32343	61,5438	6728,3	4756,5	124,069
−35	238,15	3351,50	2369,29	3,0674	32020	61,6025	6756,4	4776,3	124,187
−34	239,15	3365,46	2379,13	3,1112	31702	61,6610	6784,6	4796,2	124,305
−33	240,15	3379,43	2388,97	3,1555	31388	61,7193	6812,7	4816,0	124,422
−32	241,15	3393,41	2398,83	3,2002	31078	61,7774	6840,9	4835,9	124,539
−31	242,15	3407,40	2408,69	3,2455	30772	61,8352	6869,1	4855,8	124,656
−30	243,15	3421,39	2418,56	3,2912	30470	61,8929	6897,3	4875,7	124,772
−29	244,15	3435,39	2428,43	3,3373	30172	61,9504	6925,5	4895,6	124,888
−28	245,15	3449,40	2438,32	3,3840	29878	62,0076	6953,8	4915,5	125,004
−27	246,15	3463,41	2448,21	3,4311	29587	62,0647	6982,0	4935,4	125,119
−26	247,15	3477,43	2458,10	3,4787	29301	62,1215	7010,3	4955,4	125,233
−25	248,15	3491,46	2468,01	3,5268	29018	62,1782	7038,6	4975,4	125,347
−24	249,15	3505,50	2477,92	3,5755	28739	62,2346	7066,9	4995,3	125,461
−23	250,15	3519,54	2487,84	3,6245	28463	62,2909	7095,2	5015,3	125,575
−22	251,15	3533,59	2497,76	3,6741	28191	62,3469	7123,5	5035,3	125,688
−21	252,15	3547,65	2507,69	3,7242	27923	62,4028	7151,8	5055,4	125,800
−20	253,15	3561,71	2517,63	3,7748	27658	62,4584	7180,2	5075,4	125,912
−19	254,15	3575,78	2527,58	3,8260	27396	62,5139	7208,6	5095,4	126,024
−18	255,15	3589,85	2537,53	3,8776	27138	62,5692	7236,9	5115,5	126,136
−17	256,15	3603,93	2547,49	3,9297	26883	62,6242	7265,3	5135,6	126,247
−16	257,15	3618,02	2557,45	3,9824	26631	62,6791	7293,7	5155,7	126,357
−15	258,15	3632,11	2567,42	4,0355	26382	62,7338	7322,1	5175,8	126,468
−14	259,15	3646,21	2577,39	4,0892	26137	62,7884	7350,6	5195,9	126,578
−13	260,15	3660,32	2587,37	4,1434	25894	62,8427	7379,0	5216,0	126,687
−12	261,15	3674,43	2597,36	4,1982	25655	62,8968	7407,4	5236,1	126,796
−11	262,15	3688,55	2607,36	4,2535	25418	62,9508	7435,9	5256,3	126,905
−10	263,15	3702,67	2617,35	4,3093	25185	63,0045	7464,4	5276,4	127,013
−9	264,15	3716,80	2627,36	4,3657	24954	63,0581	7492,8	5296,6	127,121
−8	265,15	3730,94	2637,37	4,4226	24726	63,1115	7521,3	5316,8	127,229
−7	266,15	3745,08	2647,38	4,4800	24501	63,1648	7549,8	5337,0	127,336
−6	267,15	3759,22	2657,40	4,5380	24279	63,2178	7578,4	5357,2	127,443
−5	268,15	3773,37	2667,43	4,5965	24060	63,2707	7606,9	5377,4	127,550
−4	269,15	3787,53	2677,46	4,6556	23843	63,3234	7635,4	5397,6	127,656
−3	270,15	3801,69	2687,50	4,7153	23628	63,3759	7664,0	5417,8	127,762
−2	271,15	3815,85	2697,54	4,7755	23417	63,4282	7692,5	5438,1	127,868
−1	272,15	3830,02	2707,59	4,8363	23208	63,4804	7721,1	5458,3	127,973
0	273,15	3844,20	2717,64	4,8977	23001	63,5324	7749,7	5478,6	128,077
1	274,15	3858,38	2727,69	4,9596	22797	63,5842	7778,3	5498,9	128,182
2	275,15	3872,57	2737,76	5,0221	22596	63,6359	7806,9	5519,2	128,286
3	276,15	3886,76	2747,82	5,0852	22396	63,6873	7835,5	5539,4	128,390
4	277,15	3900,95	2757,89	5,1488	22200	63,7387	7864,1	5559,7	128,493

240

Hydrogen (*Continued*)

t	T	h	u	π_0	0_0	s^0	H	U	S
5	278,15	3915,15	2767,97	5,2131	22005	63,7898	7892,7	5580,1	128,596
6	279,15	3929,36	2778,05	5,2779	21813	63,8408	7921,3	5600,4	128,699
7	280,15	3943,57	2788,13	5,3433	21623	63,8916	7950,0	5620,7	128,802
8	281,15	3957,78	2798,22	5,4094	21435	63,9422	7978,6	5641,0	128,904
9	282,15	3972,00	2808,32	5,4700	21250	63,9927	8007,3	5661,4	129,005
10	283,15	3986,22	2818,41	5,5432	21067	64,0430	8036,0	5681,8	129,107
11	284,15	4000,44	2828,52	5,6110	20886	64,0932	8064,7	5702,1	129,208
12	285,15	4014,68	2838,62	5,6794	20707	64,1432	8093,3	5722,5	129,309
13	286,15	4028,91	2848,73	5,7485	20530	64,1930	8122,0	5742,9	129,409
14	287,15	4043,15	2858,85	5,8181	20355	64,2427	8150,7	5763,3	129,509
15	288,15	4057,39	2868,96	5,8884	20182	64,2922	8179,5	5783,7	129,609
16	289,15	4071,64	2879,09	5,9593	20011	64,3415	8208,2	5804,1	129,709
17	290,15	4085,89	2889,21	6,0308	19842	64,3907	8236,9	5824,5	129,808
18	291,15	4100,14	2899,34	6,1029	19675	64,4398	8265,6	5844,9	129,907
19	292,15	4114,40	2909,48	6,1757	19510	64,4887	8294,4	5865,3	130,005
20	293,15	4128,66	2919,62	6,2491	19347	64,5374	8323,1	5885,8	130,104
21	294,15	4142,93	2929,76	6,3232	19186	64,5860	8351,9	5906,2	130,201
22	295,15	4157,20	2939,90	6,3979	19026	64,6344	8380,7	5926,7	130,299
23	296,15	4171,47	2950,05	6,4732	18868	64,6827	8409,4	5947,1	130,396
24	297,15	4185,75	2960,20	6,5492	18712	64,7308	8438,2	5967,6	130,493
25	298,15	4200,03	2970,36	6,6258	18558	64,7788	8467,0	5988,1	130,590
26	299,15	4214,31	2980,52	6,7031	18406	64,8266	8495,8	6008,5	130,687
27	300,15	4228,60	2990,68	6,7810	18255	64,8743	8524,6	6029,0	130,783
28	301,15	4242,89	3000,85	6,8596	18106	64,9218	8553,4	6049,5	130,879
29	302,15	4257,18	3011,02	6,9389	17959	64,9692	8582,2	6070,0	130,974
30	303,15	4271,48	3021,19	7,0188	17813	65,0165	8611,0	6090,5	131,069
31	304,15	4285,78	3031,37	7,0994	17669	65,0636	8639,9	6111,1	131,164
32	305,15	4300,08	3041,54	7,1807	17526	65,1105	8668,7	6131,6	131,259
33	306,15	4314,39	3051,73	7,2627	17385	65,1573	8697,6	6152,1	131,353
34	307,15	4328,70	3061,91	7,3453	17246	65,2040	8726,4	6172,6	131,447
35	308,15	4343,01	3072,10	7,4286	17108	65,2505	8755,3	6193,2	131,541
36	309,15	4357,33	3082,29	7,5126	16971	65,2969	8784,1	6213,7	131,635
37	310,15	4371,65	3092,49	7,5974	16836	65,3431	8813,0	6234,3	131,728
38	311,15	4385,97	3102,68	7,6828	16703	65,3892	8841,9	6254,8	131,821
39	312,15	4400,29	3112,88	7,7689	16571	65,4352	8870,7	6275,4	131,913
40	313,15	4414,62	3123,09	7,8557	16440	65,4810	8899,6	6296,0	132,006
41	314,15	4428,95	3133,29	7,9432	16311	65,5267	8928,5	6316,5	132,098
42	315,15	4443,28	3143,50	8,0314	16183	65,5723	8957,4	6337,1	132,190
43	316,15	4457,62	3153,71	8,1203	16057	65,6177	8986,3	6357,7	132,281
44	317,15	4471,96	3163,93	8,2100	15932	65,6630	9015,2	6378,3	132,373
45	318,15	4486,30	3174,14	8,3003	15808	65,7081	9044,1	6398,9	132,464
46	319,15	4500,64	3184,36	8,3914	15685	65,7531	9073,0	6419,5	132,554
47	320,15	4514,99	3194,58	8,4832	15564	65,7980	9101,9	6440,1	132,645
48	321,15	4529,34	3204,81	8,5758	15444	65,8427	9130,9	6460,7	132,735
49	322,15	4543,69	3215,04	8,6690	15326	65,8874	9159,8	6481,3	132,825
50	323,15	4558,04	3225,27	8,7631	15208	65,9319	9188,7	6501,9	132,915
51	324,15	4572,40	3235,50	8,8578	15092	65,9762	9217,7	6522,6	133,004
52	325,15	4586,76	3245,73	8,9533	14977	66,0204	9246,6	6543,2	133,093
53	326,15	4601,12	3255,97	9,0496	14864	66,0645	9275,6	6563,8	133,182
54	327,15	4615,48	3266,21	9,1466	14751	66,1085	9304,5	6584,5	133,271
55	328,15	4629,85	3276,45	9,2443	14640	66,1524	9333,5	6605,1	133,359
56	329,15	4644,22	3286,69	9,3428	14530	66,1961	9362,5	6625,8	133,447
57	330,15	4658,59	3296,94	9,4421	14420	66,2397	9391,4	6646,4	133,535
58	331,15	4672,96	3307,19	9,5421	14313	66,2831	9420,4	6667,1	133,623
59	332,15	4687,33	3317,44	9,6429	14206	66,3265	9449,4	6687,8	133,710

Hydrogen (*Continued*)

t	T	h	u	π_0	θ_0	s^0	H	U	S
60	333,15	4701,71	3327,69	9,7445	14100	66,3697	9478,4	6708,4	133,797
61	334,15	4716,09	3337,94	9,8469	13995	66,4128	9507,4	6729,1	133,884
62	335,15	4730,47	3348,20	9,9500	13892	66,4558	9536,3	6749,8	133,971
63	336,15	4744,85	3358,46	10,053	13789	66,4986	9565,3	6770,4	134,057
64	337,15	4759,24	3368,72	10,158	13688	66,5413	9594,3	6791,1	134,143
65	338,15	4773,62	3378,98	10,264	13587	66,5839	9623,3	6811,8	134,229
66	339,15	4788,01	3389,24	10,370	13488	66,6264	9652,3	6832,5	134,315
67	340,15	4802,40	3399,51	10,477	13389	66,6688	9681,4	6853,2	134,400
C8	341,15	4816,80	3409,78	10,585	13292	66,7111	9710,4	6873,9	134,485
69	342,15	4831,19	3420,05	10,694	13195	66,7532	9739,4	6894,6	134,570
70	343,15	4845,59	3430,32	10,803	13099	66,7952	9768,4	6915,3	134,655
71	344,15	4859,98	3440,59	10,913	13005	66,8371	9797,4	6936,0	134,740
72	345,15	4874,38	3450,87	11,025	12911	66,8789	9826,5	6956,7	134,824
73	346,15	4888,79	3461,15	11,136	12818	66,9205	9855,5	6977,5	134,908
74	347,15	4903,19	3471,43	11,249	12727	66,9621	9884,5	6998,2	134,992
75	348,15	4917,59	3481,71	11,363	12636	67,0035	9913,6	7018,9	135,075
76	349,15	4932,00	3491,99	11,477	12546	67,0449	9942,6	7039,6	135,158
77	350,15	4946,41	3502,27	11,593	12456	67,0861	9971,7	7060,4	135,241
78	351,15	4960,82	3512,56	11,709	12368	67,1272	10000,7	7081,1	135,324
79	352,15	4975,23	3522,85	11,826	12281	67,1681	10029,8	7101,9	135,407
80	353,15	4989,65	3533,14	11,943	12194	67,2090	10088,5	7122,6	135,489
81	354,15	5004,06	3543,43	12,062	12108	67,2498	10087,9	7143,3	135,572
82	355,15	5018,48	3553,72	12,181	12024	67,2904	10116,9	7164,1	135,653
83	356,15	5032,90	3564,01	12,302	11939	67,3310	10146,0	7184,8	135,735
84	357,15	5047,32	3574,31	12,423	11857	67,3714	10175,1	7205,6	135,817
85	358,15	5061,74	3584,61	12,545	11774	67,4117	10204,2	7226,4	135,898
86	359,15	5076,16	3594,90	12,668	11692	67,4519	10233,2	7247,1	135,979
87	360,15	5090,58	3605,20	12,792	11611	67,4920	10262,3	7267,9	136,060
88	361,15	5105,01	3615,51	12,916	11531	67,5320	10291,4	7288,6	136,141
89	362,15	5119,44	3625,81	13,042	11452	67,5719	10320,5	7309,4	136,221
90	363,15	5133,86	3636,11	13,168	11373	67,6117	10349,6	7330,2	136,301
91	364,15	5148,29	3646,42	13,296	11295	67,6514	10378,7	7351,0	136,381
92	365,15	5162,72	3656,72	13,424	11218	67,6910	10407,7	7371,7	136,461
93	366,15	5177,16	3667,03	13,553	11141	67,7304	10436,8	7392,5	136,541
94	367,15	5191,59	3677,34	13,683	11066	67,7698	10465,9	7413,3	136,620
95	368,15	5206,02	3687,65	13,814	10991	67,8091	10495,0	7434,1	136,699
96	369,15	5220,46	3697,96	13,946	10917	67,8482	10524,1	7454,9	136,778
97	370,15	5234,90	3708,28	14,078	10843	67,8873	10553,2	7475,7	136,857
98	371,15	5249,34	3718,59	14,212	10770	67,9262	10582,3	7496,5	136,935
99	372,15	5263,78	3728,91	14,346	10698	67,9651	10611,5	7517,2	137,014
100	373,15	5278,22	3739,22	14,482	10626	68,0038	10640,6	7538,0	137,092
101	374,15	5292,66	3749,54	14,618	10555	68,0425	10669,7	7558,8	137,170
102	375,15	5307,10	3759,86	14,756	10485	68,0810	10698,8	7579,6	137,247
103	376,15	5321,55	3770,18	14,894	10415	68,1195	10727,9	7600,5	137,325
104	377,15	5335,99	3780,50	15,033	10346	68,1579	10757,0	7621,3	137,402
105	378,15	5350,44	3790,82	15,173	10278	68,1961	10786,2	7642,1	137,479
106	379,15	5364,89	3801,15	15,314	10210	68,2343	10815,3	7662,9	137,556
107	380,15	5379,34	3811,47	15,456	10143	68,2723	10844,4	7683,7	137,633
108	381,15	5393,79	3821,80	15,599	10077	68,3103	10873,5	7704,5	137,709
109	382,15	5408,24	3832,12	15,743	10011	68,3481	10902,7	7725,3	137,786
110	383,15	5422,69	3842,45	15,888	9946,0	68,3859	10931,8	7746,1	137,862
111	384,15	5437,14	3852,78	16,033	9881,3	68,4236	10960,9	7767,0	137,938
112	385,15	5451,59	3863,11	16,180	9817,2	68,4612	10990,1	7787,8	138,014
113	386,15	5466,05	3873,44	16,328	9753,6	68,4986	11019,2	7808,6	138,089
114	387,15	5480,50	3883,77	16,477	9690,6	68,5360	11048,4	7829,4	138,165

242

Hydrogen (*Continued*)

t	T	h	u	π_0	θ_0	s^0	H	U	S
115	388,15	5494,96	3894,10	16,626	9628,2	68,5733	11077,5	7850,3	138,240
116	389,15	5509,42	3904,43	16,777	9566,3	68,6105	11106,7	7871,1	138,315
117	390,15	5523,88	3914,77	16,928	9505,0	68,6476	11135,8	7891,9	138,389
118	391,15	5538,34	3925,10	17,081	9444,2	68,6846	11165,0	7912,8	138,464
119	392,15	5552,80	3935,44	17,235	9384,0	68,7216	11194,1	7933,6	138,539
120	393,15	5567,26	3945,78	17,389	9324,3	68,7584	11223,3	7954,4	138,613
121	394,15	5581,72	3956,11	17,545	9265,1	68,7951	11252,4	7975,3	138,687
122	395,15	5596,18	3966,45	17,701	9206,4	68,8318	11281,6	7996,1	138,761
123	396,15	5610,65	3976,79	17,859	9148,3	68,8683	11310,7	8017,0	138,834
124	397,15	5625,11	3987,13	18,018	9090,7	68,9048	11339,9	8037,8	138,908
125	398,15	5639,58	3997,47	18,177	9033,5	68,9412	11369,0	8058,7	138,981
126	399,15	5654,04	4007,82	18,338	8976,9	68,9775	11398,2	8079,5	139,054
127	400,15	5668,51	4018,16	18,500	8920,7	69,0137	11427,4	8100,4	139,127
128	401,15	5682,98	4028,50	18,662	8865,1	69,0498	11456,5	8121,2	139,200
129	402,15	5697,45	4038,85	18,826	8809,9	69,0858	11485,7	8142,1	139,273
130	403,15	5711,92	4049,19	18,991	8755,2	69,1217	11514,9	8162,9	139,345
131	404,15	5726,39	4059,54	19,157	8700,9	69,1576	11544,1	8183,8	139,418
132	405,15	5740,86	4069,88	19,323	8647,1	69,1933	11573,2	8204,6	139,490
133	406,15	5755,33	4080,23	19,491	8593,8	69,2290	11602,4	8225,5	139,562
134	407,15	5769,80	4090,58	19,660	8541,0	69,2646	11631,6	8246,4	139,633
135	408,15	5784,27	4100,93	19,830	8488,6	69,3001	11660,7	8267,2	139,705
136	409,15	5798,75	4111,28	20,001	8436,6	69,3355	11689,9	8288,1	139,776
137	410,15	5813,22	4121,63	20,173	8385,1	69,3709	11719,1	8308,9	139,847
138	411,15	5827,70	4131,98	20,346	8334,0	69,4061	11748,3	8329,8	139,919
139	412,15	5842,17	4142,33	20,521	8283,3	69,4413	11777,5	8350,7	139,989
140	413,15	5856,65	4152,68	20,696	8233,1	69,4764	11806,6	8371,6	140,060
141	414,15	5871,12	4163,03	20,872	8183,3	69,5114	11835,8	8392,4	140,131
142	415,15	5885,60	4173,38	21,050	8133,9	69,5463	11865,0	8413,3	140,201
143	416,15	5900,08	4183,74	21,228	8084,9	69,5811	11894,2	8434,2	140,271
144	417,15	5914,56	4194,09	21,408	8036,3	69,6159	11923,4	8455,0	140,341
145	418,15	5929,04	4204,45	21,589	7988,2	69,6505	11952,6	8475,9	140,411
146	419,15	5943,52	4214,80	21,770	7940,4	69,6851	11981,8	8496,8	140,481
147	420,15	5958,00	4225,16	21,953	7893,0	69,7196	12011,0	8517,7	140,551
148	421,15	5972,48	4235,52	22,137	7846,1	69,7540	12040,2	8538,5	140,620
149	422,15	5986,96	4245,87	22,322	7799,5	69,7884	12069,4	8559,4	140,689
150	423,15	6001,44	4256,23	22,509	7753,3	69,8226	12098,5	8580,3	140,758
151	424,15	6015,92	4266,59	22,696	7707,4	69,8568	12127,7	8601,2	140,827
152	425,15	6030,41	4276,95	22,885	7662,0	69,8909	12156,9	8622,1	140,896
153	426,15	6044,89	4287,31	23,074	7616,9	69,9250	12186,1	8643,0	140,965
154	427,15	6059,38	4297,67	23,265	7572,2	69,9589	12215,3	8663,8	141,033
155	428,15	6073,86	4308,03	23,457	7527,8	69,9928	12244,5	8684,7	141,101
156	429,15	6088,35	4318,39	23,650	7483,9	70,0266	12273,7	8705,6	141,169
157	430,15	6102,83	4328,75	23,844	7440,2	70,0603	12302,9	8726,5	141,237
158	431,15	6117,32	4339,11	24,039	7396,9	70,0939	12332,1	8747,4	141,305
159	432,15	6131,80	4349,47	24,235	7354,0	70,1275	12361,3	8768,3	141,373
160	433,15	6146,29	4359,84	24,433	7311,4	70,1610	12390,6	8789,2	141,440
161	434,15	6160,78	4370,20	24,632	7269,2	70,1944	12419,8	8810,1	141,508
162	435,15	6175,27	4380,56	24,832	7227,3	70,2277	12449,0	8831,0	141,575
163	436,15	6189,76	4390,93	25,033	7185,7	70,2610	12478,2	8851,8	141,642
164	437,15	6204,24	4401,29	25,235	7144,5	70,2942	12507,4	8872,7	141,709
165	438,15	6218,73	4411,66	25,438	7103,6	70,3273	12536,6	8893,6	141,776
166	439,15	6233,22	4422,02	25,643	7063,0	70,3603	12565,8	8914,5	141,842
167	440,15	6247,71	4432,39	25,849	7022,7	70,3933	12595,0	8935,4	141,909
168	441,15	6262,21	4442,76	26,056	6982,8	70,4261	12624,2	8956,3	141,975
169	442,15	6276,70	4453,12	26,264	6943,2	70,4590	12653,4	8977,2	142,041

Hydrogen (*Continued*)

t	T	h	u	π_0	θ_0	s^0	H	U	S
170	443,15	6291,19	4463,49	26,473	6903,8	70,4917	12682,7	8998,1	142,107
171	444,15	6305,68	4473,86	26,683	6864,8	70,5244	12711,9	9019,0	142,173
172	445,15	6320,17	4484,23	26,895	6826,1	70,5569	12741,1	9039,9	142,239
173	446,15	6334,67	4494,59	27,108	6787,7	70,5895	12770,3	9060,8	142,304
174	447,15	6349,16	4504,96	27,322	6749,6	70,6219	12799,5	9081,7	142,370
175	448,15	6363,65	4515,33	27,538	6711,8	70,6543	12828,7	9102,6	142,435
176	449,15	6378,15	4525,70	27,754	6674,3	70,6866	12858,0	9123,5	142,500
177	450,15	6392,64	4536,07	27,972	6637,1	70,7188	12887,2	9144,4	142,565
178	451,15	6407,13	4546,44	28,191	6600,2	70,7510	12916,4	9165,4	142,630
179	452,15	6421,63	4556,81	28,411	6563,5	70,7831	12945,6	9186,3	142,694
180	453,15	6436,13	4567,18	28,633	6527,2	70,8151	12974,8	9207,2	142,759
181	454,15	6450,62	4577,55	28,855	6491,1	70,8471	13004,1	9228,1	142,823
182	455,15	6465,12	4587,93	29,079	6455,3	70,8790	13033,3	9249,0	142,888
183	456,15	6479,61	4598,30	29,304	6419,7	70,9108	13062,5	9269,9	142,952
184	457,15	6494,11	4608,67	29,531	6384,5	70,9425	13091,7	9290,8	143,016
185	458,15	6508,61	4619,04	29,759	6349,5	70,9742	13121,0	9311,7	143,080
186	459,15	6523,11	4629,42	29,987	6314,8	71,0058	13150,2	9332,6	143,143
187	460,15	6537,60	4639,79	30,218	6280,3	71,0373	13179,4	9353,5	143,207
188	461,15	6552,10	4650,16	30,449	6246,1	71,0688	13208,6	9374,5	143,270
189	462,15	6566,60	4660,54	30,682	6212,2	71,1002	13237,9	9395,4	143,334
190	463,15	6581,10	4670,91	30,916	6178,5	71,1316	13267,1	9416,3	143,397
191	464,15	6595,60	4681,29	31,151	6145,1	71,1628	13296,3	9437,2	143,460
192	465,15	6610,10	4691,66	31,388	6111,9	71,1940	13325,6	9458,1	143,523
193	466,15	6624,60	4702,04	31,626	6079,0	71,2252	13354,8	9479,0	143,586
194	467,15	6639,10	4712,41	31,865	6046,3	71,2562	13384,0	9499,9	143,648
195	468,15	6653,60	4722,79	32,105	6013,8	71,2873	13413,3	9520,9	143,711
196	469,15	6668,10	4733,17	32,347	5981,6	71,3182	13442,5	9541,8	143,773
197	470,15	6682,60	4743,54	32,590	5949,7	71,3491	13471,7	9562,7	143,835
198	471,15	6697,10	4753,92	32,835	5917,9	71,3799	13501,0	9583,6	143,898
199	472,15	6711,60	4764,30	33,080	5886,5	71,4106	13530,2	9604,5	143,960
200	473,15	6726,10	4774,67	33,327	5855,2	71,4413	13559,4	9625,5	144,021
201	474,15	6740,61	4785,05	33,576	5824,2	71,4719	13588,7	9646,4	144,083
202	475,15	6755,11	4795,43	33,825	5793,4	71,5025	13617,9	9667,3	144,145
203	476,15	6769,61	4805,81	34,076	5762,8	71,5330	13647,1	9688,2	144,206
204	477,15	6784,11	4816,19	34,329	5732,5	71,5634	13676,4	9709,1	144,268
205	478,15	6798,62	4826,57	34,582	5702,4	71,5938	13705,6	9730,1	144,329
206	479,15	6813,12	4836,95	34,837	5672,5	71,6241	13734,8	9751,0	144,390
207	480,15	6827,62	4847,32	35,094	5642,8	71,6543	13764,1	9771,9	144,451
208	481,15	6842,13	4857,70	35,351	5613,3	71,6845	13793,3	9792,8	144,512
209	482,15	6856,63	4868,08	35,610	5584,0	71,7146	13822,6	9813,8	144,572
210	483,15	6871,14	4878,46	35,871	5555,0	71,7446	13851,8	9834,7	144,633
211	484,15	6885,64	4888,84	36,133	5526,2	71,7746	13881,0	9855,6	144,693
212	485,15	6900,15	4899,23	36,396	5497,5	71,8046	13910,3	9876,5	144,754
213	486,15	6914,65	4909,61	36,660	5469,1	71,8344	13939,5	9897,5	144,814
214	487,15	6929,16	4919,99	36,926	5440,9	71,8642	13968,8	9918,4	144,874
215	488,15	6943,66	4930,37	37,194	5412,9	71,8940	13998,0	9939,3	144,934
216	489,15	6958,17	4940,75	37,462	5385,1	71,9237	14027,3	9960,3	144,994
217	490,15	6972,68	4951,13	37,732	5357,5	71,9533	14056,5	9981,2	145,054
218	491,15	6987,18	4961,52	38,004	5330,1	71,9829	14085,7	10002,1	145,113
219	492,15	7001,69	4971,90	38,277	5302,8	72,0124	14115,0	10023,0	145,173
220	493,15	7016,20	4982,28	38,551	5275,8	72,0418	14144,2	10044,0	145,232
221	494,15	7030,70	4992,66	38,827	5249,0	72,0712	14173,5	10064,9	145,291
222	495,15	7045,21	5003,05	39,104	5222,3	72,1005	14202,7	10085,8	145,350
223	496,15	7059,72	5013,43	39,382	5195,9	72,1298	14232,0	10106,8	145,409
224	497,15	7074,23	5023,81	39,662	5169,6	72,1590	14261,2	10127,7	145,468

Hydrogen (*Continued*)

t	T	h	u	π_0	θ_e	s^0	H	U	S
225	498,15	7088,74	5034,20	39,943	5143,5	72,1882	14290,5	10148,6	145,527
226	499,15	7103,24	5044,58	40,226	5117,6	72,2173	14319,7	10169,6	145,586
227	500,15	7117,75	5054,97	40,510	5091,9	72,2463	14349,0	10190,5	145,644
228	501,15	7132,26	5065,35	40,796	5066,3	72,2753	14378,2	10211,4	145,703
229	502,15	7146,77	5075,74	41,083	5041,0	72,3042	14407,5	10232,4	145,761
230	503,15	7161,28	5086,12	41,372	5015,8	72,3331	14436,7	10253,3	145,819
231	504,15	7175,79	5096,51	41,662	4990,8	72,3619	14466,0	10274,3	145,877
232	505,15	7190,30	5106,89	41,953	4965,9	72,3906	14495,2	10295,2	145,935
233	506,15	7204,81	5117,28	42,246	4941,2	72,4193	14524,5	10316,1	145,993
234	507,15	7219,32	5127,67	42,541	4916,7	72,4480	14553,7	10337,1	146,051
235	508,15	7233,83	5138,05	42,836	4892,4	72,4766	14583,0	10358,0	146,108
236	509,15	7248,34	5148,44	43,134	4868,2	72,5051	14612,2	10378,9	146,166
237	510,15	7262,85	5158,83	43,433	4844,3	72,5336	14641,5	10399,9	146,223
238	511,15	7277,37	5169,21	43,733	4820,4	72,5620	14670,7	10420,8	146,281
239	512,15	7291,88	5179,60	44,035	4796,7	72,5903	14700,0	10441,8	146,338
240	513,15	7306,39	5189,99	44,338	4773,2	72,6186	14729,2	10462,7	146,395
241	514,15	7320,90	5200,37	44,643	4749,9	72,6469	14758,5	10483,6	146,452
242	515,15	7335,41	5210,76	44,949	4726,7	72,6751	14787,8	10504,6	146,509
243	516,15	7349,93	5221,15	45,257	4703,7	72,7032	14817,0	10525,5	146,565
244	517,15	7364,44	5231,54	45,566	4680,8	72,7313	14846,3	10546,5	146,622
245	518,15	7378,95	5241,93	45,877	4658,1	72,7594	14875,5	10567,4	146,679
246	519,15	7393,47	5252,32	46,189	4635,5	72,7873	14904,8	10588,4	146,735
247	520,15	7407,98	5262,71	46,503	4613,1	72,8153	14934,0	10609,3	146,791
248	521,15	7422,49	5273,10	46,818	4590,8	72,8432	14963,3	10630,2	146,847
249	522,15	7437,01	5283,49	47,135	4568,7	72,8710	14992,6	10651,2	146,904
250	523,15	7451,52	5293,88	47,454	4546,7	72,8987	15021,8	10672,1	146,960
251	524,15	7466,04	5304,27	47,774	4524,9	72,9265	15051,1	10693,1	147,015
252	525,15	7480,55	5314,66	48,095	4503,2	72,9541	15080,3	10714,0	147,071
253	526,15	7495,07	5325,05	48,418	4481,7	72,9817	15109,6	10735,0	147,127
254	527,15	7509,58	5335,44	48,743	4460,3	73,0093	15138,9	10755,9	147,182
255	528,15	7524,10	5345,83	49,069	4439,1	73,0368	15168,1	10776,9	147,238
256	529,15	7538,61	5356,22	49,397	4418,0	73,0643	15197,4	10797,8	147,293
257	530,15	7553,13	5366,61	49,726	4397,0	73,0917	15226,7	10818,8	147,348
258	531,15	7567,64	5377,00	50,057	4376,2	73,1190	15255,9	10839,7	147,404
259	532,15	7582,16	5387,40	50,390	4355,5	73,1463	15285,2	10860,7	147,459
260	533,15	7596,68	5397,79	50,724	4334,9	73,1736	15314,4	10881,6	147,514
261	534,15	7611,19	5408,18	51,060	4314,5	73,2008	15343,7	10902,6	147,568
262	535,15	7625,71	5418,57	51,397	4294,2	73,2280	15373,0	10923,5	147,623
263	536,15	7640,23	5428,97	51,736	4274,1	73,2551	15402,2	10944,5	147,678
264	537,15	7654,75	5439,36	52,076	4254,0	73,2821	15431,5	10965,4	147,732
265	538,15	7669,26	5449,75	52,418	4234,1	73,3091	15460,8	10986,4	147,787
266	539,15	7683,78	5460,15	52,762	4214,4	73,3361	15490,0	11007,3	147,841
267	540,15	7698,30	5470,54	53,107	4194,7	73,3630	15519,3	11028,3	147,895
268	541,15	7712,82	5480,94	53,454	4175,2	73,3898	15548,6	11049,2	147,949
269	542,15	7727,34	5491,33	53,803	4155,9	73,4166	15577,9	11070,2	148,004
270	543,15	7741,86	5501,73	54,153	4136,6	73,4434	15607,1	11091,2	148,057
271	544,15	7756,38	5512,12	54,505	4117,5	73,4701	15636,4	11112,1	148,111
272	545,15	7770,90	5522,52	54,858	4098,4	73,4967	15665,7	11133,1	148,165
273	546,15	7785,42	5532,91	55,213	4079,6	73,5234	15694,9	11154,0	148,219
274	547,15	7799,94	5543,31	55,570	4060,8	73,5499	15724,2	11175,0	148,272
275	548,15	7814,46	5553,71	55,928	4042,1	73,5764	15753,5	11195,9	148,326
276	549,15	7828,98	5564,10	56,288	4023,6	73,6029	15782,8	11216,9	148,379
277	550,15	7843,50	5574,50	56,650	4005,2	73,6293	15812,0	11237,9	148,432
278	551,15	7858,02	5584,90	57,014	3986,9	73,6557	15841,3	11258,8	148,485
279	552,15	7872,54	5595,29	57,379	3968,7	73,6820	15870,6	11279,8	148,539

Hydrogen (*Continued*)

l	T	h	u	π_0	θ_0	s^0	H	U	S
280	553,15	7887,07	5605,69	57,745	3950,7	73,7083	15899,9	11300,7	148,591
281	554,15	7901,59	5616,09	58,114	3932,7	73,7345	15929,1	11321,7	148,644
282	555,15	7916,11	5626,49	58,484	3914,9	73,7607	15958,4	11342,7	148,697
283	556,15	7930,63	5636,88	58,856	3897,2	73,7868	15987,7	11363,6	148,750
284	557,15	7945,16	5647,28	59,229	3879,5	73,8129	16017,0	11384,6	148,802
285	558,15	7959,68	5657,68	59,604	3862,0	73,8390	16046,2	11405,5	148,855
286	559,15	7974,20	5668,08	59,981	3844,7	73,8650	16075,5	11426,5	148,907
287	560,15	7988,73	5678,48	60,360	3827,4	73,8909	16104,8	11447,5	148,960
288	561,15	8003,25	5688,88	60,740	3810,2	73,9168	16134,1	11468,4	149,012
289	562,15	8017,78	5699,28	61,122	3793,1	73,9427	16163,4	11489,4	149,064
290	563,15	8032,30	5709,68	61,506	3776,2	73,9685	16192,6	11510,4	149,116
291	564,15	8046,83	5720,08	61,892	3759,3	73,9943	16221,9	11531,3	149,168
292	565,15	8061,35	5730,48	62,279	3742,6	74,0200	16251,2	11552,3	149,220
293	566,15	8075,88	5740,88	62,668	3725,9	74,0457	16280,5	11573,3	149,272
294	567,15	8090,40	5751,29	63,059	3709,4	74,0713	16309,8	11594,2	149,323
295	568,15	8104,93	5761,69	63,451	3692,9	74,0969	16339,1	11615,2	149,375
296	569,15	8119,46	5772,09	63,845	3676,6	74,1225	16368,3	11636,2	149,426
297	570,15	8133,98	5782,49	64,241	3660,3	74,1480	16397,6	11657,2	149,478
298	571,15	8148,51	5792,90	64,639	3644,2	74,1734	16426,9	11678,2	149,529
299	572,15	8163,04	5803,30	65,039	3628,1	74,1988	16456,2	11699,1	149,580
300	573,15	8177,56	5813,70	65,440	3612,2	74,2242	16485,5	11720,1	149,632
301	574,15	8192,09	5824,11	65,843	3596,3	74,2495	16514,8	11741,0	149,688
302	575,15	8206,62	5834,51	66,248	3580,6	74,2748	16544,1	11762,0	149,734
303	576,15	8221,15	5844,92	66,654	3564,9	74,3000	16573,3	11783,0	149,784
304	577,15	8235,68	5855,32	67,063	3549,4	74,3252	16602,6	11804,0	149,835
305	578,15	8250,21	5865,72	67,473	3533,9	74,3504	16631,9	11824,9	149,886
306	579,15	8264,74	5876,13	67,885	3518,5	74,3755	16661,2	11845,9	149,937
307	580,15	8279,27	5886,54	68,299	3503,2	74,4006	16690,5	11866,9	149,987
308	581,15	8293,80	5896,94	68,715	3488,1	74,4256	16719,8	11887,9	150,038
309	582,15	8308,33	5907,35	69,132	3473,0	74,4506	16749,1	11908,9	150,088
310	583,15	8322,86	5917,76	69,551	3457,9	74,4755	16778,4	11929,8	150,138
311	584,15	8337,39	5928,16	69,973	3443,0	74,5004	16807,7	11950,8	150,188
312	585,15	8351,92	5938,57	70,396	3428,2	74,5253	16837,0	11971,8	150,238
313	586,15	8366,46	5948,98	70,820	3413,5	74,5501	16866,3	11992,8	150,288
314	587,15	8380,99	5959,39	71,247	3398,8	74,5749	16895,6	12013,8	150,338
315	588,15	8395,52	5969,79	71,676	3384,2	74,5996	16924,9	12034,7	150,388
316	589,15	8410,05	5980,20	72,106	3369,8	74,6243	16954,2	12055,7	150,438
317	590,15	8424,59	5990,61	72,538	3355,4	74,6489	16983,5	12076,7	150,488
318	591,15	8439,12	6001,02	72,972	3341,1	74,6735	17012,8	12097,7	150,537
319	592,15	8453,66	6011,43	73,408	3326,8	74,6981	17042,1	12118,7	150,587
320	593,15	8468,19	6021,84	73,846	3312,7	74,7226	17071,4	12139,7	150,636
321	594,15	8482,73	6032,25	74,286	3298,6	74,7471	17100,7	12160,7	150,686
322	595,15	8497,26	6042,66	74,727	3284,7	74,7715	17130,0	12181,6	150,735
323	596,15	8511,80	6053,08	75,171	3270,8	74,7959	17159,3	12202,6	150,784
324	597,15	8526,33	6063,49	75,616	3257,0	74,8203	17188,6	12223,6	150,833
325	598,15	8540,87	6073,90	76,063	3243,2	74,8446	17217,9	12244,6	150,882
326	599,15	8555,41	6084,31	76,512	3229,6	74,8689	17247,2	12265,6	150,931
327	600,15	8569,94	6094,73	76,964	3216,0	74,8932	17276,5	12286,6	150,980
328	601,15	8584,48	6105,14	77,417	3202,5	74,9174	17305,8	12307,6	151,029
329	602,15	8599,02	6115,55	77,871	3189,1	74,9415	17335,1	12328,6	151,078
330	603,15	8613,56	6125,97	78,328	3175,8	74,9656	17364,4	12349,6	151,126
331	604,15	8628,10	6136,38	78,787	3162,5	74,9897	17393,7	12370,6	151,175
332	605,15	8642,64	6146,80	79,248	3149,3	75,0138	17423,0	12391,6	151,223
333	606,15	8657,18	6157,21	79,710	3136,2	75,0378	17452,4	12412,6	151,272
334	607,15	8671,72	6167,63	80,175	3123,2	75,0618	17481,7	12433,6	151,320

t	T	h	u	π_0	θ_0	s^0	H	U	S
335	608,15	8686,26	6178,04	80,641	3110,3	75,0857	17511,0	12454,6	151,368
336	609,15	8700,80	6188,46	81,110	3097,4	75,1096	17540,3	12475,6	151,416
337	610,15	8715,34	6198,88	81,580	3084,6	75,1334	17569,6	12496,6	151,464
338	611,15	8729,88	6209,30	82,053	3071,8	75,1572	17598,9	12517,6	151,512
339	612,15	8744,42	6219,71	82,527	3059,2	75,1810	17628,2	12538,6	151,560
340	613,15	8758,97	6230,13	83,003	3046,6	75,2048	17657,6	12559,6	151,608
341	614,15	8773,51	6240,55	83,482	3034,1	75,2285	17686,9	12580,6	151,656
342	615,15	8788,05	6250,97	83,962	3021,6	75,2521	17716,2	12601,6	151,704
343	616,15	8802,60	6261,39	84,444	3009,3	75,2757	17745,5	12622,6	151,751
344	617,15	8817,14	6271,81	84,929	2997,0	75,2993	17774,8	12643,6	151,799
345	618,15	8831,69	6282,23	85,415	2984,7	75,3229	17804,2	12664,6	151,846
346	619,15	8846,23	6292,65	85,903	2972,6	75,3464	17833,5	12685,6	151,894
347	620,15	8860,78	6303,07	86,394	2960,5	75,3699	17862,8	12706,6	151,941
348	621,15	8875,32	6313,49	86,886	2948,4	75,3933	17892,1	12727,6	151,988
349	622,15	8889,87	6323,92	87,380	2936,5	75,4167	17921,4	12748,6	152,036
350	623,15	8904,42	6334,34	87,877	2924,6	75,4401	17950,8	12769,6	152,083
351	624,15	8918,97	6344,76	88,375	2912,7	75,4634	17980,1	12790,7	152,130
352	625,15	8933,51	6355,19	88,876	2901,0	75,4867	18009,4	12811,7	152,177
353	626,15	8948,06	6365,61	89,378	2889,3	75,5099	18038,8	12832,7	152,223
354	627,15	8962,61	6376,04	89,883	2877,7	75,5331	18068,1	12853,7	152,270
355	628,15	8977,16	6386,46	90,389	2866,1	75,5563	18097,4	12874,7	152,317
356	629,15	8991,71	6396,89	90,898	2854,6	75,5795	18126,8	12895,7	152,364
357	630,15	9006,26	6407,31	91,409	2843,1	75,6026	18156,1	12916,8	152,410
358	631,15	9020,81	6417,74	91,922	2831,8	75,6257	18185,4	12937,8	152,457
359	632,15	9035,36	6428,17	92,437	2820,5	75,6487	18214,8	12958,8	152,503
360	633,15	9049,92	6438,59	92,953	2809,2	75,6717	18244,1	12979,8	152,550
361	634,15	9064,47	6449,02	93,473	2798,0	75,6947	18273,4	13000,8	152,596
362	635,15	9079,02	6459,45	93,994	2786,9	75,7176	18302,8	13021,9	152,642
363	636,15	9093,58	6469,88	94,517	2775,8	75,7405	18332,1	13042,9	152,688
364	637,15	9108,13	6480,31	95,042	2764,8	75,7633	18361,4	13063,9	152,734
365	638,15	9122,68	6490,74	95,570	2753,9	75,7862	18390,8	13084,9	152,780
366	639,15	9137,24	6501,17	96,099	2743,0	75,8090	18420,1	13106,0	152,826
367	640,15	9151,79	6511,60	96,631	2732,2	75,8317	18449,5	13127,0	152,872
368	641,15	9166,35	6522,03	97,165	2721,4	75,8544	18478,8	13148,0	152,918
369	642,15	9180,91	6532,47	97,701	2710,7	75,8771	18508,2	13169,1	152,964
370	643,15	9195,46	6542,90	98,239	2700,1	75,8998	18537,5	13190,1	153,009
371	644,15	9210,02	6553,33	98,779	2689,5	75,9224	18566,9	13211,1	153,055
372	645,15	9224,58	6563,77	99,321	2678,9	75,9450	18596,2	13232,2	153,101
373	646,15	9239,14	6574,20	99,866	2668,5	75,9675	18625,6	13253,2	153,146
374	647,15	9253,70	6584,64	100,41	2658,0	75,9900	18654,9	13274,2	153,191
375	648,15	9268,26	6595,07	100,96	2647,7	76,0125	18684,3	13295,3	153,237
376	649,15	9282,82	6605,51	101,51	2637,4	76,0350	18713,6	13316,3	153,282
377	650,15	9297,38	6615,94	102,06	2627,1	76,0574	18743,0	13337,3	153,327
378	651,15	9311,94	6626,38	102,62	2616,9	76,0798	18772,3	13358,4	153,372
379	652,15	9326,50	6636,82	103,17	2606,8	76,1021	18801,7	13379,4	153,417
380	653,15	9341,07	6647,26	103,73	2596,7	76,1244	18831,0	13400,5	153,462
381	654,15	9355,63	6657,70	104,30	2586,6	76,1467	18860,4	13421,5	153,507
382	655,15	9370,19	6668,14	104,86	2576,7	76,1690	18889,7	13442,6	153,552
383	656,15	9384,76	6678,58	105,43	2566,7	76,1912	18919,1	13463,6	153,597
384	657,15	9399,32	6689,02	105,99	2556,9	76,2133	18948,5	13484,7	153,642
385	658,15	9413,89	6699,46	106,57	2547,0	76,2355	18977,8	13505,7	153,686
386	659,15	9428,46	6709,90	107,14	2537,3	76,2576	19007,2	13526,8	153,731
387	660,15	9443,02	6720,34	107,71	2527,5	76,2797	19036,6	13547,8	153,775
388	661,15	9457,59	6730,79	108,29	2517,9	76,3017	19065,9	13568,9	153,820
389	662,15	9472,16	6741,23	108,87	2508,2	76,3238	19095,3	13589,9	153,864

Hydrogen (*Continued*)

t	T	h	u	π_0	θ_0	s^0	H	U	S
390	663,15	9486,73	6751,67	109,45	2498,7	76,3457	19124,7	13611,0	153,908
391	664,15	9501,30	6762,12	110,04	2489,2	76,3677	19154,0	13632,0	153,953
392	665,15	9515,87	6772,56	110,62	2479,7	76,3896	19183,4	13653,1	153,997
393	666,15	9530,44	6783,01	111,21	2470,3	76,4115	19212,8	13674,1	154,041
394	667,15	9545,01	6793,46	111,80	2460,9	76,4334	19242,2	13695,2	154,085
395	668,15	9559,58	6803,90	112,40	2451,6	76,4552	19271,5	13716,3	154,129
396	669,15	9574,15	6814,35	112,99	2442,3	76,4770	19300,9	13737,3	154,173
397	670,15	9588,72	6824,80	113,59	2433,1	76,4987	19330,3	13758,4	154,217
398	671,15	9603,30	6835,25	114,19	2423,9	76,5205	19359,7	13779,5	154,261
399	672,15	9617,87	6845,70	114,79	2414,8	76,5422	19389,1	13800,5	154,304
400	673,15	9632,45	6856,15	115,40	2405,7	76,5638	19418,4	13821,6	154,348
401	674,15	9647,02	6866,60	116,00	2396,7	76,5855	19447,8	13842,7	154,392
402	675,15	9661,60	6877,05	116,61	2387,7	76,6071	19477,2	13863,7	154,435
403	676,15	9676,18	6887,51	117,22	2378,8	76,6287	19506,6	13884,8	154,479
404	677,15	9690,75	6897,96	117,84	2369,9	76,6502	19536,0	13905,9	154,522
405	678,15	9705,33	6908,41	118,45	2361,0	76,6717	19565,4	13926,9	154,566
406	679,15	9719,91	6918,87	119,07	2352,2	76,6932	19594,8	13948,0	154,609
407	680,15	9734,49	6929,32	119,69	2343,5	76,7147	19624,1	13969,1	154,652
408	681,15	9749,07	6939,78	120,32	2334,7	76,7361	19653,5	13990,2	154,695
409	682,15	9763,65	6950,24	120,94	2326,1	76,7575	19682,9	14011,3	154,738
410	683,15	9778,23	6960,69	121,57	2317,5	76,7788	19712,3	14032,3	154,782
411	684,15	9792,82	6971,15	122,20	2308,9	76,8002	19741,7	14053,4	154,825
412	685,15	9807,40	6981,61	122,83	2300,3	76,8215	19771,1	14074,5	154,867
413	686,15	9821,98	6992,07	123,47	2291,8	76,8427	19800,5	14095,6	154,910
414	687,15	9836,57	7002,53	124,11	2283,4	76,8640	19829,9	14116,7	154,953
415	688,15	9851,15	7012,99	124,75	2275,0	76,8852	19859,3	14137,8	154,996
416	689,15	9865,74	7023,45	125,39	2266,6	76,9064	19888,7	14158,9	155,039
417	690,15	9880,32	7033,91	126,03	2258,3	76,9275	19918,1	14180,0	155,081
418	691,15	9894,91	7044,38	126,68	2250,0	76,9486	19947,5	14201,0	155,124
419	692,15	9909,50	7054,84	127,33	2241,8	76,9697	19977,0	14222,1	155,166
420	693,15	9924,09	7065,31	127,98	2233,6	76,9908	20006,4	14243,2	155,209
421	694,15	9938,68	7075,77	128,64	2225,4	77,0118	20035,8	14264,3	155,251
422	695,15	9953,27	7086,24	129,29	2217,3	77,0328	20065,2	14285,4	155,294
423	696,15	9967,86	7096,70	129,95	2209,2	77,0538	20094,6	14306,5	155,336
424	697,15	9982,45	7107,17	130,61	2201,2	77,0747	20124,0	14327,6	155,378
425	698,15	9997,04	7117,64	131,28	2193,2	77,0957	20153,4	14348,7	155,420
426	699,15	10011,64	7128,11	131,95	2185,3	77,1165	20182,9	14369,8	155,462
427	700,15	10026,23	7138,58	132,61	2177,3	77,1374	20212,3	14390,9	155,504
428	701,15	10040,82	7149,05	133,29	2169,5	77,1582	20241,7	14412,1	155,546
429	702,15	10055,42	7159,52	133,96	2161,6	77,1790	20271,1	14433,2	155,588
430	703,15	10070,02	7169,99	134,64	2153,8	77,1998	20300,5	14454,3	155,630
431	704,15	10084,61	7180,46	135,32	2146,1	77,2205	20330,0	14475,4	155,672
432	705,15	10099,21	7190,94	136,00	2138,4	77,2413	20359,4	14496,5	155,714
433	706,15	10113,81	7201,41	136,68	2130,7	77,2620	20388,8	14517,6	155,755
434	707,15	10128,41	7211,89	137,37	2123,0	77,2826	20418,3	14538,7	155,797
435	708,15	10143,01	7222,36	138,06	2115,4	77,3032	20447,7	14559,8	155,839
436	709,15	10157,61	7232,84	138,75	2107,9	77,3238	20477,1	14581,0	155,880
437	710,15	10172,21	7243,32	139,44	2100,3	77,3444	20506,6	14602,1	155,922
438	711,15	10186,81	7253,79	140,14	2092,8	77,3650	20536,0	14623,2	155,963
439	712,15	10201,42	7264,27	140,84	2085,4	77,3855	20565,4	14644,3	156,005
440	713,15	10216,02	7274,75	141,54	2077,9	77,4060	20594,9	14665,5	156,046
441	714,15	10230,63	7285,23	142,24	2070,6	77,4264	20624,3	14686,6	156,087
442	715,15	10245,23	7295,71	142,95	2063,2	77,4469	20653,8	14707,7	156,128
443	716,15	10259,84	7306,20	143,66	2055,9	77,4673	20683,2	14728,9	156,169
444	717,15	10274,45	7316,68	144,37	2048,6	77,4877	20712,7	14750,0	156,211

Hydrogen (*Continued*)

t	T	h	u	π_0	θ_0	s^0	H	U	S
445	718,15	10289,05	7327,16	145,08	2041,4	77,5080	20742,1	14771,1	156,252
446	719,15	10303,66	7337,65	145,80	2034,2	77,5284	20771,6	14792,3	156,293
447	720,15	10318,27	7348,13	146,52	2027,0	77,5487	20801,0	14813,4	156,333
448	721,15	10332,88	7358,62	147,24	2019,9	77,5689	20830,5	14834,5	156,374
449	722,15	10347,49	7369,11	147,97	2012,7	77,5892	20859,9	14855,7	156,415
450	723,15	10362,11	7379,59	148,69	2005,7	77,6094	20889,4	14876,8	156,456
451	724,15	10376,72	7390,08	149,42	1998,6	77,6296	20918,8	14898,0	156,497
452	725,15	10391,33	7400,57	150,16	1991,6	77,6498	20948,3	14919,1	156,537
453	726,15	10405,95	7411,06	150,89	1984,7	77,6699	20977,8	14940,3	156,578
454	727,15	10420,56	7421,55	151,63	1977,7	77,6900	21007,2	14961,4	156,618
455	728,15	10435,18	7432,05	152,37	1970,8	77,7101	21036,7	14982,6	156,659
456	729,15	10449,80	7442,54	153,11	1964,0	77,7302	21066,2	15003,7	156,699
457	730,15	10464,42	7453,03	153,86	1957,1	77,7502	21095,6	15024,9	156,740
458	731,15	10479,04	7463,53	154,61	1950,3	77,7702	21125,1	15046,0	156,780
459	732,15	10493,66	7474,02	155,36	1943,5	77,7902	21154,6	15067,2	156,820
460	733,15	10508,28	7484,52	156,11	1936,8	77,8102	21184,1	15088,3	156,861
461	734,15	10522,90	7495,02	156,87	1930,1	77,8301	21213,5	15109,5	156,901
462	735,15	10537,52	7505,52	157,63	1923,4	77,8500	21243,0	15130,7	156,941
463	736,15	10552,15	7516,02	158,39	1916,8	77,8699	21272,5	15151,8	156,981
464	737,15	10566,77	7526,52	159,15	1910,2	77,8897	21302,0	15173,0	157,021
465	738,15	10581,40	7537,02	159,92	1903,6	77,9095	21331,5	15194,2	157,061
466	739,15	10596,02	7547,52	160,69	1897,0	77,9293	21360,9	15215,3	157,101
467	740,15	10610,65	7558,02	161,46	1890,5	77,9491	21390,4	15236,5	157,141
468	741,15	10625,28	7568,53	162,24	1884,0	77,9689	21419,9	15257,7	157,181
469	742,15	10639,91	7579,03	163,01	1877,6	77,9886	21449,4	15278,9	157,220
470	743,15	10654,54	7589,54	163,79	1871,1	78,0083	21478,9	15300,1	157,260
471	744,15	10669,17	7600,04	164,58	1864,7	78,0280	21508,4	15321,2	157,300
472	745,15	10683,80	7610,55	165,36	1858,4	78,0476	21537,9	15342,4	157,339
473	746,15	10698,43	7621,06	166,15	1852,0	78,0672	21567,4	15363,6	157,379
474	747,15	10713,07	7631,57	166,94	1845,7	78,0868	21596,9	15384,8	157,418
475	748,15	10727,70	7642,08	167,74	1839,4	78,1064	21626,4	15406,0	157,458
476	749,15	10742,34	7652,59	168,54	1833,2	78,1260	21655,9	15427,2	157,497
477	750,15	10756,97	7663,10	169,34	1827,0	78,1455	21685,4	15448,4	157,537
478	751,15	10771,61	7673,62	170,14	1820,8	78,1650	21714,9	15469,6	157,576
479	752,15	10786,25	7684,13	170,94	1814,6	78,1845	21744,4	15490,7	157,615
480	753,15	10800,89	7694,65	171,75	1808,5	78,2039	21773,9	15511,9	157,654
481	754,15	10815,53	7705,16	172,56	1802,4	78,2233	21803,5	15533,1	157,694
482	755,15	10830,17	7715,68	173,38	1796,3	78,2427	21833,0	15554,3	157,733
483	756,15	10844,81	7726,20	174,19	1790,2	78,2621	21862,5	15575,6	157,772
484	757,15	10859,46	7736,72	175,01	1784,2	78,2815	21892,0	15596,8	157,811
485	758,15	10874,10	7747,24	175,83	1778,2	78,3008	21921,5	15618,0	157,850
486	759,15	10888,75	7757,76	176,66	1772,2	78,3201	21951,1	15639,2	157,889
487	760,15	10903,39	7768,28	177,49	1766,3	78,3394	21980,6	15660,4	157,928
488	761,15	10918,04	7778,80	178,32	1760,4	78,3586	22010,1	15681,6	157,966
489	762,15	10932,69	7789,33	179,15	1754,5	78,3779	22039,6	15702,8	158,005
490	763,15	10947,34	7799,85	179,99	1748,6	78,3971	22069,2	15724,0	158,044
491	764,15	10961,99	7810,38	180,83	1742,8	78,4163	22098,7	15745,3	158,083
492	765,15	10976,64	7820,91	181,67	1737,0	78,4354	22128,2	15766,5	158,121
493	766,15	10991,29	7831,43	182,51	1731,2	78,4546	22157,8	15787,7	158,160
494	767,15	11005,95	7841,96	183,36	1725,4	78,4737	22187,3	15808,9	158,198
495	768,15	11020,60	7852,49	184,21	1719,7	78,4928	22216,9	15830,2	158,237
496	769,15	11035,26	7863,02	185,07	1714,0	78,5118	22246,4	15851,4	158,275
497	770,15	11049,91	7873,56	185,92	1708,3	78,5309	22276,0	15872,6	158,314
498	771,15	11064,57	7884,09	186,78	1702,7	78,5499	22305,5	15893,9	158,352
499	772,15	11079,23	7894,62	187,64	1697,1	78,5689	22335,1	15915,1	158,390

Hydrogen (*Continued*)

t	T	h	u	π_0	θ_0	s^0	H	U	S
500	773,15	11093,89	7905,16	188,51	1691,5	78,5879	22364,6	15936,3	158,428
501	774,15	11108,55	7915,70	189,38	1685,9	78,6068	22394,2	15957,6	158,467
502	775,15	11123,21	7926,23	190,25	1680,3	78,6258	22423,7	15978,8	158,505
503	776,15	11137,87	7936,77	191,12	1674,8	78,6447	22453,3	16000,1	158,543
504	777,15	11152,54	7947,31	192,00	1669,3	78,6635	22482,8	16021,3	158,581
505	778,15	11167,20	7957,85	192,88	1663,8	78,6824	22512,4	16042,6	158,619
506	779,15	11181,87	7968,39	193,76	1658,4	78,7012	22542,0	16063,8	158,657
507	780,15	11196,54	7978,94	194,65	1652,9	78,7200	22571,5	16085,1	158,695
508	781,15	11211,20	7989,48	195,54	1647,5	78,7388	22601,1	16106,3	158,733
509	782,15	11225,87	8000,02	196,43	1642,2	78,7576	22630,7	16127,6	158,771
510	783,15	11240,54	8010,57	197,32	1636,8	78,7763	22660,3	16148,8	158,808
511	784,15	11255,21	8021,12	198,22	1631,5	78,7951	22689,8	16170,1	158,846
512	785,15	11269,89	8031,67	199,12	1626,2	78,8138	22719,4	16191,4	158,884
513	786,15	11284,56	8042,22	200,03	1620,9	78,8324	22749,0	16212,6	158,921
514	787,15	11299,24	8052,77	200,93	1615,6	78,8511	22778,6	16233,9	158,959
515	788,15	11313,91	8063,32	201,84	1610,4	78,8697	22808,2	16255,2	158,997
516	789,15	11328,59	8073,87	202,76	1605,2	78,8883	22837,8	16276,4	159,034
517	790,15	11343,27	8084,42	203,67	1600,0	78,9069	22867,3	16297,7	159,072
518	791,15	11357,94	8094,98	204,59	1594,8	78,9255	22896,9	16319,0	159,109
519	792,15	11372,62	8105,53	205,51	1589,6	78,9440	22926,5	16340,3	159,146
520	793,15	11387,31	8116,09	206,44	1584,5	78,9626	22956,1	16361,6	159,184
521	794,15	11401,99	8126,65	207,36	1579,4	78,9811	22985,7	16382,8	159,221
522	795,15	11416,67	8137,21	208,30	1574,3	78,9995	23015,3	16404,1	159,258
523	796,15	11431,36	8147,77	209,23	1569,3	79,0180	23044,9	16425,4	159,296
524	797,15	11446,04	8158,33	210,17	1564,2	79,0364	23074,5	16446,7	159,333
525	798,15	11460,73	8168,89	211,11	1559,2	79,0548	23104,1	16468,0	159,370
526	799,15	11475,42	8179,46	212,05	1554,2	79,0732	23133,8	16489,3	159,407
527	800,15	11490,11	8190,02	213,00	1549,3	79,0916	23163,4	16510,6	159,444
528	801,15	11504,80	8200,59	213,95	1544,3	79,1099	23193,0	16531,9	159,481
529	802,15	11519,49	8211,15	214,90	1539,4	79,1283	23222,6	16553,2	159,518
530	803,15	11534,18	8221,72	215,86	1534,5	79,1466	23252,2	16574,5	159,555
531	804,15	11548,88	8232,29	216,82	1529,6	79,1649	23281,8	16595,8	159,592
532	805,15	11563,57	8242,86	217,78	1524,7	79,1831	23311,5	16617,1	159,628
533	806,15	11578,27	8253,43	218,74	1519,9	79,2014	23341,1	16638,4	159,665
534	807,15	11592,96	8264,01	219,71	1515,1	79,2196	23370,7	16659,7	159,702
535	808,15	11607,66	8274,58	220,68	1510,3	79,2378	23400,4	16681,1	159,739
536	809,15	11622,36	8285,16	221,66	1505,5	79,2560	23430,0	16702,4	159,775
537	810,15	11637,06	8295,73	222,64	1500,7	79,2741	23459,6	16723,7	159,812
538	811,15	11651,77	8306,31	223,62	1496,0	79,2923	23489,3	16745,0	159,848
539	812,15	11666,47	8316,89	224,60	1491,3	79,3104	23518,9	16766,4	159,885
540	813,15	11681,17	8327,47	225,59	1486,6	79,3285	23548,5	16787,7	159,921
541	814,15	11695,88	8338,05	226,58	1481,9	79,3465	23578,2	16809,0	159,958
542	815,15	11710,59	8348,63	227,57	1477,2	79,3646	23607,8	16830,3	159,994
543	816,15	11725,29	8359,22	228,57	1472,6	79,3826	23637,5	16851,7	160,031
544	817,15	11740,00	8369,80	229,57	1468,0	79,4006	23667,1	16873,0	160,067
545	818,15	11754,71	8380,39	230,58	1463,4	79,4186	23696,8	16894,4	160,103
546	819,15	11769,43	8390,98	231,58	1458,8	79,4366	23726,5	16915,7	160,139
547	820,15	11784,14	8401,57	232,59	1454,2	79,4546	23756,1	16937,1	160,176
548	821,15	11798,85	8412,16	233,61	1449,7	79,4725	23785,8	16958,4	160,212
549	822,15	11813,57	8422,75	234,62	1445,1	79,4904	23815,4	16979,8	160,248
550	823,15	11828,28	8433,34	235,64	1440,6	79,5083	23845,1	17001,1	160,284
551	824,15	11843,00	8443,93	236,67	1436,2	79,5261	23874,8	17022,5	160,320
552	825,15	11857,72	8454,53	237,69	1431,7	79,5440	23904,5	17043,8	160,356
553	826,15	11872,44	8465,12	238,72	1427,2	79,5618	23934,1	17065,2	160,392
554	827,15	11887,16	8475,72	239,75	1422,8	79,5796	23963,8	17086,5	160,428

Hydrogen (*Continued*)

t	T	h	u	π_0	θ_0	s^0	H	U	S
555	828,15	11901,89	8486,32	240,79	1418,4	79,5974	23993,5	17107,9	160,464
556	829,15	11916,61	8496,92	241,83	1414,0	79,6152	24023,2	17129,3	160,499
557	830,15	11931,34	8507,52	242,87	1409,6	79,6329	24052,9	17150,6	160,535
558	831,15	11946,06	8518,12	243,92	1405,3	79,6507	24082,5	17172,0	160,571
559	832,15	11960,79	8528,73	244,97	1400,9	79,6684	24112,2	17193,4	160,607
560	833,15	11975,52	8539,33	246,02	1396,6	79,6861	24141,9	17214,8	160,642
561	834,15	11990,25	8549,94	247,08	1392,3	79,7037	24171,6	17236,2	160,678
562	835,15	12004,98	8560,54	248,14	1388,0	79,7214	24201,3	17257,5	160,714
563	836,15	12019,71	8571,15	249,20	1383,8	79,7390	24231,0	17278,9	160,749
564	837,15	12034,45	8581,76	250,27	1379,5	79,7566	24260,7	17300,3	160,785
565	838,15	12049,18	8592,37	251,34	1375,3	79,7742	24290,4	17321,7	160,820
566	839,15	12063,92	8602,99	252,41	1371,1	79,7918	24320,1	17343,1	160,855
567	840,15	12078,66	8613,60	253,49	1366,9	79,8093	24349,9	17364,5	160,891
568	841,15	12093,40	8624,21	254,57	1362,7	79,8269	24379,6	17385,9	160,926
569	842,15	12108,14	8634,83	255,65	1358,5	79,8444	24409,3	17407,3	160,962
570	843,15	12122,88	8645,45	256,74	1354,4	79,8619	24439,0	17428,7	160,997
571	844,15	12137,62	8656,07	257,83	1350,3	79,8794	24468,7	17450,1	161,032
572	845,15	12152,37	8666,69	258,92	1346,2	79,8968	24498,4	17471,5	161,067
573	846,15	12167,11	8677,31	260,02	1342,1	79,9143	24528,2	17492,9	161,102
574	847,15	12181,86	8687,93	261,12	1338,0	79,9317	24557,9	17514,3	161,137
575	848,15	12196,61	8698,56	262,22	1333,9	79,9491	24587,6	17535,8	161,173
576	849,15	12211,36	8709,18	263,33	1329,9	79,9665	24617,4	17557,2	161,208
577	850,15	12226,11	8719,81	264,44	1325,9	79,9838	24647,1	17578,6	161,243
578	851,15	12240,86	8730,44	265,56	1321,8	80,0012	24676,8	17600,0	161,278
579	852,15	12255,62	8741,07	266,67	1317,9	80,0185	24706,6	17621,5	161,312
580	853,15	12270,37	8751,70	267,79	1313,9	80,0358	24736,3	17642,9	161,347
581	854,15	12285,13	8762,33	268,92	1309,9	80,0531	24766,1	17664,3	161,382
582	855,15	12299,89	8772,96	270,05	1306,0	80,0703	24795,8	17685,8	161,417
583	856,15	12314,65	8783,60	271,18	1302,0	80,0876	24825,6	17707,2	161,452
584	857,15	12329,41	8794,23	272,31	1298,1	80,1048	24855,3	17728,6	161,487
585	858,15	12344,17	8804,87	273,45	1294,2	80,1220	24885,1	17750,1	161,521
586	859,15	12358,93	8815,51	274,60	1290,3	80,1392	24914,9	17771,5	161,556
587	860,15	12373,70	8826,15	275,74	1286,5	80,1564	24944,6	17793,0	161,591
588	861,15	12388,46	8836,79	276,89	1282,6	80,1736	24974,4	17814,4	161,625
589	862,15	12403,23	8847,43	278,04	1278,8	80,1907	25004,2	17835,9	161,660
590	863,15	12418,00	8858,08	279,20	1275,0	80,2078	25033,9	17857,4	161,694
591	864,15	12432,77	8868,73	280,36	1271,2	80,2249	25063,7	17878,8	161,729
592	865,15	12447,54	8879,37	281,52	1267,4	80,2420	25093,5	17900,3	161,763
593	866,15	12462,31	8890,02	282,69	1263,6	80,2591	25123,3	17921,7	161,797
594	867,15	12477,09	8900,67	283,86	1259,8	80,2761	25153,1	17943,2	161,832
595	868,15	12491,86	8911,32	285,04	1256,1	80,2931	25182,8	17964,7	161,866
596	869,15	12506,64	8921,97	286,22	1252,4	80,3102	25212,6	17986,2	161,900
597	870,15	12521,42	8932,63	287,40	1248,6	80,3272	25242,4	18007,6	161,935
598	871,15	12536,20	8943,28	288,58	1244,9	80,3441	25272,2	18029,1	161,969
599	872,15	12550,98	8953,94	289,77	1241,3	80,3611	25302,0	18050,6	162,003
600	873,15	12565,76	8964,60	290,96	1237,6	80,3780	25331,8	18072,1	162,037
601	874,15	12580,55	8975,26	292,16	1233,9	80,3949	25361,6	18093,6	162,071
602	875,15	12595,33	8985,92	293,36	1230,3	80,4119	25391,4	18115,1	162,105
603	876,15	12610,12	8996,58	294,56	1226,7	80,4287	25421,2	18136,6	162,140
604	877,15	12624,91	9007,25	295,77	1223,1	80,4456	25451,1	18158,1	162,174
605	878,15	12639,70	9017,91	296,98	1219,5	80,4625	25480,9	18179,6	162,207
606	879,15	12654,49	9028,58	298,20	1215,9	80,4793	25510,7	18201,1	162,241
607	880,15	12669,28	9039,25	299,42	1212,3	80,4961	25540,5	18222,6	162,275
608	881,15	12684,07	9049,92	300,64	1208,7	80,5129	25570,3	18244,1	162,309
609	882,15	12698,87	9060,59	301,86	1205,2	80,5297	25600,2	18265,6	162,343

Hydrogen (*Continued*)

t	T	h	u	π_0	θ_0	s^0	H	U	S
610	883,15	12713,67	9071,26	303,09	1201,7	80,5465	25630,0	18287,1	162,377
611	884,15	12728,46	9081,93	304,33	1198,2	80,5632	25659,8	18308,6	162,411
612	885,15	12743,26	9092,61	305,56	1194,7	80,5799	25689,7	18330,2	162,444
613	886,15	12758,06	9103,29	306,80	1191,2	80,5966	25719,5	18351,7	162,478
614	887,15	12772,87	9113,96	308,05	1187,7	80,6133	25749,3	18373,2	162,512
615	888,15	12787,67	9124,64	309,30	1184,2	80,6300	25779,2	18394,7	162,545
616	889,15	12802,48	9135,32	310,55	1180,8	80,6467	25809,0	18416,3	162,579
617	890,15	12817,28	9146,01	311,80	1177,4	80,6633	25838,9	18437,8	162,612
618	891,15	12832,09	9156,69	313,06	1173,9	80,6799	25868,7	18459,3	162,646
619	892,15	12846,90	9167,38	314,33	1170,5	80,6966	25898,6	18480,9	162,679
620	893,15	12861,71	9178,06	315,59	1167,1	80,7131	25928,4	18502,4	162,713
621	894,15	12876,53	9188,75	316,86	1163,8	80,7297	25958,3	18524,0	162,746
622	895,15	12891,34	9199,44	318,14	1160,4	80,7463	25988,2	18545,5	162,780
623	896,15	12906,16	9210,13	319,42	1157,0	80,7628	26018,0	18567,1	162,813
624	897,15	12920,97	9220,83	320,70	1153,7	80,7793	26047,9	18588,6	162,846
625	898,15	12935,79	9231,52	321,99	1150,4	80,7959	26077,8	18610,2	162,880
626	899,15	12950,61	9242,22	323,28	1147,1	80,8123	26107,7	18631,8	162,913
627	900,15	12965,43	9252,91	324,57	1143,8	80,8288	26137,5	18653,3	162,946
628	901,15	12980,26	9263,61	325,87	1140,5	80,8453	26167,4	18674,9	162,979
629	902,15	12995,08	9274,31	327,17	1137,2	80,8617	26197,3	18696,5	163,012
630	903,15	13009,91	9285,01	328,48	1133,9	80,8781	26227,2	18718,0	163,045
631	904,15	13024,73	9295,72	329,79	1130,7	80,8946	26257,1	18739,6	163,079
632	905,15	13039,56	9306,42	331,10	1127,4	80,9109	26287,0	18761,2	163,112
633	906,15	13054,39	9317,13	332,42	1124,2	80,9273	26316,9	18782,8	163,145
634	907,15	13069,23	9327,84	333,74	1121,0	80,9437	26346,8	18804,4	163,178
635	908,15	13084,06	9338,55	335,06	1117,8	80,9600	26376,7	18825,9	163,211
636	909,15	13098,90	9349,26	336,39	1114,6	80,9764	26406,6	18847,5	163,243
637	910,15	13113,73	9359,97	337,72	1111,4	80,9927	26436,5	18869,1	163,276
638	911,15	13128,57	9370,68	339,06	1108,3	81,0090	26466,4	18890,7	163,309
639	912,15	13143,41	9381,40	340,40	1105,1	81,0252	26496,3	18912,3	163,342
640	913,15	13158,25	9392,12	341,75	1102,0	81,0415	26526,2	18933,9	163,375
641	914,15	13173,09	9402,83	343,10	1098,8	81,0577	26556,2	18955,6	163,408
642	915,15	13187,94	9413,55	344,45	1095,7	81,0740	26586,1	18977,2	163,440
643	916,15	13202,78	9424,28	345,81	1092,6	81,0902	26616,0	18998,8	163,473
644	917,15	13217,63	9435,00	347,17	1089,5	81,1064	26646,0	19020,4	163,506
645	918,15	13232,48	9445,72	348,53	1086,4	81,1226	26675,9	19042,0	163,538
646	919,15	13247,33	9456,45	349,90	1083,4	81,1387	26705,8	19063,6	163,571
647	920,15	13262,18	9467,18	351,27	1080,3	81,1549	26735,8	19085,3	163,603
648	921,15	13277,04	9477,91	352,65	1077,2	81,1710	26765,7	19106,9	163,636
649	922,15	13291,89	9488,64	354,03	1074,2	81,1871	26795,7	19128,5	163,668
650	923,15	13306,75	9499,37	355,42	1071,2	81,2032	26825,6	19150,2	163,701
651	924,15	13321,61	9510,10	356,80	1068,2	81,2193	26855,6	19171,8	163,733
652	925,15	13336,47	9520,84	358,20	1065,2	81,2354	26885,5	19193,4	163,766
653	926,15	13351,33	9531,58	359,59	1062,2	81,2514	26915,5	19215,1	163,798
654	927,15	13366,19	9542,32	361,00	1059,2	81,2675	26945,4	19236,7	163,830
655	928,15	13381,06	9553,06	362,40	1056,2	81,2835	26975,4	19258,4	163,863
656	929,15	13395,92	9563,80	363,81	1053,3	81,2995	27005,4	19280,0	163,895
657	930,15	13410,79	9574,54	365,22	1050,3	81,3155	27035,3	19301,7	163,927
658	931,15	13425,66	9585,29	366,64	1047,4	81,3315	27065,3	19323,4	163,959
659	932,15	13440,53	9596,03	368,06	1044,5	81,3475	27095,3	19345,0	163,992
660	933,15	13455,40	9606,78	369,49	1041,5	81,3634	27125,3	19366,7	164,024
661	934,15	13470,28	9617,53	370,92	1038,6	81,3793	27155,3	19388,4	164,056
662	935,15	13485,15	9628,28	372,35	1035,7	81,3952	27185,3	19410,0	164,088
663	936,15	13500,03	9639,03	373,79	1032,9	81,4111	27215,2	19431,7	164,120
664	937,15	13514,91	9649,79	375,23	1030,0	81,4270	27245,2	19453,4	164,152

252

Hydrogen (*Continued*)

t	T	h	u	π_0	θ_0	s^0	H	U	S
665	938,15	13529,79	9660,54	376,68	1027,1	81,4429	27275,2	19475,1	164,184
666	939,15	13544,67	9671,30	378,13	1024,3	81,4588	27305,2	19496,8	164,216
667	940,15	13559,55	9682,06	379,59	1021,4	81,4746	27335,2	19518,5	164,248
668	941,15	13574,44	9692,82	381,05	1018,6	81,4904	27365,3	19540,1	164,280
669	942,15	13589,33	9703,58	382,51	1015,8	81,5062	27395,3	19561,8	164,312
670	943,15	13604,21	9714,35	383,98	1013,0	81,5220	27425,3	19583,5	164,344
671	944,15	13619,10	9725,11	385,45	1010,2	81,5378	27455,3	19605,2	164,375
672	945,15	13634,00	9735,88	386,92	1007,4	81,5536	27485,3	19627,0	164,407
673	946,15	13648,89	9746,65	388,41	1004,6	81,5693	27515,3	19648,7	164,439
674	947,15	13663,78	9757,42	389,89	1001,9	81,5850	27545,4	19670,4	164,471
675	948,15	13678,68	9768,19	391,38	999,14	81,6008	27575,4	19692,1	164,502
676	949,15	13693,58	9778,97	392,87	996,39	81,6165	27605,4	19713,8	164,534
677	950,15	13708,48	9789,74	394,37	993,65	81,6322	27635,5	19735,5	164,566
678	951,15	13723,38	9800,52	395,87	990,93	81,6478	27665,5	19757,3	164,597
679	952,15	13738,28	9811,30	397,38	988,21	81,6635	27695,6	19779,0	164,629
680	953,15	13753,19	9822,08	398,89	985,50	81,6791	27725,6	19800,7	164,660
681	954,15	13768,09	9832,86	400,40	982,80	81,6948	27755,7	19822,5	164,692
682	955,15	13783,00	9843,65	401,92	980,12	81,7104	27785,7	19844,2	164,723
683	956,15	13797,91	9854,43	403,44	977,44	81,7260	27815,8	19865,9	164,755
684	957,15	13812,82	9865,22	404,97	974,77	81,7416	27845,8	19887,7	164,786
685	958,15	13827,74	9876,01	406,50	972,11	81,7572	27875,9	19909,4	164,818
686	959,15	13842,65	9886,80	408,04	969,46	81,7727	27906,0	19931,2	164,849
687	960,15	13857,57	9897,59	409,58	966,82	81,7883	27936,0	19952,9	164,880
688	961,15	13872,49	9908,38	401,13	964,19	81,8038	27966,1	19974,7	164,912
689	962,15	13887,41	9919,18	412,68	961,57	81,8193	27996,2	19996,5	164,943
690	963,15	13902,33	9929,97	414,23	958,96	81,8348	28026,3	20018,2	164,974
691	964,15	13917,25	9940,77	415,79	956,36	81,8503	28056,3	20040,0	165,005
692	965,15	13932,17	9951,57	417,35	953,76	81,8658	28086,4	20061,8	165,036
693	966,15	13947,10	9962,37	418,92	951,18	81,8812	28116,5	20083,5	165,068
694	967,15	13962,03	9973,18	420,49	948,61	81,8967	28146,6	20105,3	165,099
695	968,15	13976,96	9983,98	422,06	946,04	81,9121	28176,7	20127,1	165,130
696	969,15	13991,89	9994,79	423,65	943,49	81,9275	28206,8	20148,9	165,161
697	970,15	14006,82	10005,60	425,23	940,94	81,9429	28236,9	20170,7	165,192
698	971,15	14021,76	10016,41	426,82	938,40	81,9583	28267,0	20192,5	165,223
699	972,15	14036,69	10027,22	428,41	935,87	81,9737	28297,1	20214,3	165,254
700	973,15	14051,63	10038,04	430,01	933,35	81,9890	28327,2	20236,1	165,285
701	974,15	14066,57	10048,85	431,61	930,84	82,0044	28357,4	20257,9	165,316
702	975,15	14081,51	10059,67	433,22	928,34	82,0197	28387,5	20279,7	165,347
703	976,15	14096,46	10070,49	434,83	925,85	82,0350	28417,6	20301,5	165,378
704	977,15	14111,40	10081,31	436,45	923,37	82,0503	28447,7	20323,3	165,408
705	978,15	14126,35	10092,13	438,07	920,89	82,0656	28477,9	20345,1	165,439
706	979,15	14141,29	10102,95	439,70	918,42	82,0809	28508,0	20366,9	165,470
707	980,15	14156,24	10113,78	441,33	915,97	82,0961	28538,1	20388,8	165,501
708	981,15	14171,20	10124,61	442,96	913,52	82,1114	28568,3	20410,6	165,532
709	982,15	14186,15	10135,43	444,60	911,08	82,1266	28598,4	20432,4	165,562
710	983,15	14201,10	10146,27	446,24	908,65	82,1418	28628,6	20454,3	165,593
711	984,15	14216,06	10157,10	447,89	906,22	82,1570	28658,7	20476,1	165,624
712	985,15	14231,02	10167,93	449,54	903,81	82,1722	28688,9	20497,9	165,654
713	986,15	14245,98	10178,77	451,20	901,40	82,1874	28719,0	20519,8	165,685
714	987,15	14260,94	10189,61	452,86	899,01	82,2026	28749,2	20541,6	165,715
715	988,15	14275,91	10200,45	454,53	896,62	82,2177	28779,4	20563,5	165,746
716	989,15	14290,87	10211,29	456,20	894,24	82,2329	28809,5	20585,3	165,777
717	990,15	14305,84	10222,13	457,88	891,86	82,2480	28839,7	20607,2	165,807
718	991,15	14320,81	10232,97	459,56	889,50	82,2631	28869,9	20629,1	165,837
719	992,15	14335,78	10243,82	461,24	887,14	82,2782	28900,1	20650,9	165,868

Hydrogen (*Continued*)

t	T	h	u	π_0	θ_0	s^0	H	U	S
720	993,15	14350,75	10254,67	462,93	884,80	82,2933	28930,3	20672,8	165,898
721	994,15	14365,72	10265,52	464,63	882,46	82,3083	28960,4	20694,7	165,929
722	995,15	14380,70	10276,37	466,33	880,13	82,3234	28990,6	20716,5	165,959
723	996,15	14395,68	10287,22	468,03	877,80	82,3384	29020,8	20738,4	165,989
724	997,15	14410,66	10298,08	469,74	875,49	82,3535	29051,0	20760,3	166,020
725	998,15	14425,64	10308,93	471,45	873,18	82,3685	29081,2	20782,2	166,050
726	999,15	14440,62	10319,79	473,17	870,88	82,3835	29111,4	20804,1	166,080
727	1000,15	14455,61	10330,65	474,89	868,59	82,3985	29141,6	20826,0	166,110
728	1001,15	14470,59	10341,51	476,62	866,31	82,4135	29171,8	20847,9	166,141
729	1002,15	14485,58	10352,38	478,35	864,03	82,4284	29202,1	20869,8	166,171
730	1003,15	14500,57	10363,24	480,09	861,76	82,4434	29232,3	20891,7	166,201
731	1004,15	14515,56	10374,11	481,83	859,50	82,4583	29262,5	20913,6	166,231
732	1005,15	14530,55	10384,98	483,58	857,25	82,4732	29292,7	20935,5	166,261
733	1006,15	14545,55	10395,85	485,33	855,01	82,4881	29323,0	20957,4	166,291
734	1007,15	14560,55	10406,72	487,09	852,77	82,5030	29353,2	20979,3	166,321
735	1008,15	14575,55	10417,60	488,85	850,54	82,5179	29383,4	21001,3	166,351
736	1009,15	14590,55	10428,47	490,62	848,32	82,5328	29413,7	21023,2	166,381
737	1010,15	14605,55	10439,35	492,39	846,11	82,5477	29443,9	21045,1	166,411
738	1011,15	14620,55	10450,23	494,16	843,91	82,5625	29474,2	21067,0	166,441
739	1012,15	14635,56	10461,11	495,94	841,71	82,5773	29504,4	21089,0	166,471
740	1013,15	14650,57	10472,00	497,73	839,52	82,5922	29534,7	21110,9	166,501
741	1014,15	14665,57	10482,88	499,52	837,33	82,6070	29564,9	21132,9	166,531
742	1015,15	14680,59	10493,77	501,31	835,16	82,6218	29595,2	21154,8	166,560
743	1016,15	14695,60	10504,66	503,11	832,99	82,6365	29625,4	21176,8	166,590
744	1017,15	14710,61	10515,55	504,92	830,83	82,6513	29655,7	21198,7	166,620
745	1018,15	14725,63	10526,44	506,73	828,68	82,6661	29686,0	21220,7	166,650
746	1019,15	14740,65	10537,33	508,54	826,53	82,6808	29716,3	21242,6	166,680
747	1020,15	14755,67	10548,23	510,36	824,39	82,6955	29746,5	21264,6	166,709
748	1021,15	14770,69	10559,13	512,19	822,26	82,7103	29776,8	21286,6	166,739
749	1022,15	14785,71	10570,03	514,02	820,14	82,7250	29807,1	21308,5	166,769
750	1023,15	14800,74	10580,93	515,85	818,02	82,7396	29837,4	21330,5	166,798
751	1024,15	14815,77	10591,83	517,69	815,91	82,7543	29867,7	21352,5	166,828
752	1025,15	14830,80	10602,74	519,53	813,81	82,7690	29898,0	21374,5	166,857
753	1026,15	14845,83	10613,64	521,38	811,71	82,7837	29928,3	21396,5	166,887
754	1027,15	14860,86	10624,55	523,24	809,62	82,7983	29958,6	21418,5	166,916
755	1028,15	14875,89	10635,46	525,10	807,54	82,8129	29988,9	21440,5	166,946
756	1029,15	14890,93	10646,37	526,96	805,47	82,8275	30019,2	21462,4	166,975
757	1030,15	14905,97	10657,29	528,83	803,40	82,8421	30049,5	21484,5	167,005
758	1031,15	14921,01	10668,20	530,70	801,34	82,8567	30079,9	21506,5	167,034
759	1032,15	14936,05	10679,12	532,58	799,28	82,8713	30110,2	21528,5	167,064
760	1033,15	14951,09	10690,04	534,47	797,24	82,8859	30140,5	21550,5	167,093
761	1034,15	14966,14	10700,96	536,36	795,20	82,9004	30170,8	21572,5	167,122
762	1035,15	14981,19	10711,88	538,25	793,17	82,9150	30201,2	21594,5	167,152
763	1036,15	14996,24	10722,81	540,15	791,14	82,9295	30231,5	21616,5	167,181
764	1037,15	15011,29	10733,73	542,06	789,12	82,9440	30261,9	21638,6	167,210
765	1038,15	15026,34	10744,66	543,97	787,11	82,9585	30292,2	21660,6	167,239
766	1039,15	15041,40	10755,59	545,88	785,10	82,9730	30322,6	21682,6	167,269
767	1040,15	15056,45	10766,53	547,80	783,10	82,9875	30352,9	21704,7	167,298
768	1041,15	15071,51	10777,46	549,73	781,11	83,0020	30383,3	21726,7	167,327
769	1042,15	15086,57	10788,40	551,66	779,12	83,0164	30413,6	21748,8	167,356
770	1043,15	15101,63	10799,33	553,59	777,15	83,0309	30444,0	21770,8	167,385
771	1044,15	15116,70	10810,27	555,54	775,17	83,0453	30474,4	21792,9	167,414
772	1045,15	15131,76	10821,21	557,48	773,21	83,0598	30504,7	21814,9	167,443
773	1046,15	15146,83	10832,16	559,43	771,25	83,0742	30535,1	21837,0	167,473
774	1047,15	15161,90	10843,10	561,39	769,29	83,0886	30565,5	21859,0	167,502

Hydrogen (*Continued*)

t	T	h	u	π_0	θ_0	s^0	H	U	S
775	1048,15	15176,97	10854,05	563,35	767,35	83,1029	30595,9	21881,1	167,531
776	1049,15	15192,04	10865,00	565,32	765,41	83,1173	30626,2	21903,2	167,560
777	1050,15	15207,12	10875,95	567,29	763,47	83,1317	30656,6	21925,3	167,588
778	1051,15	15222,20	10886,90	569,27	761,55	83,1460	30687,0	21947,3	167,617
779	1052,15	15237,27	10897,86	571,25	759,63	83,1604	30717,4	21969,4	167,646
780	1053,15	15252,35	10908,81	573,24	757,71	83,1747	30747,8	21991,5	167,675
781	1054,15	15267,44	10919,77	575,23	755,80	83,1890	30778,2	22013,6	167,704
782	1055,15	15282,52	10930,73	577,23	753,90	83,2033	30808,6	22035,7	167,733
783	1056,15	15297,61	10941,69	579,23	752,00	83,2176	30839,1	22057,8	167,762
784	1057,15	15312,70	10952,66	581,24	750,12	83,2319	30869,5	22079,9	167,790
785	1058,15	15327,79	10963,62	583,25	748,23	83,2461	30899,9	22102,0	167,819
786	1059,15	15342,88	10974,59	585,27	746,35	83,2604	30930,3	22124,1	167,848
787	1060,15	15357,97	10985,56	587,30	744,48	83,2746	30960,7	22146,2	167,877
788	1061,15	15373,07	10996,53	589,33	742,62	83,2889	30991,2	22168,3	167,905
789	1062,15	15388,16	11007,50	591,36	740,76	83,3031	31021,6	22190,5	167,934
790	1063,15	15403,26	11018,48	593,41	738,91	83,3173	31052,1	22212,6	167,963
791	1064,15	15418,36	11029,45	595,45	737,06	83,3315	31082,5	22234,7	167,991
792	1065,15	15433,47	11040,43	597,50	735,22	83,3457	31112,9	22256,9	168,020
793	1066,15	15448,57	11051,41	599,56	733,39	83,3599	31143,4	22279,0	168,048
794	1067,15	15463,68	11062,40	601,62	731,56	83,3740	31173,8	22301,1	168,077
795	1068,15	15478,79	11073,38	603,69	729,73	83,3882	31204,3	22323,3	168,106
796	1069,15	15493,90	11084,37	605,76	727,92	83,4023	31234,8	22345,4	168,134
797	1070,15	15509,01	11095,35	607,84	726,11	83,4165	31265,2	22367,6	168,163
798	1071,15	15524,13	11106,34	609,93	724,30	83,4306	31295,7	22389,7	168,191
799	1072,15	15539,24	11117,34	612,02	722,50	83,4447	31326,2	22411,9	168,219
800	1073,15	15554,36	11128,33	614,11	720,71	83,4588	31356,7	22434,0	168,248
801	1074,15	15569,48	11139,33	616,21	718,92	83,4729	31387,1	22456,2	168,276
802	1075,15	15584,60	11150,32	618,32	717,14	83,4869	31417,6	22478,4	168,305
803	1076,15	15599,73	11161,32	620,43	715,36	83,5010	31448,1	22500,6	168,333
804	1077,15	15614,85	11172,33	622,55	713,59	83,5150	31478,6	22522,7	168,361
805	1078,15	15629,98	11183,33	624,67	711,83	83,5291	31509,1	22544,9	168,390
806	1079,15	15645,11	11194,33	626,80	710,07	83,5431	31539,6	22567,1	168,418
807	1080,15	15660,24	11205,34	628,93	708,32	83,5571	31570,1	22589,3	168,446
808	1081,15	15675,37	11216,35	631,07	706,57	83,5711	31600,6	22611,5	168,474
809	1082,15	15690,51	11227,36	633,21	704,83	83,5851	31631,1	22633,7	168,503
810	1083,15	15705,65	11238,37	635,37	703,09	83,5991	31661,6	22655,9	168,531
811	1084,15	15720,79	11249,39	637,52	701,36	83,6131	31692,2	22678,1	168,559
812	1085,15	15735,93	11260,41	639,68	699,64	83,6270	31722,7	22700,3	168,587
813	1086,15	15751,07	11271,42	641,85	697,92	83,6410	31753,2	22722,5	168,615
814	1087,15	15766,21	11282,44	644,02	696,20	83,6549	31783,7	22744,7	168,643
815	1088,15	15781,36	11293,47	646,20	694,49	83,6688	31814,3	22767,0	168,671
816	1089,15	15796,51	11304,49	648,38	692,79	83,6827	31844,8	22789,2	168,699
817	1090,15	15811,66	11315,52	650,57	691,09	83,6967	31875,4	22811,4	168,727
818	1091,15	15826,81	11326,55	652,77	689,40	83,7105	31905,9	22833,6	168,755
819	1092,15	15841,97	11337,58	654,97	687,71	83,7244	31936,5	22855,9	168,783
820	1093,15	15857,12	11348,61	657,18	686,03	83,7383	31967,0	22878,1	168,811
821	1094,15	15872,28	11359,64	659,39	684,36	83,7522	31997,6	22900,4	168,839
822	1095,15	15887,44	11370,68	661,61	682,69	83,7660	32028,1	22922,6	168,867
823	1096,15	15902,61	11381,72	663,83	681,02	83,7798	32058,7	22944,9	168,895
824	1097,15	15917,77	11392,76	666,06	679,36	83,7937	32089,3	22967,1	168,923
825	1098,15	15932,94	11403,80	668,30	677,71	83,8075	32119,8	22989,4	168,951
826	1099,15	15948,10	11414,84	670,54	676,06	83,8213	32150,4	23011,6	168,979
827	1100,15	15963,27	11425,89	672,78	674,41	83,8351	32181,0	23033,9	169,007
828	1101,15	15978,45	11436,94	675,04	672,77	83,8489	32211,6	23056,2	169,034
829	1102,15	15993,62	11447,99	677,29	671,14	83,8627	32242,2	23078,5	169,062

Hydrogen (*Continued*)

t	T	h	u	π.	θ.	s°	H	U	S
830	1103,15	16008,80	11459.04	679,56	669,51	83,8764	32272,8	23100,7	169,090
831	1104,15	16023,97	11470,09	681,83	667,88	83,8902	32303,4	23123,0	169,118
832	1105,15	16039,15	11481,15	684,10	666,27	83,9039	32334,0	23145,3	169,145
833	1106,15	16054,34	11492,20	686,38	664,65	83,9176	32364,6	23167,6	169,173
834	1107,15	16069,52	11503,26	688,67	663,04	83,9314	32395,2	23189,9	169,201
835	1108,15	16084,70	11514,32	690,96	661,44	83,9451	32425,8	23212,2	169,228
836	1109,15	16099,89	11525,39	693,26	659,84	83,9588	32456,4	23234,5	169,256
837	1110,15	16115,08	11536,45	695,57	658,25	83,9725	32487,0	23256,8	169,283
838	1111,15	16130,27	11547,52	697,88	656,66	83,9861	32517,7	23279,1	169,311
839	1112,15	16145,47	11558,59	700,19	655,08	83,9998	32548,3	23301,4	169,339
840	1113,15	16160,66	11569,66	702,52	653,50	84,0135	32578,9	23323,7	169,366
841	1114,15	16175,86	11580,73	704,85	651,92	84,0271	32609,6	23346,1	169,394
842	1115,15	16191,06	11591,81	707,18	650,36	84,0407	32640,2	23368,4	169,421
843	1116,15	16206,26	11602,89	709,52	648,79	84,0544	32670,8	23390,7	169,449
844	1117,15	16221,46	11613,96	711,87	647,23	84,0680	32701,5	23413,1	169,476
845	1118,15	16236,67	11625,04	714,22	645,68	84,0816	32732,2	23435,4	169,503
846	1119,15	16251,88	11636,13	716,58	644,13	84,0952	32762,8	23457,7	169,531
847	1120,15	16267,09	11647,21	718,94	642,59	84,1088	32793,5	23480,1	169,558
848	1121,15	16282,30	11658,30	721,31	641,05	84,1223	32824,1	23502,4	169,586
849	1122,15	16297,51	11669,39	723,69	639,51	84,1359	32854,8	23524,8	169,613
850	1123,15	16312,72	11680,48	726,07	637,98	84,1495	32885,5	23547,1	169,640
851	1124,15	16327,94	11691,57	728,46	636,46	84,1630	32916,2	23569,5	169,668
852	1125,15	16343,16	11702,67	730,85	634,94	84,1765	32946,8	23591,9	169,695
853	1126,15	16358,38	11713,76	733,25	633,42	84,1901	32977,5	23614,2	169,722
854	1127,15	16373,61	11724,86	735,66	631,91	84,2036	33008,2	23636,6	169,749
855	1128,15	16388,83	11735,96	738,07	630,40	84,2171	33038,9	23659,0	169,777
856	1129,15	16404,06	11747,07	740,49	628,90	84,2306	33069,6	23681,4	169,804
857	1130,15	16419,29	11758,17	742,91	627,40	84,2440	33100,3	23703,8	169,831
858	1131,15	16434,52	11769,28	745,34	625,91	84,2575	33131,0	23726,2	169,858
859	1132,15	16449,75	11780,39	747,78	624,42	84,2710	33161,7	23748,5	169,885
860	1133,15	16464,98	11791,50	750,22	622,94	84,2844	33192,4	23770,9	169,912
861	1134,15	16480,22	11802,61	752,67	621,46	84,2979	33223,1	23793,3	169,939
862	1135,15	16495,46	11813,72	755,13	619,99	84,3113	33253,9	23815,8	169,967
863	1136,15	16510,70	11824,84	757,59	618,52	84,3247	33284,6	23838,2	169,994
864	1137,15	16525,94	11835,96	760,05	617,05	84,3381	33315,3	23860,6	170,021
865	1138,15	16541,19	11847,08	762,53	615,59	84,3515	33346,0	23883,0	170,048
866	1139,15	16556,44	11858,20	765,01	614,13	84,3649	33376,8	23905,4	170,075
867	1140,15	16571,68	11869,33	767,49	612,68	84,3783	33407,5	23927,8	170,102
868	1141,15	16586,94	11880,45	769,99	611,23	84,3917	33438,3	23950,3	170,129
869	1142,15	16602,19	11891,58	772,48	609,79	84,4050	33469,0	23972,7	170,155
870	1143,15	16617,44	11902,71	774,99	608,35	84,4184	33499,8	23995,2	170,182
871	1144,15	16632,70	11913,84	777,50	606,92	84,4317	33530,5	24017,6	170,209
872	1145,15	16647,96	11924,98	780,02	605,49	84,4450	33561,3	24040,0	170,236
873	1146,15	16663,22	11936,11	782,54	604,06	84,4584	33592,1	24062,5	170,263
874	1147,15	16678,48	11947,25	785,07	602,64	84,4717	33622,8	24084,9	170,290
875	1148,15	16693,75	11958,39	787,61	601,23	84,4850	33653,6	24107,4	170,317
876	1149,15	16709,01	11969,54	790,15	599,81	84,4983	33684,4	24129,9	170,343
877	1150,15	16724,28	11980,68	792,70	598,41	84,5115	33715,2	24152,3	170,370
878	1151,15	16739,55	11991,83	795,25	597,00	84,5248	33745,9	24174,8	170,397
879	1152,15	16754,83	12002,98	797,81	595,60	84,5381	33776,7	24197,3	170,424
880	1153,15	16770,10	12014,13	800,38	594,21	84,5513	33807,5	24219,8	170,450
881	1154,15	16785,38	12025,28	802,95	592,82	84,5646	33838,3	24242,2	170,477
882	1155,15	16800,66	12036,43	805,53	591,43	84,5778	33869,1	24264,7	170,504
883	1156,15	16815,94	12047,59	808,12	590,05	84,5910	33899,9	24287,2	170,530
884	1157,15	16831,22	12058,75	810,71	588,67	84,6042	33930,7	24309,7	170,557

Hydrogen (*Continued*)

t	T	h	u	π₀	θ₀	s⁰	H	U	S
				π_0	θ_0	s^0			
885	1158,15	16846,51	12069,91	813,31	587,29	84,6174	33961,5	24332,2	170,584
886	1159,15	16861,79	12081,07	815,92	585,92	84,6306	33992,4	24354,7	170,610
887	1160,15	16877,08	12092,24	818,53	584,56	84,6438	34023,2	24377,2	170,637
888	1161,15	16892,37	12103,40	821,15	583,19	84,6570	34054,0	24399,7	170,663
889	1162,15	16907,67	12114,57	823,78	581,84	84,6702	34084,8	24422,3	170,690
890	1163,15	16922,96	12125,74	826,41	580,48	84,6833	34115,7	24444,8	170,716
891	1164,15	16938,26	12136,92	829,05	579,13	84,6965	34146,5	24467,3	170,743
892	1165,15	16953,56	12148,09	831,69	577,79	84,7096	34177,4	24489,8	170,769
893	1166,15	16968,86	12159,27	834,34	576,44	84,7227	34208,2	24512,4	170,796
894	1167,15	16984,16	12170,45	837,00	575,11	84,7358	34239,1	24534,9	170,822
895	1168,15	16999,47	12181,63	839,66	573,77	84,7490	34269,9	24557,4	170,849
896	1169,15	17014,78	12192,81	842,34	572,44	84,7621	34300,8	24580,0	170,875
897	1170,15	17030,08	12204,00	845,01	571,12	84,7751	34331,6	24602,5	170,902
898	1171,15	17045,40	12215,18	847,70	569,80	84,7882	34362,5	24625,1	170,928
899	1172,15	17060,71	12226,37	850,39	568,48	84,8013	34393,4	24647,6	170,954
900	1173,15	17076,02	12237,56	853,08	567,16	84,8144	34424,2	24670,2	170,981
901	1174,15	17091,34	12248,76	855,79	565,85	84,8274	34455,1	24692,8	171,007
902	1175,15	17106,66	12259,95	858,50	564,55	84,8404	34486,0	24715,3	171,033
903	1176,15	17121,98	12271,15	861,22	563,25	84,8535	34516,9	24737,9	171,060
904	1177,15	17137,31	12282,35	863,94	561,95	84,8665	34547,8	24760,5	171,086
905	1178,15	17152,63	12293,55	866,67	560,65	84,8795	34578,7	24783,1	171,112
906	1179,15	17167,96	12304,75	869,41	559,36	84,8925	34609,6	24805,6	171,138
907	1180,15	17183,29	12315,96	872,15	558,08	84,9055	34640,5	24828,2	171,164
908	1181,15	17198,62	12327,16	874,90	556,79	84,9185	34671,4	24850,8	171,191
909	1182,15	17213,95	12338,37	877,66	555,51	84,9315	34702,3	24873,4	171,217
910	1183,15	17229,29	12349,58	880,42	554,24	84,9444	34733,2	24896,0	171,243
911	1184,15	17244,63	12360,80	883,19	552,97	84,9574	34764,1	24918,6	171,269
912	1185,15	17259,97	12372,01	885,97	551,70	84,9704	34795,1	24941,2	171,295
913	1186,15	17275,31	12383,23	888,75	550,43	84,9833	34826,0	24963,9	171,321
914	1187,15	17290,65	12394,45	891,54	549,17	84,9962	34856,9	24986,5	171,347
915	1188,15	17306,00	12405,67	894,34	547,92	85,0091	34887,9	25009,1	171,373
916	1189,15	17321,35	12416,90	897,15	546,66	85,0221	34918,8	25031,7	171,399
917	1190,15	17336,70	12428,12	899,96	545,42	85,0350	34949,7	25054,3	171,425
918	1191,15	17352,05	12439,35	902,78	544,17	85,0479	34980,7	25077,0	171,451
919	1192,15	17367,40	12450,58	905,60	542,93	85,0607	35011,6	25099,6	171,477
920	1193,15	17382,76	12461,81	908,43	541,69	85,0736	35042,6	25122,3	171,503
921	1194,15	17398,12	12473,05	911,27	540,45	85,0865	35073,6	25144,9	171,529
922	1195,15	17413,48	12484,28	914,12	539,22	85,0993	35104,5	25167,6	171,555
923	1196,15	17428,84	12495,52	916,97	538,00	85,1122	35135,5	25190,2	171,581
924	1197,15	17444,21	12506,76	919,83	536,77	85,1250	35166,5	25212,9	171,607
925	1198,15	17459,57	12518,00	922,69	535,55	85,1379	35197,5	25235,5	171,633
926	1199,15	17474,94	12529,25	925,57	534,34	85,1507	35228,4	25258,2	171,659
927	1200,15	17490,31	12540,49	928,45	533,12	85,1635	35259,4	25280,9	171,684
928	1201,15	17505,69	12551,74	931,33	531,91	85,1763	35290,4	25303,6	171,710
929	1202,15	17521,06	12562,99	934,23	530,71	85,1891	35321,4	25326,2	171,736
930	1203,15	17536,44	12574,24	937,13	529,50	85,2019	35352,4	25348,9	171,762
931	1204,15	17551,82	12585,50	940,03	528,30	85,2146	35383,4	25371,6	171,788
932	1205,15	17567,20	12596,76	942,95	527,11	85,2274	35414,4	25394,3	171,813
933	1206,15	17582,58	12608,01	945,87	525,92	85,2402	35445,4	25417,0	171,839
934	1207,15	17597,96	12619,27	948,80	524,73	85,2529	35476,4	25439,7	171,865
935	1208,15	17613,35	12630,54	951,74	523,54	85,2657	35507,5	25462,4	171,890
936	1209,15	17628,74	12641,80	954,68	522,36	85,2784	35538,5	25485,1	171,916
937	1210,15	17644,13	12653,07	957,63	521,18	85,2911	35569,5	25507,8	171,942
938	1211,15	17659,53	12664,34	960,58	520,01	85,3038	35600,5	25530,5	171,967
939	1212,15	17674,92	12675,61	963,55	518,84	85,3165	35631,6	25553,3	171,993

t	T	h	u	π_0	θ_0	s°	H	U	S
940	1213,15	17690,32	12686,88	966,52	517,67	85,3292	35662,6	25576,0	172,019
941	1214,15	17705,72	12698,16	969,50	516,50	85,3419	35693,7	25598,7	172,044
942	1215,15	17721,12	12709,44	972,48	515,34	85,3546	35724,7	25621,5	172,070
943	1216,15	17736,52	12720,71	975,48	514,18	85,3673	35755,8	25644,2	172,095
944	1217,15	17751,93	12732,00	978,47	513,03	85,3799	35786,8	25666,9	172,121
945	1218,15	17767,34	12743,28	981,48	511,88	85,3926	35817,9	25689,7	172,146
946	1219,15	17782,75	12754,57	984,50	510,73	85,4052	35849,0	25712,4	172,172
947	1220,15	17798,16	12765,85	987,52	509,58	85,4179	35880,0	25735,2	172,197
948	1221,15	17813,57	12777,14	990,54	508,44	85,4305	35911,1	25758,0	172,223
949	1222,15	17828,99	12788,43	993,58	507,30	85,4431	35942,2	25780,7	172,248
950	1223,15	17844,41	12799,73	996,62	506,17	85,4557	35973,3	25803,5	172,274
951	1224,15	17859,83	12811,02	999,67	505,04	85,4683	36004,3	25826,3	172,299
952	1225,15	17875,25	12822,32	1002,7	503,91	85,4809	36035,4	25849,0	172,324
953	1226,15	17890,68	12833,62	1005,7	502,78	85,4935	36066,5	25871,8	172,350
954	1227,15	17906,10	12844,93	1008,8	501,66	85,5061	36097,6	25894,6	1/2,375
955	1228,15	17921,53	12856,23	1011,9	500,54	85,5187	36128,7	25917,4	172,400
956	1229,15	17936,96	12867,54	1015,0	499,43	85,5312	36159,8	25940,2	172,426
957	1230,15	17952,39	12878,84	1018,1	498,32	85,5438	36190,9	25963,0	172,451
958	1231,15	17967,83	12890,15	1021,2	497,21	85,5563	36222,1	25985,8	172,476
959	1232,15	17983,27	12901,47	1024,3	496,10	85,5688	36253,2	26008,6	172,502
960	1233,15	17998,70	12912,78	1027,4	495,00	85,5814	36284,3	26031,4	172,527
961	1234,15	18014,15	12924,10	1030,5	493,90	85,5939	36315,4	26054,2	172,552
962	1235,15	18029,59	12935,42	1033,7	492,80	85,6064	36346,6	26077,0	172,577
963	1236,15	18045,03	12946,74	1036,8	491,71	85,6189	36377,7	26099,8	172,603
964	1237,15	18060,48	12958,06	1039,9	490,62	85,6314	36408,8	26122,7	172,628
965	1238,15	18075,93	12969,39	1043,1	489,53	85,6439	36440,0	26145,5	172,653
966	1239,15	18091,38	12980,71	1046,3	488,45	85,6563	36471,1	26168,3	172,678
967	1240,15	18106,84	12992,04	1049,4	487,36	85,6688	36502,3	26191,2	172,703
968	1241,15	18122,29	13003,38	1052,6	486,29	85,6813	36533,5	26214,0	172,728
969	1242,15	18137,75	13014,71	1055,8	485,21	85,6937	36564,6	26236,9	172,753
970	1243,15	18153,21	13026,04	1059,0	484,14	85,7062	36595,8	26259,7	172,778
971	1244,15	18168,67	13037,38	1062,2	483,07	85,7186	36627,0	26282,6	172,804
972	1245,15	18184,14	13048,72	1065,4	482,00	85,7310	36658,1	26305,4	172,829
973	1246,15	18199,60	13060,06	1068,6	480,94	85,7434	36689,3	26328,3	172,854
974	1247,15	18215,07	13071,41	1071,8	479,88	85,7558	36720,5	26351,2	172,879
975	1248,15	18230,54	13082,75	1075,0	478,82	85,7682	36751,7	26374,0	172,904
976	1249,15	18246,01	13094,10	1078,3	477,77	85,7806	36782,9	26396,9	172,929
977	1250,15	18261,49	13105,45	1081,5	476,72	85,7930	36814,1	26419,8	172,954
978	1251,15	18276,97	13116,81	1084,8	475,67	85,8054	36845,3	26442,7	172,979
979	1252,15	18292,44	13128,16	1088,0	474,63	85,8178	36876,5	26465,6	173,008
980	1253,15	18307,93	13139,52	1091,3	473,58	85,8301	36907,7	26488,5	173,028
981	1254,15	18323,41	13150,87	1094,6	472,54	85,8425	36938,9	26511,4	173,053
982	1255,15	18338,89	13162,24	1097,8	471,51	85,8548	36970,1	26534,3	173,078
983	1256,15	18354,38	13173,60	1101,1	470,48	85,8671	37001,3	26557,2	173,103
984	1257,15	18369,87	13184,96	1104,4	469,44	85,8795	37032,6	26580,1	173,128
985	1258,15	18385,36	13196,33	1107,7	468,42	85,8918	37063,8	26603,0	173,153
986	1259,15	18400,86	13207,70	1111,0	467,39	85,9041	37095,0	26625,9	173,178
987	1260,15	18416,35	13219,07	1114,3	466,37	85,9164	37126,3	26648,9	173,202
988	1261,15	18431,85	13230,44	1117,7	465,35	85,9287	37157,5	26671,8	173,227
989	1262,15	18447,35	13241,82	1121,0	464,33	85,9410	37188,7	26694,7	173,252
990	1263,15	18462,85	13253,20	1124,4	463,32	85,9533	37220,0	26717,7	173,277
991	1264,15	18478,35	13264,58	1127,7	462,31	85,9655	37251,3	26740,6	173,301
992	1265,15	18493,86	13275,96	1131,1	461,30	85,9778	37282,5	26763,5	173,326
993	1266,15	18509,37	13287,34	1134,4	460,30	85,9900	37313,8	26786,5	173,351
994	1267,15	18524,88	13298,73	1137,8	459,30	86,0023	37345,0	26809,4	173,375

Hydrogen (*Continued*)

t	T	h	u	π_0	θ_0	$s^\bullet$	H	U	S
995	1268,15	18540,39	13310,12	1141,2	458,30	86,0145	37376,3	26832,4	173,400
996	1269,15	18555,91	13321,51	1144,6	457,30	86,0267	37407,6	26855,4	173,425
997	1270,15	18571,42	13332,90	1148,0	456,31	86,0390	37438,9	26878,3	173,449
998	1271,15	18586,94	13344,29	1151,4	455,32	86,0512	37470,2	26901,3	173,474
999	1272,15	18602,46	13355,69	1154,8	454,33	86,0634	37501,4	26924,3	173,499
1000	1273,15	18617,99	13367,09	1158,2	453,34	86,0756	37532,7	26947,3	173,523
1001	1274,15	18633,51	13378,49	1161,6	452,36	86,0878	37564,0	26970,2	173,548
1002	1275,15	18649,04	13389,89	1165,1	451,38	86,1000	37595,3	26993,2	173,572
1003	1276,15	18664,57	13401,30	1168,5	450,40	86,1121	37626,6	27016,2	173,597
1004	1277,15	18680,10	13412,71	1172,0	449,43	86,1243	37658,0	27039,2	173,621
1005	1278,15	18695,63	13424,12	1175,4	448,45	86,1365	37689,3	27062,2	173,646
1006	1279,15	18711,17	13435,53	1178,9	447,48	86,1486	37720,6	27085,2	173,670
1007	1280,15	18726,71	13446,94	1182,4	446,52	86,1607	37751,9	27108,2	173,695
1008	1281,15	18742,25	13458,36	1185,9	445,55	86,1729	37783,2	27131,2	173,719
1009	1282,15	18757,79	13469,77	1189,3	444,59	86,1850	37814,6	27154,3	173,744
1010	1283,15	18773,33	13481,19	1192,8	443,63	86,1971	37845,9	27177,3	173,768
1011	1284,15	18788,88	13492,62	1196,4	442,68	86,2092	37877,3	27200,3	173,793
1012	1285,15	18804,43	13504,04	1199,9	441,72	86,2213	37908,6	27223,3	173,817
1013	1286,15	18819,98	13515,47	1203,4	440,77	86,2334	37939,9	27246,4	173,841
1014	1287,15	18835,53	13526,89	1206,9	439,83	86,2455	37971,3	27269,4	173,866
1015	1288,15	18851,09	13538,32	1210,5	438,88	86,2576	38002,7	27292,4	173,890
1016	1289,15	18866,64	13549,76	1214,0	437,94	86,2697	38034,0	27315,5	173,914
1017	1290,15	18882,20	13561,19	1217,6	437,00	86,2817	38065,4	27338,5	173,939
1018	1291,15	18897,76	13572,63	1221,1	436,06	86,2938	38096,8	27361,6	173,963
1019	1292,15	18913,33	13584,07	1224,7	435,12	86,3058	38128,1	27384,7	173,987
1020	1293,15	18928,89	13595,51	1228,3	434,19	86,3179	38159,5	27407,7	174,012
1021	1294,15	18944,46	13606,95	1231,9	433,26	86,3299	38190,9	27430,8	174,036
1022	1295,15	18960,03	13618,40	1235,5	432,33	86,3419	38222,3	27453,9	174,060
1023	1296,15	18975,60	13629,84	1239,1	431,41	86,3540	38253,7	27476,9	174,084
1024	1297,15	18991,17	13641,29	1242,7	430,48	86,3660	38285,1	27500,0	174,109
1025	1298,15	19006,75	13652,75	1246,3	429,56	86,3780	38316,5	27523,1	174,133
1026	1299,15	19022,33	13664,20	1249,9	428,65	86,3900	38347,9	27546,2	174,157
1027	1300,15	19037,91	13675,65	1253,6	427,73	86,4020	38379,3	27569,3	174,181
1028	1301,15	19053,49	13687,11	1257,2	426,82	86,4139	38410,7	27592,4	174,205
1029	1302,15	19069,07	13698,57	1260,9	425,91	86,4259	38442,1	27615,5	174,229
1030	1303,15	19084,66	13710,04	1264,6	425,00	86,4379	38473,5	27638,6	174,254
1031	1304,15	19100,25	13721,50	1268,2	424,10	86,4498	38505,0	27661,7	174,278
1032	1305,15	19115,84	13732,97	1271,9	423,19	86,4618	38536,4	27684,8	174,302
1033	1306,15	19131,43	13744,43	1275,6	422,29	86,4737	38567,8	27708,0	174,326
1034	1307,15	19147,03	13755,91	1279,3	421,39	86,4857	38599,3	27731,1	174,350
1035	1308,15	19162,63	13767,38	1283,0	420,50	86,4976	38630,7	27754,2	174,374
1036	1309,15	19178,23	13778,85	1286,7	419,61	86,5095	38662,2	27777,3	174,398
1037	1310,15	19193,83	13790,33	1290,4	418,72	86,5214	38693,6	27800,5	174,422
1038	1311,15	19209,43	13801,81	1294,2	417,83	86,5333	38725,1	27823,6	174,446
1039	1312,15	19225,04	13813,29	1297,9	416,94	86,5452	38756,5	27846,8	174,470
1040	1313,15	19240,64	13824,77	1301,6	416,06	86,5571	38788,0	27869,9	174,494
1041	1314,15	19256,25	13836,26	1305,4	415,18	86,5690	38819,5	27893,1	174,518
1042	1315,15	19271,87	13847,75	1309,2	414,30	86,5809	38850,9	27916,2	174,542
1043	1316,15	19287,48	13859,24	1312,9	413,42	86,5927	38882,4	27939,4	174,566
1044	1317,15	19303,10	13870,73	1316,7	412,55	86,6046	38913,9	27962,6	174,590
1045	1318,15	19318,71	13882,22	1320,5	411,68	86,6165	38945,4	27985,7	174,614
1046	1319,15	19334,34	13893,72	1324,3	410,81	86,6283	38976,9	28008,9	174,637
1047	1320,15	19349,96	13905,22	1328,1	409,94	86,6401	39008,4	28032,1	174,661
1048	1321,15	19365,58	13916,72	1331,9	409,08	86,6520	39039,9	28055,3	174,685
1049	1322,15	19381,21	13928,22	1335,8	408,21	86,6638	39071,4	28078,5	174,709

Hydrogen (*Continued*)

t	T	h	u	π_0	θ_0	s^0	H	U	S
1050	1323,15	19396,84	13939,73	1339,6	407,35	86,6756	39102,9	28101,7	174,733
1051	1324,15	19412,47	13951,23	1343,4	406,49	86,6874	39134,4	28124,9	174,757
1052	1325,15	19428,10	13962,74	1347,3	405,64	86,6992	39165,9	28148,1	174,780
1053	1326,15	19443,74	13974,25	1351,1	404,79	86,7110	39197,4	28171,3	174,804
1054	1327,15	19459,38	13985,77	1355,0	403,94	86,7228	39228,9	28194,5	174,828
1055	1328,15	19475,02	13997,28	1358,9	403,09	86,7346	39260,5	28217,7	174,852
1056	1329,15	19490,66	14008,80	1362,8	402,24	86,7464	39292,0	28240,9	174,875
1057	1330,15	19506,30	14020,32	1366,7	401,40	86,7581	39323,5	28264,1	174,899
1058	1331,15	19521,95	14031,84	1370,6	400,55	86,7699	39355,1	28287,4	174,923
1059	1332,15	19537,60	14043,37	1374,5	399,72	86,7816	39386,6	28310,6	174,947
1060	1333,15	19553,25	14054,89	1378,4	398,88	86,7934	39418,2	28333,8	174,970
1061	1334,15	19568,90	14066,42	1382,3	398,04	86,8051	39449,7	28357,1	174,994
1062	1335,15	19584,56	14077,95	1386,3	397,21	86,8169	39481,3	28380,3	175,018
1063	1336,15	19600,21	14089,48	1390,2	396,38	86,8286	39512,9	28403,6	175,041
1064	1337,15	19615,87	14101,02	1394,2	395,55	86,8403	39544,4	28426,8	175,065
1065	1338,15	19631,53	14112,56	1398,1	394,72	86,8520	39576,0	28450,1	175,088
1066	1339,15	19647,20	14124,09	1402,1	393,90	86,8637	39607,6	28473,3	175,112
1067	1340,15	19662,86	14135,64	1406,1	393,08	86,8754	39639,1	28496,6	175,136
1068	1341,15	19678,53	14147,18	1410,1	392,26	86,8871	39670,7	28519,9	175,159
1069	1342,15	19694,20	14158,72	1414,1	391,44	86,8988	39702,3	28543,1	175,183
1070	1343,15	19709,87	14170,27	1418,1	390,63	86,9104	39733,9	28566,4	175,206
1071	1344,15	19725,54	14181,82	1422,1	389,81	86,9221	39765,5	28589,7	175,230
1072	1345,15	19741,22	14193,37	1426,1	389,00	86,9338	39797,1	28613,0	175,253
1073	1346,15	19756,90	14204,93	1430,1	388,19	86,9454	39828,7	28636,3	175,277
1074	1347,15	19772,58	14216,48	1434,2	387,38	86,9570	39860,3	28659,6	175,300
1075	1348,15	19788,26	14228,04	1438,2	386,58	86,9687	39891,9	28682,9	175,324
1076	1349,15	19803,95	14239,60	1442,3	385,78	86,9803	39923,6	28706,2	175,347
1077	1350,15	19819,63	14251,16	1446,4	384,98	86,9919	39955,2	28729,5	175,371
1078	1351,15	19835,32	14262,73	1450,5	384,18	87,0036	39986,8	28752,8	175,394
1079	1352,15	19851,01	14274,30	1454,5	383,38	87,0152	40018,5	28776,1	175,417
1080	1353,15	19866,71	14285,86	1458,6	382,59	87,0268	40050,1	28799,4	175,441
1081	1354,15	19882,40	14297,44	1462,7	381,80	87,0384	40081,7	28822,8	175,464
1082	1355,15	19898,10	14309,01	1466,9	381,01	87,0499	40113,4	28846,1	175,487
1083	1356,15	19913,80	14320,58	1471,0	380,22	87,0615	40145,0	28869,4	175,511
1084	1357,15	19929,50	14332,16	1475,1	379,43	87,0731	40176,7	28892,8	175,534
1085	1358,15	19945,21	14343,74	1479,3	378,65	87,0847	40208,3	28916,1	175,557
1086	1359,15	19960,91	14355,32	1483,4	377,87	87,0962	40240,0	28939,5	175,581
1087	1360,15	19976,62	14366,91	1487,6	377,09	87,1078	40271,7	28962,8	175,604
1088	1361,15	19992,33	14378,49	1491,7	376,31	87,1193	40303,3	28986,2	175,627
1089	1362,15	20008,04	14390,08	1495,9	375,53	87,1309	40335,0	29009,5	175,651
1090	1363,15	20023,76	14401,67	1500,1	374,76	87,1424	40366,7	29032,9	175,674
1091	1364,15	20039,47	14413,26	1504,3	373,99	87,1539	40398,4	29056,3	175,697
1092	1365,15	20055,19	14424,86	1508,5	373,22	87,1654	40430,1	29079,6	175,720
1093	1366,15	20070,91	14436,46	1512,7	372,45	87,1770	40461,8	29103,0	175,744
1094	1367,15	20086,64	14448,05	1517,0	371,68	87,1885	40493,5	29126,4	175,767
1095	1368,15	20102,36	14459,66	1521,2	370,92	87,2000	40525,2	29149,8	175,790
1096	1369,15	20118,09	14471,26	1525,4	370,16	87,2114	40556,9	29173,2	175,813
1097	1370,15	20133,82	14482,86	1529,7	369,40	87,2229	40588,6	29196,6	175,836
1098	1371,15	20149,55	14494,47	1534,0	368,64	87,2344	40620,3	29220,0	175,859
1099	1372,15	20165,29	14506,08	1538,2	367,89	87,2459	40652,0	29243,4	175,882
1100	1373,15	20181,02	14517,69	1542,5	367,13	87,2573	40683,7	29266,8	175,906
1101	1374,15	20196,76	14529,31	1546,8	366,38	87,2688	40715,5	29290,2	175,929
1102	1375,15	20212,50	14540,92	1551,1	365,63	87,2803	40747,2	29313,6	175,952
1103	1376,15	20228,24	14552,54	1555,4	364,88	87,2917	40778,9	29337,0	175,975
1104	1377,15	20243,99	14564,16	1559,7	364,14	87,3031	40810,7	29360,5	175,998

t	T	h	u	π_o	θ_o	s^o	H	U	S
1105	1378,15	20259,73	14575,78	1564,1	363,39	87,3146	40842,4	29383,9	176,021
1106	1379,15	20275,48	14587,41	1568,4	362,65	87,3260	40874,2	29407,3	176,044
1107	1380,15	20291,23	14599,04	1572,7	361,91	87,3374	40905,9	29430,8	176,067
1108	1381,15	20306,99	14610,66	1577,1	361,17	87,3488	40937,7	29454,2	176,090
1109	1382,15	20322,74	14622,30	1581,5	360,44	87,3602	40969,4	29477,7	176,113
1110	1383,15	20338,50	14633,93	1585,8	359,70	87,3716	41001,2	29501,1	176,136
1111	1384,15	20354,26	14645,56	1590,2	358,97	87,3830	41033,0	29524,6	176,159
1112	1385,15	20370,02	14657,20	1594,6	358,24	87,3944	41064,7	29548,0	176,182
1113	1386,15	20385,79	14668,84	1599,0	357,51	87,4058	41096,5	29571,5	176,205
1114	1387,15	20401,55	14680,48	1603,5	356,78	87,4171	41128,3	29595,0	176,228
1115	1388,15	20417,32	14692,13	1607,9	356,06	87,4285	41160,1	29618,4	176,251
1116	1389,15	20433,09	14703,77	1612,3	355,33	87,4399	41191,9	29641,9	176,274
1117	1390,15	20448,86	14715,42	1616,8	354,61	87,4512	41223,7	29665,4	176,296
1118	1391,15	20464,64	14727,07	1621,2	353,89	87,4625	41255,5	29688,9	176,319
1119	1392,15	20480,42	14738,72	1625,7	353,17	87,4739	41287,3	29712,4	176,342
1120	1393,15	20496,19	14750,38	1630,1	352,46	87,4852	41319,1	29735,9	176,365
1121	1394,15	20511,98	14762,04	1634,6	351,74	87,4965	41350,9	29759,4	176,388
1122	1395,15	20527,76	14773,69	1639,1	351,03	87,5079	41382,7	29782,9	176,411
1123	1396,15	20543,54	14785,36	1643,6	350,32	87,5192	41414,6	29806,4	176,433
1124	1397,15	20559,33	14797,02	1648,1	349,61	87,5305	41446,4	29829,9	176,456
1125	1398,15	20575,12	14808,68	1652,6	348,91	87,5418	41478,2	29853,4	176,479
1126	1399,15	20590,91	14820,35	1657,2	348,20	87,5531	41510,0	29876,9	176,502
1127	1400,15	20606,71	14832,02	1661,7	347,50	87,5643	41541,9	29900,5	176,524
1128	1401,15	20622,50	14843,69	1666,3	346,80	87,5756	41573,7	29924,0	176,547
1129	1402,15	20638,30	14855,37	1670,8	346,10	87,5869	41605,6	29947,5	176,570
1130	1403,15	20654,10	14867,04	1675,4	345,40	87,5982	41637,4	29971,1	176,593
1131	1404,15	20669,91	14878,72	1680,0	344,70	87,6094	41669,3	29994,6	176,615
1132	1405,15	20685,71	14890,40	1684,6	344,01	87,6207	41701,2	30018,2	176,638
1133	1406,15	20701,52	14902,09	1689,2	343,32	87,6319	41733,0	30041,7	176,661
1134	1407,15	20717,33	14913,77	1693,8	342,63	87,6432	41764,9	30065,3	176,683
1135	1408,15	20733,14	14925,46	1698,4	341,94	87,6544	41796,8	30088,8	176,706
1136	1409,15	20748,95	14937,15	1703,0	341,25	87,6656	41828,6	30112,4	176,729
1137	1410,15	20764,77	14948,84	1707,7	340,56	87,6768	41860,5	30136,0	176,751
1138	1411,15	20780,58	14960,53	1712,3	339,88	87,6880	41892,4	30159,5	176,774
1139	1412,15	20796,40	14972,23	1717,0	339,20	87,6992	41924,3	30183,1	176,796
1140	1413,15	20812,23	14983,92	1721,6	338,52	87,7104	41956,2	30206,7	176,819
1141	1414,15	20828,05	14995,62	1726,3	337,84	87,7216	41988,1	30230,3	176,842
1142	1415,15	20843,88	15007,33	1731,0	337,16	87,7328	42020,0	30253,9	176,864
1143	1416,15	20859,71	15019,03	1735,7	336,49	87,7440	42051,9	30277,5	176,887
1144	1417,15	20875,54	15030,74	1740,4	335,81	87,7552	42083,8	30301,1	176,909
1145	1418,15	20891,37	15042,44	1745,1	335,14	87,7664	42115,7	30324,7	176,932
1146	1419,15	20907,20	15054,16	1749,9	334,47	87,7775	42147,7	30348,3	176,954
1147	1420,15	20923,04	15065,87	1754,6	333,80	87,7887	42179,6	30371,9	176,977
1148	1421,15	20938,88	15077,58	1759,4	333,14	87,7998	42211,5	30395,5	176,999
1149	1422,15	20954,72	15089,30	1764,1	332,47	87,8110	42243,5	30419,1	177,022
1150	1423,15	20970,56	15101,02	1768,9	331,81	87,8221	42275,4	30442,7	177,044
1151	1424,15	20986,41	15112,74	1773,7	331,15	87,8332	42307,3	30466,4	177,067
1152	1425,15	21002,26	15124,46	1778,5	330,49	87,8444	42339,3	30490,0	177,089
1153	1426,15	21018,11	15136,19	1783,3	329,83	87,8555	42371,2	30513,6	177,111
1154	1427,15	21033,96	15147,92	1788,1	329,17	87,8666	42403,2	30537,3	177,134
1155	1428,15	21049,81	15159,65	1792,9	328,52	87,8777	42435,2	30560,9	177,156
1156	1429,15	21065,67	15171,38	1797,7	327,86	87,8888	42467,1	30584,6	177,179
1157	1430,15	21081,53	15183,11	1802,6	327,21	87,8999	42499,1	30608,2	177,201
1158	1431,15	21097,39	15194,85	1807,4	326,56	87,9110	42531,1	30631,9	177,223
1159	1432,15	21113,25	15206,59	1812,3	325,91	87,9220	42563,1	30655,6	177,246

t	T	h	u	π₀	θ₀	s°	H	U	s
1160	1433,15	21129,12	15218,33	1817,1	325,26	87,9331	42595,0	30679,2	177,268
1161	1434,15	21144,99	15230,07	1822,0	324,62	87,9442	42627,0	30702,9	177,290
1162	1435,15	21160,85	15241,82	1826,9	323,98	87,9553	42659,0	30726,6	177,313
1163	1436,15	21176,73	15253,56	1831,8	323,33	87,9663	42691,0	30750,3	177,335
1164	1437,15	21192,60	15265,31	1836,7	322,69	87,9774	42723,0	30774,0	177,357
1165	1438,15	21208,48	15277,06	1841,7	322,05	87,9884	42755,0	30797,6	177,379
1166	1439,15	21224,55	15288,82	1846,6	321,42	87,9994	42787,0	30821,3	177,402
1167	1440,15	21240,23	15300,57	1851,5	320,78	88,0105	42819,0	30845,0	177,424
1168	1441,15	21256,12	15312,33	1856,5	320,15	88,0215	42851,1	30868,7	177,446
1169	1442,15	21272,00	15324,09	1861,5	319,51	88,0325	42883,1	30892,4	177,468
1170	1443,15	21287,89	15335,85	1866,4	318,88	88,0435	42915,1	30916,2	177,490
1171	1444,15	21303,77	15347,62	1871,4	318,25	88,0545	42947,1	30939,9	177,513
1172	1445,15	21319,67	15359,38	1876,4	317,63	88,0655	42979,2	30963,6	177,535
1173	1446,15	21335,56	15371,15	1881,4	317,00	88,0765	43011,2	30987,3	177,557
1174	1447,15	21351,45	15382,92	1886,5	316,37	88,0875	43043,2	31011,1	177,579
1175	1448,15	21367,35	15394,70	1891,5	315,75	88,0985	43075,3	31034,8	177,601
1176	1449,15	21383,25	15406,47	1896,5	315,13	88,1095	43107,3	31058,5	177,623
1177	1450,15	21399,15	15418,25	1901,6	314,51	88,1204	45139,4	31082,3	177,646
1178	1451,15	21415,06	15430,03	1906,6	313,89	88,1315	43171,5	31106,0	177,668
1179	1452,15	21430,96	15441,81	1911,7	313,27	88,1424	43203,5	31129,8	177,690
1180	1453,15	21446,87	15453,59	1916,8	312,66	88,1533	43235,6	31153,5	177,712
1181	1454,15	21462,78	15465,38	1921,9	312,04	88,1642	43267,7	31177,3	177,734
1182	1455,15	21478,69	15477,17	1927,0	311,43	88,1752	43299,8	31201,0	177,756
1183	1456,15	21494,61	15488,96	1932,1	310,82	88,1861	43331,8	31224,8	177,778
1184	1457,15	21510,52	15500,75	1937,2	310,21	88,1970	43363,9	31248,6	177,800
1185	1458,15	21526,44	15512,54	1942,4	309,60	88,2080	43396,0	31272,4	177,822
1186	1459,15	21542,36	15524,34	1947,5	309,00	88,2189	43428,1	31296,1	177,844
1187	1460,15	21558,29	15536,14	1952,7	308,39	88,2298	43460,2	31319,9	177,866
1188	1461,15	21574,21	15547,94	1957,8	307,79	88,2407	43492,3	31343,7	177,888
1189	1462,15	21590,14	15559,74	1963,0	307,19	88,2516	43524,4	31367,5	177,910
1190	1463,15	21606,07	15571,55	1968,2	306,59	88,2625	43556,5	31391,3	177,932
1191	1464,15	21622,00	15583,35	1973,4	305,99	88,2734	43588,7	31415,1	177,954
1192	1465,15	21637,93	15595,16	1978,6	305,39	88,2842	43620,8	31438,9	177,976
1193	1466,15	21653,87	15606,98	1983,9	304,79	88,2951	43652,9	31462,7	177,998
1194	1467,15	21669,81	15618,79	1989,1	304,20	88,3060	43685,0	31486,5	178,020
1195	1468,15	21685,75	15630,61	1994,3	303,60	88,3168	43717,2	31510,4	178,041
1196	1469,15	21701,69	15642,42	1999,6	303,01	88,3277	43749,3	31534,2	178,063
1197	1470,15	21717,63	15654,24	2004,9	302,42	88,3386	43781,4	31558,0	178,085
1198	1471,15	21733,58	15666,07	2010,1	301,83	88,3494	43813,6	31581,8	178,107
1199	1472,15	21749,53	15677,89	2015,4	301,25	88,3602	43845,7	31605,7	178,129
1200	1473,15	21765,48	15689,72	2020,7	300,66	88,3711	43877,9	31629,5	178,151
1201	1474,15	21781,43	15701,55	2026,0	300,08	88,3819	43910,1	31653,4	178,173
1202	1475,15	21797,39	15713,38	2031,4	299,49	88,3927	43942,2	31677,2	178,194
1203	1476,15	21813,35	15725,21	2036,7	298,91	88,4035	43974,4	31701,1	178,216
1204	1477,15	21829,31	15737,04	2042,0	298,33	88,4143	44006,6	31724,9	178,238
1205	1478,15	21845,27	15748,88	2047,4	297,75	88,4251	44038,8	31748,8	178,260
1206	1479,15	21861,23	15760,72	2052,8	297,17	88,4359	44070,9	31772,7	178,282
1207	1480,15	21877,20	15772,56	2058,1	296,60	88,4467	44103,1	31796,5	178,303
1208	1481,15	21893,16	15784,41	2063,5	296,02	88,4575	44135,3	31820,4	178,325
1209	1482,15	21909,13	15796,25	2068,9	295,45	88,4683	44167,5	31844,3	178,347
1210	1483,15	21925,11	15808,10	2074,3	294,88	88,4791	44199,7	31868,2	178,368
1211	1484,15	21941,08	15819,95	2079,8	294,31	88,4898	44231,9	31892,1	178,390
1212	1485,15	21957,06	15831,80	2085,2	293,74	88,5006	44264,1	31916,0	178,412
1213	1486,15	21973,04	15843,66	2090,6	293,17	88,5113	44296,3	31939,9	178,434
1214	1487,15	21989,02	15855,51	2096,1	292,60	88,5221	44328,5	31963,8	178,455

Hydrogen (*Continued*)

t	T	h	u	π_θ	θ_θ	s°	H	U	S
1215	1488,15	22005,00	15867,37	2101,6	292,04	88,5328	44360,8	31987,7	178,477
1216	1489,15	22020,99	15879,23	2107,0	291,48	88,5436	44393,0	32011,6	178,499
1217	1490,15	22036,97	15891,10	2112,5	290,91	88,5543	44425,2	32035,5	178,520
1218	1491,15	22052,96	15902,96	2118,0	290,35	88,5650	44457,5	32059,4	178,542
1219	1492,15	22068,96	15914,83	2123,5	289,79	88,5757	44489,7	32083,3	178,563
1220	1493,15	22084,95	15926,70	2129,1	289,23	88,5865	44521,9	32107,3	178,585
1221	1494,15	22100,95	15938,57	2134,6	288,68	88,5972	44554,2	32131,2	178,607
1222	1495,15	22116,94	15950,45	2140,2	288,12	88,6079	44586,4	32155,1	178,628
1223	1496,15	22132,94	15962,32	2145,7	287,57	88,6186	44618,7	32179,1	178,650
1224	1497,15	22148,95	15974,20	2151,3	287,01	88,6293	44650,9	32203,0	178,671
1225	1498,15	22164,95	15986,08	2156,9	286,46	88,6400	44683,2	32227,0	178,693
1226	1499,15	22180,96	15997,96	2162,5	285,91	88,6506	44715,5	32250,9	178,714
1227	1500,15	22196,97	16009,85	2168,1	285,36	88,6613	44747,8	32274,9	178,736
1228	1501,15	22212,98	16021,73	2173,7	284,82	88,6720	44780,0	32298,9	178,757
1229	1502,15	22228,99	16033,62	2179,3	284,27	88,6826	44812,3	32322,8	178,779
1230	1503,15	22245,01	16045,51	2184,9	283,73	88,6933	44844,6	32346,8	178,800
1231	1504,15	22261,02	16057,41	2190,6	283,18	88,7040	44876,9	32370,8	178,822
1232	1505,15	22277,04	16069,30	2196,3	282,64	88,7146	44909,2	32394,7	178,843
1233	1506,15	22293,06	16081,20	2201,9	282,10	88,7252	44941,5	32418,7	178,865
1234	1507,15	22309,09	16093,10	2207,6	281,56	88,7359	44973,8	32442,7	178,886
1235	1508,15	22325,11	16105,00	2213,3	281,02	88,7465	45006,1	32466,7	178,908
1236	1509,15	22341,14	16116,90	2219,0	280,48	88,7571	45038,4	32490,7	178,929
1237	1510,15	22357,17	16128,81	2224,7	279,95	88,7677	45070,7	32514,7	178,950
1238	1511,15	22373,21	16140,72	2230,5	279,41	88,7784	45103,0	32538,7	178,972
1239	1512,15	22389,24	16152,63	2236,2	278,88	88,7890	45135,4	32562,7	178,993
1240	1513,15	22405,28	16164,54	2242,0	278,35	88,7996	45167,7	32586,7	179,015
1241	1514,15	22421,32	16176,46	2247,7	277,82	88,8102	45200,0	32610,8	179,036
1242	1515,15	22437,36	16188,37	2253,5	277,29	88,8208	45232,4	32634,8	179,057
1243	1516,15	22453,40	16200,29	2259,3	276,76	88,8313	45264,7	32658,8	179,079
1244	1517,15	22469,45	16212,21	2265,1	276,23	88,8419	45297,1	32682,8	179,100
1245	1518,15	22485,49	16224,14	2270,9	275,71	88,8525	45329,4	32706,9	179,121
1246	1519,15	22501,54	16236,06	2276,8	275,18	88,8631	45361,8	32730,9	179,143
1247	1520,15	22517,59	16247,99	2282,6	274,66	88,8736	45394,1	32755,0	179,164
1248	1521,15	22533,65	16259,92	2288,4	274,14	88,8842	45426,5	32779,0	179,185
1249	1522,15	22549,70	16271,85	2294,3	273,62	88,8947	45458,9	32803,1	179,206
1250	1523,15	22565,76	16283,78	2300,2	273,10	88,9053	45491,2	32827,1	179,228
1251	1524,15	22581,82	16295,72	2306,1	272,58	88,9158	45523,6	32851,2	179,249
1252	1525,15	22597,89	16307,66	2312,0	272,06	88,9264	45556,0	32875,3	179,270
1253	1526,15	22613,95	16319,60	2317,9	271,55	88,9369	45588,4	32899,3	179,291
1254	1527,15	22630,02	16331,54	2323,8	271,03	88,9474	45620,8	32923,4	179,313
1255	1528,15	22646,09	16343,48	2329,7	270,52	88,9579	45653,1	32947,5	179,334
1256	1529,15	22662,16	16355,43	2335,7	270,01	88,9684	45685,5	32971,6	179,355
1257	1530,15	22678,23	16367,38	2341,6	269,50	88,9790	45717,9	32995,7	179,376
1258	1531,15	22694,30	16379,33	2347,6	268,99	88,9895	45750,4	33019,7	179,397
1259	1532,15	22710,38	16391,28	2353,6	268,48	89,0000	45782,8	33043,8	179,419
1260	1533,15	22726,46	16403,24	2359,6	267,97	89,0104	45815,2	33067,9	179,440
1261	1534,15	22742,54	16415,20	2365,6	267,46	89,0209	45847,6	33092,1	179,461
1262	1535,15	22758,63	16427,16	2371,6	266,96	89,0314	45880,0	33116,2	179,482
1263	1536,15	22774,71	16439,12	2377,6	266,45	89,0419	45912,5	33140,3	179,503
1264	1537,15	22790,80	16451,08	2383,7	265,95	89,0524	45944,9	33164,4	179,524
1265	1538,15	22806,89	16463,05	2389,7	265,45	89,0628	45977,3	33188,5	179,545
1266	1539,15	22822,98	16475,01	2395,8	264,95	89,0733	46009,8	33212,6	179,566
1267	1540,15	22839,08	16486,98	2401,9	264,45	89,0837	46042,2	33236,8	179,587
1268	1541,15	22855,17	16498,96	2408,0	263,95	89,0942	46074,7	33260,9	179,609
1269	1542,15	22871,27	16510,93	2414,1	263,46	89,1046	46107,1	33285,0	179,630

Hydrogen (*Continued*)

t	T	h	u	π₀	θ₀	s°	H	U	S
1270	1543,15	22887,37	16522,91	2420,2	262,96	89,1151	46139,6	33309,2	179,651
1271	1544,15	22903,48	16534,89	2426,3	262,47	89,1255	46172,0	33333,3	179,672
1272	1545,15	22919,58	16546,87	2432,5	261,97	89,1359	46204,5	33357,5	179,693
1273	1546,15	22935,69	16558,85	2438,6	261,48	89,1463	46237,0	33381,6	179,714
1274	1547,15	22951,80	16570,84	2444,8	260,99	89,1568	46269,5	33405,8	179,735
1275	1548,15	22967,91	16582,82	2451,0	260,50	89,1672	46301,9	33430,0	179,756
1276	1549,15	22984,02	16594,81	2457,2	260,01	89,1776	46334,4	33454,1	179,777
1277	1550,15	23000,14	16606,80	2463,4	259,53	89,1880	46366,9	33478,3	179,798
1278	1551,15	23016,26	16618,80	2469,6	259,04	89,1984	46399,4	33502,5	179,819
1279	1552,15	23032,38	16630,79	2475,8	258,55	89,2087	46431,9	33526,7	179,839
1280	1553,15	23048,50	16642,79	2482,1	258,07	89,2191	46464,4	33550,9	179,860
1281	1554,15	23064,63	16654,79	2488,3	257,59	89,2295	46496,9	33575,1	179,881
1282	1555,15	23080,75	16666,79	2494,6	257,11	89,2399	46529,4	33599,3	179,902
1283	1556,15	23096,88	16678,80	2500,9	256,63	89,2503	46561,9	33623,5	179,923
1284	1557,15	23113,01	16690,80	2507,1	256,15	89,2606	46594,4	33647,7	179,944
1285	1558,15	23129,14	16702,81	2513,5	255,67	89,2710	46627,0	33671,9	179,965
1286	1559,15	23145,28	16714,82	2519,8	255,19	89,2813	46659,5	33696,1	179,986
1287	1560,15	23161,42	16726,84	2526,1	254,71	89,2917	46692,0	33720,3	180,007
1288	1561,15	23177,56	16738,85	2532,4	254,24	89,3020	46724,6	33744,5	180,027
1289	1562,15	23193,70	16750,87	2538,8	253,77	89,3123	46757,1	33768,7	180,048
1290	1563,15	23209,84	16762,89	2545,2	253,29	89,3227	46789,6	33793,0	180,069
1291	1564,15	23225,99	16774,91	2551,5	252,82	89,3330	46822,:	33817,2	180,090
1292	1565,15	23242,13	16786,93	2557,9	252,35	89,3433	46854,7	33841,4	180,111
1293	1566,15	23258,28	16798,96	2564,3	251,88	89,3536	46887,3	33865,7	180,132
1294	1567,15	23274,44	16810,99	2570,8	251,41	89,3640	46919,9	33889,9	180,152
1295	1568,15	23290,59	16823,02	2577,2	250,94	89,3743	46952,4	33914,2	180,173
1296	1569,15	23306,75	16835,05	2583,6	250,48	89,3846	46985,0	33938,4	180,194
1297	1570,15	23322,90	16847,08	2590,1	250,01	89,3948	47017,6	33962,7	180,215
1298	1571,15	23339,07	16859,12	2596,6	249,55	89,4051	47050,2	33987,0	180,235
1299	1572,15	23355,23	16871,16	2603,0	249,09	89,4154	47082,7	34011,2	180,256
1300	1573,15	23371,39	16883,20	2609,5	248,62	89,4257	47115,3	34035,5	180,277
1301	1574,15	23387,56	16895,24	2616,0	248,16	89,4360	47147,9	34059,8	180,298
1302	1575,15	23403,73	16907,28	2622,6	247,70	89,4462	47180,5	34084,1	180,318
1303	1576,15	23419,90	16919,33	2629,1	247,24	89,4565	47213,1	34108,4	180,339
1304	1577,15	23436,07	16931,38	2635,6	246,79	89,4668	47245,7	34132,6	180,360
1305	1578,15	23452,25	16943,43	2642,2	246,33	89,4770	47278,3	34156,9	180,380
1306	1579,15	23468,43	16955,48	2648,8	245,87	89,4873	47310,9	34181,2	180,401
1307	1580,15	23484,61	16967,54	2655,4	245,42	89,4975	47343,6	34205,5	180,422
1308	1581,15	23500,79	16979,60	2662,0	244,97	89,5077	47376,2	34229,8	180,442
1309	1582,15	23516,97	16991,66	2668,6	244,51	89,5180	47408,8	34254,2	180,463
1310	1583,15	23533,16	17003,72	2675,2	244,06	89,5282	47441,4	34278,5	180,483
1311	1584,15	23549,35	17015,78	2681,8	243,61	89,5384	47474,1	34302,8	180,504
1312	1585,15	23565,54	17027,85	2688,5	243,16	89,5486	47506,7	34327,1	180,525
1313	1586,15	23581,73	17039,92	2695,2	242,71	89,5589	47539,3	34351,4	180,545
1314	1587,15	23597,92	17051,99	2701,8	242,27	89,5691	47572,0	34375,8	180,566
1315	1588,15	23614,12	17064,06	2708,5	241,82	89,5793	47604,6	34400,1	180,586
1316	1589,15	23630,32	17076,13	2715,2	241,38	89,5895	47637,3	34424,5	180,607
1317	1590,15	23646,52	17088,21	2722,0	240,93	89,5997	47670,0	34448,8	180,628
1318	1591,15	23662,72	17100,29	2728,7	240,49	89,6098	47702,6	34473,2	180,648
1319	1592,15	23678,93	17112,37	2735,4	240,05	89,6200	47735,3	34497,5	180,669
1320	1593,15	23695,13	17124,45	2742,2	239,61	89,6302	47768,0	34521,9	180,689
1321	1594,15	23711,34	17136,54	2749,0	239,16	89,6404	47800,6	34546,2	180,710
1322	1595,15	23727,56	17148,62	2755,7	238,73	89,6505	47833,3	34570,6	180,730
1323	1596,15	23743,77	17160,71	2762,5	238,29	89,6607	47866,0	34595,0	180,751
1324	1597,15	23759,99	17172,80	2769,4	237,85	89,6708	47898,7	34619,3	180,771

Hydrogen (*Continued*)

t	T	h	u	π_o	θ_o	s°	H	U	S
1325	1598,15	23776,20	17184,90	2776,2	237,41	89,6810	47931,4	34643,7	180,792
1326	1599,15	23792,42	17196,99	2783,0	236,98	89,6911	47964,1	34668,1	180,812
1327	1600,15	23808,64	17209,09	2789,9	236,55	89,7013	47996,8	34692,5	180,832
1328	1601,15	23824,87	17221,19	2796,7	236,11	89,7114	48029,5	34716,9	180,853
1329	1602,15	23841,10	17233,29	2803,6	255,68	89,7216	48062,2	34741,3	180,873
1330	1603,15	23857,32	17245,40	2810,5	235,25	89,7317	48094,9	34765,7	180,894
1331	1604,15	23873,55	17257,50	2817,4	234,82	89,7418	48127,7	34790,1	180,914
1332	1605,15	23889,79	17269,61	2824,3	234,39	89,7519	48160,4	34814,5	180,934
1333	1606,15	23906,02	17281,72	2831,3	233,96	89,7620	48193,1	34838,9	180,955
1334	1607,15	23922,26	17293,84	2838,2	233,53	89,7721	48225,8	34863,3	180,975
1335	1608,15	23938,50	17305,95	2845,2	233,11	89,7822	48258,6	34887,8	180,996
1336	1609,15	23954,74	17318,07	2852,1	232,68	89,7923	48291,3	34912,2	181,016
1337	1610,15	23970,98	17330,19	2859,1	232,26	89,8024	48324,1	34936,6	181,036
1338	1611,15	23987,23	17342,31	2866,1	231,83	89,8125	48356,8	34961,0	181,057
1339	1612,15	24003,47	17354,43	2873,1	231,41	89,8226	48389,6	34985,5	181,077
1340	1613,15	24019,72	17366,55	2880,2	230,99	89,8327	48422,3	35009,9	181,097
1341	1614,15	24035,98	17378,68	2887,2	230,57	89,8427	48455,1	35034,4	181,118
1342	1615,15	24052,23	17390,81	2894,3	230,15	89,8528	48487,9	35058,8	181,138
1343	1616,15	24068,49	17402,94	2901,3	229,73	89,8629	48520,6	35083,3	181,158
1344	1617,15	24084,74	17415,08	2908,4	229,31	89,8729	48553,4	35107,7	181,178
1345	1618,15	24101,00	17427,21	2915,5	228,90	89,8830	48586,2	35132,2	181,199
1346	1619,15	24117,27	17439,35	2922,6	228,48	89,8930	48619,0	35156,7	181,219
1347	1620,15	24133,53	17451,49	2929,8	228,07	89,9031	48651,7	35181,2	181,239
1348	1621,15	24149,80	17463,63	2936,9	227,65	89,9131	48684,5	35205,6	181,259
1349	1622,15	24166,06	17475,78	2944,0	227,24	89,9231	48717,3	35230,1	181,280
1350	1623,15	24182,34	17487,92	2951,2	226,83	89,9332	48750,1	35254,6	181,300
1351	1624,15	24198,61	17500,07	2958,4	226,42	89,9432	48782,9	35279,1	181,320
1352	1625,15	24214,88	17512,22	2965,6	226,01	89,9532	48815,8	35303,6	181,340
1353	1626,15	24231,16	17524,37	2972,8	225,60	89,9632	48848,6	35328,1	181,360
1354	1627,15	24247,44	17536,53	2980,0	225,19	89,9732	48881,4	35352,6	181,381
1355	1628,15	24263,72	17548,69	2987,3	224,78	89,9832	48914,2	35377,1	181,401
1356	1629,15	24280,00	17560,84	2994,5	224,37	89,9932	48947,0	35401,6	181,421
1357	1630,15	24296,29	17573,01	3001,8	223,97	90,0032	48979,9	35426,1	181,441
1358	1631,15	24312,58	17585,17	3009,0	223,56	90,0132	49012,7	35450,6	181,461
1359	1632,15	24328,87	17597,33	3016,3	223,16	90,0232	49045,5	35475,2	181,481
1360	1633,15	24345,16	17609,50	3023,6	222,76	90,0332	49078,4	35499,7	181,501
1361	1634,15	24361,45	17621,67	3031,0	222,36	90,0431	49111,2	35524,2	181,522
1362	1635,15	24377,75	17633,84	3038,3	221,95	90,0531	49144,1	35548,8	181,542
1363	1636,15	24394,04	17646,02	3045,7	221,55	90,0631	49176,9	35573,3	181,562
1364	1637,15	24410,35	17658,19	3053,0	221,15	90,0730	49209,8	35597,9	181,582
1365	1638,15	24426,65	17670,37	3060,4	220,76	90,0830	49242,7	35622,4	181,602
1366	1639,15	24442,95	17682,55	3067,8	220,36	90,0929	49275,5	35647,0	181,622
1367	1640,15	24459,26	17694,73	3075,2	219,96	90,1029	49308,4	35671,5	181,642
1368	1641,15	24475,57	17706,92	3082,6	219,57	90,1128	49341,3	35696,1	181,662
1369	1642,15	24491,88	17719,10	3090,0	219,17	90,1228	49374,2	35720,6	181,682
1370	1643,15	24508,19	17731,29	3097,5	218,78	90,1327	49407,0	35745,2	181,702
1371	1644,15	24524,50	17743,48	3105,0	218,38	90,1426	49439,9	35769,8	181,722
1372	1645,15	24540,82	17755,67	3112,4	217,99	90,1525	49472,8	35794,4	181,742
1373	1646,15	24557,14	17767,87	3119,9	217,60	90,1624	49505,7	35819,0	181,762
1374	1647,15	24573,46	17780,06	3127,4	217,21	90,1724	49538,6	35843,5	181,782
1375	1648,15	24589,78	17792,26	3135,0	216,82	90,1823	49571,5	35868,1	181,802
1376	1649,15	24606,11	17804,46	3142,5	216,43	90,1922	49604,4	35892,7	181,822
1377	1650,15	24622,44	17816,67	3150,0	216,04	90,2021	49637,4	35917,3	181,842
1378	1651,15	24638,77	17828,87	3157,6	215,66	90,2120	49670,3	35941,9	181,862
1379	1652,15	24655,10	17841,08	3165,2	215,27	90,2218	49703,2	35966,5	181,882

Hydrogen (*Continued*)

t	T	h	u	π.	θ.	s°	H	U	S
1380	1653,15	24671,43	17853,29	3172,8	214,89	90,2317	49736,1	35991,2	181,902
1381	1654,15	24687,77	17865,50	3180,4	214,50	90,2416	49769,1	36015,8	181,922
1382	1655,15	24704,10	17877,71	3188,0	214,12	90,2515	49802,0	36040,4	181,942
1383	1656,15	24720,44	17889,93	3195,7	213,74	90,2614	49834,9	36065,0	181,961
1384	1657,15	24736,79	17902,15	3203,3	213,35	90,2712	49867,9	36089,7	181,981
1385	1658,15	24753,13	17914,36	3211,0	212,97	90,2811	49900,8	36114,3	182,001
1386	1659,15	24769,48	17926,59	3218,7	212,59	90,2909	49933,8	36138,9	182,021
1387	1660,15	24785,82	17938,81	3226,4	212,21	90,3008	49966,7	36163,6	182,041
1388	1661,15	24802,17	17951,05	3234,1	211,83	90,3106	49999,7	36188,2	182,061
1389	1662,15	24818,53	17963,26	3241,8	211,46	90,3205	50032,7	36212,9	182,081
1390	1663,15	24834,88	17975,49	3249,5	211,08	90,3303	50065,6	36237,5	182,100
1391	1664,15	24851,24	17987,73	3257,3	210,70	90,3401	50098,6	36262,2	182,120
1392	1665,15	24867,60	17999,96	3265,1	210,33	90,3500	50131,6	36286,8	182,140
1393	1666,15	24883,96	18012,20	3272,8	209,95	90,3598	50164,6	36311,5	182,160
1394	1667,15	24900,32	18024,44	3280,6	209,58	90,3696	50197,5	36336,2	182,180
1395	1668,15	24916,68	18036,68	3288,5	209,21	90,3794	50230,5	36360,9	182,199
1396	1669,15	24933,05	18048,92	3296,3	208,84	90,3892	50263,5	36385,5	182,219
1397	1670,15	24949,42	18061,16	3304,1	208,47	90,3990	50296,5	36410,2	182,239
1398	1671,15	24965,79	18073,41	3312,0	208,10	90,4088	50329,5	36434,9	182,259
1399	1672,15	24982,16	18085,66	3319,9	207,73	90,4186	50362,5	36459,6	182,279
1400	1673,15	24998,54	18097,91	3327,8	207,36	90,4284	50395,6	36484,3	182,298
1401	1674,15	25014,92	18110,16	3335,7	206,99	90,4382	50428,6	36509,0	182,318
1402	1675,15	25031,30	18122,42	3343,6	206,62	90,4480	50461,6	36533,7	182,338
1403	1676,15	25047,68	18134,67	3351,5	206,26	90,4578	50494,6	36558,4	182,357
1404	1677,15	25064,06	18146,93	3359,5	205,89	90,4675	50527,6	36583,1	182,377
1405	1678,15	25080,45	18159,19	3367,4	205,53	90,4773	50560,7	36607,8	182,397
1406	1679,15	25096,83	18171,46	3375,4	205,16	90,4871	50593,7	36632,6	182,416
1407	1680,15	25113,22	18183,72	3383,4	204,80	90,4968	50626,8	36657,3	182,436
1408	1681,15	25129,62	18195,99	3391,4	204,44	90,5066	50659,8	36682,0	182,456
1409	1682,15	25146,01	18208,26	3399,4	204,08	90,5163	50692,8	36706,8	182,475
1410	1683,15	25162,40	18220,53	3407,5	203,72	90,5261	50725,9	36731,5	182,495
1411	1684,15	25178,80	18232,81	3415,5	203,36	90,5358	50759,0	36756,2	182,515
1412	1685,15	25195,20	18245,08	3423,6	203,00	90,5455	50792,0	36781,0	182,534
1413	1686,15	25211,61	18257,36	3431,7	202,64	90,5553	50825,1	36805,7	182,554
1414	1687,15	25228,01	18269,64	3439,8	202,28	90,5650	50858,2	36830,5	182,574
1415	1688,15	25244,42	18281,92	3447,9	201,93	90,5747	50891,2	36855,3	182,593
1416	1689,15	25260,82	18294,20	3456,1	201,57	90,5844	50924,3	36880,0	182,613
1417	1690,15	25277,23	18306,49	3464,2	201,21	90,5941	50957,4	36904,8	182,632
1418	1691,15	25293,65	18318,78	3472,4	200,86	90,6038	50990,5	36929,6	182,652
1419	1692,15	25310,06	18331,07	3480,5	200,51	90,6136	51023,6	36954,3	182,671
1420	1693,15	25326,48	18343,36	3488,7	200,15	90,6233	51056,7	36979,1	182,691
1421	1694,15	25342,90	18355,66	3496,9	199,80	90,6329	51089,8	37003,9	182,711
1422	1695,15	25359,32	18367,95	3505,2	199,45	90,6426	51122,9	37028,7	182,730
1423	1696,15	25375,74	18380,25	3513,4	199,10	90,6523	51156,0	37053,5	182,750
1424	1697,15	25392,16	18392,55	3521,7	198,75	90,6620	51189,1	37078,3	182,769
1425	1698,15	25408,59	18404,85	3529,9	198,40	90,6717	51222,2	37103,1	182,789
1426	1699,15	25425,02	18417,16	3538,2	198,05	90,6813	51255,3	37127,9	182,808
1427	1700,15	25441,45	18429,46	3546,5	197,71	90,6910	51288,4	37152,7	182,828
1428	1701,15	25457,88	18441,77	3554,8	197,36	90,7007	51321,6	37177,5	182,847
1429	1702,15	25474,32	18454,08	3563,2	197,01	90,7103	51354,7	37202,3	182,867
1430	1703,15	25490,76	18466,40	3571,5	196,67	90,7200	51387,8	37227,1	182,886
1431	1704,15	25507,20	18478,71	3579,9	196,32	90,7296	51421,0	37252,0	182,906
1432	1705,15	25523,64	18491,03	3588,3	195,98	90,7393	51454,1	37276,8	182,925
1433	1706,15	25540,08	18503,35	3596,7	195,64	90,7489	51487,3	37301,6	182,944
1434	1707,15	25556,53	18515,67	3605,1	195,30	90,7586	51520,4	37326,5	182,964
1435	1708,15	25572,97	18527,99	3613,5	194,95	90,7682	51553,6	37351,3	182,983
1436	1709,15	25589,42	18540,32	3622,0	194,61	90,7778	51586,7	37376,2	183,003
1437	1710,15	25605,87	18552,64	3630,4	194,27	90,7874	51619,9	37401,0	183,022
1438	1711,15	25622,33	18564,97	3638,9	193,93	90,7971	51653,1	37425,9	183,041
1439	1712,15	25638,78	18577,30	3647,4	193,60	90,8067	51686,2	37450,7	183,061

Hydrogen (Continued)

t	T	h	u	$\pi_\bullet$	$\theta_\bullet$	$s^\bullet$	H	U	S
1440	1713,15	25655,24	18589,64	3655,9	193,26	90,8163	51719,4	37475,6	183,080
1441	1714,15	25671,70	18601,97	3664,4	192,92	90,8259	51752,6	37500,5	183,100
1442	1715,15	25688,16	18614,31	3673,0	192,58	90,8355	51785,8	37525,3	183,119
1443	1716,15	25704,63	18626,65	3681,5	192,25	90,8451	51819,0	37550,2	183,138
1444	1717,15	25721,09	18638,99	3690,1	191,91	90,8547	51852,2	37575,1	183,158
1445	1718,15	25737,56	18651,34	3698,7	191,58	90,8643	51885,4	37600,0	183,177
1446	1719,15	25754,03	18663,68	3707,3	191,25	90,8738	51918,6	37624,9	183,196
1447	1720,15	25770,50	18676,03	3715,9	190,91	90,8834	51951,8	37649,8	183,216
1448	1721,15	25786,98	18688,38	3724,5	190,58	90,8930	51985,0	37674,6	183,235
1449	1722,15	25803,45	18700,73	3733,2	190,25	90,9026	52018,2	37699,5	183,254
1450	1723,15	25819,93	18713,08	3741,9	189,92	90,9121	52051,4	37724,5	183,273
1451	1724,15	25836,41	18725,44	3750,6	189,59	90,9217	52084,7	37749,4	183,293
1452	1725,15	25852,89	18737,80	3759,3	189,26	90,9313	52117,9	37774,3	183,312
1453	1726,15	25869,38	18750,16	3768,0	188,93	90,9408	52151,1	37799,2	183,331
1454	1727,15	25885,86	18762,52	3776,7	188,61	90,9504	52184,3	37824,1	183,350
1455	1728,15	25902,35	18774,88	3785,5	188,28	90,9599	52217,6	37849,0	183,370
1456	1729,15	25918,84	18787,25	3794,2	187,95	90,9694	52250,8	37874,0	183,389
1457	1730,15	25935,33	18799,62	3803,0	187,63	90,9790	52284,1	37898,9	183,408
1458	1731,15	25951,83	18811,99	3811,8	187,30	90,9885	52317,3	37923,8	183,427
1459	1732,15	25968,32	18824,36	3820,6	186,98	90,9980	52350,6	37948,8	183,447
1460	1733,15	25984,82	18836,73	3829,4	186,65	91,0076	52383,8	37973,7	183,466
1461	1734,15	26001,32	18849,11	3838,3	186,33	91,0171	52417,1	37998,7	183,485
1462	1735,15	26017,82	18861,49	3847,2	186,01	91,0266	52450,4	38023,6	183,504
1463	1736,15	26034,33	18873,87	3856,0	185,69	91,0361	52483,6	38048,6	183,523
1464	1737,15	26050,84	18886,25	3864,9	185,37	91,0456	52516,9	38073,5	183,542
1465	1738,15	26067,34	18898,63	3873,9	185,05	91,0551	52550,2	38098,5	183,562
1466	1739,15	26083,85	18911,02	3882,8	184,73	91,0646	52583,5	38123,5	183,581
1467	1740,15	26100,37	18923,41	3891,7	184,41	91,0741	52616,8	38148,5	183,600
1468	1741,15	26116,88	18935,80	3900,7	184,09	91,0836	52650,1	38173,4	183,619
1469	1742,15	26133,40	18948,19	3909,7	183,77	91,0931	52683,4	38198,4	183,638
1470	1743,15	26149,92	18960,58	3918,7	183,46	91,1025	52716,7	38223,4	183,657
1471	1744,15	26166,44	18972,98	3927,7	183,14	91,1120	52750,0	38248,4	183,676
1472	1745,15	26182,96	18985,38	3936,7	182,83	91,1215	52783,3	38273,4	183,695
1473	1746,15	26199,49	18997,78	3945,8	182,51	91,1309	52816,6	38298,4	183,715
1474	1747,15	26216,01	19010,18	3954,8	182,20	91,1404	52849,9	38323,4	183,734
1475	1748,15	26232,54	19022,59	3963,9	181,88	91,1499	52883,2	38348,4	183,753
1476	1749,15	26249,07	19034,99	3973,0	181,57	91,1593	52916,6	38373,4	183,772
1477	1750,15	26265,60	19047,40	3982,1	181,26	91,1688	52949,9	38398,4	183,791
1478	1751,15	26282,14	19059,81	3991,2	180,95	91,1782	52983,2	38423,4	183,810
1479	1752,15	26298,68	19072,22	4000,4	180,64	91,1877	53016,6	38448,5	183,829
1480	1753,15	26315,21	19084,64	4009,5	180,33	91,1971	53049,9	38473,5	183,848
1481	1754,15	26331,75	19097,05	4018,7	180,02	91,2065	53083,2	38498,5	183,867
1482	1755,15	26348,30	19109,47	4027,9	179,71	91,2160	53116,6	38523,6	183,886
1483	1756,15	26364,84	19121,89	4037,1	179,40	91,2254	53149,9	38548,6	183,905
1484	1757,15	26381,39	19134,32	4046,4	179,09	91,2348	53183,3	38573,6	183,924
1485	1758,15	26397,94	19146,74	4055,6	178,79	91,2442	53216,7	38598,7	183,943
1486	1759,15	26414,49	19159,17	4064,9	178,48	91,2536	53250,0	38623,7	183,962
1487	1760,15	26431,04	19171,59	4074,2	178,18	91,2630	53283,4	38648,8	183,981
1488	1761,15	26447,60	19184,03	4083,5	177,87	91,2724	53316,8	38673,8	184,000
1489	1762,15	26464,15	19196,46	4092,8	177,57	91,2818	53350,1	38698,9	184,019
1490	1763,15	26480,71	19208,89	4102,1	177,26	91,2912	53383,5	38724,0	184,038
1491	1764,15	26497,27	19221,33	4111,5	176,96	91,3006	53416,9	38749,0	184,057
1492	1765,15	26513,83	19233,77	4120,8	176,66	91,3100	53450,3	38774,1	184,075
1493	1766,15	26530,40	19246,21	4130,2	176,36	91,3194	53483,7	38799,2	184,094
1494	1767,15	26546,97	19258,65	4139,6	176,06	91,3288	53517,1	38824,3	184,113
1495	1768,15	26563,54	19271,09	4149,0	175,76	91,3381	53550,5	38849,4	184,132
1496	1769,15	26580,11	19283,54	4158,5	175,46	91,3475	53583,9	38874,5	184,151
1497	1770,15	26596,68	19295,99	4167,9	175,16	91,3569	53617,3	38899,6	184,170
1498	1771,15	26613,25	19308,44	4177,4	174,86	91,3662	53650,7	38924,7	184,189
1499	1772,15	26629,83	19320,89	4186,9	174,56	91,3756	53684,1	38949,8	184,208
1500	1773,15	26646,41	19333,35	4196,4	174,26	91,3849	53717,6	38974,9	184,227

Table III.17 Helium ($\mu = 4.0026$)

t	T	h	u	π_0	θ_0	s^0	H	U	S
—50	223,15	1213,14	749,60	3,5537	13043	26,5492	4855,7	3000,3	106,266
—49	224,15	1218,33	752,71	3,5936	12956	26,5724	4876,5	3012,8	106,359
—48	225,15	1223,52	755,83	3,6339	12870	26,5955	4897,3	3025,3	106,451
—47	226,15	1228,71	758,94	3,6743	12785	26,6186	4918,1	3037,8	106,543
—46	227,15	1233,91	762,06	3,7151	12700	26,6415	4938,8	3050,2	106,635
—45	228,15	1239,10	765,18	3,7561	12617	26,6643	4959,6	3062,7	106,726
—44	229,15	1244,29	768,29	3,7974	12534	26,6870	4980,4	3075,2	106,817
—43	230,15	1249,49	771,41	3,8390	12453	26,7096	5001,2	3087,6	106,908
—42	231,15	1254,68	774,52	3,8808	12372	26,7321	5022,0	3100,1	106,998
—41	232,15	1259,87	777,64	3,9229	12292	26,7545	5042,8	3112,6	107,088
—40	233,15	1265,07	780,76	3,9653	12213	26,7769	5063,6	3125,1	107,177
—39	234,15	1270,26	783,87	4,0080	12135	26,7991	5084,3	3137,5	107,266
—38	235,15	1275,45	786,99	4,0509	12058	26,8212	5105,1	3150,0	107,355
—37	236,15	1280,65	790,10	4,0941	11981	26,8433	5125,9	3162,5	107,443
—36	237,15	1285,84	793,22	4,1376	11905	26,8652	5146,7	3174,9	107,531
—35	238,15	1291,03	796,33	4,1813	11830	26,8871	5167,5	3187,4	107,618
—34	239,15	1296,23	799,45	4,2254	11756	26,9088	5188,3	3199,9	107,705
—33	240,15	1301,42	802,57	4,2697	11683	26,9305	5209,1	3212,4	107,792
—32	241,15	1306,61	805,68	4,3143	11610	26,9521	5229,8	3224,8	107,878
—31	242,15	1311,80	808,80	4,3591	11539	26,9736	5250,6	3237,3	107,964
—30	243,15	1317,00	811,91	4,4043	11467	26,9950	5271,4	3249,8	108,050
—29	244,15	1322,19	815,03	4,4497	11397	27,0163	5292,2	3262,2	108,135
—28	245,15	1327,38	818,15	4,4954	11327	27,0375	5313,0	3274,7	108,220
—27	246,15	1332,58	821,26	4,5414	11258	27,0586	5333,8	3287,2	108,305
—26	247,15	1337,77	824,38	4,5876	11190	27,0797	5354,6	3299,7	108,389
—25	248,15	1342,96	827,49	4,6342	11123	27,1007	5375,3	3312,1	108,473
—24	249,15	1348,16	830,61	4,6810	11056	27,1215	5396,1	3324,6	108,557
—23	250,15	1353,35	833,73	4,7281	10989	27,1423	5416,9	3337,1	108,640
—22	251,15	1358,54	836,84	4,7755	10924	27,1631	5437,7	3349,5	108,723
—21	252,15	1363,74	839,96	4,8232	10859	27,1837	5458,5	3362,0	108,805
—20	253,15	1368,93	843,07	4,8712	10795	27,2043	5479,3	3374,5	108,888
—19	254,15	1374,12	846,19	4,9194	10731	27,2247	5500,1	3387,0	108,970
—18	255,15	1379,32	849,30	4,9680	10668	27,2451	5520,8	3399,4	109,051
—17	256,15	1384,51	852,42	5,0168	10606	27,2654	5541,6	3411,9	109,133
—16	257,15	1389,70	855,54	5,0659	10544	27,2857	5562,4	3424,4	109,214
—15	258,15	1394,89	858,65	5,1153	10483	27,3058	5583,2	3436,8	109,294
—14	259,15	1400,09	861,77	5,1650	10422	27,3259	5604,0	3449,3	109,375
—13	260,15	1405,28	864,88	5,2149	10362	27,3459	5624,8	3461,8	109,455
—12	261,15	1410,47	868,00	5,2652	10302	27,3658	5645,6	3474,3	109,534
—11	262,15	1415,67	871,12	5,3157	10244	27,3857	5666,3	3486,7	109,614
—10	263,15	1420,86	874,23	5,3666	10185	27,4054	5687,1	3499,2	109,693
—9	264,15	1426,05	877,35	5,4177	10127	27,4251	5707,9	3511,7	109,772
—8	265,15	1431,25	880,46	5,4691	10070	27,4448	5728,7	3524,1	109,850
—7	266,15	1436,44	883,58	5,5208	10013	27,4643	5749,5	3536,6	109,929
—6	267,15	1441,63	886,69	5,5728	9957,8	27,4838	5770,3	3549,1	110,007
—5	268,15	1446,83	889,81	5,6251	9902,1	27,5032	5791,1	3561,6	110,084
—4	269,15	1452,02	892,93	5,6777	9847,0	27,5225	5811,9	3574,0	110,162
—3	270,15	1457,21	896,04	5,7306	9792,4	27,5418	5832,6	3586,5	110,239
—2	271,15	1462,41	899,16	5,7838	9738,2	27,5610	5853,4	3599,0	110,316
—1	272,15	1467,60	902,27	5,8373	9684,6	27,5801	5874,2	3611,4	110,392
0	273,15	1472,79	905,39	5,8910	9631,5	27,5991	5895,0	3623,9	110,468
1	274,15	1477,98	908,51	5,9451	9578,8	27,6181	5915,8	3636,4	110,544
2	275,15	1483,18	911,62	5,9995	9526,6	27,6370	5936,6	3648,9	110,620
3	276,15	1488,37	914,74	6,0541	9474,9	27,6559	5957,4	3661,3	110,695
4	277,15	1493,56	917,85	6,1091	9423,7	27,6746	5978,1	3673,8	110,770

Helium (Continued)

t	T	h	u	π_0	θ_0	s^*	H	U	S
5	278,15	1498,76	920,97	6,1643	9372,9	27,6933	5998,9	3686,3	110,845
6	279,15	1503,95	924,09	6,2199	9322,6	27,7120	6019,7	3698,7	110,920
7	280,15	1509,14	927,20	6,2758	9272,7	27,7305	6040,5	3711,2	110,994
8	281,15	1514,34	930,32	6,3319	9223,3	27,7490	6061,3	3723,7	111,068
9	282,15	1519,53	933,43	6,3884	9174,3	27,7675	6082,1	3736,2	111,142
10	283,15	1524,72	936,55	6,4451	9125,8	27,7859	6102,9	3748,6	111,216
11	284,15	1529,92	939,66	6,5022	9077,6	27,8042	6123,6	3761,1	111,289
12	285,15	1535,11	942,78	6,5595	9029,9	27,8224	6144,4	3773,6	111,362
13	286,15	1540,30	945,90	6,6172	8982,6	27,8406	6165,2	3786,0	111,435
14	287,15	1545,50	949,01	6,6752	8935,8	27,8587	6186,0	3798,5	111,507
15	288,15	1550,69	952,15	6,7334	8889,3	27,8768	6206,8	3811,0	111,580
16	289,15	1555,88	955,24	6,7920	8843,2	27,8948	6227,6	3823,5	111,652
17	290,15	1561,07	958,36	6,8509	8797,5	27,9127	6248,4	3835,9	111,723
18	291,15	1566,27	961,48	6,9101	8752,2	27,9305	6269,1	3848,4	111,795
19	292,15	1571,46	964,59	6,9695	8707,3	27,9484	6289,9	3860,9	111,866
20	293,15	1576,65	967,71	7,0293	8662,8	27,9661	6310,7	3873,3	111,937
21	294,15	1581,85	970,82	7,0894	8618,7	27,9838	6331,5	3885,8	112,008
22	295,15	1587,04	973,94	7,1498	8574,9	28,0014	6352,3	3898,3	112,078
23	296,15	1592,23	977,06	7,2106	8531,5	28,0190	6373,1	3910,8	112,149
24	297,15	1597,43	980,17	7,2716	8488,5	28,0365	6393,9	3923,2	112,219
25	298,15	1602,62	983,29	7,3329	8445,8	28,0539	6414,6	3935,7	112,289
26	299,15	1607,81	986,40	7,3946	8403,5	28,0713	6435,4	3948,2	112,358
27	300,15	1613,01	989,52	7,4565	8361,5	28,0886	6456,2	3960,6	112,428
28	301,15	1618,20	992,63	7,5188	8319,9	28,1059	6477,0	3973,1	112,497
29	302,15	1623,39	995,75	7,5813	8278,7	28,1231	6497,8	3985,6	112,566
30	303,15	1628,59	998,87	7,6442	8237,7	28,1403	6518,6	3998,1	112,634
31	304,15	1633,78	1001,98	7,7074	8197,1	28,1574	6539,4	4010,5	112,703
32	305,15	1638,97	1005,10	7,7709	8156,9	28,1744	6560,1	4023,0	112,771
33	306,15	1644,16	1008,21	7,8348	8116,9	28,1914	6580,9	4035,5	112,839
34	307,15	1649,36	1011,33	7,8989	8077,3	28,2084	6601,7	4047,9	112,907
35	308,15	1654,55	1014,45	7,9633	8038,1	28,2252	6622,5	4060,4	112,974
36	309,15	1659,74	1017,56	8,0281	7999,1	28,2421	6643,3	4072,9	113,042
37	310,15	1664,94	1020,68	8,0932	7960,4	28,2588	6664,1	4085,4	113,109
38	311,15	1670,13	1023,79	8,1586	7922,1	28,2756	6684,9	4097,8	113,176
39	312,15	1675,32	1026,91	8,2243	7884,0	28,2922	6705,6	4110,3	113,242
40	313,15	1680,52	1030,03	8,2903	7846,3	28,3088	6726,4	4122,8	113,309
41	314,15	1685,71	1033,14	8,3567	7808,9	28,3254	6747,2	4135,2	113,375
42	315,15	1690,90	1036,26	8,4233	7771,7	28,3419	6768,0	4147,7	113,441
43	316,15	1696,10	1039,37	8,4903	7734,9	28,3583	6788,8	4160,2	113,507
44	317,15	1701,29	1042,49	8,5576	7698,3	28,3747	6809,6	4172,7	113,573
45	318,15	1706,48	1045,60	8,6252	7662,1	28,3911	6830,4	4185,1	113,638
46	319,15	1711,68	1048,72	8,6932	7626,1	28,4074	6851,2	4197,6	113,703
47	320,15	1716,87	1051,84	8,7614	7590,4	28,4236	6871,9	4210,1	113,768
48	321,15	1722,06	1054,95	8,8300	7555,0	28,4398	6892,7	4222,6	113,833
49	322,15	1727,25	1058,07	8,8989	7519,8	28,4560	6913,5	4235,0	113,898
50	323,15	1732,45	1061,18	8,9681	7484,9	28,4721	6934,3	4247,5	113,962
51	324,15	1737,64	1064,30	9,0376	7450,3	28,4881	6955,1	4260,0	114,027
52	325,15	1742,83	1067,42	9,1075	7416,0	28,5041	6975,9	4272,4	114,091
53	326,15	1748,03	1070,53	9,1777	7381,9	28,5201	6996,7	4284,9	114,154
54	327,15	1753,22	1073,65	9,2482	7348,1	28,5360	7017,4	4297,4	114,218
55	328,15	1758,41	1076,76	9,3190	7314,5	28,5518	7038,2	4309,9	114,281
56	329,15	1763,61	1079,88	9,3902	7281,2	28,5676	7059,0	4322,3	114,345
57	330,15	1768,80	1082,99	9,4617	7248,1	28,5834	7079,8	4334,8	114,408
58	331,15	1773,99	1086,11	9,5335	7215,3	28,5991	7100,6	4347,3	114,471
59	332,15	1779,19	1089,23	9,6056	7182,8	28,6147	7121,4	4359,7	114,533

Helium (*Continued*)

t	T	h	u	π_0	θ_0	s^0	H	U	S
60	333,15	1784,38	1092,34	9,6781	7150,5	28,6303	7142,2	4372,2	114,596
61	334,15	1789,57	1095,46	9,7509	7118,4	28,6459	7162,9	4384,7	114,658
62	335,15	1794,77	1098,57	9,8240	7086,6	28,6614	7183,7	4397,2	114,720
63	336,15	1799,96	1101,69	9,8974	7055,0	28,6769	7204,5	4409,6	114,782
64	337,15	1805,15	1104,81	9,9712	7023,6	28,6923	7225,3	4422,1	114,844
65	338,15	1810,34	1107,92	10,045	6992,5	28,7077	7246,1	4434,6	114,905
66	339,15	1815,54	1111,04	10,119	6961,6	28,7230	7266,9	4447,0	114,967
67	340,15	1820,73	1114,15	10,194	6930,9	28,7383	7287,7	4459,5	115,028
68	341,15	1825,92	1117,27	10,269	6900,4	28,7536	7508,4	4472,0	115,089
69	342,15	1831,12	1120,39	10,345	6870,2	28,7688	7329,2	4484,5	115,150
70	343,15	1836,31	1123,50	10,420	6840,2	28,7839	7350,0	4496,9	115,211
71	344,15	1841,50	1126,62	10,496	6810,4	28,7990	7370,8	4509,4	115,271
72	345,15	1846,70	1129,73	10,573	6780,8	28,8141	7391,6	4521,9	115,331
73	346,15	1851,89	1132,85	10,650	6751,5	28,8291	7412,4	4534,3	115,391
74	347,15	1857,08	1135,96	10,727	6722,3	28,8441	7433,2	4546,8	115,451
75	348,15	1862,28	1139,08	10,804	6693,4	28,8591	7453,9	4559,3	115,511
76	349,15	1867,47	1142,20	10,882	6664,6	28,8739	7474,7	4571,3	115,571
77	350,15	1872,66	1145,31	10,960	6636,1	28,8888	7495,5	4584,2	115,630
78	351,15	1877,86	1148,43	11,038	6607,8	28,9036	7516,3	4596,7	115,690
79	352,15	1883,05	1151,54	11,117	6579,6	28,9184	7537,1	4609,2	115,749
80	353,15	1888,24	1154,66	11,196	6551,7	28,9331	7557,9	4621,6	115,808
81	354,15	1893,43	1157,78	11,276	6524,0	28,9478	7578,7	4634,1	115,866
82	355,15	1898,63	1160,89	11,355	6496,4	28,9624	7599,4	4646,6	115,925
83	356,15	1903,82	1164,01	11,436	6469,1	28,9770	7620,2	4659,1	115,983
84	357,15	1909,01	1167,12	11,516	6442,0	28,9916	7641,0	4671,5	116,042
85	358,15	1914,21	1170,24	11,597	6415,0	29,0061	7661,8	4684,0	116,100
86	359,15	1919,40	1173,36	11,678	6388,2	29,0206	7682,6	4696,5	116,158
87	360,15	1924,59	1176,47	11,759	6361,6	29,0350	7703,4	4708,9	116,216
88	361,15	1929,79	1179,59	11,841	6335,2	29,0494	7724,2	4721,4	116,273
89	362,15	1934,98	1182,70	11,923	6309,0	29,0638	7744,9	4733,9	116,331
90	363,15	1940,17	1185,82	12,006	6283,0	29,0781	7765,7	4746,4	116,388
91	364,15	1945,37	1188,93	12,089	6257,1	29,0924	7786,5	4758,8	116,445
92	365,15	1950,56	1192,05	12,172	6231,4	29,1066	7807,3	4771,3	116,502
93	366,15	1955,75	1195,17	12,255	6205,9	29,1208	7828,1	4783,8	116,559
94	367,15	1960,95	1198,28	12,339	6180,6	29,1350	7848,9	4796,2	116,616
95	368,15	1966,14	1201,40	12,423	6155,4	29,1491	7869,7	4808,7	116,672
96	369,15	1971,33	1204,51	12,508	6130,4	29,1632	7890,5	4821,2	116,729
97	370,15	1976,52	1207,63	12,593	6105,6	29,1773	7911,2	4833,7	116,785
98	371,15	1981,72	1210,75	12,678	6080,9	29,1913	7932,0	4846,1	116,841
99	372,15	1986,91	1213,86	12,764	6056,4	29,2052	7952,8	4858,6	116,897
100	373,15	1992,10	1216,98	12,849	6032,1	29,2192	7973,6	4871,1	116,953
101	374,15	1997,30	1220,09	12,936	6007,9	29,2331	7994,4	4883,5	117,008
102	375,15	2002,49	1223,21	13,022	5983,9	29,2469	8015,2	4896,0	117,064
103	376,15	2007,68	1226,33	13,109	5960,1	29,2608	8056,0	4908,5	117,119
104	377,15	2012,88	1229,44	13,197	5936,4	29,2746	8056,7	4921,0	117,174
105	378,15	2018,07	1232,56	13,284	5912,9	29,2883	8077,5	4933,4	117,229
106	379,15	2023,26	1235,67	13,372	5889,5	29,3020	8098,3	4945,9	117,284
107	380,15	2028,46	1238,79	13,461	5866,3	29,3157	8119,1	4958,4	117,339
108	381,15	2033,65	1241,90	13,549	5843,2	29,3293	8139,9	4970,8	117,394
109	382,15	2038,84	1245,02	13,638	5820,3	29,3429	8160,7	4983,3	117,448
110	383,15	2044,04	1248,14	13,728	5797,5	29,3565	8181,5	4995,8	117,502
111	384,15	2049,23	1251,25	13,817	5774,9	29,3701	8202,2	5008,3	117,557
112	385,15	2054,42	1254,37	13,908	5752,4	29,3836	8225,0	5020,7	117,611
113	386,15	2059,61	1257,48	13,998	5730,1	29,3970	8243,8	5033,2	117,665
114	387,15	2064,81	1260,60	14,089	5707,9	29,4105	8264,6	5045,7	117,718

Helium (*Continued*)

t	T	h	u	π_0	θ_0	s^0	H	U	S
115	388,15	2070,00	1263,72	14,180	5685,8	29,4238	8285,4	5058,1	117,772
116	389,15	2075,19	1266,83	14,272	5663,9	29,4372	8306,2	5070,6	117,825
117	390,15	2080,39	1269,95	14,363	5642,2	29,4505	8327,0	5083,1	117,879
118	391,15	2085,58	1273,06	14,456	5620,5	29,4638	8347,7	5095,6	117,932
119	392,15	2090,77	1276,18	14,548	5599,1	29,4771	8368,5	5108,0	117,985
120	393,15	2095,97	1279,29	14,641	5577,7	29,4903	8389,3	5120,5	118,038
121	394,15	2101,16	1282,41	14,734	5556,5	29,5035	8410,1	5133,0	118,091
122	395,15	2106,35	1285,53	14,828	5535,4	29,5167	8430,9	5145,4	118,143
123	396,15	2111,55	1288,64	14,922	5514,5	29,5298	8451,7	5157,9	118,196
124	397,15	2116,74	1291,76	15,016	5493,7	29,5429	8472,5	5170,4	118,248
125	398,15	2121,93	1294,87	15,111	5473,0	29,5559	8493,2	5182,9	118,301
126	399,15	2127,13	1297,99	15,206	5452,4	29,5690	8514,0	5195,3	118,353
127	400,15	2132,32	1301,11	15,302	5432,0	29,5820	8534,8	5207,8	118,405
128	401,15	2137,51	1304,22	15,397	5411,7	29,5949	8555,6	5220,3	118,457
129	402,15	2142,70	1307,34	15,493	5391,5	29,6079	8576,4	5232,7	118,508
130	403,15	2147,90	1310,45	15,590	5371,5	29,6208	8597,2	5245,2	118,560
131	404,15	2153,09	1313,57	15,687	5351,5	29,6336	8618,0	5257,7	118,612
132	405,15	2158,28	1316,69	15,784	5331,7	29,6465	8638,7	5270,2	118,663
133	406,15	2163,48	1319,80	15,882	5312,1	29,6593	8659,5	5282,6	118,714
134	407,15	2168,67	1322,92	15,980	5292,5	29,6720	8680,3	5295,1	118,765
135	408,15	2173,86	1326,03	16,078	5273,1	29,6848	8701,1	5307,6	118,816
136	409,15	2179,06	1329,15	16,177	5253,7	29,6975	8721,9	5320,1	118,867
137	410,15	2184,25	1332,26	16,276	5234,5	29,7101	8742,7	5332,5	118,918
138	411,15	2189,44	1335,38	16,375	5215,5	29,7228	8763,5	5345,0	118,968
139	412,15	2194,64	1338,50	16,475	5196,5	29,7354	8784,2	5357,5	119,019
140	413,15	2199,83	1341,61	16,575	5177,6	29,7480	8805,0	5369,9	119,069
141	414,15	2205,02	1344,73	16,675	5158,9	29,7606	8825,8	5382,4	119,120
142	415,15	2210,22	1347,84	16,776	5140,3	29,7731	8846,6	5394,9	119,170
143	416,15	2215,41	1350,96	16,877	5121,7	29,7856	8867,4	5407,4	119,220
144	417,15	2220,60	1354,08	16,979	5103,3	29,7980	8888,2	5419,8	119,270
145	418,15	2225,79	1357,19	17,081	5085,0	29,8105	8909,0	5432,3	119,319
146	419,15	2230,99	1360,31	17,183	5066,9	29,8229	8929,8	5444,8	119,369
147	420,15	2236,18	1363,42	17,286	5048,8	29,8352	8950,5	5457,2	119,419
148	421,15	2241,37	1366,54	17,389	5030,8	29,8476	8971,3	5469,7	119,468
149	422,15	2246,57	1369,66	17,492	5012,9	29,8599	8992,1	5482,2	119,517
150	423,15	2251,76	1372,77	17,596	4995,2	29,8722	9012,9	5494,7	119,566
151	424,15	2256,95	1375,89	17,700	4977,5	29,8845	9033,7	5507,1	119,616
152	425,15	2262,15	1379,00	17,805	4960,0	29,8967	9054,5	5519,6	119,664
153	426,15	2267,34	1382,12	17,910	4942,5	29,9089	9075,3	5532,1	119,713
154	427,15	2272,53	1385,23	18,015	4925,2	29,9211	9096,0	5544,5	119,762
155	428,15	2277,73	1388,35	18,120	4907,9	29,9332	9116,8	5557,0	119,811
156	429,15	2282,92	1391,47	18,226	4890,8	29,9453	9137,6	5569,5	119,859
157	430,15	2288,11	1394,58	18,333	4873,7	29,9574	9158,4	5582,0	119,907
158	431,15	2293,31	1397,70	18,440	4856,8	29,9695	9179,2	5594,4	119,956
159	432,15	2298,50	1400,81	18,547	4840,0	29,9815	9200,0	5606,9	120,004
160	433,15	2303,69	1403,93	18,654	4823,2	29,9935	9220,8	5619,4	120,052
161	434,15	2308,88	1407,05	18,762	4806,5	30,0055	9241,5	5631,8	120,100
162	435,15	2314,08	1410,16	18,870	4790,0	30,0174	9262,3	5644,3	120,148
163	436,15	2319,27	1413,28	18,979	4773,5	30,0293	9283,1	5656,8	120,195
164	437,15	2324,46	1416,39	19,088	4757,2	30,0412	9303,9	5669,3	120,243
165	438,15	2329,66	1419,51	19,197	4740,9	30,0531	9324,7	5681,7	120,291
166	439,15	2334,85	1422,62	19,307	4724,7	30,0649	9345,5	5694,2	120,338
167	440,15	2340,04	1425,74	19,417	4708,6	30,0767	9366,3	5706,7	120,385
168	441,15	2345,24	1428,86	19,528	4692,6	30,0885	9387,0	5719,1	120,432
169	442,15	2350,43	1431,97	19,638	4676,7	30,1003	9407,8	5731,6	120,479

Helium (*Continued*)

t	T	h	u	π_0	θ_0	s^0	H	U	S
170	443,15	2355,62	1435,09	19,750	4660,9	30,1120	9428,6	5744,1	120,526
171	444,15	2360,82	1438,20	19,861	4645,1	30,1237	9449,4	5756,6	120,573
172	445,15	2366,01	1441,32	19,973	4629,5	30,1354	9470,2	5769,0	120,620
173	446,15	2371,20	1444,44	20,086	4613,9	30,1471	9491,0	5781,5	120,667
174	447,15	2376,40	1447,55	20,198	4598,5	30,1587	9511,8	5794,0	120,713
175	448,15	2381,59	1450,67	20,311	4583,1	30,1703	9532,5	5806,4	120,760
176	449,15	2386,78	1453,78	20,425	4567,8	30,1819	9553,3	5818,9	120,806
177	450,15	2391,97	1456,90	20,539	4552,6	30,1934	9574,1	5831,4	120,852
178	451,15	2397,17	1460,02	20,653	4537,4	30,2049	9594,9	5843,9	120,898
179	452,15	2402,36	1463,13	20,768	4522,4	30,2164	9615,7	5856,3	120,944
180	453,15	2407,55	1466,25	20,883	4507,4	30,2279	9636,5	5868,8	120,990
181	454,15	2412,75	1469,36	20,998	4492,6	30,2394	9657,3	5881,3	121,036
182	455,15	2417,94	1472,48	21,114	4477,8	30,2508	9678,0	5893,7	121,082
183	456,15	2423,13	1475,59	21,230	4463,0	30,2622	9698,8	5906,2	121,127
184	457,15	2428,33	1478,71	21,347	4448,4	30,2735	9719,6	5918,7	121,173
185	458,15	2433,52	1481,83	21,464	4433,8	30,2849	9740,4	5931,2	121,218
186	459,15	2438,71	1484,94	21,581	4419,4	30,2962	9761,2	5943,6	121,264
187	460,15	2443,91	1488,06	21,699	4405,0	30,3075	9782,0	5956,1	121,309
188	461,15	2449,10	1491,17	21,817	4390,7	30,3188	9802,8	5968,6	121,354
189	462,15	2454,29	1494,29	21,935	4376,4	30,3300	9823,5	5981,0	121,399
190	463,15	2459,49	1497,41	22,154	4362,2	30,3413	9844,3	5993,5	121,444
191	464,15	2464,68	1500,52	22,173	4348,2	30,3525	9865,1	6006,0	121,489
192	465,15	2469,87	1503,64	22,293	4334,1	30,3636	9885,9	6018,5	121,533
193	466,15	2475,06	1506,75	22,413	4320,2	30,3748	9906,7	6030,9	121,578
194	467,15	2480,26	1509,87	22,533	4306,3	30,3859	9927,5	6043,4	121,623
195	468,15	2485,45	1512,99	22,654	4292,5	30,3970	9948,3	6055,9	121,667
196	469,15	2490,64	1516,10	22,775	4278,8	30,4081	9969,1	6068,3	121,711
197	470,15	2495,84	1519,22	22,897	4265,2	30,4192	9989,8	6080,8	121,756
198	471,15	2501,03	1522,33	23,019	4251,6	30,4302	10010,6	6093,3	121,800
199	472,15	2506,22	1525,45	23,141	4238,1	30,4412	10031,4	6105,8	121,844
200	473,15	2511,42	1528,56	23,264	4224,7	30,4522	10052,2	6118,2	121,888
201	474,15	2516,61	1531,68	23,387	4211,3	30,4632	10073,0	6130,7	121,932
202	475,15	2521,80	1534,80	23,510	4198,0	30,4741	10093,8	6143,2	121,976
203	476,15	2527,00	1537,91	23,634	4184,8	30,4850	10114,6	6155,6	122,019
204	477,15	2532,19	1541,03	23,759	4171,7	30,4959	10135,3	6168,1	122,063
205	478,15	2537,38	1544,14	23,883	4158,6	30,5068	10156,1	6180,6	122,106
206	479,15	2542,58	1547,26	24,008	4145,6	30,5176	10176,9	6193,1	122,150
207	480,15	2547,77	1550,38	24,134	4132,6	30,5285	10197,7	6205,5	122,193
208	481,15	2552,96	1553,49	24,260	4119,8	30,5393	10218,5	6218,0	122,236
209	482,15	2558,15	1556,61	24,386	4106,9	30,5500	10239,3	6230,5	122,280
210	483,15	2563,35	1559,72	24,513	4094,2	30,5608	10260,1	6242,9	122,323
211	484,15	2568,54	1562,84	24,640	4081,5	30,5715	10280,8	6255,4	122,366
212	485,15	2573,73	1565,96	24,767	4068,9	30,5823	10301,6	6267,9	122,409
213	486,15	2578,93	1569,07	24,895	4056,4	30,5930	10322,4	6280,4	122,451
214	487,15	2584,12	1572,19	25,023	4043,9	30,6036	10343,2	6292,8	122,494
215	488,15	2589,31	1575,30	25,152	4031,5	30,6143	10364,0	6305,3	122,537
216	489,15	2594,51	1578,42	25,281	4019,1	30,6249	10384,8	6317,8	122,579
217	490,15	2599,70	1581,53	25,410	4006,8	30,6355	10405,6	6330,2	122,622
218	491,15	2604,89	1584,65	25,540	3994,6	30,6461	10426,3	6342,7	122,664
219	492,15	2610,09	1587,77	25,670	3982,4	30,6567	10447,1	6355,2	122,706
220	493,15	2615,28	1590,88	25,801	3970,3	30,6672	10467,9	6367,7	122,749
221	494,15	2620,47	1594,00	25,932	3958,3	30,6777	10488,7	6380,1	122,791
222	495,15	2625,67	1597,11	26,063	3946,3	30,6882	10509,5	6392,6	122,833
223	496,15	2630,86	1600,23	26,195	3934,3	30,6987	10530,3	6405,1	122,875
224	497,15	2636,05	1603,35	26,327	3922,5	30,7091	10551,1	6417,6	122,916

Helium (*Continued*)

t	T	h	u	π_0	θ_0	$s^\bullet$	H	U	S
225	498,15	2641,24	1606,46	26,460	3910,7	30,7196	10571,8	6430,0	122,958
226	499,15	2646,44	1609,58	26,593	3898,9	30,7300	10592,6	6442,5	123,000
227	500,15	2651,63	1612,69	26,726	3887,2	30,7404	10613,4	6455,0	123,041
228	501,15	2656,82	1615,81	26,860	3875,6	30,7508	10634,2	6467,4	123,083
229	502,15	2662,02	1618,92	26,994	3864,0	30,7611	10655,0	6479,9	123,124
230	503,15	2667,21	1622,04	27,129	3852,5	30,7714	10675,8	6492,4	123,166
231	504,15	2672,40	1625,16	27,264	3841,1	30,7818	10696,6	6504,9	123,207
232	505,15	2677,60	1628,27	27,399	3829,7	30,7920	10717,3	6517,3	123,248
233	506,15	2682,79	1631,39	27,535	3818,3	30,8023	10738,1	6529,8	123,289
234	507,15	2687,98	1634,50	27,671	3807,0	30,8126	10758,9	6542,3	123,330
235	508,15	2693,18	1637,62	27,808	3795,8	30,8228	10779,7	6554,7	123,371
236	509,15	2698,37	1640,74	27,945	3784,6	30,8330	10800,5	6567,2	123,412
237	510,15	2703,56	1643,85	28,082	3773,5	30,8432	10821,3	6579,7	123,453
238	511,15	2708,76	1646,97	28,220	3762,4	30,8534	10842,1	6592,2	123,494
239	512,15	2713,95	1650,08	28,358	3751,4	30,8635	10862,8	6604,6	123,534
240	513,15	2719,14	1653,20	28,497	3740,5	30,8736	10883,6	6617,1	123,575
241	514,15	2724,33	1656,32	28,636	3729,6	30,8838	10904,4	6629,6	123,615
242	515,15	2729,53	1659,43	28,775	3718,7	30,8938	10925,2	6642,0	123,656
243	516,15	2734,72	1662,55	28,915	3707,9	30,9039	10946,0	6654,5	123,696
244	517,15	2739,91	1665,66	29,055	3697,1	30,9140	10966,8	6667,0	123,736
245	518,15	2745,11	1668,78	29,196	3686,5	30,9240	10987,6	6679,5	123,776
246	519,15	2750,30	1671,89	29,337	3675,8	30,9340	11008,4	6691,9	123,816
247	520,15	2755,49	1675,01	29,479	3665,2	30,9440	11029,1	6704,4	123,856
248	521,15	2760,69	1678,13	29,620	3654,7	30,9540	11049,9	6716,9	123,896
249	522,15	2765,88	1681,24	29,763	3644,2	30,9639	11070,7	6729,3	123,936
250	523,15	2771,07	1684,36	29,905	3633,7	30,9739	11091,5	6741,8	123,976
251	524,15	2776,27	1687,47	30,049	3623,3	30,9838	11112,3	6754,3	124,016
252	525,15	2781,46	1690,59	30,192	3613,0	30,9937	11133,1	6766,8	124,055
253	526,15	2786,65	1693,71	30,336	3602,7	31,0036	11153,9	6779,2	124,095
254	527,15	2791,85	1696,82	30,480	3592,4	31,0134	11174,6	6791,7	124,134
255	528,15	2797,04	1699,94	30,625	3582,2	31,0233	11195,4	6804,2	124,174
256	529,15	2802,23	1703,05	30,770	3572,1	31,0331	11216,2	6816,6	124,213
257	530,15	2807,42	1706,17	30,916	3562,0	31,0429	11237,0	6829,1	124,252
258	531,15	2812,62	1709,29	31,062	3551,9	31,0527	11257,8	6841,6	124,291
259	532,15	2817,81	1712,40	31,208	3541,9	31,0625	11278,6	6854,1	124,331
260	533,15	2823,00	1715,52	31,355	3532,0	31,0722	11299,4	6866,5	124,370
261	534,15	2828,20	1718,63	31,502	3522,1	31,0819	11320,1	6879,0	124,409
262	535,15	2833,39	1721,75	31,650	3512,2	31,0916	11340,9	6891,5	124,447
263	536,15	2838,58	1724,86	31,798	3502,4	31,1013	11361,7	6903,9	124,486
264	537,15	2843,78	1727,98	31,947	3492,6	31,1110	11382,5	6916,4	124,525
265	538,15	2848,97	1731,10	32,095	3482,9	31,1207	11403,3	6928,9	124,564
266	539,15	2854,16	1734,21	32,245	3473,2	31,1303	11424,1	6941,4	124,602
267	540,15	2859,36	1737,33	32,394	3463,5	31,1399	11444,9	6953,8	124,641
268	541,15	2864,55	1740,44	32,545	3453,9	31,1495	11465,6	6966,3	124,679
269	542,15	2869,74	1743,56	32,695	3444,4	31,1591	11486,4	6978,8	124,718
270	543,15	2874,94	1746,68	32,846	3434,9	31,1687	11507,2	6991,2	124,756
271	544,15	2880,13	1749,79	32,998	3425,4	31,1783	11528,0	7003,7	124,794
272	545,15	2885,32	1752,91	33,149	3416,0	31,1878	11548,8	7016,2	124,832
273	546,15	2890,51	1756,02	33,302	3406,6	31,1973	11569,6	7028,7	124,870
274	547,15	2895,71	1759,14	33,454	3397,3	31,2068	11590,4	7041,1	124,908
275	548,15	2900,90	1762,26	33,607	3388,0	31,2163	11611,1	7053,6	124,946
276	549,15	2906,09	1765,37	33,761	3378,7	31,2258	11631,9	7066,1	124,984
277	550,15	2911,29	1768,49	33,915	3369,5	31,2352	11652,7	7078,5	125,022
278	551,15	2916,48	1771,60	34,069	3360,4	31,2446	11673,5	7091,0	125,060
279	552,15	2921,67	1774,72	34,224	3351,2	31,2540	11694,3	7103,5	125,097

Helium (*Continued*)

t	T	h	u	π_e	θ_e	s^e	H	U	S
280	553,15	2926,87	1777,83	34,379	3342,2	31,2634	11715,1	7116,0	125,135
281	554,15	2932,06	1780,95	34,535	3333,1	31,2728	11735,9	7128,4	125,173
282	555,15	2937,25	1784,07	34,691	3324,1	31,2822	11756,6	7140,9	125,210
283	556,15	2942,45	1787,18	34,847	3315,1	31,2915	11777,4	7153,4	125,247
284	557,15	2947,64	1790,30	35,004	3306,2	31,3009	11798,2	7165,8	125,285
285	558,15	2952,83	1793,41	35,161	3297,3	31,3102	11819,0	7178,3	125,322
286	559,15	2958,03	1796,53	35,319	3288,5	31,3195	11839,8	7190,8	125,359
287	560,15	2963,22	1799,65	35,477	3279,7	31,3288	11860,6	7203,3	125,396
288	561,15	2968,41	1802,76	35,636	3270,9	31,3380	11881,4	7215,7	125,434
289	562,15	2973,60	1805,88	35,794	3262,2	31,3473	11902,1	7228,2	125,471
290	563,15	2978,80	1808,99	35,954	3253,5	31,3565	11922,9	7240,7	125,507
291	564,15	2983,99	1812,11	36,114	3244,9	31,3657	11943,7	7253,1	125,544
292	565,15	2989,18	1815,22	36,274	3236,3	31,3749	11964,5	7265,6	125,581
293	566,15	2994,38	1818,34	36,435	3227,7	31,3841	11985,3	7278,1	125,618
294	567,15	2999,57	1821,46	36,596	3219,2	31,3932	12006,1	7290,6	125,655
295	568,15	3004,76	1824,57	36,757	3210,7	31,4024	12026,9	7303,0	125,691
296	569,15	3009,96	1827,69	36,919	3202,2	31,4115	12047,7	7315,5	125,728
297	570,15	3015,15	1830,80	37,082	3193,8	31,4206	12068,4	7328,0	125,764
298	571,15	3020,34	1833,92	37,244	3185,4	31,4297	12089,2	7340,4	125,801
299	572,15	3025,54	1837,04	37,408	3177,1	31,4388	12110,0	7352,9	125,837
300	573,15	3030,73	1840,15	37,571	3168,8	31,4479	12130,8	7365,4	125,873
301	574,15	3035,92	1843,27	37,735	3160,5	31,4569	12151,6	7377,9	125,910
302	575,15	3041,12	1846,38	37,900	3152,2	31,4660	12172,4	7390,3	125,946
303	576,15	3046,31	1849,50	38,065	3144,0	31,4750	12193,2	7402,8	125,982
304	577,15	3051,50	1852,62	38,230	3135,9	31,4840	12213,9	7415,3	126,018
305	578,15	3056,69	1855,73	38,396	3127,7	31,4930	12234,7	7427,7	126,054
306	579,15	3061,89	1858,85	38,562	3119,6	31,5020	12255,5	7440,2	126,090
307	580,15	3067,08	1861,96	38,729	3111,6	31,5109	12276,3	7452,7	126,126
308	581,15	3072,27	1865,08	38,896	3103,5	31,5199	12297,1	7465,2	126,161
309	582,15	3077,47	1868,19	39,064	3095,6	31,5288	12317,9	7477,6	126,197
310	583,15	3082,66	1871,31	39,232	3087,6	31,5377	12338,7	7490,1	126,233
311	584,15	3087,85	1874,43	39,400	3079,7	31,5466	12359,4	7502,6	126,268
312	585,15	3093,05	1877,54	39,569	3071,8	31,5555	12380,2	7515,1	126,304
313	586,15	3098,24	1880,66	39,738	3063,9	31,5644	12401,0	7527,5	126,340
314	587,15	3103,43	1883,77	39,908	3056,1	31,5732	12421,8	7540,0	126,575
315	588,15	3108,63	1886,89	40,078	3048,3	31,5821	12442,6	7552,5	126,410
316	589,15	3113,82	1890,01	40,249	3040,5	31,5909	12463,4	7564,9	126,446
317	590,15	3119,01	1893,12	40,420	3032,8	31,5997	12484,2	7577,4	126,481
318	591,15	3124,21	1896,24	40,591	3025,1	31,6085	12504,9	7589,9	126,516
319	592,15	3129,40	1899,35	40,763	3017,5	31,6173	12525,7	7602,4	126,551
320	593,15	3134,59	1902,47	40,935	3009,8	31,6260	12546,5	7614,8	126,586
321	594,15	3139,78	1905,59	41,108	3002,2	31,6348	12567,3	7627,3	126,621
322	595,15	3144,98	1908,70	41,281	2994,7	31,6435	12588,1	7639,8	126,656
323	596,15	3150,17	1911,82	41,455	2987,2	31,6522	12608,9	7652,2	126,691
324	597,15	3155,36	1914,93	41,629	2979,6	31,6609	12629,7	7664,7	126,726
325	598,15	3160,56	1918,05	41,803	2972,2	31,6696	12650,4	7677,2	126,761
326	599,15	3165,75	1921,16	41,978	2964,7	31,6783	12671,2	7689,7	126,796
327	600,15	3170,94	1924,28	42,154	2957,3	31,6869	12692,0	7702,1	126,830
328	601,15	3176,14	1927,40	42,330	2950,0	31,6956	12712,8	7714,6	126,865
329	602,15	3181,33	1930,51	42,506	2942,6	31,7042	12733,6	7727,1	126,899
330	603,15	3186,52	1933,63	42,682	2935,3	31,7128	12754,4	7739,5	126,934
331	604,15	3191,72	1936,74	42,860	2928,0	31,7214	12775,2	7752,0	126,968
332	605,15	3196,91	1939,86	43,037	2920,8	31,7300	12795,9	7764,5	127,003
333	606,15	3202,10	1942,98	43,215	2913,5	31,7386	12816,7	7777,0	127,037
334	607,15	3207,30	1946,09	43,394	2906,3	31,7472	12837,5	7789,4	127,071

Helium (*Continued*)

t	T	h	u	π_0	θ_0	s^0	H	U	S
335	608,15	3212,49	1949,21	43,573	2899,2	31,7557	12858,3	7801,9	127,105
336	609,15	3217,68	1952,32	43,752	2892,0	31,7642	12879,1	7814,4	127,140
337	610,15	3222,87	1955,44	43,932	2884,9	31,7728	12899,9	7826,8	127,174
338	611,15	3228,07	1958,55	44,112	2877,9	31,7813	12920,7	7839,3	127,208
339	612,15	3233,26	1961,67	44,293	2870,8	31,7898	12941,4	7851,8	127,242
340	613,15	3238,45	1964,79	44,474	2863,8	31,7982	12962,2	7864,3	127,276
341	614,15	3243,65	1967,90	44,655	2856,8	31,8067	12983,0	7876,7	127,309
342	615,15	3248,84	1971,02	44,837	2849,8	31,8151	13003,8	7889,2	127,343
343	616,15	3254,03	1974,13	45,020	2842,9	31,8236	13024,6	7901,7	127,377
344	617,15	3259,23	1977,25	45,203	2836,0	31,8320	13045,4	7914,1	127,411
345	618,15	3264,42	1980,37	45,386	2829,1	31,8404	13066,2	7926,6	127,444
346	619,15	3269,61	1983,48	45,570	2822,3	31,8488	13087,0	7939,1	127,478
347	620,15	3274,81	1986,60	45,754	2815,4	31,8572	13107,7	127,512
348	621,15	3280,00	1989,71	45,939	2808,6	31,8656	13128,5	7964,0	127,545
349	622,15	3285,19	1992,83	46,124	2801,9	31,8739	13149,3	7976,5	127,579
350	623,15	3290,39	1995,95	46,309	2795,1	31,8822	13170,1	7989,0	127,612
351	624,15	3295,58	1999,06	46,495	2788,4	31,8906	13190,9	8001,4	127,645
352	625,15	3300,77	2002,18	46,682	2781,7	31,8989	13211,7	8013,9	127,678
353	626,15	3305,96	2005,29	46,869	2775,1	31,9072	13232,5	8026,4	127,712
354	627,15	3311,16	2008,41	47,056	2768,4	31,9155	13253,2	8038,9	127,745
355	628,15	3316,35	2011,52	47,244	2761,8	31,9238	13274,0	8051,3	127,778
356	629,15	3321,54	2014,64	47,432	2755,2	31,9320	13294,8	8063,8	127,811
357	630,15	3326,74	2017,76	47,621	2748,7	31,9403	13315,6	8076,3	127,844
358	631,15	3331,93	2020,87	47,810	2742,1	31,9485	13336,4	8088,7	127,877
359	632,15	3337,12	2023,99	48,000	2735,6	31,9567	13357,2	8101,2	127,910
360	633,15	3342,32	2027,10	48,190	2729,2	31,9649	13378,0	8113,7	127,943
361	634,15	3347,51	2030,22	48,380	2722,7	31,9731	13398,7	8126,2	127,976
362	635,15	3352,70	2033,34	48,571	2716,3	31,9813	13419,5	8138,6	128,008
363	636,15	3357,90	2036,45	48,762	2709,9	31,9895	13440,3	8151,1	128,041
364	637,15	3363,09	2039,57	48,954	2703,5	31,9976	13461,1	8163,6	128,074
365	638,15	3368,28	2042,68	49,147	2697,2	32,0058	13481,9	8176,0	128,106
366	639,15	3373,48	2045,80	49,339	2690,8	32,0139	13502,7	8188,5	128,139
367	640,15	3378,67	2048,92	49,533	2684,5	32,0220	13523,5	8201,0	128,171
368	641,15	3383,86	2052,03	49,726	2678,2	32,0301	13544,2	8213,5	128,204
369	642,15	3389,05	2055,15	49,920	2672,0	32,0382	13565,0	8225,9	128,236
370	643,15	3394,25	2058,26	50,115	2665,8	32,0463	13585,8	8238,4	128,269
371	644,15	3399,44	2061,38	50,310	2659,6	32,0544	13606,6	8250,9	128,301
372	645,15	3404,63	2064,49	50,506	2653,4	32,0624	13627,4	8263,3	128,333
373	646,15	3409,83	2067,61	50,701	2647,2	32,0705	13648,2	8275,8	128,365
374	647,15	3415,02	2070,73	50,898	2641,1	32,0785	13669,0	8288,3	128,397
375	648,15	3420,21	2073,84	51,095	2635,0	32,0865	13689,7	8300,8	128,430
376	649,15	3425,41	2076,96	51,292	2628,9	32,0945	13710,5	8313,2	128,462
377	650,15	3430,60	2080,07	51,490	2622,8	32,1025	13731,3	8325,7	128,494
378	651,15	3435,79	2083,19	51,688	2616,8	32,1105	13752,1	8338,2	128,525
379	652,15	3440,99	2086,31	51,887	2610,8	32,1185	13772,9	8350,6	128,557
380	653,15	3446,18	2089,42	52,086	2604,8	32,1264	13793,7	8363,1	128,589
381	654,15	3451,37	2092,54	52,285	2598,8	32,1344	13814,5	8375,6	128,621
382	655,15	3456,57	2095,65	52,485	2592,9	32,1423	13835,2	8388,1	128,653
383	656,15	3461,76	2098,77	52,686	2586,9	32,1502	13856,0	8400,5	128,684
384	657,15	3466,95	2101,89	52,887	2581,0	32,1581	13876,8	8413,0	128,716
385	658,15	3472,14	2105,00	53,088	2575,1	32,1660	13897,6	8425,5	128,748
386	659,15	3477,34	2108,12	53,290	2569,3	32,1739	13918,4	8437,9	128,779
387	660,15	3482,53	2111,23	53,493	2563,5	32,1818	13939,2	8450,4	128,811
388	661,15	3487,72	2114,35	53,695	2557,6	32,1896	13960,0	8462,9	128,842
389	662,15	3492,92	2117,46	53,899	2551,8	32,1975	13980,7	8475,4	128,874

Helium (*Continued*)

t	T	h	u	π₀	θ₀	s⁰	H	U	S
390	663,15	3498,11	2120,58	54,102	2546,1	32,2053	14001,5	8487,8	128,905
391	664,15	3503,30	2123,70	54,307	2540,3	32,2132	14022,3	8500,3	128,936
392	665,15	3508,50	2126,81	54,511	2534,6	32,2210	14043,1	8512,8	128,968
393	666,15	3513,69	2129,93	54,716	2528,9	32,2288	14063,9	8525,2	128,999
394	667,15	3518,88	2133,04	54,922	2523,2	32,2366	14084,7	8537,7	129,030
395	668,15	3524,08	2136,16	55,128	2517,6	32,2443	14105,5	8550,2	129,061
396	669,15	3529,27	2139,28	55,335	2511,9	32,2521	14126,3	8562,7	129,092
397	670,15	3534,46	2142,39	55,541	2506,3	32,2599	14147,0	8575,1	129,123
398	671,15	3539,66	2145,51	55,749	2500,7	32,2676	14167,8	8587,6	129,154
399	672,15	3544,85	2148,62	55,957	2495,1	32,2753	14188,6	8600,1	129,185
400	673,15	3550,04	2151,74	56,165	2489,6	32,2831	14209,4	8612,6	129,216
401	674,15	3555,23	2154,85	56,374	2484,0	32,2908	14230,2	8625,0	129,247
402	675,15	3560,43	2157,97	56,583	2478,5	32,2985	14251,0	8637,5	129,278
403	676,15	3565,62	2161,09	56,793	2473,0	32,3062	14271,8	8650,0	129,309
404	677,15	3570,81	2164,20	57,003	2467,5	32,3138	14292,5	8662,4	129,339
405	678,15	3576,01	2167,32	57,214	2462,1	32,3215	14313,3	8674,9	129,370
406	679,15	3581,20	2170,43	57,425	2456,6	32,3291	14334,1	8687,4	129,401
407	680,15	3586,39	2173,55	57,637	2451,2	32,3368	14354,9	8699,9	129,431
408	681,15	3591,59	2176,67	57,849	2445,8	32,3444	14375,7	8712,3	129,462
409	682,15	3596,78	2179,78	58,061	2440,4	32,3520	14396,5	8724,8	129,492
410	683,15	3601,97	2182,90	58,274	2435,1	32,3596	14417,3	8737,3	129,523
411	684,15	3607,17	2186,01	58,488	2429,8	32,3672	14438,0	8749,7	129,553
412	685,15	3612,36	2189,13	58,702	2424,4	32,3748	14458,8	8762,2	129,583
413	686,15	3617,55	2192,25	58,916	2419,1	32,3824	14479,6	8774,7	129,614
414	687,15	3622,75	2195,36	59,131	2413,9	32,3900	14500,4	8787,2	129,644
415	688,15	3627,94	2198,48	59,347	2408,6	32,3975	14521,2	8799,6	129,674
416	689,15	3633,13	2201,59	59,562	2403,4	32,4050	14542,0	8812,1	129,704
417	690,15	3638,32	2204,71	59,779	2398,1	32,4126	14562,8	8824,6	129,735
418	691,15	3643,52	2207,82	59,995	2392,9	32,4201	14583,5	8837,0	129,765
419	692,15	3648,71	2210,94	60,213	2387,7	32,4276	14604,3	8849,5	129,795
420	693,15	3653,90	2214,06	60,430	2382,6	32,4351	14625,1	8862,0	129,825
421	694,15	3659,10	2217,17	60,649	2377,4	32,4426	14645,9	8874,5	129,855
422	695,15	3664,29	2220,29	60,867	2372,3	32,4501	14666,7	8886,9	129,885
423	696,15	3669,48	2223,40	61,086	2367,2	32,4575	14687,5	8899,4	129,915
424	697,15	3674,68	2226,52	61,306	2362,1	32,4650	14708,3	8911,9	129,944
425	698,15	3679,87	2229,64	61,526	2357,0	32,4724	14729,0	8924,3	129,974
426	699,15	3685,06	2232,75	61,747	2352,0	32,4799	14749,8	8936,8	130,004
427	700,15	3690,26	2235,87	61,968	2346,9	32,4873	14770,6	8949,3	130,034
428	701,15	3695,45	2238,98	62,189	2341,9	32,4947	14791,4	8961,8	130,063
429	702,15	3700,64	2242,10	62,411	2336,9	32,5021	14812,2	8974,2	130,093
430	703,15	3705,84	2245,22	62,634	2331,9	32,5095	14833,0	8986,7	130,122
431	704,15	3711,03	2248,33	62,857	2327,0	32,5169	14853,8	8999,2	130,152
432	705,15	3716,22	2251,45	63,080	2322,0	32,5242	14874,5	9011,6	130,182
433	706,15	3721,41	2254,56	63,304	2317,1	32,5316	14895,3	9024,1	130,211
434	707,15	3726,61	2257,68	63,528	2312,2	32,5389	14916,1	9036,6	130,240
435	708,15	3731,80	2260,79	63,753	2307,3	32,5463	14936,9	9049,1	130,270
436	709,15	3736,99	2263,91	63,978	2302,4	32,5536	14957,7	9061,5	130,299
437	710,15	3742,19	2267,03	64,204	2297,5	32,5609	14978,5	9074,0	130,328
438	711,15	3747,38	2270,14	64,430	2292,7	32,5682	14999,3	9086,5	130,358
439	712,15	3752,57	2273,26	64,657	2287,9	32,5755	15020,0	9098,9	130,387
440	713,15	3757,77	2276,37	64,884	2283,1	32,5828	15040,8	9111,4	130,416
441	714,15	3762,96	2279,49	65,112	2278,3	32,5901	15061,6	9123,9	130,445
442	715,15	3768,15	2282,61	65,340	2273,5	32,5974	15082,4	9136,4	130,474
443	716,15	3773,35	2285,72	65,569	2268,7	32,6046	15103,2	9148,8	130,503
444	717,15	3778,54	2288,84	65,798	2264,0	32,6119	15124,0	9161,3	130,532

276

Helium (*Continued*)

t	T	h	u	π₀	θ₀	s°	H	U	S
445	718,15	3783,73	2291,95	66,028	2259,3	32,6191	15144,8	9173,8	130,561
446	719,15	3788,93	2295,07	66,258	2254,5	32,6263	15165,6	9186,2	130,590
447	720,15	3794,12	2298,19	66,488	2249,9	32,6335	15186,3	9198,7	130,619
448	721,15	3799,31	2301,30	66,719	2245,2	32,6408	15207,1	9211,2	130,648
449	722,15	3804,50	2304,42	66,951	2240,5	32,6480	15227,9	9223,7	130,677
450	723,15	3809,70	2307,53	67,183	2235,9	32,6551	15248,7	9236,1	130,705
451	724,15	3814,89	2310,65	67,415	2231,2	32,6623	15269,5	9248,6	130,734
452	725,15	3820,08	2313,76	67,648	2226,6	32,6695	15290,3	9261,1	130,763
453	726,15	3825,28	2316,88	67,882	2222,0	32,6766	15311,1	9273,5	130,792
454	727,15	3830,47	2320,00	68,116	2217,4	32,6838	15331,8	9286,0	130,820
455	728,15	3835,66	2323,11	68,350	2212,9	32,6909	15352,6	9298,5	130,849
456	729,15	3840,86	2326,23	68,585	2208,3	32,6980	15373,4	9311,0	130,877
457	730,15	3846,05	2329,34	68,821	2203,8	32,7052	15394,2	9323,4	130,906
458	731,15	3851,24	2332,46	69,056	2199,3	32,7123	15415,0	9335,9	130,934
459	732,15	3856,44	2335,58	69,293	2194,8	32,7194	15435,8	9348,4	130,963
460	733,15	3861,63	2338,69	69,530	2190,3	32,7265	15456,6	9360,8	130,991
461	734,15	3866,82	2341,81	69,767	2185,8	32,7335	15477,3	9373,3	131,019
462	735,15	3872,02	2344,92	70,005	2181,3	32,7406	15498,1	9385,8	131,048
463	736,15	3877,21	2348,04	70,243	2176,9	32,7477	15518,9	9398,3	131,076
464	737,15	3882,40	2351,15	70,482	2172,5	32,7547	15539,7	9410,7	131,104
465	738,15	3887,59	2354,27	70,721	2168,1	32,7618	15560,5	9423,2	131,132
466	739,15	3892,79	2357,39	70,961	2163,7	32,7688	15581,3	9435,7	131,160
467	740,15	3897,98	2360,50	71,201	2159,3	32,7758	15602,1	9448,1	131,188
468	741,15	3903,17	2363,62	71,442	2154,9	32,7828	15622,8	9460,6	131,217
469	742,15	3908,37	2366,73	71,683	2150,6	32,7898	15643,6	9473,1	131,245
470	743,15	3913,56	2369,85	71,925	2146,2	32,7968	15664,4	9485,6	131,273
471	744,15	3918,75	2372,97	72,167	2141,9	32,8038	15685,2	9498,0	131,300
472	745,15	3923,95	2376,08	72,410	2137,6	32,8108	15706,0	9510,5	131,328
473	746,15	3929,14	2379,20	72,653	2133,3	32,8177	15726,8	9523,0	131,356
474	747,15	3934,33	2382,31	72,897	2129,0	32,8247	15747,6	9535,4	131,384
475	748,15	3939,53	2385,43	73,141	2124,7	32,8316	15768,3	9547,9	131,412
476	749,15	3944,72	2388,55	73,386	2120,5	32,8386	15789,1	9560,4	131,440
477	750,15	3949,91	2391,66	73,631	2116,2	32,8455	15809,9	9572,9	131,467
478	751,15	3955,11	2394,78	73,876	2112,0	32,8524	15830,7	9585,3	131,495
479	752,15	3960,30	2397,89	74,122	2107,8	32,8593	15851,5	9597,8	131,523
480	753,15	3965,49	2401,01	74,369	2103,6	32,8662	15872,3	9610,3	131,550
481	754,15	3970,68	2404,12	74,616	2099,4	32,8731	15893,1	9622,7	131,578
482	755,15	3975,88	2407,24	74,864	2095,3	32,8800	15913,8	9635,2	131,605
483	756,15	3981,07	2410,36	75,112	2091,1	32,8869	15934,6	9647,7	131,633
484	757,15	3986,26	2413,47	75,360	2087,0	32,8937	15955,4	9660,2	131,660
485	758,15	3991,46	2416,59	75,609	2082,8	32,9006	15976,2	9672,6	131,688
486	759,15	3996,65	2419,70	75,859	2078,7	32,9074	15997,0	9685,1	131,715
487	760,15	4001,84	2422,82	76,109	2074,6	32,9143	16017,8	9697,6	131,743
488	761,15	4007,04	2425,94	76,360	2070,5	32,9211	16038,6	9710,1	131,770
489	762,15	4012,23	2429,05	76,611	2066,5	32,9279	16059,3	9722,5	131,797
490	763,15	4017,42	2432,17	76,862	2062,4	32,9347	16080,1	9735,0	131,825
491	764,15	4022,62	2435,28	77,114	2058,4	32,9415	16100,9	9747,5	131,852
492	765,15	4027,81	2438,40	77,367	2054,3	32,9483	16121,7	9759,9	131,879
493	766,15	4033,00	2441,52	77,620	2050,3	32,9551	16142,5	9772,4	131,906
494	767,15	4038,20	2444,63	77,873	2046,3	32,9619	16163,3	9784,9	131,933
495	768,15	4043,39	2447,75	78,127	2042,3	32,9686	16184,1	9797,4	131,960
496	769,15	4048,58	2450,86	78,382	2038,3	32,9754	16204,9	9809,8	131,987
497	770,15	4053,77	2453,98	78,637	2034,3	32,9821	16225,6	9822,3	132,014
498	771,15	4058,97	2457,09	78,892	2030,4	32,9889	16246,4	9834,8	132,041
499	772,15	4064,16	2460,21	79,149	2026,4	32,9956	16267,2	9847,2	132,068

Helium (*Continued*)

t	T	h	u	π₀	θ₀	s⁰	H	U	S
500	773,15	4069,35	2463,33	79,405	2022,5	33,0023	16288,0	9859,7	132,095
501	774,55	4074,55	2466,44	79,662	2018,6	33,0090	16308,8	9872,2	132,122
502	775,15	4079,74	2469,56	79,920	2014,7	33,0157	16329,6	9884,7	132,149
503	776,15	4084,93	2472,67	80,178	2010,8	33,0224	16350,4	9897,1	132,176
504	777,15	4090,13	2475,79	80,436	2006,9	33,0291	16371,1	9909,6	132,202
505	778,15	4095,32	2478,91	80,695	2003,1	33,0358	16391,9	9922,1	132,229
506	779,15	4100,51	2482,02	80,955	1999,2	33,0425	16412,7	9934,5	132,256
507	780,15	4105,71	2485,14	81,215	1995,4	33,0491	16433,5	9947,0	132,282
508	781,15	4110,90	2488,25	81,475	1991,5	33,0558	16454,3	9959,5	132,309
509	782,15	4116,09	2491,37	81,736	1987,7	33,0624	16475,1	9972,0	132,336
510	783,15	4121,29	2494,49	81,998	1983,9	33,0691	16495,9	9984,4	132,362
511	784,15	4126,48	2497,60	82,260	1980,1	33,0757	16516,6	9996,9	132,389
512	785,15	4131,67	2500,72	82,522	1976,3	33,0823	16537,4	10009,4	132,415
513	786,15	4136,86	2503,83	82,785	1972,6	33,0889	16558,2	10021,8	132,442
514	787,15	4142,06	2506,95	83,049	1968,8	33,0955	16579,0	10034,3	132,468
515	788,15	4147,25	2510,06	83,313	1965,1	33,1021	16599,8	10046,8	132,495
516	789,15	4152,44	2513,18	83,577	1961,3	33,1087	16620,6	10059,3	132,521
517	790,15	4157,64	2516,30	83,842	1957,6	33,1153	16641,4	10071,7	132,547
518	791,15	4162,83	2519,41	84,108	1953,9	33,1218	16662,1	10084,2	132,574
519	792,15	4168,02	2522,53	84,374	1950,2	33,1284	16682,9	10096,7	132,600
520	793,15	4173,22	2525,64	84,640	1946,5	33,1350	16703,7	10109,1	132,626
521	794,15	4178,41	2528,76	84,907	1942,8	33,1415	16724,5	10121,6	132,652
522	795,15	4183,60	2531,88	85,175	1939,2	33,1480	16745,3	10134,1	132,678
523	796,15	4188,80	2534,99	85,443	1935,5	33,1546	16766,1	10146,6	132,704
524	797,15	4193,99	2538,11	85,711	1931,9	33,1611	16786,9	10159,0	132,731
525	798,15	4199,18	2541,22	85,981	1928,2	33,1676	16807,6	10171,5	132,757
526	799,15	4204,38	2544,34	86,250	1924,6	33,1741	16828,4	10184,0	132,783
527	800,15	4209,57	2547,45	86,520	1921,0	33,1806	16849,2	10196,4	132,809
528	801,15	4214,76	2550,57	86,791	1917,4	33,1871	16870,0	10208,9	132,835
529	802,15	4219,95	2553,69	87,062	1913,8	33,1936	16890,8	10221,4	132,861
530	803,15	4225,15	2556,80	87,333	1910,3	33,2000	16911,6	10233,9	152,886
531	804,15	4230,34	2559,92	87,606	1906,7	33,2065	16932,4	10246,3	132,912
532	805,15	4235,53	2563,03	87,878	1908,1	33,2129	16953,1	10258,8	132,938
533	806,15	4240,73	2566,15	88,151	1899,6	33,2194	16973,9	10271,3	132,964
534	807,15	4245,92	2569,27	88,425	1896,1	33,2258	16994,7	10283,7	132,990
535	808,15	4251,11	2572,38	88,699	1892,6	33,2323	17015,5	10296,2	133,015
536	809,15	4256,31	2575,50	88,974	1889,0	33,2387	17036,3	10308,7	133,041
537	810,15	4261,50	2578,61	89,249	1885,6	33,2451	17057,1	10321,2	133,067
538	811,15	4266,69	2581,73	89,524	1882,1	33,2515	17077,9	10333,6	133,092
539	812,15	4271,89	2584,85	89,801	1878,6	33,2579	17098,6	10346,1	133,118
540	813,15	4277,08	2587,96	90,077	1875,1	33,2643	17119,4	10358,6	133,144
541	814,15	4282,27	2591,08	90,355	1871,7	33,2707	17140,2	10371,0	133,169
542	815,15	4287,47	2594,19	90,632	1868,2	33,2770	17161,0	10383,5	133,195
543	816,15	4292,66	2597,31	90,910	1864,8	33,2834	17181,8	10396,0	133,220
544	817,15	4297,85	2600,42	91,189	1861,4	33,2898	17202,6	10408,5	133,246
545	818,15	4303,04	2603,54	91,468	1858,0	33,2961	17223,4	10420,9	133,271
546	819,15	4308,24	2606,66	91.748	1854,6	33,3025	17244,2	10433,4	133,296
547	820,15	4313,43	2609,77	92,028	1851,2	33,3088	17264,9	10445,9	133,322
548	821,15	4318,62	2612,89	92,309	1847,8	33,3151	17285,7	10458,3	133,347
549	822,15	4323,82	2616,00	92,591	1844,4	33,3214	17306,5	10470,8	133,372
550	823,15	4329,01	2619,12	92,872	1841,1	33,3278	17327,3	10483,3	133,398
551	824,15	4334,20	2622,24	93,155	1837,7	33,3341	17348,1	10495,8	133,423
552	825,15	4339,40	2625,35	93,437	1834,4	33,3404	17368,9	10508,2	133,448
553	826,15	4344,59	2628,47	93,721	1831,0	33,3467	17389,7	10520,7	133,477
554	827,15	4349,78	2631,58	94,005	1827,3	33,3529	17410,4	10533,2	133,498

t	T	h	u	π_0	θ_0	$s^\bullet$	H	U	S
555	828,15	4354,98	2634,70	94,289	1824,4	33,3592	17431,2	10545,6	133,524
556	829,15	4360,17	2637,82	94,574	1821,1	33,3655	17452,0	10558,1	133,549
557	830,15	4365,36	2640,93	94,859	1817,8	33,3717	17472,8	10570,6	133,574
558	831,15	4370,56	2644,05	95,145	1814,5	33,3780	17493,6	10583,1	133,599
559	832,15	4375,75	2647,16	95,432	1811,3	33,3842	17514,4	10595,5	133,624
560	833,15	4380,94	2650,28	95,719	1808,0	33,3905	17535,2	10608,0	133,649
561	834,15	4386,13	2653,39	96,006	1804,8	33,3967	17555,9	10620,5	133,674
562	835,15	4391,33	2656,51	96,294	1801,5	33,4029	17576,7	10632,9	133,699
563	836,15	4396,52	2659,63	96,583	1798,3	33,4091	17597,5	10645,4	133,723
564	837,15	4401,71	2662,74	96,872	1795,1	33,4153	17618,3	10657,9	133,748
565	838,15	4406,91	2665,86	97,161	1791,9	33,4215	17639,1	10670,4	133,773
566	839,15	4412,10	2668,97	97,451	1788,7	33,4277	17659,9	10682,8	133,798
567	840,15	4417,29	2672,09	97,742	1785,5	33,4339	17680,7	10695,3	133,823
568	841,15	4422,49	2675,21	98,033	1782,3	33,4401	17701,4	10707,8	133,847
569	842,15	4427,68	2678,32	98,325	1779,1	33,4463	17722,2	10720,2	133,872
570	843,15	4432,87	2681,44	98,617	1775,9	33,4524	17743,0	10732,7	133,897
571	844,15	4438,07	2684,55	98,909	1772,8	33,4586	17763,8	10745,2	133,921
572	845,15	4443,26	2687,67	99,203	1769,6	33,4647	17784,6	10757,7	133,946
573	846,15	4448,45	2690,78	99,496	1766,5	33,4709	17805,4	10770,1	133,971
574	847,15	4453,65	2693,90	99,791	1763,4	33,4770	17826,2	10782,6	133,995
575	848,15	4458,84	2697,02	100,08	1760,3	33,4831	17846,9	10795,1	134,020
576	849,15	4464,03	2700,13	100,38	1757,2	33,4893	17867,7	10807,6	134,044
577	850,15	4469,22	2703,25	100,67	1754,1	33,4954	17888,5	10820,0	134,069
578	851,15	4474,42	2706,36	100,97	1751,0	33,5015	17909,3	10832,5	134,093
579	852,15	4479,61	2709,48	101,27	1747,9	33,5076	17930,1	10845,0	134,117
580	853,15	4484,80	2712,60	101,56	1744,8	33,5137	17950,9	10857,4	134,142
581	854,15	4490,00	2715,71	101,86	1741,7	33,5197	17971,7	10869,9	134,166
582	855,15	4495,19	2718,83	102,16	1738,7	33,5258	17992,4	10882,4	134,190
583	856,15	4500,38	2721,94	102,46	1735,6	33,5319	18013,2	10894,9	134,215
584	857,15	4505,58	2725,06	102,76	1732,6	33,5379	18034,0	10907,3	134,239
585	858,15	4510,77	2728,18	103,06	1729,6	33,5440	18054,8	10919,8	134,263
586	859,15	4515,96	2731,29	103,36	1726,6	33,5501	18075,6	10932,3	134,287
587	860,15	4521,16	2734,41	103,66	1723,6	33,5561	18096,4	10944,7	134,312
588	861,15	4526,35	2737,52	103,96	1720,6	33,5621	18117,2	10957,2	134,336
589	862,15	4531,54	2740,64	104,26	1717,6	33,5682	18137,9	10969,7	134,360
590	863,15	4536,74	2743,75	104,56	1714,6	33,5742	18158,7	10982,2	134,384
591	864,15	4541,93	2746,87	104,87	1711,6	33,5802	18179,5	10994,6	134,408
592	865,15	4547,12	2749,99	105,17	1708,6	33,5862	18200,3	11007,1	134,432
593	866,15	4552,31	2753,10	105,48	1705,7	33,5922	18221,1	11019,6	134,456
594	867,15	4557,51	2756,22	105,78	1702,7	33,5982	18241,9	11032,0	134,480
595	868,15	4562,70	2759,33	106,09	1699,8	33,6042	18262,7	11044,5	134,504
596	869,15	4567,89	2762,45	106,39	1696,8	33,6101	18283,5	11057,0	134,528
597	870,15	4573,09	2765,57	106,70	1693,9	33,6161	18304,2	11069,5	134,552
598	871,15	4578,28	2768,68	107,00	1691,0	33,6221	18325,0	11081,9	134,576
599	872,15	4583,47	2771,80	107,31	1688,1	33,6280	18345,8	11094,4	134,600
600	873,15	4588,67	2774,91	107,62	1685,2	33,6340	18366,6	11106,9	134,623
601	874,15	4593,86	2778,03	107,93	1682,3	33,6399	18387,4	11119,3	134,647
602	875,15	4599,05	2781,15	108,24	1679,4	33,6459	18408,2	11131,8	134,671
603	876,15	4604,25	2784,26	108,55	1676,6	33,6518	18429,0	11144,3	134,695
604	877,15	4609,44	2787,38	108,86	1673,7	33,6577	18449,7	11156,8	134,718
605	878,15	4614,63	2790,49	109,17	1670,8	33,6636	18470,5	11169,2	134,742
606	879,15	4619,83	2793,61	109,48	1668,0	33,6696	18491,3	11181,7	134,766
607	880,15	4625,02	2796,72	109,79	1665,1	33,6755	18512,1	11194,2	134,789
608	881,15	4630,21	2799,84	110,10	1662,3	33,6814	18532,9	11206,6	134,813
609	882,15	4635,40	2802,96	110,41	1659,5	33,6872	18553,7	11219,1	134,837

Helium (*Continued*)

t	T	h	u	π₀	θ₀	s°	H	U	S
610	883,15	4640,60	2806,07	110,73	1656,7	33,6931	18574,5	11231,6	134,860
611	884,15	4645,79	2809,19	111,04	1653,9	33,6990	18595,2	11244,1	134,884
612	885,15	4650,98	2812,30	111,36	1651,0	33,7049	18616,0	11256,5	134,907
613	886,15	4656,18	2815,42	111,67	1648,3	33,7107	18636,8	11269,0	134,931
614	887,15	4661,37	2818,54	111,99	1645,5	33,7166	18657,6	11281,5	134,954
615	888,15	4666,56	2821,65	112,30	1642,7	33,7224	18678,4	11293,9	134,977
616	889,15	4671,76	2824,77	112,62	1639,9	33,7283	18699,2	11306,4	135,001
617	890,15	4676,95	2827,88	112,94	1637,2	33,7341	18720,0	11318,9	135,024
618	891,15	4682,14	2831,00	113,25	1634,4	33,7400	18740,7	11331,4	135,048
619	892,15	4687,34	2834,12	113,57	1631,7	33,7458	18761,5	11343,8	135,071
620	893,15	4692,53	2837,23	113,89	1628,9	33,7516	18782,3	11356,3	135,094
621	894,15	4697,72	2840,35	114,21	1626,2	33,7574	18803,1	11368,8	135,117
622	895,15	4702,92	2843,46	114,53	1623,5	33,7632	18823,9	11381,2	135,141
623	896,15	4708,11	2846,58	114,85	1620,7	33,7690	18844,7	11393,7	135,164
624	897,15	4713,30	2849,69	115,17	1618,0	33,7748	18865,5	11406,2	135,187
625	898,15	4718,49	2852,81	115,49	1615,3	33,7806	18886,2	11418,7	135,210
626	899,15	4723,69	2855,93	115,81	1612,6	33,7864	18907,0	11431,1	135,233
627	900,15	4728,88	2859,04	116,13	1610,0	33,7921	18927,8	11443,6	135,256
628	901,15	4734,07	2862,16	116,46	1607,3	33,7979	18948,6	11456,1	135,280
629	902,15	4739,27	2865,27	116,78	1604,6	33,8037	18969,4	11468,5	135,303
630	903,15	4744,46	2868,39	117,10	1601,9	33,8094	18990,2	11481,0	135,326
631	904,15	4749,65	2871,51	117,43	1599,3	33,8152	19011,0	11493,5	135,349
632	905,15	4754,85	2874,62	117,75	1596,6	33,8209	19031,7	11506,0	135,372
633	906,15	4760,04	2877,74	118,08	1594,0	33,8266	19052,5	11518,4	135,395
634	907,15	4765,23	2880,85	118,41	1591,4	33,8324	19073,3	11530,9	135,417
635	908,15	4770,43	2883,97	118,73	1588,7	33,8381	19094,1	11543,4	135,440
636	909,15	4775,62	2887,08	119,06	1586,1	33,8438	19114,9	11555,8	135,463
637	910,15	4780,81	2890,20	119,39	1583,5	33,8495	19135,7	11568,3	135,486
638	911,15	4786,01	2893,32	119,71	1580,9	33,8552	19156,5	11580,8	135,509
639	912,15	4791,20	2896,43	120,04	1578,3	33,8609	19177,2	11593,3	135,532
640	913,15	4796,39	2899,55	120,37	1575,7	33,8666	19198,0	11605,7	135,554
641	914,15	4801,58	2902,66	120,70	1573,1	33,8723	19218,8	11618,2	135,577
642	915,15	4806,78	2905,78	121,03	1570,5	33,8780	19239,6	11630,7	135,600
643	916,15	4811,97	2908,90	121,36	1568,0	33,8836	19260,4	11643,1	135,623
644	917,15	4817,16	2912,01	121,70	1565,4	33,8893	19281,2	11655,6	135,645
645	918,15	4822,36	2915,13	122,03	1562,8	33,8950	19302,0	11668,1	135,668
646	919,15	4827,55	2918,24	122,36	1560,3	33,9006	19322,8	11680,6	135,691
647	920,15	4832,74	2921,36	122,69	1557,7	33,9063	19343,5	11693,0	135,713
648	921,15	4837,94	2924,48	123,03	1555,2	33,9119	19364,3	11705,5	135,736
649	922,15	4843,13	2927,59	123,36	1552,7	33,9175	19385,1	11718,0	135,758
650	923,15	4848,32	2930,71	123,70	1550,2	33,9232	19405,9	11730,4	135,781
651	924,15	4853,52	2933,82	124,03	1547,6	33,9288	19426,7	11742,9	135,803
652	925,15	4858,71	2936,94	124,37	1545,1	33,9344	19447,5	11755,4	135,826
653	926,15	4863,90	2940,05	124,70	1542,6	33,9400	19468,3	11767,9	135,848
654	927,15	4869,10	2943,17	125,04	1540,1	33,9456	19489,0	11780,3	135,871
655	928,15	4874,29	2946,29	125,38	1537,6	33,9512	19509,8	11792,8	135,893
656	929,15	4879,48	2949,40	125,72	1535,2	33,9568	19530,6	11805,3	135,916
657	930,15	4884,67	2952,52	126,05	1532,7	33,9624	19551,4	11817,7	135,938
658	931,15	4889,87	2955,63	126,39	1530,2	33,9680	19572,2	11830,2	135,960
659	932,15	4895,06	2958,75	126,73	1527,8	33,9736	19593,0	11842,7	135,983
660	933,15	4900,25	2961,87	127,07	1525,3	33,9791	19613,8	11855,2	136,005
661	934,15	4905,45	2964,98	127,41	1522,9	33,9847	19634,5	11867,6	136,027
662	935,15	4910,64	2968,10	127,75	1520,4	33,9902	19655,3	11880,1	136,049
663	936,15	4915,83	2971,21	128,10	1518,0	33,9958	19676,1	11892,6	136,072
664	937,15	4921,03	2974,33	128,44	1515,6	34,0013	19696,9	11905,1	136,094

Helium (*Continued*)

t	T	h	u	π_0	θ_0	s^0	H	U	S
665	938,15	4926,22	2977,45	128,78	1513,1	34,0069	19717,7	11917,5	136,116
666	939,15	4931,41	2980,56	129,13	1510,7	34,0124	19738,5	11930,0	136,138
667	940,15	4936,61	2983,68	129,47	1508,3	34,0179	19759,3	11942,5	136,160
668	941,15	4941,80	2986,79	129,81	1505,9	34,0235	19780,0	11954,9	136,182
669	942,15	4946,99	2989,91	130,16	1503,5	34,0290	19800,8	11967,4	136,204
670	943,15	4952,19	2993,02	130,50	1501,1	34,0345	19821,6	11979,9	136,226
671	944,15	4957,38	2996,14	130,85	1498,7	34,0400	19842,4	11992,4	136,248
672	945,15	4962,57	2999,26	131,20	1496,3	34,0455	19863,2	12004,8	136,270
673	946,15	4967,76	3002,37	131,55	1494,0	34,0510	19884,0	12017,3	136,292
674	947,15	4972,96	3005,49	131,89	1491,6	34,0565	19904,8	12029,8	136,314
675	948,15	4978,15	3008,60	132,24	1489,3	34,0619	19925,5	12042,2	136,336
676	949,15	4983,34	3011,72	132,59	1486,9	34,0674	19946,3	12054,7	136,358
677	950,15	4988,54	3014,84	132,94	1484,6	34,0729	19967,1	12067,2	136,380
678	951,15	4993,73	3017,95	133,29	1482,2	34,0783	19987,9	12079,7	136,402
679	952,15	4998,92	3021,07	133,64	1479,9	34,0838	20008,7	12092,1	136,424
680	953,15	5004,12	3024,18	133,99	1477,5	34,0892	20029,5	12104,6	136,446
681	954,15	5009,31	3027,30	134,34	1475,2	34,0947	20050,3	12117,1	136,467
682	955,15	5014,50	3030,42	134,70	1472,9	34,1001	20071,0	12129,5	136,489
683	956,15	5019,70	3033,53	135,05	1470,6	34,1056	20091,8	12142,0	136,511
684	957,15	5024,89	3036,65	135,40	1468,3	34,1110	20112,6	12154,5	136,533
685	958,15	5030,08	3039,76	135,76	1466,0	34,1164	20133,4	12167,0	136,554
686	959,15	5035,28	3042,88	136,11	1463,7	34,1218	20154,2	12179,4	136,576
687	960,15	5040,47	3045,99	136,47	1461,4	34,1272	20175,0	12191,9	136,598
688	961,15	5045,66	3049,11	136,82	1459,1	34,1327	20195,8	12204,4	136,619
689	962,15	5050,85	3052,23	137,18	1456,9	34,1381	20216,5	12216,8	136,641
690	963,15	5056,05	3055,34	137,53	1454,6	34,1434	20237,3	12229,3	136,663
691	964,15	5061,24	3058,46	137,89	1452,3	34,1488	20258,1	12241,8	136,684
692	965,15	5066,43	3061,57	138,25	1450,1	34,1542	20278,9	12254,3	136,706
693	966,15	5071,63	3064,69	138,61	1447,8	34,1596	20299,7	12266,7	136,727
694	967,15	5076,82	3067,81	138,97	1445,6	34,1650	20320,5	12279,2	136,749
695	968,15	5082,01	3070,92	139,33	1443,3	34,1703	20341,3	12291,7	136,770
696	969,15	5087,21	3074,04	139,69	1441,1	34,1757	20362,1	12304,1	136,792
697	970,15	5092,40	3077,15	140,05	1438,9	34,1811	20382,8	12316,6	136,813
698	971,15	5097,59	3080,27	140,41	1436,7	34,1864	20403,6	12329,1	136,835
699	972,15	5102,79	3083,38	140,77	1434,4	34,1917	20424,4	12341,6	136,856
700	973,15	5107,98	3086,50	141,13	1432,2	34,1971	20445,2	12354,0	136,877
701	974,15	5113,17	3089,62	141,49	1430,0	34,2024	20466,0	12366,5	136,899
702	975,15	5118,37	3092,73	141,86	1427,8	34,2078	20486,8	12379,0	136,920
703	976,15	5123,56	3095,85	142,22	1425,6	34,2131	20507,6	12391,4	136,941
704	977,15	5128,75	3098,96	142,59	1423,4	34,2184	20528,3	12403,9	136,963
705	978,15	5133,94	3102,08	142,95	1421,3	34,2237	20549,1	12416,4	136,984
706	979,15	5139,14	3105,20	143,32	1419,1	34,2290	20569,9	12428,9	137,005
707	980,15	5144,33	3108,31	143,68	1416,9	34,2343	20590,7	12441,3	137,026
708	981,15	5149,52	3111,43	144,05	1414,8	34,2396	20611,5	12453,8	137,047
709	982,15	5154,72	3114,54	144,42	1412,6	34,2449	20632,3	12466,3	137,069
710	983,15	5159,91	3117,66	144,79	1410,4	34,2502	20653,1	12478,7	137,090
711	984,15	5165,10	3120,78	145,15	1408,3	34,2555	20673,8	12491,2	137,111
712	985,15	5170,30	3123,89	145,52	1406,1	34,2607	20694,6	12503,7	137,132
713	986,15	5175,49	3127,01	145,89	1404,0	34,2660	20715,4	12516,2	137,153
714	987,15	5180,68	3130,12	146,26	1401,9	34,2713	20736,2	12528,6	137,174
715	988,15	5185,88	3133,24	146,63	1399,7	34,2765	20757,0	12541,1	137,195
716	989,15	5191,07	3136,35	147,00	1397,6	34,2818	20777,8	12553,6	137,216
717	990,15	5196,26	3139,47	147,38	1395,5	34,2870	20798,6	12566,0	137,237
718	991,15	5201,46	3142,59	147,75	1393,4	34,2923	20819,3	12578,5	137,258
719	992,15	5206,65	3145,70	148,12	1391,3	34,2975	20840,1	12591,0	137,279

t	T	h	u	π_o	θ_o	s^o	H	U	S
720	993,15	5211,84	3148,82	148,50	1389,2	34,3027	20860,9	12603,5	137,300
721	994,15	5217,03	3151,93	148,87	1387,1	34,3080	20881,7	12615,9	137,321
722	995,15	5222,23	3155,05	149,24	1585,0	34,3132	20902,5	12628,4	137,342
723	996,15	5227,42	3158,17	149,62	1382,9	34,3184	20923,3	12640,9	137,363
724	997,15	5232,61	3161,28	150,00	1380,8	34,3236	20944,1	12653,3	137,384
725	998,15	5237,81	3164,40	150,37	1378,8	34,3288	20964,8	12665,8	137,405
726	999,15	5243,00	3167,51	150,75	1376,7	34,3340	20985,6	12678,3	137,425
727	1000,15	5248,19	3170,63	151,13	1374,6	34,3392	21006,4	12690,8	137,446
728	1001,15	5253,39	3173,75	151,50	1372,6	34,3444	21027,2	12703,2	137,467
729	1002,15	5258,58	3176,86	151,88	1370,5	34,3496	21048,0	12715,7	137,488
730	1003,15	5263,77	3179,98	152,26	1368,5	34,3548	21068,8	12728,2	137,508
731	1004,15	5268,97	3183,09	152,64	1366,4	34,3599	21089,6	12740,6	137,529
732	1005,15	5274,16	3186,21	153,02	1364,4	34,3651	21110,3	12753,1	137,550
733	1006,15	5279,35	3189,32	153,40	1362,3	34,3703	21131,1	12765,6	137,570
734	1007,15	5284,55	3192,44	153,78	1360,3	34,3754	21151,9	12778,1	137,591
735	1008,15	5289,74	3195,56	154,17	1358,3	34,3806	21172,7	12790,5	137,612
736	1009,15	5291,93	3198,67	154,55	1356,3	34,3857	21193,5	12803,0	137,632
737	1010,15	5300,12	3201,79	154,93	1354,3	34,3909	21214,3	12815,5	137,653
738	1011,15	5305,32	3204,90	155,32	1352,3	34,3960	21235,1	12827,9	137,673
739	1012,15	5310,51	3208,02	155,70	1350,3	34,4011	21255,8	12840,4	137,694
740	1013,15	5315,70	3211,14	156,09	1348,3	34,4063	21276,6	12852,9	137,715
741	1014,15	5320,90	3214,25	156,47	1346,3	34,4114	21297,4	12865,4	137,735
742	1015,15	5326,09	3217,37	156,86	1344,3	34,4165	21318,2	12877,8	137,756
743	1016,15	5331,28	3220,48	157,24	1342,3	34,4216	21339,0	12890,3	137,776
744	1017,15	5336,48	3223,60	157,63	1340,3	34,4267	21359,8	12902,8	137,796
745	1018,15	5341,67	3226,71	158,02	1338,3	34,4318	21380,6	12915,2	137,817
746	1019,15	5346,86	3229,83	158,41	1336,4	34,4369	21401,4	12927,7	137,837
747	1020,15	5352,06	3232,95	158,80	1334,4	34,4420	21422,1	12940,2	137,858
748	1021,15	5357,25	3236,06	159,18	1332,4	34,4471	21442,9	12952,7	137,878
749	1022,15	5362,44	3239,18	159,57	1330,5	34,4522	21463,7	12965,1	137,898
750	1023,15	5367,64	3242,29	159,97	1328,5	34,4573	21484,5	12977,6	137,919
751	1024,15	5372,83	3245,41	160,36	1326,6	34,4624	21505,3	12990,1	137,939
752	1025,15	5378,02	3248,53	160,75	1324,6	34,4674	21526,1	13002,6	137,959
753	1026,15	5383,21	3251,64	161,14	1322,7	34,4725	21546,9	13015,0	137,980
754	1027,15	5388,41	3254,76	161,53	1320,8	34,4775	21567,6	13027,5	138,000
755	1028,15	5393,60	3257,87	161,93	1318,9	34,4826	21588,4	13040,0	138,020
756	1029,15	5398,79	3260,99	162,32	1316,9	34,4876	21609,2	13052,4	138,040
757	1030,15	5403,99	3264,11	162,72	1315,0	34,4927	21630,0	13064,9	138,060
758	1031,15	5409,18	3267,22	163,11	1313,1	34,4977	21650,8	13077,4	138,081
759	1032,15	5414,37	3270,34	163,51	1311,2	34,5028	21671,6	13089,9	138,101
760	1033,15	5419,57	3273,45	163,90	1309,3	34,5078	21692,4	13102,3	138,121
761	1034,15	5424,76	3276,57	164,30	1307,4	34,5128	21713,1	13114,8	138,141
762	1035,15	5429,95	3279,68	164,70	1305,5	34,5178	21733,9	13127,3	138,161
763	1036,15	5435,15	3282,80	165,10	1303,6	34,5228	21754,7	13139,7	138,181
764	1037,15	5440,34	3285,92	165,49	1301,7	34,5279	21775,5	13152,2	138,201
765	1038,15	5445,53	3289,03	165,89	1299,8	34,5329	21796,3	13164,7	138,221
766	1039,15	5450,73	3292,15	166,29	1298,0	34,5379	21817,1	13177,2	138,241
767	1040,15	5455,92	3295,26	166,69	1296,1	34,5429	21837,9	13189,6	138,261
768	1041,15	5461,11	3298,38	167,09	1294,2	34,5478	21858,6	13202,1	138,281
769	1042,15	5466,30	3301,50	167,50	1292,4	34,5528	21879,4	13214,6	138,301
770	1043,15	5471,50	3304,61	167,90	1290,5	34,5578	21900,2	13227,0	138,321
771	1044,15	5476,69	3307,73	168,30	1288,7	34,5628	21921,0	13239,5	138,341
772	1045,15	5481,88	3310,84	168,70	1286,8	34,5678	21941,8	13252,0	138,361
773	1046,15	5487,08	3313,96	169,11	1285,0	34,5727	21962,6	13264,5	138,381
774	1047,15	5492,27	3317,08	169,51	1283,1	34,5777	21983,4	13276,9	138,401

Helium (Continued)

t	T	h	u	π_e	θ_e	s^o	H	U	S
775	1048,15	5497,46	3320,19	169,92	1281,3	34,5826	22004,1	13289,4	138,420
776	1049,15	5502,66	3323,31	170,32	1279,5	34,5876	22024,9	13301,9	138,440
777	1050,15	5507,85	3326,42	170,73	1277,6	34,5925	22045,7	13314,3	138,460
778	1051,15	5513,04	3329,54	171,14	1275,8	34,5975	22066,5	13326,8	138,480
779	1052,15	5518,24	3332,65	171,54	1274,0	34,6024	22087,3	13339,3	138,500
780	1053,15	5523,43	3335,77	171,95	1272,2	34,6074	22108,1	13351,8	138,519
781	1054,15	5528,62	3338,89	172,36	1270,4	34,6123	22128,9	13364,2	138,539
782	1055,15	5533,81	3342,00	172,77	1268,6	34,6172	22149,6	13376,7	138,559
783	1056,15	5539,01	3345,12	173,18	1266,8	34,6221	22170,4	13389,2	138,579
784	1057,15	5544,20	3348,23	173,59	1265,0	34,6270	22191,2	13401,6	138,598
785	1058,15	5549,39	3351,35	174,00	1263,2	34,6320	22212,0	13414,1	138,618
786	1059,15	5554,59	3354,47	174,41	1261,4	34,6369	22232,8	13426,6	138,638
787	1060,15	5559,78	3357,58	174,82	1259,6	34,6418	22253,6	13439,1	138,657
788	1061,15	5564,97	3360,70	175,23	1257,8	34,6467	22274,4	13451,5	138,677
789	1062,15	5570,17	3363,81	175,65	1256,0	34,6516	22295,1	13464,0	138,696
790	1063,15	5575,36	3366,93	176,06	1254,3	34,6564	22315,9	13476,5	138,716
791	1064,15	5580,55	3370,05	176,48	1252,5	34,6613	22336,7	13488,9	138,735
792	1065,15	5585,75	3373,16	176,89	1250,7	34,6662	22357,5	13501,4	138,755
793	1066,15	5590,94	3376,28	177,31	1249,0	34,6711	22378,3	13513,9	138,774
794	1067,15	5596,13	3379,39	177,72	1247,2	34,6759	22399,1	13526,4	138,794
795	1068,15	5601,33	3382,51	178,14	1245,5	34,6808	22419,9	13538,8	138,813
796	1069,15	5606,52	3385,62	178,56	1243,7	34,6857	22440,7	13551,3	138,833
797	1070,15	5611,71	3388,74	178,97	1242,0	34,6905	22461,4	13563,8	138,852
798	1071,15	5616,90	3391,86	179,39	1240,2	34,6954	22482,2	13576,2	138,872
799	1072,15	5622,10	3394,97	179,81	1238,5	34,7002	22503,0	13588,7	138,891
800	1073,15	5627,29	3398,09	180,23	1236,8	34,7051	22523,8	13601,2	138,910
801	1074,15	5632,48	3401,20	180,65	1235,0	34,7099	22544,6	13613,7	138,930
802	1075,15	5637,68	3404,32	181,07	1233,3	34,7147	22565,4	13626,1	138,949
803	1076,15	5642,87	3407,44	181,49	1231,6	34,7196	22586,2	13638,6	138,968
804	1077,15	5648,06	3410,55	181,92	1229,9	34,7244	22606,9	13651,1	138,988
805	1078,15	5653,26	3413,67	182,34	1228,2	34,7292	22627,7	13663,5	139,007
806	1079,15	5658,45	3416,78	182,76	1226,5	34,7340	22648,5	13676,0	139,026
807	1080,15	5663,64	3419,90	183,18	1224,8	34,7388	22669,3	13688,5	139,046
808	1081,15	5668,84	3423,01	183,61	1223,1	34,7436	22690,1	13701,0	139,065
809	1082,15	5674,03	3426,13	184,03	1221,4	34,7484	22710,9	13713,4	139,084
810	1083,15	5679,22	3429,25	184,46	1219,7	34,7532	22731,7	13725,9	139,103
811	1084,15	5684,42	3432,36	184,89	1218,0	34,7580	22752,4	13738,4	139,122
812	1085,15	5689,61	3435,48	185,31	1216,3	34,7628	22773,2	13750,8	139,142
813	1086,15	5694,80	3438,59	185,74	1214,6	34,7676	22794,0	13763,3	139,161
814	1087,15	5699,99	3441,71	186,17	1213,0	34,7724	22814,8	13775,8	139,180
815	1088,15	5705,19	3444,83	186,60	1211,3	34,7771	22835,6	13788,3	139,199
816	1089,15	5710,38	3447,94	187,02	1209,6	34,7819	22856,4	13800,7	139,218
817	1090,15	5715,57	3451,06	187,45	1208,0	34,7867	22877,2	13813,2	139,237
818	1091,15	5720,77	3454,17	187,88	1206,3	34,7914	22897,9	13825,7	139,256
819	1092,15	5725,96	3457,29	188,32	1204,6	34,7962	22918,7	13838,1	139,275
820	1093,15	5731,15	3460,41	188,75	1203,0	34,8009	22939,5	13850,6	139,294
821	1094,15	5736,35	3463,52	189,18	1201,3	34,8057	22960,3	13863,1	139,313
822	1095,15	5741,54	3466,64	189,61	1199,7	34,8104	22981,1	13875,6	139,332
823	1096,15	5746,73	3469,75	190,04	1198,1	34,8152	23001,9	13888,0	139,351
824	1097,15	5751,93	3472,87	190,48	1196,4	34,8199	23022,7	13900,5	139,370
825	1098,15	5757,12	3475,98	190,91	1194,8	34,8246	23043,4	13913,0	139,389
826	1099,15	5762,31	3479,10	191,35	1193,1	34,8294	23064,2	13925,4	139,408
827	1100,15	5767,51	3482,22	191,78	1191,5	34,8341	23085,0	13937,9	139,427
828	1101,15	5772,70	3485,33	192,22	1189,9	34,8388	23105,8	13950,4	139,446
829	1102,15	5777,89	3488,45	192,66	1188,3	34,8435	23126,6	13962,9	139,465

Helium (*Continued*)

t	T	h	u	π_0	θ_0	s^0	H	U	S
830	1103,15	5783,08	3491,56	193,09	1186,7	34,8482	23147,4	13975,3	139,484
831	1104,15	5788,28	3494,68	193,53	1185,1	34,8529	23168,2	13987,8	139,502
832	1105,15	5793,47	3497,80	193,97	1183,4	34,8576	23188,9	14000,3	139,521
833	1106,15	5798,66	3500,91	194,41	1181,8	34,8623	23209,7	14012,7	139,540
834	1107,15	5803,86	3504,03	194,85	1180,2	34,8670	23230,5	14025,2	139,559
835	1108,15	5809,05	3507,14	195,29	1178,6	34,8717	23251,3	14037,7	139,578
836	1109,15	5814,24	3510,26	195,73	1177,0	34,8764	23272,1	14050,2	139,596
837	1110,15	5819,44	3513,38	196,17	1175,5	34,8811	23292,9	14062,6	139,615
838	1111,15	5824,63	3516,49	196,61	1173,9	34,8858	23313,7	14075,1	139,634
839	1112,15	5829,82	3519,61	197,06	1172,3	34,8904	23334,4	14087,6	139,652
840	1113,15	5835,02	3522,72	197,50	1170,7	34,8951	23355,2	14100,1	139,671
841	1114,15	5840,21	3525,84	197,94	1169,1	34,8998	23376,0	14112,5	139,690
842	1115,15	5845,40	3528,95	198,39	1167,6	34,9044	23396,8	14125,0	139,708
843	1116,15	5850,60	3532,07	198,83	1166,0	34,9091	23417,6	14137,5	139,727
844	1117,15	5855,79	3535,19	199,28	1164,4	34,9137	23438,4	14149,9	139,746
845	1118,15	5860,98	3538,30	199,72	1162,9	34,9184	23459,2	14162,4	139,764
846	1119,15	5866,17	3541,42	200,17	1161,3	34,9230	23480,0	14174,9	139,783
847	1120,15	5871,37	3544,53	200,62	1159,8	34,9277	23500,7	14187,4	139,801
848	1121,15	5876,56	3547,65	201,07	1158,2	34,9323	23521,5	14199,8	139,820
849	1122,15	5881,75	3550,77	201,51	1156,7	34,9369	23542,3	14212,3	139,839
850	1123,15	5886,95	3553,88	201,96	1155,1	34,9415	23563,1	14224,8	139,857
851	1124,15	5892,14	3557,00	202,41	1153,6	34,9462	23583,9	14237,2	139,876
852	1125,15	5897,33	3560,11	202,86	1152,0	34,9508	23604,7	14249,7	139,894
853	1126,15	5902,53	3563,23	203,32	1150,5	34,9554	23625,5	14262,2	139,912
854	1127,15	5907,72	3566,35	203,77	1149,0	34,9600	23646,2	14274,7	139,931
855	1128,15	5912,91	3569,46	204,22	1147,4	34,9646	23667,0	14287,1	139,949
856	1129,15	5918,11	3572,58	204,67	1145,9	34,9692	23687,8	14299,6	139,968
857	1130,15	5923,30	3575,69	205,13	1144,4	34,9738	23708,6	14312,1	139,986
858	1131,15	5928,49	3578,81	205,58	1142,9	34,9784	23729,4	14324,5	140,005
859	1132,15	5933,69	3581,92	206,03	1141,4	34,9830	23750,2	14337,0	140,023
860	1133,15	5938,88	3585,04	206,49	1139,9	34,9876	23771,0	14349,5	140,041
861	1134,15	5944,07	3588,16	206,95	1138,3	34,9922	23791,7	14362,0	140,060
862	1135,15	5949,26	3591,27	207,40	1136,8	34,9967	23812,5	14374,4	140,078
863	1136,15	5954,46	3594,39	207,86	1135,3	35,0013	23833,3	14386,9	140,096
864	1137,15	5959,65	3597,50	208,32	1133,8	35,0059	23854,1	14399,4	140,115
865	1138,15	5964,84	3600,62	208,78	1132,3	35,0104	23874,9	14411,8	140,133
866	1139,15	5970,04	3603,74	209,23	1130,9	35,0150	23895,7	14424,3	140,151
867	1140,15	5975,23	3606,85	209,69	1129,4	35,0196	23916,5	14436,8	140,169
868	1141,15	5980,42	3609,97	210,15	1127,9	35,0241	23937,2	14449,3	140,188
869	1142,15	5985,62	3613,08	210,61	1126,4	35,0287	23958,0	14461,7	140,206
870	1143,15	5990,81	3616,20	211,08	1124,9	35,0332	23978,8	14474,2	140,224
871	1144,15	5996,00	3619,31	211,54	1123,5	35,0377	23999,6	14486,7	140,242
872	1145,15	6001,20	3622,43	212,00	1122,0	35,0423	24020,4	14499,1	140,260
873	1146,15	6006,39	3625,55	212,46	1120,5	35,0468	24041,2	14511,6	140,278
874	1147,15	6011,58	3628,66	212,93	1119,0	35,0513	24062,0	14524,1	140,297
875	1148,15	6016,78	3631,78	213,39	1117,6	35,0559	24082,7	14536,6	140,315
876	1149,15	6021,97	3634,89	213,86	1116,1	35,0604	24103,5	14549,0	140,333
877	1150,15	6027,16	3638,01	214,32	1114,7	35,0649	24124,3	14561,5	140,351
878	1151,15	6032,35	3641,13	214,79	1113,2	35,0694	24145,1	14574,0	140,369
879	1152,15	6037,55	3644,24	215,25	1111,8	35,0739	24165,9	14586,4	140,387
880	1153,15	6042,74	3647,36	215,72	1110,3	35,0784	24186,7	14598,9	140,405
881	1154,15	6047,93	3650,47	216,19	1108,9	35,0829	24207,5	14611,4	140,423
882	1155,15	6053,13	3653,59	216,66	1107,4	35,0874	24228,2	14623,9	140,441
883	1156,15	6058,32	3656,71	217,13	1106,0	35,0919	24249,0	14636,3	140,459
884	1157,15	6063,51	3659,82	217,60	1104,6	35,0964	24269,8	14648,8	140,477

Helium (*Continued*)

t	T	h	u	$\pi_\bullet$	$\theta_\bullet$	s°	H	U	S
885	1158,15	6068,71	3662,94	218,07	1103,1	35,1009	24290,6	14661,3	140,495
886	1159,15	6073,90	3666,05	218,54	1101,7	35,1054	24311,4	14673,7	140,513
887	1160,15	6079,09	3669,17	219,01	1100,3	35,1099	24332,2	14686,2	140,531
888	1161,15	6084,29	3672,28	219,48	1098,9	35,1143	24353,0	14698,7	140,549
889	1162,15	6089,48	3675,40	219,96	1097,5	35,1188	24373,7	14711,2	140,567
890	1163,15	6094,67	3678,52	220,43	1096,0	35,1233	24394,5	14723,6	140,584
891	1164,15	6099,87	3681,63	220,90	1094,6	35,1277	24415,3	14736,1	140,602
892	1165,15	6105,06	3684,75	221,38	1093,2	35,1322	24436,1	14748,6	140,620
893	1166,15	6110,25	3687,86	221,85	1091,8	35,1367	24456,9	14761,0	140,638
894	1167,15	6115,44	3690,98	222,33	1090,4	35,1411	24477,7	14773,5	140,656
895	1168,15	6120,64	3694,10	222,81	1089,0	35,1456	24498,5	14786,0	140,674
896	1169,15	6125,85	3697,21	223,28	1087,6	35,1500	24519,3	14798,5	140,691
897	1170,15	6131,02	3700,33	223,76	1086,2	35,1544	24540,0	14810,9	140,709
898	1171,15	6136,22	3703,44	224,24	1084,8	35,1589	24560,8	14823,4	140,727
899	1172,15	6141,41	3706,56	224,72	1083,4	35,1633	24581,6	14835,9	140,745
900	1173,15	6146,60	3709,68	225,20	1082,1	35,1677	24602,4	14848,3	140,762
901	1174,15	6151,80	3712,79	225,68	1080,7	35,1722	24623,2	14860,8	140,780
902	1175,15	6156,99	3715,91	226,16	1079,3	35,1766	24644,0	14873,3	140,798
903	1176,15	6162,18	3719,02	226,64	1077,9	35,1810	24664,8	14885,8	140,815
904	1177,15	6167,38	3722,14	227,12	1076,5	35,1854	24685,5	14898,2	140,833
905	1178,15	6172,57	3725,25	227,61	1075,2	35,1898	24706,3	14910,7	140,851
906	1179,15	6177,76	3728,37	228,09	1073,8	35,1942	24727,1	14923,2	140,868
907	1180,15	6182,96	3731,49	228,57	1072,4	35,1986	24747,9	14935,6	140,886
908	1181,15	6188,15	3734,60	229,06	1071,1	35,2030	24768,7	14948,1	140,904
909	1182,15	6193,34	3737,72	229,54	1069,7	35,2074	24789,5	14960,6	140,921
910	1183,15	6198,53	3740,83	230,03	1068,4	35,2118	24810,3	14973,1	140,939
911	1184,15	6203,73	3743,95	230,51	1067,0	35,2162	24831,0	14985,5	140,956
912	1185,15	6208,92	3747,07	231,00	1065,7	35,2206	24851,8	14998,0	140,974
913	1186,15	6214,11	3750,18	231,49	1064,3	35,2250	24872,6	15010,5	140,991
914	1187,15	6219,31	3753,30	231,98	1063,0	35,2293	24893,4	15022,9	141,009
915	1188,15	6224,50	3756,41	232,47	1061,6	35,2337	24914,2	15035,4	141,026
916	1189,15	6229,69	3759,53	232,96	1060,3	35,2381	24935,0	15047,9	141,044
917	1190,15	6234,89	3762,64	233,45	1059,0	35,2424	24955,8	15060,4	141,061
918	1191,15	6240,08	3765,76	233,94	1057,6	35,2468	24976,5	15072,8	141,079
919	1192,15	6245,27	3768,88	234,43	1056,3	35,2512	24997,3	15085,3	141,096
920	1193,15	6250,47	3771,99	234,92	1055,0	35,2555	25018,1	15097,8	141,114
921	1194,15	6255,66	3775,11	235,41	1053,6	35,2599	25038,9	15110,2	141,131
922	1195,15	6260,85	3778,22	235,91	1052,3	35,2642	25059,7	15122,7	141,149
923	1196,15	6266,05	3781,34	236,40	1051,0	35,2686	25080,5	15135,2	141,166
924	1197,15	6271,24	3784,46	236,89	1049,7	35,2729	25101,3	15147,7	141,183
925	1198,15	6276,43	3787,57	237,39	1048,4	35,2772	25122,0	15160,1	141,201
926	1199,15	6281,62	3790,69	237,88	1047,1	35,2816	25142,8	15172,6	141,218
927	1200,15	6286,82	3793,80	238,38	1045,7	35,2859	25163,6	15185,1	141,235
928	1201,15	6292,01	3796,92	238,88	1044,4	35,2902	25184,4	15197,6	141,253
929	1202,15	6297,20	3800,04	239,37	1043,1	35,2945	25205,2	15210,0	141,270
930	1203,15	6302,40	3803,15	239,87	1041,8	35,2989	25226,0	15222,5	141,287
931	1204,15	6307,59	3806,27	240,37	1040,5	35,3032	25246,8	15235,0	141,304
932	1205,15	6312,78	3809,38	240,87	1039,2	35,3075	25267,5	15247,4	141,322
933	1206,15	6317,98	3812,50	241,37	1037,9	35,3118	25288,3	15259,9	141,339
934	1207,15	6323,17	3815,61	241,87	1036,7	35,3161	25309,1	15272,4	141,356
935	1208,15	6328,36	3818,73	242,37	1035,4	35,3204	25329,9	15284,9	141,373
936	1209,15	6333,56	3821,85	242,87	1034,1	35,3247	25350,7	15297,3	141,391
937	1210,15	6338,75	3824,96	243,38	1032,8	35,3290	25371,5	15309,8	141,408
938	1211,15	6343,94	3828,08	243,88	1031,5	35,3333	25392,3	15322,3	141,425
939	1212,15	6349,14	3831,19	244,38	1030,3	35,3376	25413,0	15334,7	141,442

Helium (*Continued*)

t	T	h	u	π_0	θ_0	s^0	H	U	S
940	1213,15	6354,33	3834,31	244,89	1029,0	35,3418	25433,8	15347,2	141,459
941	1214,15	6359,52	3837,43	245,39	1027,7	35,3461	25454,6	15359,7	141,476
942	1215,15	6364,71	3840,54	245,90	1026,4	35,3504	25475,4	15372,2	141,494
943	1216,15	6369,91	3843,66	246,41	1025,2	35,3547	25496,2	15384,6	141,511
944	1217,15	6375,10	3846,77	246,91	1023,9	35,3589	25517,0	15397,1	141,528
945	1218,15	6380,29	3849,89	247,42	1022,6	35,3632	25537,8	15409,6	141,545
946	1219,15	6385,49	3853,01	247,93	1021,4	35,3675	25558,6	15422,0	141,562
947	1220,15	6390,68	3856,12	248,44	1020,1	35,3717	25579,3	15434,5	141,579
948	1221,15	6395,87	3859,24	248,95	1018,9	35,3760	25600,1	15447,0	141,596
949	1222,15	6401,07	3862,35	249,46	1017,6	35,3802	25620,9	15459,5	141,613
950	1223,15	6406,26	3865,47	249,97	1016,4	35,3845	25641,7	15471,9	141,630
951	1224,15	6411,45	3868,58	250,48	1015,1	35,3887	25662,5	15484,4	141,647
952	1225,15	6416,65	3871,70	250,99	1013,9	35,3930	25683,3	15496,9	141,664
953	1226,15	6421,84	3874,82	251,50	1012,7	35,3972	25704,1	15509,3	141,681
954	1227,15	6427,03	3877,93	252,01	1011,4	35,4014	25724,8	15521,8	141,698
955	1228,15	6432,23	3881,05	252,53	1010,2	35,4057	25745,6	15534,3	141,715
956	1229,15	6437,42	3884,16	253,04	1009,0	35,4099	25766,4	15546,8	141,732
957	1230,15	6442,61	3887,28	253,56	1007,7	35,4141	25787,2	15559,2	141,749
958	1231,15	6447,80	3890,40	254,07	1006,5	35,4183	25808,0	15571,7	141,765
959	1232,15	6453,00	3893,51	254,59	1005,3	35,4225	25828,8	15584,2	141,782
960	1233,15	6458,19	3896,63	255,11	1004,0	35,4268	25849,6	15596,6	141,799
961	1234,15	6463,38	3899,74	255,62	1002,8	35,4310	25870,3	15609,1	141,816
962	1235,15	6468,58	3902,86	256,14	1001,6	35,4352	25891,1	15621,6	141,833
963	1236,15	6473,77	3905,98	256,66	1000,4	35,4394	25911,9	15634,1	141,850
964	1237,15	6478,96	3909,09	257,18	999,22	35,4436	25932,7	15646,5	141,866
965	1238,15	6484,16	3912,21	257,70	998,01	35,4478	25953,5	15659,0	141,883
966	1239,15	6489,35	3915,32	258,22	996,81	35,4520	25974,3	15671,5	141,900
967	1240,15	6494,54	3918,44	258,74	995,60	35,4562	25995,1	15683,9	141,917
968	1241,15	6499,74	3921,55	259,26	994,40	35,4603	26015,8	15696,4	141,934
969	1242,15	6504,93	3924,67	259,79	993,20	35,4645	26036,6	15708,9	141,950
970	1243,15	6510,12	3927,79	260,31	992,00	35,4687	26057,4	15721,4	141,967
971	1244,15	6515,32	3930,90	260,83	990,80	35,4729	26078,2	15733,8	141,984
972	1245,15	6520,51	3934,02	261,36	989,61	35,4771	26099,0	15746,3	142,000
973	1246,15	6525,70	3937,13	261,88	988,42	35,4812	26119,8	15758,8	142,017
974	1247,15	6530,89	3940,25	262,41	987,23	35,4854	26140,6	15771,2	142,034
975	1248,15	6536,09	3943,37	262,94	986,04	35,4896	26161,3	15783,7	142,050
976	1249,15	6541,28	3946,48	263,46	984,86	35,4937	26182,1	15796,2	142,067
977	1250,15	6546,47	3949,60	263,99	983,68	35,4979	26202,9	15808,7	142,084
978	1251,15	6551,67	3952,71	264,52	982,50	35,5020	26223,7	15821,1	142,100
979	1252,15	6556,86	3955,83	265,05	981,32	35,5062	26244,5	15833,6	142,117
980	1253,15	6562,05	3958,94	265,58	980,15	35,5103	26265,3	15846,1	142,134
981	1254,15	6567,25	3962,06	266,11	978,98	35,5145	26286,1	15858,5	142,150
982	1255,15	6572,44	3965,18	266,64	977,81	35,5186	26306,8	15871,0	142,167
983	1256,15	6577,63	3968,29	267,17	976,64	35,5227	26327,6	15883,5	142,183
984	1257,15	6582,83	3971,41	267,70	975,47	35,5269	26348,4	15896,0	142,200
985	1258,15	6588,02	3974,52	268,23	974,31	35,5310	26369,2	15908,4	142,216
986	1259,15	6593,21	3977,64	268,77	973,15	35,5351	26390,0	15920,9	142,233
987	1260,15	6598,41	3980,76	269,30	971,99	35,5392	26410,8	15933,4	142,249
988	1261,15	6603,60	3983,87	269,84	970,84	35,5434	26431,6	15945,8	142,266
989	1262,15	6608,79	3986,99	270,37	969,68	35,5475	26452,3	15958,3	142,282
990	1263,15	6613,98	3990,10	270,91	968,53	35,5516	26473,1	15970,8	142,299
991	1264,15	6619,18	3993,22	271,44	967,38	35,5557	26493,9	15983,3	142,315
992	1265,15	6624,37	3996,34	271,98	966,24	35,5598	26514,7	15995,7	142,332
993	1266,15	6629,56	3999,45	272,52	965,09	35,5639	26535,5	16008,2	142,348
994	1267,15	6634,76	4002,57	273,06	963,95	35,5680	26556,3	16020,7	142,365

Helium (*Continued*)

t	T	h	u	π_0	θ_0	s^0	H	U	S
995	1268,15	6639,95	4005,68	273,60	962,81	35,5721	26577,1	16033,1	142,381
996	1269,15	6645,14	4008,80	274,14	961,67	35,5762	26597,9	16045,6	142,397
997	1270,15	6650,34	4011,91	274,68	960,54	35,5803	26618,6	16058,1	142,414
998	1271,15	6655,53	4015,03	275,22	959,40	35,5844	26639,4	16070,6	142,430
999	1272,15	6660,72	4018,15	275,76	958,27	35,5885	26660,2	16083,0	142,446
1000	1273,15	6665,92	4021,26	276,30	957,14	35,5925	26681,0	16095,5	142,463
1001	1274,15	6671,11	4024,38	276,84	956,02	35,5966	26701,8	16108,0	142,479
1002	1275,15	6676,30	4027,49	277,39	954,89	35,6007	26722,6	16120,4	142,495
1003	1276,15	6681,50	4030,61	277,93	953,77	35,6048	26743,4	16132,9	142,512
1004	1277,15	6686,69	4033,73	278,48	952,65	35,6088	26764,1	16145,4	142,528
1005	1278,15	6691,88	4036,84	279,02	951,53	35,6129	26784,9	16157,9	142,544
1006	1279,15	6697,07	4039,96	279,57	950,42	35,6170	26805,7	16170,3	142,560
1007	1280,15	6702,27	4043,07	280,11	949,30	35,6210	26826,5	16182,8	142,577
1008	1281,15	6707,46	4046,19	280,66	948,19	35,6251	26847,3	16195,3	142,593
1009	1282,15	6712,65	4049,31	281,21	947,08	35,6291	26868,1	16207,7	142,609
1010	1283,15	6717,85	4052,42	281,76	945,98	35,6332	26888,9	16220,2	142,625
1011	1284,15	6723,04	4055,54	282,31	944,87	35,6372	26909,6	16232,7	142,642
1012	1285,15	6728,23	4058,65	282,86	943,77	35,6413	26930,4	16245,2	142,658
1013	1286,15	6733,43	4061,77	283,41	942,67	35,6453	26951,2	16257,6	142,674
1014	1287,15	6738,62	4064,88	283,96	941,57	35,6493	26972,0	16270,1	142,690
1015	1288,15	6743,81	4068,00	284,51	940,47	35,6534	26992,8	16282,6	142,706
1016	1289,15	6749,01	4071,12	285,06	939,38	35,6574	27013,6	16295,1	142,722
1017	1290,15	6754,20	4074,23	285,62	938,29	35,6614	27034,4	16307,5	142,738
1018	1291,15	6759,39	4077,35	286,17	937,20	35,6654	27055,1	16320,0	142,755
1019	1292,15	6764,59	4080,46	286,72	936,11	35,6695	27075,9	16332,5	142,771
1020	1293,15	6769,78	4083,58	287,28	935,02	35,6735	27096,7	16344,9	142,787
1021	1294,15	6774,97	4086,70	287,84	933,94	35,6775	27117,5	16357,4	142,803
1022	1295,15	6780,16	4089,81	288,39	932,86	35,6815	27138,3	16369,9	142,819
1023	1296,15	6785,36	4092,93	288,95	931,78	35,6855	27159,1	16382,4	142,835
1024	1297,15	6790,55	4096,04	289,51	930,70	35,6895	27179,9	16394,8	142,851
1025	1298,15	6795,74	4099,16	290,06	929,63	35,6935	27200,6	16407,3	142,867
1026	1299,15	6800,94	4102,28	290,62	928,55	35,6975	27221,4	16419,8	142,883
1027	1300,15	6806,13	4105,39	291,18	927,48	35,7015	27242,2	16432,2	142,899
1028	1301,15	6811,32	4108,51	291,74	926,41	35,7055	27263,0	16444,7	142,915
1029	1302,15	6816,52	4111,62	292,30	925,35	35,7095	27283,8	16457,2	142,931
1030	1303,15	6821,71	4114,74	292,87	924,28	35,7135	27304,6	16469,7	142,947
1031	1304,15	6826,90	4117,85	293,43	923,22	35,7175	27325,4	16482,1	142,963
1032	1305,15	6832,10	4120,97	293,99	922,16	35,7215	27346,1	16494,6	142,979
1033	1306,15	6837,29	4124,09	294,55	921,10	35,7254	27366,9	16507,1	142,995
1034	1307,15	6842,48	4127,20	295,12	920,04	35,7294	27387,7	16519,5	143,011
1035	1308,15	6847,68	4130,32	295,68	918,99	35,7334	27408,5	16532,0	143,026
1036	1309,15	6852,87	4133,43	296,25	917,94	35,7373	27429,3	16544,5	143,042
1037	1310,15	6858,06	4136,55	296,81	916,88	35,7413	27450,1	16557,0	143,058
1038	1311,15	6863,25	4139,67	297,38	915,84	35,7453	27470,9	16569,4	143,074
1039	1312,15	6868,45	4142,78	297,95	914,79	35,7492	27491,6	16581,9	143,090
1040	1313,15	6873,64	4145,90	298,52	913,74	35,7532	27512,4	16594,4	143,106
1041	1314,15	6878,83	4149,01	299,09	912,70	35,7571	27533,2	16606,8	143,122
1042	1315,15	6884,03	4152,13	299,65	911,66	35,7611	27554,0	16619,3	143,137
1043	1316,15	6889,22	4155,24	300,22	910,62	35,7650	27574,8	16631,8	143,153
1044	1317,15	6894,41	4158,36	300,80	909,59	35,7690	27595,6	16644,3	143,169
1045	1318,15	6899,61	4161,48	301,37	908,55	35,7729	27616,4	16656,7	143,185
1046	1319,15	6904,80	4164,59	301,94	907,52	35,7769	27637,2	16669,2	143,200
1047	1320,15	6909,99	4167,71	302,51	906,49	35,7808	27657,9	16681,7	143,216
1048	1321,15	6915,19	4170,82	303,08	905,46	35,7847	27678,7	16694,1	143,232
1049	1322,15	6920,38	4173,94	303,66	904,43	35,7887	27699,5	16706,6	143,248

Helium (*Continued*)

t	T	h	u	π_0	θ_0	s^0	H	U	S
1050	1323,15	6925,57	4177,06	304,23	903,41	35,7926	27720,3	16719,1	143,263
1051	1324,15	6930,77	4180,17	304,81	902,38	35,7965	27741,1	16731,6	143,279
1052	1325,15	6935,96	4183,29	305,38	901,36	35,8004	27761,9	16744,0	143,295
1053	1326,15	6941,15	4186,40	305,96	900,34	35,8043	27782,7	16756,5	143,310
1054	1327,15	6946,34	4189,52	306,54	899,32	35,8083	27803,4	16769,0	143,326
1055	1328,15	6951,54	4192,64	307,12	898,31	35,8122	27824,2	16781,4	143,342
1056	1329,15	6956,73	4195,75	307,69	897,30	35,8161	27845,0	16793,9	143,357
1057	1330,15	6961,92	4198,87	308,27	896,28	35,8200	27865,8	16806,4	143,373
1058	1331,15	6967,12	4201,98	308,85	895,27	35,8239	27886,6	16818,9	143,389
1059	1332,15	6972,31	4205,10	309,43	894,27	35,8278	27907,4	16831,3	143,404
1060	1333,15	6977,50	4208,21	310,01	893,26	35,8317	27928,2	16843,8	143,420
1061	1334,15	6982,70	4211,33	310,60	892,26	35,8356	27948,9	16856,3	143,435
1062	1335,15	6987,89	4214,45	311,18	891,25	35,8395	27969,7	16868,7	143,451
1063	1336,15	6993,08	4217,56	311,76	890,25	35,8434	27990,5	16881,2	143,467
1064	1337,15	6998,28	4220,68	312,34	889,25	35,8472	28011,3	16893,7	143,482
1065	1338,15	7003,47	4223,79	312,93	888,26	35,8511	28032,1	16906,2	143,498
1066	1339,15	7008,66	4226,91	313,51	887,26	35,8550	28052,9	16918,6	143,513
1067	1340,15	7013,86	4230,03	314,10	886,27	35,8589	28073,7	16931,1	143,529
1068	1341,15	7019,05	4233,14	314,69	885,28	35,8628	28094,4	16943,6	143,514
1069	1342,15	7024,24	4236,26	315,27	884,29	35,8666	28115,2	16956,0	143,560
1070	1343,15	7029,43	4239,37	315,86	883,30	35,8705	28136,0	16968,5	143,575
1071	1344,15	7034,63	4242,49	316,45	882,32	35,8744	28156,8	16981,0	143,591
1072	1345,15	7039,82	4245,61	317,04	881,33	35,8782	28177,6	16993,5	143,606
1073	1346,15	7045,01	4248,72	317,63	880,35	35,8821	28198,4	17005,9	143,622
1074	1347,15	7050,21	4251,84	318,22	879,37	35,8859	28219,2	17018,4	143,637
1075	1348,15	7055,40	4254,95	318,81	878,39	35,8898	28239,9	17030,9	143,652
1076	1349,15	7060,59	4258,07	319,40	877,42	35,8936	28260,7	17043,3	143,668
1077	1350,15	7065,79	4261,18	319,99	876,44	35,8975	28281,5	17055,8	143,683
1078	1351,15	7070,98	4264,30	320,58	875,47	35,9013	28302,3	17068,3	143,699
1079	1352,15	7076,17	4267,42	321,18	874,50	35,9052	28323,1	17080,8	143,714
1080	1353,15	7081,37	4270,53	321,77	873,53	35,9090	28343,9	17093,2	143,729
1081	1354,15	7086,56	4273,65	322,37	872,56	35,9129	28364,7	17105,7	143,745
1082	1355,15	7091,75	4276,76	322,96	871,60	35,9167	28385,4	17118,2	143,760
1083	1356,15	7096,95	4279,88	323,56	870,63	35,9205	28406,2	17130,6	143,775
1084	1357,15	7102,14	4283,00	324,16	869,67	35,9243	28427,0	17143,1	143,791
1085	1358,15	7107,33	4286,11	324,75	868,71	35,9282	28447,8	17155,6	143,806
1086	1359,15	7112,52	4289,23	325,35	867,75	35,9320	28468,6	17168,1	143,821
1087	1360,15	7117,72	4292,34	325,95	866,79	35,9358	28489,4	17180,5	143,837
1088	1361,15	7122,91	4295,46	326,55	865,84	35,9396	28510,2	17193,0	143,852
1089	1362,15	7128,10	4298,58	327,15	864,89	35,9434	28530,9	17205,5	143,867
1090	1363,15	7133,30	4301,69	327,75	863,93	35,9473	28551,7	17217,9	143,882
1091	1364,15	7138,49	4304,81	328,35	862,98	35,9511	28572,5	17230,4	143,898
1092	1365,15	7143,68	4307,92	328,95	862,04	35,9549	28593,3	17242,9	143,913
1093	1366,15	7148,88	4311,04	329,56	861,09	35,9587	28614,1	17255,4	143,928
1094	1367,15	7154,07	4314,15	330,16	860,15	35,9625	28634,9	17267,8	143,943
1095	1368,15	7159,26	4317,27	330,76	859,20	35,9663	28655,7	17280,3	143,959
1096	1369,15	7164,46	4320,39	331,37	858,26	35,9701	28676,5	17292,8	143,974
1097	1370,15	7169,65	4323,50	331,97	857,32	35,9738	28697,2	17305,2	143,989
1098	1371,15	7174,84	4326,62	332,58	856,38	35,9776	28718,0	17317,7	144,004
1099	1372,15	7180,04	4329,73	333,19	855,45	35,9814	28738,8	17330,2	144,019
1100	1373,15	7185,23	4332,85	333,79	854,51	35,9852	28759,6	17342,7	144,034
1101	1374,15	7190,42	4335,97	334,40	853,58	35,9890	28780,4	17355,1	144,050
1102	1375,15	7195,61	4339,08	335,01	852,65	35,9928	28801,2	17367,6	144,065
1103	1376,15	7200,81	4342,20	335,62	851,72	35,9965	28822,0	17380,1	144,080
1104	1377,15	7206,00	4345,31	336,23	850,79	36,0003	28842,7	17392,6	144,095

Helium (*Continued*)

t	T	h	u	π_0	θ_0	s^0	H	U	S
1105	1378,15	7211,19	4348,43	336,84	849,87	36,0041	28863,5	17405,0	144,110
1106	1379,15	7216,39	4351,54	337,45	848,94	36,0079	28884,3	17417,5	144,125
1107	1380,15	7221,58	4354,66	338,06	848,02	36,0116	28905,1	17430,0	144,140
1108	1381,15	7226,77	4357,78	338,68	847,10	36,0154	28925,9	17442,4	144,155
1109	1382,15	7231,97	4360,89	339,29	846,18	36,0191	28946,7	17454,9	144,170
1110	1383,15	7237,16	4364,01	339,90	845,26	36,0229	28967,5	17467,4	144,185
1111	1384,15	7242,35	4367,12	340,52	844,35	36,0266	28988,2	17479,9	144,200
1112	1385,15	7247,55	4370,24	341,13	843,43	36,0304	29009,0	17492,3	144,215
1113	1386,15	7252,74	4373,36	341,75	842,52	36,0341	29029,8	17504,8	144,230
1114	1387,15	7257,93	4376,47	342,37	841,61	36,0379	29050,6	17517,3	144,245
1115	1388,15	7263,13	4379,59	342,98	840,70	36,0416	29071,4	17529,7	144,260
1116	1389,15	7268,32	4382,70	343,60	839,79	36,0454	29092,2	17542,2	144,275
1117	1390,15	7273,51	4385,82	344,22	838,89	36,0491	29113,0	17554,7	144,290
1118	1391,15	7278,70	4388,94	344,84	837,98	36,0528	29133,7	17567,2	144,305
1119	1392,15	7283,90	4392,05	345,46	837,08	36,0566	29154,5	17579,6	144,320
1120	1393,15	7289,09	4395,17	346,08	836,18	36,0603	29175,3	17592,1	144,335
1121	1394,15	7294,28	4398,28	346,70	835,28	36,0640	29196,1	17604,6	144,350
1122	1395,15	7299,48	4401,40	347,33	834,38	36,0678	29216,9	17617,0	144,365
1123	1396,15	7304,67	4404,51	347,95	833,49	36,0715	29237,7	17629,5	144,380
1124	1397,15	7309,86	4407,63	348,57	832,59	36,0752	29258,5	17642,0	144,395
1125	1398,15	7315,06	4410,75	349,20	831,70	36,0789	29279,2	17654,5	144,409
1126	1399,15	7320,25	4413,86	349,82	830,81	36,0826	29300,0	17666,9	144,424
1127	1400,15	7325,44	4416,98	350,45	829,92	36,0863	29320,8	17679,4	144,439
1128	1401,15	7330,64	4420,09	351,07	829,03	36,0900	29341,6	17691,9	144,454
1129	1402,15	7335,83	4423,21	351,70	828,14	36,0937	29362,4	17704,3	144,469
1130	1403,15	7341,02	4426,33	352,33	827,26	36,0974	29383,2	17716,8	144,484
1131	1404,15	7346,22	4429,44	352,95	826,37	36,1011	29404,0	17729,5	144,498
1132	1405,15	7351,41	4432,56	353,58	825,49	36,1048	29424,7	17741,8	144,513
1133	1406,15	7356,60	4435,67	354,21	824,61	36,1085	29445,5	17754,2	144,528
1134	1407,15	7361,79	4438,79	354,84	823,73	36,1122	29466,3	17766,7	144,543
1135	1408,15	7366,99	4441,91	355,47	822,85	36,1159	29487,1	17779,2	144,558
1136	1409,15	7372,18	4445,02	356,10	821,98	36,1196	29507,9	17791,6	144,572
1137	1410,15	7377,37	4448,14	356,74	821,10	36,1233	29528,7	17804,1	144,587
1138	1411,15	7382,57	4451,25	357,37	820,23	36,1270	29549,5	17816,6	144,602
1139	1412,15	7387,76	4454,37	358,00	819,36	36,1306	29570,2	17829,1	144,617
1140	1413,15	7392,95	4457,48	358,64	818,49	36,1343	29591,0	17841,5	144,631
1141	1414,15	7398,15	4460,60	359,27	817,62	36,1380	29611,8	17854,0	144,646
1142	1415,15	7403,34	4463,72	359,91	816,76	36,1417	29632,6	17866,5	144,661
1143	1416,15	7408,53	4466,83	360,54	815,89	36,1453	29653,4	17878,9	144,675
1144	1417,15	7413,73	4469,95	361,18	815,03	36,1490	29674,2	17891,4	144,690
1145	1418,15	7418,92	4473,06	361,82	814,17	36,1527	29695,0	17903,9	144,705
1146	1419,15	7424,11	4476,18	362,46	813,31	36,1563	29715,8	17916,4	144,719
1147	1420,15	7429,31	4479,30	363,09	812,45	36,1600	29736,5	17928,8	144,734
1148	1421,15	7434,50	4482,41	363,73	811,59	36,1636	29757,3	17941,3	144,749
1149	1422,15	7439,69	4485,53	364,37	810,73	36,1673	29778,1	17953,8	144,763
1150	1423,15	7444,88	4488,64	365,02	809,88	36,1709	29798,9	17966,2	144,778
1151	1424,15	7450,08	4491,76	365,66	809,03	36,1746	29819,7	17978,7	144,792
1152	1425,15	7455,27	4494,87	366,30	808,17	36,1782	29840,5	17991,2	144,807
1153	1426,15	7460,46	4497,99	366,94	807,32	36,1819	29861,3	18003,7	144,822
1154	1427,15	7465,66	4501,11	367,59	806,48	36,1855	29882,0	18016,1	144,836
1155	1428,15	7470,85	4504,22	368,23	805,63	36,1892	29902,8	18028,6	144,851
1156	1429,15	7476,04	4507,34	368,87	804,78	36,1928	29923,6	18041,1	144,865
1157	1430,15	7481,24	4510,45	369,52	803,94	36,1964	29944,4	18053,5	144,880
1158	1431,15	7486,43	4513,57	370,17	803,10	36,2001	29965,2	18066,0	144,894
1159	1432,15	7491,62	4516,69	370,81	802,26	36,2037	29986,0	18078,5	144,909

t	T	h	u	π_0	θ_0	s^0	H	U	S
1160	1433,15	7496,82	4519,80	371,46	801,42	36,2073	30006,8	18091,0	144,923
1161	1434,15	7502,01	4522,92	372,11	800,58	36,2109	30027,5	18103,4	144,938
1162	1435,15	7507,20	4526,03	372,76	799,74	36,2145	30048,3	18115,9	144,952
1163	1436,15	7512,40	4529,15	373,41	798,91	36,2182	30069,1	18128,4	144,967
1164	1437,15	7517,59	4532,27	374,06	798,07	36,2218	30089,9	18140,8	144,981
1165	1438,15	7522,78	4535,38	374,71	797,24	36,2254	30110,7	18153,3	144,996
1166	1439,15	7527,97	4538,50	375,36	796,41	36,2290	30131,5	18165,8	145,010
1167	1440,15	7533,17	4541,61	376,01	795,58	36,2326	30152,3	18178,3	145,025
1168	1441,15	7538,36	4544,73	376,67	794,75	36,2362	30173,0	18190,7	145,039
1169	1442,15	7543,55	4547,84	377,32	793,93	36,2398	30193,8	18203,2	145,053
1170	1443,15	7548,75	4550,96	377,98	793,10	36,2434	30214,6	18215,7	145,068
1171	1444,15	7553,94	4554,08	378,63	792,28	36,2470	30235,4	18228,1	145,082
1172	1445,15	7559,13	4557,19	379,29	791,46	36,2506	30256,2	18240,6	145,097
1173	1446,15	7564,33	4560,31	379,94	790,64	36,2542	30277,0	18253,1	145,111
1174	1447,15	7569,52	4563,42	380,60	789,82	36,2578	30297,8	18265,6	145,125
1175	1448,15	7574,71	4566,54	381,26	789,00	36,2614	30318,5	18278,0	145,140
1176	1449,15	7579,91	4569,66	381,92	788,18	36,2650	30339,3	18290,5	145,154
1177	1450,15	7585,10	4572,77	382,58	787,37	36,2685	30360,1	18303,0	145,168
1178	1451,15	7590,29	4575,89	383,24	786,55	36,2721	30380,9	18315,4	145,183
1179	1452,15	7595,49	4579,00	383,90	785,74	36,2757	30401,7	18327,9	145,197
1180	1453,15	7600,68	4582,12	384,56	784,93	36,2793	30422,5	18340,4	145,211
1181	1454,15	7605,87	4585,24	385,22	784,12	36,2828	30443,3	18352,9	145,226
1182	1455,15	7611,06	4588,35	385,88	783,31	36,2864	30464,0	18365,3	145,240
1183	1456,15	7616,26	4591,47	386,55	782,50	36,2900	30484,8	18377,8	145,254
1184	1457,15	7621,45	4594,58	387,21	781,70	36,2936	30505,6	18390,3	145,269
1185	1458,15	7626,64	4597,70	387,87	780,90	36,2971	30526,4	18402,7	145,283
1186	1459,15	7631,84	4600,81	388,54	780,09	36,3007	30547,2	18415,2	145,297
1187	1460,15	7637,03	4603,93	389,21	779,29	36,3042	30568,0	18427,7	145,311
1188	1461,15	7642,22	4607,05	389,87	778,49	36,3078	30588,8	18440,2	145,326
1189	1462,15	7647,42	4610,16	390,54	777,69	36,3113	30609,5	18452,6	145,340
1190	1463,15	7652,61	4613,28	391,21	776,90	36,3149	30630,3	18465,1	145,354
1191	1464,15	7657,80	4616,39	391,88	776,10	36,3184	30651,1	18477,6	145,368
1192	1465,15	7663,00	4619,51	392,55	775,31	36,3220	30671,9	18490,1	145,382
1193	1466,15	7668,19	4622,63	393,22	774,51	36,3255	30692,7	18502,5	145,397
1194	1467,15	7673,38	4625,74	393,89	773,72	36,3291	30713,5	18515,0	145,411
1195	1468,15	7678,58	4628,86	394,56	772,93	36,3326	30734,3	18527,5	145,425
1196	1469,15	7683,77	4631,97	395,23	772,14	36,3361	30755,1	18539,9	145,439
1197	1470,15	7688,96	4635,09	395,90	771,35	36,3397	30775,8	18552,4	145,453
1198	1471,15	7694,15	4638,21	396,58	770,57	36,3432	30796,6	18564,9	145,467
1199	1472,15	7699,35	4641,32	397,25	769,78	36,3467	30817,4	18577,4	145,481
1200	1473,15	7704,54	4644,44	397,93	769,00	36,3503	30838,2	18589,8	145,496
1201	1474,15	7709,73	4647,55	398,60	768,22	36,3538	30859,0	18602,3	145,510
1202	1475,15	7714,93	4650,67	399,28	767,44	36,3573	30879,8	18614,8	145,524
1203	1476,15	7720,12	4653,78	399,96	766,66	36,3608	30900,6	18627,2	145,538
1204	1477,15	7725,31	4656,90	400,63	765,88	36,3643	30921,3	18639,7	145,552
1205	1478,15	7730,51	4660,02	401,31	765,10	36,3679	30942,1	18652,2	145,566
1206	1479,15	7735,70	4663,13	401,99	764,32	36,3714	30962,9	18664,7	145,580
1207	1480,15	7740,89	4666,25	402,67	763,55	36,3749	30983,7	18677,1	145,594
1208	1481,15	7746,09	4669,36	403,35	762,78	36,3784	31004,5	18689,6	145,608
1209	1482,15	7751,28	4672,48	404,03	762,00	36,3819	31025,3	18702,1	145,622
1210	1483,15	7756,47	4675,60	404,71	761,23	36,3854	31046,1	18714,5	145,636
1211	1484,15	7761,67	4678,71	405,40	760,47	36,3889	31066,8	18727,0	145,650
1212	1485,15	7766,86	4681,83	406,08	759,70	36,3924	31087,6	18739,5	145,664
1213	1486,15	7772,05	4684,94	406,76	758,93	36,3959	31108,4	18752,0	145,678
1214	1487,15	7777,24	4688,06	407,45	758,17	36,3994	31129,2	18764,4	145,692

Helium (*Continued*)

i	T	h	u	π_0	θ_0	s^0	H	U	S
1215	1488,15	7782,44	4691,17	408,13	757,40	36,4029	31150,0	18776,9	145,706
1216	1489,15	7787,63	4694,29	408,82	756,64	36,4064	31170,8	18789,4	145,720
1217	1490,15	7792,82	4697,41	409,51	755,88	36,4098	31191,6	18801,8	145,734
1218	1491,15	7798,02	4700,52	410,19	755,12	36,4133	31212,3	18814,3	145,748
1219	1492,15	7803,21	4703,64	410,88	754,36	36,4168	31233,1	18826,8	145,762
1220	1493,15	7808,40	4706,75	411,57	753,60	36,4203	31253,9	18839,3	145,776
1221	1494,15	7813,60	4709,87	412,26	752,84	36,4238	31274,7	18851,7	145,790
1222	1495,15	7818,79	4712,99	412,95	752,09	36,4272	31295,5	18864,2	145,804
1223	1496,15	7823,98	4716,10	413,64	751,33	36,4307	31316,3	18876,7	145,818
1224	1497,15	7829,18	4719,22	414,33	750,58	36,4342	31337,1	18889,1	145,831
1225	1498,15	7834,37	4722,33	415,02	749,83	36,4377	31357,8	18901,6	145,845
1226	1499,15	7839,56	4725,45	415,72	749,08	36,4411	31378,6	18914,1	145,859
1227	1500,15	7844,76	4728,57	416,41	748,33	36,4446	31399,4	18926,6	145,873
1228	1501,15	7849,95	4731,68	417,10	747,58	36,4480	31420,2	18939,0	145,887
1229	1502,15	7855,14	4734,80	417,80	746,84	36,4515	31441,0	18951,5	145,901
1230	1503,15	7860,33	4737,91	418,50	746,09	36,4550	31461,8	18964,0	145,915
1231	1504,15	7865,53	4741,03	419,19	745,35	36,4584	31482,6	18976,4	145,928
1232	1505,15	7870,72	4744,14	419,89	744,61	36,4619	31503,3	18988,9	145,942
1233	1506,15	7875,91	4747,26	420,59	743,86	36,4653	31524,1	19001,4	145,956
1234	1507,15	7881,11	4750,38	421,29	743,12	36,4688	31544,9	19013,9	145,970
1235	1508,15	7886,30	4753,49	421,98	742,38	36,4722	31565,7	19026,3	145,984
1236	1509,15	7891,49	4756,61	422,68	741,65	36,4756	31586,5	19038,8	145,997
1237	1510,15	7896,69	4759,72	423,38	740,91	36,4791	31607,3	19051,3	146,011
1238	1511,15	7901,88	4762,84	424,09	740,18	36,4825	31628,1	19063,7	146,025
1239	1512,15	7907,07	4765,96	424,79	739,44	36,4860	31648,8	19076,2	146,039
1240	1513,15	7912,27	4769,07	425,49	738,71	36,4894	31669,6	19088,7	146,052
1241	1514,15	7917,46	4772,19	426,19	737,98	36,4928	31690,4	19101,2	146,066
1242	1515,15	7922,65	4775,30	426,90	737,25	36,4962	31711,2	19113,6	146,080
1243	1516,15	7927,85	4778,42	427,60	736,52	36,4997	31732,0	19126,1	146,094
1244	1517,15	7933,04	4781,54	428,31	735,79	36,5031	31752,8	19138,6	146,107
1245	1518,15	7938,23	4784,65	429,01	735,06	36,5065	31773,6	19151,0	146,121
1246	1519,15	7943,42	4787,77	429,72	734,34	36,5099	31794,4	19163,5	146,135
1247	1520,15	7948,62	4790,88	430,43	733,61	36,5134	31815,1	19176,0	146,148
1248	1521,15	7953,81	4794,00	431,14	732,89	36,5168	31835,9	19188,5	146,162
1249	1522,15	7959,00	4797,11	431,85	732,17	36,5202	31856,7	19200,9	146,176
1250	1523,15	7964,20	4800,23	432,56	731,45	36,5236	31877,5	19213,4	146,189
1251	1524,15	7969,39	4803,35	433,27	730,73	36,5270	31898,3	19225,9	146,203
1252	1525,15	7974,58	4806,46	433,98	730,01	36,5304	31919,1	19238,3	146,217
1253	1526,15	7979,78	4809,58	434,69	729,29	36,5338	31939,9	19250,8	146,230
1254	1527,15	7984,97	4812,69	435,40	728,57	36,5372	31960,6	19263,3	146,244
1255	1528,15	7990,16	4815,81	436,11	727,86	36,5406	31981,4	19275,8	146,257
1256	1529,15	7995,36	4818,93	436,83	727,14	36,5440	32002,2	19288,2	146,271
1257	1530,15	8000,55	4822,04	437,54	726,43	36,5474	32023,0	19300,7	146,285
1258	1531,15	8005,74	4825,16	438,26	725,72	36,5508	32043,8	19313,2	146,298
1259	1532,15	8010,94	4828,27	438,97	725,01	36,5542	32064,6	19325,6	146,312
1260	1533,15	8016,13	4831,39	439,69	724,30	36,5576	32085,4	19338,1	146,325
1261	1534,15	8021,32	4834,51	440,41	723,59	36,5610	32106,1	19350,6	146,339
1262	1535,15	8026,51	4837,62	441,13	722,89	36,5643	32126,9	19363,1	146,352
1263	1536,15	8031,71	4840,74	441,84	722,18	36,5677	32147,7	19375,5	146,366
1264	1537,15	8036,90	4843,85	442,56	721,48	36,5711	32168,5	19388,0	146,380
1265	1538,15	8042,09	4846,97	443,28	720,77	36,5745	32189,3	19400,5	146,393
1266	1539,15	8047,29	4850,08	444,00	720,07	36,5779	32210,1	19412,9	146,407
1267	1540,15	8052,48	4853,20	444,73	719,37	36,5812	32230,9	19425,4	146,420
1268	1541,15	8057,67	4856,32	445,45	718,67	36,5846	32251,6	19437,9	146,434
1269	1542,15	8062,87	4859,43	446,17	717,97	36,5880	32272,4	19450,4	146,447

Helium (*Continued*)

t	T	h	u	π_0	θ_0	s^0	H	U	S
1270	1543,15	8068,06	4862,55	446,90	717,27	36,5913	32293,2	19462,8	146,461
1271	1544,15	8073,25	4865,66	447,62	716,58	36,5947	32314,0	19475,3	146,474
1272	1545,15	8078,45	4868,78	448,34	715,88	36,5981	32334,8	19487,8	146,487
1273	1546,15	8083,64	4871,90	449,07	715,19	36,6014	32355,6	19500,2	146,501
1274	1547,15	8088,83	4875,01	449,80	714,49	36,6048	32376,4	19512,7	146,514
1275	1548,15	8094,03	4878,13	450,52	713,80	36,6081	32397,1	19525,2	146,528
1276	1549,15	8099,22	4881,24	451,25	713,11	36,6115	32417,9	19537,7	146,541
1277	1550,15	8104,41	4884,36	451,98	712,42	36,6148	32438,7	19550,1	146,555
1278	1551,15	8109,60	4887,47	452,71	711,73	36,6182	32459,5	19562,6	146,568
1279	1552,15	8114,80	4890,59	453,44	711,04	36,6215	32480,3	19575,1	146,581
1280	1553,15	8119,99	4893,71	454,17	710,36	36,6249	32501,1	19587,6	146,595
1281	1554,15	8125,18	4896,82	454,90	709,67	36,6282	32521,9	19600,0	146,608
1282	1555,15	8130,38	4899,94	455,63	708,99	36,6316	32542,6	19612,5	146,622
1283	1556,15	8135,57	4903,05	456,37	708,30	36,6349	32563,4	19625,0	146,635
1284	1557,15	8140,76	4906,17	457,10	707,62	36,6382	32584,2	19637,4	146,648
1285	1558,15	8145,96	4909,29	457,83	706,94	36,6416	32605,0	19649,9	146,662
1286	1559,15	8151,15	4912,40	458,57	706,26	36,6449	32625,8	19662,4	146,675
1287	1560,15	8156,34	4915,52	459,31	705,58	36,6482	32646,6	19674,9	146,688
1288	1561,15	8161,54	4918,63	460,04	704,90	36,6516	32667,4	19687,3	146,702
1289	1562,15	8166,73	4921,75	460,78	704,23	36,6549	32688,1	19699,8	146,715
1290	1563,15	8171,92	4924,87	461,52	703,55	36,6582	32708,9	19712,3	146,728
1291	1564,15	8177,12	4927,98	462,25	702,88	36,6615	32729,7	19724,7	146,741
1292	1565,15	8182,31	4931,10	462,99	702,20	36,6649	32750,5	19737,2	146,755
1293	1566,15	8187,50	4934,21	463,73	701,53	36,6682	32771,3	19749,7	146,768
1294	1567,15	8192,69	4937,33	464,47	700,86	36,6715	32792,1	19762,2	146,781
1295	1568,15	8197,89	4940,44	465,22	700,19	36,6748	32812,9	19774,6	146,795
1296	1569,15	8203,08	4943,56	465,96	699,52	36,6781	32833,7	19787,1	146,808
1297	1570,15	8208,27	4946,68	466,70	698,85	36,6814	32854,4	19799,6	146,821
1298	1571,15	8213,47	4949,79	467,44	698,18	36,6847	32875,2	19812,0	146,834
1299	1572,15	8218,66	4952,91	468,19	697,52	36,6880	32896,0	19824,5	146,848
1300	1573,15	8223,85	4956,02	468,93	696,85	36,6913	32916,8	19837,0	146,861
1301	1574,15	8229,05	4959,14	469,68	696,19	36,6946	32937,6	19849,5	146,874
1302	1575,15	8234,24	4962,26	470,43	695,53	36,6979	32958,4	19861,9	146,887
1303	1576,15	8239,43	4965,37	471,17	694,86	36,7012	32979,2	19874,4	146,900
1304	1577,15	8244,63	4968,49	471,92	694,20	36,7045	32999,9	19886,9	146,914
1305	1578,15	8249,82	4971,60	472,67	693,54	36,7078	33020,7	19899,3	146,927
1306	1579,15	8255,01	4974,72	473,42	692,88	36,7111	33041,5	19911,8	146,940
1307	1580,15	8260,21	4977,84	474,17	692,23	36,7144	33062,3	19924,3	146,953
1308	1581,15	8265,40	4980,95	474,92	691,57	36,7177	33083,1	19936,8	146,966
1309	1582,15	8270,59	4984,07	475,67	690,91	36,7210	33103,9	19949,2	146,979
1310	1583,15	8275,78	4987,18	476,42	690,26	36,7242	33124,7	19961,7	146,992
1311	1584,15	8280,98	4990,30	477,17	689,61	36,7275	33145,4	19974,2	147,006
1312	1585,15	8286,17	4993,41	477,93	688,95	36,7308	33166,2	19986,6	147,019
1313	1586,15	8291,36	4996,53	478,68	688,30	36,7341	33187,0	19999,1	147,032
1314	1587,15	8296,56	4999,65	479,44	687,65	36,7373	33207,8	20011,6	147,045
1315	1588,15	8301,75	5002,76	480,19	687,00	36,7406	33228,6	20024,1	147,058
1316	1589,15	8306,94	5005,88	480,95	686,35	36,7439	33249,4	20036,5	147,071
1317	1590,15	8312,14	5008,99	481,70	685,71	36,7471	33270,2	20049,0	147,084
1318	1591,15	8317,33	5012,11	482,46	685,06	36,7504	33290,9	20061,5	147,097
1319	1592,15	8322,52	5015,23	483,22	684,42	36,7537	33311,7	20073,9	147,110
1320	1593,15	8327,72	5018,34	483,98	683,77	36,7569	33332,5	20086,4	147,123
1321	1594,15	8332,91	5021,46	484,74	683,13	36,7602	33353,3	20098,9	147,136
1322	1595,15	8338,10	5024,57	485,50	682,49	36,7635	33374,1	20111,4	147,149
1323	1596,15	8343,30	5027,69	486,26	681,84	36,7667	33394,9	20123,8	147,162
1324	1597,15	8348,49	5030,80	487,02	681,20	36,7700	33415,7	20136,3	147,175

Helium (*Continued*)

t	T	h	u	π_0	θ_0	s^0	H	U	S
1325	1598,15	8353,68	5033,92	487,79	680,56	36,7732	33436,4	20148,8	147,188
1326	1599,15	8358,87	5037,04	488,55	679,93	36,7765	33457,2	20161,2	147,201
1327	1600,15	8364,07	5040,15	489,31	679,29	36,7797	33478,0	20173,7	147,214
1328	1601,15	8369,26	5043,27	490,08	678,65	36,7829	33498,8	20186,2	147,227
1329	1602,15	8374,45	5046,38	490,84	678,02	36,7862	33519,6	20198,7	147,240
1330	1603,15	8379,65	5049,50	491,61	677,38	36,7894	33540,4	20211,1	147,253
1331	1604,15	8384,84	5052,62	492,38	676,75	36,7927	33561,2	20223,6	147,266
1332	1605,15	8390,03	5055,73	493,15	676,12	36,7959	33581,9	20236,1	147,279
1333	1606,15	8395,23	5058,85	493,91	675,49	36,7991	33602,7	20248,5	147,292
1334	1607,15	8400,42	5061,96	494,68	674,86	36,8024	33623,5	20261,0	147,305
1335	1608,15	8405,61	5065,08	495,45	674,23	36,8056	33644,3	20273,5	147,318
1336	1609,15	8410,81	5068,20	496,22	673,60	36,8088	33665,1	20286,0	147,331
1337	1610,15	8416,00	5071,31	496,99	672,97	36,8121	33685,9	20298,4	147,344
1338	1611,15	8421,19	5074,43	497,77	672,34	36,8153	33706,7	20310,9	147,357
1339	1612,15	8426,39	5077,54	498,54	671,72	36,8185	33727,4	20323,4	147,370
1340	1613,15	8431,58	5080,66	499,31	671,09	36,8217	33748,2	20335,8	147,383
1341	1614,15	8436,77	5083,77	500,09	670,47	36,8249	33769,0	20348,3	147,396
1342	1615,15	8441,96	5086,89	500,86	669,85	36,8282	33789,8	20360,8	147,408
1343	1616,15	8447,16	5090,01	501,64	669,23	36,8314	33810,6	20373,3	147,421
1344	1617,15	8452,35	5093,12	502,41	668,61	36,8346	33831,4	20385,7	147,434
1345	1618,15	8457,54	5096,24	503,19	667,99	36,8378	33852,2	20398,2	147,447
1346	1619,15	8462,74	5099,35	503,97	667,37	36,8410	33873,0	20410,7	147,460
1347	1620,15	8467,93	5102,47	504,75	666,75	36,8442	33893,7	20423,1	147,473
1348	1621,15	8473,12	5105,59	505,53	666,13	36,8474	33914,5	20435,6	147,485
1349	1622,15	8478,32	5108,70	506,31	665,52	36,8506	33935,3	20448,1	147,498
1350	1623,15	8483,51	5111,82	507,09	664,90	36,8538	33956,1	20460,6	147,511
1351	1624,15	8488,70	5114,93	507,87	664,29	36,8570	33976,9	20473,0	147,524
1352	1625,15	8493,90	5118,05	508,65	663,68	36,8602	33997,7	20485,5	147,537
1353	1626,15	8499,09	5121,17	509,43	663,06	36,8634	34018,5	20498,0	147,549
1354	1627,15	8504,28	5124,28	510,22	662,45	36,8666	34039,2	20510,4	147,562
1355	1628,15	8509,48	5127,40	511,00	661,84	36,8698	34060,0	20522,9	147,575
1356	1629,15	8514,67	5130,51	511,79	661,23	36,8730	34080,8	20535,4	147,588
1357	1630,15	8519,86	5133,63	512,57	660,62	36,8762	34101,6	20547,9	147,601
1358	1631,15	8525,05	5136,74	513,36	660,02	36,8793	34122,4	20560,3	147,613
1359	1632,15	8530,25	5139,86	514,15	659,41	36,8825	34143,2	20572,8	147,626
1360	1633,15	8535,44	5142,98	514,93	658,80	36,8857	34164,0	20585,3	147,639
1361	1634,15	8540,63	5146,09	515,72	658,20	36,8889	34184,7	20597,7	147,651
1362	1635,15	8545,83	5149,21	516,51	657,60	36,8921	34205,5	20610,2	147,664
1363	1636,15	8551,02	5152,32	517,30	656,99	36,8952	34226,3	20622,7	147,677
1364	1637,15	8556,21	5155,44	518,09	656,39	36,8984	34247,1	20635,2	147,690
1365	1638,15	8561,41	5158,56	518,88	655,79	36,9016	34267,9	20647,6	147,702
1366	1639,15	8566,60	5161,67	519,68	655,19	36,9048	34288,7	20660,1	147,715
1367	1640,15	8571,79	5164,79	520,47	654,59	36,9079	34309,5	20672,6	147,728
1368	1641,15	8576,99	5167,90	521,26	653,99	36,9111	34330,2	20685,0	147,740
1369	1642,15	8582,18	5171,02	522,06	653,40	36,9143	34351,0	20697,5	147,753
1370	1643,15	8587,37	5174,14	522,85	652,80	36,9174	34371,8	20710,0	147,766
1371	1644,15	8592,57	5177,25	523,65	652,20	36,9206	34392,6	20722,5	147,778
1372	1645,15	8597,76	5180,37	524,44	651,61	36,9237	34413,4	20734,9	147,791
1373	1646,15	8602,95	5183,48	525,24	651,02	36,9269	34434,2	20747,4	147,804
1374	1647,15	8608,14	5186,60	526,04	650,42	36,9300	34455,0	20759,9	147,816
1375	1648,15	8613,34	5189,71	526,84	649,83	36,9332	34475,7	20772,4	147,829
1376	1649,15	8618,53	5192,83	527,64	649,24	36,9363	34496,5	20784,8	147,841
1377	1650,15	8623,72	5195,95	528,44	648,65	36,9395	34517,3	20797,3	147,854
1378	1651,15	8628,92	5199,06	529,24	648,06	36,9426	34538,1	20809,8	147,867
1379	1652,15	8634,11	5202,18	530,04	647,47	36,9458	34558,9	20822,2	147,879

Helium (*Continued*)

t	T	h	u	π_0	θ_0	s^0	H	U	S
1380	1653,15	8639,30	5205,29	530,84	646,89	36,9489	34579,7	20834,7	147,892
1381	1654,15	8644,50	5208,41	531,65	646,30	36,9521	34600,5	20847,2	147,904
1382	1655,15	8649,69	5211,53	532,45	645,71	36,9552	34621,2	20859,7	147,917
1383	1656,15	8654,88	5214,64	533,26	645,13	36,9583	34642,0	20872,1	147,929
1384	1657,15	8660.08	5217,76	534,06	644,54	36,9615	34662,8	20884,6	147,942
1385	1658,15	8665,27	5220,87	534,87	643,96	36,9646	34683,6	20897,1	147,955
1386	1659,15	8670,46	5223,99	535,67	643,38	36,9677	34704,4	20909,5	147,967
1387	1660,15	8675,66	5227,10	536,48	642,80	36,9709	34725,2	20922,0	147,980
1388	1661,15	8680,85	5230,22	537,29	642,22	36,9740	34746,0	20934,5	147,992
1389	1662,15	8686,04	5233,34	538,10	641,64	36,9771	34766,7	20947,0	148,005
1390	1663,15	8691,23	5236,45	538,91	641,06	36,9802	34787,5	20959,4	148,017
1391	1664,15	8696,43	5239,57	539,72	640,48	36,9834	34808,3	20971,9	148,030
1392	1665,15	8701,62	5242,68	540,53	639,91	36,9865	34829,1	20984,4	148,042
1393	1666,15	8706,81	5245,80	541,34	639,33	36,9896	34849,9	20996,8	148,055
1394	1667,15	8712,01	5248,92	542,15	638,75	36,9927	34870,7	21009,3	148,067
1395	1668,15	8717,20	5252,03	542,97	638,18	36,9958	34891,5	21021,8	148,080
1396	1669,15	8722,39	5255,15	543,78	637,61	36,9989	34912,3	21034,3	148,092
1397	1670,15	8727,59	5258,26	544,60	637,03	37,0021	34933,0	21046,7	148,104
1398	1671,15	8732,78	5261,38	545,41	636,46	37,0052	34953,8	21059,2	148,117
1399	1672,15	8737,97	5264,50	546,23	635,89	37,0083	34974,6	21071,7	148,129
1400	1673,15	8743,17	5267,61	547,05	635,32	37,0114	34995,4	21084,1	148,142
1401	1674,15	8748,36	5270,73	547,86	634,75	37,0145	35016,2	21096,6	148,154
1402	1675,15	8753,55	5273,84	548,68	634,18	37,0176	35037,0	21109,1	148,167
1403	1676,15	8758,75	5276,96	549,50	633,62	37,0207	35057,8	21121,6	148,179
1404	1677,15	8763,94	5280,07	550,32	633,05	37,0238	35078,5	21134,0	148,191
1405	1678,15	8769,13	5283,19	551,14	632,48	37,0269	35099,3	21146,5	148,204
1406	1679,15	8774,32	5286,31	551,96	631,92	37,0300	35120,1	21159,0	148,216
1407	1680,15	8779,52	5289,42	552,79	631,35	37,0331	35140,9	21171,4	148,229
1408	1681,15	8784,71	5292,54	553,61	630,79	37,0361	35161,7	21183,9	148,241
1409	1682,15	8789,90	5295,65	554,43	630,23	37,0392	35182,5	21196,4	148,253
1410	1683,15	8795,10	5298,77	555,26	629,67	37,0423	35203,3	21208,9	148,266
1411	1684,15	8800,29	5301,89	556,08	629,11	37,0454	35224,0	21221,3	148,278
1412	1685,15	8805,48	5305,00	556,91	628,55	37,0485	35244,8	21233,8	148,290
1413	1686,15	8810,68	5308,12	557,73	627,99	37,0516	35265,6	21246,3	148,303
1414	1687,15	8815,87	5311,23	558,56	627,43	37,0546	35286,4	21258,7	148,315
1415	1688,15	8821,06	5314,35	559,39	626,87	37,0577	35307,2	21271,2	148,327
1416	1689,15	8826,26	5317,47	560,22	626,32	37,0608	35328,0	21283,7	148,340
1417	1690,15	8831,45	5320,58	561,05	625,76	37,0639	35348,8	21296,2	148,352
1418	1691,15	8836,64	5323,70	561,88	625,20	37,0669	35369,5	21308,6	148,364
1419	1692,15	8841,84	5326,81	562,71	624,65	37,0700	35390,3	21321,1	148,376
1420	1693,15	8847,03	5329,93	563,54	624,10	37,0731	35411,1	21333,6	148,389
1421	1694,15	8852,22	5333,04	564,37	623,54	37,0761	35431,9	21346,0	148,401
1422	1695,15	8857,41	5336,16	565,21	622,99	37,0792	35452,7	21358,5	148,413
1423	1696,15	8862,61	5339,28	566,04	622,44	37,0823	35473,5	21371,0	148,426
1424	1697,15	8867,80	5342,39	566,87	621,89	37,0853	35494,3	21383,5	148,438
1425	1698,15	8872,99	5345,51	567,71	621,34	37,0884	35515,0	21395,9	148,450
1426	1699,15	8878,19	5348,62	568,55	620,79	37,0915	35535,8	21408,4	148,462
1427	1700,15	8883,38	5351,74	569,38	620,25	37,0945	35556,6	21420,9	148,474
1428	1701,15	8888,57	5354,86	570,22	619,70	37,0976	35577,4	21433,3	148,487
1429	1702,15	8893,77	5357,97	571,06	619,15	37,1006	35598,2	21445,8	148,499
1430	1703,15	8898,96	5361,09	571,90	618,61	37,1037	35619,0	21458,3	148,511
1431	1704,15	8904,15	5364,20	572,74	618,06	37,1067	35639,8	21470,8	148,523
1432	1705,15	8909,35	5367,32	573,58	617,52	37,1098	35660,5	21483,2	148,536
1433	1706,15	8914,54	5370,44	574,42	616,98	37,1128	35681,3	21495,7	148,548
1434	1707,15	8919,73	5373,55	575,26	616,44	37,1158	35702,1	21508,2	148,560
1435	1708,15	8924,93	5376,67	576,10	615,89	37,1189	35722,9	21520,6	148,572
1436	1709,15	8930,12	5379,78	576,95	615,35	37,1219	35743,7	21533,1	148,584
1437	1710,15	8935,31	5382,90	577,79	614,81	37,1250	35764,5	21545,6	148,596
1438	1711,15	8940,50	5386,01	578,64	614,28	37,1280	35785,3	21558,1	148,609
1439	1712,15	8945,70	5389,13	579,48	613,74	37,1310	35806,0	21570,5	148,621

Helium (*Continued*)

t	τ	h	u	π₀	θ₀	s⁰	H	U	S
1440	1713,15	8950,89	5392,25	580,33	613,20	37,1341	35826,8	21583,0	148,633
1441	1714,15	8956,08	5395,36	581,18	612,66	37,1371	35847,6	21595,5	148,645
1442	1715,15	8961,28	5398,48	582,03	612,13	37,1401	35868,4	21607,9	148,657
1443	1716,15	8966,47	5401,59	582,87	611,59	37,1432	35889,2	21620,4	148,669
1444	1717,15	8971,66	5404,71	583,72	611,06	37,1462	35910,0	21632,9	148,681
1445	1718,15	8976,86	5407,83	584,57	610,53	37,1492	35930,8	21645,4	148,693
1446	1719,15	8982,05	5410,94	585,42	609,99	37,1522	35951,6	21657,8	148,705
1447	1720,15	8987,24	5414,06	586,28	609,46	37,1552	35972,3	21670,3	148,718
1448	1721,15	8992,44	5417,17	587,13	608,93	37,1583	35993,1	21682,8	148,730
1449	1722,15	8997,63	5420,29	587,98	608,40	37,1613	36013,9	21695,2	148,742
1450	1723,15	9002,82	5423,40	588,84	607,87	37,1643	36034,7	21707,7	148,754
1451	1724,15	9008,02	5426,52	589,69	607,34	37,1673	36055,5	21720,2	148,766
1452	1725,15	9013,21	5429,64	590,55	606,81	37,1703	36076,3	21732,7	148,778
1453	1726,15	9018,40	5432,75	591,40	606,29	37,1733	36097,1	21745,1	148,790
1454	1727,15	9023,59	5435,87	592,26	605,76	37,1763	36117,8	21,57,6	148,802
1455	1728,15	9028,79	5438,98	593,12	605,23	37,1793	36138,6	21770,1	148,814
1456	1729,15	9033,98	5442,10	593,98	604,71	37,1823	36159,4	21782,5	148,826
1457	1730,15	9039,17	5445,22	594,83	604,18	37,1853	36180,2	21795 0	148,838
1458	1731,15	9044,37	5448,33	595,69	603,66	37,1883	36201,0	21807,5	148,850
1459	1732,15	9049,56	5451,45	596,55	603,14	37,1913	36221,8	21820,0	148,862
1460	1733,15	9054,75	5454,56	597,42	602,62	37,1943	36242,6	21832,4	148,874
1461	1734,15	9059,95	5457,68	598,28	602,10	37,1973	36263,3	21844,9	148,886
1462	1735,15	9065,14	5460,80	599,14	601,58	37,2003	36284,1	21857,4	148,898
1463	1736,15	9070,33	5463,91	600,00	601,06	37,2033	36304,9	21869,9	148,910
1464	1737,15	9075,53	5467,03	600,87	600,54	37,2063	36325,7	21882,3	148,922
1465	1738,15	9080,72	5470,14	601,73	600,02	37,2093	36346,5	21894,8	148,934
1466	1739,15	9085,91	5473,26	602,60	599,50	37,2123	36367,3	21907,3	148,946
1467	1740,15	9091,11	5476,37	6 3,47	598,98	37,2153	36388,1	21919,7	148,958
1468	1741,15	9096,30	5479,49	604,33	598,47	37,2183	36408,8	21932,2	148,970
1469	1742,15	9101,49	5482,61	605,20	597,95	37,2212	36429,6	21944,7	148,982
1470	1743,15	9106,68	5485,72	606,07	597,44	37,2242	36450,4	21957,2	148,994
1471	1744,15	9111,88	5488,84	606,94	596,92	37,2272	36471,2	21969,6	149,006
1472	1745,15	9117,07	5491,95	607,81	596,41	37,2302	36492,0	21982,1	149,017
1473	1746,15	9122,26	5495,07	608,68	595,90	37,2331	36512,8	21994,6	149,029
1474	1747,15	9127,46	5498,19	609,55	595,39	37,2361	36533,6	22007,0	149,041
1475	1748,15	9132,65	5501,30	610,43	594,88	37,2391	36554,3	22019,5	149,053
1476	1749,15	9137,84	5504,42	611,30	594,37	37,2421	36575,1	22032,0	149,065
1477	1750,15	9143,04	5507,53	612,17	593,86	37,2450	36595,9	22044,5	149,077
1478	1751,15	9148,23	5510,65	613,05	593,35	37,2480	36616,7	22056,9	149,089
1479	1752,15	9153,42	5513,77	613,92	592,84	37,2510	36637,5	22069,4	149,101
1480	1753,15	9158,62	5516,88	614,80	592,33	37,2539	36658,3	22081,9	149,113
1481	1754,15	9163,81	5520,00	615,68	591,83	37,2569	36679,1	22094,3	149,124
1482	1755,15	9169,00	5523,11	616,56	591,32	37,2598	36699,8	22106,8	149,136
1483	1756,15	9174,20	5526,23	617,43	590,82	37,2628	36720,6	22119,3	149,148
1484	1757,15	9179,39	5529,34	618,31	590,31	37,2658	36741,4	22131,8	149,160
1485	1758,15	9184,58	5532,46	619,19	589,81	37,2687	36762,2	22144,2	149,172
1486	1759,15	9189,77	5535,58	620,07	589,31	37,2717	36783,0	22156,7	149,184
1487	1760,15	9194,97	5538,69	620,96	588,80	37,2746	36803,8	22169,2	149,195
1488	1761,15	9200,16	5541,81	621,84	588,30	37,2776	36824,6	22181,6	149,207
1489	1762,15	9205,35	5544,92	622,72	587,80	37,2805	36845,3	22194,1	149,219
1490	1763,15	9210,55	5548,04	623,61	587,30	37,2835	36866,1	22206,6	149,231
1491	1764,15	9215,74	5551,16	624,49	586,80	37,2864	36886,9	22219,1	149,243
1492	1765,15	9220,93	5554,27	625,38	586,30	37,2893	36907,7	22231,5	149,254
1493	1766,15	9226,13	5557,39	626,26	585,81	37,2923	36928,5	22244,0	149,266
1494	1767,15	9231,32	5560,50	627,15	585,31	37,2952	36949,3	22256,5	149,278
1495	1768,15	9236,51	5563,62	628,04	584,81	37,2982	36970,1	22268,9	149,290
1496	1769,15	9241,71	5566,73	628,92	584,32	37,3011	36990,9	22281,4	149,301
1497	1770,15	9246,90	5569,85	629,81	583,82	37,3040	37011,6	22293,9	149,313
1498	1771,15	9252,09	5572,97	630,70	583,33	37,3070	37032,4	22306,4	149,325
1499	1772,15	9257,29	5576,08	631,59	582,83	37,3099	37053,2	22318,8	149,337
1500	1773,15	9262,48	5579,20	632,49	582,34	37,3128	37074,0	22331,3	149,348